Lecture Notes in Mechanical Engineering

Lecture Notes in Mechanical Engineering (LNME) publishes the latest developments in Mechanical Engineering—quickly, informally and with high quality. Original research reported in proceedings and post-proceedings represents the core of LNME. Volumes published in LNME embrace all aspects, subfields and new challenges of mechanical engineering. Topics in the series include:

- Engineering Design
- Machinery and Machine Elements
- Mechanical Structures and Stress Analysis
- Automotive Engineering
- Engine Technology
- Aerospace Technology and Astronautics
- Nanotechnology and Microengineering
- Control, Robotics, Mechatronics
- MEMS
- Theoretical and Applied Mechanics
- Dynamical Systems, Control
- Fluid Mechanics
- Engineering Thermodynamics, Heat and Mass Transfer
- Manufacturing
- Precision Engineering, Instrumentation, Measurement
- Materials Engineering
- Tribology and Surface Technology

To submit a proposal or request further information, please contact the Springer Editor of your location:

China: Dr. Mengchu Huang at mengchu.huang@springer.com
India: Priya Vyas at priya.vyas@springer.com
Rest of Asia, Australia, New Zealand: Swati Meherishi at swati.meherishi@springer.com
All other countries: Dr. Leontina Di Cecco at Leontina.dicecco@springer.com

To submit a proposal for a monograph, please check our Springer Tracts in Mechanical Engineering at http://www.springer.com/series/11693 or contact Leontina.dicecco@springer.com

Indexed by SCOPUS. The books of the series are submitted for indexing to Web of Science.

More information about this series at http://www.springer.com/series/11236

Shripad Revankar · Swarnendu Sen ·
Debjyoti Sahu
Editors

Proceedings of International Conference on Thermofluids

KIIT Thermo 2020

Springer

Editors
Shripad Revankar
Purdue University
West Lafayette, IN, USA

Swarnendu Sen
Jadavpur University
Kolkata, West Bengal, India

Debjyoti Sahu
Kalinga Institute of Industrial Technology
Bhubaneswar, Odisha, India

ISSN 2195-4356 ISSN 2195-4364 (electronic)
Lecture Notes in Mechanical Engineering
ISBN 978-981-15-7830-4 ISBN 978-981-15-7831-1 (eBook)
https://doi.org/10.1007/978-981-15-7831-1

This Springer imprint is published by the registered company Springer Nature Singapore Pte Ltd.
The registered company address is: 152 Beach Road, #21-01/04 Gateway East, Singapore 189721, Singapore

Preface

This International Conference on Thermofluids (KIIT Thermo 2020) was held in the School of Mechanical Engineering inside the thriving campus of KIIT Deemed to be University, Bhubaneswar, India, during 23–24 January 2020. With the passion for quality education, visionary educationalist Dr. Achyuta Samanta established KIIT Deemed to be University, formerly Kalinga Institute of Industrial Technology, which is a co-educational autonomous university located at Bhubaneswar in the state of Odisha, India. KIIT was established in 1992 as an Industrial Training Institution (ITI) which has grown to become KIIT Deemed to be University in 2004. It was one of the youngest institutions to be recognized as a Deemed University in India. All the academic programmes are accredited by NAAC of University Grants Commission of India and NBA as per Washington Accord of All India Council for Technical Education, which are benchmarks of quality education. NAAC (Government of India appointed agency to evaluate universities) has conferred KIIT with the highest grade "A" with a CGPA of 3.36/4. KIIT Deemed to be University recently achieved the tag of Institution of Eminence (IoE) by Ministry of Human Resource Development, Government of India. KIIT School of Mechanical Engineering, established in the year 1997, produces graduates, postgraduates and researches who can meet the needs of the industry which require learning attitude, exposure to latest technology and research. Recent consultancy and core research areas of the school include thermofluids and renewable energy, material processing technology, cleaner manufacturing technology, development of the racing car, quality engineering and management. Residual stresses in fusion-welded structure and surface finish optimization by high-pressure impingement cooling are the other areas of interest. Research and development activities of the school are supported by private companies and Government of India organizations like ARDB, BRNS, AICTE and DST.

The International Conference on Thermofluids (KIIT Thermo 2020) provided an ideal platform and brought together the researchers, scientists, engineers, industrial experts, scholars and students to share and widen their knowledge on theoretical, numerical and experimental developments in the fields of thermodynamics, combustion, heat transfer and fluid mechanics. This conference offered excellent opportunities for the participants to have a direct exchange of ideas and experiences, to mine potential research problems and to forge research relations alongside finding partners for future collaborations. The conference has invited eminent speakers from the industry and academia for delivering keynote lectures and plenary talks.

A promising volume of research has been carried out in the area of thermofluids recently. The niche of thermofluids ranges from power plant cycle analysis, thermal management, flow visualization to HVAC (heating, ventilation and air-conditioning) analysis. Researchers in the field of thermofluids view various streams of thermal and fluid engineering through a holistic approach. Given the gamut of engineering challenges pertaining to thermofluids that the industry is currently faced with, a holistic effort involving and transcending thermal and fluid engineering is essential. KIIT Thermo 2020 proceedings span over 11 topical tracks, which are well balanced in content and manageable in terms of the number of contributions and create an adequate discussion space for the interesting topics. There were 45 oral presentations and about 15 poster presentations by participants which brought great opportunity to share their recent research experience among each other graciously.

Efforts taken by peer reviewers contributed to improve the quality of manuscripts, and their constructive critical comments, improvements and corrections to the authors are gratefully appreciated. We on behalf of the organizers, Prof. A. K. Sahoo, Conference Chair, Prof. P. C. Mishra, Conference Co-chair and Dr. Gyan Sagar Sinha, Convener are very much grateful to the international/national advisory committee, session chairs, student volunteers and administrative assistants from the institute management who selflessly contributed to the success of this conference. Also, we are thankful to all the authors who submitted papers, because of which the conference became a successful one. It was the quality of their presentations and their passion to communicate with the other participants that really made this conference a great success. Last but not least, we are thankful for the gracious support of Springer for believing us in every step of our journey towards success. Their cooperation was not only the strength but also an inspiration for the organizers.

Edited by:
Prof. Shripad Revankar
Prof. Swarnendu Sen
Dr. Debjyoti Sahu

Bhubaneswar, India

Dr. Debjyoti Sahu
Secretary, KIIT Thermo 2020
debjyoti.sahufme@kiit.ac.in
debjyothi@gmail.com

Contents

About the Editors

Dr. Shripad Revankar is a Professor of Nuclear Engineering and Director of Multiphase and Fuel Cell Research Laboratory in the School of Nuclear Engineering at Purdue University, West Lafayette, Indiana. He received his B.S., M.S. and Ph.D. in Physics from Karnatak University, India, and M.Eng. in Nuclear Engineering from McMaster University, Canada. He has worked as a Postdoctoral Researcher at Lawrence Berkeley Laboratory and at the Nuclear Engineering Department, University of California, Berkeley, from 1984 to 1987. Then, in August 1987, he joined the School of Nuclear Engineering as a Visiting Assistant Professor. Since then, he has been serving SNE, currently as a Professor. He also serves as BK21 Plus Visiting Professor in the Division of Advanced Nuclear Engineering at Pohang University of Science and Technology (POSTECH), South Korea. Dr. Revankar has published over 375 peer-reviewed technical articles in archival scientific journals and conference proceedings and author/co-author of three recent books: Advances in Nuclear Fuels, InTech publisher 2012, Fuel Cells-Principles, Design, and Analysis, CRC Press 2014, and Storage and Hybridization of Nuclear Energy: Techno-economic Integration of Renewable and Nuclear Energy, Academic Press, recently published in November 2018. He has chaired 25 M.S. thesis, and 12 Ph.D. thesis and has served on 90 M.S. and Ph.D. thesis committees. He has mentored over 35 Visiting Scholars and Postdoctoral Researchers. He has presented over 110 invited seminars. Dr. Revankar has served as research and educational Consultant to academia, national laboratories and industries in Canada, China, Hong Kong, India, South Korea, and USA. He is Chief Editor of Frontier in Energy – Nuclear Energy – and Chief Editor of the International Journal of Magnetism & Nuclear Science. He is also on editorial boards of other six international journals including Heat Transfer Engineering, Journal of Thermodynamics, Nuclear Engineering and Technology, and has served as Guest Editor for Nuclear Engineering and Design. Dr. Revankar is a Life Member of American Nuclear Society (ANS), American Society of Mechanical Engineer (ASME), American Institute of Chemical Engineers (AIChE), Korean Nuclear Society, (KNS), and Indian Society for Heat and Mass Transfer (ISHMT). He is also a Member of American Society for Engineering Education (ASEE),

Electrochemical Society (ECS) and American Association for Advancement of Science (AAAS). He was Chair of Thermal Hydraulics Division of ANS in 2007–2008, Chair of ASME K-13 Committee on Heat Transfer in Multiphase Systems in 2009-2011, Executive Member of the AIChE Transport and Energy Processes Division in 2006–2009, and Chair of the ASEE Nuclear and Radiological Division in 2008–2009. He is currently Vice Chair and Chair Elect Nuclear Engineering Division of ASME. He was elected as a Fellow of ASME in 2008, Fellow of ANS in 2015 and Fellow of AIChE in 2017.

Prof. (Dr.) Swarnendu Sen has been involved in teaching and research since 1988 at Jadavpur University, Kolkata. He is serving this esteemed institute as a Professor since 2007. Prior to academics, he worked for HCL Ltd. He was a Visiting Research Fellow in the University of Illinois at Chicago in 2002–2003 and a Visiting Scholar at Virginia Tech in 2006. He received DAAD fellowship from Germany in 2011 as well as in 2013. His research area includes combustion dynamics, transport processes in nanofluids and carbon nanostructure synthesis. Prof. Sen has guided more than 15 Ph.D.s and more than 150 papers and his h-index is 35. He is one among the pioneers in the field of natural circulation loop. He is a Member of Indian Society for Heating Refrigerating and Air-conditioning Engineers, Associate Member of The Institution of Engineers (India), Life Member of The Combustion Institute, Indian Section, and Life Member of Indian Society of Heat and Mass Transfer.

Dr. Debjyoti Sahu currently serves as an Assistant Professor at the School of Mechanical Engineering, KIIT Deemed to be University, Bhubaneswar. Dr. Sahu obtained his B.E. in Automobile Engineering from Malnad College of Engineering, Hassan, and M.Tech. in Energy Systems Engineering from BVB College of Engineering and Technology Hubli, Karnataka. He obtained his Ph.D. in Energy Technology from Indian Institute of Technology Guwahati, India. He has taught various subjects in UG and PG levels at the Department of Mechanical Engineering, Amrita Vishwa Vidyapeetham, Bengaluru, during 2008–2018. Now he teaches Fuels and Emission, Automobile Engineering and IC Engine and Gas Turbines at KIIT Deemed to be University, Bhubaneswar. He has been involved in several experimental research including hydrogen adsorption, metal organic frameworks, refrigeration and bio-digestion. His work on portable bio-digestion is mentioned in an article published by ENSIA, the most popular environment magazine. He has published several papers on refrigeration, adsorption and IC engine performance. Dr. Sahu is a Life Member of Indian Science Congress and Institute of Engineers, India.

Fluid Dynamics

Large Eddy Simulation Modeling in 2D Lid-Driven Cavity

Banamali Dalai and Manas Kumar Laha

Abstract Primitive variable formulation of Navier–Stokes equation is solved in a lid-driven cavity using SIMPLE algorithm. The equations are discretized using finite volume method in a staggered grid mesh. The turbulence flow phenomena are observed in a 2D lid-driven cavity using large eddy simulation modeling of flow. Smagorinsky model is applied for the formulation of eddy viscosity. The solution is obtained upto maximum Reynolds number 4500 using grid sizes 66 × 66, 81 × 81, 101 × 101 and 121 × 121. A comparison has been drawn between turbulence model and without turbulence model. The study of comparison between two models consists of velocity profiles along the center of the cavity, location of primary and secondary eddies, velocity vector plot and pressure contours.

Keywords SIMPLE · FVM · Staggered grid · LES · Smagorinsky model

1 Introduction

Two-dimensional lid-driven cavity is a square cavity having all the walls except top wall which is rigidly fixed and stationary as shown in Fig. 1. The top wall is allowed to move toward right or left with non-dimensional speed unity. Non-dimensional length of each side of the wall is unity. Initially, the cavity is filled with fluid which is at rest. As the lid starts moving toward right with uniform velocity unity, the fluid flow in the cavity is set up. As the Reynolds number gradually increases, the number of primary and secondary eddies increases in the cavity. The large eddy which appears at Re = 0.00001 is called primary eddy. With increase of Reynolds number, levels of corner eddies gradually grow in the cavity [1, 2].

B. Dalai (✉)
Faculty, Mechanical Engineering, College of Engineering and Technology, Bhubaneswar, Biju Patnaik University of Technology, Rourkela, Odisha, India
e-mail: banamali.2000@gmail.com

M. K. Laha
Faculty, Aerospace Engineering, Indian Institute of Technology, Kharagpur, West Bengal, India
e-mail: mlaha@aero.iitkgp.ac.in

S. Revankar et al. (eds.), *Proceedings of International Conference on Thermofluids*, Lecture Notes in Mechanical Engineering,
https://doi.org/10.1007/978-981-15-7831-1_1

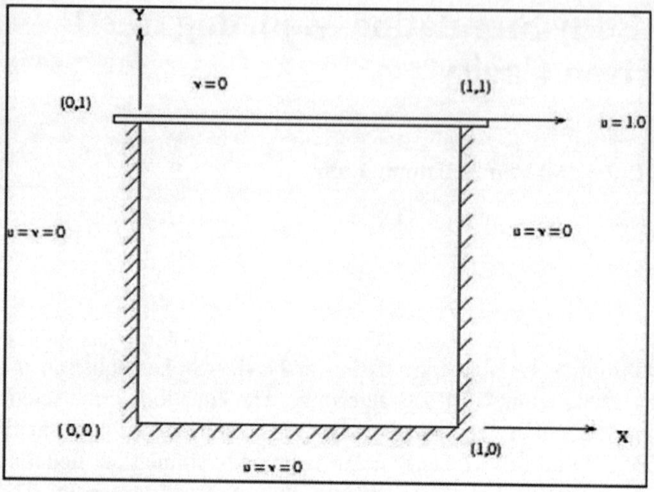

Fig. 1 Schematic diagram of a lid-driven cavity

According to the appearance of the eddies, these are named as secondary, tertiary and so on. The present work contains large eddy simulation modeling of Navier–Stokes equation in the lid-driven cavity to study the effect of chaos and turbulence at higher Reynolds number flow. Primitive variable form of Navier-Stokes equation is preferred than the stream function-vorticity form because of simplicity of incorporating the eddy viscosity term in the governing equation. Few literatures regarding the chaos effect in the cavity are discussed below.

Ghia et al. studied the primitive variable formulation of Navier–Stokes equation using multigrid method in a lid-driven cavity [3]. They were successful to solve the N–S equation upto Re = 10,000 in the grid sizes 129 × 129 and 257 × 257. Chen solved the large eddy simulation modeling of stream function-vorticity form of Navier–Stokes equation using lattice Boltzmann method [1]. It is reported that the flow is unsteady at Re = 20,000 and at Re = 50,000, and the flow becomes turbulent. Two and three-dimensional lid-driven cavities using primitive variable formulation of Navier–Stokes equation are already reported [2]. They implemented large eddy simulation modeling of Navier–Stokes equation in three-dimensional lid-driven cavities. They found that the flow is unsteady at Re = 3000 and steady upto Re = 2000. They found the presence of Taylor–Gortler-like vortices in the cavity after Re = 3000. Peng et al. studied the transition of flow from laminar to turbulent in a lid-driven cavity using Navier–Stokes equations [4]. They discretized the convection terms by a seventh-order accurate upwind scheme and the diffusion terms by a sixth-order accurate central difference scheme. The rest of the terms were discretized by second-order accurate central difference scheme. They obtained the solution using MAC numerical method. The steady-state flow became unstable via Hopf. bifurcation at Re = 7402, 7694 and 7704 in the grid size 100 × 100, 150 × 150 and 200 × 200, respectively. This type of behavior was called branch-I Period-1 pattern by them.

The second branch exhibited more complicated behavior including quasi-periodic and period-doubling behavior via Hopf. bifurcations in the Reynolds number range from 10,300 to 11,000. They called it branch-II period-2 pattern. Bruneau and Saad studied the transition of flow from laminar to turbulent in the lid-driven cavity using Navier–Stokes equations in two-dimensional lid-driven cavity [5]. They carried out linear stability analysis by the method of Lyapunov exponents and determined the critical Reynolds number based on a Hopf. bifurcation. The numerical simulation was based on finite difference discretization on uniform staggered grid. The discretized equation was solved using multigrid solver with cell-by-cell relaxation procedure. They observed first critical Reynolds number at 8000, this being the point where the solution loses stability. It becomes periodic at Re = 8050 with non-dimensional fundamental frequency 0.45.

From the literature, it is observed that the large eddy simulation modeling in primitive variable formulation of N–S equation in a two-dimensional lid-driven cavity will be quite useful. So, the objective of this study is to find out the steady-state solution in two-dimensional lid-driven cavity by application of large eddy simulation modeling with Smagorinsky model of eddy viscosity. The region of unsteadiness and the effect of turbulence in the cavity will be studied.

2 Formulation of the Problem

Two-dimensional incompressible Navier–Stokes equation in dimensional form after passing through the filter is:

$$\frac{\partial \bar{u}_i}{\partial t} + \frac{\partial}{\partial x_j}(\bar{u}_i \bar{u}_j) = -\frac{1}{\rho}\frac{\partial \bar{p}}{\partial x_i} + 2\frac{\partial}{\partial x_j}\{(\nu + \nu_t)\overline{S}_{ij}\} \tag{1}$$

and the continuity equation is expressed as:

$$\frac{\partial u_i}{\partial x_i} = 0 \tag{2}$$

where i, j are indices along x- and y-directions which are denoted by $i, j = 1$ or 2. $\bar{u}_i$ is the filtered velocity component along i direction. ν and ν_t are the molecular and eddy viscosity, respectively. If ν_t is zero, then the LES model becomes primitive variable formulation of Navier–Stokes equation without large eddy simulation modeling.

Where $\overline{S}_{ij} = \frac{1}{2}\left(\frac{\partial \bar{u}_i}{\partial x_j} + \frac{\partial \bar{u}_j}{\partial x_i}\right)$ and $\nu_{t_{ij}} = (C_s\Delta)^2\sqrt{2\overline{S}_{ij}\overline{S}_{ij}}$ are called sub-grid scale tensor and eddy viscosity respectively in which C_s is called Smagorinsky constant and Δ is the sub-grid length scale which is computed by taking square root of the product of sub-grid length scales along x- and y-directions ($\Delta = (\Delta_1\Delta_2)^{1/2}$); Δ_1 and Δ_2 are called sub-grid length scales along x- and y-directions, respectively. The value of C_s is 0.18 calculated by Smagorinsky (Leisure) but the value 0.10–0.12 as shown by Jordan et al. and Leisure is sufficient in computation [6, 7]. In this computation, the value of C_s is 0.10. The tensor product is evaluated in this manner:

$\overline{S}_{ij}\overline{S}_{ij} = \begin{bmatrix} S_{11} & S_{12} \\ S_{21} & S_{22} \end{bmatrix}\begin{bmatrix} S_{11} & S_{12} \\ S_{21} & S_{22} \end{bmatrix}$ and the eddy viscosity is evaluated as: $\nu_{t_{ij}} = \begin{bmatrix} \nu_{t_{11}} & \nu_{t_{12}} \\ \nu_{t_{21}} & \nu_{t_{22}} \end{bmatrix}$, where

$$\nu_{t_{11}} = (C_s\Delta)^2\sqrt{2(S_{11}S_{11} + S_{12}S_{21})},\ \nu_{t_{12}} = (C_s\Delta)^2\sqrt{2(S_{11}S_{12} + S_{12}S_{22})},$$
$$\nu_{t_{22}} = (C_s\Delta)^2\sqrt{2(S_{21}S_{12} + S_{22}S_{22})} \tag{3}$$

2.1 Numerical Method

The non-dimensional Navier–Stokes equation (Eqs. 1 and 2) in x- and y-directions is:

$$\frac{\partial \bar{u}}{\partial t} + \frac{\partial \bar{u}^2}{\partial x} + \frac{\partial \bar{u}\bar{v}}{\partial y} = -\frac{\partial \bar{p}}{\partial x} + \left(\frac{2}{\mathrm{Re}} + 2\nu_{t_{11}}\right)\frac{\partial^2 \bar{u}}{\partial x^2} + \left(\frac{1}{\mathrm{Re}} + \nu_{t_{12}}\right)\left(\frac{\partial^2 \bar{u}}{\partial y^2} + \frac{\partial^2 \bar{v}}{\partial x \partial y}\right)$$
$$+ 2\overline{S}_{11}\frac{\partial \nu_{t_{11}}}{\partial x} + 2\overline{S}_{12}\frac{\partial \nu_{t12}}{\partial y} \tag{4}$$

$$\frac{\partial \bar{v}}{\partial t} + \frac{\partial \bar{u}\bar{v}}{\partial x} + \frac{\partial \bar{v}^2}{\partial y} = -\frac{\partial \bar{p}}{\partial y} + \left(\frac{1}{\mathrm{Re}} + \nu_{t_{12}}\right)\left(\frac{\partial^2 \bar{v}}{\partial x^2} + \frac{\partial^2 \bar{u}}{\partial x \partial y}\right) + \left(\frac{2}{\mathrm{Re}} + 2\nu_{t_{22}}\right)\frac{\partial^2 \bar{v}}{\partial y^2}$$
$$+ 2\overline{S}_{21}\frac{\partial \nu_{t_{21}}}{\partial x} + 2\overline{S}_{22}\frac{\partial \nu_{t_{22}}}{\partial y} \tag{5}$$

and the continuity equation is:

$$\frac{\partial \bar{u}}{\partial x} + \frac{\partial \bar{v}}{\partial y} = 0 \tag{6}$$

where Re is the Reynolds number and $\bar{p}$ is the filtered pressure. The boundary conditions for the two-dimensional lid-driven cavity are:

$$\text{At } x = 0,\ \bar{u} = \bar{v} = 0;\ \text{at } y = 0,\ \bar{u} = \bar{v} = 0;\ \text{at } x = 1,\ \bar{u} = \bar{v} = 0$$
$$\text{and at } y = 1,\ \bar{u} = 1,\ \bar{v} = 0 \tag{7}$$

Equations (4) and (5) are discretized using second-order accurate central difference scheme in a finite volume staggered grid mesh, and the solution is obtained following SIMPLE method [8]. Time derivative terms are discretized using first-order accurate forward difference scheme. Initial condition is that fluid flow starts from the rest, i.e., properties of fluid are at atmospheric condition and velocity is zero in the computational domain at time $t = 0$. The solution is obtained using time marching explicit scheme in Eqs. (4) and (5). Equations(4) and (5) along with boundary conditions are iterated in each time step till the desired convergence. The time step 0.001 is used for iteration at all Reynolds numbers.

2.2 *Convergence Details*

Convergence criterion in LES modeling for two-dimensional lid-driven cavity is the satisfaction of continuity Eq. (6) upto the order 10^{-18}. First, converged solution is obtained at Re = 1000. Then, this solution is assumed as initial value for subsequent higher Reynolds number solution. This procedure is continued upto Re = 4600. The grid sizes used for this solution are 81 × 81, 101 × 101 and 121 × 121 in a range of Reynolds numbers from 1000 to 15,000.

2.3 *Validation Study*

The computed results in the grid size 121 × 121 are validated with the results of Ghia et al. in the grid size 129 × 129 as shown in Fig. 2 [3]. Computed velocity profiles using large eddy simulation modeling and without large eddy simulation modeling match very closely with Ghia et al. [3].

3 Results and Discussion

The converged solution is obtained upto Re = 4600 which is the steady-state solution. The solution at Re = 4800 is unsteady in the grid sizes 66 × 66, 81 × 81, 101 × 101 and 121 × 121. The pressure contour in the lid-driven cavity shows that pressure distribution is uniform at the central region, whereas at the lower corners, the pressure is very weak. On the other hand, at the top right corner of the cavity, the pressure is high and variation is large which can be observed from the pressure contour plots as shown in Fig. 3. Fluctuation of pressure is observed at the central portion and nearby regions of it at Re = 7500. Gradually, that fluctuation of pressure has spread to all portion of the cavity at higher Reynolds number.

Velocity vector plot in Fig. 4 shows that vortex flow is set up in the cavity with central large primary eddy which rotates clockwise. As the Reynolds number increases, the radius of the primary eddy becomes large and ring-like structure. At Re = 5000, the right lower corner eddies fluctuate and those indicate early presence

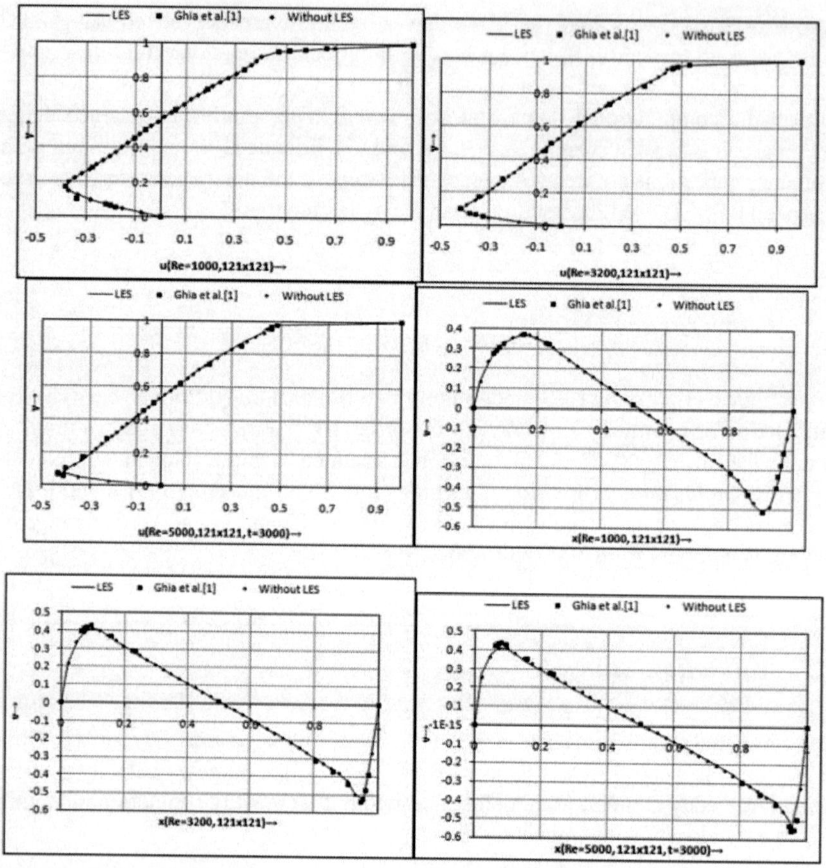

Fig. 2 *u* and *v*-velocity profiles along the center of the cavity at different Reynolds numbers

of turbulence. Chaotic character of flow becomes more prominent on the secondary eddies at Re = 7500. At Re = 10,000, the chaotic nature of flow is observed at the lower left corner as well as upper left corner of the cavity as seen from Fig. 5.

Figure 6 shows that *u*, *v*, *p* and continuity equation converge monotonically to the desired level at a point (0.374, 0.2479) using large eddy simulation modeling at Re = 5000, whereas Fig. 7 shows that those do not converge to the desired level at the same point and same Reynolds number when large eddy simulation modeling is applied on the Navier–Stokes equation.

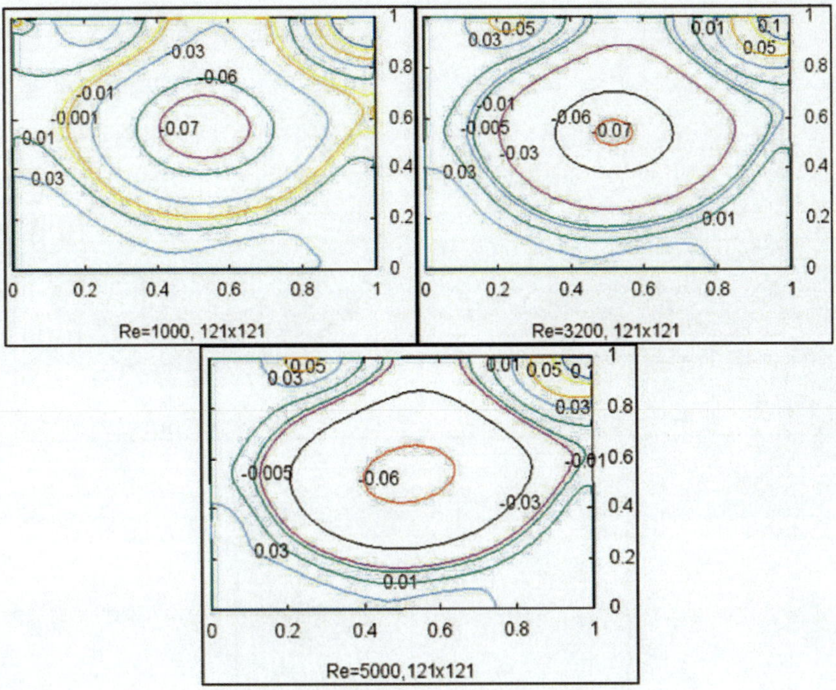

Fig. 3 Pressure contours at different Reynolds numbers

4 Conclusion

The large eddy simulation modeling of Navier–Stokes equation in 2D lid-driven cavity is studied with application of Smagorinsky model in the eddy viscosity. It was observed that the flow is steady up to Re = 4500, and after that, the unsteadiness of flow is observed. At Re = 5000, the lower right corner of the cavity captures the unsteadiness and subsequent higher Reynolds numbers become chaotic in nature at that region. The chaotic flow has spread to the lower left corner and upper left corner of the cavity at higher Reynolds number of the flow.

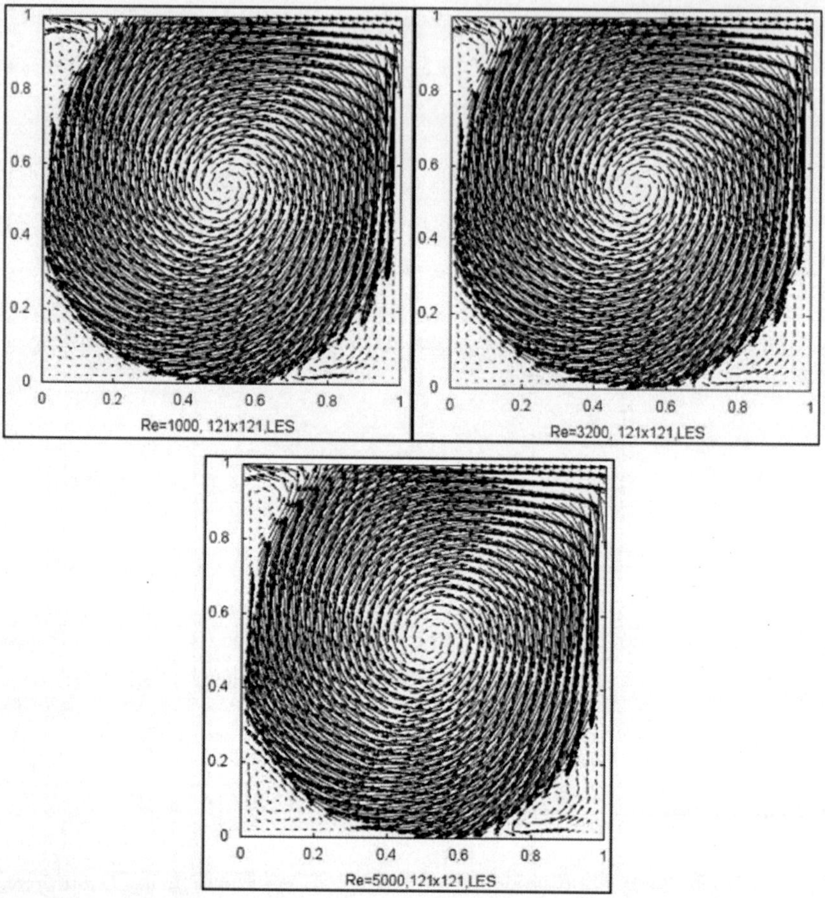

Fig. 4 Velocity vector plots at different Reynolds numbers using LES

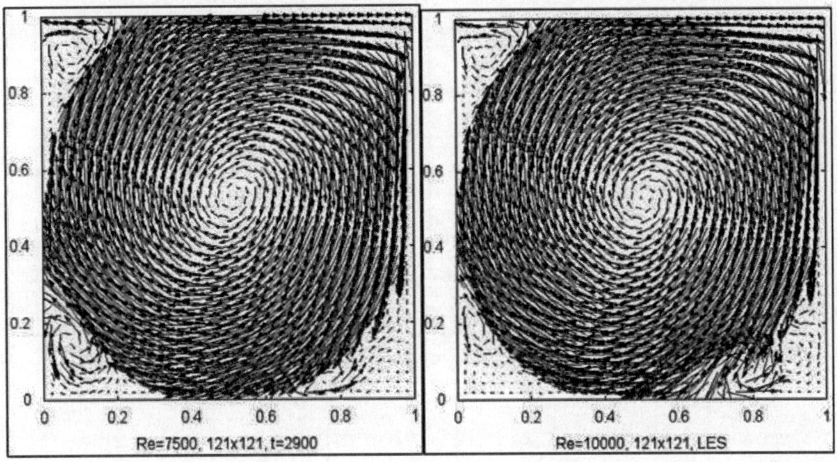

Fig. 5 Velocity vector plots at Re = 7500 and 10,000 using LES

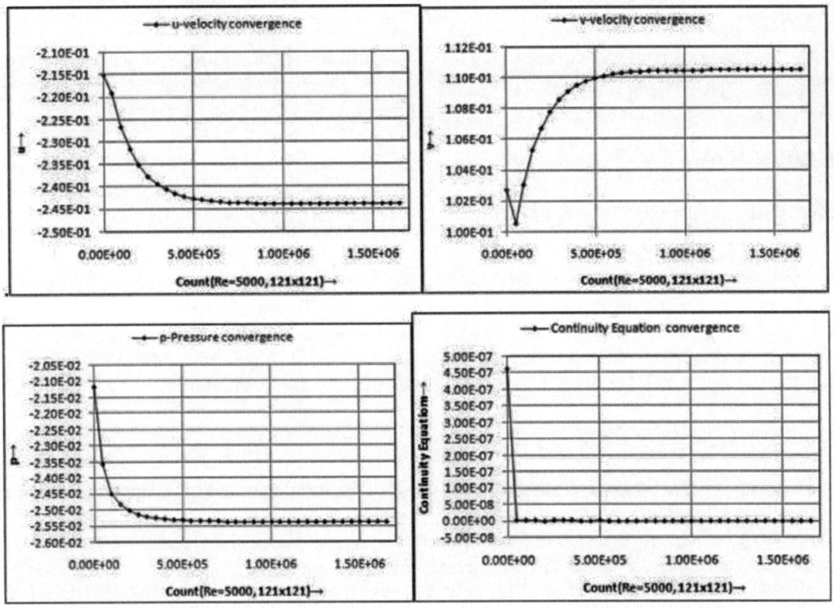

Fig. 6 Convergence of u, v, p and continuity equation at a point (0.374, 0.2479) for Re = 5000 without LES model

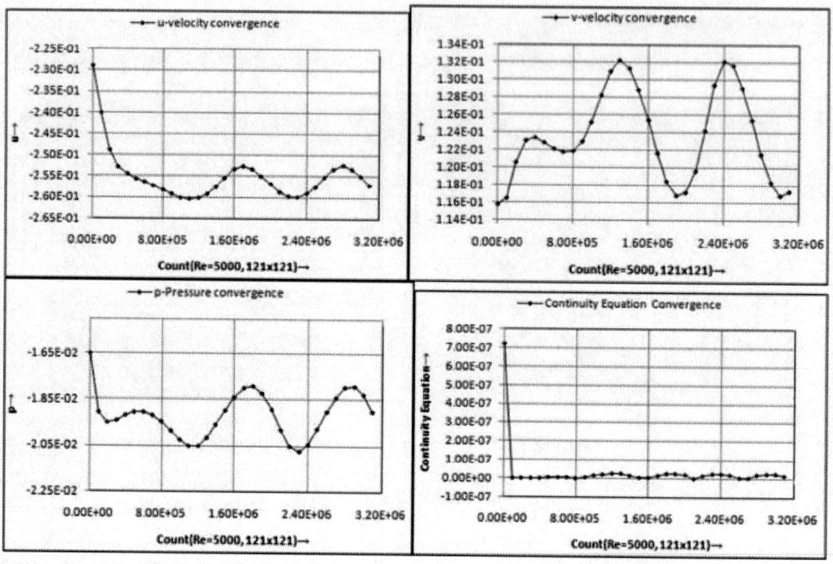

Fig. 7 *u*, *v*, *p* and continuity equation profile at a point (0.374, 0.2479) for Re = 5000 using LES model

References

1. Chen S (2009) A large-eddy-based lattice Boltzmann model for turbulent flow simulation. Appl Math Comput 215:591–598
2. Dalai B, Laha MK (2014) Incompressible viscous flows in two and three dimensional lid-driven cavities. PhD thesis dissertation, Aerospace Engineering, IIT Kharagpur
3. Ghia U, Ghia KN, Shin CT (1982) High Reynolds number solutions for incompressible flow using the Navier–Stokes equations and a multigrid method. J Comput Phys 48:387–411
4. Peng YF, Shiau YH, Hwang RR (2003) Transition in a 2D lid-driven cavity flow. Comput Fluids 32:337–352
5. Bruneau CH, Saad M (2006) The 2D lid-driven cavity problem revisited. Comput Fluids 35:326–348
6. Lesieur M (1997) Large eddy simulations, Chap XII. In: Turbulence in fluids. Kluwer Academic Publishers, London/Boston/Dordrecht, pp 375–406
7. Jordan SA, Raghab SA (1994) On the unsteady and turbulent characteristics of the three-dimensional shear driven cavity flow. J Fluids Eng 116:439
8. Patankar SV (1981) Calculation of the flow field. In: Numerical heat transfer and fluid flow, 6.7, the SIMPLE algorithm. Hemisphere Publishing Corporation, p 126

Numerical Study on Flow Over Two Inline Triangular Cylinders—Influence of Gap Ratio on Vortex Shedding

Manu Sivan and S. Ajith Kumar

Abstract In the present study, numerical simulation was carried out for the flow around the two inline equilateral triangular cylinders at different gap ratios (G/D) for various Reynolds numbers (Re). The numerical investigations are carried out for $75 \leq \text{Re} \leq 200$ and $0 \leq G/D \leq 2.5$. The governing differential equations for laminar incompressible flow are discretized using finite volume method and PISO algorithm is employed to convert PDE to algebraic equation using the open-source software package. The effect of G/D ratio and Re on the non-dimensional shedding frequency Strouhal number (St) is investigated with the aid of vorticity plot and velocity profiles.

Keywords Triangular cylinder · Gap ratio · Strouhal number · Reynolds number

Nomenclature

Re	Reynolds number
G/D	Gap ratio
St	Strouhal number
C_l	Coefficient of lift

1 Introduction

Flow past multiple cylinders has got plenty of industrial relevance compared to the single cylinder configuration, specifically for triangular cylinders. The scope and applications of such a configuration have wide spectrum of applications in design of heat exchangers, nuclear power plants cooling towers, offshore structures, chimney

M. Sivan (✉) · S. Ajith Kumar
Department of Mechanical Engineering, Amrita Vishwa Vidyapeetham, Amritapuri, India
e-mail: manusivan@am.amrita.edu

S. Revankar et al. (eds.), *Proceedings of International Conference on Thermofluids*, Lecture Notes in Mechanical Engineering,
https://doi.org/10.1007/978-981-15-7831-1_2

stacks, and power transmission cables, etc. [1–7]. The flow patterns and the wake behind the cylinder have undergone spurious modifications due to the strong wake interactions of multiple cylinders [1–6, 8–16]. Due to its enormous importance to engineering applications and because of complex flow patterns, many researchers have studied both experimentally and numerically flow across an array of cylinders in different configurations. Some of such studies are briefly discussed below.

Flow over multiple cylinders has gained significant attention and numerous experimental and numerical studies have been done for inline, side by side, and staggered configurations. Detailed review on flow over circular cylinders has been by done by Zdravkovich [1] and Sumner [2]. Experimental studies on flow over circular cylinders in pair can be seen in the works of Igarashi [3], Bearman and Wadcock [8], and Xu and Zhou [9]. Igarashi [3] analysed flow dynamics in the wake of two circular cylinders in inline configuration for the Reynolds number in the subcritical range at different spacing ratio. Six different flow patterns were identified and concluded that distance between cylinders has a strong dependence on flow patterns. Bearman and Wadcock [8] analyzed the proximity interference of the side by side placed circular cylinder at Re = 25,000. Flow patterns similar for the case of the single cylinder were observed when the spacing between the cylinders is small and when the spacing is greater than one diameter, the cylinder shed vortices separately. Xu and Zhou [9] studied the effect of changing the gap ratio and Re on Strouhal number for the case flow over inline circular cylinders. Based on St–Re relation and gap ratios, the flow patterns were classified into different regimes.

Numerical study on the dynamics of flow over multiple circular cylinders can be found in the works of Li et al., Meneghini and Satara [4], Mittal and Prasanth [11] and Liu et al. [12]. Li et al. observed that when gap ratio is less than 3, the vortex shedding is from downstream cylinder only, but shedding occurs from both cylinders for higher gap ratios. A 2-D FEM study is done by Meneghini and Satara [4] to understand the flow interference of two cylinders of circular cross section placed in side by side and inline configurations at Re = 200. The value of average drag coefficient changes from negative value to positive value for the downstream cylinder, when the gap ratio becomes greater than three times the diameter of the cylinder. Mittal and Prasanth [11] analyzed the flow over two equally sized circular cylinders arranged in staggered as well as inline arrangement at Re 100. For spacing ratio of $5.5D$ for which the analysis is carried out, the authors observed that the cylinder placed upstream behaves characteristics similar to the case of flow over a single cylinder qualitatively.

Studies on flow over multiple cylinders having square cross section of different arrangements have got good attention by researchers [5–7, 13]. An experimental study on flow over two cylinders having square cross section arranged in side by side manner with transverse spacing ratio from 1.02 to 6 at Re = 47,000 is done by Alam et al. [5]. Four distinct flow regimes were observed and the physical aspects in each regimes are investigated in detail. In the experimental work of flow over inline square cylinders, Yen et al. [13] observed three different flow field characteristic modes, single, reattached, and binary modes, respectively. The Re and gap ratio effects over the flow patterns at the downstream of cylinders were also investigated by the authors. Flows over multiple inline square cylinders are numerically studied by Sarwar et al. [7]. The study is conducted for two, three, and four cylinders with different spacing ratios. For all cases depending on Re, three different states of flow were observed, steady state, transitional state, and unsteady state. For each state, the flow dynamics and corresponding aerodynamic coefficients are also different.

In comparison with that of circular and square cylinders, literatures on flow past multiple triangular cylinders are very less investigated [14–16]. The interaction of two freely rotating triangular cylinders arranged inline at Re = 200 is numerically investigated by Wang et al. [14]. The effect of varying the spacing between the cylinders on the dynamic behavior of the cylinders and vortical structure of the flow were thoroughly investigated. Flow over two inline triangular cylinders which oscillate transversely in a uniform flow at Re = 100 is numerically investigated by Zhai et al. [15]. The study targets on two particular spacing between the cylinders corresponding to vortex formation regime and vortex suppression regime, respectively. The deposition and dispersion of aerosols of varying diameters over two triangular cylinders in inline configuration are conducted by Ghafouri et al. [16].

In the present study, flow over a pair of equilateral triangular cylinders of same dimension was analyzed for different gap ratio and Reynolds number (Re). The Strouhal number, St, is defined as non-dimensional shedding frequency ($St = fD/U\infty$). The variation in the St due to placing a cylinder in the downstream is thoroughly studied.

2 Numerical Method

The non-dimensional incompressible governing (continuity and Navier–Stokes) equations in the vector form are.

$$\nabla \cdot \mathbf{V} = 0 \tag{1}$$

$$\frac{\partial \mathbf{V}}{\partial t} + (\mathbf{V} \cdot \nabla)\mathbf{V} = -\nabla p + \frac{1}{\mathrm{Re}} \nabla^2 \mathbf{V} \tag{2}$$

The governing Eqs. (1) and (2) are discretized using finite volume method [17–21]. The present study covers the transient flow over equilateral triangular cylinders in inline arrangement and the PISO algorithm is employed to convert PDE to a system of algebraic equations, are solved simultaneously. We made use of the open-source code OpenFOAM [22] for this purpose adopting its well-known transient solver pisoFoam.

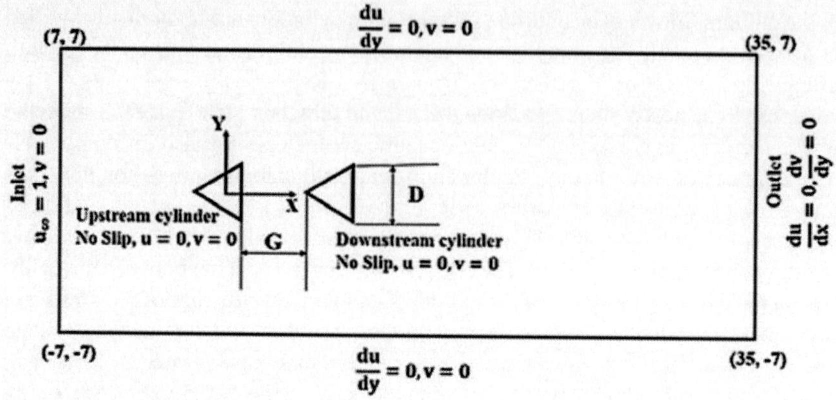

Fig. 1 Computational domain

2.1 Computational Domain

A schematic diagram representing the computational domain used for the present study is shown in Fig. 1. The boundary conditions used in the present study are indicated in Fig. 1. Two equilateral triangles of same side length (D) are used and the distance from the base of the upstream cylinder and to the apex of the down-stream cylinder as shown in Fig. 1 is taken as 'G'. The centroid of upstream cylinder is taken as origin.

2.2 Grid Independence Test

A grid independence test is conducted to find an optimum grid for the analysis of flow past two triangular cylinders in inline arrangement as shown in Fig. 1. Computational domain along with the boundary conditions and other notations is shown in Fig. 1. The simulations are carried out for Re $=$ 100 and $G/D = 1.0$. Five different cases consist of different number of quadrilateral cells were analyzed. The obtained results are tabulated in Table 1. It is found from this grid independence test that for $N =$

Table 1 Grid independency test

Number of cells (N)	St
2144	0.1220
8150	0.1300
32,600	**0.1330**
73,350	0.1333
130,400	0.1340

Table 2 Validation of the OpenFOAM code for flow over the circular cylinder at Re = 100

Literature	St	Reference
Williamson (1996)	0.164	[17]
Le et al. (2006)	0.160	[18]
Russell and Wang (2003)	0.169	[19]
Mittal and Raghuvanshi (2001)	0.168	[20]
Present study	**0.165**	

32,600 and above, the St is not changing considerably and hence we selected N = 32,600 (shown in bold letters in Table 1) as the optimum value for the rest of the numerical predictions.

2.3 *Validation*

The OpenFOAM code is validated for the case of flow over single circular cylinder at Re = 100, and the St is calculated and compared with the literature in Table 2 (Values in bold letters shows the present study value). A good agreement of the calculated value of St can been seen in the literature with a maximum deviation of less than 3.5%.

3 Results and Discussions

The numerical analysis is carried for the flow over two inline equilateral triangular cylinders with apex facing the flow as shown in Fig. 1. The numerical investigations are performed for $75 \leq \text{Re} \leq 200$ and $0 \leq G/D \leq 2.5$ for the optimum grid obtained after grid independent test using the pisoFoam solver. The code is validated for the case of flow over single circular cylinder at Re = 100 [17–20]. The effect of Re and G/D ratio on the vortex shedding frequency of the upstream cylinder when a secondary cylinder is introduced in the wake is discussed in the preceding section.

Figure 2 represents the variation of the non-dimensional shedding frequency, St with Re for different values of G/D ratios. In Fig. 2, we also made an attempt to compare the plots with a single cylinder configuration. It is already established that as Re increases, St increases for flow over triangular cylinder [21]. In the present study, we observed that the same trend can be extended to flow over apex facing multiple triangular cylinder as well.

The temporal variation of the coefficient of lift C_l of the upstream cylinder is shown in Fig. 3 for various G/D ratios at Re = 100 for the same range of time. It is clear from the plots that the upstream cylinder experiences an oscillatory motion due to periodic oscillations generated, all having different amplitudes and frequencies of shedding. Figure 4 represents the power spectrum density of the coefficient of lift

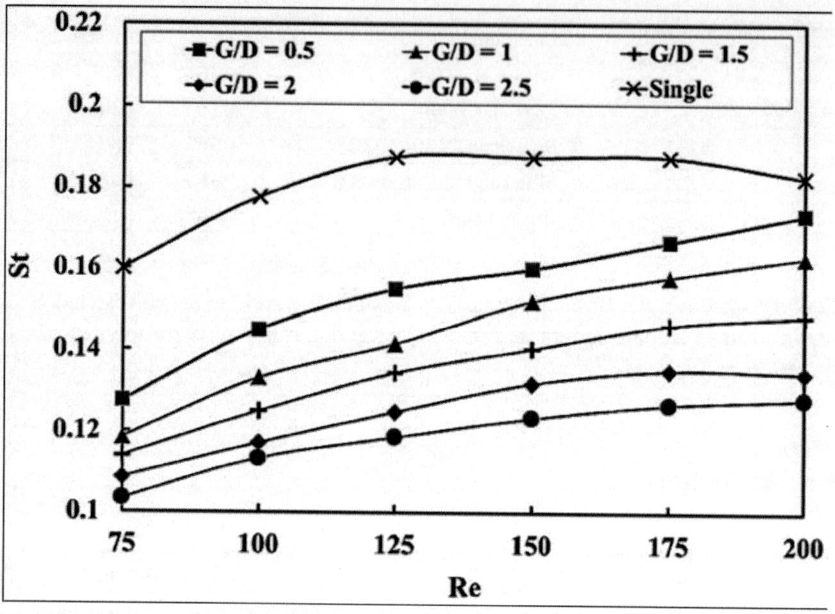

Fig. 2 Variation of St with Re for different *G*/*D* ratios

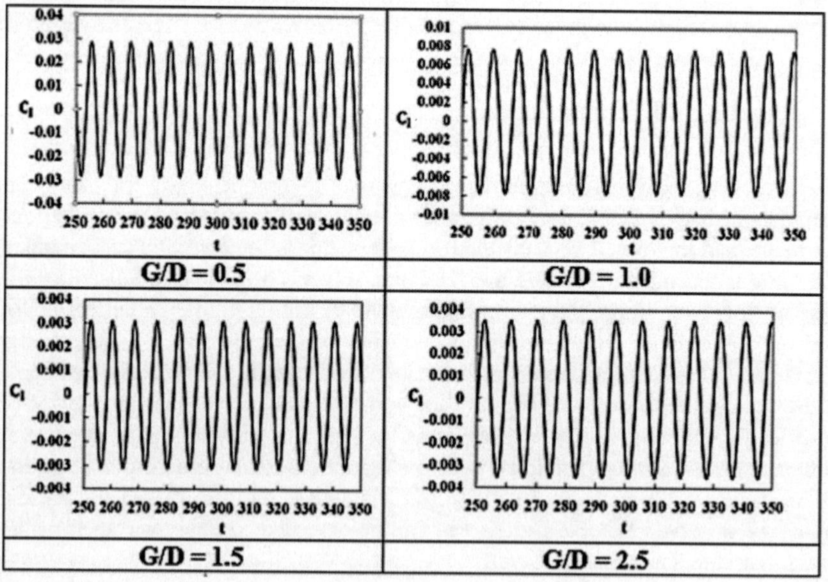

Fig. 3 C_l versus time plot for different *G*/*D* ratio

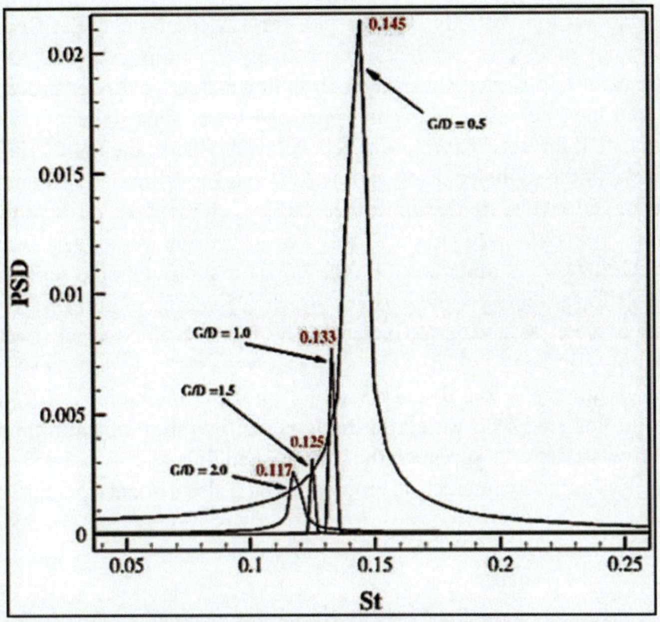

Fig. 4 PSD of lift coefficient at Re = 100 for different *G*/*D* ratios

(C_l) oscillations which provides the information about vortex shedding frequency (f). The St of different gap ratios is indicated in the plot itself. It can be seen in Fig. 2 that the St decreases with gap ratio, *G*/*D*. An attempt is made to establish the reason for this.

Figure 5 shows the variation of St with *G*/*D* ratio for different values of Re. As discussed, a decreasing trend of St is observed as *G*/*D* increases for a given Re. We have used the vorticity plots and velocity profiles in the next paragraph to substantiate the results obtained.

Fig. 5 Variation of St with *G*/*D* ratios for different Re

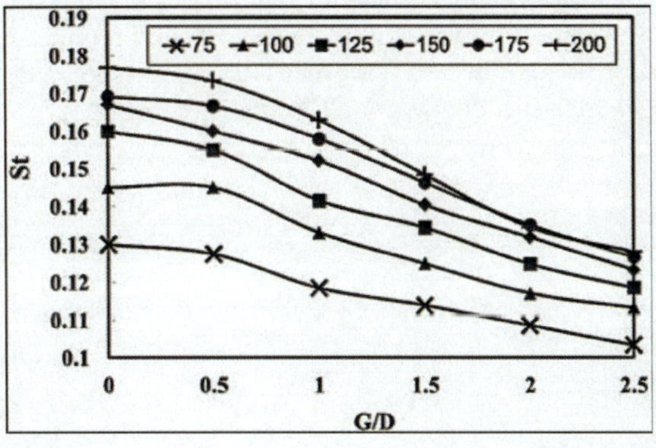

Figure 6 shows the vorticity contour plot for a shedding cycle at Re = 100 at three *G*/*D* ratios of 0.5, 1.5, and 2.5, respectively. It is observed that as the *G*/*D* ratio increases, the separated shear layer from downstream cylinder becomes more stabilized and the interaction between upper and lower shear layers is suppressed and consequently, the vortex formation gets delayed. Hence, the number of vortices shedding behind the cylinder decreases as *G*/*D* ratio is increased. Figure 7 shows the time averaged *x*-velocity distribution at $x/D = 7$ in the wake of the downstream cylinder for three *G*/*D* ratio 0.5, 1.5, and 2.5. The mean velocity is found to be maximum for $G/D = 0.5$ and minimum for 2.5. This shows that the separated shear layer acceleration is more for the case of $G/D = 0.5$, which in turn results in faster detachment of vortices and is attributed to the increase in St when *G*/*D* ratio is decreased.

Strouhal frequency is found suppressed for all the combinations of Re and *G*/*D* ratios used in this study. So we claim that this configuration of multiple cylinders is an effective strategy to suppress the vortex shedding and associated structural damages in fluid structure interaction problems which also evident from the literatures [1–3, 10].

4 Conclusions

In the present work, flow past two equilateral triangular cylinders in inline arrangement for $75 \geq \text{Re} \geq 200$ and $0 \geq G/D \geq 2.5$ is conducted. The effect of placing a cylinder in the downstream on the St of the upstream cylinder is studied numerically. A PISO algorithm is employed for numerical solution of the 2D incompressible laminar flow around the equilateral triangular cylinders. For flow over single triangular cylinder, St increases with Re, similar trend is also observed for flow past multiple triangular cylinders. The Strouhal number for the two cylinder configuration is found to be less than that of the flow over single cylinder cases, for all the Re and *G*/*D* ratios considered in this study. Hence, placing a cylinder in the wake of the triangular cylinder can be treated as an effective vortex shedding suppression mechanism and has got wide applications in fluid–structure interaction problems. It is evident from vorticity plots and velocity profiles that the shear layer interaction decreases and decelerates, when *G*/*D* ratio is increased and is attributed to suppression of vortex shedding when *G*/*D* ratio is increased.

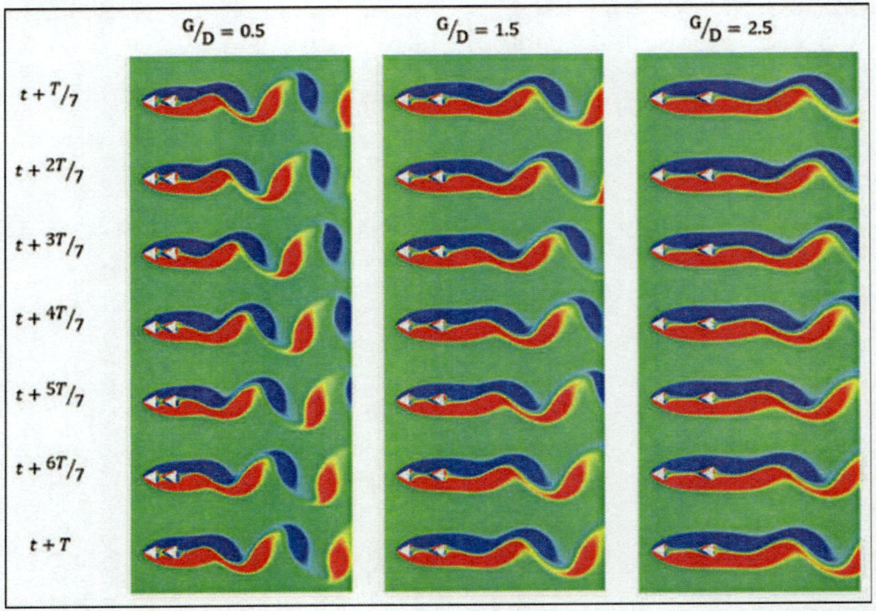

Fig. 6 Vorticity contours for one shedding cycle at different *G*/*D* ratio and Re = 100

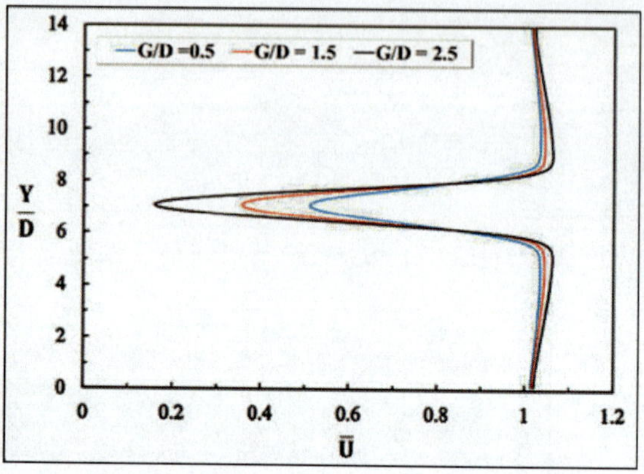

Fig. 7 Variation of time averaged *x*-velocity at $x/D = 7$ in the wake of the downstream cylinder

References

1. Zdravkovich MM (1977) Review of flow interference between two circular cylinders in various arrangements. J Fluids Eng (ASME) 99:618–633
2. Sumner D (2010) Two circular cylinders in cross-flow: a review. J Fluids Struct 26:849–899
3. Igarashi T (1981) Characteristics of the flow around two circular cylinders arranged in tandem (1st report). Bull JSME 24:323–331
4. Meneghini R, Satara F (2001) Numerical simulation of flow interference between two circular cylinders in tandem and side-by-side arrangements. J Fluids Struct 15:327–350
5. Alam MM, Zhou Y, Wang XW (2011) The wake of two side-by-side square cylinders. J Fluid Mech 669:432–471
6. Kim MK, Kim DK, Yoon SH, Lee DHJ (2008) Measurements of the flow fields around two square cylinders in a tandem arrangement. J Mech Sci Technol 22:397–407
7. Sarwar WA, Islam SUL, Faiz L, Rahman H (2018) Numerical investigation of transitions in flow states and variation in aerodynamic forces for flow around square cylinders arranged inline. Chin J Aeronaut 31:2111–2123
8. Bearman PW, Wadcock AJ (1973) The interaction between a pair of circular cylinders normal to a stream. J Fluid Mech 61:499–511
9. Xu G, Zhou Y (2004) Strouhal numbers in the wake of two inline cylinders. Exp Fluids 37:248–256
10. Li J, Chambarel A, Donneaud M, Martin R (1991) Numerical study of laminar flow past one and two circular cylinders. Comput Fluids 19:155–170
11. Mittal S, Prasanth TK (2009) Vortex-induced vibration of two circular cylinders at low Reynolds number. J Fluids Struct 25:731–741
12. Liu M-M, Lu L, Teng B, Zhao M, Tang G-Q (2014) Re-examination of laminar flow over twin circular cylinders in tandem arrangement. Fluid Dyn Res 46
13. Yen SC, San KC, Chuang TH (2008) Interactions of tandem square cylinders at low Reynolds numbers. Exp Therm Fluid Sci 32:927–938
14. Wang S, Zhu L, Xing Z, He GW (2011) Flow past two freely rotatable triangular cylinders in tandem arrangement. J Fluids Eng 133:081202
15. Zhai Q, Wang HK, Gong LY (2017) Numerical study of flow past two transversely oscillating triangular cylinders in tandem at low Reynolds number. J Appl Fluid Mech 10:1247–1260
16. Ghafouri S, Alizadeh M, Seyyedi SM, Hassanzadeh AH, Ganji DD (2015) Deposition and dispersion of aerosols over triangular cylinders in a two-dimensional channel; effect of cylinder location and arrangement. J Mol Liq 206:228–238

17. Williamson CHK (1996) Vortex dynamics in the cylinder wake. Annu Rev Fluid Mech 28:477–539
18. Le DV, Khoo BC, Peraire J (2006) An immersed interface method for viscous incompressible flows involving rigid and flexible boundaries. J Comput Phys 220:109–138
19. Russell D, Wang JZ (2003) A cartesian grid method for modeling multiple moving objects in 2D incompressible viscous flow. J Comput Phys 191:177–205
20. Mittal S, Raghuvanshi A (2001) Control of vortex shedding behind circular cylinder for flows at low Reynolds numbers. Int J Numer Methods Fluids 35:421–447
21. De AK, Dalal A (2006) Numerical simulation of unconfined flow past a triangular cylinder. Int J Numer Methods Fluids 52(7):801–821
22. OpenFoam homepage. https://openfoam.org/

Fluid–Structure Interaction Modelling of Physiological Loading-Induced Canalicular Fluid Motion in Osteocyte Network

Rakesh Kumar, Abhishek Kumar Tiwari, Dharmendra Tripathi, Niti Nipun Sharma, and Milan Khadiya

Abstract The present study develops a fluid–structure Interaction (FSI) model to characterize the physiological loading-induced canalicular fluid motion in an osteocyte network within the bone. The effect of poromechanical properties on the canalicular fluid motion is also studied. The outcomes indicate that fluid motion in the network varies with gait events and fluid motion occurs from compressive to tensile strain environment. This work will provide a better understanding of osteocyte mechanical environment and biochemical communication regulating the bone's adaptation which is although tedious to be explored using experimental techniques.

Keywords Osteocyte · Canaliculi · Fluid motion · Fluid–structure interaction

1 Introduction

Mechanical loading is important to maintain the desired bone health [1, 2]. It is noticed that exogenous mechanical loading helps in the recovery of bone health when subjected to disorders such as osteoporosis [3], bone/muscle disuse, or osteogenesis Imperfecta [4] (OI). Bone response to loading-induced mechanical environment is typically coordinated by bone cells namely osteocytes [5]. Osteocytes sense the loading-induced mechanical environment and accordingly sends the biochemical signals to other bone cells, namely osteoblasts and osteoclasts to initiate bone addition

R. Kumar · N. N. Sharma · M. Khadiya
Department of Mechanical Engineering, Manipal University Jaipur, Jaipur, Rajasthan 303007, India

A. K. Tiwari (✉)
Department of Applied Mechanics, Motilal Nehru National Institute of Technology Allahabad, Prayagraj, Uttar Pradesh 211004, India
e-mail: aktiwari@mnnit.ac.in

D. Tripathi
Department of Mathematics, National Institute of Technology, Uttarakhand, Srinagar 246174, India

S. Revankar et al. (eds.), *Proceedings of International Conference on Thermofluids*, Lecture Notes in Mechanical Engineering,
https://doi.org/10.1007/978-981-15-7831-1_3

or removal process so that a certain level of bone health is maintained [6, 7]. Osteocytes embedded in the lacunar space form a sensory and communication network within the bone. The osteocyte cell processes passing through fluid filled canalicular spaces connect the remote osteocytes and form the network. Analytical studies have shown that bone matrix surrounding the osteocyte lacuna deforms under physiological loading. This generates pressure difference which allows the interstitial fluid to flow in the canalicular spaces [8, 9]. Fluid motion across the canalicular network acts as a medium of communication for neighboring cells. Fluid flow develops shear stress which deforms osteocyte cell body. In vitro studies have repeatedly reported that fluid shear is sensed by the osteocytes and transduce this mechanical signal into biochemical signals required for bone remodelling activities [10, 11]. It is noticed that bone experiences dynamic physiological loading during daily activities such as physical exercise and walking [12]. Physiological loading maintains a certain level of fluid flow within the lacunar space to regulate the osteo-activities. The characterization of loading-induced canalicular fluid motion in the osteocyte network is a challenge. There are experimental studies in the literature where fluorescent labelling has been used to visualize the fluid environment of osteocyte cells [13]. However, the experimental quantification of fluid environment surrounding the osteocyte is cumbersome. As bone structure deformation interacts and drives the canalicular fluid flow, multiphysics analysis may be useful to characterize the fluid flow behavior across the canalicular network. There are studies which have used mathematical and computational approaches to study the behavior of fluid motion within the bone [14, 15]. Biot's theory of poroelasticity [16] is usually applied to compute the loading-derived pressure gradient and fluid motion in lacunar-canalicular spaces. These studies presented idealized model of osteocyte canaliculi to compute fluid flow and fluid shear within the bone; however, fluid flows across the canalicular network and this aspect has not been incorporated. In recent time, a multiphysics modelling technique, namely fluid–structure interaction (FSI) modelling has been useful in characterizing the loading-induced fluid environment of osteocyte. This analysis has presented important findings in the literature. For example, FSI model explained that osteocyte cell body experiences hydrodynamic pressure, whereas the cell process is subjected to shear stress. Similarly, McNamara et al. [17] explained that osteocyte projection amplifies the strain environment of osteocytes. Geometry of pericellular space also affect the fluid velocity. Nevertheless, there is hardly any FSI model which explained how canalicular fluid distributes in the canalicular network as a result loading. This gap is the objective for the presented article. Accordingly, this study attempts to characterize the canalicular fluid motion pattern with the osteocyte cell network distributed in the osteon. FSI model in combination with poroelasticity is developed to study the fluid motion. The present work also studies the effect of loading pattern as well as the effect of poromechanical properties such as porosity and permeability. To serve this purpose, fluid motion is studied in a tissue subjected to brittle bone disease, namely osteogenesis imperfecta which has a different loading pattern [18] and poromechanical properties [19]. The canalicular fluid motion in the osteocyte network of an osteon subjected to physiological loading is computed. Fluid motion at different fraction of gait cycle is studied and a comparison is made between the fluid flow pattern in healthy and OI bones. Ultimately, this work will provide a better understanding on canalicular fluid motion and biomechanical regulation of bone remodeling which is indeed required to design and develop the biomechanical strategies to improve the bone health.

2 Material and Methods

2.1 The Osteon and Osteocyte Network Model

A three-dimensional model of osteon [20] is developed as shown in Fig. 1a. Osteon is idealized as a hollow cylinder with Harvesian canal [21]. Osteocytes are embedded and distributed within the osteon as shown in Fig. 1a. Each osteocyte is connected to other osteocytes circumferentially and radially with cylindrical canaliculi of in the osteonal cross-section, and longitudinally along the osteon length. Nevertheless, canaliculi or cell processes are randomly oriented within the osteon however the three

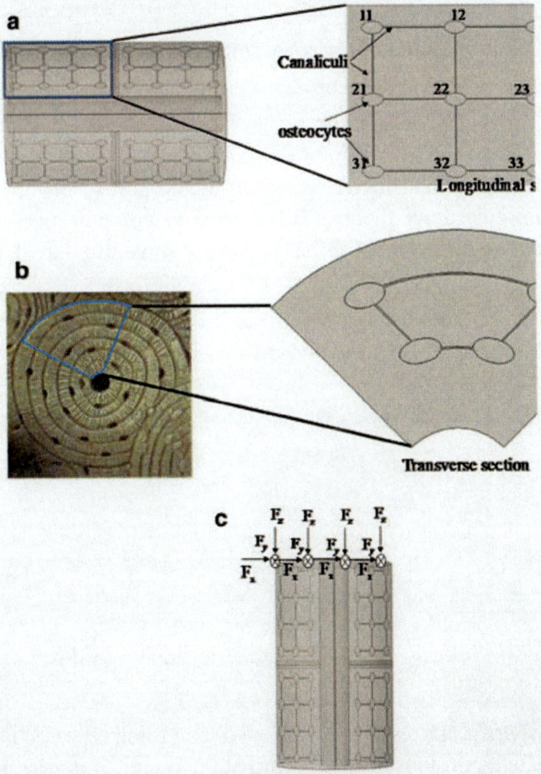

Fig. 1 Three-dimensional model of osteon **a** longitudinal section, **b** transverse section and **c** loading configuration of osteon

directions mentioned above represent the idealized directions along which the canaliculi are typically oriented and fluid flow may occur. Osteocytes are modelled in form of ellipse with a major axis and minor axis. These osteocytes are connected to each other through a micro-channel. Osteon is assumed as a poroelastic and transversely elastic material. Poromechanical properties are assigned to the osteon in accordance with healthy and OI bones as mentioned in the literature. Gait loading specifically ground reaction force data available for healthy [22] and OI [18] bone reported in the literature are used to define the loading on osteon (Fig. 2). After applying the boundary condition, we use default physics-controlled mesh for meshing the osteon model. The mesh grid setting is set normal in which tetrahedral element is used. A fluid–structure interaction module available in COMSOL Multiphysics is used to simulate the interface between the solid osteon deformation and canalicular fluid motion. COMSOL uses a time-dependent solver within the geometry to modify the boundary of solid domain for a few time iteration and repeats N times until solution converges. Fluid motion under the effect of applied loading pattern is studied (Fig. 2). A region of interest is chosen in the osteon cross section and fluid motion is investigated between the two osteocytes reservoirs. The models are meshed before performing the computation.

2.2 *Governing Equations*

2.2.1 Poroelasticity

Osteon is considered as a deformable porous material with interstitial fluid phase. Poroelasticity theory [16] is used to compute pressure gradients. The governing equations used in COMSOL during simulation are defined as follow [23]:

$$-\nabla \cdot \sigma = F_{\mathrm{V}},\ \sigma = s \tag{1}$$

$$s - s_{\mathrm{o}} = C : (\varepsilon - \varepsilon_{\mathrm{o}} - \varepsilon_{\mathrm{inl}}) - \alpha p I \tag{2}$$

$$\varepsilon = \frac{1}{2}\left((\nabla\delta)^{\mathrm{T}} + \nabla\delta\right) \tag{3}$$

$$\rho S \frac{\partial p}{\partial t} + \nabla \cdot (\rho u) = Q - \rho\alpha \frac{\partial \varepsilon_{\mathrm{vol}}}{\partial t} \tag{4}$$

$$u = -\frac{k}{\mu}\nabla p \tag{5}$$

where σ is stress tensor, F_{V} is volume force vector, ρ is density, u is Darcy's velocity, ε is strain tensor, δ is deformation vector, p is pressure, α is Biot's constant, k is

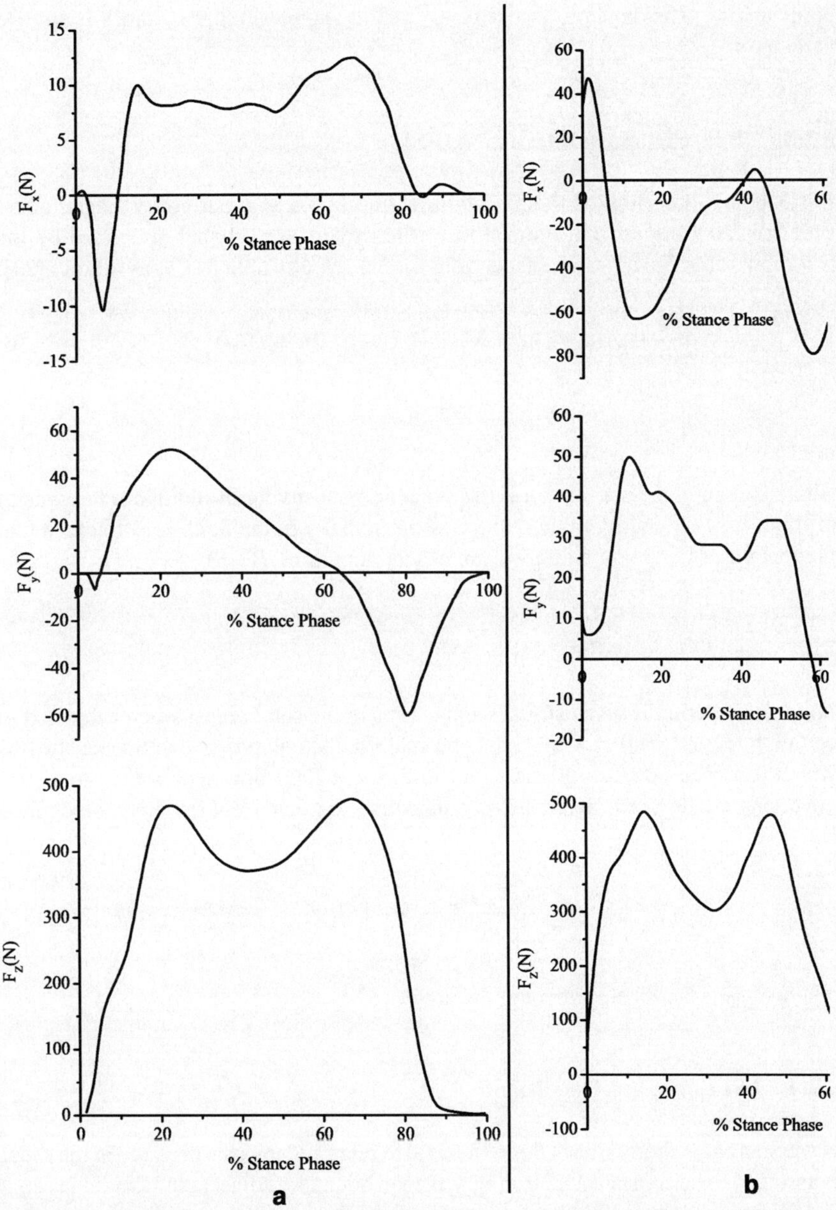

Fig. 2 Forces in *x*, *y* and *z* directions from gait analysis recorded for **a** healthy normal bone [22] and **b** OI patient bone [18]

permeability, μ is dynamic viscosity, Q is fluid source/sink-term, and S is storage parameter.

2.2.2 Fluid–Structure Interaction (FSI)

FSI couple's fluid flow with solid deformation by using an arbitrary Lagrangian–Eulerian (ALE) approach. Fluid flow in the canalicular channel is defined by the incompressible Navier–Stokes equations for the velocity field,u (u,v) as follows [23]:

$$\rho\frac{\partial u}{\partial t} + \rho(u \cdot \nabla)u = \nabla \cdot \left[-pI + \mu(\nabla u + (\nabla u)^{\mathrm{T}})\right] + F \tag{6}$$

$$\rho\nabla \cdot u = 0 \tag{7}$$

where p denotes the pressure, μ is the dynamic viscosity for the fluid, ρ is the density of fluid, I is the identity matrix, and u is the velocity of the fluid. It is assumed that gravitation or other volume forces are absent, therefore, $F = 0$.

2.2.3 Solid Deformation

Solid deformation induces stresses and strains in the solid object when subjected to certain forces or loading. Object may be subjected to compressive or tensile stresses which form the pressure gradient and initiate the fluid flow within the bone. The governing equation for deformation connecting the fluid flow force can be defined as:

$$\rho\frac{\partial^2 \delta_{\mathrm{solid}}}{\partial t^2} - \nabla \cdot \sigma = F_{\mathrm{v}} \tag{8}$$

where δ_{solid} is the displacement of the solid.

2.2.4 FSI Boundary Conditions

Haversian canal dimensions allows the fluid to relax and thus the pressure in the canal is assumed to be constant. Accordingly, osteon boundary adjacent to Harvesian canal is assumed impermeable. No macroscopic fluid flow is allowed across this surface. In addition to this, mainly two boundary conditions are applied:

1. Fluid velocities are defined in term of the solid boundary deformation rate.

$$u_{\text{fluid}} = \dot{\delta}_{\text{solid}}$$
$$\dot{\delta}_{\text{solid}} = \frac{\partial \delta_{\text{solid}}}{\partial t} \tag{9}$$

2. Forces on solid boundary and fluid velocity can be related using the following relation:

$$F_{\text{T}} = \left[-pI + \mu(\nabla u + (\nabla u)^{\text{T}})\right] \tag{10}$$

The model of osteon with osteocyte network is coupled with poroelastic and FSI governing equations to compute the loading-induced fluid flow. Fluid motion is finally studied in the canalicular network in longitudinal and transverse planes of the osteon at 20, 40, 60, and 80% of stance phase of the gait cycle. Osteocytes positions are also numbered to discuss the results obtained in the network as shown in Fig. 1.

3 Result and Discussion

FSI analysis explains the fluid flow pattern around the lacunar-canalicular architecture. Fluid flow patterns are studied at 20, 40, 60, and 80% of the stance phase. The stance phase includes initial contact, loading response, mid stance, terminal, and pre-swing phase. Fluid flow around the osteocyte lacunae and in the canaliculi are shown in Figs. 3 and 4. It can be observed that fluid confines within the lacuna and acts as the reservoir. Figure 3 explains the fluid motion in the canaliculi connecting osteocyte locations 11–21, 12–22, 13–23, and 14–24 which are oriented in the transverse direction of the osteon. Figure 4 explains the fluid motion in the canaliculi connecting osteocyte locations 11–12, 21–22, 31–32, and 41–42 which are oriented in longitudinal direction of the osteon. Fluid flow varies with loading cycle. In healthy bone, fluid motion is high at 20 and 60% of the gait cycle, whereas it is relatively low at 40 and 80% of the cycle (Fig. 5). In OI bone which is brittle and porous, fluid motion does not vary much with loading cycle. Fluid motion is also high in healthy bone as compared to OI bone. This indicates that an increase in the porosity level of solid bone matrix may reduce the pressure gradient and hence the fluid velocity within the canaliculi. Fluid motion is relatively high in the canaliculi oriented in the longitudinal direction adjacent to the osteon boundary where the loading is applied, whereas it decreases near the osteon boundary where displacement is constrained (Fig. 4). Fluid flow is high near the lacunae adjacent to outer permeable surface of the osteon, e.g., 11, 12, 13, and 14 which decreases while moving towards the lacunae near the Harversian canals, e.g., 21, 22, 23, and 24 (Figs. 5 and 6). This is also due to the fact the outer wall of osteon is defined more permeable as compared to the inner wall and hence the fluid motion in the lacunae near to this wall is high. It indicates that poromechanical characteristics, e.g., permeability and porosity play important

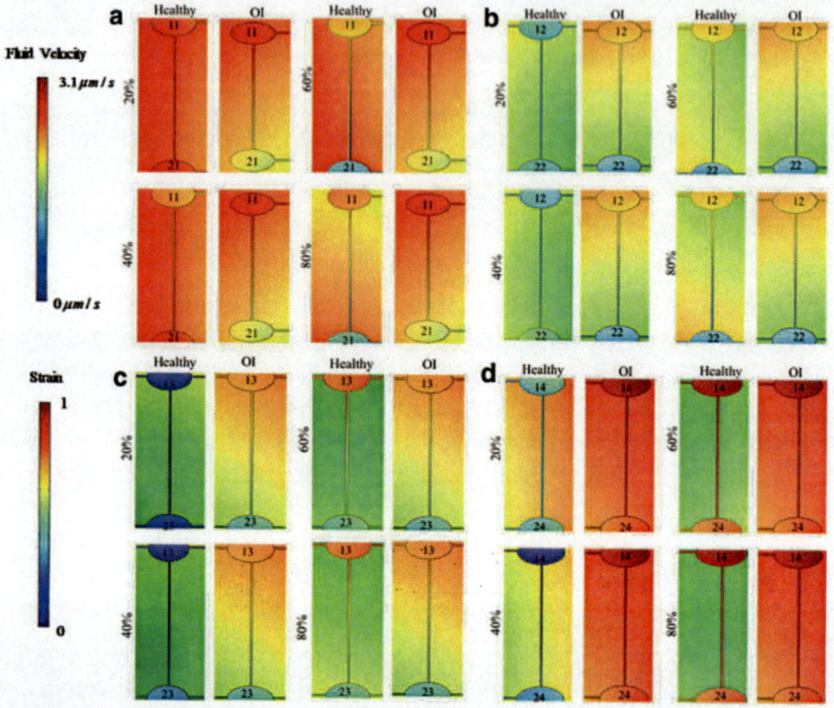

Fig. 3 Fluid motion (μm/s) in the canaliculi connecting osteocyte locations **a** 11–21, **b** 12–22, **c** 13–23, and **d** 14–24 at 20, 40, 60, and 80% of the gait cycle

role in deciding the desired mechanical stimuli, e.g., fluid motion for osteogenic or remodelling activities.

In addition to this, fluid velocity profile has also been plotted between the canalicular walls (Figs. 5 and 6). Fluid motion is laminar and maximum near the center of the canalicular walls. In addition to this, fluid velocity profile is parabolic across the canalicular walls. Fluid velocity is zero at the wall and satisfies no slip condition (Figs. 5 and 6). Fluid velocity profiles also vary with gait cycle. Nevertheless, fluid velocity profiles do not change much with gait cycle in the case of porous and brittle OI bone (Figs. 5 and 6).

Fluid motion has also been studied in a sector of osteon's circular cross section as shown in Fig. 7. Fluid velocity radially increases and is high near the outer wall of the osteon as outer wall of the osteon is considered permeable (Fig. 7). Thus, aforesaid findings strongly depend on the applied boundary conditions.

The work presented here does not considers the variation in permeability of osteon along its wall while defining the boundary conditions. This model also does not consider vascular porosities. Nevertheless, these limitations will be addressed with a more robust model in the future work. It is however important to note that fluid motion does not change during normal physiological loading in OI bone [24]. Therefore,

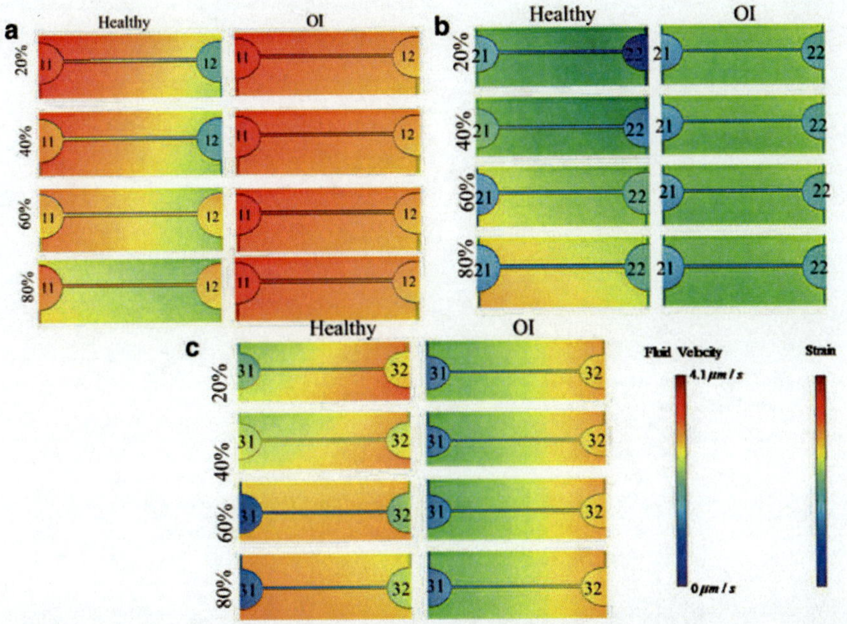

Fig. 4 Fluid motion (μm/s) in the canaliculi connecting osteocyte locations **a** 11–12, **b** 21–22, **c** 31–32 at 20, 40, 60, and 80% of the gait cycle

bone cell may experience a static mechanical environment, e.g., constant fluid shear. In such case, bone cell will not relax, and it may adversely affect the mechano-sensitivity. This may will ultimately lead to delayed remodeling activities.

4 Conclusion

The study concludes that effect of ground reaction forces on bone are maximum at 20 and 80% of stance phase of gait cycle in normal bone case and at 20, 40, and 60% of stance phase in diseased OI bone case. This is mainly because ground reaction forces are higher in gait cycle near these points. It was found that volumetric displacement is much higher in diseased OI bone than normal bone as it is very brittle.

All the above findings show that canalicular fluid is driven by loading-induced pressure. Canalicular fluid motion significantly vary with gait cycle. Thus, there is scope to enhance the fluid motion within the bone using well designed biomechanical therapies such as exercises. This will ultimately increase the osteogenic activities within the bone and may be useful in improving the bone health. In contrast, a different strategy may be required to enhance the fluid motion within a bone with disorder, e.g., osteoporosis or OI.

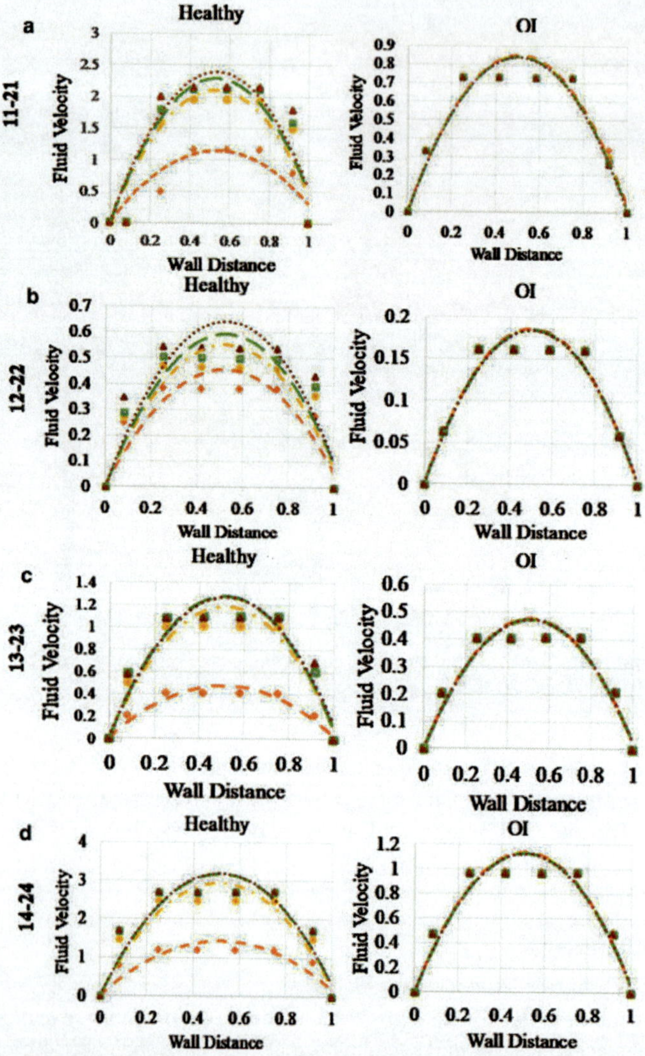

Fig. 5 Velocity profiles (μm/s) in the canaliculi connecting the osteocyte locations **a** 11–21, **b** 12–22, **c** 13–23, and **d** 14–24 for 20, 40, 60, and 80% of the gait cycle

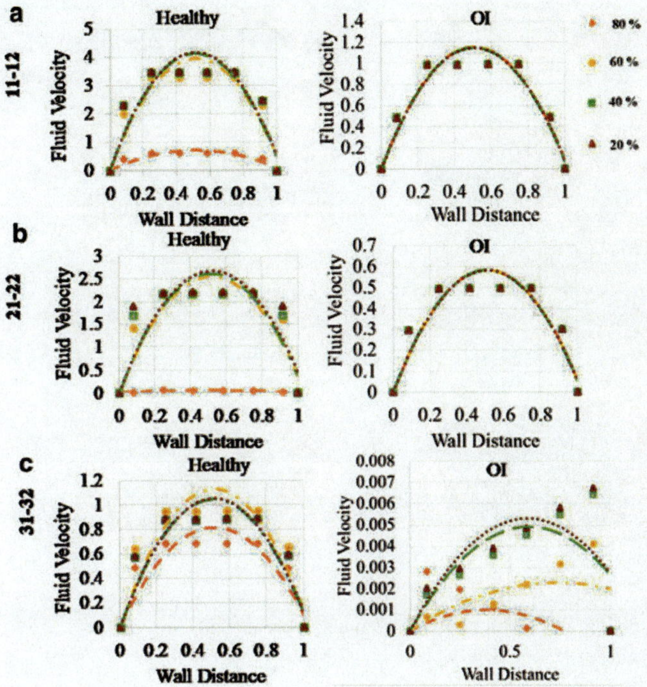

Fig. 6 Velocity profiles in the canaliculi connecting the osteocyte locations **a** 11–12, **b** 21–22, **c** 31–32 for 20, 40, 60, and 80% of the gait cycle

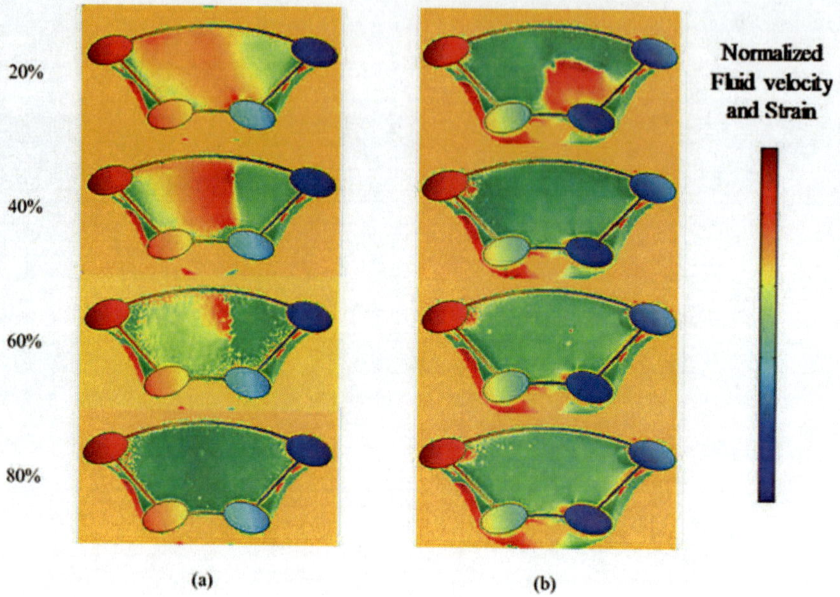

Fig. 7 Fluid motion in a sector of osteon's circular cross section

References

1. Shackelford LC (2004) Resistance exercise as a countermeasure to disuse-induced bone loss. J Appl Physiol 97:119–129. https://doi.org/10.1152/japplphysiol.00741.2003
2. Rubin C, Recker R, Cullen D, Ryaby J, McCabe J, McLeod K (2003) Prevention of post-menopausal bone loss by a low-magnitude, high-frequency mechanical stimuli: a clinical trial assessing compliance, efficacy, and safety. J Bone Miner Res 19:343–351. https://doi.org/10.1359/JBMR.0301251
3. Lau RY, Guo X (2011) A review on current osteoporosis research: with special focus on disuse bone loss. J Osteoporos 2011:1–6. https://doi.org/10.4061/2011/293808
4. Wu X-G, Chen WY (2013) A hollow osteon model for examining its poroelastic behaviors: mathematically modeling an osteon with different boundary cases. Eur J Mech A-Solids 40:34–49
5. Klein-Nulend J, Van der Plas A, Semeins C, Ajubi N, Frangos J, Nijweide P, Burger E (1995) Sensitivity of osteocytes to biomechanical stress in vitro. FASEB J 9:441–445
6. You J, Yellowley C, Donahue H, Zhang Y, Chen Q, Jacobs C (2000) Substrate deformation levels associated with routine physical activity are less stimulatory to bone cells relative to loading-induced oscillatory fluid flow. J Biomech Eng 122:387–393
7. Owan I, Burr DB, Turner CH, Qiu J, Tu Y, Onyia JE, Duncan RL (1997) Mechanotransduction in bone: osteoblasts are more responsive to fluid forces than mechanical strain. Am J Physiol-Cell Physiol 273:C810–C815
8. Weinbaum S, Cowin S, Zeng Y (1994) A model for the excitation of osteocytes by mechanical loading-induced bone fluid shear stresses. J Biomech 27:339–360
9. Han Y, Cowin SC, Schaffler MB, Weinbaum S (2004) Mechanotransduction and strain amplification in osteocyte cell processes. Proc Natl Acad Sci 101:16689–16694

10. Bacabac RG, Smit TH, Mullender MG, Dijcks SJ, Van Loon JJ, Klein-Nulend J (2004) Nitric oxide production by bone cells is fluid shear stress rate dependent. Biochem Biophys Res Commun 315:823–829
11. Klein-Nulend J, Bakker AD, Bacabac RG, Vatsa A, Weinbaum S (2013) Mechanosensation and transduction in osteocytes. Bone 54:182–190
12. Palombaro KM (2005) Effects of walking-only interventions on bone mineral density at various skeletal sites: a meta-analysis. J Geriatr Phys Ther 28:102–107
13. Price C, Zhou X, Li W, Wang L (2011) Real-time measurement of solute transport within the lacunar-canalicular system of mechanically loaded bone: direct evidence for load-induced fluid flow. J Bone Miner Res 26:277–285
14. Tiwari AK, Prasad J (2016) Computer modelling of bone's adaptation: the role of normal strain, shear strain and fluid flow. Biomech Model Mechanobiol 1–16. https://doi.org/10.1007/s10237-016-0824-z
15. Tiwari AK, Kumar R, Tripathi D, Badhyal S (2018) In silico modeling of bone adaptation to rest-inserted loading: strain energy density versus fluid flow as stimulus. J Theor Biol 446:110–127
16. Biot MA (1941) General theory of three-dimensional consolidation. J Appl Phys 12:155–164
17. Verbruggen SW, Vaughan TJ, McNamara LM (2014) Fluid flow in the osteocyte mechanical environment: a fluid–structure interaction approach. Biomech Model Mechanobiol 13:85–97. https://doi.org/10.1007/s10237-013-0487-y
18. Fritz JM, Guan Y, Wang M, Smith PA, Harris GF (2009) A fracture risk assessment model of the femur in children with osteogenesis imperfecta (OI) during gait. Med Eng Phys 31:1043–1048
19. Jameson JR (2014) Characterization of bone material properties and microstructure in osteogenesis imperfecta/brittle bone disease
20. Rho J-Y, Kuhn-Spearing L, Zioupos P (1998) Mechanical properties and the hierarchical structure of bone. Med Eng Phys 20:92–102
21. Greenwald AS, Boden SD, Goldberg VM, Khan Y, Laurencin CT, Rosier RN (2001) Bone-graft substitutes: facts, fictions, and applications. JBJS 83:98–103
22. Giakas G, Baltzopoulos V, Dangerfield PH, Dorgan JC, Dalmira S (1996) Comparison of gait patterns between healthy and scoliotic patients using time and frequency domain analysis of ground reaction forces. Spine 21:2235–2242
23. Holzbecher E (2013) Poroelasticity benchmarking for FEM on analytical solutions. In: Excerpt from the proceedings of the COMSOL conference, Rotterdam, pp 1–7
24. Shrivas NV, Tiwari AK, Kumar R, Tripathi D, Sharma VR (2018) Investigation on loading-induced fluid flow in osteogenesis imperfecta bone. In: ASME 2018 5th joint US-European fluids engineering division summer meeting. American society of mechanical engineers digital collection

Generation of Temperature Profile by Artificial Neural Network in Flow of Non-Newtonian Third Grade Fluid Through Two Parallel Plates

Vijay Kumar Mishra, Sumanta Chaudhuri, Jitendra K. Patel, and Arnab Sengupta

Abstract Generation of temperature profile of third grade fluid in flow through parallel plates maintained at uniform heat flux is reported. Exact solution of heat transfer problem with third grade fluid is very difficult due to high level of nonlinearity. Viscous dissipation considerations make it even more difficult. In the present problem, least square method (LSM) has been employed to solve the governing equations. The velocity and temperature profile computed by LSM is used to train the artificial neural network (ANN). Once ANN is trained, it is able to give temperature profile corresponding to any velocity profile fed into the ANN. This work demonstrates the scope of ANN use to solve these types of complicated and problem of practical use. Artificial neural network (ANN) is employed in the solving of the problem with scaled conjugate gradient (SCG) as training algorithm.

Keywords Third grade fluid · Artificial neural network · Scaled conjugate gradient · Parameter estimations

Nomenclature

A	Third grade fluid parameter
Ac	Cross-sectional area
$A_1, A_2, A_3, \dots A_n$	Kinematic tensor
a_0, a_2, a_4, a_6, a_8	Constants
Br	Brinkman number
$b_0, b_2, b_4, b_6, b_8, b_{10}, b_{12}$	Constants
C_P	Specific heat at constant pressure
$c_1, c_2, \dots$	Constants
c_i	ith constant

V. K. Mishra · S. Chaudhuri (✉) · J. K. Patel · A. Sengupta
School of Mechanical Engineering, Kalinga Institute of Industrial Technology (KIIT) Deemed to be University, Bhubaneswar 751024, India
e-mail: sumanta.chaudhurifme@kiit.ac.in

S. Revankar et al. (eds.), *Proceedings of International Conference on Thermofluids*, Lecture Notes in Mechanical Engineering,
https://doi.org/10.1007/978-981-15-7831-1_4

D	Differential operator
D/Dt	Material derivative
f	Body force per unit volume
g	Function
h	Half depth of channel
k_{th}	Thermal conductivity of the fluid
L	Length of the channel
l_1, l_2	Constants
N	Non-dimensional pressure gradient
Nu	Nusselt number
p^*	Dimensional pressure
q	Heat flux ratio
q_1, q_1	Heat fluxes at lower and upper walls
R	Residual
S	Sum of square of residual
T^*	Dimensional temperature
T^*_{m}	Bulk mean temperature
$T^*_{\mathrm{w_l}}$	Temperature of the lower wall
u	Non-dimensional velocity along axial direction
u_{N}	Non-dimensional velocity for Newtonian fluid
u^*	Dimensional velocity along axial direction
u_0	Average velocity
V^*	Velocity vector
$v, \tilde{v}$	Functions
x, y, z	Non-dimensional coordinate
x^*, y^*, z^*	Dimensional coordinates
w_i	*i*th weight function

Greek Symbols

α_1, α_2	Material constants
β	Constant
$\beta_1, \beta_2, \beta_3, \ldots$	Material constants
ρ	Density of the fluid
μ	Dynamic viscosity of the fluid
Θ	Non-dimensional temperature
Θ_{N}	Non-dimensional temperature for Newtonian fluid
Φ_i	Base function
τ	Stress

1 Introduction

Non-Newtonian fluids are used in various industries such as: paper and pulp, polymer processing, and food processing. There are different equipments like pump, heat exchanger, mixer, etc., which are used in these industries. For good performance of these devices, accurate analysis of fluid flow and heat transfer problem is desired. Governing equations involving non-Newtonian fluids are difficult to solve due to the fact that these fluids do not obey linear stress–strain relation, as is done by Newtonian fluids. Different types of non-Newtonian models are developed, depending upon the rheological behavior of the fluid. Some of the most commonly model are: power law fluid [1–3], Sisko fluid [4–7], third grade fluid [4], Oldroid fluid, Casson fluid, etc. Ability of considering shear thinning and thickening phenomena by third grade fluid model makes this model suitable for the present study [8–13]. The governing equations in the present study are solved by least square method (LSM), a semi-analytical method [14–18], widely employed by the researchers.

Various industrial problems involve optimization techniques for the search of better performance of any thermal device or for design considerations such as in inverse problems. Various optimization methods are commonly used for solving these types of problems, such as: genetic algorithm, pattern search algorithm, and simulated annealing. But the use of artificial neural network (ANN) to solve these problems is new and attracted various new researches [19–21]. Without requiring prior knowledge of the relationships of the parameters involved in the problem, ANN is able to solve various complex problems by learning from the examples through iterations. Capabilities of ANN to handle nonlinear relationships, uncertainties and noisy data, make it more versatile for complicated problems.

In the present work, a single and direct procedure for estimating the temperature profile from semi-analytically generated velocity profile in flow between two parallel plates using ANN is developed. The relevant governing equations are first of all solved by LSM, and then the velocity and the temperature profile are obtained. The velocity and the temperature profiles are then used to train the ANN. Once the ANN is trained, a velocity profile is fed into the ANN model, the ANN gives the corresponding temperature profile.

2 Formulation

Schematic of the present problem is shown in Fig. 1a. A third grade fluid flows between two parallel plates, subjected to uniform heat flux from both upper and lower plate's sides. The flow is assumed to be: hydro-dynamically and thermally fully developed, steady state, laminar, incompressible and fluid properties are constant. The height (H) and width (W) are small as compared to length (L). The flow is governed by the following equations:

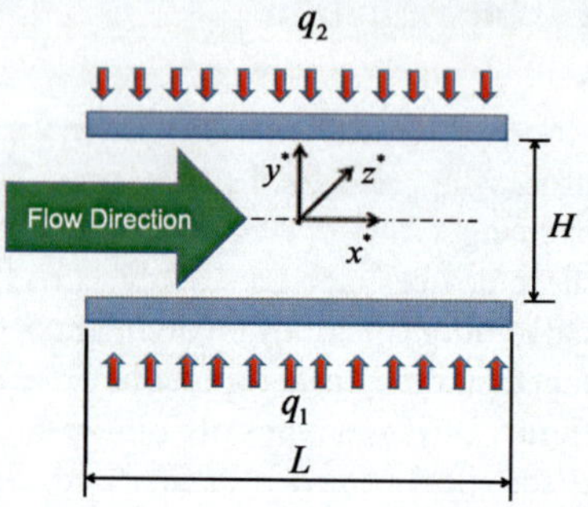

Fig. 1 Schematic of the present problem to be addressed

Continuity Equation

$$\nabla \cdot V^* = 0 \tag{1}$$

where V^* is the velocity vector.

Momentum Equation

$$\rho \frac{DV^*}{Dt} = \nabla \cdot \tau + f \tag{2}$$

Body forces are assumed to be negligible.

In the momentum equation, the stress tensor varies linearly with the strain rate tensor for a Newtonian fluid. But in case of non-Newtonian fluids, the stress tensor behaves nonlinearly with the change in strain rate tensor. For a third grade fluid, the shear stress and shear strain rate equation is given by Eq. (3), and constants used can be expressed by Eqs. (4) and (5).

$$\tau = -pI + \mu A_1 + \alpha_1 A_2 + \alpha_2 A_1^2 + \beta_1 A_3 + \beta_2 (A_1 A_2 + A_2 A_1) + \beta_3 (\mathrm{tr} A_1^2) A_1 \tag{3}$$

$$A_1 = (\mathrm{grad} V^*) + (\mathrm{grad} V^*)^{\mathrm{Transpose}} \tag{4}$$

$$A_n = \frac{\mathrm{d}A_{n-1}}{\mathrm{d}t} + A_{n-1}(\mathrm{grad} V^*) + (\mathrm{grad} V^*)^{\mathrm{Transpose}} A_{n-1}, \quad n = 1, 2, 3 \tag{5}$$

Energy conservation equation:

$$\rho c_{\mathrm{p}} \frac{\mathrm{d}T^*}{\mathrm{d}t} = \tau : \mathrm{grad}(V^*) - \nabla \cdot (-k_{\mathrm{th}} \nabla T^*) \tag{6}$$

Assumption of hydro-dynamically fully developed flow condition gives rise to existence of only axial velocity component u^* (along the x^* axis) which depends on y^* alone. This leads to zero values for the other two velocity components along the vertical and transverse direction. Thus, the velocity vector becomes:

$$v^* = \left[u^*(y^*), 0, 0\right] \tag{7}$$

Momentum equations after simplifications:

$$\frac{\partial p^*}{\partial x^*} = \mu \frac{\mathrm{d}^2 u^*}{\mathrm{d}y^{*2}} + 2(\beta_2 + \beta_3) \frac{\mathrm{d}}{\mathrm{d}y^*}\left(\frac{\mathrm{d}u^*}{\mathrm{d}y^*}\right)^3 \tag{8}$$

$$\frac{\partial p^*}{\partial y^*} = (2\alpha_1 + \alpha_2) \frac{\mathrm{d}}{\mathrm{d}y^*}\left(\frac{\mathrm{d}u^*}{\mathrm{d}y^*}\right)^2 \tag{9}$$

$$\frac{\partial p^*}{\partial z^*} = 0 \tag{10}$$

In Eq. (8), the pressure gradient in the x^* direction will be a constant. If assumed velocity field from Eq. (7) is substituted in the energy equation given in Eq. (6), it simplifies to:

$$\rho c_{\mathrm{p}} u^* \frac{\partial T^*}{\partial x^*} = k_{\mathrm{th}}\left(\frac{\partial^2 T^*}{\partial x^{*2}} + \frac{\partial^2 T^*}{\partial y^{*2}}\right) + \mu\left(\frac{\mathrm{d}u^*}{\mathrm{d}y^*}\right)^2 + 2(\beta_2 + \beta_3)\left(\frac{\mathrm{d}u^*}{\mathrm{d}y^*}\right)^4 \tag{11}$$

where k_{th} is the thermal conductivity of the fluid and viscous dissipation is represented by second term. Whereas, the third term accounts for the effect of non-Newtonian parameters. Zero values for β_2 and β_3 in Eq. (11) gives the corresponding situation for Newtonian fluid flow. Under thermally fully developed condition, the temperature profile depends only on y^*, and can be written as:

$$\theta = \frac{T^* - T^*_{\mathrm{w_l}}}{T^*_{\mathrm{m}} - T^*_{\mathrm{w_l}}}$$

where T^*_{m} is bulk temperature of the fluid and $T^*_{\mathrm{w_l}}$ is the temperature of the lower wall. Due to constant wall heat flux, the expression for non-dimensional temperature along the axial direction becomes:

$$\frac{\partial T^*}{\partial x^*} = \frac{\mathrm{d}T^*_{\mathrm{m}}}{\mathrm{d}x^*} = \text{const}, \quad \frac{\partial^2 T^*}{\partial x^{*2}} = 0 \tag{12}$$

Substitution of the relations of Eq. (12) into Eq. (11) gives

$$\rho c_p u^* \frac{\partial T^*}{\partial x^*} = k_{th}\left(\frac{\partial^2 T^*}{\partial y^{*2}}\right) + \mu\left(\frac{du^*}{dy^*}\right)^2 + 2(\beta_2 + \beta_3)\left(\frac{du^*}{dy^*}\right)^4 \quad (13)$$

Boundary conditions: For velocity field, no-slip condition at both the plates are used

$$u*(-h) = 0,\ u*(h) = 0 \quad (14)$$

Both the plates are subjected to constant heat flux

$$k_{th}\left(\frac{\partial T^*}{\partial y^*}\right)_{-h} = \mp q_1,\ k_{th}\left(\frac{\partial T^*}{\partial y^*}\right)_{h} = \pm q_2 \quad (15)$$

For the lower wall, −ve sign is taken in Eq. (15), since the temperature increases with the decrease in the vertical (y^*) coordinate. But, for the upper wall, +ve sign is taken for heating, as the temperature increases with the increase of the y^* coordinate. A non-dimensional temperature scale is considered, because of the absence of the knowledge of the plate temperature, and is expressed as follows:

$$\theta = \frac{T^* - T^*_{w_1}}{q_1 h / k_{th}} \quad (16)$$

where q_1 is the uniform and constant heat flux at the lower plate. Non-dimensionalization of the momentum and energy equation can be carried out with the help of the following non-dimensional variables are used:

$$u = \frac{u^*}{u_0},\ y = \frac{y^*}{h} \quad (17)$$

Thus, the momentum and the energy equation in non-dimensional form are:

$$\frac{d^2u}{dy^2} + 6A\left(\frac{du}{dy}\right)^2 \frac{d^2u}{dy^2} = N \quad (18)$$

$$\beta u = \frac{d^2\theta}{dy^2} + \mathrm{Br}\left(\frac{du}{dy}\right)^2 + 2A\mathrm{Br}\left(\frac{du}{dy}\right)^4 \quad (19)$$

$$A = \frac{\beta_2 + \beta_3}{\mu}\left(\frac{u_0}{h}\right)^2,\ \mathrm{Br} = \frac{\mu u_0^2}{q_1 h},\ u_0 = -\frac{1}{N}\frac{dp^*}{dx^*}\frac{h^2}{\mu},\ \beta = \frac{\rho c_p u_0}{q_1}\frac{dT^*}{dx^*} \quad (20)$$

Here, u_0 is average velocity.

The boundary conditions for momentum equation can be written in non-dimensional form as:

$$u(-1) = 0, \ u(1) = 0 \tag{21}$$

One easy boundary condition for such problems could be symmetry condition, i.e., applying boundary condition at the center line. But the selection of base function for velocity in the least square method will become difficult in such scenario.

The energy equation is subjected to the following non-dimensional boundary conditions:

$$\theta(-1) = 0 \tag{22}$$

$$\frac{d\theta}{dx}(-1) = \mp 1 \tag{23}$$

$$\frac{d\theta}{dx}(1) = \pm \frac{q_2}{q_1} = q \tag{24}$$

For highly nonlinear equations such as Eqs. (18) and (19), it is difficult to obtain analytical solutions in closed forms. Analytical solutions can be used to validate experimental/numerical solution, also it helps to explain the physics of the problem. Thus, even approximate analytical or semi-analytical solutions are also useful. Motivated by this fact, various researchers constantly put efforts to come up with new and advanced analytical or semi-analytical techniques such as: homotopy perturbation method (HPM) [22], homotopy analysis method (HAM) [23], Adomian decomposition method (ADM) [24], and least square method (LSM). These methods are explored to solve various engineering problems. In the present work, LSM has been employed as it has some advantages compared to others techniques. Construction of homotopy is required in HAM and HPM, transformation is required in DTM [25], but LSM is free from such hurdles. LSM implementation is easy and described in brief in the next section.

3 Results and Discussion

The governing equations are solved by LSM, and the results are validated with the exact solution by Danish et al. [9] in Fig. 2a. In present result, Ha = 0 is considered for validation to remove the effect of magnetic field. Two different values of the third grade fluid parameter A (denoted as β in the study of Danish et al. [9]) are used for validation, and the agreement is found to be good. Also, the present method is validated for Newtonian fluid as shown in Fig. 2b.

Obtaining the velocity and the temperature profile for different values of parameter of third grade fluid constitutes the semi-analytical way of the study. Artificial neural network (ANN) approach constitutes the second part of the study. In the second part, the problem temperature and the velocity profile obtained by LSM are used to train the ANN by scaled conjugate gradient (SCG) method. Once the ANN is trained, an

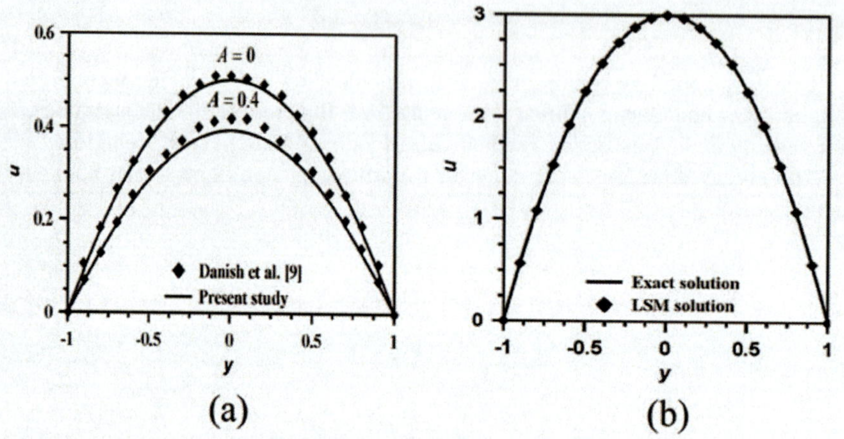

Fig. 2 Validation of present result with exact solution **a** comparison of velocity profile for third grade fluid, **b** comparison of velocity profile for Newtonian fluid

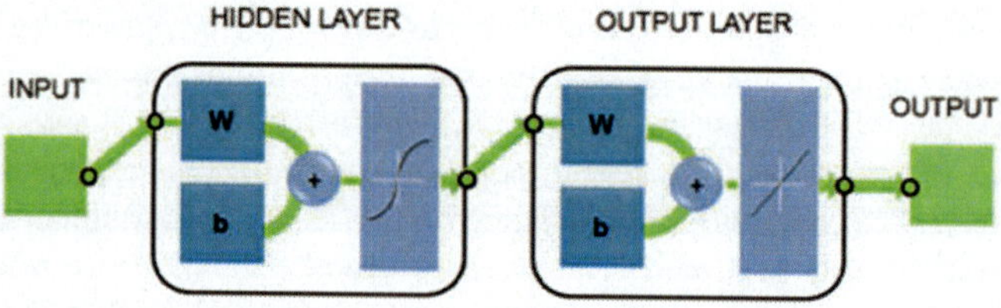

Fig. 3 ANN model used for estimation of parameter in inverse analysis

unknown velocity profile (computed by LSM method) is fed into the ANN, and the ANN as shown in Fig. 3 gives the corresponding temperature profile as output.

In the training of ANN, the velocity and the temperature profiles for 15 different values of the parameter of third grade fluid are fed. The velocity profiles are used as inputs and the temperature profiles are used as target output in the ANN. During the training of ANN, the data fed into ANN are split into three parts: training, validation, and test.

Figure 4 shows the histogram of the input data used in the ANN model. Large amount of data falling on zero error line indicates good learning of the ANN. While significant amount of data lying away from zero error line indicates that the ANN model is not over trained. This means that the ANN model can handle new types of cases as well. In the performance graph of Fig. 5, it is observed that the error is continuously decreasing.

Late occurring of best validation indicates over fitting case, and early occurring indicates under fitting. Both are undesirable for a good ANN model. Regression analysis of the overall ANN model is presented in Fig. 6.

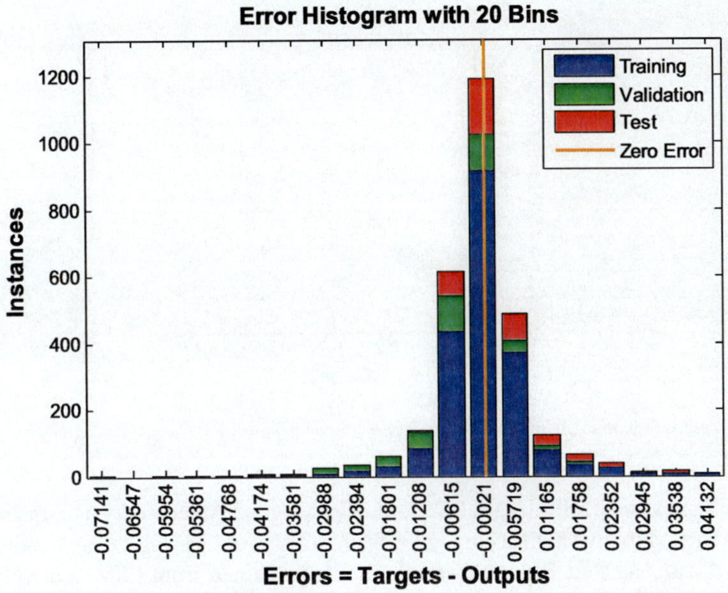

Fig. 4 Histogram of input data in ANN model

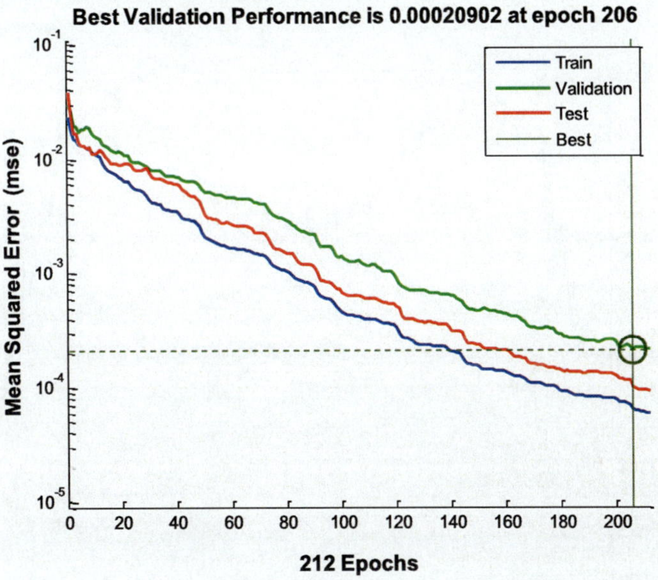

Fig. 5 Performance curves in ANN model with SCG algorithm

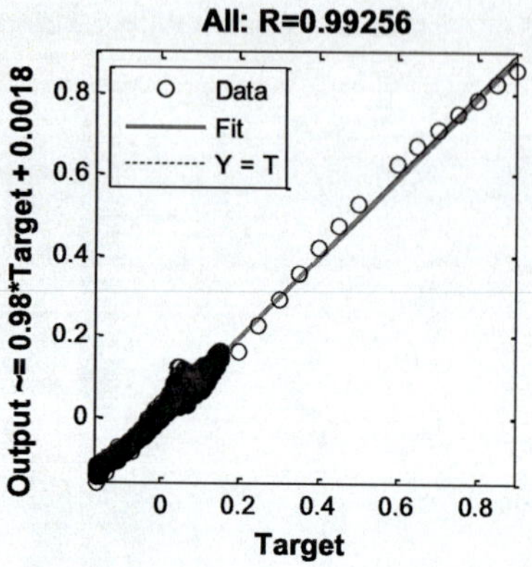

Fig. 6 Regression analysis of the ANN model

The regression coefficient of 0.99256 indicates good ANN model, i.e., the model is well trained to give high accuracy for the familiar cases as well as can handle new types of cases as well. The temperature profiles obtained from LSM and ANN are compared in Fig. 7.

The ANN is able to give the temperature profile with high accuracy in 0.01 s. The zig-zag nature of the curve for ANN depends upon various factors such as: number of

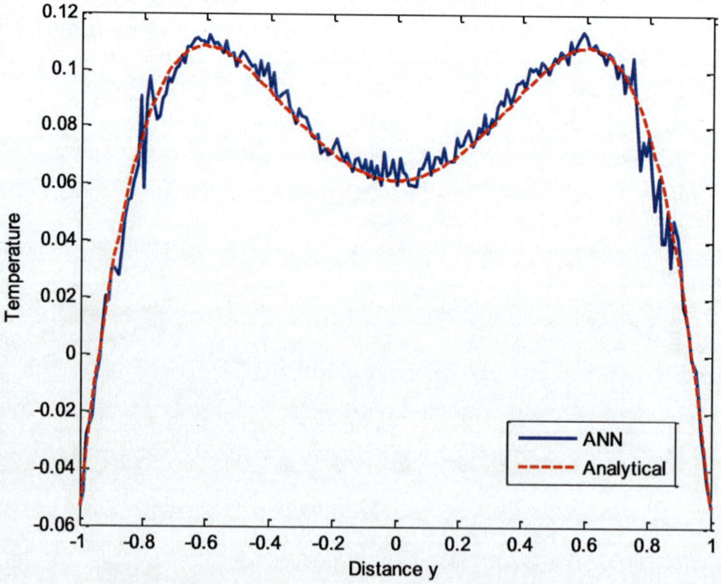

Fig. 7 Comparison of temperature profile obtained from the ANN model with that obtained from LSM

neurons, number of test cases, and distribution of input data for training, validation, and testing.

4 Conclusions

Temperature profile for third grade fluid flowing through two parallel plates was computed by using artificial neural network (ANN). Solving the governing equations by least square method (LSM) constituted the first part of the problem. The velocity and temperature profile obtained by least square method (LSM) was used in the artificial neural network (ANN) model. Scaled conjugate gradient (SCG) was employed as training algorithm in the artificial neural network (ANN) model. Temperature profile was generated by artificial neural network (ANN) for a given velocity profile obtained by least square method (LSM). Scaled conjugate gradient (SCG) is suitable for large problem as it is more memory efficient. Artificial neural network (ANN) with scaled conjugate gradient (SCG) algorithm is found to be suitable for present type of problems, as it takes moderate time with high accuracy of estimation.

References

1. Al Mukahal FHH, Wilson SK, Duffy BR (2015) A rivulet of a power-law fluid with constant width draining down a slowly varying substrate. J Non-Newton Fluid Mech 224:30–39
2. Jalil M, Asghar S (2013) Flow of power-law fluid over a stretching surface: a lie group analysis. Int J Non-Linear Mech 48:65–71
3. Tso CP, Sheela FJ, Hung YM (2010) Viscous dissipation effects of power-law fluid flow within parallel plates with constant heat fluxes. J Non-Newton Fluid Mech 165:625–630
4. Sisko AW (1958) The flow of lubricating greases. Ind Eng Chem Res 50:1789–1799
5. Bhatti MM, Zeeshan A, Ellahi R (2016) Endoscope analysis on peristaltic blood flow of Sisko fluid with Titanium magneto-nano particles. Comput Biol Med 78:29–41
6. Khan M, Munuwar S, Abbasbandy S (2010) Steady flow and heat transfer of Sisko fluid in annular pipe. Int J Heat Mass Transf 53:1290–1297
7. Wang L, Jian Y, Liu Q, Li F, Chan L (2016) Electromagnetohydrodynamic flow and heat transfer of third grade fluids between two micro-parallel plates. Colloids Surf A 494:87–94
8. Siddiqui AM, Zeb A, Ghori QK, Benharbit AM (2008) Homotopy perturbation method for heat transfer flow of a third grade fluid between parallel plates. Chaos, Solitons Fractals 36:182–192
9. Danish M, Kumar S, Kumar S (2012) Exact analytical solutions for the Poiseuille and Couette–Poiseuille flow of third grade fluid between parallel plates. Commun Non Linear Sci 17:1089–1097
10. Ebaid A (2014) Remarks on the homotopy perturbation method for the peristaltic flow of Jeffrey fluid with nano-particles in an asymmetric channel. Comput Math Appl 68:77–85
11. Sheela-Fransisca J, Tso CP (2009) Viscous dissipation effects on parallel plates with constant heat flux boundary conditions. Int Commun Heat Mass Transfer 36:249–254
12. Chakraborty R, Dey R, Chakraborty S (2013) Thermal characteristics of electromagnetohydrodynamic flows in narrow channels with viscous dissipation and Joule heating under constant wall heat flux. Int J Heat Mass Transf 67:1151–1162

13. Barisik M, Yazicioglu AG, Setin C, Kakak S (2015) Analytical solution of thermally developing microtube heat transfer including axial heat conduction, viscous dissipation, and rarefaction effects. Int Commun Heat Mass Transfer 67:81–88
14. Ozisik MN (1993) Heat conduction, 2nd edn. Wiley, USA
15. Hatami M, Sheikholeslami M, Ganji DD (2014) Laminar flow and heat transfer of nano fluid between contracting and rotation disks by least square method. Powder Technol 253:769–779
16. Hatami M, Ganji DD (2013) Heat transfer and flow analysis for SA-TiO_2 non-Newtonian nano fluid passing through the porous media between two co-axial cylinders. J Mol Liq 188:155–161
17. Pourmehran O, Rahimi-Gorji M, Gorji-Bandpy M, Ganji DD (2015) Analytical investigation of squeezing unsteady nanofluid flow between parallel plates by LSM and CM. Alex Eng J 54:17–26
18. Fakour M, Vahabzadhe A, Ganji DD, Hatami M (2015) Analytical study of micro polar fluid flow and heat transfer in a channel with permeable walls. J Mol Liq 204:198–204
19. Dumek V, Druckmuller M, Raudensk, Woodbury KA (1993) Novel approaches to the ICHP: neural networks and expert systems, inverse problems in engineering: theory and practice. In: Proceedings of the first international conference on inverse problems in engineering, ASME no. I00357, pp 275–282
20. Jambunathan K, Hartle S, Ashforth-Frost S, Fontama VN (1996) Evaluating convective heat transfer coefficients using neural networks. Int J Heat Mass Transf 39:2329–2332
21. Sablani SS (2001) A neural network approach for non-iterative calculation of heat transfer coefficient in fluid-particle systems. Chem Eng Process 40:363–369
22. Yun Y, Temuer C (2015) Application of the homotopy perturbation method for the large deflection problem of a circular plate. Appl Math Model 39:1308–1316
23. Liao S (2003) On the analytic solution of magnetohydrodynamic flows of non-Newtonian fluids over a stretching sheet. J Fluid Mech 488:189–212
24. Singla RK, Das R (2015) Adomian decomposition method for a stepped fin with all temperature dependent mode of heat transfer. Int J Heat Mass Transf 82:447–459
25. Yaghoobi H, Torabi M (2011) The application of differential transformation method to nonlinear equations arising in heat transfer. Int Commun Heat Mass Transfer 38:815–820

Fluid Flow and Heat Transfer

Fluid Flow and Heat Transfer Characteristics for an Impinging Jet with Various Angles of Inclination of Impingement Surface

Jublee John Mili, Tanmoy Mondal, and Akshoy Ranjan Paul

Abstract Present study involves the thermofluidic characteristics of a two-dimensional impinging isothermal jet for various angles of inclination of impinging surface. The RANS (Reynolds-averaged Navier–Stokes) equations based on the standard k–ε turbulence model has been used to predict the turbulent flow and heat transfer fields. The angle of inclination of the impinging surface is varied in the range $\alpha = 0°$–$10°$, the Reynolds number is considered to be Re = 11,500 and the impinging surface is subjected to the constant wall heat flux. According to the present computational results, the impingement point, a location where the jet strikes on the bottom wall after issuing from the nozzle, gradually shifts towards the right side of the jet center-line with the increase of angle of inclination. The skin friction coefficient decreases but the pressure coefficient increases with increasing the angle of inclination. The Nusselt number, a measure of rate of convective heat transfer, decreases drastically as the angle of inclination increases from $\alpha = 0°$ to $\alpha = 10°$.

Keywords Impinging jet · Heat transfer · Impingement surface · k–ε turbulence model

1 Introduction

Jet impingement techniques are widely used in heat transfer cooling ranging from thermal management of electronic gadgets to the cooling of rocket launching stations. Although many experimental and numerical studies are found in literature on impinging jets, most of the studies are limited to the flow characteristics of impinging jet on a flat surface [1–3]. A few studies are available in the area of heat transfer for an impinging jet on an inclined plane. Seyedein et al. [1] numerically investigated two-dimensional flow field and heat transfer of a turbulent slot jet [1]. Pramanik et al. [4] investigated the fluid flow of a confined impinging jet slot [4]. Yang et al. [5]

J. J. Mili (✉) · T. Mondal · A. R. Paul
Department of Applied Mechanics, Motilal Nehru National Institute of Technology Allahabad, Prayagraj, Uttar Pradesh 211004, India
e-mail: nampozonmili@gmail.com

S. Revankar et al. (eds.), *Proceedings of International Conference on Thermofluids*, Lecture Notes in Mechanical Engineering,
https://doi.org/10.1007/978-981-15-7831-1_5

numerically studied variations of jet exit at Reynolds number (Re = 5920–23,700), dimensionless jet to-surface distance ($h/w = 0.5$–12), dimensionless jet width ($w/x =$ 0.033–0.05) and the heat flux varying between 1663 and 5663 W/m^2 [5]. Exhaustive numerical study of impinging jets has also been carried out on non-inclined surfaces or curved surfaces with either heated or non-heated impingement surface but not on varying inclined surfaces or with high constant heat flux impingement surface [6–9]. The present study, therefore, focuses on both the fluid flow and heat transfer characteristics of impinging jet on inclined plane. In the present study, both flow and heat transfer characteristics are studied in order to investigate the effects of flow field due to heat transfer of the impinging jet.

2 Mathematical Formulation

For the present study, the Reynolds number was set at 11,500 and the angle of inclination (α) was varied from $\alpha = 0°$ to $\alpha = 10°$. The jet width (w) was 4 mm and h/w ratio is 2 for $\alpha = 0°$ and height from jet inlet to impingement surface increases (h) with the increase in angle of inclination. A schematic diagram is shown in Fig. 1.

Here, the flow is assumed to be steady, two dimensional and incompressible. The body force is neglected. Flow of fluid and heat transfer behaviour is assumed to be constant. The Reynolds averaged Navier–Stokes equations (RANS) based on two equations standard $k - \varepsilon$ is used to predict the turbulent flow. The dimensional form of the governing equations is as represented through Eqs. (1)–(5).

Continuity equation:

$$\frac{\partial U_j}{\partial x_j} = 0 \tag{1}$$

Momentum equation:

$$\frac{\partial (U_i U_j)}{\partial x_j} = -\frac{\partial P}{\partial x_i} - \frac{\partial \tau_{ij}}{\partial x_j} + \frac{\partial \left(-\overline{u_i' u_j'}\right)}{\partial x_j} \tag{2}$$

where $\tau_{ij} = -\mu\left(\frac{\partial U_i}{\partial x_j} + \frac{\partial U_j}{\partial x_i}\right)$.

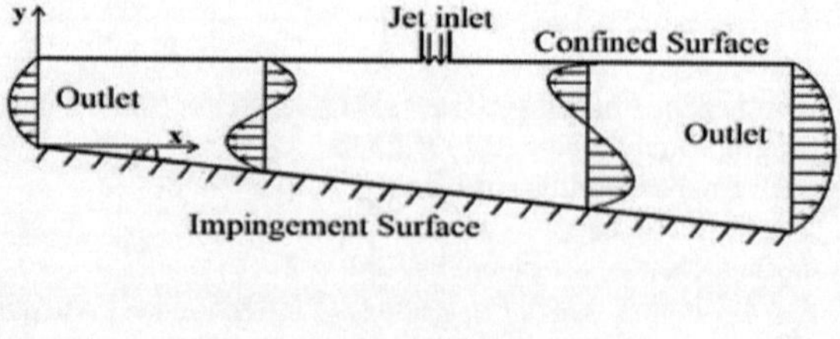

Fig. 1 Schematic diagram of an impinging jet on an inclined surface

Energy equation:

$$C_{\mu}\frac{\partial(u_jT)}{\partial x_j}=\lambda\frac{\partial}{\partial x_j}\left(\frac{\partial u_i}{\partial x_j}+\frac{\partial u_j}{\partial x_i}\right)-\frac{\partial Pu_i}{\partial x_i}-\tau_{ij}\frac{\partial u_j}{\partial x_j} \tag{3}$$

Turbulent kinetic energy (k) transport equation:

$$\frac{\partial(kU_i)}{\partial x_i}=\frac{\partial}{\partial x_i}\left(\frac{\upsilon_t}{\sigma_k}\frac{\partial k}{\partial x_i}\right)+\upsilon_t\left(\frac{\partial U_i}{\partial x_j}+\frac{\partial U_j}{\partial x_i}\right)\frac{\partial U_i}{\partial x_j}-\varepsilon \tag{4}$$

Dissipation (ε) transport equation:

$$\frac{\partial(\varepsilon U_i)}{\partial x_i}=\frac{\partial}{\partial x_i}\left(\frac{\upsilon_t}{\sigma_\varepsilon}\frac{\partial \varepsilon}{\partial x_i}\right)+C_{1\varepsilon}\frac{\varepsilon}{k}P-C_{2\varepsilon}\frac{\varepsilon^2}{k} \tag{5}$$

where P is the generation of k and is given by, $P=\upsilon_t\left(\frac{\partial U_i}{\partial x_j}+\frac{\partial U_j}{\partial x_i}\right)\frac{\partial U_i}{\partial x_j}$

turbulent viscosity, $\upsilon_t=C_\mu\frac{k^2}{\varepsilon}$.

The constant for the above equation are $C_1=1.44,\ C_2=1.92,\ C_\mu=0.09,\ \sigma_\varepsilon=1.30$ and $\sigma_k=1.00$ [10].

2.1 Numerical Schemes and Boundary Conditions

The set of Eqs. (1)–(5) are computationally solved by using commercial software ANSYS Fluent. The Finite Volume method is used for the spatial discretization of the RANS equations. The SIMPLE (Semi-implicit method for pressured linked equations) algorithm is used for coupling the pressure and velocities. First-order upwind scheme is used for the discretization of convective terms and the central difference scheme is used for the discretization of viscous diffusion terms of the momentum and turbulent transport equations.

At inlet, u and v velocities are considered as 50 m/s and 0 m/s, respectively. The turbulent kinetic energy (k) and rate of dissipation equation of turbulent kinetic energy (ε) at the inlet are set as $k=1.5uI^2$ (where I is the inlet turbulence intensity and is considered to be 0.05) and $\varepsilon=C_\mu\frac{k^{3/2}}{l}$ (where l is the turbulent length scale and is considered to be 0.07). The pressure p at inlet is set as 1 atmospheric. At the exit, the outlet boundary condition is considered for all the variables. No-slip ($u=0\,\mathrm{m/s}$) and no-penetration conditions ($v=0\,\mathrm{m/s}$) are considered for both solid walls. For temperature boundary conditions, an adiabatic confined wall is considered and the impingement wall is subjected to a constant wall heat flux i.e. $q''=10^2\,\mathrm{W/m^2}$. The temperature at the inlet and exit is maintained at ambient temperature i.e. $T_\infty=300$ K.

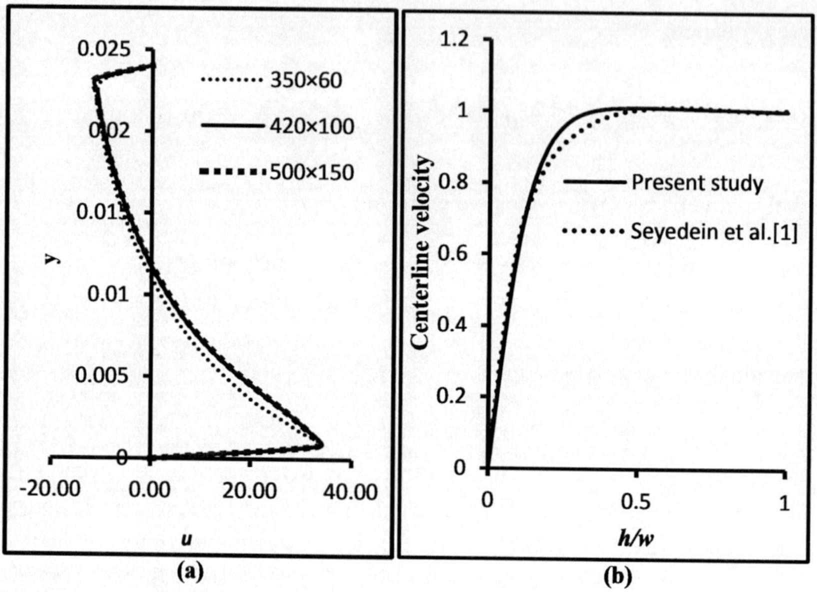

Fig. 2 (**a**) Grid independence test result; (**b**) validation of present computational result

2.2 Grid Independence Test and Validation

A non-uniform grid is generated for the present computational domain. In order to carry out the grid independence test, three different grid densities are used i.e. 350 × 60, 420 × 120 to 500 × 150. The grid independence test is shown in Fig. 2a for the u velocity profile at $x = 195$ mm for $\alpha = 0°$. As can be seen, the solutions remain almost same if the grid density is changed from 420 × 100 to 500 × 150. Therefore, for the remainder of the present study is conducted using the grid density of 420 × 100.

To validate the present computational results, the non-dimensional jet centre-line velocity computed by the present numerical method can be compared with the experimental values reported by Seyedein et al. in Fig. 2(b) for Reynolds number Re = 9900, angle of inclination of impingement surface $\alpha = 0°$ and ratio of distance between the jet inlet to the impingement wall and the jet width $h/w = 7.5$ [1]. As seen, the present computational results agree well with the corresponding experimental results in all the region except there is some deviation in the region close to the plane of the jet centre-line. This could be attributed to the bending of jet in either direction of jet centre-line.

3 Results and Discussion

Figure 3(a), (b) shows the streamline plots in the region close to the jet inlet for $\alpha = 0°$ and 10°, respectively. After issuing from the nozzle, the jet impinges at the impingement point (ip) on the bottom wall. As a result, two counter-rotating vortices, one on the right side and other on the left side of the jet centerline, are formed close to the jet inlet. These two vortices about the jet centerline remain symmetric for $\alpha =$ 0°, but asymmetric for $\alpha = 10°$. In case of $\alpha = 10°$, the vortex formed on the right side of the jet centerline is found to be larger than that formed on the left side. The reason is the confinement effect created by the two walls is greater on the left side of the jet centerline. Due to this asymmetry in the confinement effect, the impingement point is located little downstream of the jet centerline for $\alpha = 10°$, whereas in case of $\alpha = 0°$, the impingement point is formed exactly at point where the jet centerline intersect the bottom wall.

The variations of skin friction coefficient (C_f) and the pressure coefficient (C_p) at the impingement wall are shown in Fig. 4(a), (b), respectively, for different angle of inclination of the impingement surface i.e. $\alpha = 0°$, 5° and 10°. As can be seen from Fig. 4, wherever the skin friction is maximum, and the pressure coefficient is minimum there or vice versa. At the impingement point the skin friction coefficient becomes minimum i.e. $C_f = 0$ because at the impingement point the streamwise velocity $u = 0$. The two peaks appearing in Fig. 4a indicate the reattachment of the jet on the top solid wall after issuing from the jet inlet. As seen from Fig. 4a, with the increase of the angle of inclination, the skin friction coefficient gradually decreases. This is because, the momentum of jet decreases with the increase of the confinement effect between the two solid walls. A similar trend of variation is noticed for the pressure coefficient as the angle of inclination α decreases. At the impingement point, the pressure coefficient is maximum since the static pressure is maximum there.

The effects of flow field on the heat transfer results are shown in Fig. 5(a), (b) in terms of variations of bottom wall temperature and local Nusselt number along the bottom wall, respectively. As noticed, the wall temperature T_w is minimum at the impingement point; however, at this location the value of local Nusselt number Nu is maximum. This is due to the fact of cold jet impinging on a heated wall. With

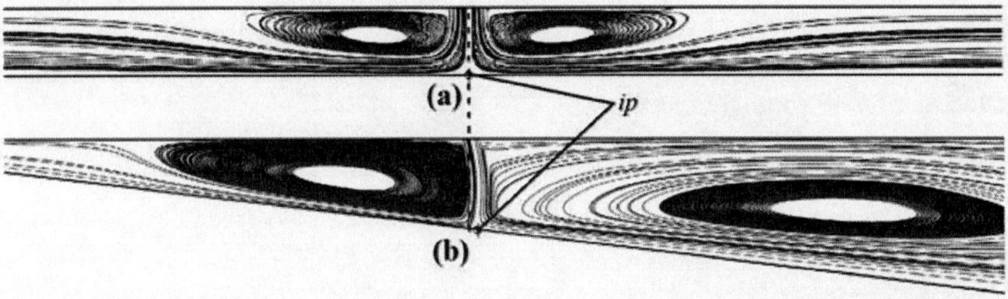

Fig. 3 Steamline plots for **(a)** $\alpha = 0°$ and **(b)** $\alpha = 10°$

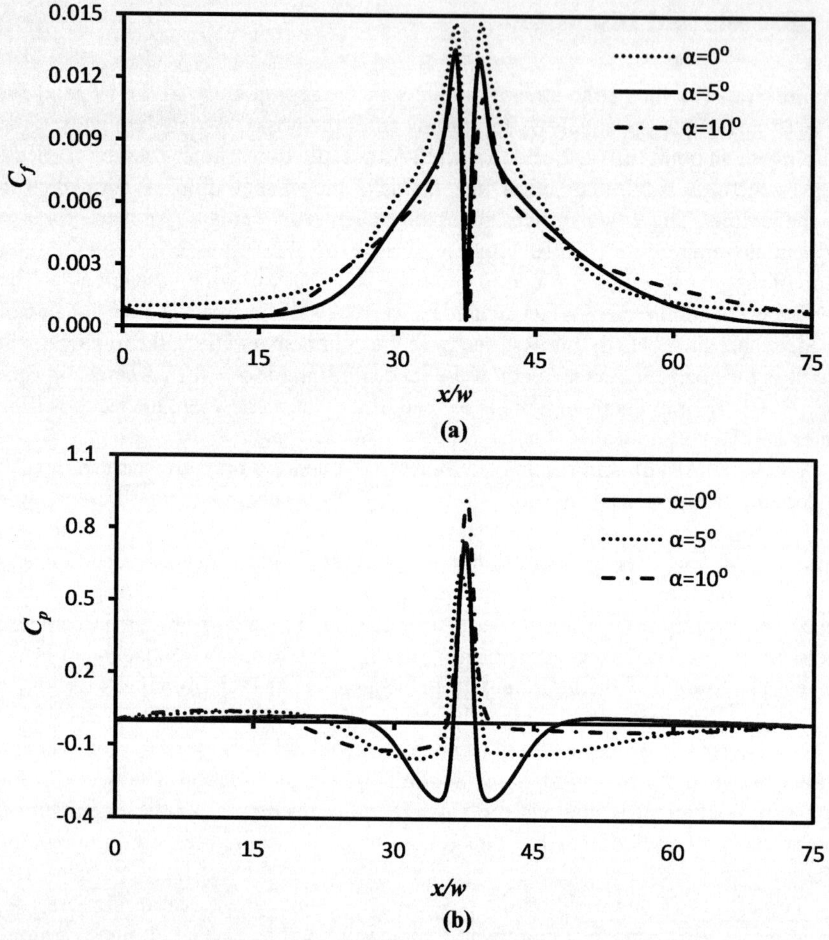

Fig. 4 (**a**) Distribution of skin friction coefficient; (**b**) the variation of pressure coefficient

the downstream distance, the wall temperature increase but local Nusselt number decreases, indicating development of thermal boundary layer. It can also be noticed from Fig. 5(b) that the Nusselt number decreases as the increases from $\alpha = 0°$ to 10°. Therefore, the convective heat transfer rate decreases with the increase of angle of inclination of impingement surface.

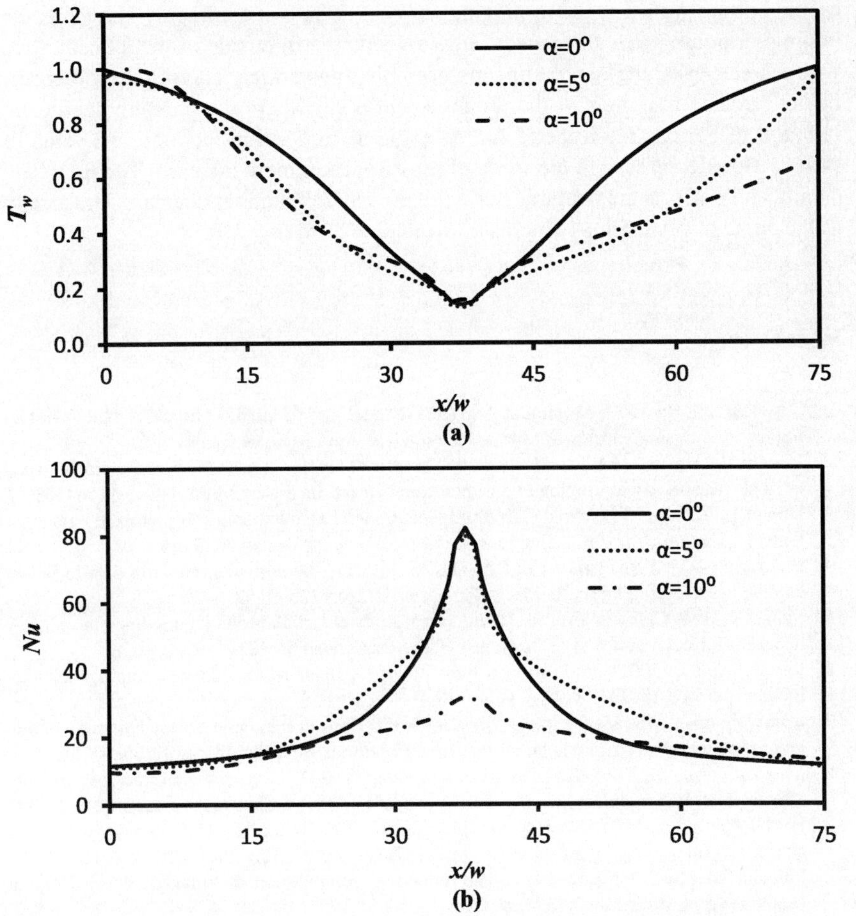

Fig. 5 (**a**) Variation of temperature and (**b**) variation of Nusselt number

4 Conclusions

In the present study, the fluid flow and heat transfer characteristics of an impinging isothermal jet have been studied for various angles of inclination of the impingement surface in the range of 0°–10°. The Reynolds averaged Navier–Stokes (RANS) equations based on standard $k - \varepsilon$ model has been solved to obtain the numerical results. The Reynolds number based on the jet width of 4 mm is considered as Re = 11,500 and the bottom wall (i.e. the impinging surface) is maintained at constant heat flux condition.

The present computational results indicate that the flow field remains symmetric about the jet centerline at $\alpha = 0°$; however, with the increase of angle of inclination, the flow field becomes asymmetric. The location at which the jet impinges on the bottom wall (i.e. the impinging surface), which is commonly known as the impinging point, gradually moves more towards the right side of the jet centerline with increasing of angle of inclination of impinging surface. The skin friction coefficient remains minimum at the impingement point where the pressure coefficient is found to be maximum there. As the angle of inclination increase, the value of skin friction decreases but the value of pressure coefficient increase. The flow field significantly affects the heat transfer field: the Nusselt number decreases drastically as the angle of inclination is increased from $\alpha = 0°$ to $\alpha = 10°$.

References

1. Seyedein SH, Hasan M, Mujumdar AS (1994) Modeling of a single confined turbulent slot jet impingement using various k–ε turbulence models. Appl Math Model 18:526–537
2. Beitelmal AH, Saad MA, Patel CD (2000) The effect of inclination on the heat transfer between a flat surface and an impinging two-dimensional air jet. Int J Heat Fluid Flow 21:156–163
3. Dong LL, Leung CW, Cheung CS (2002) Heat transfer characteristics of premixed butane/air flame jet impinging on an inclined flat surface. Heat Mass Transf 39:19–26
4. Pramanik S, Madhusudana Achari A, Das MK (2012) Numerical simulation of a turbulent confined slot impinging jet. Ind Eng Chem Res 51(26):9153–9163
5. Yang YT, Wei TC, Wang YH (2011) Numerical study of turbulent slot jet impingement cooling on a semi-circular concave surface. Int J Heat Mass Transf 54:482–489
6. Roy S, Patel P (2003) Study of heat transfer for a pair of rectangular jets impinging on an inclined surface. Int J Heat Mass Transf 46:411–425
7. Ramezanpur A, Mirzaee I, Firth D, Shirvani H (2007) A numerical heat transfer study of slot jet impinging on an inclined plate. Int J Numer Methods Heat Fluid Flow 17:661–676
8. Ying Z, Guiping L, Xueqin B, Lizhan B, Dongsheng W (2017) Experimental study of curvature effects on jet impingement heat transfer on concave surfaces. Chin J Aeronaut 30(2):586–594
9. Hu G, Zhang L (2007) Experimental and numerical study on heat transfer with impinging circular jet on a convex hemispherical surface. Heat Transfer Eng 28(12):1008–1016
10. Launder BE, Spalding DB (1974) The numerical computation of turbulent flows. Comput Methods Appl Mech Eng 3:269–289

Implementation of Improved Wall Function for Buffer Sub-layer in OpenFOAM

R. Lakshman, Jha Rahul Binod, and Ranjan Basak

Abstract In computational fluid dynamics the value of y^+ is a defining parameter in the accuracy of the result. In most of the available CFD tools, the computation near the wall has been done based on the concept of two zonal equation (viscous and log-law region). In this study, a new wall function has been implemented in the open-source CFD tool OpenFOAM for obtaining the result of near-wall properties. This newly implemented wall function is based on three-zonal equations compared to the two zonal equations used in standard wall function. A comparative analysis of the newly implemented wall function with the standard wall function has been conducted and the results are discussed. The analysis is conducted for the fluid flow over the backward-facing step. Comparative analysis was also conducted by keeping the value of y^+ in viscous as well as in buffer region.

Keywords Wall function · CFD · OpenFOAM

1 Introduction

Turbulence modeling uses numerical models to predict the variation of fluid properties at different positions across the fluid flow [1]. These numerical models use some empirical equation called wall function to satisfy the physics involved in fluid flow in the region near to the wall. When we deal with the fluid flow across a narrow channel the channel boundaries have a great impact on the variation of fluid properties. For no-slip condition, there is rapid variation in fluid property in the near-wall region, therefore, the gradient of properties is large near-wall which requires very fine meshing of the near-wall area thereby, increasing the computational cost. In order to reduce the computational effort, wall function has been introduced which works like a bridge in the intermediate region between the wall and the fully developed turbulent region. The near-wall flow can be divided into three sublayers [2] based on

R. Lakshman (✉) · J. R. Binod · R. Basak
Department of Mechanical Engineering, National Institute of Technology Sikkim, Ravangla, South Sikkim, India
e-mail: lakshmanr03@gmail.com

S. Revankar et al. (eds.), *Proceedings of International Conference on Thermofluids*, Lecture Notes in Mechanical Engineering,
https://doi.org/10.1007/978-981-15-7831-1_6

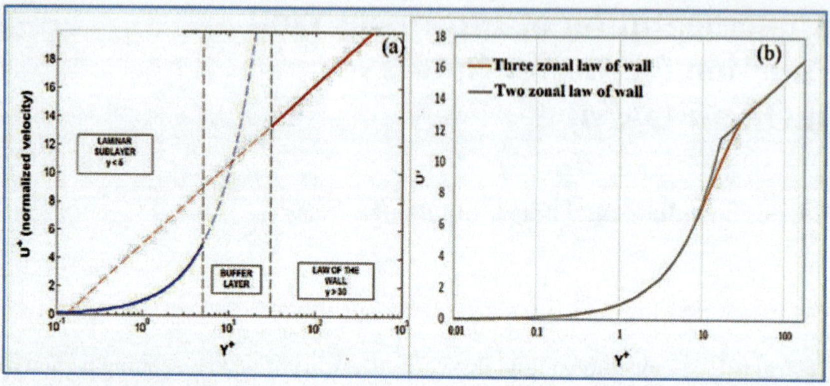

Fig. 1 **a** Curve representing law of wall, **b** comparison between three-zonal and conventional two zonal wall function

a dimensionless number y^+ namely, viscous sublayer ($y^+ < 5$) fully turbulent sublayer ($y^+ > 30$) and buffer sublayer ($5 < y^+ < 30$).

In standard wall functions available in CFD tools are based on two zonal equations without considering buffer region [3–5]. In our study, we have tried to establish an empirical relation between y^+ and U^+ for the buffer region using the boundary condition proposed by Chmielewski and Gieras [6]. Due to the ease of accessibility to the source code, OpenFOAM is having greater attraction among the researchers [7, 8]. Here, in this study, a new wall function called "nutBufferWallFunction" is developed by modifying the standard wall function (nutkWallFunction) available in OpenFOAM [9].

2 Theory

2.1 *Standard Wall Function*

The standard wall function, given in most of the computational tools are based on the division of near-wall region in two sublayers namely viscous sublayer ($y^+ < 11.225$) and log-law sublayer ($y^+ > 11.225$) as shown in Fig. 1a. The relation between U^+ and y^+ in viscous sublayer and log-law sublayer is shown in Eqs. (1) and (2), respectively;

$$U^+ = y^+ \tag{1}$$

$$U^+ = \frac{1}{K}\ln\left(Ey^+\right) \tag{2}$$

$$y^+ = \frac{y * U_\tau}{\vartheta};\ U^+ = \frac{U}{U_\tau} \tag{3}$$

where,

U—mean velocity, K—von Karman constant (= 0.41), E—empirical constant (= 9.793), U_τ—friction velocity, y—distance from wall to first cell centre, ϑ—kinematic viscosity.

2.2 *Implementation of New Wall Function*

The major drawback in standard wall function is the lack of clarity in the buffer sublayer. In this study, a new wall function is implemented based on the boundary condition used by Chmielewski and Gieras [6]. The curve showing the relation between U^+ and y^+ for standard and the new wall function is shown in Fig. 1.

Considering $\emptyset_1$, $\emptyset_2$ and $\emptyset_3$ as a function representing the value of U^+ in viscous, buffer and log-law region, respectively, the boundary conditions applied are shown below

$$[\emptyset_2]_{y_1^+} = [\emptyset_1]_{y_1^+} \tag{4}$$

$$[\emptyset_2]_{y_2^+} = [\emptyset_3]_{y_2^+} \tag{5}$$

$$\frac{\mathrm{d}\emptyset_2}{\mathrm{d}y^+}_{y_1^+} = \frac{\mathrm{d}\emptyset_1}{\mathrm{d}y^+}_{y_1^+} \tag{6}$$

$$\frac{\mathrm{d}\emptyset_2}{\mathrm{d}y^+}_{y_2^+} = \frac{\mathrm{d}\emptyset_3}{\mathrm{d}y^+}_{y_2^+} \tag{7}$$

Since there are four boundary conditions, $\emptyset_2$ can be expressed by third-order polynomial:

$$\emptyset_2(y^+) = a_0 + a_1 y^+ + a_2(y^+)^2 + a_3(y^+)^3 \tag{8}$$

Calculating function derivative:

$$\frac{\mathrm{d}\emptyset_2}{\mathrm{d}y^+} = a_1 + 2a_2(y^+) + 3a_3(y^+)^2 \tag{9}$$

Calculating derivative from $\emptyset_1(y^+)$ and $\emptyset_3(y^+)$:

$$\frac{d\emptyset_1}{dy^+} = 1 \tag{10}$$

$$\frac{d\emptyset_3}{dy^+} = \frac{1}{ky^+} \tag{11}$$

The system of equations can be shown in matrix form as Eq. (12):

$$\begin{bmatrix} 1 & y_1^+ & (y_1^+)^2 & (y_1^+)^3 \\ 1 & y_2^+ & (y_2^+)^2 & (y_2^+)^3 \\ 0 & 1 & 2y_1^+ & 3(y_1^+)^2 \\ 0 & 1 & 2y_2^+ & 3(y_2^+)^2 \end{bmatrix} \cdot \begin{bmatrix} a_0 \\ a_1 \\ a_2 \\ a_3 \end{bmatrix} = \begin{bmatrix} y_1^+ \\ \frac{1}{k}\ln(Ey^+) \\ 1 \\ \frac{1}{ky^+} \end{bmatrix} \tag{12}$$

The value of the constants obtained by solving the matrix (12) are;

$$a_0 = -1.092 \qquad a_1 = 1.4519$$

$$a_2 = -0.0496 \qquad a_3 = 0.000596$$

Thus, Eq. (8) can be written as

$$\emptyset_2(y^+) = -1.092 + 1.4519y^+ \pm 0.0496(y^+)^2 + 0.000596(y^+)^3 \tag{13}$$

Thus, the relation between U^+ and Y^+ in all the three sublayer can be shown as below:

$$\left.\begin{array}{l} \text{For viscous sublayer } (y^+ < 5) \quad U^+ = y^+ \\ \text{For buffer sublayer } (5 < y^+ < 30); \\ \quad U^+ = -1.092 + 1.4519y^+ - 0.0496(y^+)^2 + 0.000596(y^+)^3 \\ \text{For turbulent sublayer } (y^+ > 30) \quad U^+ = \frac{1}{k}\ln(Ey^+) \end{array}\right\} \tag{14}$$

2.3 *New Eddy Viscosity Expression for Buffer Sublayer*

An effective expression for eddy viscosity provides a better prediction of near-wall properties. Equation (14) is used for the derivation of new expression for eddy viscosity and the same is coded in OpenFOAM source code. The expression for the eddy viscosity ϑ_τ used in the source code is derived as follows.

The expression for wall shear stress can be written as:

$$\tau_{\omega} = \rho u_{\tau} \tag{15}$$

where,

$$u_{\tau} = \frac{U}{U+} \tag{16}$$

Using Eqs. (13) and (16) in Eq. (15)

$$\tau_{\omega} = \rho u_{\tau}\frac{U}{U+} = \frac{\rho u_{\tau}(U_{c} - U_{w})}{-1.092 + 1.4519y+ - 0.0496(y+)^{2} + 0.000596(y+)^{3}} \tag{17}$$

where, U is the velocity of first cell centre near the wall and in the direction of flow. Wall shear stress τ_{ω} for turbulent flow can also be written as;

$$\tau_{\omega} = \vartheta_{\text{eff}}\frac{(U_{c} - U_{w})}{y} \tag{18}$$

Comparing Eqs. (17) and (18),

$$\vartheta_{\text{eff}} = \vartheta + \vartheta_{\tau} = \frac{y^{+}\vartheta}{-1.092 + 1.4519y+ - 0.0496(y+)^{2} + 0.000596(y+)^{3}} \tag{19}$$

From this equation, the turbulent viscosity can be obtained as;

$$\vartheta_{\tau} = \vartheta\left(\frac{y^{+}}{-1.092 + 1.4519y+ - 0.0496(y+)^{2} + 0.000596(y+)^{3}} - 1\right) \tag{20}$$

A new wall function namely nutBufferWallFunction has been implemented in OpenFOAM library by altering the standard ϑ_{τ} wall function. In nutBufferWallFunction Eq. (20) is used for obtaining eddy viscosity in buffer region.

3 Numerical Setup

The geometry of the backward-facing step used in the present study is shown in Fig. 2a. For the analysis two different mesh whose y^{+} value coming in the viscous region ($y^{+} \approx 2$) and buffer region ($y^{+} \approx 10$) has been taken. As inlet boundary condition velocity equal to 44.2 m/s and as an outlet boundary condition, a constant zero pressure is given. Proper turbulence properties have been chosen by using the inlet velocity, the dimension of the geometry and kinematic viscosity (0.0000156 m^{2}/s). The OpenFOAM provided solver SimpleFOAM is used for simulation. For the

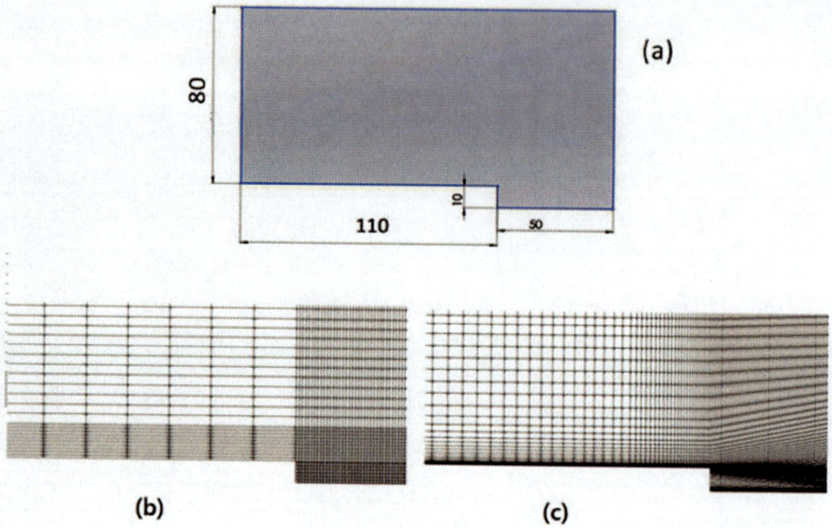

Fig. 2 **a** Dimensions of backward-facing step used for simulation (in mm); **b** mesh having $y^+ \approx 10$ (buffer sublayer); **c** mesh with $y^+ \approx 2$ (viscous sublayer)

numerical simulation, the residual value of 10^{-4} has been taken as convergence criteria for pressure and 10^{-6} for other parameters.

4 Results and Discussion

A comparative analysis of the newly implemented nutBufferWallFunction with standard wall function has been carried out on the basis of backward-facing step experimental data from Driver's and Seegmiller's research [10]. Due to the availability of experimental data, the results of the coefficient of friction (cf) value is been used for the comparative analysis. All three conventional RANS turbulence modeling has been used for this study.

From Fig. 3, it is clear that both the experimental and simulation result almost agrees with each other when the y^+ value comes in the viscous region but when the y^+ value is in the buffer region the result deteriorates. Figure 3 shows the deviation of result from the experimental data is very high when the y^+ value lies in the buffer region.

In the present study, a comparative analysis using the newly implemented nutBufferWallFunction with standard wall function has been conducted taking y^+ in the buffer region. Figure 4 shows the comparison of results obtained while using standard wall function and nutBufferWallfunction by using K-Epsilon turbulence

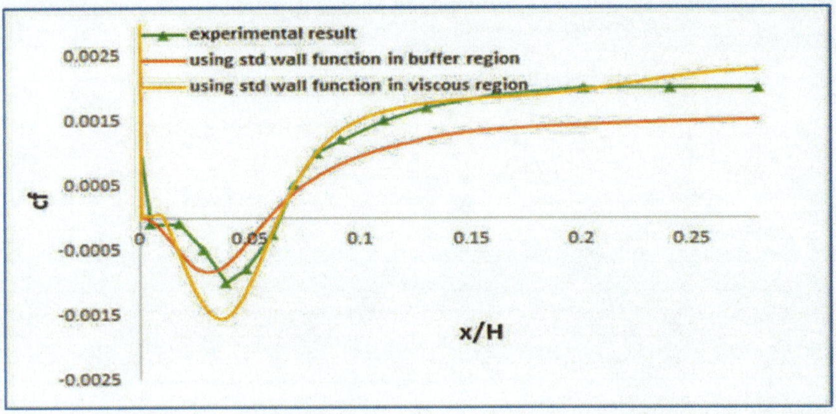

Fig. 3 Comparison of cf value along lower wall obtained by simulation and experimental data for y^+ in viscous region

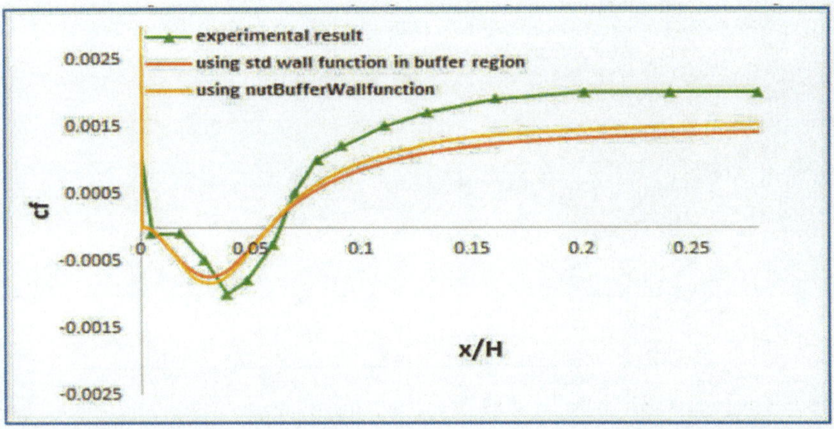

Fig. 4 Comparison of cf value for standard and nutBufferWallFunction with experimental data using RANS k ε model

model. From the result, it is evident that the results obtained by using nutBufferWallfunction are close to the experimental values compared to the one using standard wall function.

Figures 5 and 6 shows the results obtained while using the k ω turbulence model and k ω SST turbulence model, respectively. The value of cf predicted by both the model shows wobbling since they are very sensitive in the buffer region. Even in these models, the results obtained by using nutBufferWallfunction are superior over the one obtained by using standard wall function. From the comparative study, it is evident that the newly implemented wall function is providing better results compared to the

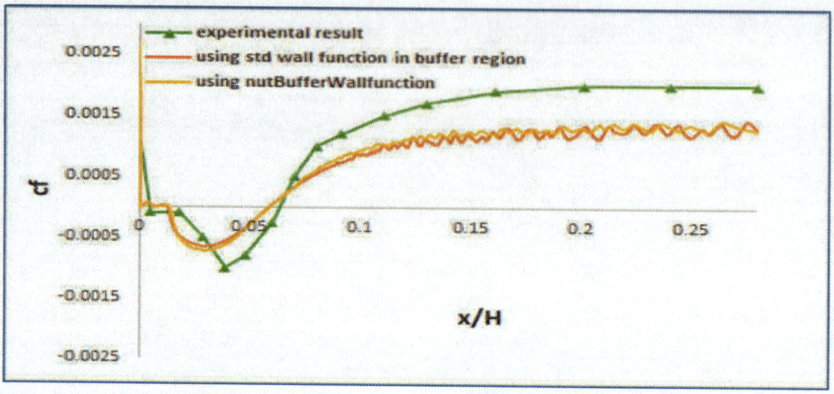

Fig. 5 Comparison of cf value for standard and nutBufferWallFunction with experimental data using RANS $k\ \omega$ model

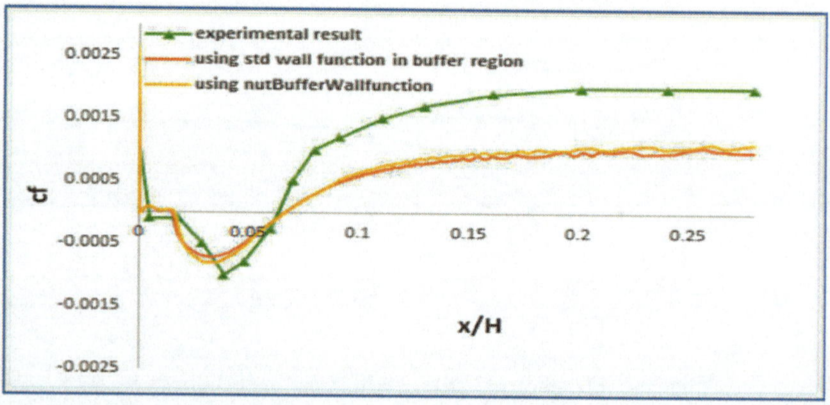

Fig. 6 Comparison of cf value for standard and nutBufferWallFunction with experimental data using RANS $k\ \omega$ SST model

standard wall function available in CFD code. The same trend can be seen in all the RANS turbulence models used for this study.

5 Conclusion

A new wall function namely nutBufferWallFunction has been implemented in OpenFOAM source code. The study has been conducted for backward-facing step experiment using different RANS turbulence models. The value of coefficient of friction along the lower wall has been taken for the analytical study. From the comparative analysis, it is found that the newly implemented wall function is providing better results compared to the available standard wall function. This nutBufferWallFunction can be used for obtaining a better result for computational grid having y^+ value in the buffer region. The proposed wall function can be useful for getting accurate simulation result for complex geometries.

References

1. Hinze JO (1975) Turbulence, 2nd edn. McGraw Hill, New York
2. Van Driest ER (1956) On turbulent flow near a wall. J Aeronaut Sci 23(11):1007–1011
3. Šarić S, Basara B, Žunič Z (2017) Advanced near-wall modeling for engine heat transfer. Int J Heat Fluid Flow 1(63):205–211
4. Joshi A, Assam A, Nived MR, Eswaran V (2019) A generalised wall function including compressibility and pressure-gradient terms for the Spalart–Allmaras turbulence model. J Turbul 20(10):626–660
5. Li W, Zhang C, Chen T, Min J, Sénéchal D, Mimouni S (2019) A unified wall function for wall condensation modelling in containment multi-component flows. Nucl Eng Des 1(348):24–36
6. Chmielewski M, Gieras M (2013) Three-zonal wall function for k-ε turbulence models. Comput Methods Sci Technol 19(2):107–114
7. Lakshman R, Basak R (2018) Analysis of transformed fifth order polynomial curve for the contraction of wind tunnel by using OpenFOAM. IOP Conf Ser Mater Sci Eng 377(1):012048
8. Greenshields CJ (2015) OpenFOAM—the open source CFD toolbox—user guide. OpenFOAM Foundation Ltd.
9. Liu F (2016) A thorough description of how wall functions are implemented in OpenFOAM. In: Proceedings of CFD with OpenSource software, pp 1–33
10. Driver DM, Seegmiller HL (1985) Features of a reattaching turbulent shear layer in divergent channelflow. AIAA J 23(2):163–171

Numerical Simulation of Low Reynolds Number Gusty Flow Past Two Side-By-Side Circular Cylinders

Yagneshkumar A. Joshi, Deep Pandya, Rameshkumar Bhoraniya, and Atal Bihari Harichandan

Abstract The fluid flow past circular cylinders is still being investigated to understand the complex phenomenon involved. There are several applications in the area of structural and mechanical designs where cylinders in the side-by-side arrangement are being used. In some of the applications, cylinder array has to experience flow with different gust frequencies naturally while some of the applications may have advantages by introducing artificial gust frequencies in the flow. The proposed work represents simulation results of low Reynolds number (Re = 100) gusty flow past two circular cylinders in a side-by-side arrangement with three different cases of different gust frequencies and with three different gaps between cylinders. Extended flux reconstruction-based two-dimensional fully explicit Navier–Stokes solver is used for the numerical computation with a collocated grid comprising triangular mesh which solves the complete Navier–Stokes equation in the physical plane itself. To clearly understand the flow features associated with wake pattern streamlines and vorticity contours are evaluated. The Strouhal number is computed to derive the vortex shedding frequency. The variations in flow properties for three gusty flow cases with three different values of transverse gap are investigated and concluded.

Keywords Gusty flow · Low Reynolds number · Cylinder

Y. A. Joshi (✉) · D. Pandya
Marwadi University (MU), Rajkot, Gujarat, India
e-mail: eryagnesh@gmail.com

R. Bhoraniya
Department of Mechanical Engineering, Marwadi University (MU), Rajkot, Gujarat, India

A. B. Harichandan
School of Mechanical Engineering, Kalinga Institute of Industrial Technology (KIIT) Deemed to be University, Bhubaneswar, Odisha 751024, India

S. Revankar et al. (eds.), *Proceedings of International Conference on Thermofluids*, Lecture Notes in Mechanical Engineering,
https://doi.org/10.1007/978-981-15-7831-1_7

1 Introduction

The vortex shedding is very interesting flow feature observed for flow past bluff bodies. Study of the effects of vortex shedding is significant in aerodynamics and wind engineering. Whenever the flow past bluff bodies like cylinder, due to the unsteady pressure distribution behind the cylinders, vortices shed continuously from each side of cylinder and cylinder is subjected to fluctuating aerodynamic forces. This alternate shedding of vortices forms a wake pattern called Von Karman vortex street in the downstream and the frequency of vortex shedding which is defined as Strouhal number determines the structural vibration that body experience. The investigation of this structural vibration becomes very significant in many engineering practices like singing, flutter and galloping of overhead power lines, tall towers and antenna and bridge construction due to the wind gust. Also, the mentioned structure involves several cylinders in side-by-side array. The literatures reviewed show that lots of numerical and experimental works are reported for incompressible flow past cylinder arrangements but the previous work has mainly focused on uniform flow (Table 1).

The gusty flow past circular cylindrical structures in different arrangements are yet not been dealt with in depth. Such study can be very helpful for the structural

Table 1 Literature review

Author	Year	Important observations
Harichandan and Roy [1]	2010	They introduced explicit consistent flux reconstruction (CFR) scheme for unstaggered grid has simulated uniform flow over single as well as multiple cylinders. They found that when circular cylinders are side-by-side arranged, the intervention effect is stronger for small transverse gap
Bryja [2]	2009	Author carried out stochastic analysis of suspension bridge under the wind gust
Golubev and Nguyen [3]	2010	They investigated gust-airfoil response to analyze viscous effect with high accuracy
Afgan et al. [4]	2013	They numerically investigated flow over flat plate with uniform inlet as well as oblique incident gust flow at Re 750 with varying amplitude and time period
Zhan et al. [5]	2017	They investigated the behavior of power extraction performance on flapping foil at Reynolds number of 1100 with gusty flow. Higher efficiency of power extraction is achieved by stronger fluctuation of gust amplitude
Rana et al. [6]	2017	They numerically investigated the gust response of rotating circular cylinder at low Reynolds number to study combined effect of rotation of cylinder and gust impulse
Parekh et al. [7]	2018	They studied the gusty flow past single cylinder. They reported variations of coefficient of lift and drag and flow separation point with variation in the angular frequency of the gust (ω)

design as well as mechanical flow problems. These motivated authors for the present work.

2 Methodology

In this paper, gust flow past side-by-side array of two cylinders is numerically investigated with different gust frequency and different transverse gap between cylinders at low Reynolds number of 100. Navier–Stokes solver developed by Harichandan and Roy [1] for two-dimensional incompressible flow past object is adopted and extended its application for side-by-side arrangement of circular cylinder for the numerical computation. The streamlines and the vorticity contours are evaluated for the clear understanding of flow features associated with wake pattern. The Strouhal number is computed to derive the vortex shedding frequency.

2.1 FVM Solver and Validation

The solver is based on finite volume method (FVM) used of collocated grid comprising triangular mesh. Key feature of the solver is flux reconstruction-based fully explicit scheme that solves the complete Navier–Stokes equation in the physical plane itself. The inlet uniform velocity of domain is perturbed with a velocity component represented as, $v_g = I\cos(kx - \omega t)$ where I is intensity ($I = I \times U_\infty$) of gust relative to the mean flow, k is gust wave number, ω represents the angular gust frequency and U_∞ stands for inlet velocity. The solution is based on temporal property and not on spatial property.

Linear gust frequency can be defined as $f = \omega/2\pi$ while the gust frequency imposed can be represented as $f_i = \omega D/2U_\infty$.

Figure 1 represents the vorticity contour plot, the streamlines plot and the evolution of lift and drag force coefficients for flow past a single circular cylinder at Re 100. A data plotting software tool "Tecplot" is used to plot results obtained by used solver.

The solver is validated by comparing the numerical results obtained from solver for uniform flow past single circular cylinder at Re = 100 with the result available in open literature as shown in Table 2.

The results obtained through solver is in good agreement with open literature results; hence, we can say that the used solver is reliable for the numerical investigation of incompressible 2D flows at low value of Re.

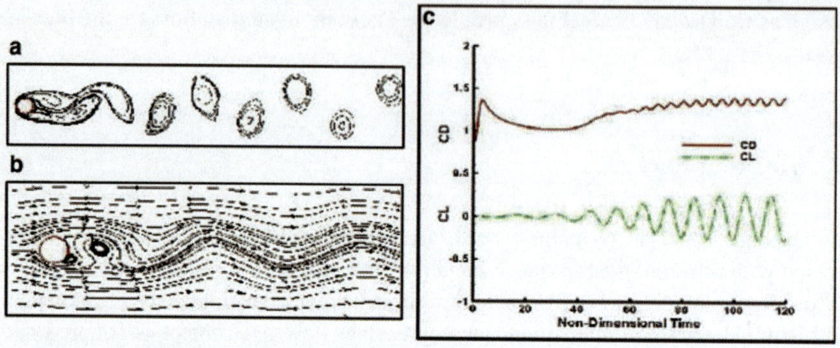

Fig. 1 **a–c** Vorticity contour, streamlines and lift and drag force coefficients plot

Table 2 Lift coefficient and drag coefficient for flow past single cylinder at Re = 100

Parameters	Drag coefficient (CD)	Lift coefficient (CL)
Reynolds number	100	100
Meneghini [8]	1.370 ± 0.010	–
Ding et al. [9]	1.356 ± 0.010	±0.287
Braza et al. [10]	1.364 ± 0.015	±0.25
Present results	1.359 ± 0.010	±0.251

2.2 *Flow Past Two Side-By-Side Circular Cylinders*

In this paper, gust flow past side-by-side array of two cylinders is numerically investigated with different gust frequency and different transverse gap between cylinders at low Reynolds number of 100. Figure 2 shows the physical flow problem where two circular cylinders of equal diameter D are configured in side-by-side arrangement to the direction of flow. The gust source is generated by superimposing v—velocity component perturbation of sinusoidal wave in the upstream direction of cylinders. "T" represents the transverse gap that is the center-to-center distance between two cylinders. Rectangular domain of $23D \times 35D$ is considered for computation. Upper and lower wall are away at a distance of 10 times the cylinder diameter from the nearest cylinder center. The distance of upper and lower wall of flow domain is same from corresponding cylinders. The inlet and the outlet wall are at a distance of 10 times and 20 times the cylinder diameter from the nearest cylinder center, respectively.

Figure 3 shows the unstructured grid comprising of triangular mesh. Investigation has been carried out for three different transverse gap of cylinders $3D$, $2D$ and $1.5D$, respectively with three different gust frequencies 0.2π, 0.5π and 1.0π.

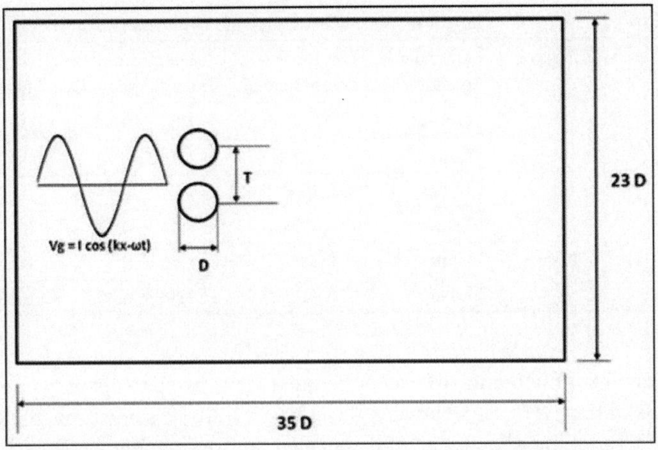

Fig. 2 Physical problem definition

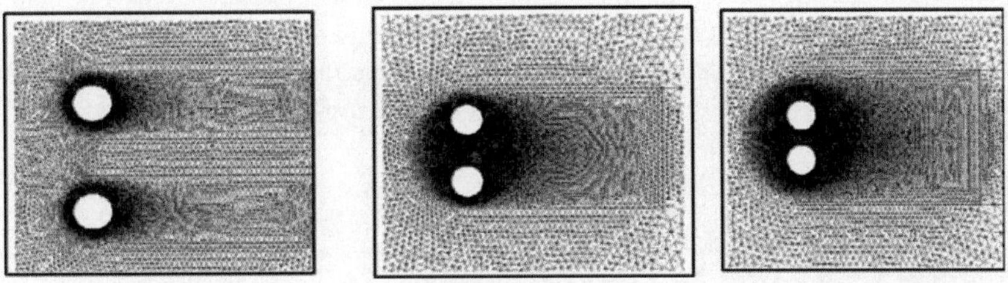

Transverse-Gap 3×D   Transverse-Gap 2×D   Transverse Gap 1.5×D

Fig. 3 Triangular meshing around cylinders with different values of *T*

3 Results and Discussion

Result data obtained for combination of three different gust frequencies and three different transverse gap and hence total nine cases are simulated and analyzed. The data obtained is evaluated in terms of force coefficients and Strouhal number. The force coefficients CL and CD are analyzed through open-source "Tecplot" software while the Strouhal number is calculated through fast Fourier transform (FFT) analysis using open-source "Origin360" software.

3.1 CASE A: Angular Gust Frequency 0.2π

Referred to Table 3, it is observed that gap flow slightly deflects toward upper cylinder and consequently wake recirculation is observed downstream of lower cylinder which

Table 3 Flow parameters for gusty flow past two side-by-side cylinders (Re = 100, $\omega = 0.2\pi$)

Gust freq.	Trans. gap T	Cylinder position	Lift force coefficient CL	Drag force coefficient CD	Strouhal number St
$\omega = 0.2\pi$	$T = 3D$	Upper	0.1 ± 0.67	1.04 ± 0.18	0.10
		Lower	-0.1 ± 0.8	1.46 ± 0.26	0.10
	$T = 2D$	Upper	0.19 ± 0.63	1.01 ± 0.15	0.10
		Lower	-0.09 ± 0.63	1.41 ± 0.31	0.10
	$T = 1.5D$	Upper	0.35 ± 0.68	1.16 ± 0.3	0.10
		Lower	-0.41 ± 0.73	1.58 ± 0.23	0.10

experiences higher fluctuation of lift coefficient as compared to upper cylinder for all three values of transverse gap between the cylinders. At the same time, the values of CL are positive for upper cylinder while negative for lower cylinder. The CL is lesser for $T = 3D$ and increasing in nature with decrease in the gap to $T = 2D$ and $T = 1.5D$. It is also observed that the values of CD for all three transverse gap are lesser for upper cylinder and higher for lower cylinder and increases with decrease in transverse gap. One narrow wake pattern is formed immediately behind the lower cylinder and one wide wake pattern is observed at some distance downstream of upper cylinder. The deflected gap flow oscillates randomly that result in flip-flopping pattern of wake. The Strouhal number is in between 0.09 and 0.1.

3.2 *CASE B: Angular Gust Frequency 0.5π*

Referred to Table 4 for the case of $\omega = 0.5\pi$, similar nature of gap flow which is slightly deflects toward upper cylinder and consequently wake recirculation is observed downstream of lower cylinder which experiences higher fluctuation of lift coefficient as compared to upper cylinder for $T = 3D$ and $T = 2D$, but it is reversed in nature for $T = 1.5D$. The values of CL are positive for upper cylinder while negative for lower cylinder. The fluctuation of CL is higher for $T = 3D$ for upper cylinder.

Table 4 Flow parameters for gusty flow past two side-by-side cylinders (Re = 100, $\omega = 0.5\pi$)

Gust freq.	Trans. gap T	Cylinder position	Lift force coefficient CL	Drag force coefficient CD	Strouhal number St
$\omega = 0.5\pi$	$T = 3D$	Upper	0.01 ± 0.3	1 ± 0.08	0.22
		Lower	-0.13 ± 0.31	1.27 ± 0.07	0.22
	$T = 2D$	Upper	0.19 ± 0.17	1.01 ± 0.05	0.25
		Lower	-0.3 ± 0.11	1.4 ± 0.07	0.25
	$T = 1.5D$	Upper	0.39 ± 0.13	1.03 ± 0.05	0.25
		Lower	-0.43 ± 0.18	1.43 ± 0.04	0.25

Table 5 Flow parameters for gusty flow past two side-by-side cylinders (Re = 100, $\omega = 1.0\pi$)

Gust freq.	Trans. gap T	Cylinder position	Lift force coefficient CL	Drag force coefficient CD	Strouhal number St
$\omega = 1.0\pi$	$T = 3D$	Upper	0.03 ± 0.13	1.05 ± 0.06	0.15
		Lower	$- 0.11 \pm 0.15$	1.35 ± 0.05	0.16
	$T = 2D$	Upper	0.23 ± 0.13	1.05 ± 0.06	0.19
		Lower	$- 0.2 \pm 0.15$	1.47 ± 0.07	0.15
	$T = 1.5D$	Upper	0.43 ± 0.06	1.09 ± 0.04	0.16
		Lower	$- 0.46 \pm 0.16$	1.48 ± 0.05	0.17

The values of CD for all three transverse gap are lesser for compared to in case of $\omega = 0.2\pi$. The Strouhal number is in between around 0.240 and 0.254 which is higher in compare to in case of $\omega = 0.2\pi$.

3.3 *CASE C: Angular Gust Frequency 1π*

Referred to Table 5 for $T = 3D$, lower cylinder experiences higher fluctuation while for $T = 2D$ and $T = 1.5D$ upper cylinder experiences higher fluctuation of lift coefficient. The value of CL increases with decrease in the transverse gap between the cylinders. It is also observed that the values of CD for transverse gap are higher for upper cylinder for $T = 3D$ and $T = 2D$ while it is lesser for upper cylinder for $T = 1.5D$. The Strouhal number is in between 0.148 and 0.161 which is lesser in compare to $\omega = 0.5\pi$ but it is higher than that for $\omega = 0.2\pi$.

4 Conclusions

The Strouhal number (St) indicates rapidly subsequent vortex shedding on downstream of cylinders. The investigated value of St, for all three constant values of transverse gap ($T = 3D$, $2D$, & $1.5D$) between cylinders in side-by-side arrangement, is found lesser for angular gust frequency (ω) = 0.2π, increased for $\omega = 0.5\pi$ and again decreased for $\omega = 1\pi$. When the similar flow with similar arrangement is investigated with three different considered values of T, keeping value of ω constant, it is found that the transverse gap is not affecting significantly on the Strouhal number for Re 100.

The lift coefficient (CL) for lower cylinder is negative, while it is positive for upper cylinder in all the considered cases for different values of ω and T. The drag coefficient (CD) is lesser for upper cylinder than lower cylinder. The CD increases with decrease in transverse gap and it increases with increase in gust angular frequency for considered cases of ω and T.

References

1. Harichandan AB, Roy A (2010) Numerical investigation of low Reynolds number flow past two and three circular cylinders using unstructured grid CFR scheme. Int J Heat Fluid Flow 31(2):154–171
2. Bryja D (2009) Stochastic response analysis of suspension bridge under gusty wind with time-dependent mean velocity. Arch Civ Mech Eng 9(2):15–38
3. Golubev VV, Nguyen L (2010) High-accuracy low-Re simulations of airfoil-gust and airfoil-vortex interactions, pp 1–19
4. Afgan I, Benhamadouche S, Han X, Sagaut P, Laurence D (2013) Flow over a flat plate with uniform inlet and incident coherent gusts. J Fluid Mech 720:457–485
5. Zhan J, Xu B, Wu J, Wu J (2017) Power extraction performance of a semi-activated flapping foil in gusty flow. J Bionic Eng 14(1):99–110
6. Rana K, Manzoor S, Sheikh NA, Ali M, Ali HM (2017) Gust response of a rotating circular cylinder in the vortex suppression regime. Int J Heat Mass Transf 115:763–776
7. Parekh CJ, Roy A, Harichandan AB (2018) Numerical simulation of incompressible gusty flow past a circular cylinder. Alex Eng J 57(4):3321–3332
8. Meneghini FSJR (2001) Numerical simulation of flow interference between two circular cylinders in tandem and side-by-side arrangements. J Fluids Struct 15:327–350
9. Ding H, Shu C, Yeo KS, Xu D (2007) Numerical simulation of flows around two circular cylinders by mesh-free least square-based finite difference methods. Int J Numer Methods Fluids 53:305–332
10. Braza M, Chassaing P, Ha Minh H (1986) Numerical study and physical analysis of the pressure and velocity fields in the near wake of a circular cylinder. J Fluid Mech 165:79–130

Comparative Analysis of Inlet Boundary Conditions for Atmospheric Boundary Layer Simulation Using OpenFOAM

R. Lakshman, Nitin Pal, and Ranjan Basak

Abstract The atmospheric boundary layer has been extensively studied by using computational fluid dynamics. The simulation of the atmospheric boundary layer has been conducted either by solving RANS equation or by LES simulation. LES simulation provides higher accuracy in the result but with a very high computational cost. Due to this higher cost in computational effort, RANS turbulence modelling has been preferred by most of the researchers. In Computational fluid dynamics, the role of inlet conditions is more important for any simulation problem. In this study, a comparative analysis has been conducted using open source CFD tool OpenFoam for different inlet boundary conditions for the atmospheric boundary layer. Horizontal homogeneity is a very keen factor while simulating the atmospheric boundary layer. The analysis has been conducted by taking the inlet and outlet profiles of velocity, turbulent dissipation rate, and turbulent kinetic energy. Analysis with different turbulence model constants was also conducted and the results are discussed.

Keywords Atmospheric boundary layer simulation · OpenFOAM · Horizontal homogeneity · CFD · RANS

1 Introduction

For the simulation of the atmospheric boundary layer, horizontal homogeneity is a very important parameter. A zero streamwise gradient for all the variables refers to horizontal homogeneity. The atmospheric boundary layer having horizontal homogeneity is called an equilibrium atmospheric boundary layer. An improper boundary condition can affect the equilibrium atmospheric boundary layer [1]. The significance of simulating equilibrium atmospheric boundary layer has been highlighted by several researchers [2, 3]. Careful selection of inlet boundary conditions is required to

R. Lakshman (✉) · N. Pal · R. Basak
Department of Mechanical Engineering, National Institute of Technology Sikkim, Ravangla, South Sikkim, India
e-mail: lakshmanr03@gmail.com

S. Revankar et al. (eds.), *Proceedings of International Conference on Thermofluids*, Lecture Notes in Mechanical Engineering,
https://doi.org/10.1007/978-981-15-7831-1_8

achieve a horizontally homogeneous ABL profile. Altering the wall function parameters have done some slight improvement to the level of homogeneity of the atmospheric boundary layer [4]. Richard and Hoxey suggested an inlet boundary condition for the mean velocity and turbulence variables. It is considered as one of the most important accomplishments and is widely used in the RANS atmospheric boundary layer simulation [5]. One of the drawbacks of this model is the constant value allocated for turbulence kinetic energy which is different from real case scenario. Yang et al. has further investigated the simulation of atmospheric boundary layer and based on the assumption of turbulence equilibrium, an expression for turbulence kinetic energy has been derived [6]. Other researchers also have done credible work on the simulation of atmospheric boundary layer [7, 8].

In this study, the RANS k–ε turbulence model has been used to compare the result of ABL simulation using inlet boundary condition proposed by Richard and Hoxey and Yang et al. Due to the flexibility in computation, open-source CFD tool OpenFOAM has been attracted to many researchers [9, 10, 11]. Here, in this study, OpenFOAM is used for the simulation of the atmospheric boundary layer [12].

2 Theory

Transformation of the k–ε model transport equations are given by

$$\frac{\partial(\rho k)}{\partial t}+\frac{\partial(\rho k u_j)}{\partial x_j}=\frac{\partial(\rho k)}{\partial x_j}\left[\left(\mu+\frac{\mu_t}{\sigma_k}\right)\frac{\partial k}{\partial x_j}\right]+P_k-\rho\varepsilon \tag{1}$$

$$\frac{\partial(\rho\varepsilon)}{\partial t}+\frac{\partial(\rho\varepsilon u_j)}{\partial x_j}=\frac{\partial(\rho k)}{\partial x_j}\left[\left(\mu+\frac{\mu_t}{\sigma_\varepsilon}\right)\frac{\partial\varepsilon}{\partial x_j}\right]+C_{1\varepsilon}\frac{\varepsilon}{k}P_k-C_{2\varepsilon}\rho\frac{\varepsilon^2}{k} \tag{2}$$

where ρ is the density of fluid, k, ε and μ_t are the turbulent kinetic energy, turbulence dissipation rate, and eddy viscosity respectively and P_k is the production of turbulent kinetic energy. In k–ε turbulence model Eddy viscosity can be shown as Eq. (3).

$$\mu_t=\rho C_\mu\frac{k^2}{\varepsilon} \tag{3}$$

Yang et al. has proposed an inlet boundary condition for atmospheric boundary layer flow by taking the assumption of steady, incompressible and horizontally homogeneous flow. Horizontal homogeneity indicates that u, k, and ε are the invariant with the computational domain of the geometry and adjacent coordinates x, y, and only alter with height above ground (z). Applying all these assumptions and considering highly turbulent flows ($\mu_t \gg \mu$), in Eqs. (1) and (2). The transport equation for turbulence kinetic energy can simplified as

$$k(z+z_o)\frac{\partial k}{\partial z} = \text{const.} \tag{4}$$

Assuming the rate of dissipation equal to rate of turbulent kinetic energy, the turbulent dissipation rate, ε can be simplified as Eq. (5). Equation (6) represents the mean velocity profile based on logarithmic law.

$$\varepsilon = C_\mu^{1/2} k(z)\frac{\partial u}{\partial z} \tag{5}$$

$$u = \frac{u_*}{k}\ln\left(\frac{z+z_o}{z_o}\right) \tag{6}$$

After obtaining the solution of Eq. (4), a simple linear transformation of the constants are performed to obtain the expression of turbulent kinetic energy as Eq. (7). Considering the relationship ε with k and u, and by using Eqs. (5) and (6), the expression for ε is shown in the Eq. (8).

$$k = \frac{u_*^2}{\sqrt{C_\mu}}\sqrt{C_1 \cdot \ln\left(\frac{z+z_o}{z_o}\right) + C_2} \tag{7}$$

$$\varepsilon = \frac{u_*^3}{k(z+z_0)}\sqrt{C_1 \cdot \ln\left(\frac{z+z_o}{z_o}\right) + C_2} \tag{8}$$

Richard and Hoxey have also proposed an inlet boundary condition for atmospheric boundary layer simulation. They have taken the assumption of steady incompressible 2-dimensional flow, zero vertical velocity and a constant pressure and shear stress for obtaining the inlet boundary condition for k–ε turbulence model. From Reynolds Average Navier–stokes transport equation for k–ε turbulence model and the above-mentioned assumptions, Richard and Hoxey came with the following expression for inlet boundary condition.

$$u = \frac{u_*}{k}\ln\left(\frac{z+z_o}{z_o}\right) \tag{9}$$

$$k = \frac{u_*^2}{\sqrt{C_\mu}} \tag{10}$$

$$\varepsilon = \frac{u_*^3}{k(z+z_0)} \tag{11}$$

where K is the von Karman constant, z_o is the aerodynamic roughness dimension, u_* is the friction velocity and C_μ is the turbulence model constant.

In this study along with the standard turbulence model constant, a modified turbulence model constants where also used based on the previous studies. In this paper, the

Table 1 The turbulence model constants used for STD and ALT model

Model	C_μ	σ_k	σ_ε	$C_{1\varepsilon}$	$C_{2\varepsilon}$
STD model	0.09	1.0	1.3	1.44	1.90
ALT model	0.028	1.67	2.51	1.5	1.92

simulation with standard turbulence model constants are named as STD model and the modified one is named as ALT model. The values of turbulence model constants for both cases used in this paper are listed in Table 1.

3 Numerical Model

The geometry used for this study is taken similar to the TJ-1 wind tunnel model. The details of the TJ-1 wind tunnel test section is shown in Table 2 and the geometry created for the analysis is shown in Fig. 1. The analysis has been conducted in the open-source CFD tool OpenFOAM. OpenFOAM provided SimpleFOAM solver is used for the simulation. Inlet condition similar to TJ-1 wind tunnel test has been taken for the simulation. Constant zero pressure has been given at the outlet. Structured hexahedral Meshing is done using snappyHexMesh utility. A total of 217,440 cells with an average non-orthogonality of 9.736 has been selected for the simulation.

Table 2 Details of TJ-1 wind tunnel test section

Element	Dimension (m)
Width	1.8
Height	1.8
Length	12

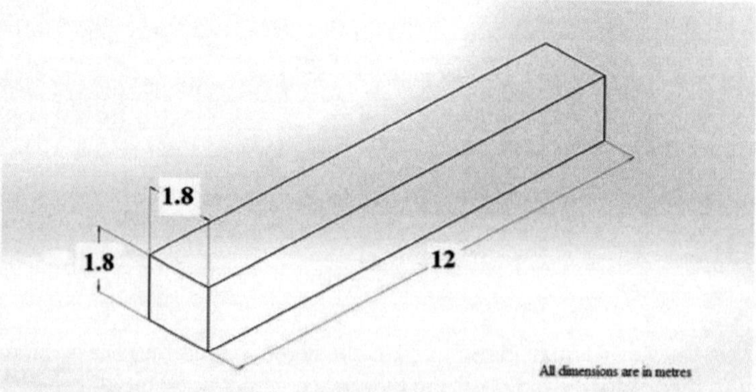

Fig. 1 The dimensions of the geometry used for the simulation

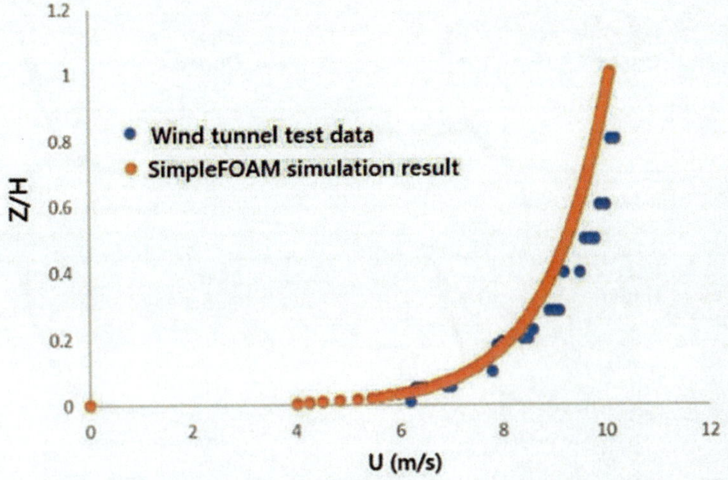

Fig. 2 Comparison of velocity profile at the outlet

Figure 2 shows the validation results of simpleFOAM solver with the experimental data. Figure 2 shows good agreement with the experimental result.

4 Result and Discussion

In the study, the inlet boundary conditions provided by Richard and Hoxey is been compared with the one proposed by Yang et al. The open-source computational fluid dynamics tool OpenFOAM is used for the analysis. The study has been done by taking two models, named as STD model and ALT model. Here we compare the horizontal homogeneity for the velocity (u), turbulence kinetic energy (k) and dissipation rate (ε) by using the inlet conditions provided by Richard and Hoxey and Yang et al.

From Fig. 3a, it is clear that by using the inlet condition provided by Richard and Hoxey the velocity profile is sustained throughout the domain in the case of ALT model. For STD model, there exist some deviation of the profile at outlet compared to inlet. This shows the ALT model is suitable for predicting the velocity profile for atmospheric boundary layer simulation. The results obtained by using the Yang et al. boundary condition is also similar to Richard and Hoxey (Fig. 3b).

Poor sustainability of turbulent kinetic energy is observed when using Richard and Hoxey inlet condition (Fig. 4a), for both STD and ALT model. STD model provided slightly better result compared to ALT model. The irregularity is mainly due to the constant value of turbulent kinetic energy provided at the inlet. Yang's inlet boundary condition provides better horizontal homogeneity compared to the one provided by Richard and Hoxey. Compared to STD model, ALT model is giving

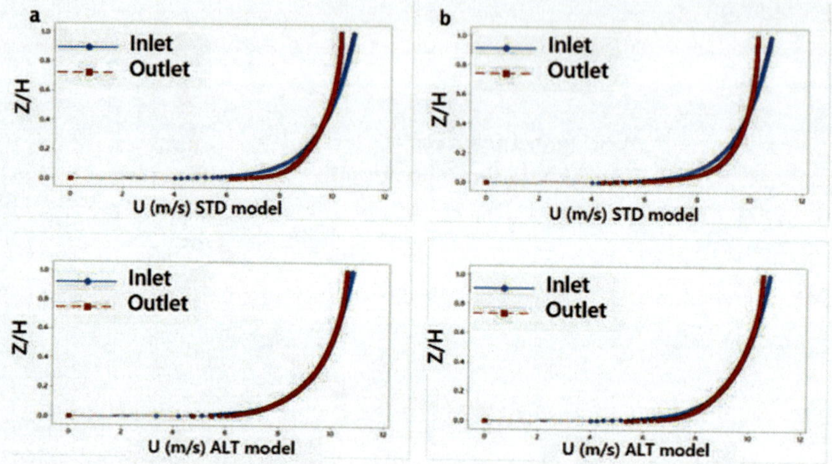

Fig. 3 **a** The comparison of predicted outlet velocity with the inlet velocity by using STD and ALT model using Richard and Hoxey inlet boundary condition; **b** using Yang et al. inlet boundary condition

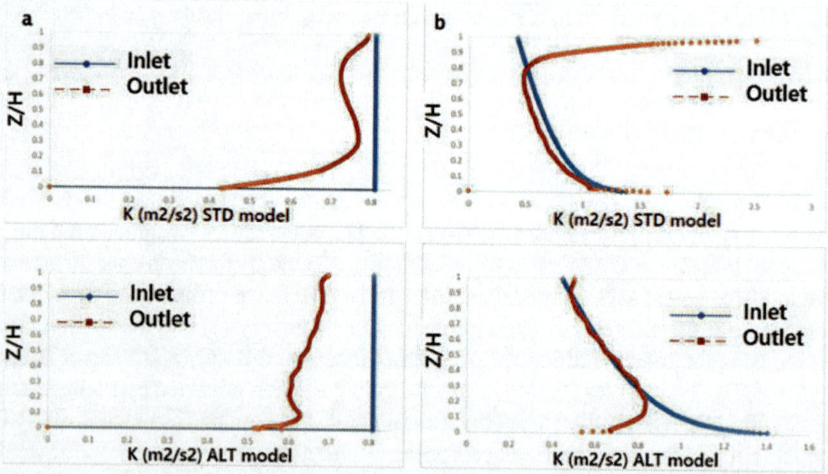

Fig. 4 **a** The comparison of predicted outlet TKE with the inlet TKE by using STD and ALT model using Richard and Hoxey inlet boundary condition; **b** using Yang et al. inlet boundary condition

better results for Yang et al. boundary condition. But even in ALT model there is some loss in sustainability of TKE is seen up to ¼th height of the domain. This is one of the drawbacks seen in ALT model.

From Fig. 5b, it is seen that Yang et al. boundary condition with STD model is providing better horizontal homogeneity for turbulence dissipation rate. The profile

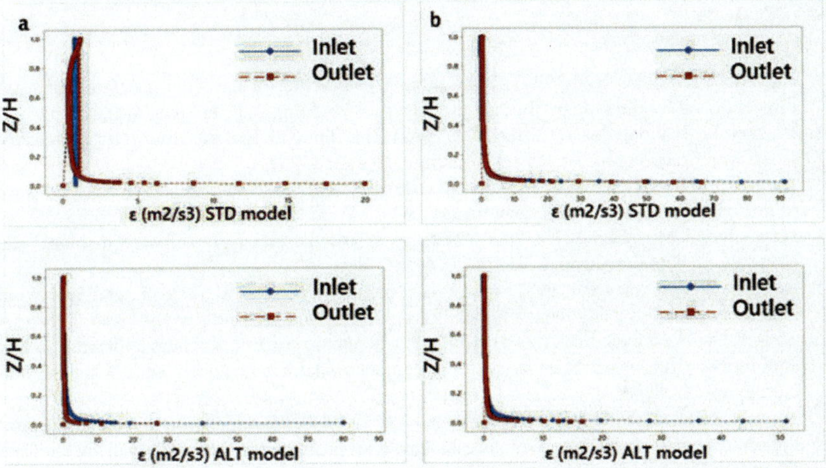

Fig. 5 **a** The comparison of predicted outlet turbulence dissipation rate with the inlet TKE by using STD and ALT model using Richard and Hoxey inlet boundary condition; **b** using Yang et al. inlet boundary condition

obtained by using Richard and Hoxey boundary condition cannot be preferred for the simulation of atmospheric boundary layer (Fig. 5a). Compared to the STD model, the Richard and Hoxey boundary condition with ALT model gives better performance.

5 Conclusion

In the CFD simulations of the atmospheric boundary layer, horizontal homogeneity of the profile plays an important role in the consistency of the results. A comparative analysis of the inlet boundary condition for the atmospheric boundary layer simulation proposed by Richard and Hoxey and Yang et al. has been conducted by using OpenFOAM. An analysis with different turbulence model constants (STD model and ALT model) were also conducted. From the simulation results, it is observed that the velocity profile and the turbulence dissipation rate profile obtained by using both inlet condition are giving satisfactory result, whereas for the Richard and Hoxey turbulent kinetic energy profile is giving very unsatisfactory result. Yang et al. inlet profile for turbulent kinetic energy is giving reasonable horizontal homogeneity with the cost of some deflection near the ground. From all the results, it is observed that ALT model is giving better horizontal homogeneity compared to STD model.

References

1. Richards PJ, Younis BA (1990) Comments on "prediction of the wind-generated pressure distribution around buildings" by E.H. Mathews. J Wind Eng Ind Aerodyn 34(1):107–110
2. Blocken B, Stathopoulos T, Carmeliet J (2007) CFD simulation of the atmospheric boundary layer: wall function problems. Atmos Environ 41(2):238–252
3. Hargreaves DM, Wright NG (2007) On the use of the k–ε model in commercial CFD software to model the neutral atmospheric boundary layer. J Wind Eng Ind Aerodyn 95(5):355–369
4. Hargreaves DM, Wright NG (2006) The use of commercial CFD software to model the atmospheric boundary layer. CWE 14(2006):797–800
5. Richards PJ, Hoxey RP (1993) Appropriate boundary conditions for computational wind engineering models using the k-ϵ turbulence model. J Wind Eng Ind Aerodyn 1(46):145–153
6. Yang Y, Gu M, Chen S, Jin X (2009) New inflow boundary conditions for modelling the neutral equilibrium atmospheric boundary layer in computational wind engineering. J Wind Eng Ind Aerodyn 97(2):88–95
7. Toparlar Y, Blocken B, Maiheu B, van Heijst G (2019) CFD simulation of the near-neutral atmospheric boundary layer: new temperature inlet profile consistent with wall functions. J Wind Eng Ind Aerodyn 1(191):91–102
8. Han Y, Stoellinger M, Naughton J (2016) Large eddy simulation for atmospheric boundary layer flow over flat and complex terrains. J Phys Conf Ser 753(3):032044
9. Lakshman R, Basak R (2018) Analysis of transformed fifth order polynomial curve for the contraction of wind tunnel by using OpenFOAM. IOP Conf Ser Mater Sci Eng 377(1):012048
10. Lakshman R, Basak R (2020) Analysis of transformed sixth-order polynomial for the contraction wall profile by using OpenFOAM. In: Recent advances in theoretical, applied, computational and experimental mechanics 2020. Springer, Singapore, pp. 133–144
11. Peralta C, Nugusse H, Kokilavani SP, Schmidt J, Stoevesandt B (2014) Validation of the simpleFoam (RANS) solver for the atmospheric boundary layer in complex terrain. In: ITM web of conferences 2014, vol 2. EDP Sciences, p 01002
12. Greenshields CJ (2015) OpenFOAM—the open source CFD toolbox—user guide. OpenFOAM Foundation Ltd.

Parametric Analysis of Coupled Thermal Hydraulic Instabilities in Forced Flow Channel Using Reduced-Order Three-Zone Model

Daya Shankar, Harabindu Debnath, and Indira Kar

Abstract One of the significant issues of supercritical water reactor (SCWR) is the variation of coolant density along its axial direction of flow, which further creates a stability issue in the reactor. This causes the dynamic instability in the coolant channel. Present work is focused to study the parametric effects on the stability of the system using a simplified three-zone lumped parameter model by considering the neutronics coupled with thermal hydraulics. The coolant channel is divided into three zones, namely heavy fluid region, light fluid region, and intermediate heavy and light mixture fluid region. The interface of these hypothetical regions is separated by time dependence boundaries. A set of governing equations are solved for the individual zones that provide another set of algebraic and ODEs, which has coolant enthalpy and the length of each boundary as a primary variable. The US reference design of SCWR has been considered at 25 MPa as the operating pressure. Increasing the length of the coolant channel and hydraulic diameter can destabilize the system. Increasing inlet orifice and decreasing in the exit orifice coefficient have a stabilizing effect on the system.

Keywords Supercritical water reactor · Coolant channel · Inlet orifice · Lumped parameter model

Nomenclature

Area (m^2)	A
Absolute temperature (K)	T
$\beta\, \gamma$ (–)	σ
Core length (m)	L
Density (kg m^{-3})	ρ
Delayed neutron fraction (–)	β

D. Shankar (✉) · H. Debnath · I. Kar
Department of Mechanical Engineering, ICFAI University Tripura, Agartala, India
e-mail: iitgdaya@gmail.com

S. Revankar et al. (eds.), *Proceedings of International Conference on Thermofluids*, Lecture Notes in Mechanical Engineering,
https://doi.org/10.1007/978-981-15-7831-1_9

Enthalpy (kJ kg^{-1})	h
Euler number (–)	N_{Eu}
Feedback reactivity (–)	k
Froude number (–)	N_{Fr}
Friction factor	f
Gravitational acceleration (m s^{-2})	g
Heat flux (kW m^{-2})	$q'o'$
Heat capacity of fuel rod (kJ K^{-1})	C_f
Heat transfer coefficient (kW m^{-2} K^{-1})	α
Heated perimeter (m)	π_h
Hydraulic diameter (m)	D_h
Isobaric specific heat (kJ kg^{-1} K^{-1})	C_p
Mass flux (kg m^{-2} s^{-1})	G
Number of fuel rods (–)	n_f
Neutron generation time (–)	γ
Pseudosubcooling number	N_{SPC}
Precursor density (cm^{-3})	C
Power (kW)	P
Pressure (N m^{-2})	p
Restriction coefficient (m^{-1})	K
Space coordinate (m)	z
Time (s)	t
Transcritical phase-change number (–)	N_{TPC}
Volumetric expansion coefficient (K^{-1})	β

1 Introduction

Supercritical water is characterized with good heat transport characteristics and expansion of substantial volumetric near the point of pseudocritical, identifying itself as a good coolant for modern nuclear reactors. While the temperature of higher cycle promises enhanced thermal efficiency, elimination of several bulky parts, like the steam separator, dryer, and recirculation channels, offers compact design and economically competitive structure. The absence of different phase changes removes the associated constraint with the critical heat flux. Such systems exhibit complicated stability behaviour and may often opt to get large amplitude oscillations with little change in the operating conditions. Functioning in the unstable regime is highly undesirable, and it is better to be avoided, which gives separate thermohydraulic and power fluctuations, sometimes followed by structural vibrations. This makes it necessary to get a simplified perception about the functioning procedure of systems in forced and natural flow situation, with mainly eye on increasing the rate of flow and coefficient of heat transfer.

Vast property variation experienced by the supercritical fluid near the pseudocritical point is the main reason of inducing instability of density wave in the SCWR core, as the density variation in the heated channel can change substantially with little change in flow rate, input power. Therefore, many researchers have attempted to decode the corresponding stability behaviour, taking both experimental and theoretical/computational approaches. Yi et al. [1] studied the stability characteristics of a once-through core of SCWR under a pulse perturbation by calculating system decay ratio. Zhao et al. [2] took a 3-zoned model to approximate the changes of density with respect to enthalpy. To analyse thermal hydraulic stability of SCWR, they determined some novel dimensionless groups referring US design. A stability code for frequency domain (SCWRSA) was developed by Yang [3], making work for a repetition of solution statement, to get the steady-state flow distribution under boundary condition of constant total rate of flow and similar pressure drop in parallel channels. Ambrosini [4] studied the stability of single heated channel with same inlet as well as outlet pressures. A standard$k$–ε turbulence model was employed by Ampomah-Amoako and Ambrosini [5] in a CFD package to predict the instability of heated supercritical channels and a sub-channel fuel assembly. Hou et al. [6] analysed the dynamic characteristics for the fast spectrum zone a parallel channel system. Decent number of research studies focussing on the linear and transient analyses of supercritical loops which is naturally circulated can also be found in the literature [7–9]. Most of them discussed the role of computational aspects, such as nodalization, time spacing, and discretization scheme, on the stability prediction. Quite a few experimental works on natural circulation-based system can also be found [10, 11]. In fact, a scrupulous review of the available literature indicates towards a larger volume of research towards such buoyancy-drive systems, mostly because of their intricate stability behaviour in a simple configuration and lower ranges of flow rates and driving forces.

Forced flow in supercritical channels is more relevant to large-scale applications. They experience larger variation in properties across the core and are more prone towards unstable oscillations. So characteristics of heat transportation of those systems, giving the possible overview reduced strength of heat transfer and exploring the buoyancy effect and acceleration of flow, have been an object of many research studies [12, 13]. But the number of research works looking more into the stability aspects, with a detailed treatment of the property variation, is really scarce. Some database shows the adoption of both linear and nonlinear techniques, with each having its own advantage over the other. While the linear approach can promptly predict the location of stability threshold, exact nature of the temporal variation can be obtained only through the nonlinear one. So, they can be viewed to complement each other, and a proper stability analysis can employ a suitable combination of both of them.

Full solution of the conservation equations can be extremely time-taking, even with the most advanced computational resources. That is also not necessary if the focus is on the identification of the stability threshold and to get an overview of the system response on the side of the stability map. The present study, therefore, aims towards the development of a simplified-order model and employs the same for both linear stability prediction and transient analyses. Basic conservation equations are converted to a set of algebraic and ordinary differential equations by adopting a lumped parameter approach, which is detailed in the next section. Developed set of equations is employed for a comprehensive parametric appraisal of the stability analyses of referring US design of SCWR, with the objective being the identification of a safer combination of geometric and operating variables.

2 Mathematical Model Development

In the present research study, a reduced-order model is considered, where a vertical heater channel is separated into three regions (Fig. 1) following Zhang et al. [14]. The regions are divided from each other by time-dependent boundaries. They can be viewed as (1) liquid-like heavier fluid region with constant density, (2) mixture region experiencing continuous change in density with enthalpy, and (3) vapour-like lighter fluid with constant density.

2.1 Conservation Equations

1-D versions of corresponding momentum, mass, and energy conservation equations can be summarized as

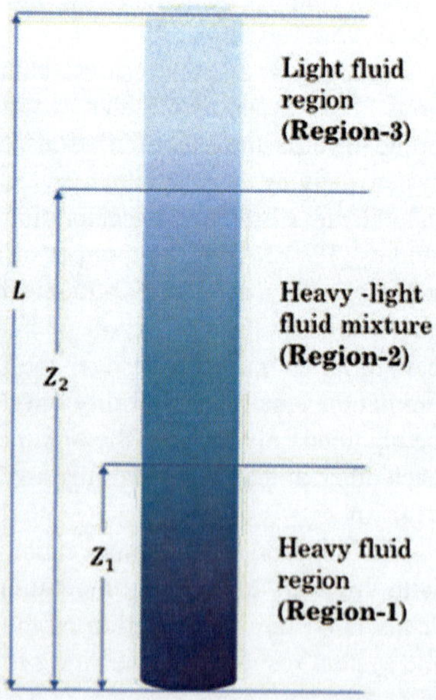

Fig. 1 Schematic view of the three zones of the heated channel

$$\frac{\partial \rho}{\partial t} + \frac{\partial G}{\partial z} = 0 \tag{1}$$

$$\frac{\partial G}{\partial t} + \frac{\partial}{\partial z}\left(\frac{G^2}{\rho}\right) = -\frac{\partial p}{\partial z} - \rho g - \left[\frac{f}{D_{\mathrm{h}}} + 2K_{\mathrm{i}}\delta(z) + 2K_{\mathrm{o}}\delta(z-L)\right]\frac{G^2}{2\rho} \tag{2}$$

$$\frac{\partial}{\partial t}(\rho h) + \frac{\partial}{\partial z}(Gh) = \left(\frac{\pi_{\mathrm{h}}}{A}\right)q''(t) \tag{3}$$

Here, heat loss to the surrounding and viscous dissipation effects are neglected. The following definitions are used to non-dimensionalize the conservation equations.

$$z^* = \frac{z}{L_{\mathrm{c}}},\quad t^* = \frac{tG_{\mathrm{c}}}{\rho_{\mathrm{pc}}L_{\mathrm{c}}},\quad G^* = \frac{G}{G_{\mathrm{c}}},\quad \rho^* = \frac{\rho}{\rho_{\mathrm{pc}}},$$

$$p^* = \frac{(p-p_{\mathrm{r}})\rho_{\mathrm{pc}}}{G_{\mathrm{c}}^2},\quad T^* = \frac{T-T_{\mathrm{pc}}}{T_{\mathrm{pc}}},$$

$$h^* = \left(h - h_{\mathrm{pc}}\right)\frac{\beta_{\mathrm{pc}}}{C_{\mathrm{p\cdot pc}}},\quad N_{\mathrm{Fr}} = \frac{G_{\mathrm{c}}^2}{\rho_{\mathrm{pc}}^2 g L_{\mathrm{c}}},\quad N_{\mathrm{TPC}} = \frac{q_0''\pi_{\mathrm{h}}L_{\mathrm{c}}}{AG_{\mathrm{c}}}\frac{\beta_{\mathrm{pc}}}{C_{\mathrm{p\cdot pc}}},$$

$$N_{\mathrm{Eu}} = \frac{fL_{\mathrm{c}}}{2D_{\mathrm{h}}} \tag{4}$$

Similarly, the non-dimensional forms of the conservation equations can be presented as

$$\frac{\partial \rho^*}{\partial t^*} + \frac{\partial G^*}{\partial z^*} = 0 \tag{5}$$

$$\frac{\partial G^*}{\partial t^*} + \frac{\partial}{\partial z^*}\left(\frac{G^{*2}}{\rho^*}\right) = -\frac{\partial p^*}{\partial z^*} - \frac{\rho^*-1}{N_{\mathrm{Fr}}} - \left[N_{\mathrm{Eu}} + K_{\mathrm{i}}\delta^*(z^*) + K_{\mathrm{o}}\delta^*(z^*-1)\right]\frac{G^{*2}}{\rho^*} \tag{6}$$

$$\frac{\partial}{\partial t^*}(\rho^* h^*) + \frac{\partial}{\partial z^*}(G^* h^*) = N_{\mathrm{TPC}} f(t^*) \tag{7}$$

Here, the definition of N_{TPC} is related to reference level of power q_0'' and the function (t), which in turn is dependent on the neutronics.

2.2 *Profiles of Density and Enthalpy*

The state of equation for each of the identified regions is obtained by fitting separate piecewise linear functions with IAPWS data for water at 25 MPa (Fig. 2). Corresponding equations are given below.

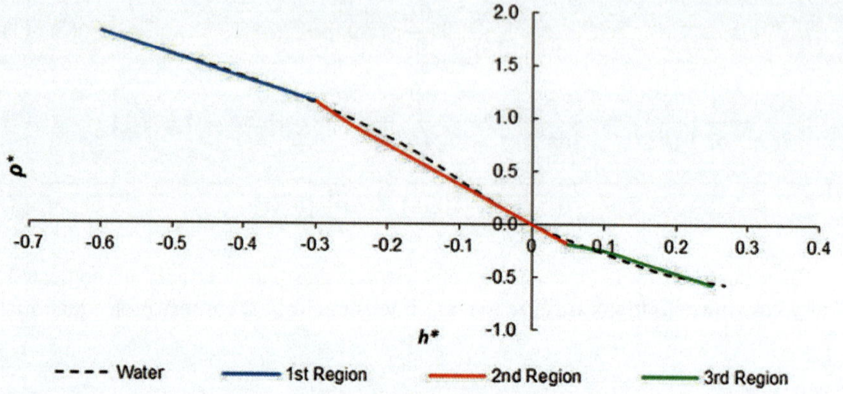

Fig. 2 Fitting of equation of state with IAPWS data for water at 25 MPa

$$\begin{aligned} \rho^* &= -2.21h^* + 0.49 && \text{for } -0.6 < h^* \le -0.3 \\ \rho^* &= -3.7168h^* + 0.0485 && \text{for } -0.3 < h^* \le 0.05 \\ \rho^* &= -2.11h^* - 0.0363 && \text{for } -0.05 \le h^* \end{aligned} \tag{8}$$

Linear variation in axial enthalpy profile is considered for each of the zone as shown below.

$$\begin{aligned} h^* &= \left(h_1^* - h_i^*\right)\frac{z^*}{z_1^*} + h_i^* \\ h^* &= \left[\frac{h_1^* - h_2^*}{z_1^* - z_2^*}\right]\left(z^* - z_1^*\right) + h_1^* \\ h^* &= \left[\frac{h_2^* - h_o^*}{z_2^* - 1}\right]\left(z^* - 1\right) + h_o^* \end{aligned} \tag{9}$$

Here, z_1^* and z_2^* represent the location of the separating boundaries, whereas h_1^* and h_2^* are respective fluid enthalpies.

2.3 *Model of Lumped Parameter*

The non-dimensional versions of the conservation Eqs. (5)–(7) are integrated separately over the three regions, while incorporating the abovementioned enthalpy and density profiles. Adding of the mass and energy conservation equations for the first zone ($z^* \le z_1^*$) yields

$$\frac{dz_1^*}{dt^*} = \frac{G_i^* - G_1^*}{A_1} \quad G_1^* = \frac{N_{\text{TPCM}} z_1^* A_1 + G_i^* h_i^* A_1 - G_i^* B_1}{h_1^* - B_1} \tag{10}$$

Following the similar procedure for the second zone ($z_1^* < z^* \leq z_2^*$),

$$\frac{dz_2^*}{dt^*} = \frac{G_1^* - G_2^* - D_2 \frac{dz_1^*}{dt^*}}{C_2}$$

$$G_2^* = \frac{N_{\mathrm{TPCM}} C_2 \left(z_2^* - z_1^*\right) C_1 - G_1^* \left(h_1^* C_2 - A_2\right) + \left(D_2 A_2 - B_2 C_2\right) \frac{dz_1^*}{dt^*}}{\left(h_2^* C_2 - A_2\right)} \tag{11}$$

Finally, repeating the procedure for the third zone ($z_2^* < z^* \leq 1$),

$$\frac{dh_o^*}{dt^*} = \frac{G_2^* - G_o^* - A_3 \frac{dz_2^*}{dt^*}}{B_3}$$

$$G_o^* = \frac{\left(B_4 - h_2^* B_3\right) G_2^* + \left(B_3 A_4 - A_3 B_4\right) \frac{dz_2^*}{dt^*} - B_3 N_{\mathrm{TPCM}} \left(1 - z_2^*\right)}{\left(B_4 - B_3 h_o^*\right)} \tag{12}$$

All mentioned coefficients in the above equations are mere combinations of the state variables and associated constants. The newly obtained set of equations are solved for constant mass flow rate boundary condition. Accordingly, the solution of the momentum conservation equation is not required, thereby identifying six state variables as z_1^*, z_2^*, G_1^*, G_2^*, G_o^* and h_o^*, in order to characterize the thermal hydraulic model.

2.4 Model of Fuel Rod Dynamic and Neutronics

The coupling between the thermal hydraulic characteristics and the nuclear reactions can be simulated by introducing the heat transfer through the fuel rods. Corresponding dimensionless equation can be written as

$$\frac{\partial}{\partial t^*}\left(\rho^* h^*\right)_j + \frac{\partial}{\partial z}\left(G^* h^*\right)_j = N_j'\left(T_f^* - T_j^*\right) \tag{13}$$

Here, h_j^* represents the average enthalpy of the jth node, whereas T_j^* is the corresponding average coolant temperature. The power generated inside the code is directly reliant on the neutron density. Non-dimensional conservation equations for the core power and precursor density can be summarized as

$$\frac{dP^*}{dt^*} = \frac{k - \beta}{\gamma} P^* + \sigma C^* \tag{14}$$

$$\frac{dC^*}{dt^*} = \frac{\beta}{\gamma} P^* - \sigma C^* \tag{15}$$

Here, the non-dimensional power P^* determined the nature of $f(t^*)$ introduced in Eq. (7).

3 Results and Discussion

It is already mentioned that the present analysis is based on US reference design of SCWR. Here, the dynamic simulation and the linear stability results are characterized in terms of N_{TPC} and N_{SPC}. In Fig. 3, sample stability map is presented. To understand the nature instability, transient simulation has been done under some operating conditions. In Fig. 3, three coloured points show such cases. For each such case, a small perturbation is inflicted on one of the state variables, and similarly, relating profiles of all the variables are taken over a suitably long interval. In Fig. 3, the green point corresponds to a stable system, and it is linked a temporal variation in z_1 which is shown in Fig. 4. It is proved that the oscillations moderate earlier and the profile moves with a near-constant value. Similarly, in stable fixed-point phase, portrait gives a converging spiral. For a system in the region of the limited stable boundary, the temporal development of z_1 is shown in Fig. 5, which is shown in Fig. 3 marked as a blue point. Hardly any recognizable changes are to be observed in the maximum displacement of oscillation over the specified time span, and it shows limit cycle which is stable. In addition to some stages of initial development, similarly phase shows the shape of a perfect circle. An unstable state corresponds by red point in Fig. 3. The variation in z_1 is shown in Fig. 6. The oscillation amplitude is rapidly growing which is clearly visible. The portrait on the state-space plane is separating away from a fixed stable point, exhibiting a spiral and signifying a typically unstable system.

Figure 3 represents the stability maps of the channel at three different lengths. For a given inlet temperature, with the increase in channel length, neutral stability

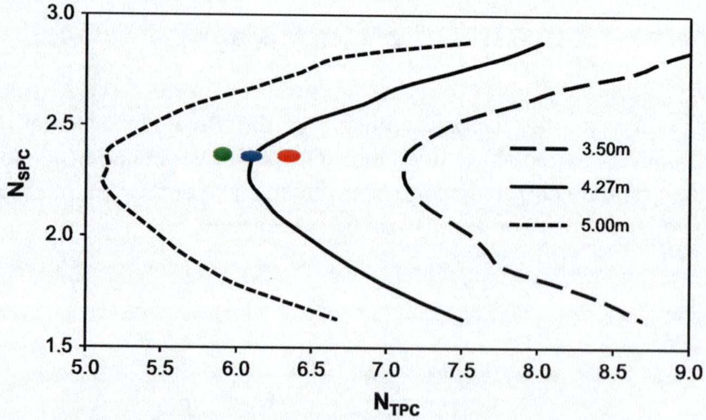

Fig. 3 Stability map for three different channel lengths

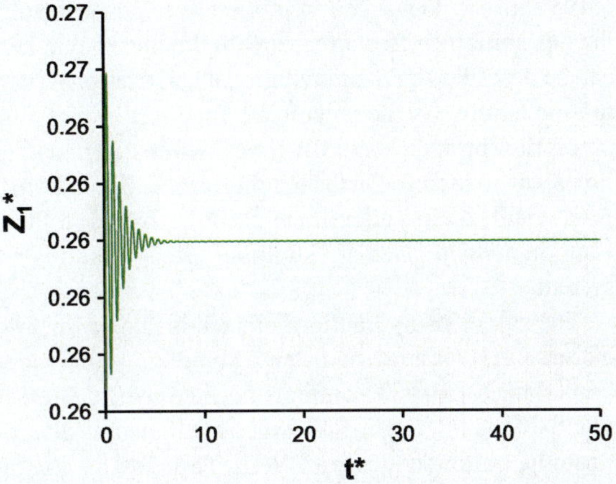

Fig. 4 Transient variation in the first zone boundary at $N_{SPC} = 6.0$ and $N_{TPC} = 2.24$

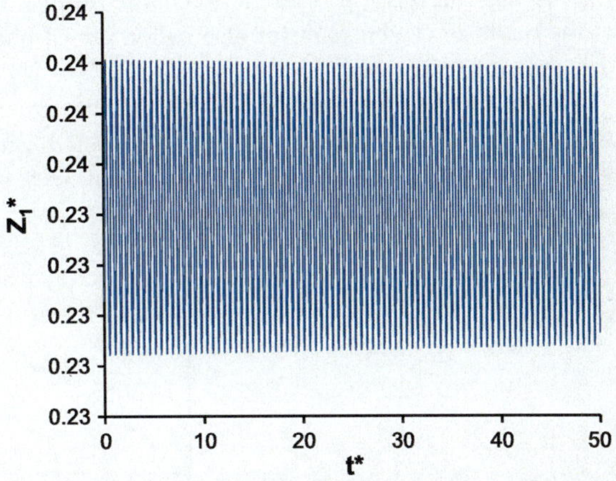

Fig. 5 Transient variation in the first zone boundary at $N_{SPC} = 6.13$ and $N_{TPC} = 2.24$

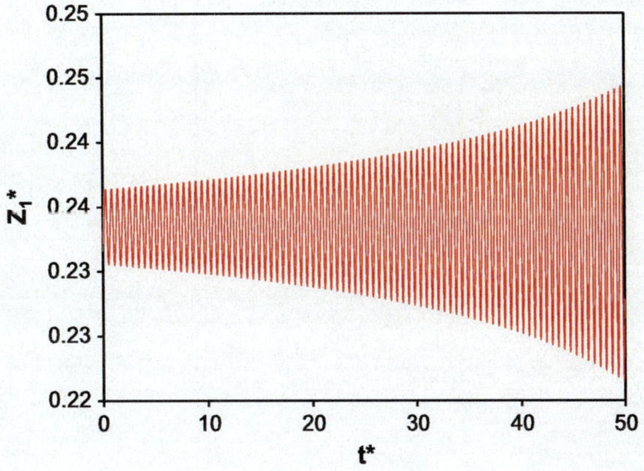

Fig. 6 Transient variation in the first zone boundary at $N_{SPC}. = 6.28$ and $N_{TPC} = 2.24$

curve shifts in the direction of the lower power level. Therefore, it is observed that the destabilizing effect creates with the increase in heated length. This phenomenon can be described by the gravitational and frictional pressure reduction in the equation of momentum. On the other hand, for a stated inlet temperature and power, frictional pressure drop reduces to 0.4 times, while the gravitational component becomes 1.8 times, for an increase in the length from 3.50 to 4.27 m, for identical rate of mass flow. As a result, the pseudocritical point is shifted in the direction of the inlet channel. Friction always gives a restraining effect which is weakening destabilizing of the system.

The effect of hydraulic diameter is shown in Fig. 7. It is found that with the increase in hydraulic diameter, a stabilizing influence, despite reduction in the frictional forces, exists. The effect of orifice coefficients at the inlet and outlet is shown in Fig. 8. In the inlet, rising of orifice coefficient is always found to stabilize the system strongly, quite identical to BWRs. Drawing an analogy with two-phase systems, it can be argued that across the higher density zone any increase in the pressure loss increases the in-phase pressure loss. On the reverse, the effect of a flow resistance at the outlet of the channel (or also called out-of-phase pressure loss) is strongly destabilizing.

On the stability threshold, the effects of negative enthalpy reactivity coefficient and fuel time constant are shown in Figs. 9 and 10, respectively, while increase in the first is found to destabilize the system, and later is found to have the opposing influence.

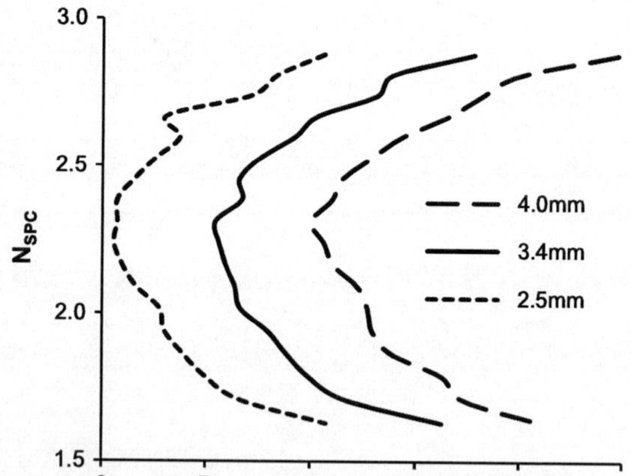

Fig. 7 Effect of channel diameter on stability map

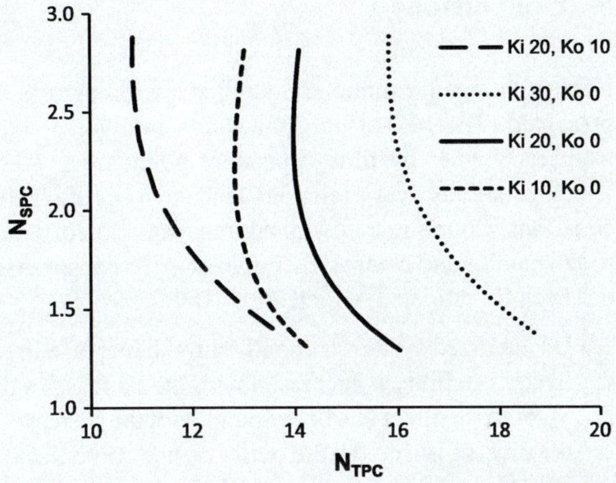

Fig. 8 Effect of orifice coefficients on stability map

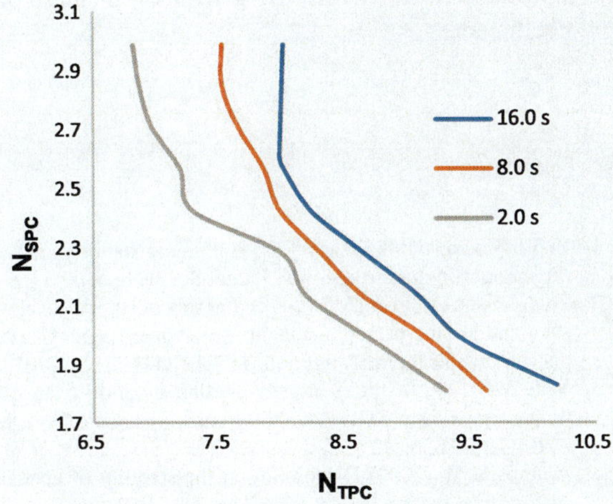

Fig. 9 Effect of fuel time constant on stability map

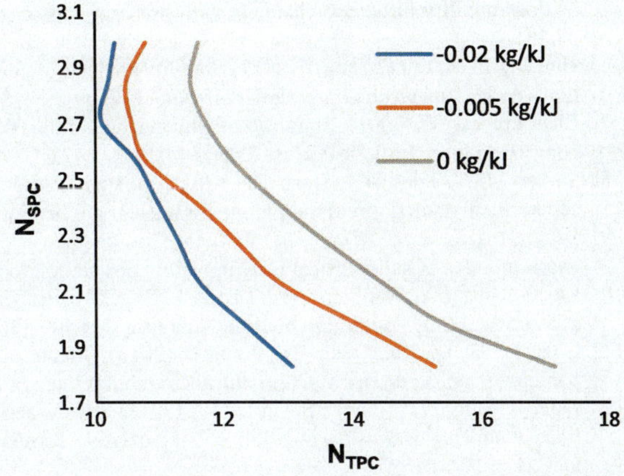

Fig. 10 Effect of enthalpy reactivity coefficient on stability map

4 Conclusions

Here, a lumped parameter model of a three-zone supercritical flow channel is presented. The heated section, which is divided into three regions, is separated from each other by time-dependent boundaries. The integration form of conservation equations over each zone leads to a set of algebraic and ordinary differential equations. Coupling that with adopted equation of state and suitable models for fuel rod dynamics and neutronics, effect of different parameters on the stability threshold has been investigated, which are as follows:

1. Destabilized system is found with the increase in heated channel length.
2. System stability is increased with the increase in the hydraulic diameter.
3. Stabilizing effect has been found with the increase in orifice coefficient, whereas the reverse is true for the outlet orifice coefficient.
4. Reactor stability decreases with the increase in fuel time constant.
5. Reactor stability is increasing with the increase in negative enthalpy reactivity coefficient.

References

1. Yi TT, Koshizuka S, Oka Y (2004) A linear stability analysis of supercritical water reactors, (I): thermal-hydraulic stability. J Nucl Sci Technol 41:1166–1175
2. Zhao J, Saha P, Kazimi MS (2005) Stability of supercritical water-cooled reactor during steady-state and sliding pressure start-up. In: Proceedings of 11th international topical meeting on nuclear reactor thermal-hydraulics, NURETH-11, pp 1–21
3. Yang WS (2005) Initial implementation of multi-channel thermal-hydraulics capability in frequency domain SCWR stability analysis code SCWRSA. Argonne National Laboratory, ANL-GenIV-056
4. Ambrosini W (2009) Discussion on the stability of heated channels with different fluids at supercritical pressures. Nucl Eng Des 239:2952–3296
5. Ampomah-Amoako E, Ambrosini W (2013) Developing a CFD methodology for the analysis of flow stability in heated channels with fluids at supercritical pressures. Ann Nucl Energy 54:251–262
6. Hou D, Lin M, Liu P, Yang Y (2011) Stability analysis of parallel-channel systems with forced flows under supercritical pressure. Ann Nucl Energy 38:2386–2396
7. Chatoorgoon V (2001) Stability of supercritical fluid flow in a single-channel natural-convection loop. Int J Heat Mass Transf 44:1963–1972
8. Sarkar MKS, Tilak AK, Basu DN (2014) A state-of-the-art review of recent advances in supercritical natural circulation loops for nuclear applications. Ann Nucl Energy 73:250–263
9. Sarkar MKS, Basu DN (2017) Influence of geometric parameters on thermal hydraulic characteristics of supercritical CO_2 in natural circulation loop. Ned Eng Des 324:402–415
10. Liu G, Huang Y, Wang J, Lv F, Leung LHK (2016) Experiments on the basic behavior of supercritical CO_2 natural circulation. Nucl Eng Des 300:376–383
11. Chen L, Zhang XR, Deng BL, Jiang B (2013) Effects of inclination angle and operation parameters on supercritical CO_2 natural circulation loop. Nucl Eng Des 265:895–908
12. Jäger W, Hugo V, Espinoza S, Hurtado A (2011) Review and proposal for heat transfer predictions at supercritical water conditions using existing correlations and experiments. Nucl Eng Des 241:2184–2203

13. Ruspini LC, Marcel CP, Clausse A (2014) Two-phase flow instabilities: a review. Int J Heat Mass Transf 71:521–548
14. Zhang Y, Li H, Li L, Wang T, Zhang Q, Lei X (2015) A new model for studying the density wave instabilities of supercritical water flows in tubes. Appl Therm Eng 75:397–409

Numerical Simulation of a Single-Pass Parallel Flow Solar Air Heater with Circular Fins Using S2S Radiation Model

Praveen Alok, Sai Charan Teja Javvadi, Pavan Kumar Konchada, and G. Raam Dheep

Abstract Solar air heater (SAH) is a heat exchanger device that converts solar radiation into thermal energy, using solar thermal technology (energy from the sun is trapped and absorbed to heat the air). The conventional SAH is inefficient due to poor heat transfer characteristics from the absorber. In this present study, a parallel flow SAH consisting of circular fins attached to the absorber is designed to enhance the heat transfer rate. Computational fluid dynamics (CFD)-based numerical simulation is carried out in ANSYS Fluent using the k–ε turbulence model with enhanced wall treatment for different mass flow rates and different solar radiation intensities. The simulation output is validated using experimental results which are available in the literature. The heat transfer characteristics, absorber, and outlet air temperature are shown using contours. The mean deviation in the outlet air temperature predicted by the simulation is found to be ~7%. The simulation and experimental results conclude that surface-to-surface radiation model can accurately predict the fluid flow behavior inside the SAH.

Keywords Solar air heater · Parallel flow · S2S model · Conjugate heat transfer

Nomenclature

ρ	Density, kg/m^3
k	Turbulent kinetic energy, m^2/s^2
ε	Dissipation rate
P	Pressure
u	Velocity, m/s

P. Alok · S. C. T. Javvadi · P. K. Konchada
CADFEM India, Hyderabad 500082, India

G. R. Dheep (✉)
Department of Power Electronics and Electrical Engineering, Lovely Professional University, Punjab 144411, India
e-mail: raamdheep@gmail.com

S. Revankar et al. (eds.), *Proceedings of International Conference on Thermofluids*, Lecture Notes in Mechanical Engineering,
https://doi.org/10.1007/978-981-15-7831-1_10

T Temperature, K
μ Dynamic viscosity, kg/m s
Γ Molecular diffusivities, μ/Pr
Γ_t Thermal diffusivities, μ_t/Pr_t
K Thermal conductivity, W/mK

1 Introduction

Solar air heater is a heat exchanger device which converts the incident solar radiation into thermal energy. The absorber inside the SAH receives solar radiation, converts to heat, and transfers it to the air medium. The temperature of the outlet air ranges from 70 to 110 °C. SAHs are used for space heating, drying agricultural food products, and industrial applications [1]. The conventional flat plate solar air heaters are usually inefficient due to reduced heat transfer from absorber to air. Many techniques are implemented to enhance the efficiency of solar air heater. The characteristics of heat transfer can be improved either by modifying the airflow path or geometries of absorber plate. The air is passed either between absorber and glazing (glass) or between absorber and base plate as single-pass or double-pass path. Depending upon the airflow path, the resting time of air inside the SAH duct rises the air temperature which in turn increases the efficiency of the system. Increasing the effective surface area to enhance heat transfer and in turn convective heat transfer coefficient, using extended surface of absorber with different geometries such as fins, ribs, and wire mesh is found to be another effective method of enhancing the heat transfer rate [2]. The influence of different shapes, operating parameters, and geometrical shapes of absorber on thermal and hydraulic performance of SAH duct is being studied by numerous researchers with experimental analysis and computational fluid dynamics simulations. Prasad and Mullick designed a SAH duct with artificial roughness using wires with smaller diameter of 0.84 mm to enhance heat transfer coefficient for relative roughness with height and pitch of 0.019 and 12.7 mm, respectively [3]. Liu et al. studied the heat transfer coefficient of SAH using extended surfaces on the absorber [4]. The author found that extended surface protrudes beyond the sublayer which results in pressure drop without increasing heat transfer coefficient. It is found that the artificial roughness with smaller roughness height, pitch, and Reynolds number is more efficient. Researchers also studied the thermal performance of SAH's with the geometries such as V, W, L, arc, semicircular, triangular, rectangular, sawtooth, traverse-shaped ribs, using experimental and numerical simulations [5]. The detailed literature review on SAH shows that only limited studies are carried out on thermal performance of circular fins and double pass-based SAH. The main objective of this present work is to investigate the thermal performance of single-pass parallel flow SAH with circular fins attached above the absorber plate, using numerical simulations. It is observed in detail on the performance characteristics of SAH using CFD simulations. The radiation model used for simulating the solar air heater has rarely been reported in any other work. Recently, Rajarajeswari et al. simulated the solar air heater using Rosseland model [6]. But the feasibility and accuracy of other radiation models in predicting the fluid flow behavior inside the solar air heater are unknown. The present work incorporates on the feasibility of surface-to-surface radiation model (S2S Model) in predicting the temperature variation of absorber and most importantly air. The results obtained are compared with the experimental results of Raam Dheep and Sreekumar [7] and numerical simulation results of Rajarajeswari et al. [6]. The flow behavior, temperature distribution of air inside the SAH's duct, and outlet air temperature are studied with respect to different mass flow rates and solar radiation intensities.

2 CFD Analysis

2.1 *Modeling and Meshing*

Modeling of the solar air heater is done in SpaceClaim based on the experimental work done by Raam Dheep and Sreekumar [7]. Raam Dheep and Sreekumar have already done an extensive work in evaluating the performance of SAH for different mass flow rates and solar intensities [7]. The dimensions of solar air heater are taken from the work done by Raam Dheep and Sreekumar [7]. After modeling, the design in geometry preprocessor, ANSYS SpaceClaim, the model is been imported to ANSYS Workbench for meshing purpose. Figures 1 and 2 show the schematic diagram of SAH along with its outer and cross-sectional views with dimensions. Length and diameter of longitudinal fins are 2000 mm and 9 mm, respectively. Mesh independence study has been performed to check the dependence of mesh on the result sensitivity of

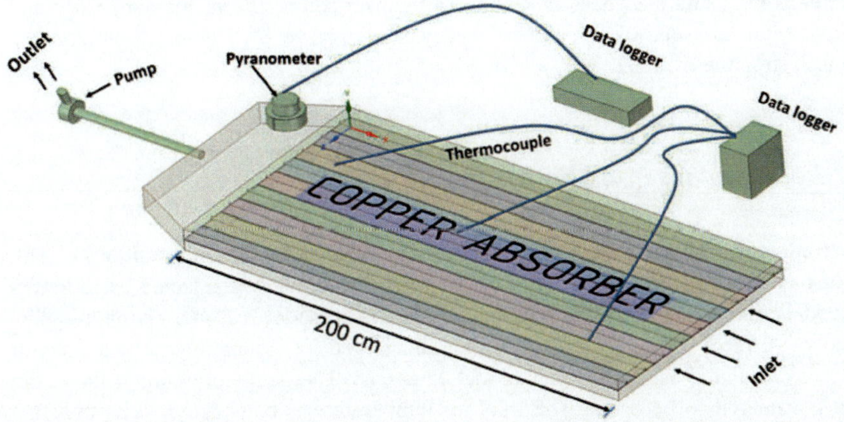

Fig. 1 Schematic diagram of solar air heater

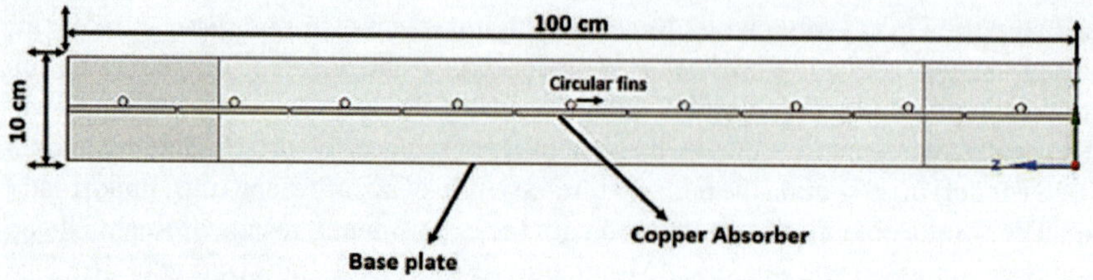

Fig. 2 Cross-sectional view of solar air heaters

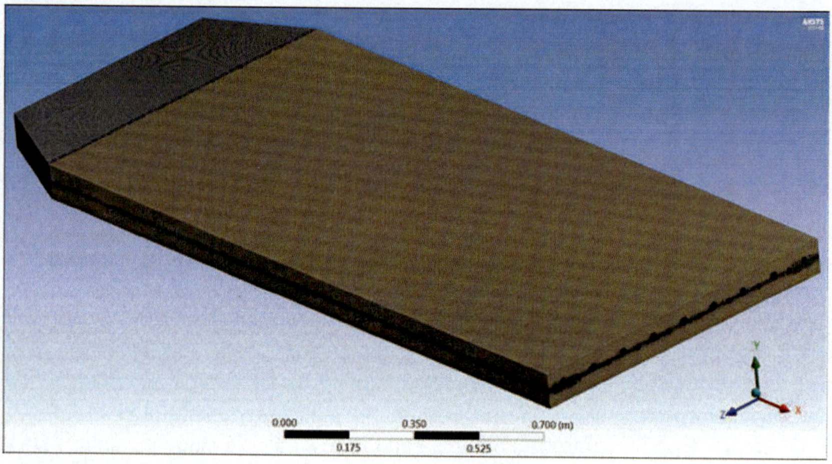

Fig. 3 Fine mesh structure (hexahedral + tetrahedral mesh)

the simulation. All the simulations are performed with ~5 million elements. Aspect ratio is <15, and skewness is <0.95. The convergence criteria for continuity and momentum are set to $1e^{-04}$, and for energy is set to be $1e^{-06}$. Figure 3 shows the fine mesh structure of SAH.

2.2 *Boundary Conditions*

An attempt has been made to simulate the input solar radiation using ANSYS Fluent. Surface to surface (S2S) along with solar tracing models is selected as radiation models, and k–ε realizable enhanced wall treatment model is used for simulating the flow [8]. Table 1 shows the assumptions considered for simulations. The density of air is calculated using ideal gas model equations. The gauge pressure at the outlet is considered to be 0 bar. The inlet air temperature is considered to be constant throughout simulation.

Table 1 Assumptions and considerations

Parameters	Values
Solar absorptivity of copper absorber	0.95
Solar transmissivity of glass	0.9
Solar absorptivity of base plate	0.8

Table 2 Analysis parameters and boundary conditions

Parameters	Values/boundary conditions/model type
Radiation model	Solar ray tracing (solar radiation) + surface-to-surface model (internal radiation)
Pressure–velocity coupling	SIMPLE scheme
Side walls heat transfer	Adiabatic
Density of air	Ideal gas equation
Turbulence model	k–ε enhanced wall treatment model
Outlet pressure	P_{atm}
Inlet fluid temperature	35 °C (constant)
Global position	Latitude 11.9°, longitude 79.8°, time zone +5.3 GMT
Inlet mass flux/mass flow rate	90 kg/h m^2 (0.05 kg/s), 75 kg/h m^2 (0.0416 kg/s), 60 kg/h m^2 (0.033 kg/s)
Gravity direction	Negative Y axis

To simulate the solar radiation, solar ray tracing algorithm is used. Solar ray tracing uses multiple positioning vectors and two illumination parameters to set solar radiation intensities to represent solar load. Absorptivity and transmissivity are defined for all the solid material. Glass is modeled as semitransparent object while other solid components are modeled as opaque. Steady-state simulations are performed for different mass flow rates and for different time intervals. The simulation results are compared with those of experimental results of Raam Dheep and Sreekumar [7]. Table 2 shows simulation parameters and boundary conditions, used for simulating SAH.

2.3 Governing Equations

Continuity, momentum, and energy equations are a set of basic governing equations to analyze the flow behavior within the domain. The flow phenomenon in a conventional SAH duct is governed by these continuity, momentum, and energy equations as written below. In the Cartesian tensor system, these equations can be written as [9];

Continuity Equation:

$$\frac{\partial}{\partial x_i}(\rho u_i) = 0 \tag{1}$$

Momentum Equation:

$$\frac{\partial}{\partial x_i}(\rho u_i u_j) = -\frac{\partial P}{\partial x_i} + \frac{\partial}{\partial x_j}\left[\mu\left(\frac{\partial u_i}{\partial x_j} + \frac{\partial u_j}{\partial x_i}\right)\right] + \frac{\partial}{\partial x_j}\left(-\rho\overline{u_i' u_j'}\right) \tag{2}$$

Energy Equation:

$$\frac{\partial}{\partial x_i}(\rho u_i T) = \frac{\partial}{\partial x_j}\left((\Gamma + \Gamma_t)\frac{\partial T}{\partial x_J}\right) \tag{3}$$

2.4 Turbulence Modeling

According to the idea of Reynolds decomposition, the dependent variables of the system (as per governing Navier–Stokes equations) will be decomposed into mean and fluctuating parts. The fluctuating parts are solved by using turbulence models. To simulate the effects of turbulence, realizable k–ε turbulence with enhanced wall treatment model is used. Enhanced wall treatment is considered as a near-wall modeling method which uses two-layer approach to predict the flow at the viscous sub-layer region and turbulent region. It can resolve viscous sub-layer if the mesh is fine enough. For conjugate heat transfer problems, resolving viscous sub-layer is important to accurately predict the heat transfer between the walls and fluid. For $\mathrm{Re}_y < 200$, one equation model of Wolfstein is used. For $\mathrm{Re}_y > 200$, two equation k–ε models are used [10]. Enhanced wall treatment model has the ability to resolve both the viscous sub-layer and turbulent region which is not possible using standard and scalable wall functions approach.

3 Results and Discussions

Solar radiation intensity data is taken from experiment done by Raam Dheep and Sreekumar [7]. In order to validate the numerical results with those of experiment results, the same dimensions of SAH are considered for simulation. It is ensured to keep $Y+$ value for all the simulation to be below 5. This is due to the reason that the simulation model involves calculations of heat transfer which requires $Y+$ value to be below 5 to capture the viscous sub-layer. Figures 4 and 7 shows the absorber temperature for different mass flow rates 0.05 kg/s, 0.041 kg/and 0.033 kg/s, respectively, at 10 a.m. and 12 p.m. The temperature of absorber increases with the decrease in mass flow rates or mass flux. This is due to the reduction of heat transfer coefficient

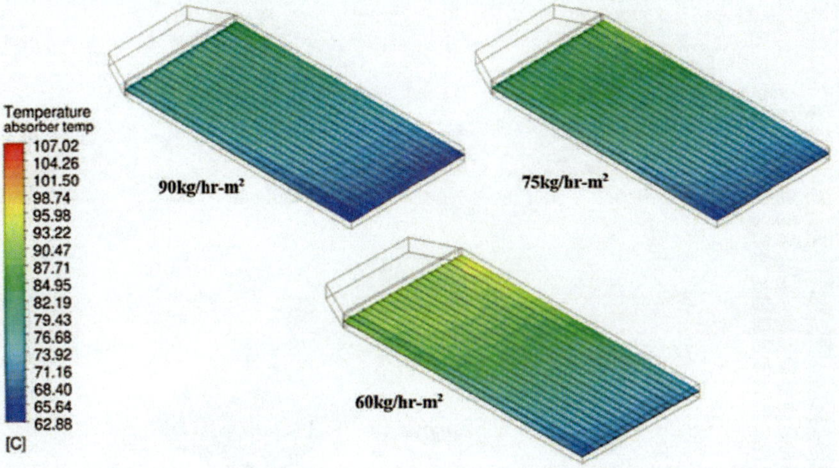

Fig. 4 Absorber temperature distribution for 0.05 kg/s (top left), 0.041 kg/s (top right), and 0.033 kg/s (bottom middle) at 10 a.m.

values and decrease in Reynolds number. Figure 5 shows the comparison of average absorber temperature obtained from the simulation with the average temperature obtained from the experiment at 10 a.m. and 12 p.m. The experiment encompasses on recording the absorber temperature at 4 different locations (at multiple locations with regular intervals on the absorber plate at 40 cm, 80, 120, 160 cm from the inlet positioning) at along the length of the absorber. Average temperature data of 4 different probe locations, obtained from experimental results were compared to simulation results. Experimental results show that absorber temperature increases with the decrease in mass flux or mass flow rates. The same behavior is also observed in the simulation. However, the simulation results overpredicted the absorber temperature results by ~16%. The deviation is primarily due to the fact that better and finer tuning of solar beam direction is required. In experiment, solar radiation is unevenly

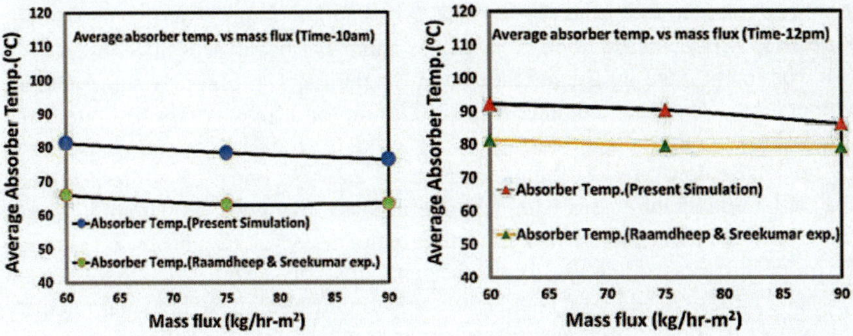

Fig. 5 Average absorber temperature versus mass flux at 10 a.m. (left) and at 12 p.m. (right)

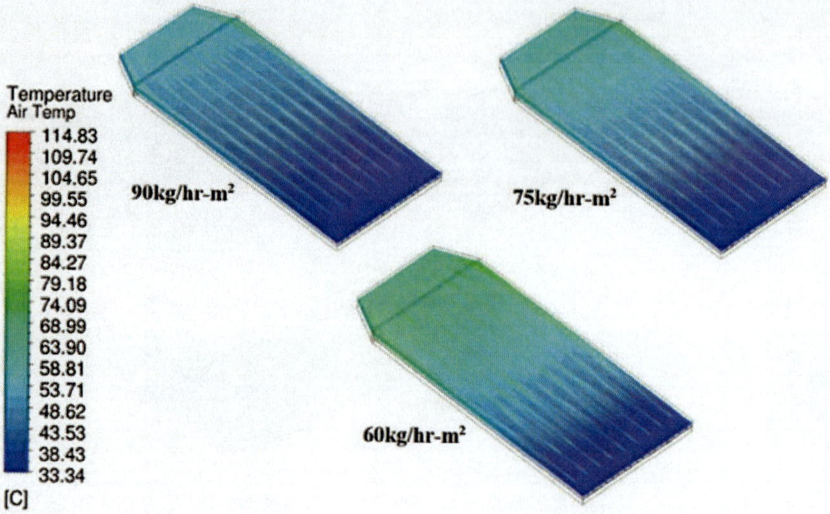

Fig. 6 Air temperature distribution along a plane for 0.05 kg/s (top left), 0.041 kg/s (top right), and 0.033 kg/s (bottom middle) at 10 a.m.

distributed and is more concentrated near to middle portion of SAH. This forms as the reason for the highest temperature value of the absorber in experiment at the middle portion of SAH [7]. Moreover, finetuning of solar transmissivity of glass and solar absorptivity of copper plates is also required in accordance with the experiment.

Figure 6 shows the air temperature distribution along a plane at 10 a.m. for mass flow rates 0.05 kg/s, 0.041 kg/s and 0.033 kg/s respectively. The air temperature increases from inlet towards outlet. Figure 9 shows the comparison of outlet temperature obtained from simulation to the experiment done at 10 a.m. [7]. The simulation outlet temperature for 90 (0.05 kg/s) kg/h m^2, 75 (0.041 kg/s) kg/h m^2, and 60 (0.033 kg/s) kg/h m^2 mass flow rates at 10 a.m. are 52.1 °C, 56.1 °C, and 61.5 °C, respectively. The average deviation obtained from experiments in predicting the outlet temperature at 10 a.m. is found to be ~2.24%. The outlet temperature is highest for lowest mass flow rate and vice versa. The simulation results show that air temperature rises with decrease in mass flow rates and same behavior is also observed in the experiment done by Raam Dheep and Sreekumar [7]. Figure 8 shows the air temperature distribution along the plane at 12 p.m. for different mass flow rates. The simulation outlet temperature for 90 (0.05 kg/s), 75 (0.041 kg/s), 60 (0.033 kg/s) kg/h m^2 mass flow rates at 12 p.m. are 57.8 °C, 62.3 °C, and 69.2 °C, respectively. The outlet temperature rises with decrease in mass flow rates, which are in accordance with the experiment. Also, the outlet temperatures at 12 p.m. are higher than the outlet temperatures at 10 a.m. for the same flow rates. This is due to the reason that solar intensity at 12 p.m. is more as compared to 10 a.m. Higher solar intensity led to more heating of solar absorber which in turn increases the outlet temperature.

Also, the average absorber temperature obtained from the simulation at 12 p.m. for 90, 75, 60 kg/h m^2 mass flow rates is ~12 °C higher than average absorber temperature at 10 a.m. The average absorber temperature obtained in experiment at 12 p.m. is ~15 °C higher than the average absorber temperature obtained at 10 a.m. Both experiment and simulation results almost showed the same behavior in the average absorber temperature rise with respect to the increase in solar radiation intensity (Figs. 7, 8 and 9).

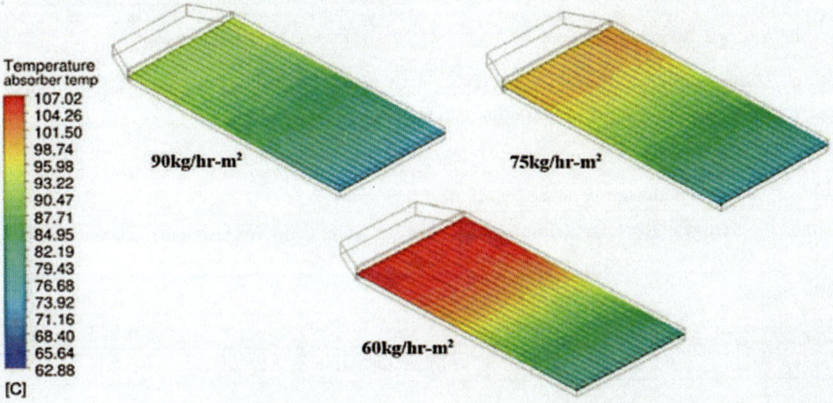

Fig. 7 Absorber temperature distribution for 0.05 kg/s (top left), 0.041 kg/s (top right), and 0.033 kg/s (bottom middle) at 12 p.m.

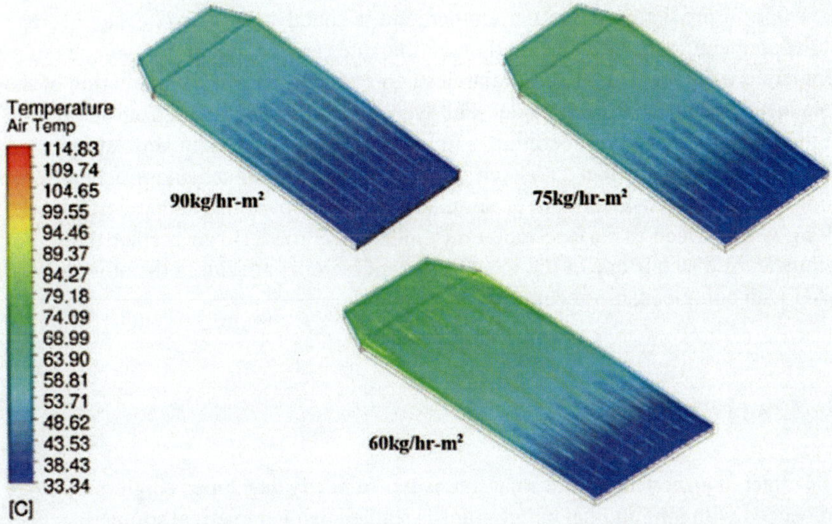

Fig. 8 Air temperature distribution along a plane for 0.05 kg/s (top left), 0.041 kg/s (top right), and 0.033 kg/s (bottom middle) at 12 p.m.

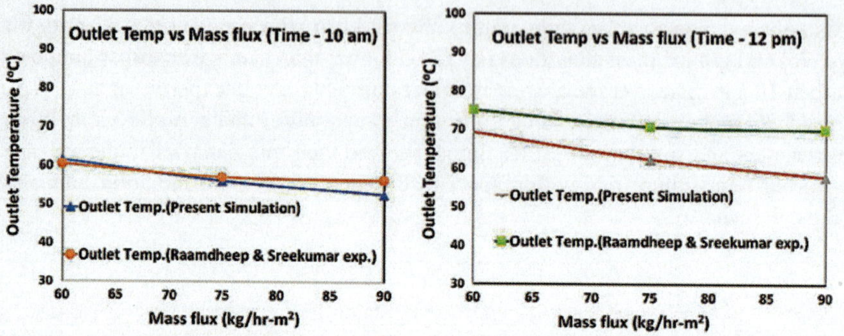

Fig. 9 Average outlet temperature versus mass flux at 10 a.m. (left) and at 12 p.m. (right). Outlet temperature comparison from the experimental results

Table 3 Average outlet temperature for all mass flow rates

Time	Average outlet temp. (simulation) (°C)	Average outlet temp. (experiment) (°C)	Deviation (%)
10 a.m.	56.7	58	2.24
12 p.m.	63.1	71.8	12.11
		Mean deviation = 7.17%	

Table 3 shows the deviation of outlet temperature predicted by simulation from experiment results at different time intervals. The average deviation is found to be ~2.2% and ~12% for 10 a.m. and 12 p.m., respectively. The overall average deviation in predicting the outlet temperature is found to be as ~7%. The average outlet temperature at 12 p.m. for all mass flow rates is higher than the average outlet temperature obtained at 10 a.m. Rajarajeswari et al. did numerical simulation of the solar heater using Rosseland model. The average deviation obtained from simulation in predicting the outlet temperature is found be ~14%. The present simulation using S2S model predicted the outlet temperature of air with much more accuracy as compared to Rosseland model The simulation results proved that solar tracing model along with surface-to-surface radiation model (S2S model) can predict the outlet temperature which is one of the most important factors in deciding the efficiency of SAH with commendable accuracy.

4 Conclusions

The effect of airflow rates and solar intensities of newly developed single-pass SAH embedded with longitudinal circular fins is studied using numerical simulation. S2S model along with solar ray tracing algorithm is used to simulate effect of solar intensity on the temperature of air and absorber for different mass flow rates and solar intensities. The simulation results showed that the outlet air temperature and absorber temperature increase with decrease in mass flow rates. The simulation also exhibited that absorber and outlet temperatures rise with the increase in solar radiation intensity. The simulation results predicted the outlet air temperature with mean deviation of ~7% from the experimental results. Surface-to-surface model can be used to predict the outlet temperature of air with commendable accuracy and hence can be used as an alternative to experiment to envisage the performance and efficiency of solar air heater.

Acknowledgements The authors are profoundly grateful to CADFEM India and ANSYS Inc.

References

1. Singh I, Singh S (2018) CFD analysis of solar air heater duct having square wave profiled transverse ribs as roughness elements. Sol Energy 162:442–453
2. Kumar R, Goel V, Kumar A, Khurana S, Singh P, Bopche SB (2018) Numerical investigation of heat transfer and friction factor in ribbed triangular duct solar air heater using computational fluid dynamics (CFD). J Mech Sci Technol 32(1):399–404
3. Prasad K, Mullick SC (1985) Heat transfer characteristics of a solar air heater used for drying purposes. Appl Energy 13:83–93
4. Ye-Di Liu L, Diaz A, Suryanarayana NV (1984) Heat transfer enhancement in air heating fiat-plate solar collectors. J Sol Energy Eng 106:358–363
5. Kumar R, Geol V, Kumar A (2017) A parametric study of the 2D model of solar air heater with elliptical rib roughness using CFD. J Mech Sci Technol 31(2):959–964
6. Rajarajeswari K, Alok P, Sreekumar A (2018) Simulation and experimental investigation of fluid flow in porous and non-porous solar air heaters. Sol Energy 171:258–270
7. Raam Dheep G, Sreekumar A (2019) Experimental studies on energy and exergy analysis of a single-pass parallel flow solar air heater. ASME J Sol Energy Eng 142(1):011003. https://doi.org/10.1115/1.4044127
8. ANSYS fluent user Guide, release 19.2
9. Yadav AS, Bhagoria JL (2013) Heat transfer and fluid flow analysis of solar air heater: a review of CFD approach. Renew Sustain Energy Rev 23:60–79
10. https://www.afs.enea.it/project/neptunius/docs/fluent/html/th/node101.htm

High-Speed Flow

Parametric Study of Blast Wave Formation in a Shock Tube

Sachin Pullil, N. Vaibhav, R. Sanjay, and S. R. Nagaraja

Abstract Blast waves are formed when there is a sudden release of energy from a point in space, like, from an explosion. These waves propagate through air, come into contact with structures or human beings, and proceed to cause fatal or irreversible damage. Many methods have been developed to attenuate blast waves. However, to conduct physical experiments, the use of real explosives is for the risk of collateral damage. Thus, shock tubes are utilized to create blast waves within a confined region so that they can be studied safely. When the diaphragm bursts, a shock wave is formed in the driven section of the tube and, eventually, the pressure profile of the wave changes to that of a blast wave. The paper focuses on a parametric study of the variation of driver section length and pressure in the formation of blast waves in a shock tube. The distance at which the blast wave forms increases with increasing driver section pressure and length. The experimental results show that the pressure wave closely matches with that of Friedlander waveform at higher driver pressures.

Keywords Blast wave · Shock tube · Friedlander wave

1 Introduction

Shock/blast waves are used in many innovative industrial [1] and biological medical applications [2, 3]. The blast profile follows a distinct wave profile of that of Friedlander equation. Dewey [4] have compared the blast wave profile to that of the Friedlander equation and determined the various ranges of the blastwave profile in terms of distance from source to that of Friedlander waveform. Many methods are used to create shock/blast waves. The oldest form of creating a blast wave is through detonation. However, a blast wave can also be created by various other methods. One method is described by Courtney et al. [5], where a shock tube filled with a mixture of oxygen and acetylene in driver section is used. On ignition, it creates a pressure

S. Pullil · N. Vaibhav · R. Sanjay · S. R. Nagaraja (✉)
Department of Mechanical Engineering, Amrita School of Engineering, Amrita Vishwa Vidyapeetham, Bengaluru, India
e-mail: sr_nagaraja@blr.amrita.edu

S. Revankar et al. (eds.), *Proceedings of International Conference on Thermofluids*, Lecture Notes in Mechanical Engineering,
https://doi.org/10.1007/978-981-15-7831-1_11

difference between driver section and driven section, hence creating a blast wave in the driven section. One more method of creating a blast wave profile is discussed by Courtney and Courtney [6], where the blast wave profile is created by using a firearm rifle to increase the pressure and using the barrel as the shock tube. The difference in pressure causes the formation of blast wave profile. An alternative method to create a blast wave on a small scale, as put forward by Ram and Sadot [7], is by using an exploding wire. The wire is subject to high current and, in turn, a large amount of heat is generated, which leads to vapourizing of a liquid, which is also accompanied by a blast wave. A familiar method used to create a blast wave profile is to increase the length of the driven section of the shock tube as mentioned by Chandra et al. [8], where the blast wave profile becomes more prominent over the length of the shock tube (driven length). Sandhu et al. [9] have shown that the blast wave profile depends on the driver pressure as well as the driven length of the shock tube; larger the driver pressures, the longer will be the distance of formation of blast wave profile over the driven tube. Susheel Kumar and Nagaraja [10] have studied the effect of shock tube diameter on the profile of the shock wave. They have compared shock tubes of different diameters. The major difference between the simulation process and the experimental process to create the blast wave profile in a shock tube is that, in experimentation, there is energy loss due to friction between the pipe wall and the wave created, which is discussed in the paper by Ngomo et al. [11], where the pressure of the blast wave formed over the length of the shock tube decreases due to friction forces. In this paper, the conventional method to produce a blast wave in a shock tube is explored by varying the parameters: driver section length and pressure. The shock tube method provides a cheap and safe alternative to creating a blast wave. In this paper, the distance at which a blast wave is formed is studied for different driver and driven section lengths and diver pressures using finite element software COMSOL multiphysics. The results are verified using a hand-operated mini shock tube.

2 Numerical Simulations

Numerical simulations were conducted using COMSOL multiphysics. The geometry of the shock tube was modelled and boundary conditions such as initial pressure and temperature were set. The simulations were run and the pressure profiles at different points along the driven section of the shock tube were studied. The waveform for pressure was compared with the corresponding Friedlander waveform to check if the pressure profile fits that of a blast wave. When the pressure profile at a point becomes similar to the Friedlander waveform, the location of that point along the driven section is considered to be when the blast wave is fully formed within the shock tube. Finally, the distance at which the blast wave forms for each simulation is tabulated and analyzed.

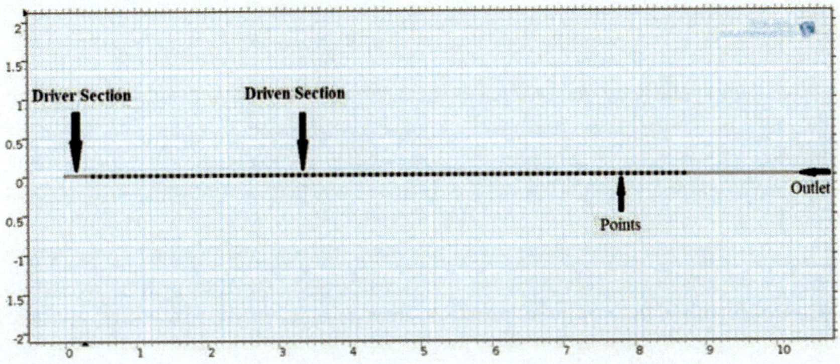

Fig. 1 Geometry of shock tube for simulation with 0.3 m driver section length

2.1 Geometry

The geometry of the shock tube consists of two rectangles to represent the driver and driven sections. In the simulations, driver section length was varied from 0.3 to 1.5 m with steps of 0.3 m—for a total of five different lengths. The driven section length was varied from 10 to 20 m. The criteria for the driven section length is that location in the section at which the blast wave forms must be before the end of the tube. So, the driven section length is always kept high to ensure blast wave forms within the tube. Finally, points were added along the driven section of the shock tube at a uniform interval of 0.1 m in order to inspect the pressure profile at that location within the shock tube. Figure 1 shows the geometry for one particular simulation where the driver section length was 0.3 m and the driven section length was 10 m. The axes represent distance in meters.

2.2 Boundary Conditions

The pressure inside the driver section of the shock tube was varied from 5 to 25 bar at intervals of 5 bar. This allowed for a total of five different driver section pressures for each driver section length. As such, there were 25 simulations in total. The driven section pressure was always maintained at 1 bar. The temperature throughout the shock tube was kept constant at 300 K. Finally, the right wall of the driven section was made to be an outlet. This was done in order to prevent the shock from rebounding off the wall and disrupting the pressure readings at the concerned points.

2.3 Simulation Results and Analysis

Figure 2 shows the pressure profiles at various points along the driven section for a shock tube of 0.6 m driver length and 25 bar driver pressure. For points closer to the diaphragm, there is a high pressure that is sustained for a while, and then the pressure decreases gradually below 1 bar before it rises once again. In the simulation results, as shown in the figure, it can be seen that for points that are closer to the diaphragm, there is a sudden jump in pressure to a high value that is sustained for 1–2 ms before it decreases gradually. This matches the description of a shock wave. For points that were further down the driven section, there is a sudden rise to a high pressure which is then immediately followed by a steady decline in pressure. This matches the description of a blast wave. It can be seen that initially the pressure profile is that of a shock wave, and then, it gradually changes to that of a blast wave. In order to locate the point at which the pressure profile of the shock wave changes to that of a blast wave, the profile at each point was plotted with its corresponding Friedlander waveform. If the pressure profile matches its Friedlander waveform, then it can be classified as a blast wave, and that particular point is considered as the point along the shock tube at which a blast wave is formed. The plotting software used to plot the pressure profile and the Friedlander waveform was MATLAB.

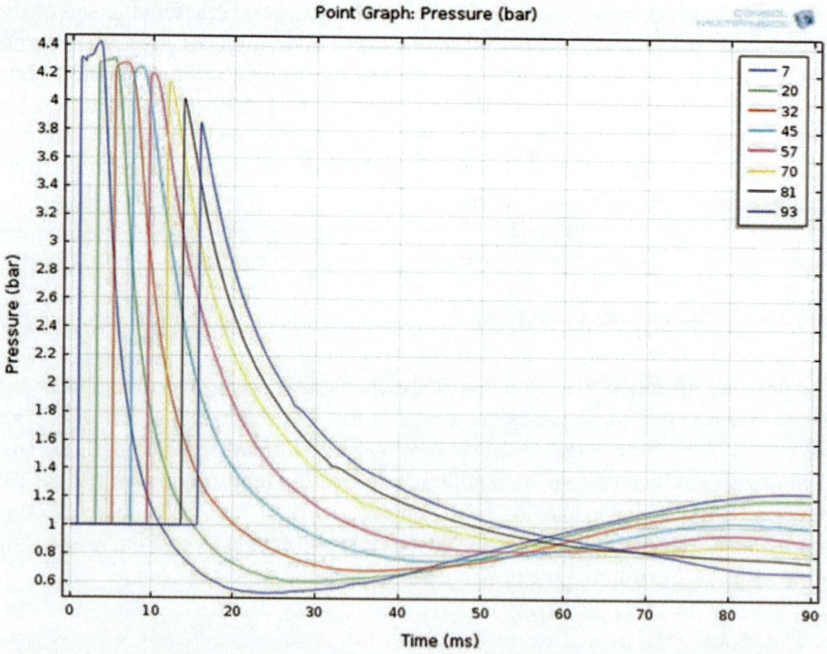

Fig. 2 Pressure profiles for different points in the driven section for a driver section of 0.6 m length at 25 bar pressure

Figure 3 shows the simulated pressure profiles against corresponding Friedlander waveform. It is seen that the profile agrees closely with the Friedlander waveform and so it can be classified as blast wave. For a shock tube of driver section length and pressure of 0.6 m and 25 bar, the blast wave forms at 3.2 m away from the diaphragm. Similarly, for a shock tube of driver section length and pressure of 0.3 m and 5 bar, the blast wave forms at 1.6 m away from the diaphragm. Analysis was conducted for all 25 cases and the distance at which the blast wave forms is noted. Figure 4 shows the distance at which the blast wave forms for different driver pressure and driver lengths. It can be observed that for the same driver length, increased driver pressure results in increased distance along the driven section of the tube at which the blast wave forms. Also, for the same driver pressure, increased driver length results in increased distance along the driven section of the tube at which the blast wave forms. Another observation that is made is that, for a particular driver section length, the length along the driven section at which blast wave forms seems to achieve a near-constant value at higher driver pressures. For a 0.6 m driver length section, for driver pressures of 20 bar and 25 bar, the lengths along the driven section of the tube at which blast wave forms is 6.9 m and 7.0 m, respectively. This could imply that at higher pressures, the location along the driven length at which the shock wave becomes a blast wave may converge to a specific point. The same pattern is observed in all the other driver lengths considered for the output readings of 20 and 25 bar.

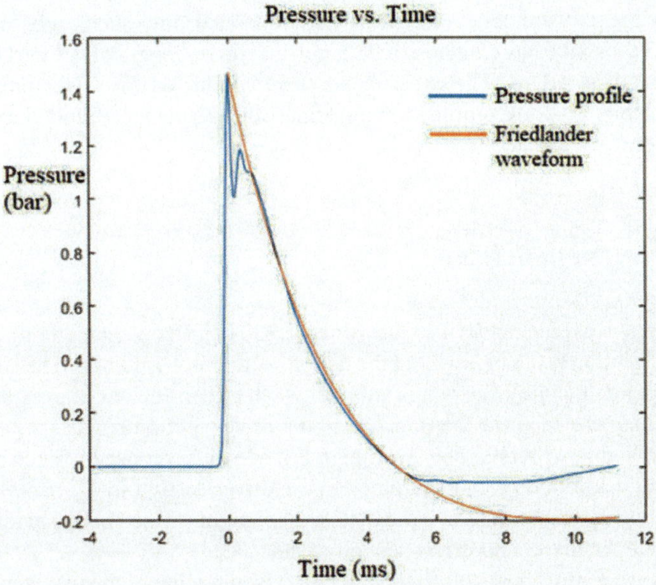

Fig. 3 Pressure profile at 1.6 m away from the diaphragm in the driven section, for a driver section length and pressure of 0.3 m and 5 bar, and corresponding Friedlander waveform

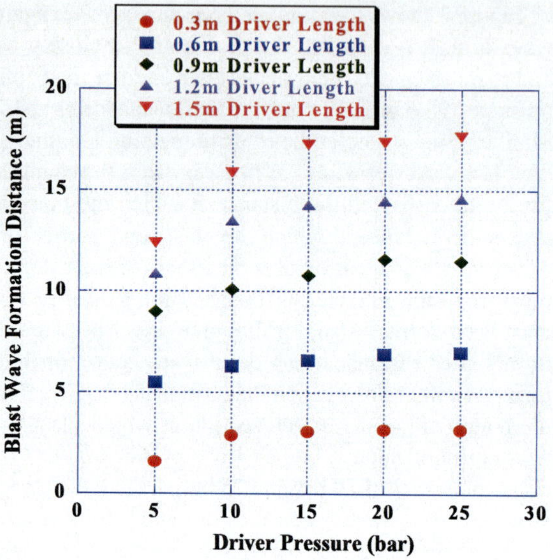

Fig. 4 The distance at which blast wave forms for different driver pressures and driver lengths

3 Experimental Section

The experiments were conducted using hand-operated mini shock tube developed [12]. The shock tube has diameter of 26.5 mm and driver length of 0.3 m. The driven section length is 4.1 m. The experiments were conducted for driver pressures of 4.5 and 7.5 bar. Pressure profiles were obtained and results were plotted against the Friedlander wave forms.

3.1 Experimental Setup

The shock tube used is made of stainless steel SS304 and has a diameter of 26.5 mm. The driver section is 0.3 m long and the driven section is 4.1 m long. The shock tube is hand-operated and makes use of a lever which, when rotated, moves the piston forward along the tube till the pressure in the driver section reaches a peak value when the diaphragm bursts. The diaphragm is made of tracer paper. When one sheet is used, the maximum pressure attained in the driver section is 4.5 bar. When two sheets are used, the pressure is 7.5 bar. The diaphragm is fixed between the flanges that separate the driver and driver section of the tube.

The pressure sensors are plush mounted in driven section of the tube and a digital storage oscilloscope is used to store the data. Holes were drilled in the driver section of the tube and nuts were brazed on each hole to seat the sensor and screw it in place. For every reading, one sensor was placed at 1.635 m from the diaphragm. This was the reference point and called "Point 0". Excluding this point, there were 11 points

Fig. 5 Experimental setup

at nearly even intervals along the driven section of the tube. The second sensor was placed at the required points along the shock tube. Figure 5 shows the complete experimental setup.

4 Results and Discussion

The experiments were conducted for driver pressures of 4.5 and 7.5 bar. Pressure readings were taken for every point from 1 to 11 and plotted. Figure 6 shows the experimental results for 4.5 bar driver pressure and pressure profile for points 6, 7 and 9, which are 2.835, 3.035 and 3.435 from the diaphragm, respectively. The pressure readings only for these points are shown in order to facilitate analysis. From the graph, it is seen that at point 6 and 7, there is a sustained high pressure for almost 2 ms before the pressure declines. At point 9, there is a peak pressure for an instant which then begins to decline. From the graphs, it is seen that at points 6 and 7, the high pressure is sustained for a while before it slowly falls. This pressure profile closely matches that of a shock wave. For point 9, the pressure attains a peak value and then begins to decline. This description matches that of a blast pressure profile. However, to check if it is truly a blast wave, the profile is plotted with its corresponding Friedlander waveform, as shown in Fig. 7. The experiment is repeated for a driver pressure of 7.5 bar. The blast wave and the corresponding Friedlander wave form are as shown in Fig. 8. When compared to the pressure profile for the same point under 4.5 bar driver section pressure, the pressure profile for 7.5 bar fits the Friedlander waveform more closely.

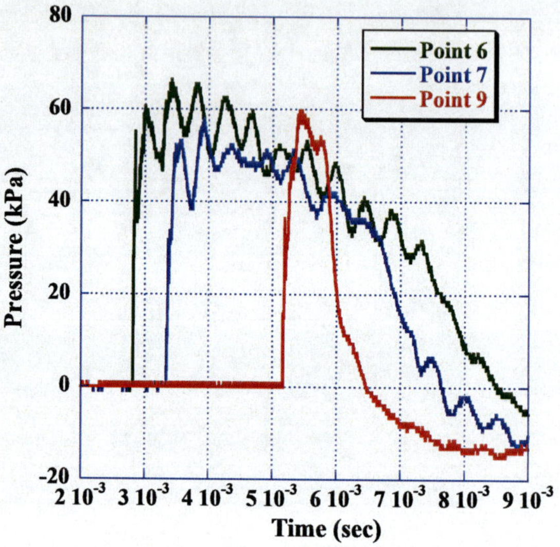

Fig. 6 Pressure versus time readings for 4.5 bar driver pressure at points 0, 2, 7 and 11 along the driven section

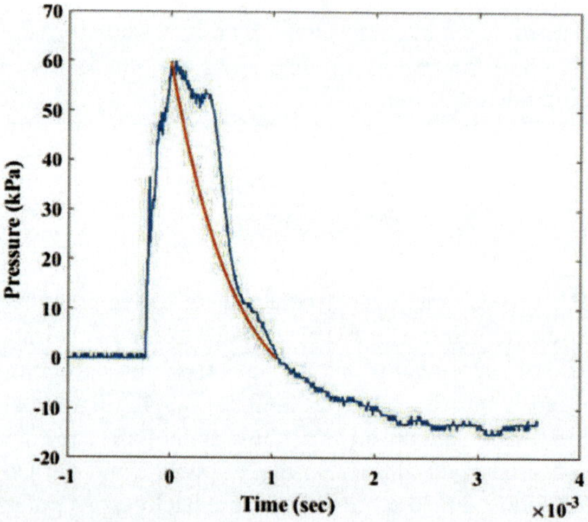

Fig. 7 Pressure plot for 4.5 bar driver pressure and corresponding Friedlander wave form

5 Conclusion

It is observed that the distance at which the blast wave forms in the shock tube increases with increase in driver pressure and driver length. It is also found that the distance along the driven section of the shock tube at which the blast wave pressure profile forms tends to converge at higher pressures. This could be indicative of an upper limit to the blast formation distance. It is also observed that smoother and more accurate pressure wave is obtained at driver pressures higher than 20 bar—regardless

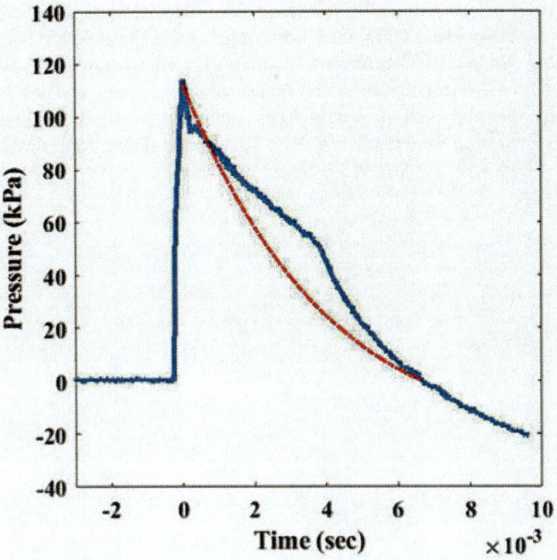

Fig. 8 Pressure plot for 7.5 bar driver pressure and corresponding Friedlander wave form

of driver length. From experimental results, it is seen that the pressure wave obtained conforms more closely to the Friedlander waveform as the driver pressure increases.

References

1. Nagaraja SR, Prasad JK, Jagadeesh G (2013) Theoretical-experimental study of shock wave assisted metal forming process using diaphragmless shock tube. Proc IMechE, Part G: J Aerosp Eng 226:1534–1543
2. Jagadeesh G, Divya Prakash G, Rakesh SG, Allam US, Gopala Krishna M, Eswarappa SM, Chakravortty D (2011) Needleless vaccine delivery using micro-shock waves. Clin Vacc Immunol 18:539–545
3. Nagaraja SR, Rakesh SG, Prasad JK, Barhai PK, Jagadeesh G (2012) Investigations on micro-blast wave assisted metal foil forming for biomedical applications. Int J Mech Sci 61:1–7
4. Dewey JM (2010) The shape of the blast wave: studies of the friedlander equation. Mcmillen & Associates, Department of Physics & Astronomy, University Of Victoria, BC Canada
5. Courtney AC, Andrusiv LP, Courtney MW (2012) Oxy-acetylene driven laboratory scale shock tubes for studying blast wave effects. Rev Sci Instrum 83:045111
6. Courtney MW, Courtney AC (2011) A Table-top Blast Driven Shock Tube. Review of Scientific Instruments. https://doi.org/10.1063/1.3518970
7. Ram O, Sadot O (2012) Implementation of the exploding wire technique to study blast-wave–structure interaction. Exp Fluids 53:1335–1345
8. Chandra N, Ganpule S, Kleinschmit NN, Feng R, Holmberg AD, Sundaramurthy A, Selvan V, Alai A (2012) Evolution of blast wave profiles in simulated air blasts experiment and computational modelling. Shock Waves 22:403–415
9. Sandhu IS, Sharma A, Prince, Singh MK, Kumari R, Alegaonkar PS, Saroha DR (2017) Study of blast wave pressure modification through rubber foam. In: 11th International Symposium on Plasticity and Impact Mechanics, Procedia Engineering, vol. 173, pp 570–576

10. Susheel Kumar BP, Nagaraja SR (2015) Comparative study of micro and macro shock tube flows using CFD. Int J Appl Eng Res 10(78):249–254
11. Ngomo D, Chaudhuri A, Chinnayya A, Hadjadj A (2010) Numerical study of shock propagation and attenuation in narrow tubes including friction and heat losses. Comput Fluids 39:1711–1721
12. Nambiar GK, Sriram M, Dharanidhar M, Nair PG, Nagaraja SR (2017) Design and fabrication of mini Shock tube. In: IOP Conference series: Material Science and Engineering, p. 012025

Intermittent Afterburner Engagement Leading to Single Engine Landing in a Typical Bypass Military Aero Engine

Saroj Kumar Muduli, Subrata Kumar Rout, Benudhar Sahoo, and P. C. Mishra

Abstract Thrust of an aero engine is augmented with the help of afterburners which are useful during takeoff and tactical maneuvers. It provides higher thrust without increase in engine size and helps to take off with maximum weapon load. Failure of afterburner engagement influences takeoff and/or abandon mission requirements. The current case studies are deliberated with respect to a low bypass turbofan engine having a thrust to weight ratio of 8:1 with afterburner on. The ignition of afterburner is by hot streak method where fuel is supplied for a short duration of time into the gas stream of the core engine just upstream of the turbine. The paper highlights the ignition system and its problem during afterburner engagement leading to single engine landings in twin engine aircraft configuration. It also highlights importance of various systems during afterburning ignition and the deficiencies in fuel nozzles in its manufacturing and assembly configurations leading to intermittent afterburner engagement thus affecting the reliability of aircraft/aero engine operation.

Keywords Afterburner · Exhaust fairing · Flame holder · Spray nozzle · Hot streak ignition

1 Introduction

Basic thrust of an aero engine is augmented by using an afterburner which is used during takeoff, climb and combat performance of aircraft. It augments higher thrust without increasing size of the aero engine which in turn increases the weight and frontal area of the aircraft. This is beneficial in military applications carrying out combat operations. A turbofan engine with multi-zone afterburning is used to achieve higher thrust along with aerial refueling. It allows fighter aircraft to take off with

S. K. Muduli (✉) · S. K. Rout · B. Sahoo
RCMA (Koraput), CEMILAC, DRDO, Koraput, Odisha, India
e-mail: sk.muduli@cemilac.drdo.in

P. C. Mishra
KIIT Deemed University, Bhubaneswar, Odisha, India

S. Revankar et al. (eds.), *Proceedings of International Conference on Thermofluids*, Lecture Notes in Mechanical Engineering,
https://doi.org/10.1007/978-981-15-7831-1_12

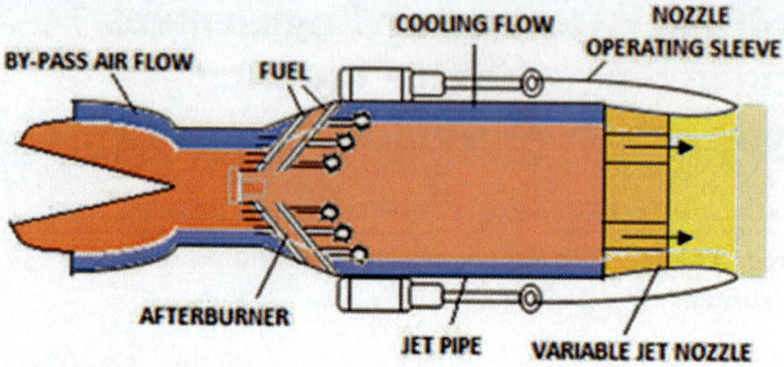

Fig. 1 Configuration of afterburner

maximum weapon load from relatively short airfields, followed by a climb to altitude [1]. Increased power could be obtained by increasing the air mass flow, but this would increase the frontal area and hence weight and specific fuel consumption. Thus, afterburning is an appropriate method of thrust augmentation for short duration. Afterburner utilizes the unburnt oxygen in the exhaust gas by supplying and burning fuel in the jet pipe. The resultant increase in the temperature of the exhaust gas gives an increased velocity of the jet leaving the propelling nozzle and therefore increases the engine thrust. To work under all operating conditions, an afterburning jet pipe is fitted with a variable area propelling nozzle [2].

1.1 Principle

Afterburner produces increased power with the same size of the aero engine. It increases thrust by burning additional fuel in jet pipe, and the gas thus produced is accelerated to higher velocity. Normally, the diameter of the jet pipe remains in the closed position(minimum) during non-reheat operation, but it opens to an exit diameter (beyond minimum) suitable to accommodate the increase in volume of the gas formed due to combustion of additional fuel. Successful afterburning is accompanied with enhanced thrust felt in the form of a jerk by the pilot during aircraft in operation. The detail principle is shown in Fig. 1.

1.2 Objective

Afterburning is a process in which fuel gets injected into the hot exhaust gases to have additional thrust for a short duration required during takeoff, climb, cruise and combat maneuvers [2]. In case of failure of afterburner engagement or of intermittent engagement, the pilot has to abandon the flight operation in a single engine configuration or land with single engine after switching off the defective engine in a twin engine aircraft. The amount of risk involved is phenomenal as safe landing depends upon speed of aircraft, altitude and power available to the aero engine. The basic objective in these two case studies is to find out the reason of the failure of afterburning system leading to single engine landing and highlight the lapses in the system if any contributed from manufacturing of the components to operation at different flight envelopes.

2 Literature Review

The cases cited in the paper are practical problems encountered during operation of a bypass military aero engine and their investigations. Though, exact cases of this nature are not available in the literature; however, researchers have been continuously working in the areas of afterburner design, development and operation. A few relevant literatures available on afterburner design are placed in Table 1.

3 Afterburner Assembly

The afterburner comprises exhaust mixer, transition section, casing with heat shield and exhaust fairing.

3.1 Exhaust Mixer

The purpose of the exhaust mixer is to mix the main gas duct stream with bypass air flow before transition section. The exhaust mixer carries many pockets for mounting of thermocouples, swirl nozzle of "flame path" system, pressure probes, etc.

3.2 Transition Section

Transition section provides steady burning of fuel in the afterburner. It consists of heat shield, flame stabilization section, fuel manifolds and casing to carry these components [8–10].

Table 1 Relevant literature on afterburner design

Authors	Year	Major highlights	Ref.
Anand R, Lokesharun D, Rajkumar S and Kirubakaran R	2017	Highlighted about application of "NUMECA" software for the three dimensional analysis in afterburner by carrying out cold flow analysis	[2]
H C Karcher and Otto T Kreusser, (US patent)	1959	This is related to jet propulsion of aircraft and more particularly to the augmentation of thrust of reaction engines	[3]
Paul C. Mangan, Wayne E. Kohman	1959	This patent described about "hot streak" ignition arrangement for afterburner fuel control system relates to gas turbine power plants	[4]
T. L. Schilling	1960	This patent related to turbo machines and in particular to a fuel metering valve for use to ignite the reheat fuel in an afterburner during reheat operation	[5]
T. E. Kuhn, H. C. Mongia, T. W. Bruce and E. Buchanan	1982	Worked on afterburner design methodology oriented toward large turbofan engines. Described a design technique based on the use of semi analytical methods along with a multidimensional aerothermal analysis	[6]
Houshang B. Ebrahima	2006	Provided an overview of military aircraft gas turbine reheat performance issues including design, history, operation, combustion instabilities and ignition. Discussed about the progress of CFD and its utilization to support the design of augmenters (reheat systems)	[7]

3.3 *Exhaust Fairing*

It is designed to decrease the exhaust gas energy losses. The exhaust fairing perforations are intended to decrease afterburner intermittent burning [12–14].

3.4 *Fuel Manifolds and Flame Holders*

Fuel is delivered through jets of manifolds consisting of one or more concentric rings located downstream of the low pressure turbine. Flame holders and exhaust fairings are placed further downstream to decrease the exhaust gas energy losses to have a stable flame in the reheat regime (Fig. 2).

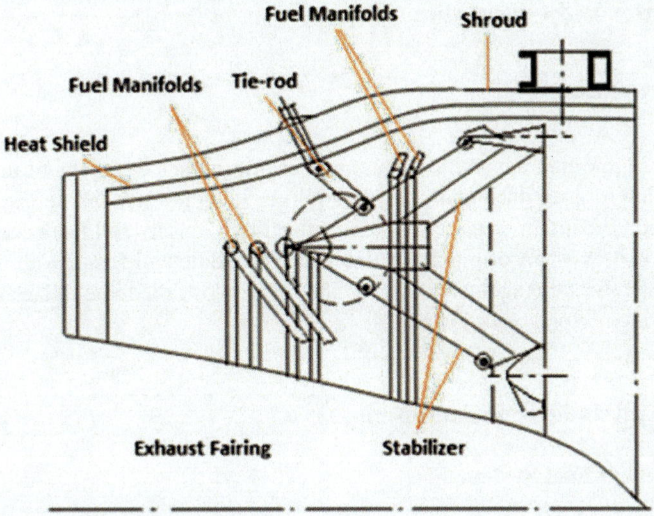

Fig. 2 Transition sectional view of afterburner assembly [11]

3.5 Hot Streak Method of Ignition

Ignition of the afterburner fuel is achieved using several different methods such as (i) hot streak method, (ii) spark or arc ignition method and (iii) the pilot burner method. In the present case, hot streak method is in use for ignition wherein fuel is injected in the main combustor for a short period which is taken away toward afterburning zone through the turbine sections for further combustion [9] (Fig. 3).

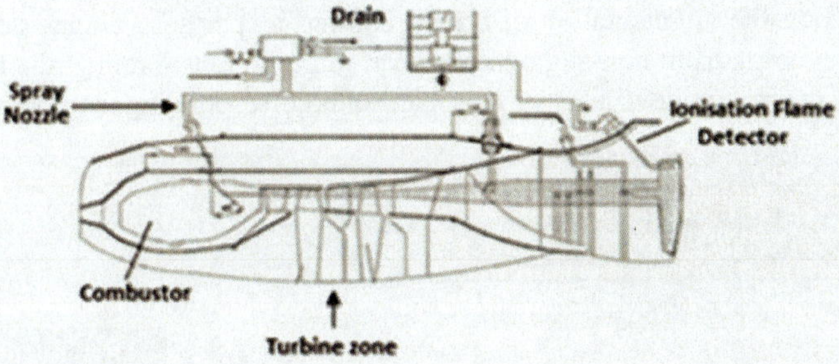

Fig. 3 Representative sketch of hot streak ignition [11]

4 Results and Discussion

4.1 Case Study—I

The aero engine had experienced intermittent afterburner engagement in air with no repetition of snag during aircraft ground run or engine standalone test bed run. Rectification including electrical system and various sensors and feedback systems were checked as a part of regular maintenance schedule and found with no abnormality. The reheat engagement could be intermittent or failed due to lack of any of the following reasons placed in Fig. 4.

4.1.1 Analysis of Cause and Effect

4.1.2 Reheat Fuel System

Reheat fuel system is intended to supply metered fuel to the spray bar of the afterburner fuel manifold nozzles [15, 16]. It does the following functions:

a. Lights up the afterburner
b. Supplies desired quantity of fuel in the afterburner combustion chamber at various ratings
c. Ensures engine acceleration and deceleration control under augmented thrust conditions.

4.1.3 Reheat Fuel Regulation

There are multiple reheat fuel manifolds placed concentric to each other. The number depends upon the augmentation of thrust required as per aero engine design. As the throttle moves from non-augmented power rating to reheat rating, the fuel from reheat fuel pump supplies fuel to various manifolds corresponding to the set pressure.

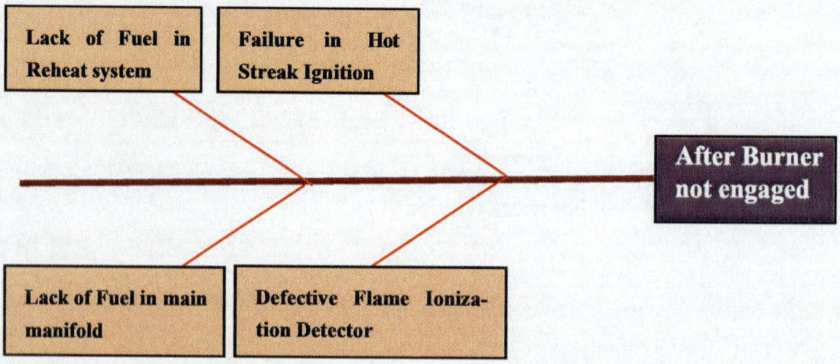

Fig. 4 Cause and effect diagram

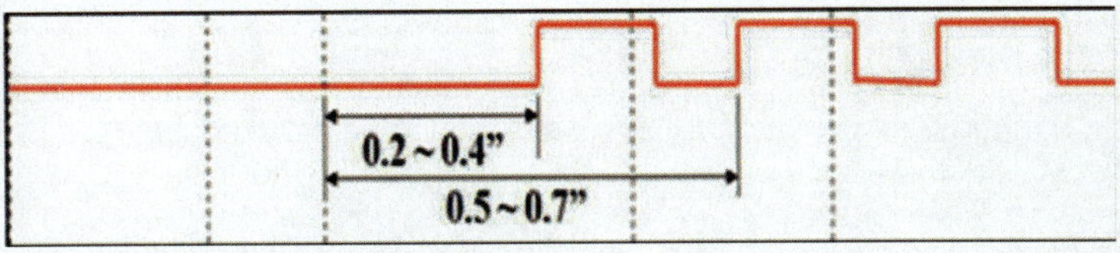

Fig. 5 Command for hot streak light up [11]

Similarly, the cutoff valves are provided with rubber seals to ensure air tightness of the manifold during non-augmented power ratings.

4.1.4 Ionization Flame Detector

The flame detector is designed to detect the presence of flame in the afterburner and sends the signal to engine control unit that afterburner has started operating. Two detectors are connected in parallel are used in the system providing reliable output of signals at different reheat operation modes. One produces a signal on the presence of flame in the afterburner area and the other on the afterburner outer stabilizer area.

4.1.5 Operation of Hot Streak Ignition System

Fuel for the *A/B* ignition system is supplied to flow metering device from afterburner fuel pump. The fuel pressure depends on the engine operating mode. The fuel supplied to the spray nozzle is injected in the main combustion chamber, ignites and is carried away by a jet stream behind LP turbine. There the jet flame initiated by the spray nozzle is re-enforced by fuel injected by the swirling type nozzle. The flame ignites in the afterburner fuel supplied by afterburner starting manifold (Fig. 5).

In 0.2 to 0.4 s, electromagnetic valve gets energized by the command from engine electronic controller. In case afterburner not started in 0.5–0.7 s in response to the command, solenoid valve turns on to supply second pulses of fuel. Within the full cycle of system operation, three pulses of fuel are supplied to nozzle. Upon fuel in afterburner is ignited, the same is sensed by ionization flame detector and fed to engine controller to continue with afterburner as shown in Fig. 3.

4.1.6 Snag Analysis

The aero engine was mounted on the engine test bed for confirmation. Fuel pressure of afterburner light up unit and that of manifolds were checked. Systems such as jet nozzle control, movement of guide vane mechanism were checked and found confirmed to the requirement. The aero engine was stripped, and the hot streak

Fig. 6 Spray nozzle and its direction of flow

initiation jet was checked. Projection of fuel jet was not in order which has led to a scattered fuel jet instead of a beamed one shown in Fig. 6.

4.1.7 Root Cause

- Lesser projection of jet supplying fuel for hot streak from combustion chamber.
- Presence of large chamfer at the jet exit & high surface roughness of the jet exit area during manufacturing.

4.1.8 Recommendations

- Projection of jet supplying fuel for initiation of hot streak toward mid value instead of lower value of the design band during assembly followed by jet stream as per standard.
- Roughness, chamfer of the jet including direction of the jet assembled to main combustion chamber needs to be ensured (Fig. 7).

4.2 Case Study—II

The aircraft was in a practice sortie wherein delay in reheat engagement at takeoff and reheat not engaging in air at an altitude of 3 km was reported. The snag could not be rectified at operating base with the laid down maintenance procedures. The aero engine was withdrawn to the repair agency for proper investigation and rectification of the defect.

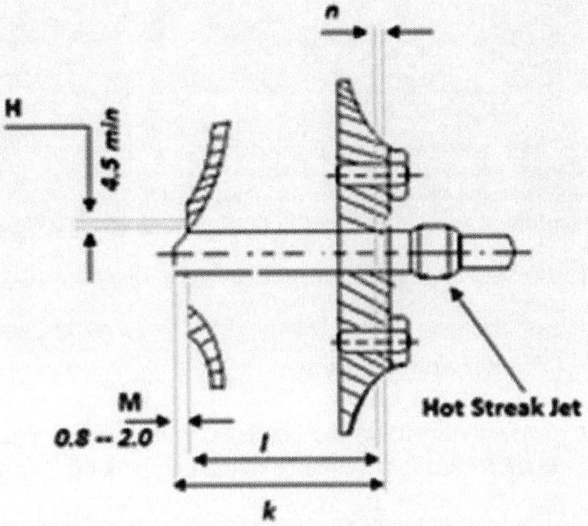

Fig. 7 Projection of hot streak jet [11]

4.2.1 Snag Analysis

Serviceability of various fuel aggregates during afterburning, electrical systems and ionization flame sensors was checked and found in order. Afterburner was not getting engaged was confirmed during confirmation test of the aero engine on test bed. However, afterburner was getting engaged during gradual movement of the throttle from max to reheat mode during confirmation test on ground. Aero engine was stripped and analyzed as per cause and effect placed for case study—I as shown in Fig. 4. It was observed that the swirling type nozzle sprayer found blocked partially. The swirling type nozzle supplies fuel from flow metering device to the jet stream behind low pressure turbine, augmenting the jet flame created by the spray nozzle. The nozzle jet is positioned opposite to the jet stream direction (Refer to Fig. 3).

Flow check of the swirl type nozzle on test rig was carried out at the preset pressure and 50% higher than the preset value. The fuel flow was not up to the rated value. Flow check of the other systems such as spray nozzle and afterburner fuel distributor was satisfactory. On further investigation, it was found that the blockage in the swirl type nozzle was due to the presence of metallic burrs at the jet and formation of carbon shoots inside the jet. Presence of metallic burrs inside the jet could be due to error in surface finish of the jet manufacturing (Fig. 8).

4.2.2 Root Cause

- Poor surface finish of the jet hole leading to deposit of carbon soots over a period of engine operation.

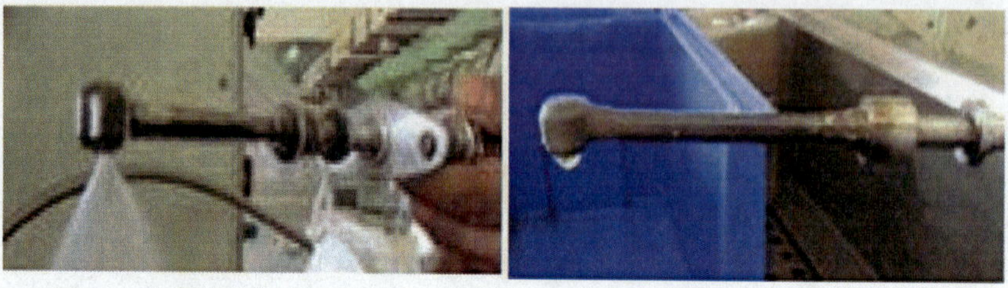

Fig. 8 View of the partially blocked jet and flow of Swirl Type Nozzle

4.2.3 Recommendation

- Surface finish of the jet hole with correct diameter to be ensured during manufacturing acceptance of the component.

The recommendations worked out during investigation such as implementation of correct projection of the spray nozzle jet with proper chamfer and surface finish at the jet exit in case study—I and polishing of the swirl type nozzle jet diameter without any metal burrs or projections and surface finish as specified in drawing (Fig. 9) in case study—II were implemented. In order to overcome repetitiveness, the same was recorded in the form of quality control instructions. No such cases are reported after careful implementation of the recommendations discussed.

5 Conclusion

This paper highlights two different case studies wherein the aircraft has been landed with single engine due to the non-performance of afterburner system. These case studies are live cases where the authors are involved during the snag analysis and find out the exact cause of failure of afterburning system in operation. Each component in the system needs to function efficiently to have successful operation of aero engines as found not in order in positioning of spray nozzle inside main combustor from where hot streak ignition is initiated in case study—I and lack of fuel supply by spray type nozzle to support hot streak ignition in case study—II. Malfunction of any component can lead to failure of mission and loss of life. Hence, utmost care must be exercised during manufacturing, assembly and testing of such aeroengine components.

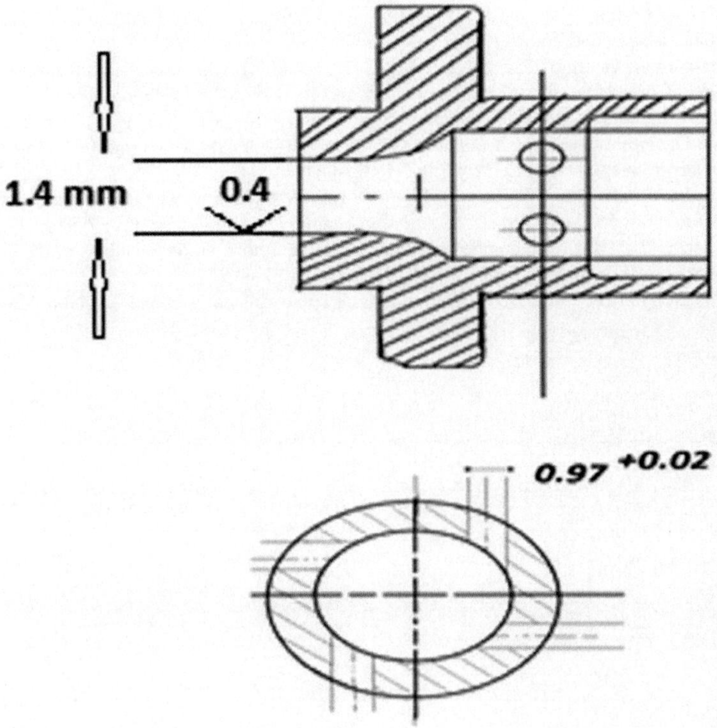

Fig. 9 Dimension of jet of Swirl Type Nozzle [11]

References

1. MacIsaac B, Langton R (2011) Gas turbine propulsion systems, 1st edn. Wiley, New York
2. Anand R, Lokesharun D, Rajkumar S, Kirubakaran R (2017) 3D CFD analysis in an afterburner using NUMECA. IJAREM. ISSN: 2456-2033, pp 20–37
3. Karcher HC, Kreusser OT (1959) Augmenter type afterburner for jet propelled aircraft. US patent US2882679A, 21 Apr 1959
4. Mangan PC, Kohman WE (1959) Hot streak ignition arrangement for afterburner fuel control system. US patent US2892307A, 30 June 1959
5. Schilling TL (1960) Afterburner ignition system. US patent US2936586A, 17 May 1960
6. Kuhn TE, Mongia HC, Bruce TW, Buchana E (1982) Small turbine engine augmenter design methodology. In: 18th AIAA/SAE/ASME joint propulsion conference, Cleveland, Ohio, USA
7. Ebrahimi HB (2006) Overview of gas turbine augmenter design, operation, combustion oscillation. In: 42nd AIAA/ASME/SAE/ASEE joint propulsion conference, 9–12 July 2006
8. Lovett JA, Brogan TP, Philippona D (2004) Development needs for advanced afterburner designs. In: Proceedings of 40th AIAA/ASME/SAE/ASEE joint propulsion conference and exhibit, Florida, USA
9. Oates GC (1985) Aerothermodynamics of aircraft engine components. AIAA education series, Pandora Books
10. Mattingly JD (2006) Elements of propulsion, gas turbines and rockets, 2nd edn. American Institute of Aeronautics and Astronautics, Virginia

11. Engine Operation Manual of a Low Bypass Turbofan Engine, Book1, Part1. Hindustan Aeronautics Limited, India (1996)
12. Gohardani A, Georgios S, Doulgeris RS (2010) Challenges of future aircraft propulsion: a review of distributed propulsion technology and its potential application for the all-electric commercial aircraft. Prog Aerosp Sci 47(5):369–391
13. Turan O, Aydın H, Hikmet Karakoc T, Midilli A (2014) First law approach of a low bypass turbofan engine. J Autom Control Eng 2(1):62–66
14. Millhouse PT, Kramer SC, King PI, Mykytka EF (2000) Identifying optimal fan compressor pressure ratios for the mixed-stream turbofan engine. J Propul Power 16(1):79–86
15. Bennett HW (1983) Aero engine development for the future. In: Proceedings of the Institute of Mechanical Engineers, Series A, vol 197(July), pp 149–157
16. Wilde GL (1978) Future large civil turbofans and powerplants. Aeronaut J 82(July):281–299

Numerical Simulation of Interaction of Blast Wave Generated from Cannon with Wall at Different Pressure Ratio

Ansab Khan, Abhishek Kundu, and Akshoy Ranjan Paul

Abstract A computational fluid dynamic method has been applied to simulate the pressure blast from a cannon and to investigate the pressure distribution on a wall located in front of the cannon. By solving this problem, peak value of pressure on wall has been calculated for different pressure ratio of driver and driven section in the cannon. By taking the maximum value of pressure at a particular pressure ratio as factor of safety, we have to invent a blast wave resisting jacket which will be helpful for blast wave explosion. The Advection Upstream Splitting Method (AUSM) scheme with third-order Monotone Upstream centered Scheme for Conservation Laws (MUSCL) approach were used for solving the unsteady, axisymmetric Navier–Stokes equation. Density-based solver was selected for present simulation due to its high-speed compressibility. The standard *K*-epsilon model with standard wall function is used for this simulation which is highly suitable for wall phenomenon problems. At the initial condition of this problem, the driver section of cannon is given to high pressure and driven section is at atmospheric pressure. After firing from cannon, flow of air take place from driver section to driven section and at advance time air flows out from the cannon. At exit of the cannon blast wave formation takes place which move in forward direction and after some time it strikes to the wall and move in backward direction. The computational result indicated that on increasing the pressure ratio in cannon causes the multiple fluctuations in pressure on the wall. The increase in pressure ratio also increases the peak pressure on wall.

Keywords Blast wave · Navier–stokes equations · Spherical wave-wall interaction · Computational fluid dynamics

A. Khan (✉) · A. Kundu · A. R. Paul
Department of Applied Mechanics, Motilal Nehru National Institute of Technology Allahabad, Prayagraj, Uttar Pradesh, India
e-mail: ansab325@gmail.com

S. Revankar et al. (eds.), *Proceedings of International Conference on Thermofluids*, Lecture Notes in Mechanical Engineering,
https://doi.org/10.1007/978-981-15-7831-1_13

1 Introduction

Generation of spherical blast wave at the exit of the canon takes place due to high pressure in the driver section and at comparatively lower pressure in the driven section. If a wall is located in front of the cannon, then this blast wave strikes the wall and gets reflected in backward direction. At the time of striking of blast wave, the pressure on wall increases to peak value. Sathish et al. [1] investigate the pressure distribution along with the different positions away from the cannon axis. Sathish et al. [1] patches very high pressure of 3450 atm and high temperature of 2500 K in driver section and atmospheric pressure and temperature in driven section to compute the pressure variation at positions away from the axis. Kundu et al. [2] reports the interaction of vortex generated from a shock tube with a wall. The vortex ring impinges on a wall kept at 300 mm from the open end of shock tube. They used the pressure ratio of 10:1 in the 165 mm of driver section and 1200 mm of driven section of shock tube. Murugan et al. [3] and Jialing et al. [4] examined the blast wave interacting with three different objects of cone, circular disc and sphere with diaphragm ratio of 56. They observed that the load acting on sphere was more as compared to conical objects. Kontis et al. [5] examine the interaction of shock wave induced vortices with flat plat and perforated plate. They observed that a region of strong flow development is generated near the wall surface due to flow interaction of reflected waves. This flow behaviour causes the generation of multiple pressure fluctuations on the wall. In the present study, we have to investigate the pressure variation on wall kept in front of the cannon at different pressure ratio of 10, 20, 30 and 40. Sathish et al. [1] used the Spalart–Allmaras model for simulation but we use standard K-epsilon, standard wall function model because this model is highly suitable for wall phenomenon problem. Many researchers [6–8] examined the vortex ring formation at exit of tube and interaction of these vortices with wall. The present study shows the impinging of vortices with wall at different pressure ratio.

2 Numerical Methodology

2.1 Governing Equations

Navier–Stokes equations are the governing equations of Computational Fluid Dynamics. The governing equations for two dimensional, unsteady, incompressible flows for blast wave simulation are continuity equation (Eq. 1), momentum conservation equation (Eq. 2) and energy conservation equation (Eq. 3). These equations in differential form are written as

Continuity Equation

$$\frac{D\rho}{Dt} + \rho\frac{\partial U_i}{\partial x_i} = 0 \tag{1}$$

Momentum Equation

$$\rho\frac{\partial U_j}{\partial t} + \rho U_i\frac{\partial U_j}{\partial x_i} = -\frac{\partial P}{\partial x_j} - \frac{\partial \tau_{ij}}{\partial x_i} + \rho g_j \tag{2}$$

Energy Equation

$$\rho c_\mu\frac{\partial T}{\partial t} + \rho c_\mu U_i\frac{\partial T}{\partial x_i} = -P\frac{\partial U_i}{\partial x_i} + \lambda\frac{\partial^2 T}{\partial x_i^2} - \tau_{ij}\frac{\partial U_j}{\partial x_i} \tag{3}$$

2.2 *Computational Domain and Boundary Conditions*

The computational domain for the interaction of cannon generated blast wave with wall is shown in Fig. 1. The length of cannon is 200 mm such that driver and driven section of cannon are equal lengths of 100 mm each. The wall is kept at 100 mm from the exit of the cannon. The diameter of cannon are 10 mm. The driven section and the domain outside the cannon is kept at atmospheric conditions initially with pressure of 1 atm and temperature of 300 K. The pressure and temperature of driver section are patched very high for four different conditions. This problem has been solved for four different pressure of driver section of 10, 20, 30 and 40 atm. The temperature of driver section for each case was calculated from Rankine–Huguniot equations. The following table represents the initial conditions of four cases to carry out the present simulations (Table 1).

The boundary conditions of problem are shown in Fig. 1.

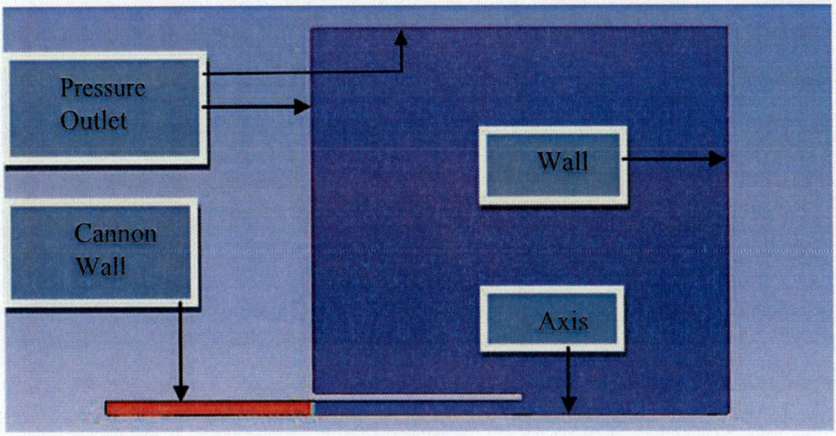

Fig. 1 Computational domain and boundary conditions

Table 1 Initial conditions used for present simulation

Test cases	*P*1 (atm)	*T*1 (K)	*P*2 (atm)	*T*2 (K)
Case 1	10	786.4	1	300
Case 2	20	1289.3	1	300
Case 3	30	1782.2	1	300
Case 4	40	2285.1	1	300

The atmospheric domain in transverse direction away from the axis of cannon is extended to a length of 267 mm. The problem is simulated for half portion of the domain due to symmetry nature so the computing effort of full domain can be reduced. The meshing of the computational domain was done for the simulation of problem.

3 Results and Discussion

3.1 Validation

The present simulation has validated with Sathish et al. [1]. They evaluated the pressure distribution at a point away from the axis. In this section, the effect of high pressure (3450 atm) and various initial conditions were compared with Sathish et al. [1]. The domain they have taken is different from present problem. They have taken 300 mm cannon in which 100 mm is patched to very high pressure of 3450 atm and high temperature of 2500 K initially. Their domain is extended to 600 mm in axial direction from the exit of cannon. The next 200 mm of cannon and domain outside the cannon is at atmospheric condition initially. The comparison was done by evaluating the pressure distribution at $X = 300$ mm, $Y = 100$ mm which is shown in Fig. 2.

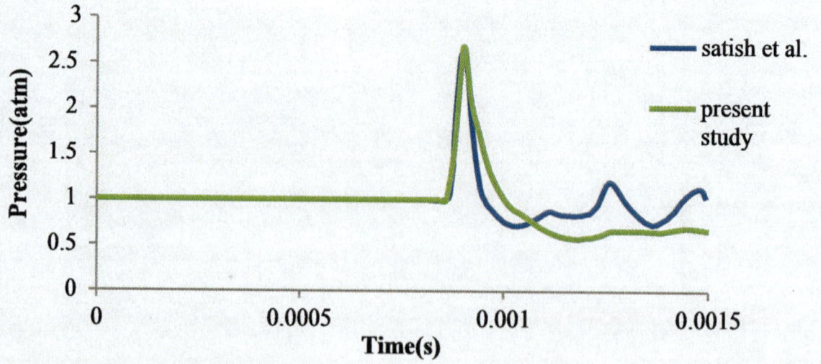

Fig. 2 Validations of result with [1]

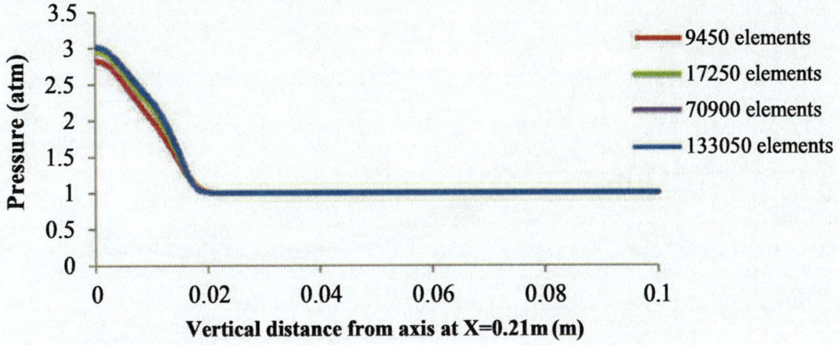

Fig. 3 Grid independency test

The result of the present study shows some deviation in pressure distribution after attaining peak pressure. The peak value of pressure of both cases is the same as shown in Fig. 3.The cause of deviation is dynamic mesh use by Sathish et al. (2011) [1] whereas present study use static mesh.

3.2 Grid Independency Test

GIT was done to make the results independent from the number of elements. This was done by evaluating the pressure on line at ($X = 0.21$ m, $Y = 0$) to ($X = 0.21$ m, $Y = 0.1$ m) at time of 0.18 ms. The result was evaluated by increasing the number of elements and observed the pressure variation. The result of GIT is shown in Fig. 3 in which the graph become similar after 70,900 elements. So, the present study is simulated at 70,900 elements.

3.3 Pressure Distributions on Wall

The objective of this study is to investigate the pressure distributions on wall located in front of the cannon. In this problem, initially, driver section is patched to high pressure and temperature and driven section is at atmospheric conditions. After firing from cannon, flow of gas takes place from driver section to driven section due to pressure gradients and finally blast wave formation takes place at the exit of cannon. The pressure on wall remains atmospheric until the blast wave from cannon reaches to wall. When the blast wave strike to wall, pressure on wall increases to peak value and then back to atmospheric pressure after blast wave get reflected from wall. This problem is solved for four cases and for each case pressure distribution on wall is shown in Fig. 4.

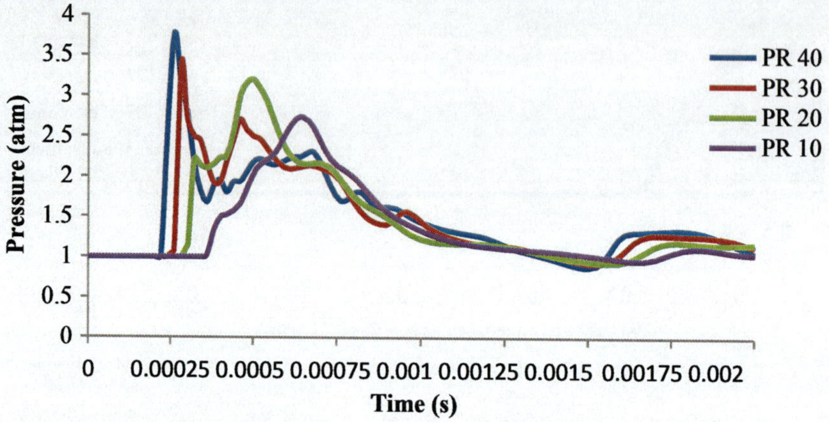

Fig. 4 Pressure variation on wall

Table 2 Result for different cases

Test case	Pressure ratio	Peak pressure on wall (atm)	Time of peak pressure (ms)
Case1	10	2.74	0.64
Case2	20	3.2	0.5
Case3	30	3.4	0.28
Case4	40	3.8	0.26

The pressure distribution on wall shows that multiple fluctuations in pressure at higher pressure ratio. The graph shows that peak pressure on wall increases on increasing the pressure ratio. It also shows that the time at which pressure attains peak value decreases on increasing the pressure ratio. The pressure fluctuations on wall increases on increasing the pressure ratio (Table 2).

In the present work, interaction of a blast wave from the canon exit with a nearby wall has been studied at different pressure ratios. Major aim of this study is to calculate the load at the wall due to the interaction of a blast wave. This study is useful for designing of a blast wave mitigating or attenuating structure which can rescue the larger commercial building or military vehicles from the larger terrorist attack across the globe.

4 Conclusion

The pressure distribution on wall due to pressure blast from cannon has been simulated by using CFD software ANSYS FLUENT. The Advection Upstream Splitting Method (AUSM) scheme with third-order Monotone Upstream centered Scheme for

Conservation Laws (MUSCL) approach were used to solve the unsteady, axisymmetric, compressible Navier–Stokes equations. The four cases of pressure ratio in cannon were solved to investigate the pressure distribution on wall. The result shows that on increase in pressure ratio in cannon, causes multiple pressure fluctuations on wall as well as increase in the peak pressure on wall. The time of attaining peak pressure decreases on increasing the pressure ratio.

References

1. Xavier S (2011) Numerical analysis of gun barrel pressure blast using dynamic mesh adaption. Dissertations and thesis, p 258
2. Kundu A, De S, Thangadurai M, Dora CL, Das D (2016) Numerical visualization of shock tube-generated vortex–wall interaction using a fifth-order upwind scheme. J Vis 19:667–678
3. Dey S, Murugan T, Chatterjee D (2018) Numerical visualization of blast wave interacting with objects. J Appl Fluid Mech 11(5):1201–1206
4. Jialing JE (1999) Numerical simulation of shock (blast) wave interaction with bodies. Commun Nonlin Sci Numer Simul 4(1):1–7
5. Kontis K, An R, Zare-Behtash H, Kounadis D (2008) Head-on collision of shock wave induced vortices with solid and perforated walls. Phys Fluids 20:016104
6. Mariani R, Kontis K, Gongora-Orozco N (2013) Head on collisions of compressible vortex rings on a smooth solid surface. Shock Waves 23:381–398
7. Minota T, Nishida M, Lee MG (1997) Shock formation by compressible vortex ring impinging on a wall. Fluid Dyn Res 21:1–17
8. Murugan T, Das D (2012) Experimental study on a compressible vortex ring in collision with a wall. J Vis 15:321–332 (2012)

Review on Hypersonic SSTO Engine (HSE) with Variable Diffusers, Double Annular Combustors and N_2O-Oxygen Enhancer

Vedant Gupta, S. S. Godara, and B. Tripathi

Abstract This paper details a novel concept of an engine capable of flying on a single-sage-to-orbit (SSTO) aircraft. Leveraging the positive facets of existing technologies of extremely high-speed flight regime, and in an attempt to improve on the deficiencies of the same, the proposed engine will be made to operate in an envelope of vastly varying flight parameters. All factors will be optimized for the continuously changing free stream conditions. In the study, it is found that the fixed geometries of diffusers and nozzles lead to unwanted results which are addressed by the HSE. Another novelty introduced is of having an oxygen enhancer (not oxidizer) in the form of N_2O which will be used to increase level of O_2 in the flowing mass. Subsequently, an enhanced version of a generic nitrogen generating system (NGS) is employed to cut emissions. The resulting engine will be capable of reaching high hypersonic Mach numbers from ground up using it radical concentric annuli, inner one of which will house an afterburning turbojet and outer one a scramjet with its own combustion section.

Keywords Hypersonic engine · Diffuser · Annular combustor

Abbreviations

SSTO	Single-stage-to-orbit
N_2O	Nitrous oxide
HSE	Hypersonic SSTO engine
NGS	Nitrogen generating system

V. Gupta
Mechanical Engineering Department (Aeronautical Engineering), University Departments, Rajasthan Technical University, Kota, Rajasthan, India

S. S. Godara (✉) · B. Tripathi
Mechanical Engineering Department, University Departments, Rajasthan Technical University, Kota, Rajasthan, India
e-mail: ssgodara@rtu.ac.in

S. Revankar et al. (eds.), *Proceedings of International Conference on Thermofluids*, Lecture Notes in Mechanical Engineering,
https://doi.org/10.1007/978-981-15-7831-1_14

RLV	Reusable launch vehicle
TSTO	Two stage to orbit
RTO	Ramjet takeover
RBCC	Rocket-based combined cycle
SCCREAM	Simulated combined-cycle rocket engine analysis module
TBCC	Turbine-based combined cycle
TRCC	Turbo rocket combined cycle
ATREX	Air turbo ramjet engine with expander cycle
LACE	Liquid air cycle engine
C-D	Convergent-divergent
SABRE	Synergistic air-breathing rocket engine
TE	Turbine engine
OA	Outer annulus
IA	Inner annulus
STO	Scramjet takeover
DC	Deflector cone
ORG	Oxygen rich gas
EAC	Enhancer-assisted combustion
NSS	Nitrogen separation system
TETO	Turbine engine takeover
FCOC	Fuel cooled oil cooler
ASM	Air separation module
I_{sp}	Specific impulse
CFD	Computational fluid dynamics

1 Introduction

SSTO is a separate class of airplanes that can be used for payload delivery into orbit using a single stage without disbursing any hardware from the aircraft. This puts it aloof from the traditional multi-stage rocketry [1]. The main idea behind the intermittent research that has taken place from the 1960s is the reduction in costs which result from expending the engines, fuel tanks and all the other expensive hardware on-board. Companies like Lockheed Martin, Boeing, NASA [2], McDonnell Douglas, ARCA Space, etc., have all invested heavily into the concept. As an alternative approach, organizations like SpaceX and Blue Origins have been successful in introducing reusable rockets. Albeit the idea is a major success and a revolution in space technology, many factors limit the reusability and drive up the refurbishment costs. What the space community yearns for is a system that can be used again for thousands of times and a potential for multiple orbital deliveries in a single day. SSTOs provide this platform but due to the umpteen complications for a propulsion system, a working SSTO has never been completely tested. Since air-breathing propulsion alone cannot power an aircraft to orbit, a rocket engine has been thus incorporated into all concepts, or at times even a separate rocket engine has been carried on aircraft to help them attain speeds where ramjet takeover (RTO) can occur. However, with the advance of material technology, we now can build with an engine that offers a high specific impulse and longevity of an air-breathing engine, enormous thrust of rocket engines, reusability, efficiency and safety [3]. HSE provides the perfect propulsion system with air being breathed at all times and an oxygen enhancer (not oxidizer) added only during some critical times of flight. It is a comprehensive engine satisfying all flight requirements. The main objectives of this study are to establish a concept for a propulsion platform capable of flying an SSTO aircraft which will have great technical and economic vitality by providing a medium to conduct satellite insertions and space missions at a fraction of the current cost.

2 Previous Work

To collect valuable data and create an all-inclusive design, fit to be used for the development of hypersonic SSTO engine literatures are studied thoroughly. This also helps in recognizing important supporting systems for the single-stage-to-orbit (SSTO) aircraft engine (Table 1).

Table 1 Concepts for SSTO propulsion proposed till now [4, 5]

Name	Inception	Main elements	Concept	Research work if any
RBCC [6]	Late 1990s	Ducted rocket, scramjet, ramjet	Common gas path for all three with variable inlet	SCCREAM [7, 8]
TBCC [9]	Early 1990s	TE, ramjet, scramjet	Switchable flow path for TE and high-speed unit [10]	Aerojet Rocketdyne for SR-72. ISTAR Engine for X-43B
TRCC [11]	Derived designs came up in late 1950s	Gas generator, afterburning TE, C-D exhaust nozzle	Independent combustion gas drives compressor. Exhaust gas mix with fan discharge and combust before a C-D nozzle	Pratt and Whitney J-58 ATREX
LACE [12]	Early 1960s	Rocket engine	Gathers part of LOX from atmosphere	Marquardt testbed
SABRE [1]	1988	Heat exchanger, rocket engine	LACE operation while in atmosphere. Resorts to using stored LOX for space flight	Being built and tested by Reaction Engines Ltd., UK

3 HSE Design Operation

3.1 Design

Since the engine has to deal with intertwined complexities of gas dynamics, shock waves, efficient stagnation temperatures and static pressures of the discharge, reduction of spillage drag, etc., the design has to be corroborative of the many factors. The main idea is to give the engine versatility for different regimes and envelopes. Following are the major design features explained.

Nacelle: HSE will have a design housing dual concentric annulus as shown in Fig. 1. Simply put, it will follow a pipe-in-pipe configuration. It may sound like a generic turbofan engine design, but HSE will have a cross section for the outer annulus (OA) to be much larger than that for the inner annulus (IA), unlike the case in the former. The IA will carry an afterburning turbojet engine capable of propelling the aircraft to speeds of Mach 3 to Mach 3.5 [13]. The role of propelling after that will be handed to the OA which will act as the scramjet. This will be called scramjet takeover (STO) [14].

Inlet and Variable Diffusers: HSE will feature separate variable area diffusers for both the annuli. Similar to variable nozzles of high-speed jet engines, the diffusers here will be varying the Turkey Feather arrangement operated by hydraulic actuators to vary the flow dynamics. After STO, the turbojet will be taken out of action by totally covering the IA. The Turkey Feathers on the IA will completely close by making a cone which is referred to as deflector cone (DC). Feathers will be linked with zero tolerance to prevent any pressure loss [15]. The feature of a variable diffuser not only enables selective operation of the annuli, but also plays a role in the efficiency by allowing for desirable placing of the shockwaves.

DC: The cone formed by the feathers of the IA will have translation abilities which is instrumental in setting up of the desired position of the external and internal

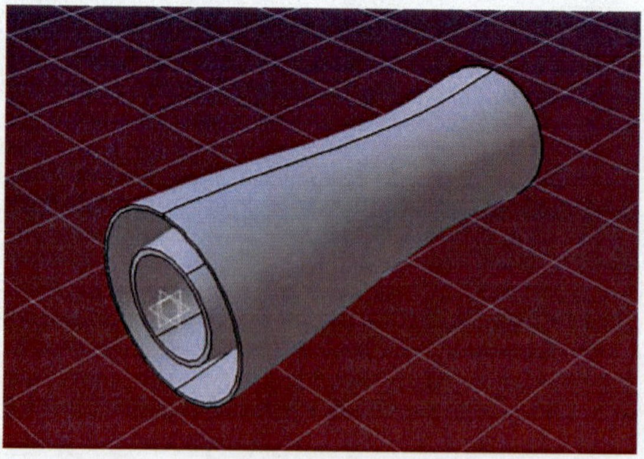

Fig. 1 Concentric annuli

Fig. 2 Approximation of the exterior geometry

shockwaves for ram air compression. It is the most critical design feature that governs the compression efficiency and consequently the combustion (Fig. 2).

Combustors: HSE will have a combustion chamber in both of the annuli. After STO, the OA will carry out the combustion of the ram air with fuel. Both of the combustors will be in tandem and not one on top of another. This will avoid excessive heating of the metal interface. More about the combustion procedure is explained in the "operations" section.

Nozzles: The nozzles of both annuli will be independently operated like the diffusers. They too will be controlled by the Turkey Feather arrangement. Since the exit mass flow is a detrimental criterion for engine efficiency, allowing the nozzles to change their area is very important for the HSE. The combined operation of variable nozzles and diffusers will lead to a largely effective operation [7].

Cowling: The outer skin of the engine will protect the engine's exterior and will house the multiple ducts and lines. The generic inlet cowl will be absent from this design to accommodate for the movement of the diffuser feathers [12].

3.2 Operations

Below given are the important procedures and operations for functioning of the HSE.

Take-Off and Climb: Here, the Turkey Feathers of the OA and would cover it by forming the lowest drag causing slope with the open IA feathers. The capture area will be mediated to cause the least spillage drag. The turbojet will have the capacity of reaching Mach 3.

STO: Now, the OA will open up and the IA feathers will close forming the DC which will also have translation abilities as discussed above and will move to perfectly position the frontal external shock, the diffuser oblique shocks and terminating normal shock. The ram air will be compressed and combusted with the fuel.

Combustion and Oxygen Enhancing: During the turbojet operation, there will be the naturally aspirated combustion of fuel with compressed air taking place. The combustion will need to be accentuated for these particular stages of flight:

- After STO, when the Mach number is high but not particularly high enough for scramjet propulsion
- When it begins to reach higher altitudes where oxygen is very little in quantity.

It is required to upscale the conditions or else incomplete or no combustion will take place. Concepts mentioned above take help of liquidized oxidizers or resort to direct combustion of rocket propellants in a separate cycle for producing thrust. HSE uses nitrous oxide (N_2O), not as oxidizer but as an oxygen enhancer. N_2O undergoes the following reaction (1) when temperature of 577 °C is supplied to it:

$$2N_2O \rightarrow O_2 + 2N_2 \quad (1)$$

This oxygen produced can be harnessed to allow for better combustion after mixing with the air. Direct combustion of fuels with oxidizers is rather difficult to control in jet engine environments and is more suited to conventional rocketry. It also requires thicker walls made of stronger and heavier materials to sustain the higher pressures. It leads to very high average emissions per flight. In case of N_2O, a direct interaction of itself and hydrocarbon fuel leads to emission of large quantities of other harmful oxides of nitrogen which are substantial contributors of greenhouse effects. What HSE does is pre-mix the incoming air with oxygen rich gas (ORG) obtained by allowing for dissipation of N_2O. Then, on reaching the combustion section, enhancer-assisted combustion (EAC) will take place. It is important to note that EAC occurs only during the above-mentioned events. Once the scramjet can sustain combustion by air alone, supply of ORG stops. To further reduce the possible emissions in the combustion, a system called nitrogen separation system (NSS) will be added to the engine. NSS will reduce the amount of N_2 from the ORG leading to cleaner combustion. Also, since N_2 is partially inert, its presence in the combustion mixture can heavily reduce the efficiency; although it may be used for other purposes. More about NSS has been explained in the following section.

Descent and Landing: After the payload insertion, the aircraft then re-enters the atmosphere with both the engines not operating and at an angle allowing for maximum deceleration depending on the design and material of the aircraft. The IA will be closed as the dynamic pressures and temperatures would possible damage the delicate turbine blades. The OA, however, would remain open with the setting of the diffuser and the DC will be such that to create shockwaves at the best possible locations which would produce the greatest pressure drag which will be useful in slowing down the vehicle. The nozzle too will be calibrated such that to complement the above. At the altitude of the turbine engine takeover (TETO), the OA will close and the IA will open for turbojet-assisted descent and landing.

3.3 Important Systems

The following are the most important systems of the HSE.

Variable Diffuser and Nozzle Control: Variable geometry of diffusers and nozzles will allow for total flow control authority. It is imperative to have the above in favour of varying external conditions. Hydraulically controlled interlinked feathers will give the variable area characteristics.

N_2O Delivery System (NDS): Pressurized liquid N_2O will be kept in a separate tank and consumed based on the need. It needs to be pre-heated to its disassociation temperature and only then can it be filtered for pure O_2. This is achieved by the simple regenerative principle of heat. Regular turbofan and turbojet engines use hot lube oil to heat the cold fuel to ease combustion, and consequently, the oil gets cooled up too. This happens in a device called fuel cooled oil cooler (FCOC). The high-speed air encountered by the aircraft can heat the skin to very high temperatures, and this heat can be used by the NDS to increase its temperature all the while cooling the exposed areas of the aircraft. The obtained N_2O can now be sent to the NSS.

Nitrogen Separation System (NSS): All jetliners today are required to be equipped with a fuel tank inerting system or a tank blanketing system, wherein mainly the centre tank is fed with N_2-rich air to create a comparatively inert atmosphere inside. This lowers chances of ignition fuel vapours (due to any spark-producing system) which can be catastrophic. The core part of this system is an air separation module (ASM) as shown in Fig. 3, wherein hollow fibres filter out large portions of O_2 from air, rendering a N_2-rich gas. Although the small percentage of O_2 still in the air will not be able to overwhelm the inert N_2. HSE will use a similar advanced system which will completely separate the two gases obtained from N_2O dissociation. The O_2 will be routed to the pre-mixer in the OA and will go on performing combustion. The scavenged N_2 can serve numerous purposes in the aircraft from tank blanketing,

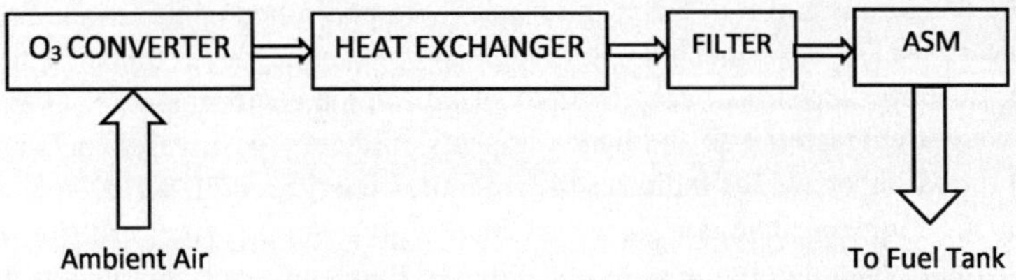

Fig. 3 NGS in commercial aircrafts

Table 2 Comparison of enhancer contenders

Parameter	N_2O (nitrous oxide)	H_2O_2 (hydrogen peroxide)	LOX (liquid oxygen)	N_2H_4 (hydrazine)
Density	1.22 g/cc	1.05 g/cc	1.14 g/cc	1.02 g/cc
Freezing point	−90.86 °C	−0.43 °C	218.79 °C	2 °C
$\Delta_F h_{298}$	+82.05 kJ/mol	−187.8 kJ/mol	0 kJ/mol	+50.63 kJ/mol
I_{sp}	170 s	161 s	500 s (with LH_2)	230 s

skin cooling, cooling of leading edges of wings, cooling of cowlings, cooling of DC, blanketing of nozzle inner surfaces, etc.

Boundary Layer Isolation System: Interaction of boundary layer and shock waves can lead to lot of inefficiencies and cause drag [16]. The boundary layer at the OA diffuser can hence be sucked off from the scramjet inlet and isolated. Since it is relatively cooler, it can be used for absorbing heat from critical surfaces or devices in or outside the engine.

3.4 *Selection of Oxygen Enhancer*

Table 2 shows a comparative study between various contenders for the role of O_2 enhancer in brief that was carried out. The below mentioned were the most critical when selecting the enhancer, while many other were considered for the same.

Higher density of fuel will ensure more mass stored for a fixed volume and thus is a crucial parameter. N_2O with the highest density among the contenders will be ideal to this regard. The freezing point is a critical factor because of low temperatures at high altitudes. N_2O offers a large enough negative value to disallow it from freezing to a solid. LOX, however, offers a value that renders it practically un-freezable. Due the highest I_{sp}, the combustion is more conducive for the environment of a rocket engine where the main requirement is of the net thrust and having the highest exit velocity. The environment of HSE which houses important and delicate machinery all around the gas path, it is arguable that having LOX as the enhancer or a direct oxidizer can be more dangerous and it is better to have a more controlled combustion. Engines prior have used LOX as a fuel but it is seen that more weight goes as thicker walls and heavier materials have to be employed to sustain in the environment. N_2O ensures lower weight characteristics and hence impacts efficiency positively. The standard heat of formation or $\Delta_f H_{298}$ indicates the amount of energy required to dissociate the compound. A high positive $\Delta_f H_{298}$ means that energy required to break the formation bonds is more than that required to make it, and thus giving a desired control over the dissociation and combustibility. For a rocket engine, the same would make it tougher to conduct the reaction; but since the HSE is air breathing and it is desirous to dissipate the unwanted excess heat from the air for a useful action, using N_2O as oxygen enhancer is beneficial. One may debate on using hydrazine for its inherent role as mono-propellant, but the reaction will also produce high levels of toxic gases which harm the environment and also lead to solid depositions on the inside of OA driving up refurbishment costs. Above all parameters, safety is the primary. N_2O gives the best combustion and thrust while giving excellent safety in storage and overall operation ability. The trade-off between high safety and a mediocre I_{sp} is always accepted, especially at high altitudes, where the latter is not such a main criterion. Exposition to even some large quantities of this gas to humans (here, pilot) leads to no adversity.

On the basis of its overall ability to provide controlled combustibility, a decent I_{sp}, a manageable freezing point, better safety, easy storage and not so harsh environmental effects than any of the contenders, and N_2O appears as the best possible solution for oxygen enhancing of the combustion. Previously, it has been used as propellant for rocket engines which testifies its acceptance in the space industry.

4 Development of HSE

Developing the HSE idea with complete systems and subsystems and further study feature centric feasibility:

- Increasing know-how of the topic and solving theoretical equations of gas dynamics, heat transfer and aero-thermodynamics, concerning the design.
- Studying the surrounding systems and their effects on performance.
- Selecting suitable turbojet for HSE and analysing its characteristics.

Developing a pedantic CAD model with majority features and studying of all possible flow properties in different flight regimes using CFD, while improving the former [17]:

- Designing a 3D model incorporating results of theoretical analysis carried out in the previous step.
- Initial conceptual understanding to be developed by 2D analysis and recognizing the probable work areas.
- Carrying CFD analysis in 3D-based with feed-ins based on the data obtained from the 2D analysis.
- Analysing the performance of compressor, turbine and combustion chamber of the turbojet in HSE from model of selected engine and improvising them to meet our needs.
- Undertaking heat transfer analysis for understanding the concept applied to various implementations in flight, like heating of N_2O till dissociation, cooling of ram air, blanketing of nozzles, etc.
- Creating the final 3D model.
- Creating a scaled-down 3D printed model based on the design obtained in the previous step.

- Printing the different models of the inlets and nozzles for various conditions.
- Testing of the model in subsonic, supersonic and hypersonic wind tunnels and collection valuable data [18].
- Implementation of final design tweaks to the engine and the associated systems and introduce them in the CAD model.

Thus, the paper also tries to develop the idea of operating the engine sustainably, which is often overlooked in development of such machines citing the magnitude of end result and other inherent complexities. The concept is laid for further accentuation and design development.

5 Conclusion

This paper overviews SSTO flight, an upcoming endeavour of aerospace engineering. It discusses the various propulsion technologies that promise to bring SSTO to practice and points out the challenges related to the same. In order to try and alleviate those, the authors propose a new engine which can give sustained hypersonic flight right from standstill. Isolating the concerns of materials, integration with aircraft and fuel, focus has been laid on the conceptualization, operation and optimization. The engine is given versatility to enable desired flight characteristics.

References

1. Tang M, Hamilton BA, Chase RL (2008) Paper, the quest for hypersonic flight with air breathing propulsion. In: AIAA international space planes & hypersonic systems and technologies conference (2008)
2. Lamorte N, Friedmann PP, Dalle DJ, Torrez SM (2015) Paper, uncertainty propagation in integrated airframe-propulsion system analysis for hypersonic vehicles. J Propul Power (2015)
3. McClinton CR (2008) Paper, high speed/hypersonic aircraft technology development. NATO OTAN
4. Adams N, Schranner FS (2011) Paper, CFD aided development of an experimental setup to investigate internal supersonic combustion. AIAA Aerospace Science Meeting
5. Bulman MJ, Siebenhaar A (2011) Paper, combined cycle propulsion: aerojet innovations for practical hypersonic vehicles. In: AIAA international space planes & hypersonic systems and technologies conference
6. Olds JR, Bradford J, Charania A, Ledsinnger L (1999) Paper, hyperion: an SSTO vision vehicle concept utilizing RBCC. AIAA Annu J (1999)
7. Huang W, Yan L, Tan J (2014) Paper, survey on the mode transition technique in combined cycle propulsion systems. Aerosp Sci Technol
8. Olds J, Bradford JE (1997) Paper, SCCREAM (simulated combined-cycle rocket engine analysis module): a conceptual RBCC engine design tool. AIAA J
9. Sanders BW, Weir LJ (2008) Paper, aerodynamic design of a dual-flow Mach 7 hypersonic inlet system for a turbine based combined cycle hypersonic propulsion system. NASA Technical Reports

10. Bhattrai S, Tang H (2013) Paper, comparative performance analysis of combined-cycle pulse detonation turbofan engines (PDTEs). Sci Direct-Propul Power Res
11. Reddy DR, Blankson I (2010) Paper, emerging air-breathing propulsion technologies. Encyclopedia of aerospace engineering. Willey, New York
12. Cikanke HA, Bartolotta PA, Klem MD (2007) Review paper, NASA hypersonic propulsion—overview of progress from 1995 to 2005
13. Hank JM, Murphy JS, Mutzman RC (2008) Paper, the X-51 A Scramjet Engine Flight Demonstration Program. AIAA international space planes & hypersonic systems and technologies conference
14. Paniagua G, Szokol S (2008) Paper, contrarotating turbine aerodesign for an advanced hypersonic propulsion system. J Propul Power 24(6):1269–1277
15. Rathakrishnan E (2013) Gas dynamics, 5th edn. PHI Learning
16. Mattingly JD (2017) Elements of gas turbine propulsion. Tata McGraw Hill Edition
17. McDaniel J, Chelliah H, Goyne CP (2009) Paper, US National Centre for Hypersonic Combined Cycle Propulsion: an overview. In: 16th AIAA/DLR/DGLR international space planes & hypersonic systems & technology conference
18. Tapasvi V, Satyanarayana Gupta M, Kumarswamy T (2015) Paper, designing and simulating compressible flow in a nozzle. Int J Eng Adv Technol

Modified Method of Characteristics for Analysing Cold Flow in Bell-Type Rocket Nozzle

S. Panigrahi, P. S. Maity, Gyan Sagar Sinha, D. Dangi, and Atal Bihari Harichandan

Abstract To achieve higher thrust force, numerical investigation of flow behaviour of bell-type rocket nozzle has been carried out in the present research work. Analysis of thermodynamic properties is performed by using ANSYS Fluent software. K-ω shear stress transport model has been used to study the turbulent components of thermodynamic properties. Overall performance gain of a bell-type nozzle and nozzle geometric modifications are discussed. The nozzle geometric modifications are highlighted by executing a C-code with concepts of the method of characteristics (MOC) and modified method of characteristics (MMOC). With modified geometry as per the modified method of characteristics (MMOC), the present numerical study has ensured the fully expanded cold flow in nozzle. The results obtained by the method of characteristics (MOC) are compared with the calculations of the thrust force for the modified geometry.

Keywords Nozzle · Shock separation · Restricted shock separation · Fully expanded flow · MOC

Nomenclature

M	Mach number
$\vartheta(M)$	Prandtl–Meyer function
ϑ	Prandtl–Meyer angle
θ	Angle made with respect to streamline coordinate axis
μ	Mach angle with respect to characteristic lines

S. Panigrahi · P. S. Maity · G. S. Sinha · A. B. Harichandan (✉)
Department of Aerospace Engineering, Kalinga Institute of Industrial Technology (KIIT) Deemed to be University, Bhubaneswar, Odisha, India 751024
e-mail: harichandan.iitkgp@gmail.com

D. Dangi
Department of Mechanical Engineering, Marwadi University, Rajkot, India 360003

S. Revankar et al. (eds.), *Proceedings of International Conference on Thermofluids*, Lecture Notes in Mechanical Engineering,
https://doi.org/10.1007/978-981-15-7831-1_15

1 Introduction

Due to an increase in demand of satellite throughout the world along with rocket launching programs, engineers are more concerned about efficiency of the rocket nozzles. The bell-shaped nozzles are considered the most [1]. But due to insufficient expansion of fluids, improper mixing and insufficient combustion always lead to lower performance of rocket nozzles than the theoretical values. Figure 1 depicts all different causes of loss in rocket nozzles. A fully-expanded flow at the exit of divergence section of a rocket launcher vehicle is an essential condition. Due to the fluid flow, under expansion or over expansion flows are experienced. Side load increases on wall surface of nozzle due to insufficient expansion, thus reducing the performance and affects the thrust force.

The rocket engine can be damaged if such dynamic loads are not considered carefully. Such conditions are often not considered in the design stage. To develop the geometry of bell-type rocket nozzle, Kbab et al. [2] have used MOC. To achieve Mach number 3.2, they have created the bell-type nozzle and achieved a maximum average of Mach 3.18 by using k-ω SST turbulence model. By considering mechanical and thermal loads existing the wall of nozzle, Hosterst et al. [3] have performed the numerical analysis on dual bell nozzle. In an axisymmetric over-expanded thrust-optimized contour nozzle, Link et al. [4] have carried out a numerical investigation of transient flows. Flow separations are observed that results in lateral loads experienced by nozzles during both start-up and shutdown operations. On conical-type nozzle, Natta-et-al. [5] carried out numerical study and investigated the supersonic flow by the geometry created by MOC. By adopting the enthalpy (H), temperature (T) and perfect gas model, Zebbiche et al. [6] ensured that the Mach number and the stagnation temperature to be lower than the error between HT and PG model. The behaviour of flow inside the nozzle of VOLVO S7 was analysed by Ostlund et al. [7]. Flow separation and shock wave occurrence in the nozzle are shown as the basic phenomena in the result. By implementing MOC at low and high expansion ratios,

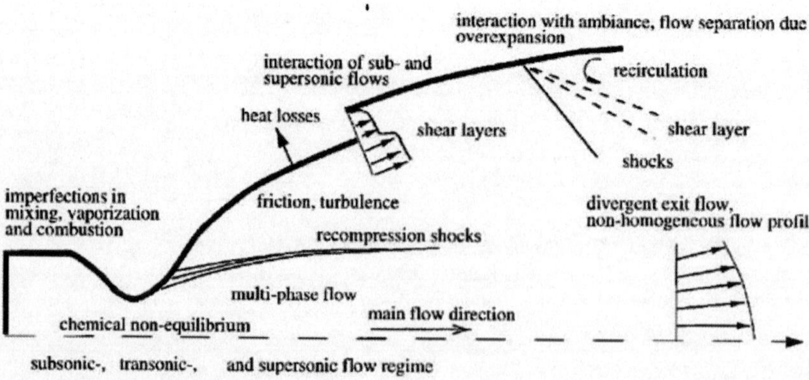

Fig. 1 Flow behaviour in rocket nozzle

Harefirld et al. [8] developed a mathematical model. By MOC analysis, Prince [9] considered a 3D quasi-situation by variable laminar thickness. The smallest span of axisymmetric Laval nozzle which is analogous to simple conical-type was determined by the construction of hypersonic shock tunnel respectively by Dumitrescu [10]. To identify shock, at different wall and inlet pressure, experimental investigation of cold flow in conical nozzle, thrust-optimized nozzle, and bell nozzle were done by Welle et al. [11]. To create supersonic wall, contour 2D (SSL) straight sonic line and (CSL) minimum length of nozzle for exit Mach number of 2,4, and 6 MOC were used by Argrow et al. [12]. Here, the influence of inlet geometry in sub-sonic and supersonic flow field on sonic-line shape, its location and flow angle are determined by the author. By using MOC, Benton et al. [13] developed geometry of nozzle at Mach 6 for air, Mach 13.5, and Mach 17 for nitrogen. The oblique shock increases near the inflection point if the Mach number increases from Mach 6 to Mach 17 were observed. Hence, using MOC gives the best nozzle geometry up to Mach 8. By varying the Reynolds number, Louisos et al. [14] numerically investigated thrust production in 2D supersonic bell micro-nozzle at different configuration (100, 80, and 60%). 60% bell configuration that exhibits the greatest efficiency was finally concluded. Shock associated noise from imperfectly expanded supersonic jet-like nozzle at Mach number 1.67 was carried out experimentally and theoretically by Tam et al. [15]. Side-loads and correlative flow physics of a thrust-optimized parabolic nozzle during fire in the hole staging event, Jia et al. [16] investigated numerically. Main properties like amplitude, phase position, and frequency are analysed. MOC and MMOC are proposed and studied to improve the performance of the rocket nozzle. For lower ambient pressure than the nozzle exit pressure, the flow becomes under-expanded but changes to over-expanded flow for higher ambient pressure than the nozzle exit pressure. Hence, the main objective is to achieve fully expanded flow inside the nozzle with higher thrust force.

2 Governing Equations

The following major equations are considered for present numerical analysis.

$$\left(1-\frac{u^2}{a^2}\right)\frac{\partial u}{\partial x}+\left(1-\frac{v^2}{a^2}\right)\frac{\partial v}{\partial y}-\frac{2uv}{a^2}\frac{\partial v}{\partial y}=0 \tag{1}$$

which can be simplified as shown in Eqs. 2 and 3;

$$\frac{\partial u}{\partial x}=\frac{\frac{2uv}{a^2}\frac{\partial u}{\partial y}-\left(1-\frac{v^2}{a^2}\right)\frac{\partial v}{\partial y}}{\left(1-\frac{u^2}{a^2}\right)} \tag{2}$$

Table 1 Grid independent study for Mach number 3.0

Maximum face size of the element	Total number of elements	Exit Mach number
9e−004	15,690	2.92
8e−004	19,884	2.91
7e−004	25,762	2.91
6e−004	35,152	2.90
5e−004	50,181	2.89
4e−004	78,390	2.89

$$\frac{\partial y}{\partial x} = \frac{\frac{-uv}{a^2} \pm \sqrt{\left(\frac{u^2+v^2}{a^2} - 1\right)}}{\left(1 - \frac{u^2}{a^2}\right)} \tag{3}$$

Equation (3) defines the characteristic curves in the physical *xy*-space. The under-root term is:

$$\frac{u^2 + v^2}{a^2} - 1 = \frac{V^2}{a^2} - 1 = M^2 - 1 \tag{4}$$

Furthermore,

$$\mathrm{d}\theta = \mp\sqrt{M^2 - 1}\frac{\mathrm{d}V}{V} \tag{5}$$

3 Grid Independence Test

For the analysis, ANSYS Fluent has been used. By using density-based solver, 2D compressible flow has been analysed. For calculation of fluctuating part of flow properties, energy equation has been considered and K-ω turbulence model is used. By MOC, grid independence test has been carried out for the bell-type nozzle geometry at Mach number $(M) = 3.0$. Maximum face size of the element in every simulation is chosen to be 5e–004 by authors based on the grid independence test (Table 1).

4 Results and Discussion

By using MOC, the geometry of bell-type convergent-divergent nozzle has been created. To obtain the dimensions, a C-program is developed from which the profile of the nozzle is generated. By using ANSYS Fluent, numerical simulations on the generated nozzle profile are obtained. For Mach numbers beyond 3.2, MMOC customizes the geometry created by MOC for better performance of rocket nozzle. Detailed analysis has to be done of numerical estimations for the variation in thermodynamic properties like Mach number, pressure, velocity, and density for the geometries developed by MOC and MMOC. Authors considered an ideal bell-type nozzle with designed exit Mach number of 3.2 for the base case. The average Mach number at the exit has been observed to be 3.17 by implementing the technique of MOC using ANSYS Fluent software. This was validated with the numerical results of Kbab et al. [2] in which the exit Mach number of 3.18 has been obtained for similar operating conditions. With that, authors carried out the present simulations for exit Mach numbers 3.25 and 3.50 with MOC and MMOC. For the geometry developed through MOC with $M = 3.25$, grid independence test has been carried out that ensured a very fine mesh with more than 50,000 elements for better performance of the nozzle. However, for the geometries created by MMOC, authors considered the same element spacing to obtain the computational grids at the mentioned Mach numbers.

Figure 2 presents the thermodynamic properties like Mach number, pressure, and velocity for geometry developed through MOC and MMOC at $M = 3.25$. From pressure contours, the fluid flow continuously converges from inlet to throat section and it expands from throat section to divergent section, even the stagnation pressure continuously decreases along the length of nozzle is being observed. However, the flow velocity increases from convergent to divergent section of nozzle as flow progresses. Average Mach number and velocity are obtained to be 3.13 and 620.65 m/s, respectively, for geometry developed by MOC. But, it highlights the flow details in the nozzle with average pressure of 103,955.24 Pa corresponding to same geometry.

Case 1: For M = 3.25

The numerically obtained average pressure for the geometry developed through MMOC is almost comparable with the atmospheric pressure that ensures fully-expanded flow in the nozzle. The average Mach number and average velocity are noticed to be higher than the results obtained for MOC generated geometry. The percentage difference of results obtained through MOC and MMOC approaches is highlighted in Table 2.

Case 2: For M = 3.50

Figure 3 represents difference in thermodynamic properties like Mach number, pressure, and velocity for geometries developed through MOC and MMOC at $M = 3.50$. The fluid flow continuously converges from inlet to throat section followed by continuous expansion from throat to outlet of nozzle. The average Mach number and average velocity are obtained to be 3.36 and 633.21 m/s, respectively. The under-expanded flow obtained with geometry developed by MOC is further modified through MMOC, and fully-developed flow has been achieved in the nozzle. Table 3 highlights the difference in thermodynamic properties for geometry developed through MOC and MMOC.

Case 1: For M = 3.25

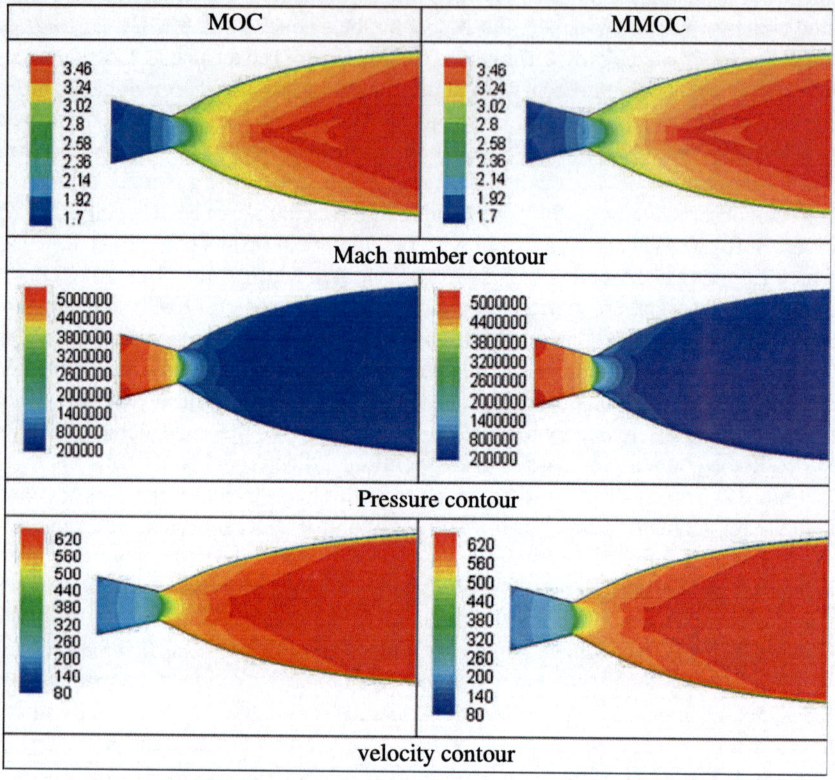

Fig. 2 Mach number, pressure, and velocity for geometries developed at $M = 3.25$

Table 2 Difference of thermodynamic properties through MOC and MMOC approaches

Property (at nozzle exit)	MOC	MMOC (1.69% *Y*-axis increment)	% Variation between MOC and MMOC
Mach number	3.13	3.15	0.638
Static Pressure(pa)	103,955.24	101,322.48	2.53
Flow Velocity(m/s)	620.65	622.66	0.324
Density(kg/m^3)	3.60	3.53	1.94
Thrust(N)	125.5500	128.1125	2.49

5 Conclusion

Numerical simulations on the nozzle profiles generated by MOC and MMOC are performed by using ANSYS Fluent software. Under-expanded flow has been experienced for nozzle geometry generated by MOC. However, fully expanded flow with

Case 2: For M = 3.50

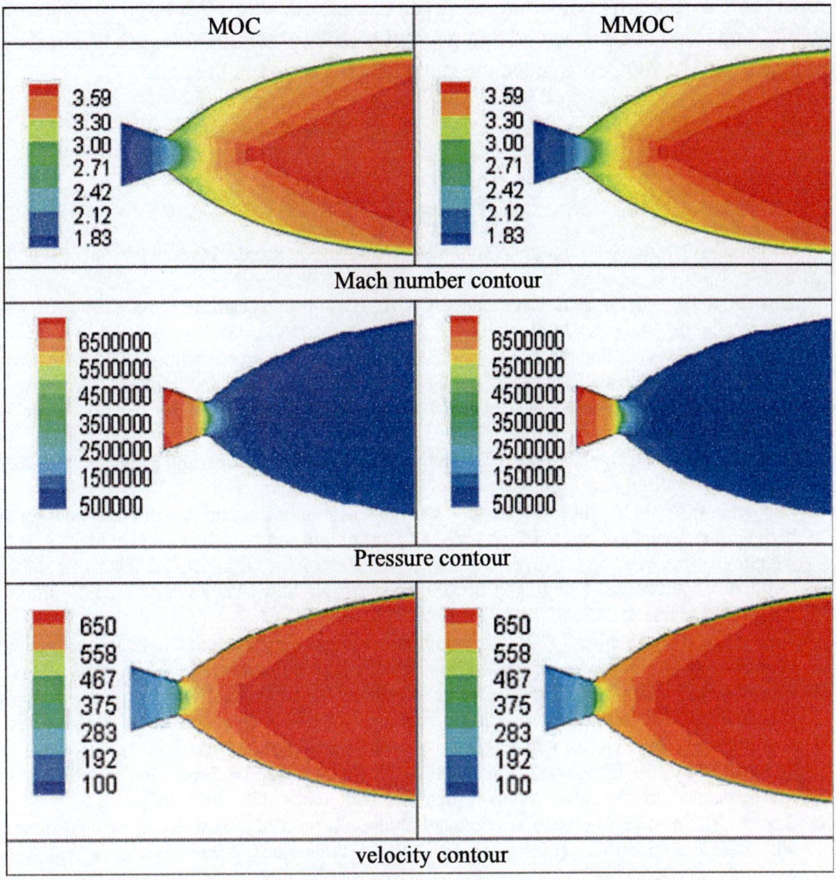

Fig. 3 Mach number, pressure, and velocity for geometries developed $M = 3.50$

Table 3 Difference in thermodynamic properties through MOC and MMOC approaches

Property (at nozzle exit)	MOC	MMOC (2.065% Y-axis increment)	% Variation between MOC and MMOC
Mach number	3.36	3.39	0.89
Static pressure (pa)	104,672.074	101,326.48	3.19
Flow velocity (m/s)	633.21	636.54	0.52
Density (kg/m^3)	3.99	3.90	2.25
Thrust (N)	231.59	238.56	3.00

almost 100% expansion is established for geometry generated by MMOC. Also, thrust force and Mach number are observed to increase with MMOC as compared to MOC. It has also been observed that geometry having minimum length of nozzle is obtained with MMOC to achieve the same Mach number at the exit.

References

1. Hagemann G, Immich H, Dumnov D (1998) Advance rocket nozzle. J Propul Power 14(5):620–634
2. Kbab H, Sellam M, Lagab L, Hamitouche T (2016) Design and performance evaluation of dual bell nozzle. Int J Eng Res Methodol. https://doi.org/10.1016/j.actaastro.2016.10.015
3. Hosters N, Klaus M, Behr M, Reimerdes HG (2013) Application of partitioned field approach to transient aero-thermal problem in rocket nozzle. Comput Fluids 88:795–803
4. Lijo V, Kim HD, Setoguchi T, Matsuo S (2010) Numerical simulation of transient flow in rocket propulsion nozzle. Int J Heat Fluid Flow 31:409–407
5. Natta P, Kumar VR, Rao YVH (2012) Flow analysis of rocket nozzle using computational fluid dynamics. Int J Eng Res Appl 2(5):1226–1235
6. Zebbiche T, Youbi Z (2006) Supersonic two dimensional minimum length nozzle design at high temperature: application of air. In: 42nd joint propulsion conference and exhibit, AIAA, p 4599
7. Ostlund J, Muhammad-Klingmann B (2005) Supersonic flow separation with application to rocket engine nozzle. ASME. 58:143–177
8. Hartfield Jr R, Burkhalter J (2015) A complete and robust approach to axisymmetric method of characteristics for nozzle design. In: 51st joint propulsion conference, AIAA, p 4217
9. Prince Jr D (1982) The method of characteristics for supersonic flow analysis—a fresh perspective. In: 3rd thermophysics, fluids, plasma and heat transfer conference, AIAA, p 996
10. Dumitrescu L (1975) Minimum length axisymmetric laval nozzle. AAIA 13(4):520–521
11. Welle R, Hardy B, Murdock J, Majamaki A, Hawkins G (2003) Separation instabilities in over-expanded nozzles. In: 39th joint propulsion conference and exhibit, AIAA, p 5239
12. Argrow B, Emanuel G (1989) A computational analysis of the transonic flow field of two dimensional minimum length nozzles. In: 20th fluid dynamics, plasma dynamics and lasers conference, AIAA, p 1822
13. Benton J, Perkins J (1990) Limitations of the methods of characteristics when applied to axisymmetric hypersonic nozzle design. In: 28th Aerospace Sciences Meeting, AIAA, p 0192.
14. Louisos W, Hitt D (2008) Numerical studies of thrust production in 2-D supersonic bell micronozzles. In: 44th joint propulsion conference and exhibit, AIAA, p 5233
15. Tam C, Tanna H (1982) Shock associated noise of supersonic jets from convergent-divergent nozzles. J Sound Vibr 81(3):337–358
16. Jia R, Jiang Z, Zhang W (2016) Transient three-dimensional side-loads analysis of a thrust-optimized parabolic nozzle during staging. Acta Astronaut 122:137–145

Aerodynamics

Performance Analysis of NACA4412 Airfoil with Gurney Flap

Ankit Kumar, Pooja Chaubdar, Gyan Sagar Sinha, and Atal Bihari Harichandan

Abstract Gurney flap has been used to increase lift in varied types of wings used in aerial vehicles. It is also preferred as it increases pressure on the pressure side of the airfoil thus increasing the lift. Gurney flaps delay the onset of boundary layer separation in fluid flows. In the present research, analysis is performed on NACA4412 airfoil attached with gurney flap of different lengths with respect to chord lengths at 0°, 4° and 8° angles of attack. Ground effect phenomenon is also done at different height regime for different height to chord ratio through computational fluid dynamics (CFD). Analysis is carried out and coefficient of lift, drag, airfoil performance and pressure distribution is studied. In gurney flap, the lift force increases with decreasing ground clearance due to pressure region created under the airfoil. Phenomena of gurney flap are applicable for increase in coefficient of lift during the takeoff of the aircraft.

Keywords Gurney flap · Ground clearance · Pressure

1 Introduction

Gurney flap, otherwise, known as wicker bill, is a micro-tab or strip after trailing edge on airfoil's pressure side. It escalates the aerodynamics of the airfoil same as that of an advanced designed airfoil. This device was initially used in race cars to enhance the down force for lateral traction which is needed during high velocity turns. The gurney flap helps in keeping the boundary layer attached up to trailing edge increasing lift coefficient.

Myose et al. [1] have investigated the gurney flap effect on 2D airfoils, 3D wings and a reflection plane. NACA0011 and GS(W)-2 airfoils were used during single element airfoil part. It was found that when gurney flaps were attached, the performance improved on the trailing edge element. In 3D wings, performance improved when gurney flaps were connected inboard rather than outboard. Jefferey et al. [2]

A. Kumar · P. Chaubdar · G. S. Sinha · A. B. Harichandan (✉)
School of Mechanical Engineering, KIIT Deemed University, Bhubaneswar, Odisha 751024, India
e-mail: harichandan.iitkgp@gmail.com

S. Revankar et al. (eds.), *Proceedings of International Conference on Thermofluids*, Lecture Notes in Mechanical Engineering,
https://doi.org/10.1007/978-981-15-7831-1_16

examined single element wing attached with gurney flap and calculated pressure, force and the velocity by laser Doppler anemometry (LDA). Its spectral analysis indicated that wake contains a von-Karman vortex sheet of alternate shed vortex which is verified through the help of smoke visualization downstream of gurney flap. The trailing edge suction increases because of shedding in the airfoil and the flow at the trailing edge is decelerated on the pressure surface due to upstream face of the device. Cole et al. [3] have investigated the applications of gurney flap at different height regimes and chord-wise locations of various airfoils. The aerodynamic effects were studied in subsonic, low-turbulence wind tunnel. The performance of the each gurney flap has been calculated by variation in maximum coefficient of lift of the airfoil. Ahmed et al. [4] have calculated the flow performance of NACA4412 airfoil experimentally with low turbulence and moving ground simulation at Reynolds no. 3×10^5 with different angle of attacks from 0–10 degree and the ground clearance of trailing edge varying from 5 to 100% of chord. It was observed that lift decreased with ground clearance up to angle of attack of 4 degrees and increased for high angles due to high pressure on lower surface. The drag increased near to the ground for the angle of attack (AOA) investigated majorly due to the modification of the lower surface pressure distribution. Webb et al. [5] performed a wind tunnel test to observe the effect of gurney flap on airfoil. The tests were performed at several different angles of attack with three different gurney flap lengths of 1, 2.5 and 5% of the airfoil chord length. After a result, they found that there is a major increase in lift coefficient and a decrease in coefficient of drag when a gurney flap is installed. In both airfoil and wing, with gurney flap attached, a significant increment in coefficient of lift and decrement in coefficient of drag was observed. Aramendia [6] has studied the implementation of gurney flap in DU91W (2)250 airfoil to find optimal gurney flap length to increase the performance of the airfoil. They found that gurney flap lengths between 0.25%C and 0.75%C, the C_l/C_d ratio for positive angle of attack is higher. Camocardi et al. [7] did an experiment in wind tunnel on an airfoil with movable gurney flap attached. The up-down movement of the flap was done through electromagnetism. For gurney flap attached airfoil, the coefficient of lift for fixed is more than the movable gurney flap. Maughmer [8] has investigated the aerodynamics of gurney flaps experimentally in subsonic wind tunnels at Reynolds no. 10^6. It was found that the coefficient of lift can be incremented by 30% by use of gurney flaps. The increased lift in airfoil with gurney flaps attached to them increases the lower surface pressure upstream of the flap location. Chand [9] has performed an aerodynamics analysis of flow over a multi-element airfoil and compared to the standard NACA airfoils. Flaps and slats have been used at different angles of attack have been computationally analyzed. The results showed that multi-element airfoils perform better than standard NACA airfoil. Li [10] has experimentally measured the wake profiles and surface pressure distributions and determined the effects of gurney flaps on NACA 0012 airfoil. The gurney flap attached airfoil amplified the coefficient of lift and the highest C_l/C_d ratio was achieved when the flap height was 2% of chord length. Jang [11] has investigated NACA 4412 airfoil with gurney flap attached numerically and compared with experimental results. It was found that gurney flap increases the C_l with a small increase in C_d. The gurney flap attached to airfoil also increases loading on the entire length of the airfoil and an excessive loading is observed at trailing edge. Graham [12] has studied the performance of thickness of gurney flap on airfoil experimentally in an open-circuit subsonic wind tunnel. The increase in lift due to flap increase with height of the flap and decreased with its thickness. The aspect ratio of gurney flap had direct correlation with change in lift at 0° angle of attack. Jain [13] has investigated the effects of gurney flap on NACA 0012 airfoil aerodynamics numerically for different height regimes at different positions from trailing edge. It was found that gurney flaps with height to chord ratio (H/C) of 0.015 can be installed perpendicular from the trailing edge to increase maximum C_l. Storms [14] experimentally measured surface pressure distributions and wake profiles for NACA 4412 airfoil attached with gurney flap. The delayed boundary layer resulted in C_l enhancement by 0.34. Dam [15] did gurney flap experiments on wigs and airfoil in wind tunnel at 10^6 Reynolds number. It was found that span-wise alteration with flap geometry enhanced lift without increase in drag comparatively.

2 Governing Equations and Numerical Considerations

Flow over airfoil is computed by Navier–Stokes equation neglecting the source terms:

$$\frac{\partial u}{\partial t} + u \cdot \nabla u = -\frac{\nabla p}{\rho} + v\nabla^2 u \tag{1}$$

where v, u, p and ρ are kinematic viscosity, velocity, pressure and density of fluid, respectively.

Figure 1 shows the computational domain and the close-up view of the mesh around an NACA4412 airfoil with gurney flap at 0° angle of attack and H/C 0.4. The present study has been validated with the results of Justin et al. [16] for analyzing the aerodynamic performance of NACA 4412 airfoil with ground effect phenomenon. In the present research work, authors have chosen similar airfoil for analysis. Grid independent study has been carried out using ANSYS Fluent at height to chord ratio

Fig.1 Computational domain and close-up view of NACA4412 airfoil with gurney flap at 0° A0A and H/C 0.4

Table 1 Grid independence study

Grid no	No. of cells	Coefficient of lift	Error (%)
Grid 1	104,989	0.3911	7.4
Grid 2	203,515	0.4160	6.3
Grid 3	380,666	0.4267	2.6

of 0.4 at 0° angle of attack and the lift coefficient values for three-different meshed are presented in Table 1.

Based on the above study, grid 3 has been chosen for present research work at different H/C ratio and different angles of attack. The present grid has 400 nodes on suction and the pressure surface of airfoil, 300 on inlet and outlet of the rectangular domain and 400 on top and bottom of the computational domain. The gurney flap considered for investigating ground effect phenomenon is of 0.5, 1, 1.5, 2 and 2.5% of chord length. K-omega SST turbulence model was used for investigation. Green-Gauss node-based gradient method is used for its high accuracy in aerodynamic flows. No slip condition at suction surface and the pressure surface of airfoil with symmetry condition at top and bottom, velocity inlet at inlet and pressure outlet at outlet of the computational domain are the assigned boundary conditions.

3 Code Validation

The coefficient of lift and drag at various height regimes are considered for comparing the present results with the experimental and numerical results of Justin et al. [16] and are shown below.

Figures 2 and 3 show the variation in coefficient of lift at different heights at 0° and 2° angles of attack, respectively, whereas Figs. 4 and 5 show the variation in coefficient of drag at different heights at 0° and 2° angles of attack, respectively. The difference between the present numerical results for coefficients of lift and drag and the experimental results of Justin et al. [16] has been found to be less than 25%. Thus,

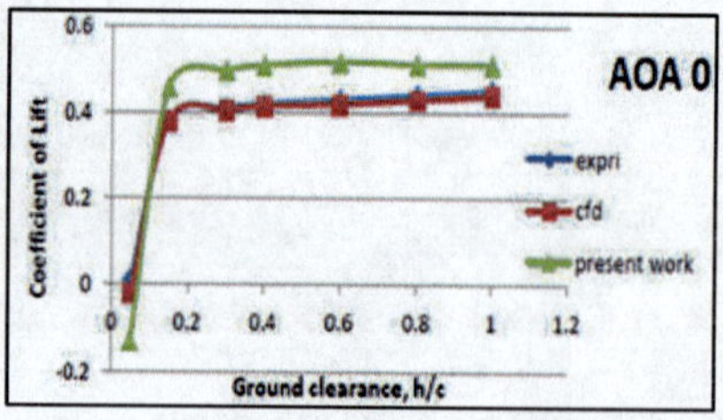

H/C ratio	Difference
0.05	8%
0.15	19%
0.3	18%
0.4	20%
0.6	21%
0.8	23%
1	23%

Fig. 2 Lift coefficient at different height regimes and 0° angle of attack

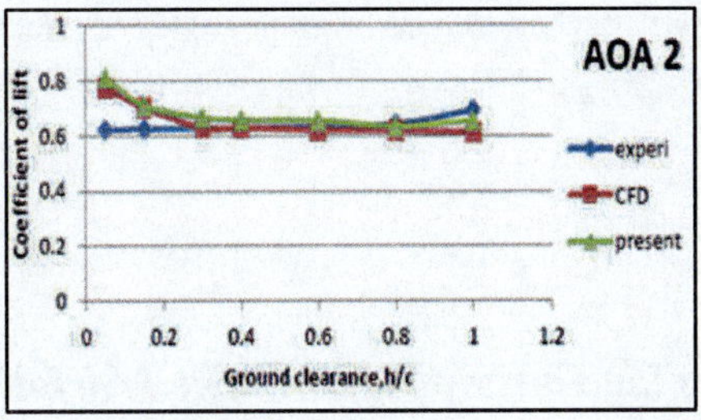

H/C ratio	Difference
0.05	23.36%
0.15	11.90%
0.3	5.63%
0.4	4.52%
0.6	3.38%
0.8	1.45%
1	5.62

Fig. 3 Lift coefficient at different height regimes and 2° angle of attack

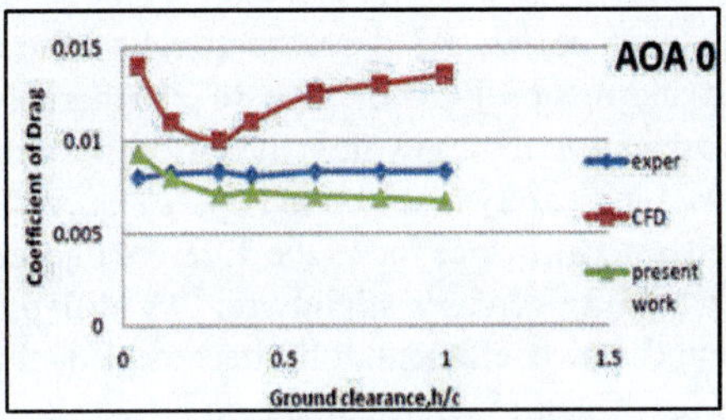

H/C ratio	Difference
0.05	13.97%
0.15	2.43%
0.3	14.4%
0.4	11.11%
0.6	15.66%
0.8	16.86%
1	19.27%

Fig. 4 Drag coefficient at different height regimes and 0° angle of attack

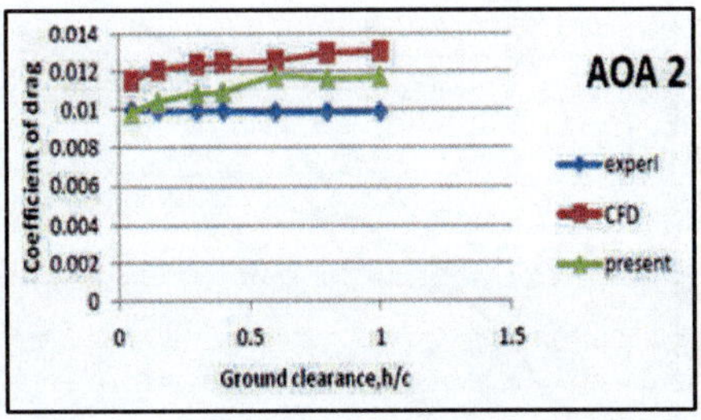

H/C ratio	Difference
0.05	7.62%
0.15	5.12%
0.3	3.36%
0.4	1.70%
0.6	8.73%
0.8	14%
1	18.60%

Fig. 5 Drag coefficient at different height regimes and 2° angle of attack

the present numerical model is considered further for the performance analysis of NACA 4412 airfoil with a gurney flap attached to it at different angles of attack with different H/C ratio.

4 Results and Discussion

4.1 Aircraft Performance of the Gurney Flap Attached Airfoil

Numerical simulations are performed for the flow over NACA4412 airfoil attached with gurney flap at 30 m/s. At different angles of attack like 0°, 4° and 8°, the coefficient of lift has been observed to increases with increase in length of gurney flap as shown in Fig. 6.

Whereas, the coefficient of drag does not always decrease affecting the airfoil performance. So, the aerodynamic performance does not increase considerably with increasing flap length. At 1.5% chord length of gurney flap, the airfoil performance increases but then decrease after that significantly is shown in Fig. 7. The coefficient of lift at 8° angle of attack gives a better performance for all the different flap lengths. At 4° angle of attack, the coefficient of lift decreases slightly for 2% of flap length and then increases. For 1.5% flap length, the coefficient of lift becomes equal for all angles of attack and then decreases for 4° and 8° angles of attack.

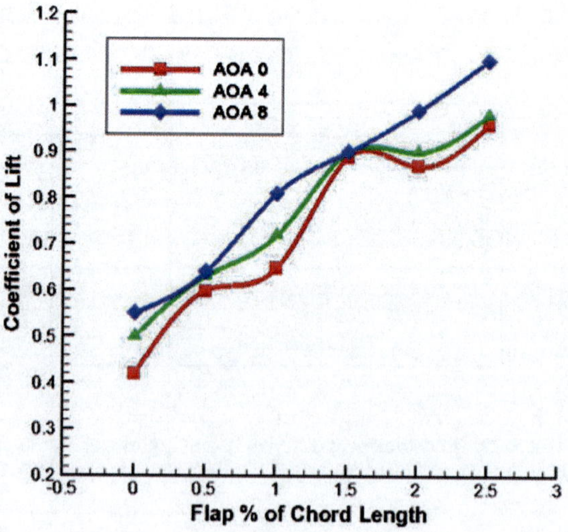

Fig. 6 Coefficient of lift (C_l) at different flap lengths

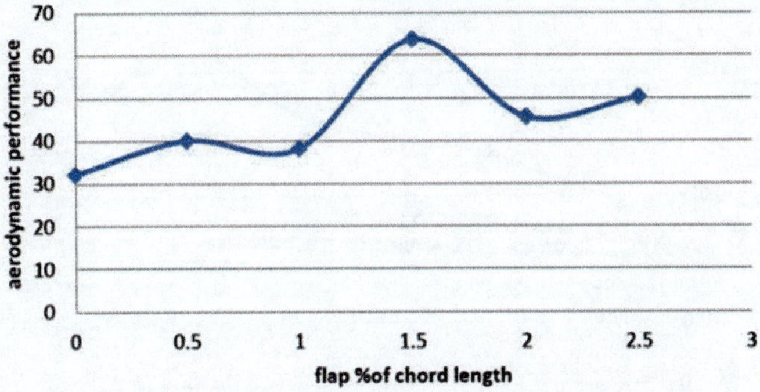

Fig. 7 Aerodynamic performance (C_l/C_d) at different flap length at 0° angle of attack

4.2 *Ground Clearance Effect*

Figures 8 and 9 show the coefficients of lift and drag at different height to chord ratio for various gurney flap lengths attached to NACA4412 airfoil. The coefficient of lift decreases with increase in height of ground clearance at a given gurney flap length. Whereas, the coefficient of drag increases slightly for increasing height to chord ratio for 0.5, 1, 1.5 and 2% of chord of gurney flap but for 2.5%C gurney flap length, the drag coefficient suddenly increases after chord to height ratio becomes greater than equal to 0.8.

Figures 10 and 11 show the lift and drag coefficients for a different height to chord ratio at 4° and 8° angles of attack, respectively. As the mass flow rate and the height under the airfoil decreases, the pressure on pressure side of the airfoil starts developing. The increasing pressure increases the coefficient of lift, and drag

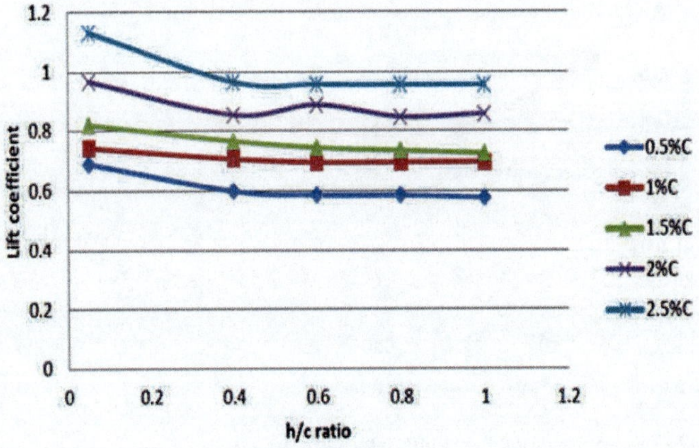

Fig. 8 C_l at different height regimes for different flap lengths

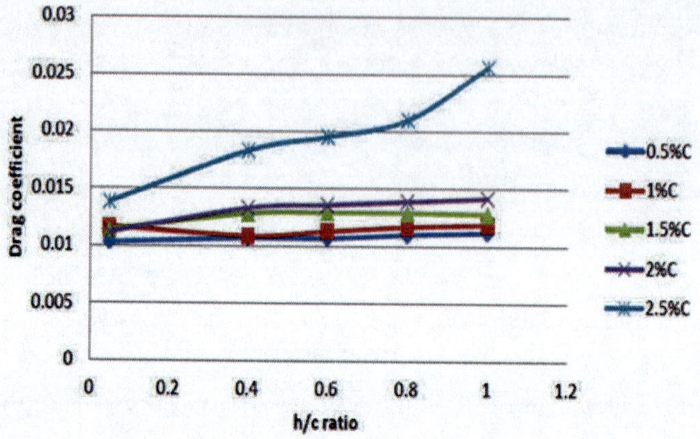

Fig. 9 C_d at different height regimes for different flap lengths

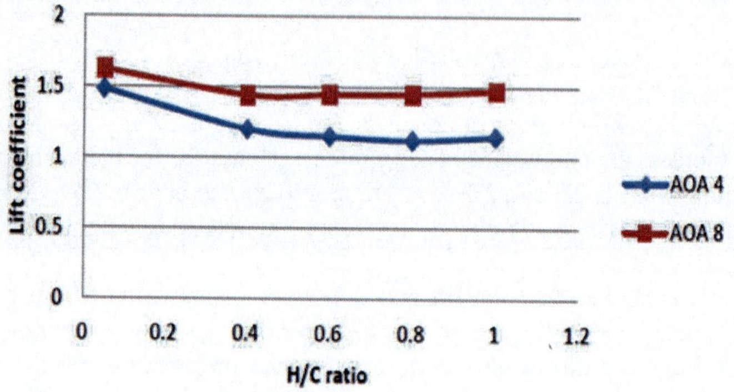

Fig. 10 C_l at various height regime at different AOA

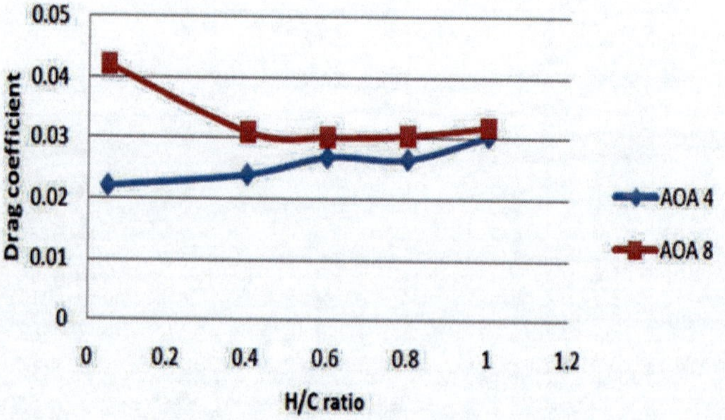

Fig. 11 C_d at various height regime at different AOA

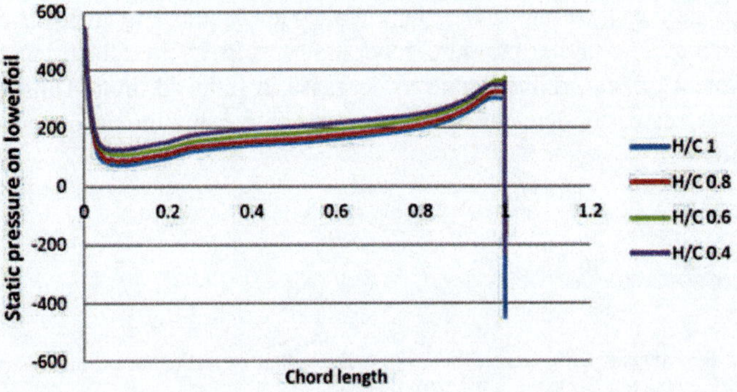

Fig. 12 Pressure on lower surface of the airfoil at different height regimes

starts decreasing due to decrease in induced drag on the airfoil. The coefficient of lift though starts decreasing at both the angle of attack slightly but is still higher than the analysis when ground clearance effect is not considered corresponding to specific angle of attack.

4.3 Pressure Distribution

The pressure distribution on suction side and pressure side of the airfoil at different height regimes differ as shown in Fig. 12. At different height regimes of the gurney flap attached to the airfoil, static pressure developed on the lower surface increases from 0% of chord length up to 1% chord length and the static pressure then decreases at flap length of 1% chord length itself. With the increased pressure on the pressure side of the airfoil, the lift force increases making the performance better.

5 Conclusions

In the present research work, NACA4412 airfoil is analyzed using ANSYS Fluent attached with a gurney flap with and without considering ground effect. At a particular angle of attack, the use of gurney flap increases the lifting force on the airfoil. The length of gurney flap considered in present work is 0.5, 1, 1.5, 2 and 2.5% of chord length at 0°, 4° and 8°. Air velocity of 30 m/s and K-omega turbulence model is considered for the investigation. Green-Gauss node-based gradient method has been used for its high effectiveness in aerodynamic studies. The increasing length of the flap results an increasing coefficient of lift but the aerodynamics performance is not always better. At 1.5%C, the airfoil attached with gurney flap gives the best performance. The force performance of the airfoil increases with ground effect.

Pressure begins to generate under the airfoil with decrease in mass flow rate and height under it. The higher pressure generation results in overall lift increment and the coefficient of drag reduced due to decrease in induced drag. Thus, the airfoil performance increases. The gurney flap helps in increasing the lift during takeoff of the aircraft.

References

1. Myose R, Papadakis M, Heron I (1998) Gurney flap experiments on airfoils, wings, and reflection plane model. J Aircraft 35(2):206–215
2. Jeffrey D, Zhang X, Hurst W (2000) Aerodynamics of Gurney flaps on a single-element high-lift wing. J Aircraft 37(2):295–304
3. Cole A, Vieria O, Coder G, Premi A, Maughmer D (2013) Experimental Investigation into the effect of Gurney flaps on various airfoils. J Aircraft 50(4)
4. Ahmed R, Takasaki T, Kohama Y (2007) Aerodynamics of a NACA4412 airfoil in ground effect. J Aircraft 45(1). https://doi.org/10.2514/1.23872
5. Webb J, Higgenbotham H, Liebshutz D, Potts D, Tondreau E, Ashworth J (2001) Analysis of Gurney flap effects on a NACA0012 airfoil/wing section. In: AIAA-2001–2483, A01–31037. 19th applied aerodynamics conference, 11–14 June 2001
6. Aramendia U, Fernandez-Gamiz E, Zulueta A, Saenez A, Teso-Fz-Betono D (2019) Parametric study of a Gurney flap implementation in a DU91W(2)250 airfoil. Energies 12:294. https://doi.org/10.3390/en12020294
7. Camocardi ME, Leo JMD, Delnero JS, Lerner JLC (2011) Experimental study of a NACA 4412 airfoil with movable Gurney. In: Flap 49th AIAA Aerospace Sciences Meeting including the New Horizons Forum and Aerospace Exposition AIAA 2011–1309
8. Maughmer MD, Bramesfeld G (2008) Experimental investigation of Gurney flaps. J Aircraft 45(6)
9. Chand DV, Sriram R, Kumar DU (2016) Aerodynamic analysis of multi element airfoil. Int J Sci Res Publ 6(7). ISSN 2550–3153
10. Li Y, Wang J, Zhang P (2002) Effects of Gurney flaps on a NACA 0012 airfoil. Flow Turbulence Combust 68:27–39
11. Jang CS, Ross JC, Cummings RM (1998) Numerical investigation of an airfoil with a Gurney flap. Aircraft Des 1(2):75–88
12. Graham M, Muradian A, Traub LW (2018) Experimental study on the effect of gurney flap thickness on airfoil performance. J Aircraft. https://doi.org/10.2514/1.C034547
13. Jain S, Sitaram N, Krishnaswamy S (2015) Computational investigations of the effects of gurney flap on airfoil aerodynamics. International Scholarly Research Notices. Hindawi. Article ID 402358, DOI: 1155/2015/402358
14. Storms BL, Jang CS (1994) Lift enhancement of an airfoil using a gurney flap and vortex generators. J Aircraft 31(1)
15. Dam CPV, Yen DT (1999) Gurney flap experiments on airfoil and wings. J Aircraft 36(2) 484–486. Engineering Notes
16. Smith JL, Graham HZ, Smith JE (2008) The validation of an airfoil in the ground effect regime using 2-D CFD analysis. In: 26th AIAA aerodynamics measurement technology and ground testing conference, AIAA 2008–4262

Effect of Ground Clearance and Air Temperature on Drag and Lift for NACA 2412 Airfoil

Prakash Ghose and Rishitosh Ranjan

Abstract The flow behavior over a NACA 2412 airfoil is studied computationally. The pressure distribution and wall shear stresses over the airfoil were obtained through computational modeling. Total lift and drag forces were calculated by considering both friction and pressure effect over the airfoil. Simulation was carried out by changing the ground clearance from 0.1 to 1.5 m. Along with that the effect of wind temperature and velocity on drag and lift force was also studied at zero angle of attack. It has been observed that both the drag and lift force increases nonlinearly with linear variation of velocity. The air temperature does not affect the drag and lift significantly. It was also observed that a higher lift is obtained on the lower surface when the airfoil is away from the ground. However, if the ground clearance becomes very less, a negative lift has been observed. This is due to the narrow region of flow that increases the air velocity leads to create a lower pressure region at the bottom part of the air flow.

Keywords NACA airfoil · Lift and drag · Ground clearance

1 Introduction

Since the successful flying of aircraft by Wright brothers, a continuous effort put by various researchers to improve the lift-to-drag ratio. An increase in lift-to-drag ratio means more aerodynamically efficient flying. During lift up and landing when the aeroplane wings are closed to the ground, an increase in lift-to-drag ratio is obtained. Due to its high aerodynamic efficiency within a considerable speed, many research works have been carried out to utilize the advantages of ground effect. Ahmed and Sharma [1], Luo and Chen [2], Ying et al. [3] have studied the ground effect on drag, lift and pressure distribution over the NACA 0015 series airfoil. Since NACA 0015 is a symmetrical wing, a significant change in angle of attack only causes a considerable

P. Ghose (✉) · R. Ranjan
School of Mechanical Engineering, KIIT Deemed To Be University, Bhubaneswar, Odisha 751024, India
e-mail: pghosefme@kiit.ac.in

S. Revankar et al. (eds.), *Proceedings of International Conference on Thermofluids*, Lecture Notes in Mechanical Engineering,
https://doi.org/10.1007/978-981-15-7831-1_17

drag-lift variation. Zhigang and Wei [4] have included the effect of engine exhaust over the wing to study the ground effect on drag-lift ratio through pressure distribution around the airfoil. Yang et al. [5] and Lee et al. [6] have investigated the effect of ground clearance for a wavy ground. Jung et al. [7] have studied the ground effect of different NACA series airfoil at various angle of attack. The work performed by Wei et al. [8] focused on the impact of ground clearance on the flow separation stability over and around the airfoil. Lee et al. [9] have investigated the ground proximity impact on aerodynamic requirement for an oscillating airfoil. From the literature, it has been observed that most of the studies are performed for symmetrical airfoil at various angle of attack. But the air temperature also may affect the drag and lift significantly due to the change in physical properties of air such as viscosity and density. The significant difference of drag and lift may be observed between too cold area (closer to earth pole where the temperature falls less than −50 °C) and too hot areas (desert-like Sahara or Australian desert, where temperature may rise more than 50 °C in summer season).

In this work, the effect of ground clearance and the effect of air temperature on drag and lift have been studied through computational modeling for NACA 2412 airfoil. This computational work is concerned with computation of the drag and lift on a stationary airfoil exposed to different velocities of air flow. However, the variation of angle of attack is not performed and kept as zero degree.

2 Computational Modeling

2.1 *Profile Generation for Four-Digit NACA2412 Airfoil*

The airfoils in the NACA four-digit airfoil having four numeric of significant meaning. The first number is the maximum camber height in percentage of the chord length. The second one represents the position of the maximum camber from the leading edge in one-tenth percentage of chord length. The last two numbers designate the maximum depth of the airfoil in percentage of the chord length (Fig. 1).

The upper and lower coordinates for the surface are calculated by the following equations:

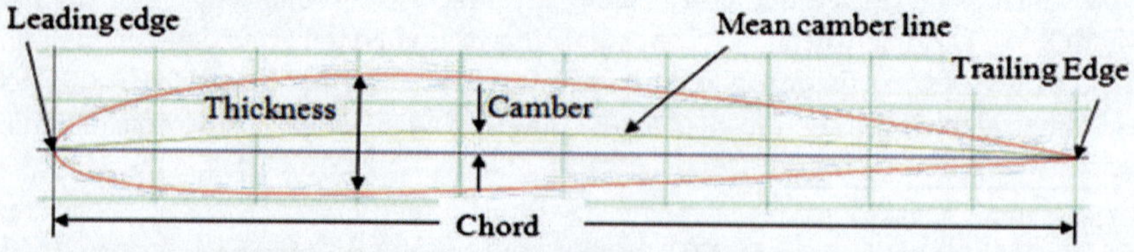

Fig. 1 NACA airfoil profile with various nomenclature

$$x_t = x - y_{af} \sin\theta \tag{1}$$

$$y_t = y_c - y_{af} \cos\theta \tag{2}$$

$$x_b = x - y_{af} \sin\theta \tag{3}$$

$$y_b = y_c - y_{af} \cos\theta \tag{4}$$

where x is the position along the chord, y_c is the mean camber line, y_{af} is the corresponding thickness distribution at different axial positions and θ is the angle between two corresponding point starting from leading edge. The thickness variation for the wing sections is expressed as:

$$\pm y_{af} = 5t\left(0.2969\sqrt{x} - 0.126x - 0.3516x^2 + 0.2843x^3 - 0.1015x^4\right) \tag{5}$$

The mean camber line is calculated as;

$$y_c = m/p^2\left(2px - x^2\right) \tag{6}$$

$$y_c = \left[m/\left(1 - p^2\right)\right]\left[\left(1 - 2p + 2px - x^2\right)\right] \tag{7}$$

where m is the highest camber height and p is the maximum camber position from leading edge. The camber line y_c as shown in Eq. (6) is used for the front of the airfoil up to $x < p$, where the camber line is calculated for the rest of the part through Eq. (7). In this work, total eighty points for both top and bottom surface are generated by using the mathematical formula discussed and imported to ANSYS workbench [10] in order to create the NACA 2412 airfoil profile.

2.2 *Method of Solution*

The 2D airfoil profile surrounded by a rectangular fluid domain is solved in Eulerian frame. The flow is assumed to be incompressible. Time averaged conservation of mass and momentum (RANS) equations are solved. In order to evaluate, the Reynolds stresses present in RANS equations, and standard two equation k-ε turbulent model [11] is used. The wall ahead of leading edge on fluid domain is used as velocity inlet boundary and all remain three walls are taken as pressure out let boundary. The airfoil wall is considered as no slip wall. SIMPLE pressure velocity coupling method along with second-order upwind scheme for spatial discretization of various governing equations is used. The converging limit has been set as 10^{-4} for various governing equations; especially, converging limit for energy equation it is used as

10^{-6}. The entire problem is solved by using the in-built code of commercial CFD tool ANSYS Fluent 14.0 [12].

2.3 Various Parameters Used in Simulation

The summation of the forces that acts over the airfoil surface and normal to the air flow direction is the lift force and the sum of the forces that acts parallel to the flow is evaluated as drag force. Both viscous and pressure effects on drag and lift calculation are considered for analysis.

To study the temperature effect on lift and drag in this cold flow simulation, the properties such as density and viscosity have been calculated at different temperature and used it as properties of air in material tab ANSYS Fluent as given in Table 1.

Static pressure coefficient is a non-dimensional parameter, which represents the pressure distribution around the airfoil. Mathematically, it is expressed as;

$$C_p = \frac{\text{static pressure} - \text{static pressure at inlet}}{\text{Dynamic pressure}} = \frac{p - p_\infty}{0.5\ \rho U_\infty^2} \tag{8}$$

Total lift force is the sum of pressure lift and friction lift. The pressure drag is calculated as;

$$L = \oint (p \times \mathrm{d}s) \cos \emptyset + \oint (\tau \times \mathrm{d}s) \sin \emptyset \tag{9}$$

where p is the pressure and τ is the wall shear stress on mesh line on the airfoil wall, ds is the mesh length and $\emptyset$ is the angle of mesh line to the x-direction. Similarly, the total drag force is calculated as;

$$D = \oint (p \times \mathrm{d}s) \sin \emptyset + \oint (\tau \times \mathrm{d}s) \cos \emptyset \tag{10}$$

Table 1 Thermophysical properties of air at various temperature

Temperature (°C)	Viscosity (N s/m^2)	Density (kg/m^3)
20	1.836924×10^{-5}	1.195
30	1.8860763×10^{-5}	1.148
40	1.9344628×10^{-5}	1.099
50	$1.98211248 \times 10^{-5}$	1.047

3 Results and Discussion

In the table, the lift and drag force are given for different sizes of NACA 2142 airfoil at different velocities by keeping the angle of attack as zero. Through the comparison between experiment performed by Rashed et al. [13] and current simulation, it has been observed that the comparatively higher drag and lift is obtained with larger wing area (Chord length × Wing span) exposed to higher air velocity (Table 2). However, the drag and lift coefficients obtained from simulated result are at same order with the experimentation.

3.1 Effect of Velocity on Drag and Lift

From Fig. 2a, it has been depicted that the total drag as well as the viscous and pressure drag are increasing nonlinearly with leaner variation of air velocity. Similarly, from Fig. 2b, it has been observed that the lift force is also increasing nonlinearly with leaner variation of air velocity. However, it is a well-known phenomenon that drag and lift forces vary with the square of the free stream velocity. The lift force due to viscous effect is predominated by the lift generated due to pressure. Therefore, due

Table 2 Reported result from selected literature [13]

Author	Chord length (mm)	Wing span (mm)	Air velocity (m/s)	Drag force (N/m of wing span)	Lift Force (N/m of wing span)	Coefficient of drag	Coefficient of lift
Rashed et al. (experiment)	200	120	4.5	0.133	0.516	0.054	0.21
Simulation	1000	-	13.8	2.7	37	0.023	0.32

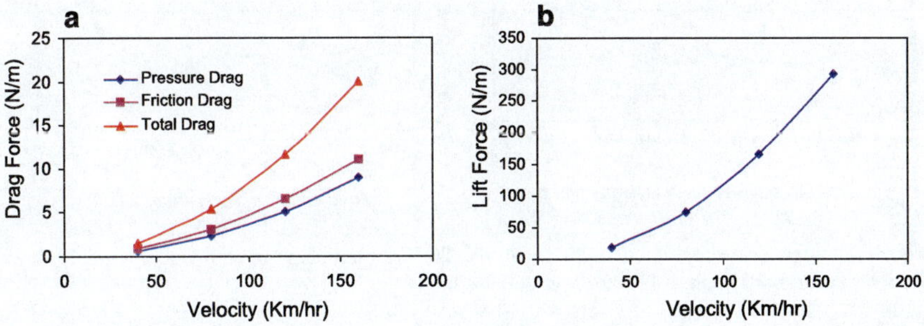

Fig. 2 **a** Drag force and **b** lift force on the aerofoil at 20 °C air temperature with 1 m ground clearance at a different air velocity

to the very negligible effect imparted by viscous has not been discussed here. It also has been observed that the lift force is around 15 times than that of drag force at any velocity.

In Fig. 3a–d, it is shown that the static pressure coefficient throughout the bottom surface of the airfoil is always greater than upper side surface of the airfoil irrespective of air velocity. Therefore, the lower surface is also called pressure side and upper portion known as suction side. However, since the viscous effect is negligible as compared to pressure effect for lift building, the net lift force can be represented by the static pressure coefficient area in between lower and upper surface. From this Fig. 3a–d, it has been seen that the net area under the lower and upper surface increases with increasing the velocity. Therefore, the lift force increases with increasing in velocity that has been clearly as shown in Fig. 2b. Moreover, it also observed that for higher inlet velocity the pressure coefficient distribution along both top and bottom surface became less, since it follow the Bernoulli's law that as velocity increases pressure decreases. For low inlet velocity, the reverse case has been obtained.

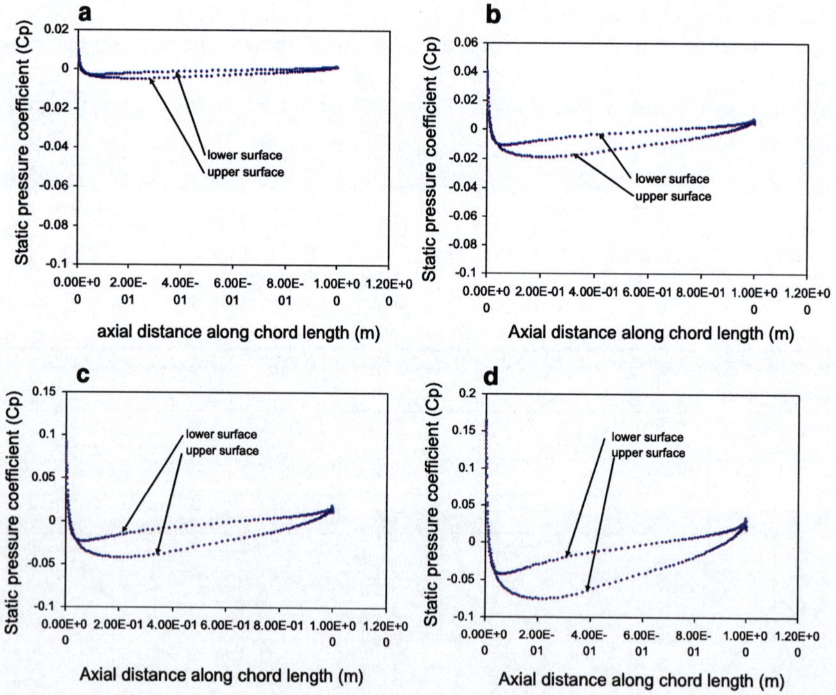

Fig. 3 Static pressure coefficient of the aerofoil at 20 °C air temperature with 1 m ground clearance for **a** 40 km/h, **b** 80 km/h, **c** 120 km/h, **d** 160 km/h

3.2 *Effect of Temperature on Drag and Lift*

The pressure field over the airfoil at various air temperature such as 20, 30, 40 and 50 °C is shown in Fig. 4a–d. However, it has been noticed from Fig. 5a, b that the air temperature does not significantly affect the drag and lift within the given temperature range. In Table 3, the values of drag and lift force are also given. Both drag and lift

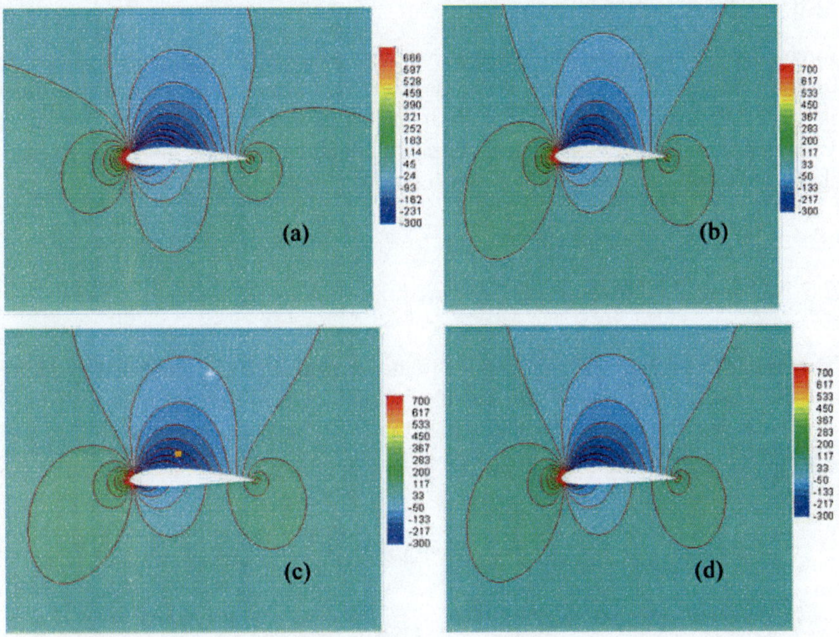

Fig. 4 Pressure contour around the aerofoil with 120 km/h of jet velocity and 1 m ground clearance at **a** 20 °C, **b** 30 °C, **c** 40 °C, **d** 50 °C

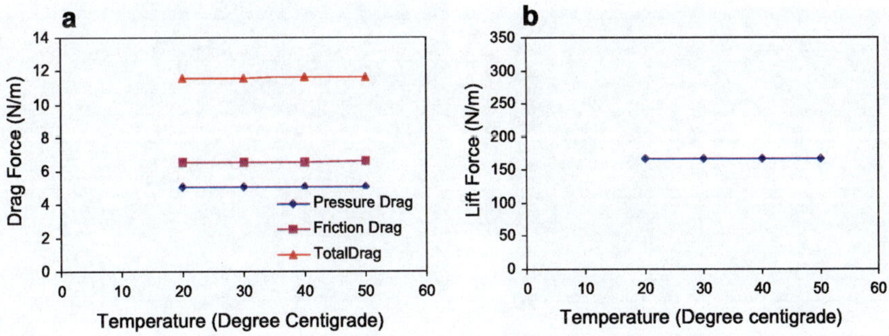

Fig. 5 **a** Drag force and **b** lift force on the aerofoil at 120 km/h with 1 m ground clearance at different air temperature

Table 3 Drag and lift force at various temperature

Temperature (°C)	Pressure drag (N/m)	Friction drag (N/m)	Total drag (N/m)	Total lift (N/m)
20	5.052	6.472	11.524	165.619
30	5.055	6.497	11.552	166.538
40	5.058	6.521	11.579	166.433
50	5.061	6.544	11.605	166.331

force directly depends on wall pressure and wall shear stress. Wall pressure depends on the velocity of air. Since velocity is same in all temperature condition, there is no change in pressure. But wall shear stress is affected by viscosity. The small variation of temperature (20–50 °C) does not affect the viscosity much. As a result, the temperature variation of air has negligible effect on both drag and lift.

3.3 *Ground Clearance Effect on Drag and Lift*

The ground clearance effect on drag and lift has been studied at a constant air velocity and constant air temperature. The pressure field contour has been given through Fig. 6 a–d. From Fig. 7a, b, it has been observed that with the ground clearance 0.5, 1 and 1.5 m, both drag and lift are almost same and both the drag and lift force does not

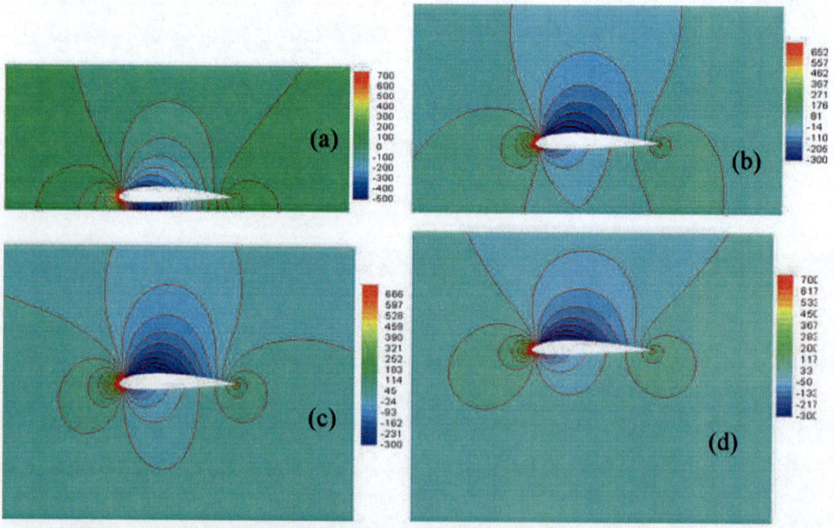

Fig. 6 Pressure contour around the aero foil with 120 km/h of jet velocity and 200 °C air temperature at ground clearance of **a** 0.1 m, **b** 0.5 m, **c** 1 m, **d** 1.5 m

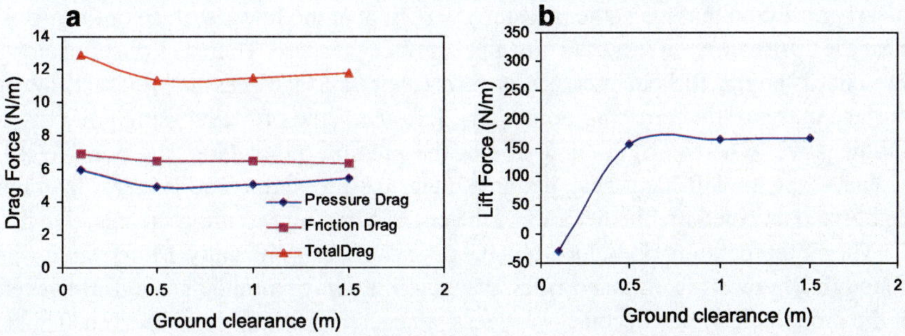

Fig. 7 **a** Drag force and **b** lift force on the aerofoil at 120 km/h and with 20 °C air temperature for different ground clearance

affected by ground clearance. But from Fig. 7b, an undesirable negative lift has been observed with the ground clearance of 0.1. Since ground clearance decreases, the space in between lower surface of airfoil and the ground is much squeezed, so that the velocity vicinity to lower airfoil surface increases by following the Bernoulli's law. As a result, pressure at the lower surface decreases. From Fig. 8a, it has been

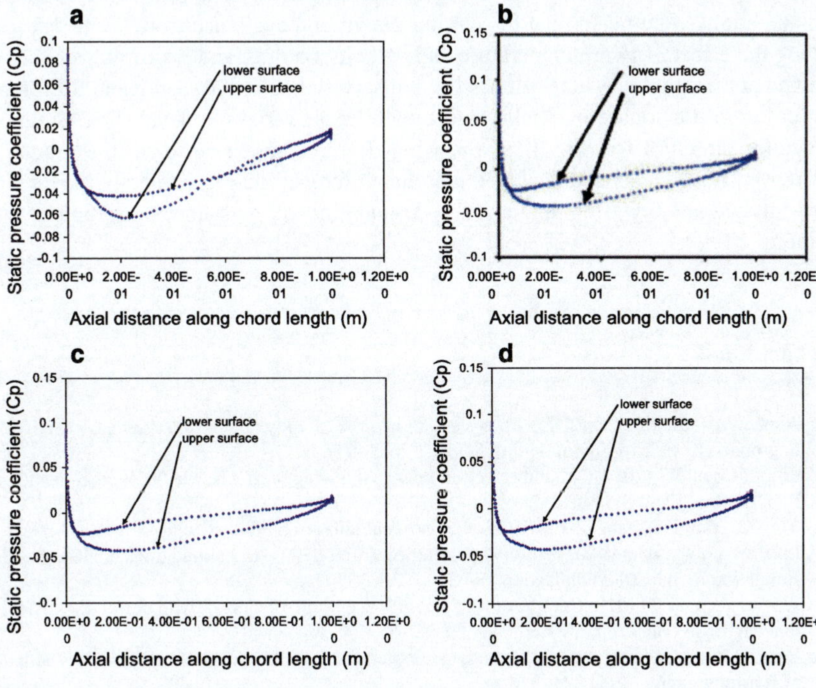

Fig. 8 Static pressure coefficient of the aerofoil with 120 km/h of jet velocity and 200 °C air temperature at ground clearance of **a** 0.1 m, **b** 0.5 m, **c** 1 m, **d** 1.5 m

clearly observed that the static pressure coefficient at the lower surface of the airfoil near the upstream even less than the upper surface pressure. There is a position along the chord length, the net pressure is zero (upper surface pressure line and lower surface pressure line crossing point). The left-hand side of the crossing point gives the negative, whereas right-hand side of the pressure point gives the positive lift. However, as the left-hand side, net area under the pressure curve is larger than the right-hand side net are, the net area of the entire curve representing the negative lift.

The air temperature does not affect the drag and lift significantly. Moreover, it was found that high lift is obtained when the airfoil is away from the ground. However, if the ground clearance becomes very less, a negative lift has been observed. This is due to the narrow region of flow that increases the air velocity leads to create a lower pressure region at the bottom part of the air flow.

4 Conclusion

The flow behavior over a NACA 2412 airfoil has been studied computationally by using commercial CFD software ANSYS Fluent in laboratory. The pressure distribution and wall shear stresses on the airfoil wall surface are obtained. Lift and drag forces were calculated over the surface by considering both friction and form drag/lift. Computations were performed by varying the ground clearance from 0.1 to 1.5 m. Along the effect of wind temperature and velocity on drag and lift force were also studied at zero angle of attack. It has been observed that both the drag and lift force increases with a square parabolic nature with linear variation of velocity. Ground clearance largely affect the lift if the wing becomes much closer to the ground. Moreover, it also has been observed that the air temperature has a negligible effect on both drag and lift. Both drag and lift vary nonlinearly with linear variation in air velocity.

References

1. Ahmed MR, Sharma SD (2005) An investigation on the aerodynamics of a symmetrical airfoil in ground effect. Exp Thermal Fluid Sci 29(6):633–647
2. Luo S, Chen YS (2012) Ground effect on flow past a wing with a NACA0015 cross-section. Exp Thermal Fluid Sci 40:18–28
3. Ying C, Yang W, Yang Z (2010) Ground viscous effect on stall of wing in ground effect. In: Proceedings of 3rd international conference on modeling and simulation, modeling and simulation in industrial application, vol 3, pp 230–233
4. Zhigang Y, Wei Y (2010) Complex flow for wing-in-ground effect craft with power augmented ram engine in cruise. Chin J Aeronaut 23(1):1–8
5. Yang W, Ying CJ, Yang Z (2010) Aerodynamic study of WIG craft near curved ground. Journal of Hydrodynamics 22(5):371–376
6. Lee T, Tremblay-Dionne V (2018) Experimental investigation of the aerodynamics and flowfield of a NACA 0015 airfoil over a wavy ground. J Fluids Eng 140(7):071202.

7. Jung J, Yoon HS, Chun HH, Hung PA, Elsamni OA (2012) Mean flow characteristics of two-dimensional wings in ground effect. Int J Naval Archit Ocean Eng 4(2):151–161
8. Wei H, Yu P, Li LKB (2018) Ground effects on the stability of separated flow around a NACA 4415 airfoil at low Reynolds numbers. Aerosp Sci Technol 72:63–76
9. Lee T, Majeed A, Siddiqui B, Tremblay Dionne V (2018) Impact of ground proximity on the aerodynamic properties of an unsteady airfoil. Part G: J Aerosp Eng 232(10):1814–1830
10. Lawrence, K.L.: ANSYS workbench tutorial release 14. SDC Publications
11. Sharma R, Bansal RJ (2013) CFD simulation for flow over passenger car using tail plates for aerodynamic drag reduction. J Mech Civil Eng 7(5):28–35
12. Fluent A (2014) 14.5. User's and theory guide. Canonsburg, Pennsylvania, USA: ANSYS. Inc.
13. Ahammad R, Hasan MK, Rahman M, Chakraborty M (2015) Performance test of Naca 2412 airfoil. In: International conference on mechanical engineering and renewable energy Bangladesh

Aerodynamic Characteristics of a Square Cylinder: Effect of Dissimilar Leading Edges

Akhil S. Nair, S. Ajith Kumar, K. Arun Kumar, and R. Ajith Kumar

Abstract This paper presents the results of flow over square cylinder with dissimilar leading edges such as corner recession and corner arc at a Reynolds number 200. The leading edges are modified in such a way that the dimensions of the bottom leading edge are kept fixed and that of top leading edge is varied in the predefined proportions or cut ratios between 0 and 1 with respect to bottom edge. This study reveals that the flow over square cylinder with dissimilar leading edges shows an inverse relationship between Strouhal number and coefficient of drag for cut ratios less than 0.5 but beyond 0.5, both these parameters register an increasing trend.

Keywords Bluff body · Leading edges · Strouhal number · Corner modification

Nomenclature

C_{d}	Coefficient of drag
St	Strouhal number
ρ	Density
U_{∞}	Free stream velocity
D	Characteristic length
μ	Dynamic viscosity
U_{s}	Velocity of vortices relative to the free stream

A. S. Nair (✉) · S. A. Kumar · K. A. Kumar · R. A. Kumar
Department of Mechanical Engineering, Amrita Vishwa Vidyapeetham, Amritapuri, India
e-mail: akhilsnair181295@gmail.com

S. Revankar et al. (eds.), *Proceedings of International Conference on Thermofluids*, Lecture Notes in Mechanical Engineering,
https://doi.org/10.1007/978-981-15-7831-1_18

1 Introduction

The study of aerodynamic characteristics of bluff body has got many relevant practical applications. The past studies mainly focused on the general aerodynamic characteristics like coefficient of drag (C_d), Strouhal number (St), etc. The drag which is undesirable inevitably occurs in practical applications. But, it can be nullified or at least controlled to a certain extent by active and passive methods. In large erected buildings, structures like bridges, the effect of pressure drag is more pronounced which in fact could result in structural failure. Therefore, proper analysis of the same at various conditions of fluid–structure interaction is of great importance in the design of such structures. In building structures, the most important shape considered is the square shape. This is mainly because of the space management and easiness in construction. Even though it has these advantages, there is a possible chance for the prismatic body (square shape) to experience flow-induced vibration (FIV) under certain flow conditions. One of the key factors which decides the intensity of FIV is the body geometry and the 'attitude' of the body to the flow [1]. Hence, study of the body geometry and its effects on the aerodynamic characteristics is of primal importance to achieve passive control of the fluid forces acting on the body and the consequent FIV. Previous studies report some interesting results on flow around corner modified square cylinders as described below.

Tamura et al. [2] found that the wake behind a square cylinder becomes narrower when the corner is chamfered, which indeed contribute more to the drag reduction at high Reynolds numbers compared to other corner modifications. Hu et al. [3] investigated the near wake characteristics of square cylinders with different corner radii experimentally. He reported that, when the corner radius ratio increases from 0 to 0.5, the maximum vorticity of the vortex attenuates, and the circulation associated with vortices drops progressively by 50%. Carassale et al. [4] conducted numerical and experimental analysis of a square cylinder with corner modifications and reported that rounded corners reduce critical angle of incidence at which the flow reattaches on to the lateral side of the square cylinder. Kawai [5] analyzed the effect of corner modifications on square sectioned tall buildings and reported that corner roundness is the best method to reduce aerodynamic instabilities compared to chamfered and recessed corners. Kumar et al. [6] studied the near wake characteristics of transversely oscillating square-section cylinders with dissimilar leading edges and concluded that corner radius ratio had a major role in the vortex formation and shedding mechanisms. He et al. [7] observed that the vortices form further away from the cylinders for the cut-corner case of square cylinder. After the corners are cut, a pair of recirculation flow is formed in the cut-out region, which acts as a smooth virtual shape that facilitates the flow to go around the corners and thus, the separation is suppressed, and the drag is reduced. Miran and Sohn [8] conducted test with variation in corner radius ratio (corner radius to cylinder diameter, R/D, where R is the corner radius and D is the cylinder diameter). The Strouhal number increases rapidly at first and then it gradually rises. The minimum values of the mean drag coefficient and the RMS value of lift coefficient were found around $R/D = 0.2$.

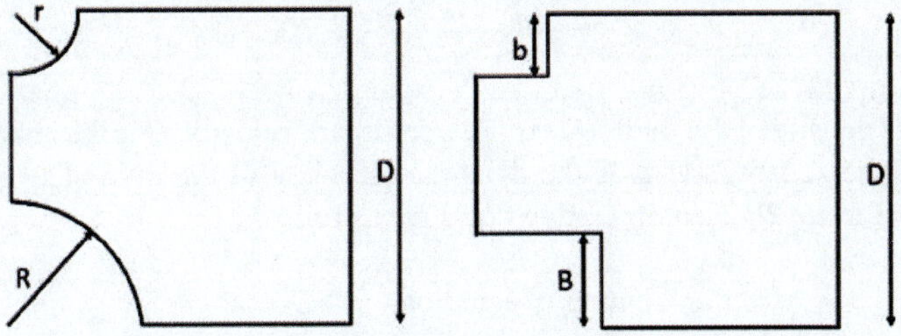

(a) Arced cut cylinder (b) recessed cylinder

Fig. 1 Schematic pictures of modified square cylinders

Literature shows that the numerical study on the effect of dissimilar leading edges such as square cut and arced cut on the aerodynamic characteristics of square cylinder has not done yet.

For the present study, square cylinder is modified at the leading edges with shapes such as arced cut and recession as shown in Fig. 1. The variation is such a way that, the bottom leading edges are kept constant and the top leading edges are varied for different cut ratios.

2 Computational Method and Validation

Two-dimensional flow domain of a square cylinder is modeled and meshed in Gambit preprocessing software, and the flow around this computational domain (Fig. 2) is analyzed using ANSYS Fluent 2019 [9]. The dimensions used for the computational domain and the boundary conditions used are mentioned in the figure. The fluid taken for the analysis is air with a flow velocity of 0.087 m/s corresponds to a Reynolds number,

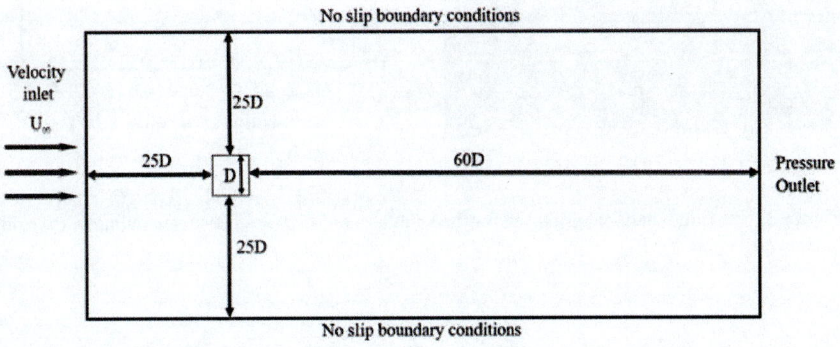

Fig. 2 Schematic representation of the computational domain

$$\mathrm{Re} = \frac{\rho \times U_\infty \times D}{\mu} = 200.$$

where ρ is density of the fluid, U_∞ is the free stream velocity, D is the characteristic length of the square, and μ is the dynamic viscosity of the fluid. The governing equations for the 2D incompressible flows are as follows,

$$\text{Continuity equation} \quad \nabla \cdot U = 0 \tag{1}$$

$$\text{Momentum equation} \quad \rho(\partial U/\partial t) + \rho U(\nabla \cdot U) = -\nabla P + \mu(\nabla^2 U) \tag{2}$$

The PISO algorithm is employed for the pressure velocity coupling to get time accurate solutions [10]. For pressure and momentum discretization, second-order upwind schemes are used. Second-order implicit scheme is applied for temporal discretization. Time step given for the computation is 0.01 to satisfy the Courant–Friedrichs–Lewy (CFL) criterion for stability of the scheme.

The grid independency study (Fig. 3) has been carried out for various grid sizes and the variation of C_d for different grid sizes has been analyzed. The mesh size of 44,020 is found optimum. Any increase in the grid size beyond this value (~44,020) does not change the value of C_d appreciably. Hence, this grid size is selected for the remaining computations in the rest of the paper.

The validation for the flow over square cylinder has been carried out for Re = 200, and for checking the corner modification, the simulation for flow over square cylinders with recessed, rounded, and chamfered leading edges has been conducted with same Reynolds number (Table 1). The percentage error hence obtained when simulated is less than 5% compared to that of the previous results.

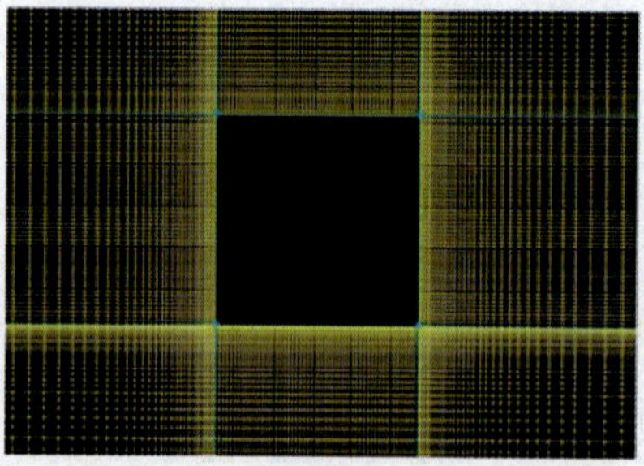

(a)

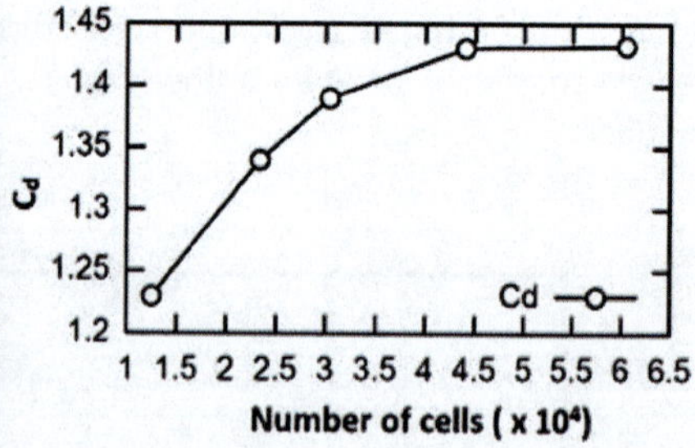

(b)

Figure 3. (a) Computational grid used for the simulation of flow around square cylinder, (b) grid independence test results

Table 1 Validation of square cylinder with modified leading edges for Re = 200

Sl. no.	Data	Ambreen et al. [11]		Present results	
		C_d	St	C_d	St
1	Uncut square cylinder	1.42	0.157	1.435	0.153
2	With corner recession	1.26	0.17	1.3	0.166
3	With rounded corners	1.23	0.165	1.26	0.159
4	With chamfered corners	1.22	0.17	1.25	0.166

3 Results and Discussion

A numerical study is conducted on the flow over square cylinder for two corner modifications, viz., corner arc and corner recession. The numerical investigations are carried out for different cut ratios between 0 and 1 for a representative Reynolds number of 200. The coefficient of drag (Cd) and Strouhal number (St) are evaluated for both the configurations and are plotted for different cut ratios.

3.1 Effect of Corner Recession on Cut Ratio (b/B)

The flow over modified square cylinder with corner recession is simulated. The C_d and S_t are plotted (Fig. 4a) against different b/B ratios. It is observed that the variation of C_d decreases and St increases till $b/B = 0.5$ like any other conventional bluff body. But beyond $b/B = 0.5$, both these parameters register an increasing trend which is not seen (not reported) for conventional bluff bodies with similar leading edges. Figure 4b shows variation of ($C_d \times$ St) for different b/B ratios. It can be seen that the ($C_d \times$ St) is almost constant for b/B ratios up to 0.5. However, both C_d, St and consequently, ($C_d \times$ St) increase with the increase in b/B ratio afterwards. As could be seen in Fig. 5, for $0 \leq b/B \leq 0.25$, a recirculation region develops in the vicinity of the leading square cuts and eventually leads to decrease in C_d.

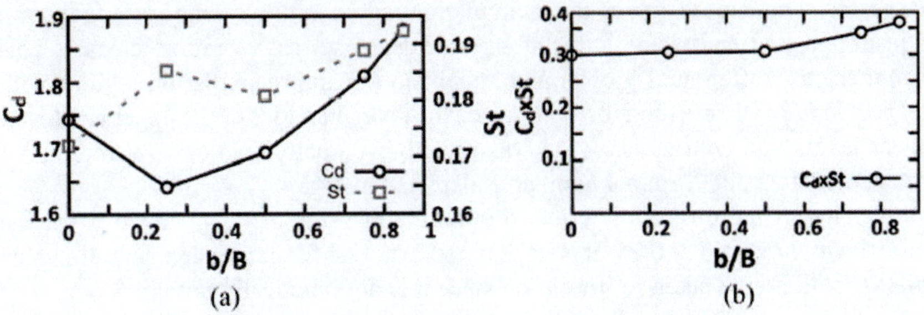

Fig. 4 **(a)** C_d and St versus b/B ratio, **(b)** $C_d \times$ St versus b/B ratio for square cut cylinder

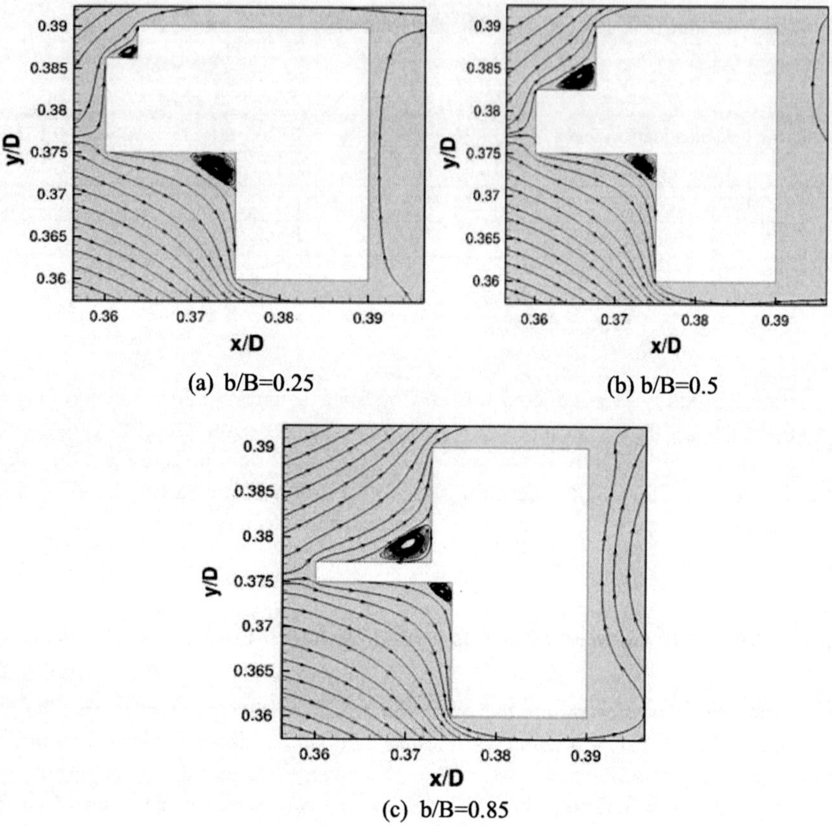

Fig. 5 Streamline contours of flow past square cylinder with corner recession for various ratios (0.25, 0.5, and 0.85)

The streamline contours of the flow past corner recessed square is given in Fig. 5. It could be noticed that, the recirculation region (bubble) formed in the upper corner cut is minimum for $b/B = 0.25$ (Fig. 5a) and it grows to larger size as b/B ratio increases, whereas the size of the recirculation region in the lower corner cut shows a just opposite trend wherein the bubble size decreases as b/B ratio increases. This simultaneous differential growth and attenuation of the recirculation regions with respect to b/B ratio are thought to influence the characteristics of the upper and lower shear layers and consequently the wake width eventually leading to variations of drag coefficient and Strouhal number as depicted in Fig. 4.

Among all the b/B ratios tested, the minimum value of C_d obtained is for the model with b/B ratio of 0.25. It is to be noted that, like for corner arc, cut ratio value of 1.0 could not be taken for recession since it is not practically possible.

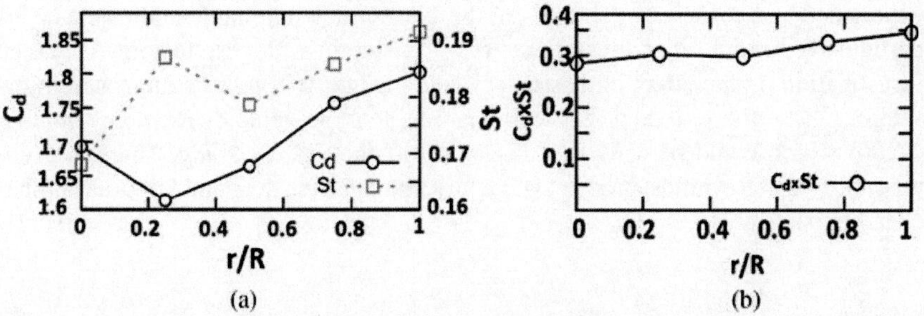

Fig. 6 **(a)** C_d and St versus cut ratio, **(b)** $C_d \times$ St versus cut ratio for arc cut cylinder

3.2 *Effect of Arc Cut on Cut Ratio* (r/R)

The C_d and St for modified square cylinder with arc cut are plotted for various cut ratios (r/R) and is shown in Fig. 6a. It can be seen in Fig. 6 that the trends of C_d, St, and $C_d \times$ St are somewhat similar to the case of corner recession. That is, till $r/R = 0.5$, C_d and St exhibit an inverse relationship with each other further to which both these parameters increase with increase in r/R ratio, and correspondingly $C_d \times$ St increases. For these trends observed (Fig. 6), presumably similar flow physics as explained in Sect. 3.1 (refer to Fig. 5a–c) is thought to be applicable. Quite interestingly, similar to the case of corner recession, the minimum C_d is experienced at $r/R = 0.25$.

It is also worthy to mention here that, the increasing trend $C_d \times$ St for values of b/B and r/R ratios greater than 0.5 is something unique not reported before, to authors knowledge. The said product ($C_d \times$ St) is known to be a function of vortex spacing ratio (b/a, where 'a' is the spacing between two adjacent vortices in a row and 'b' is the spacing between two vortices in upper and lower rows) and velocity ratio, U_s/U_o (where U_s is the velocity of vortices relative to the free stream and U_∞ is the free stream velocity) as given below

$$C_d \cdot S_t = \frac{4}{\pi}\left(1 - \frac{U_s}{U_\infty}\right)\left(\frac{U_s}{U_\infty}\right)^2\left[\mathrm{Coth}^2\left(\frac{\pi b}{a}\right) + \left(\frac{U_\infty}{U_S} - 2\right)\frac{\pi b}{a}\mathrm{Coth}\left(\frac{\pi b}{a}\right)\right] \tag{3}$$

In the light of this equation [12], it could be inferred that, the trends depicted in Figs. 4b and 6b reflect changes in the vortex spacing ratio and velocity ratio of the resulting wake due to corner modification of the square cylinder (recession and arc) for corner cut ratio greater than 0.5. This is primarily due to the changes in the flow structure particularly the changes which upper and lower shear layers undergo, as mentioned before.

As the results of the current study indicate, the, increasing trend of the product ($C_d \times$ St) for cut ratios greater than 0.5 is counter-intuitive and not known reported for bluff bodies with similar leading edges as mentioned earlier. This generates an academic curiosity as to further investigate the subtle changes the shear layers experience and consequent changes in the vortex shedding mode and mechanisms particularly for cut ratios greater than 0.5. Considering all these factors, a detailed investigation on the effect of dissimilar leading edges will be communicated in the future. Furthermore, the current results are at a single value of Reynolds number (~200) which could be seen as a limitation of the present study. There is every possibility that Reynolds number can greatly influence the drag and Strouhal number characteristics [11].

4 Conclusions

The numerical study of flow over simple square cylinder and modified square cylinder has been conducted and compared with literature. The present study, effect of dissimilar leading edges for a square cylinder on the aerodynamic characteristics such as coefficient of drag and Strouhal number has been investigated at a low Reynolds number value of 200. The results led to following conclusions.

- Both arced cut and square cut (recession) cylinders exhibit similar overall trends of drag and Strouhal number.
- Till cut ratio (arc and recession) of 0.5, drag coefficient decreases whereas Strouhal number increases displaying an inverse relationship with each other.
- For cut ratios greater than 0.5, C_d and St show an increasing behavior which is unseen for conventional bluff bodies with similar leading edges.
- For both cases of corner recession and corner arc, the lowest C_d value obtained is at cut ratio = 0.25.

References

1. Blevins RD (1990) Flow induced vibrations, 2nd edn. Van Nostrand Reinhold, New York, p 1977
2. Tamura T, Miyagi T, Kitagishi T (1998) Numerical prediction of unsteady pressures on a square cylinder with various corner shapes. J Wind Eng Ind Aerodyn 531–542
3. Hu JC, Zhou Y, Dalton C (2004) Effects of the corner radius on the near wake of a square prism. Exp Fluids 106–118
4. Carassale L, Freda A, Marre-Brunenghi M (2014) Experimental investigation on the aerodynamic behavior of square cylinders with rounded corners. J Fluids Struct 195–204
5. Kawai H (1998) Effect of corner modifications on aeroelastic instabilities of tall buildings. J Wind Eng Ind Aerodyn 719–729
6. Kumar RA, Arunkumar K, Hariprasad CM (2015) Effect of dissimilar leading edges on the flow structures around a square cylinder. J Pressure Vessel Technol 1–10 (2015)
7. He GS, Li N, Wang JJ (2014) Drag reduction of square cylinders with cut-corners at the front edges. Exp Fluids 1–11 (2014)
8. Miran S, Sohn CH (2014) Effect of rounded corners on two degree of freedom naturally oscillating square cylinder. Int J Numer Methods Heat Fluid Flow 2355–2374
9. ANSYS, Inc. (2019) ANSYS fluent user's guide, Release 2019 R1

10. Barton IE (1998) Comparison of simple- and Piso-type algorithms for transient flows. Int J Numer Methods Fluids 459–483
11. Ambreen T, Kim MH (2018) Flow and heat transfer characteristics over a square cylinder with corner modifications. Int J Heat Mass Transfer 50–57
12. Obasaju ED (1983) An investigation of the effects of incidence on the flow around a square section cylinder. Aeronaut Q 243–259

Numerical Investigation of Various Wake Patterns in Flow Past Two Side-By-Side Cylinders

P. Pandey, D. Singh, P. Das, D. Ghosh, and Atal Bihari Harichandan

Abstract In the present research work, an incompressible unsteady viscous two-dimensional finite volume Navier–Stokes solver is developed to investigate flow past two-dimensional geometries. Two-dimensional analysis of flow past any arbitrary body geometry can lead to comprehensive parametric study so that the complete flow details can be captured. In the present solver, the full Navier–Stokes equations have been solved numerically in the physical plane itself by considering "Consistent Flux Reconstruction" technique on a collocated unstructured mesh comprising of triangular cells. A C-code has been developed to estimate the primitive variables and to analyze the flow characteristics. The flow characteristics around a single unconfined circular cylinder along with two circular cylinders in side-by-side arrangement at various gap ratios have been considered for the investigation. Streamlines and vorticity contours are plotted to visualize instantaneous flow field. Lift and drag coefficients and Strouhal number are computed in order to characterize the flow. Different types of wake patterns, like in-phase or anti-phase synchronized wake patterns and deflected wake patterns, are observed based on various gap ratios between cylinders. For a certain Reynolds number, the vortex shedding frequency from either of the cylinders in a side-by-side arrangement has been estimated to be identical.

Keywords Finite volume method · Incompressible N-S solver · Unstructured grid · Unconfined circular cylinder · Strouhal number · Gap ratio

1 Introduction

Flow past bluff bodies are always of great interest for many researchers over the years. Flow past circular cylinders are more widely investigated because of its practical importance in many areas surrounding engineering and aerodynamics such as cooling systems for power plants, transmission lines, heat exchanger tubes, earth

P. Pandey · D. Singh · P. Das · D. Ghosh · A. B. Harichandan (✉)
Department of Aerospace Engineering, KIIT Deemed University (IoE), Bhubaneswar, Odisha 751024, India
e-mail: harichandan.iitkgp@gmail.com

S. Revankar et al. (eds.), *Proceedings of International Conference on Thermofluids*, Lecture Notes in Mechanical Engineering,
https://doi.org/10.1007/978-981-15-7831-1_19

extended structures, bridges, etc. Numerical investigation of unsteady flow past circular cylinder is complex since it encompasses very subtle physical and dynamics phenomenon like flow separation, recirculation, vortex shedding, etc. When these geometries are exposed to the fluid flow, they enable flow impel vibration which may lead to failure of structure. So to excluding such conditions and for much more effective designing, it becomes necessary to study and gain detailed features of the flow field. Furthermore, it becomes more complicated to investigate flow over multiple circular cylinders as compared to the single circular cylinder due to the dynamically interactions between vortices shed from cylinder and Von-Karman vortex street that appears in the region of wake.

Many research works have been carried out on this topic by previous researchers proposing various numerical schemes. Wu [1] used local domain free discretization and immersed boundary method (LDFD-IBM), Berrone et al. [2] applied adpative finite element method (AFEM) for numerical investigation. Roy and Bandyopadhyay [3] proposed finite explicit scheme for face centre flux reconstruction. Flow around multiple cylinders with different arrangement was experimentally observed by Zdravkovich [4] and Williamson [5]. Williamson [5] observed periodic and synchronized characteristics of vortex shedding at the larger gap around the cylinders in the consecutive arrangement. Zdravkovich [4] observed irregular vibrations for varying transverse gap between cylinders ($1.1D < T < 2.2D$) where D is cylinder diameter and T is the transverse gap between cylinders. He reported flip-flopping patterns of the fluid flow at the transverse gap of 1.5D. Further, Liu et al. [6] investigated flow characteristics and wake pattern past side-by-side cylinders and concluded that the patterns strongly depends upon transverse gap in between the cylinders as compared to Reynolds number. Harichandan and Roy [7] also reported in-phase or anti-phase synchronized wake pattern at larger gap and flip-flopping pattern at smaller gap. In the present investigation, 2D Navier-Stoker Solver is developed for the incompressible viscous flow passing through single and multiple cylinders. The flow variables are calculated using a scheme that is fully explicit in nature. Collocated unstructured grid approach with consistent flux reconstruction is applied for solving the momentum equations. Pressure-Poisson equation is used for the pressure–velocity coupling. For the simplicity and better understanding, the flow over an unconfined circular cylinder is investigated and its results are taken as reference to analyze the interaction of side-by-side cylinders.

2 Governing Equations

The two-dimensional continuity and momentum equations for the flow of incompressible viscous fluid in conservative and non-dimensional form are considered for the present problem.

Mass conservation equation:

$$\frac{\partial U}{\partial x} + \frac{\partial V}{\partial y} = 0 \tag{1}$$

Momentum equation:

X-direction: $$\frac{\partial U}{\partial t} + \frac{\partial (U^2)}{\partial x} + \frac{\partial (UV)}{\partial y} = -\frac{\partial P}{\partial x} + \frac{1}{\mathrm{Re}} \times \left(\frac{\partial^2 U}{\partial x^2} + \frac{\partial^2 U}{\partial y^2}\right) \tag{2}$$

Y-direction: $$\frac{\partial V}{\partial t} + \frac{\partial (UV)}{\partial x} + \frac{\partial (V^2)}{\partial y} = -\frac{\partial P}{\partial y} + \frac{1}{\mathrm{Re}} \times \left(\frac{\partial^2 V}{\partial x^2} + \frac{\partial^2 V}{\partial y^2}\right) \tag{3}$$

where U and V denotes the velocity components in the directions X- and Y respectively, Re is the Reynolds number, P is the ratio of pressure and density, and t is the non-dimensional time.

3 Problem Description

The geometrical description of two side-by-side cylinder is presented in the following diagram. 'D' denotes the diameter of cylinders, 'G' indicates the distance between the two centres of two consecutive cylinders in transverse direction and the computation domain adopted is rectangular $35D \times 23D$ (Fig. 1).

3.1 *Initial and Boundary Conditions*

In accordance of elliptic–parabolic behaviour of the governing equations, the solution marches with time and the spatial disturbance can exist along any direction. An uniform free-stream condition for pressure and velocity is defined for every cell.

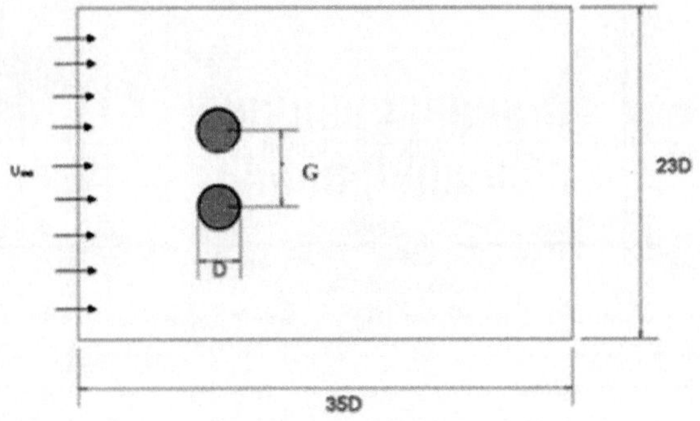

Fig. 1 Geometrical depiction of flow domain

Around the body structure no-slip condition is applied and Dirichlet boundary condition is used at outer boundaries. The top, bottom, inlet, and outlet boundaries are defined considerably away from the cylinders.

4 Results and Discussion

Further proceeding study, the described scheme is rectified for the simulation of flow past single circular cylinder at Re = 100 and Re = 200 and for the simulation of the flow parameters of two face-to-face circular cylinders of Re = 100 for varying transverse distances at the cylinders. In order to analyse the flow features around cylinders and in the wake region, the vorticity contours and streamlines are highlighted. Coefficients of lift and coefficient of drag with relation Strouhal number which characterize the flow are also calculated and in comparison with other literatures for the validation of the present solver.

4.1 Flow Past a Single Circular Cylinder

The unsteady flow is simulated on unstructured triangular mesh. A proper grid independence test is carried out with different numbers of the nodes on the body surface that prompted authors to consider a computation domain with 13,220 nodes and 26,144 cells for the present simulations having 160 nodes on the body surface. The non-dimensional time step $t = 10^{-3}$ s used in the computation. The convergence criterion of Pressure-Poisson equation is considered to be less than 10^{-6}.

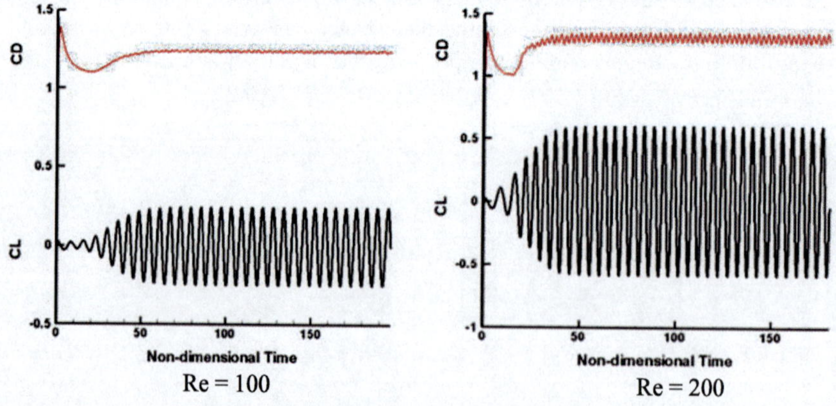

Fig. 2 Force coefficients at Re = 100 and 200

Table 1 Flow parameters at Re = 100 and 200

Parameters	Lift coefficient (C_L)		Drag coefficient (C_D)		Strouhal number (St)	
	Re = 100	Re = 200	Re = 100	Re = 200	Re = 100	Re = 200
Ding et al. [10]	±0.287	±0.659	1.356 ± 0.010	1.348 ± 0.05	0.166	0.196
Harichandan and Roy [7]	±0.278	±0.602	1.352 ± 0.010	1.320 ± 0.05	0.161	0.192
Present result	±0.260	±0.607	1.345 ± 0.010	1.312 ± 0.05	0.163	0.194

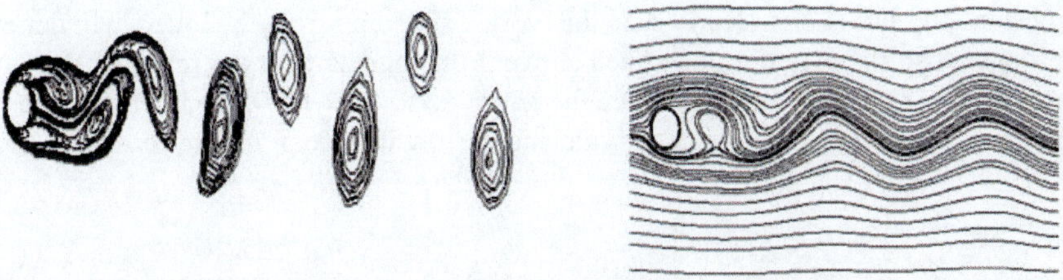

Fig. 3 Vorticity and streamline contours for past a single cylinder at Re = 100

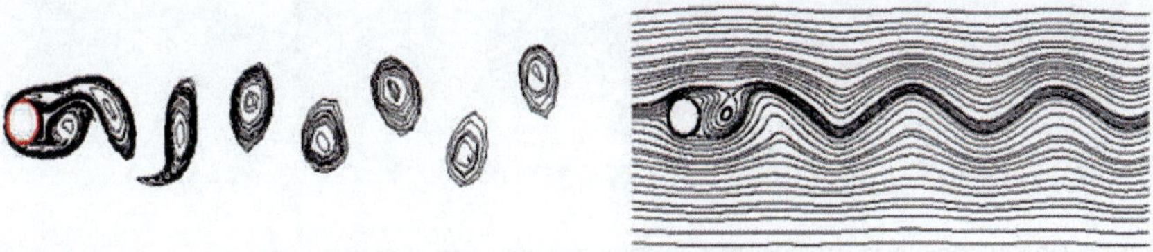

Fig. 4 Vorticity and streamline contours for past a single cylinder at Re = 200

The unsteady variation of force coefficients is presented in Fig. 2 for Re = 100 and. The force coefficients along with Strouhal number are quantified in Table 1.

The vorticity and streamlines contours for flow past single circular cylinder are shown in for Re = 100 and Re = 200 in Figs. 3 and 4 respectively.

4.2 Flow Across Cylinders at Same Time

Investigation of flow across multiple cylinders of circular cross sections increases the complexity with respect to a single circular cylinder because of the dynamic interplay between vortices shed from cylinder and Von-Karman vortex street that appears in the region of wake. In light of present investigation, simulation for the numerical values has been carried out for three different cases of side-by-side arrangement of two circular cylinders for the different gap ratios where value of '*G*' was 3*D*, 2*D* and 1.5*D*. The unsteady flow characteristic at Re = 100 is visualized on mesh with triangular cross sections as shown in Fig. 5.

The unsteady flow behaviour past side-by-side cylinders corresponding to different '*G*' values are visualized. Figures 6 and 7 show different contours of vorticity and streamline at various transverse gap for Re = 100, respectively. There are two Karman vortex streets observed for *G* = 3*D* case which are expected for the transverse gap more than 2D according to literatures. The flow characteristics indicate that vortex shedding occurs in the test cylinders approximately at same time, merges and at some distance, these merging stops. For *G* = 2*D*, bias and reversed pattern can be seen as we put the cylinders in the closer in the space the interface effect become substantially large. It is clearly seen that vortex shedding from the lower cylinder is stronger than the cylinder positioned above it that agrees with the result of Hesam and Navid [8]. As we further reduce the gap ratio to *G* = 1.5*D*, the interface effect become more severe. In this case, wide and narrow wake regions are formed in the

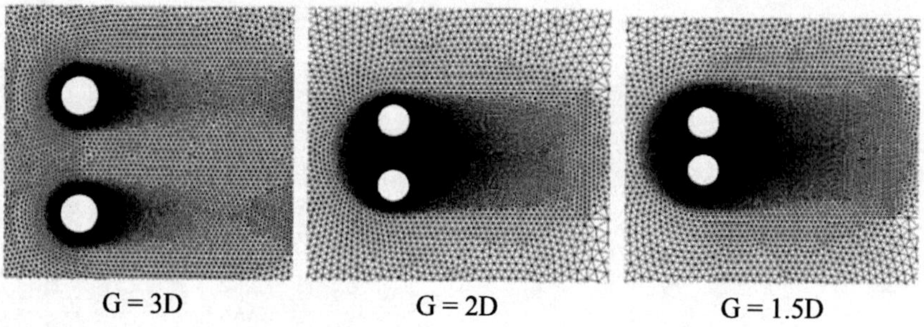

Fig. 5 Unstructured grid for side-by-side cylinders with varied transverse gaps

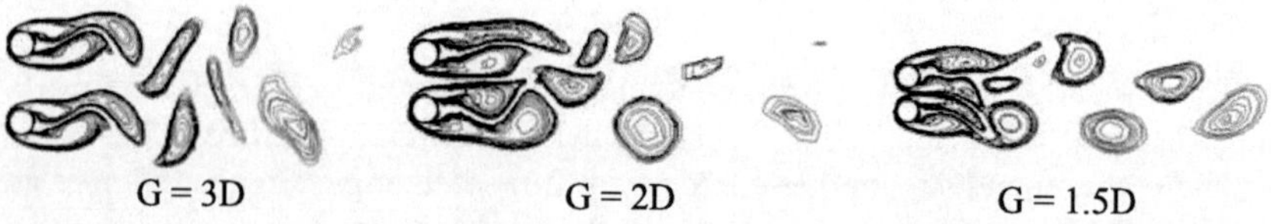

Fig. 6 Vorticity contours at Re = 100

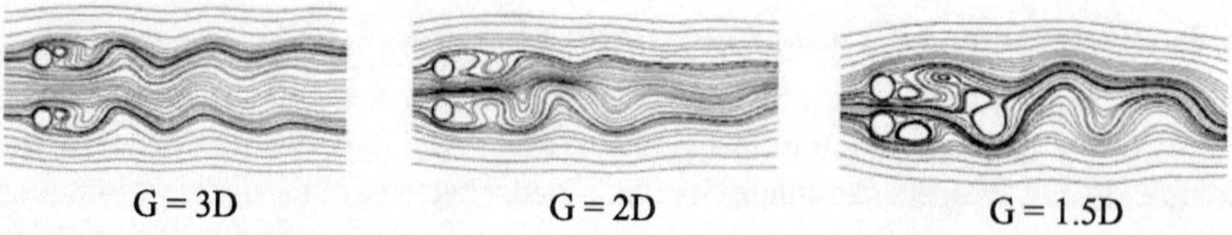

Fig. 7 Streamline contours at Re = 100

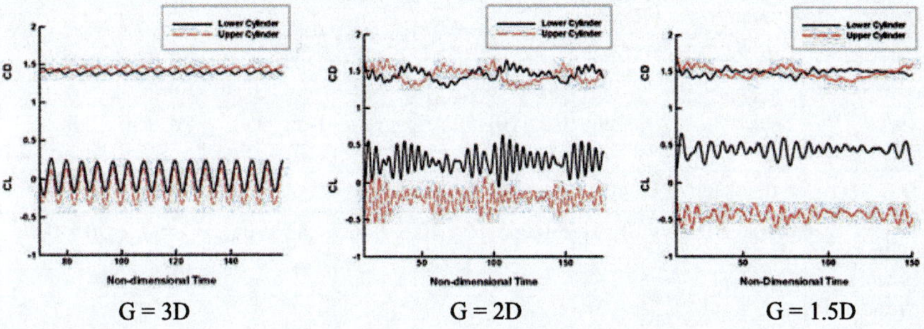

Fig. 8 Unsteady variation of force coefficients at Re = 100

rear end of the cylinders which cause the flip-flop wake pattern. This type of diagonal and reverse pattern can be seen while observing in the investigation of Wu [1].

The temporal variation of force coefficients are presented in Fig. 8 for all the cases of side-by-side arrangement at Re = 100. CL denotes the lift coefficient and CD denotes the drag coefficient. The results for both the cylinders is plotted in one graph which makes it more suitable for the comparison. It is seen that the temporal change in the coefficient of lift and drag for $G = 3D$ is synchronized with time which suggest that the fully periodic nature of vortex shedding. For $G = 2D$, very irregular flip-flopping wake patterns are obtained. As we reduce the transverse distance between cylinder to $G = 1.5D$, the fluctuation becomes very unsteady. The flow is deflected to the cylinder experiencing higher frequency which agrees with the result of Xu et al. [9].

The values of flow parameters for the present as well as the result from other researchers are presented in Table 2. For lower cylinder, lift coefficient is positive whereas for upper cylinder, it is negative. Reasonably good agreement has been obtained between the present results and results available from literature.

5 Conclusion

An incompressible 2D Navier–Stokes solver has been developed using finite volume approach. The flow is laminar, unsteady, and viscous in nature. The flow characteristics at Re = 100 and 200 around single unconfined circular cylinder has been considered for validation. The good agreement of present results with others confirms the accuracy and convergence. Different wake patterns are noticed in the downstream of cylinders based on varied transverse gap ratio between them. For larger transverse gap, $G = 3D$, synchronized patterns (in-phase or out-phase) are noticed. With decrease of gap ratio ($G = 2D$), very irregular, deflected, and unsteady flow patterns are obtained. Whereas for closer transverse gap, ($G = 1.5D$), flip-flop, and biased wake patterns are noticed.

Table 2 Flow parameters at Re = 100

Parameters	Lift coefficient (C_L)		Drag coefficient (C_D)		Strouhal number (St)	
	Present	Ding et al. [10]	Present	Ding et al. [10]	Present	Ding et al. [10]
$G = 3D$ Upper cylinder	-0.103 ± 0.03	-0.131 ± 0.253	1.38 ± 0.03	1.56 ± 0.03	0.180	0.182
Lower cylinder	0.103 ± 0.03	0.131 ± 0.253	1.40 ± 0.03	1.56 ± 0.03	0.180	0.182
$G = 2D$ Upper cylinder	−0.23	–	1.45	–	–	–
Lower cylinder	0.30	–	1.44	–	–	–
$G = 1.5D$ Upper Cylinder	−0.43	−0.46	1.47	1.53	–	-
Lower cylinder	0.45	0.47	1.46	1.51	–	–

References

1. Wu YL (2017) Numerical simulation of flows past multiple cylinders using the hybrid local domain free discretization and immersed boundary method. In: Ocean engineering, vol 141, pp 477–492. ISSN 0029-8018
2. Berrone S, Garbero V, Marro M (2011) Numerical simulation of low-Reynolds number flows past rectangular cylinders based on adaptive finite element and finite volume methods. Comput Fluids 40:92–112
3. Roy A, Bandyopadhyay G (2006) A finite volume method for viscous incompressible flows using a consistent flux reconstruction scheme. Int J Numer Methods Fluids 52:297–319
4. Zdravkovich MM (1985) Flow induced oscillations of two interfering circular cylinders. J Sound Vib 101:511–521
5. Williamson CHK (1985) Evolution of a single wake behind a pair of bluff bodies. J Fluid Mech 159:1–18
6. Shu C, Liu NY, Chew YT (2007) A novel immersed boundary velocity correction-lattice Boltzmann method and its application to simulate flow past a circular cylinder. J Comput Phys 226:1607–1622
7. Harichandan A, Roy A (2010) Numerical investigation of low Reynolds number flow past two and three circular cylinders using unstructured grid CFR scheme. Int J Heat Fluid Flow 31:154–171
8. Hesam SM, Navid N (2011) Numerical simulation of flow over two side-by-side circular cylinders. J Hydrodyn B 23(6):792–805. ISSN 1001-6058
9. Xu SJ, Zhou Y, So RMC (2003) Reynolds number effects on flow structure behind two side-by-side cylinders. Phys Fluids 15(5):1214–1219
10. Ding H, Shu C, Yeo KS, Xu D (2007) Numerical simulation of flows around two circular cylinders by mesh-free least square-based finite difference methods. Int J Numer Meth Fluids 53:305–332

Multiphase Flow and Phase Change Material

Effects of Joules Heating and Soret Number on Double-Diffusive Mixed Convection Flow in an Enclosed Square Cavity

A. Satheesh, C. G. Mohan, P. Padmanathan, and Agarwal Anmol Pallav Kumar

Abstract In this paper, a double-diffusive mixed convection flow in an enclosed cavity with the effects of Soret number (Sr) and Joule's heating parameter (J) is presented. Finite volume approach-based differed quadratic upstream interpolation for convective kinematics (QUICK) scheme and central difference scheme are used to resolve the respective convective and diffusive fluxes. The results are presented in terms of temperature, concentration and stream line contours by varying the buoyancy ratio ($-100 \leq N \leq 100$), Soret number (Sr = 0 and 1) and Joules number ($J = 0$ and 1). The present results are validated with the reported literature. Results show that the temperature is uniformly distributed with the absence of Soret number and a bulk fluid diffusion takes place inside the cavity once the Soret effect is induced. The effects of Joules heating on average Nu and Sh are negligible for Sr = 0, and it has a significant variation with Soret number.

Keywords Double-diffusion · Mixed convection · Soret number · Joules heating · Buoyancy ratio

Nomenclature

Bo	Magnetic induction (Tesla)
c	Concentration (kg/m^3)
C	Dimensionless concentration
D	Mass diffusivity
Gr	Solutal Grashof number
Gr_T	Thermal Grashof number
Ha	Hartmann number
g	Gravitational acceleration (m/s^2)

A. Satheesh (✉) · C. G. Mohan · P. Padmanathan · A. A. P. Kumar
School of Mechanical Engineering, Vellore Institute of Technology, Vellore, Tamilnadu 632014, India
e-mail: satheesh.a@vit.ac.in

S. Revankar et al. (eds.), *Proceedings of International Conference on Thermofluids*, Lecture Notes in Mechanical Engineering,
https://doi.org/10.1007/978-981-15-7831-1_20

J	Joules heating parameter
k_T	Thermal conductivity (W/mK)
Le	Lewis number
N	Buoyancy ratio
Nu	Nusselt number
P	Dimensionless pressure
Pr	Prandtl number
Re	Reynolds number
Ri	Richardson number
Sc	Schmidt number
Sr	Soret number
Sh	Sherwood number
T	Dimensionless temperature
U, V	Dimensionless velocity components along x- and y-axes
X, Y	Dimensionless Cartesian coordinates

Greek Symbols

β	Expansion coefficient
θ	Dimensional temperature (K)
μ	Dynamic viscosity (N s/m^2)
ν	Kinematic viscosity (m^2/s)
ρ	Density (kg/m^3)
α	Thermal diffusivity (m^2/s)
σ_e	Electrical conductivity (S/m)

Subscripts

C	Cold, concentration
f	Fluid
H	Hot
T	Thermal
o	Initial

1 Introduction

Double-diffusion mixed convective flow with externally applied forces is the dual effect of temperature and concentration gradients in which heat and mass transfer occur simultaneously. In cavity studies, this is an important field of research which was extensively investigated because of various engineering applications in which they are often encountered, such as renewable energy sectors, applications in electronic systems, heat exchangers [1–6], Various studies and wide research on this topic can improve knowledge of fluid dynamics, thermodynamics and crystal growth in microgravity, and these results are applied in mass transport, including heat and energy transfer in Earth's interior, atmosphere and oceans sciences. Also, the Soret effect is a phenomenon in which mixtures of bulk moving particles where the different types of particle exhibit different responses to the force of a temperature gradient. It is observed from the literature that many investigations have been carried out in the past few decades.

Hyun and Lee [1] have numerically studied using a rectangular cavity with opposing ($N < 0$) and cooperating ($N > 0$) temperature and concentration gradients. The time-dependent transport equations are solved for predicting the nature of the flow field. Based on this work, many studies have been focused on the buoyancy flow investigations. The importance of Soret (Sr) and Dufour (Du) effects is gained attention and applied these concepts on the mixed, free and forced convection flows. A steady laminar boundary layer flow over a vertical flat plate is studied by Kafoussias and Williams [2]. Teamah and Maghlany [3] have investigated the double-diffusive mixed convection flow under the combined effects of diffusion. The local and average Nusselt and Sherwood numbers are reported for various factors. Results show that the heat and mass transfer rate is increased with decrease in Richardson number (Ri). Whereas, the mass transfer alone found to increase with Lewis number (Le). However, they did not study Sr and Du effects. Pal and Mondal [4] have consolidated impacts of Sr and Du on blended convective stream along an extending sheet surrounded in saturated permeable medium using shooting method. They found that the thermal gradient increases with increase in Sr and Du parameter. For higher values of Sr, there is an enhancement in the fixation dispersion with formation of concentration peak. Wang et al. [5, 6] have analysed the influence of Sr and Du on double-diffusive convection inside a horizontal enclosure. With decrease in aspect ratio, the oscillatory convection changes from periodic to steady state. Including Sr, the effects of Hall current and Joules heating have been presented by Reddy and Chamkha [7]. They investigated the magnetohydrodynamic stream of a Newtonian fluid in a vertical direct soaked permeable medium in nearness of Hall current, Joule heating and Soret impact. It is observed that there is no change in the effects on velocity and temperature when compared to previously mentioned research works, but the effect of Soret number on concentration is noticeable. Furthermore, the fluid layer friction makes on account of a resistive constrain called as Lorentz force. Hence, with increase in magnetic parameter, the concentration decreases and the temperature increases. By incorporating Joule heating effect, the flow velocities, concentration and temperature profiles are increased. LBM is used for numerical simulation of double-diffusive convection inside a vertical enclosure in presence of Soret and Dufour effect by Ren and Chan [8]. The average Nu and Sh are increased with higher Ra (10^5), Pr (0.7, 1.0), Le (2.0, 30), Sr (0.1) and Du. However, the average values of Nu and Sh decrease with increase in the aspect ratio. Nithyadevi and Yang

[9] have studied the Soret and Dufour effects, these effects are more important when dual diffusion gradients are large. Recently, Mohan and Satheesh [10] have studied the effect of Soret number in an enclosed, steady, incompressible square cavity. They concluded that the influence of Sr on thermal diffusion is insignificant for low Ri. Also, it is found to increase significantly with higher Ri. It is observed from the literature that there are many research work was reported by considering only the effects of Soret and Dufour numbers on double-diffusive mixed convection flows [2, 4–6, 9–15] and the effect of Soret number in magnetic fluids flows [7, 14, 16, 17]. However, very limited work was proposed by considering the Joules heating parameter [18, 19]. Hence, in the present study, the above two important and significant parameters, such as, Soret number and Joules heating parameter are used to investigate the double-diffusive mixed convection flow in an enclosed square cavity with different buoyancy ratios.

2 Simple Model Including Quantity Demand

The steady state, 2D, incompressible flow in a four-sided cavity are shown in Fig. 1. Left vertical wall is maintained with high temperature (T_{H}) and concentration (C_{H}), and the right vertical wall is low temperature (T_{C}) and concentration (C_{C}). Both the top and bottom walls are adiabatic. The top wall is moving with a uniform velocity (U) in positive X-direction, and other walls remain stationary.

The fluid properties except density remain constant throughout the study. The density is calculated from Boussinesq approximation as given in the following Eq. (1).

$$\rho = \rho_o[1 - \beta_{\mathrm{T}}(\theta - \theta_{\mathrm{c}}) - \beta_{\mathrm{C}}(c - c_{\mathrm{c}})]. \tag{1}$$

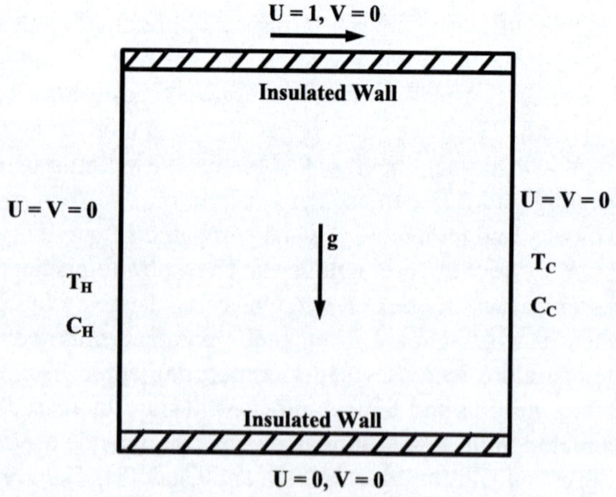

Fig. 1 Schematic diagram of the problem

where

$$\beta_T = -\frac{1}{\rho_o}\left(\frac{\partial \rho}{\partial \theta}\right)_{P,c} \quad \beta_C = -\frac{1}{\rho_o}\left(\frac{\partial \rho}{\partial c}\right)_{P,\theta} \tag{2}$$

After invoking Boussinesq approximation in the governing equation and incorporating the Soret and Joule's heating parameters with considering negligible viscous dissipation, the following set of non-dimensional equations (Eqs. 3–7) are obtained.

$$\frac{\partial U}{\partial X} + \frac{\partial V}{\partial Y} = 0 \tag{3}$$

$$U\frac{\partial U}{\partial X} + V\frac{\partial U}{\partial Y} = -\frac{\partial P}{\partial X} + \frac{1}{\mathrm{Re}}\left(\frac{\partial^2 U}{\partial X^2} + \frac{\partial^2 U}{\partial Y^2}\right) \tag{4}$$

$$U\frac{\partial V}{\partial X} + V\frac{\partial V}{\partial Y} = -\frac{\partial P}{\partial Y} + \frac{1}{\mathrm{Re}}\left(\frac{\partial^2 V}{\partial X^2} + \frac{\partial^2 V}{\partial Y^2}\right) + \mathrm{Ri}(T + \mathrm{NC}) + \frac{\mathrm{Ha}^2}{\mathrm{Re}}V \tag{5}$$

$$U\frac{\partial T}{\partial X} + V\frac{\partial T}{\partial Y} = \frac{1}{\mathrm{RePr}}\left(\frac{\partial^2 T}{\partial X^2} + \frac{\partial^2 T}{\partial Y^2}\right) + \mathrm{JV}^2 \tag{6}$$

$$U\frac{\partial C}{\partial X} + V\frac{\partial C}{\partial Y} = \frac{1}{\mathrm{ReSc}}\left(\frac{\partial^2 C}{\partial X^2} + \frac{\partial^2 C}{\partial Y^2}\right) + \frac{Sr}{Re}\left(\frac{\partial^2 T}{\partial X^2} + \frac{\partial^2 T}{\partial Y^2}\right) \tag{7}$$

The local and average Nu number and Sh number of the hot wall of the cavity is determined by the following Eqs. (8) and (9).

$$\mathrm{Nu} = -\left(\frac{\partial T}{\partial x}\right)_{x=0} \quad \mathrm{Nu_{avg}} = -\frac{1}{L}\int_0^L \mathrm{Nu}\, dy \tag{8}$$

$$\mathrm{Sh} = -\left(\frac{\partial C}{\partial x}\right)_{x=0} \quad \mathrm{Sh_{avg}} = -\frac{1}{L}\int_0^L \mathrm{Sh}\, dy \tag{9}$$

3 Methodology

A double-diffusive mixed convection flow in an enclosed square cavity with the effects of Soret number (Sr) and Joule's heating parameter (J) is presented in this paper. FVM-based SIMPLE algorithm is used to obtain numerical solutions. The discretized equations are solved using line-by-line procedure, combining the tri-diagonal matrix algorithm (TDMA). The iterative procedure is repeated until the convergence criteria of the relative error between the successive iterations are fixed as 10^{-10}. The convergence criterion less than 10^{-10} does not cause any appreciable change in the final results.

4 Results and Discussion

The grid independent study for mid-plane horizontal velocity is shown in Fig. 2. Different grids sizes of 51 × 51, 81 × 81, 101 × 101 and 121 × 121 are investigated. It is found that there is no considerable change in velocity after increasing the grid size from 101 × 101. Hence, analysis is carried out using 101 × 101 grid sizes for further investigations.

In Fig. 3, the temperature, concentration and stream line contours from Teamah and Maghlany [3] are validated with the present study. Also, a comparison of average Nusselt number for Le = 1.0; Ri = 0.01 and Re = 100 with Al-Amiri et al. [20] is presented in Table 1. It is observed from these numerical results that the present FVM code has good agreement with the reported literature.

Figure 4 shows the different contours for Le = 1.0, Sr = 1.0 without considering J. When $N = -100$, the temperature is extremely high at the bottom left corner of the wall and keeps decreasing steadily and is lowest at the top right corner of the cavity. The temperature is seen to be reasonable towards the centre of the cavity and has value around 0.55. As it move towards the top right corner of the cavity, the temperature reaches extremely low as 0.1. For increase in buoyancy ratio, it can be observed that the heat is being transferred to the top left corner of the cavity, whereas the temperature is seen to be minimum at the bottom at right corner of the cavity for $N = 0$. For positive buoyancy ratio values, the temperature is highest at the top left

Fig. 2 Grid independent test

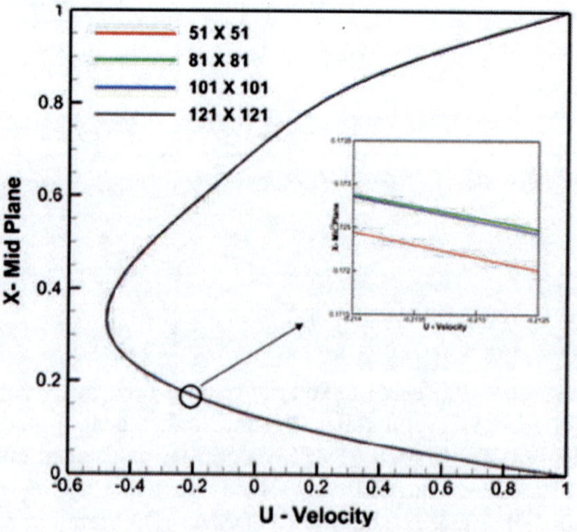

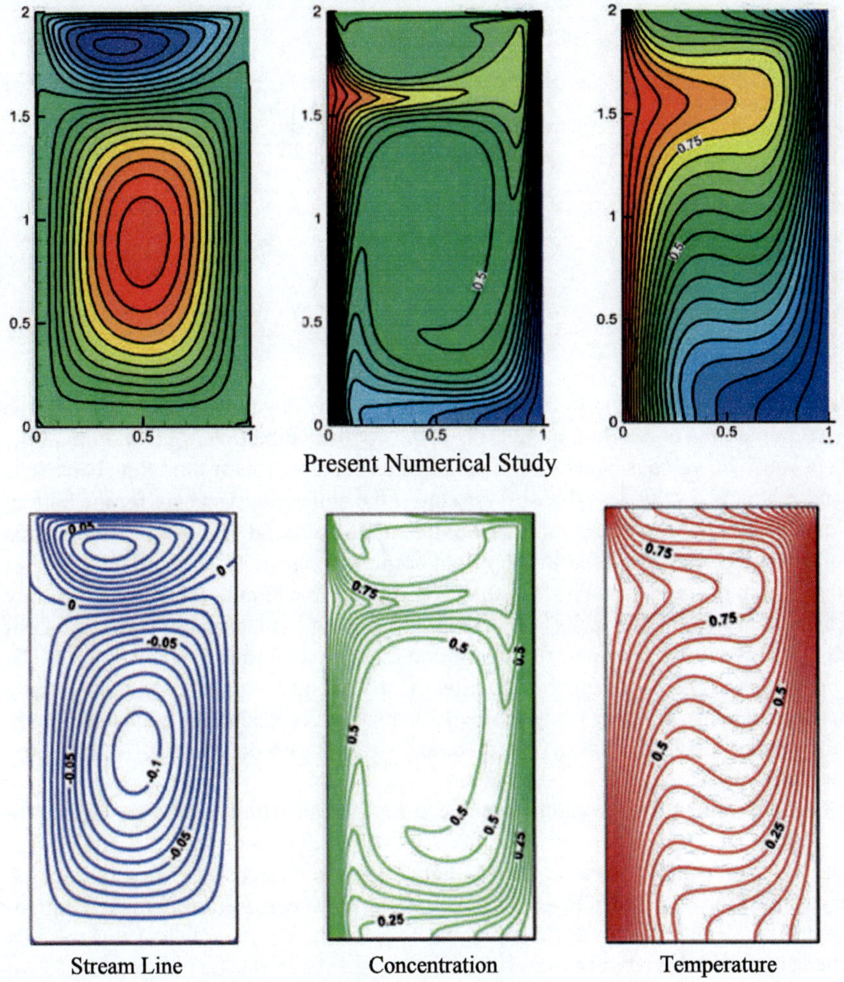

Fig. 3 Validation with Teamah and Maghlany (2010) work

corner of the cavity and the temperature is lower towards the right bottom corner of the cavity. However, the temperature is neither maximum nor minimum towards the middle of the cavity is about 0.5. For $N = -100$, the concentration of fluid particles is maximum and greater than 1.0 (around 1.2) towards the bottom of the cavity and keeps on gradually decreasing towards the cavity. For $N = 0$, the concentration of fluid particles shift to the left side of the cavity and is maximum on the left and is seen to be minimum at the right side of the cavity.

The concentration of the fluid particles is around 0.4 towards the middle of the cavity. For $N = 50$ and 100, the concentration is maximum at the top end of the cavity and is greater than unity, whereas it keeps decreasing on moving downwards

Table 1 Boundary conditions applied for present numerical study

At top wall	$U = 1.0,\ V = 0$	$\frac{\partial T}{\partial y} = \frac{\partial C}{\partial y} = 0$	
At bottom wall	$U = V = 0$	$\frac{\partial T}{\partial y} = \frac{\partial C}{\partial y} = 0$	
At left wall	$U = V = 0$		$T = T_H$ and $C = C_H$
At right wall	$U = V = 0$		$T = T_C$ and $C = C_C$

and is less than 0, towards the bottom of the cavity. For $N = -100$ and -50, the streamline flow of the fluid particles is in the opposite direction of the motion of the top wall. The value is negative which indicates the direction of fluid flow behaviour and is having a value less than 1.0 for almost the entire cavity except for the bottom left part. For $N = -50$, the value of the streamline is around 0.11 towards the middle of the cavity and is 0.01 at the top right corner of the cavity and around -0.05 at the bottom of the cavity. For $N = 0$, the fluid flows in a circular direction uniformly distributed with in the cavity, but it keeps on varying in magnitude, however, and it is 0.06 towards the centre of the cavity, and around 0.005 towards the bottom of the cavity. For $N = 50$ and 100, the direction of the fluid flow is positive and thus having a magnitude greater than 1.0. For $N = 100$, the value is around 0.1 and is maximum at the right top of the cavity, i.e. 0.18, and again it is reduced to 0.1 towards the bottom of the cavity.

Figure 5 shows the variation of temperature, concentration and stream line for Le = 1.0, Sr = 1.0 with considering J. For $N = -100$, the temperature is maximum at the bottom left corner of the cavity and gradually decreases on moving towards the right side. For $N = -50$, the temperature is gradually decreased from the left bottom corner of the cavity towards the top layer of the cavity. For $N = 0$, the value towards the centre of the cavity becomes 0.5 while it is slowly started decreasing to 0.25 on moving downwards and then to 0.05 towards the right bottom of the cavity. For $N =$ 50 and 100, the temperature is maximum at the top left corner of the cavity, whereas it is lowest towards the bottom right corner of the cavity and its value is around 0.1 as seen in the contours and towards the lower side of the cavity its value is around 0.35 and increases on moving up the cavity. On observing the concentration behaviour for $N = -100$ and -50, the concentration of the fluid is maximum and its value is more than 1.0 towards the bottom of the cavity, whereas its value is around 0.6 towards the centre of the cavity and decreases on moving up further. For $N = 0$, the concentration of fluid particles is maximum towards the left side of the cavity and low towards the right side. For $N = 50$ and 100, the concentration is highest at the top and least at the bottom side of the cavity. For streamline contours at $N = -100$ and -50, the velocity of the fluid is maximum towards the topside and it keeps decreasing on moving down the cavity. For $N = 0$, the fluid flows in a circular direction throughout the cavity and

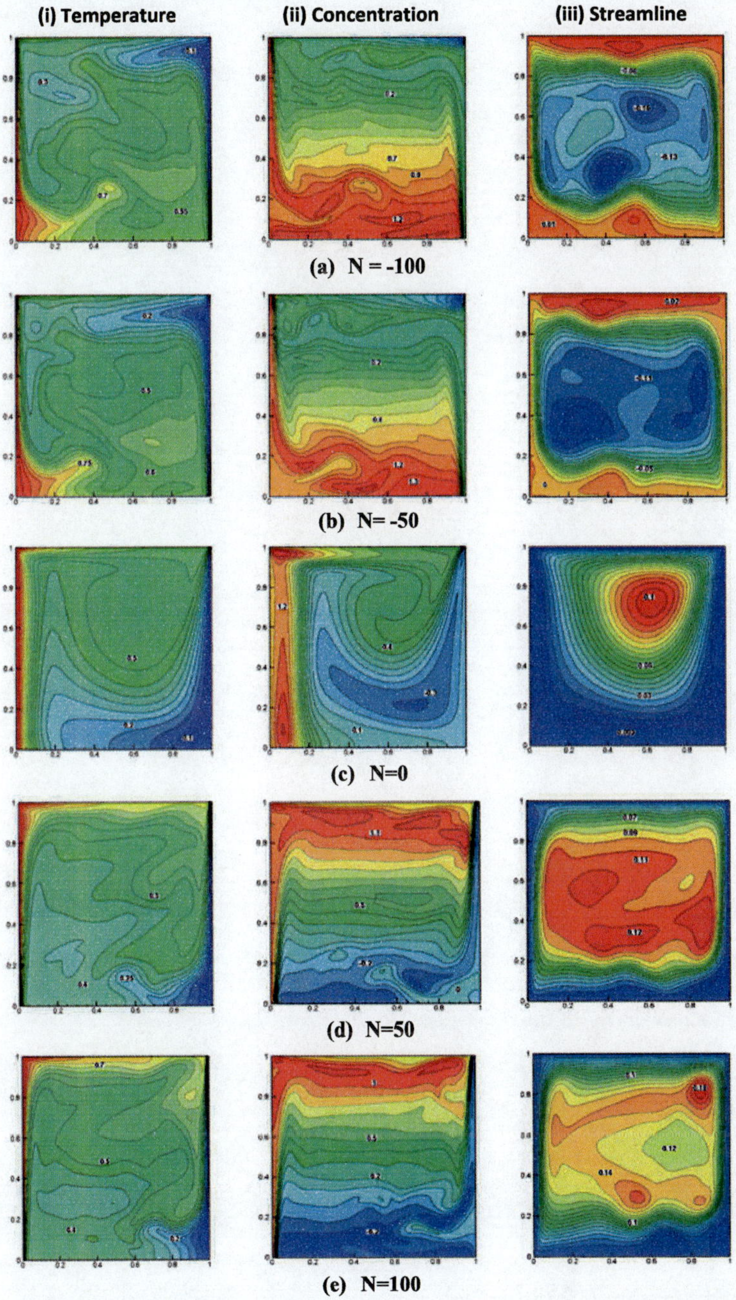

Fig. 4 Effect of buoyancy ratio on (i) temperature, (ii) concentration and (iii) stream line contours for Le = 1.0, Sr = 1.0 and J = 0.0

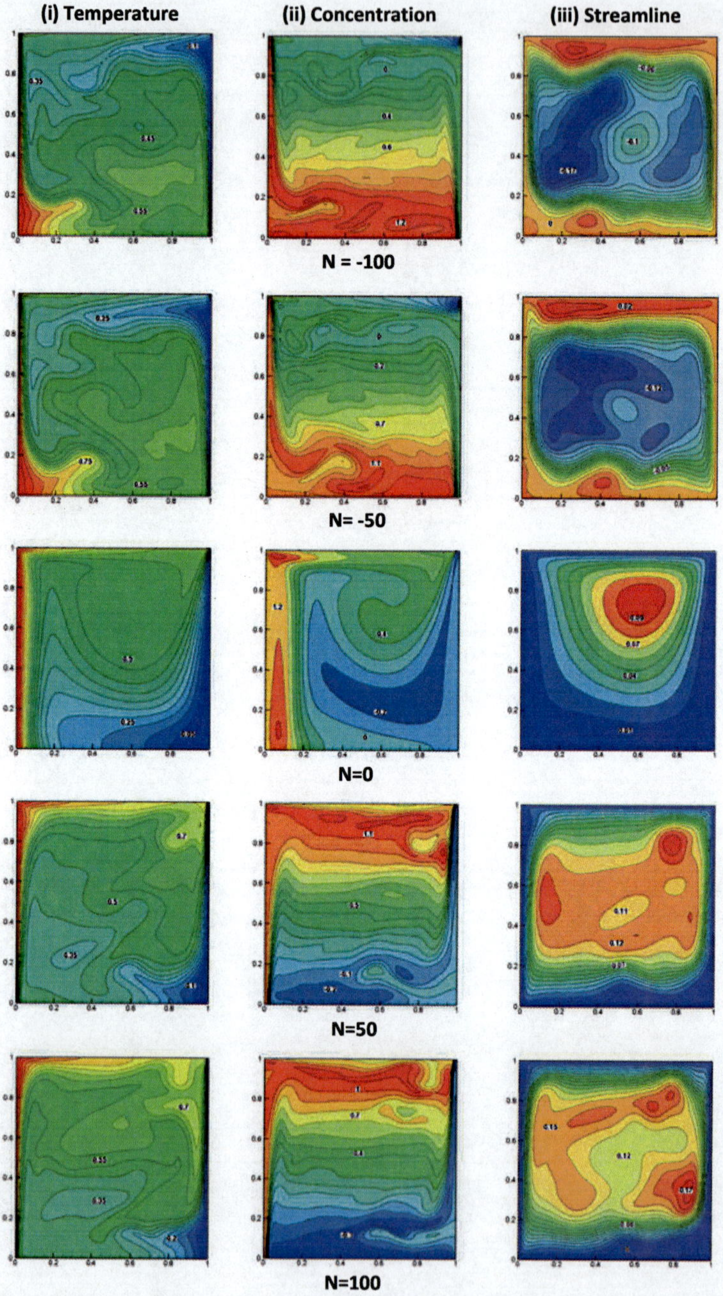

Fig. 5 Effect of buoyancy ratio on (i) temperature, (ii) concentration and (iii) stream line contours for Le = 1.0, Sr = 1.0 and J = 1.0

its value is 0.09 towards the upper part and decreases along the walls of the cavity and is 0.04 towards the centre of the cavity and 0.01 at lower side of the cavity. For $N = 50$ and 100, the velocity towards the middle of the cavity and multiple loops have been formed each having different velocity and streamline values.

The variation in temperature is observed in the graph at horizontal mid-plane for values of different buoyancy ratios (as shown in Fig. 6). It is observed for the buoyancy ratio values, i.e. $N = -100$, $N = -50$ and $N = 0$, the temperature starts decreasing on moving from left to right as we keep on gradually increasing the distance from horizontal mid-plane and opposite behaviour is seen for the positive values of buoyancy ratios, i.e. when $N = 50$ and $N = 100$. As we introduce the Soret effect, for non-zero buoyancy ratios, the linearity of the curve changes to the abnormal zig-zag curves when Soret effect is present. The ups and downs in the curve indicate [sudden rise and fall of temperature or] temperature variation and it suggests

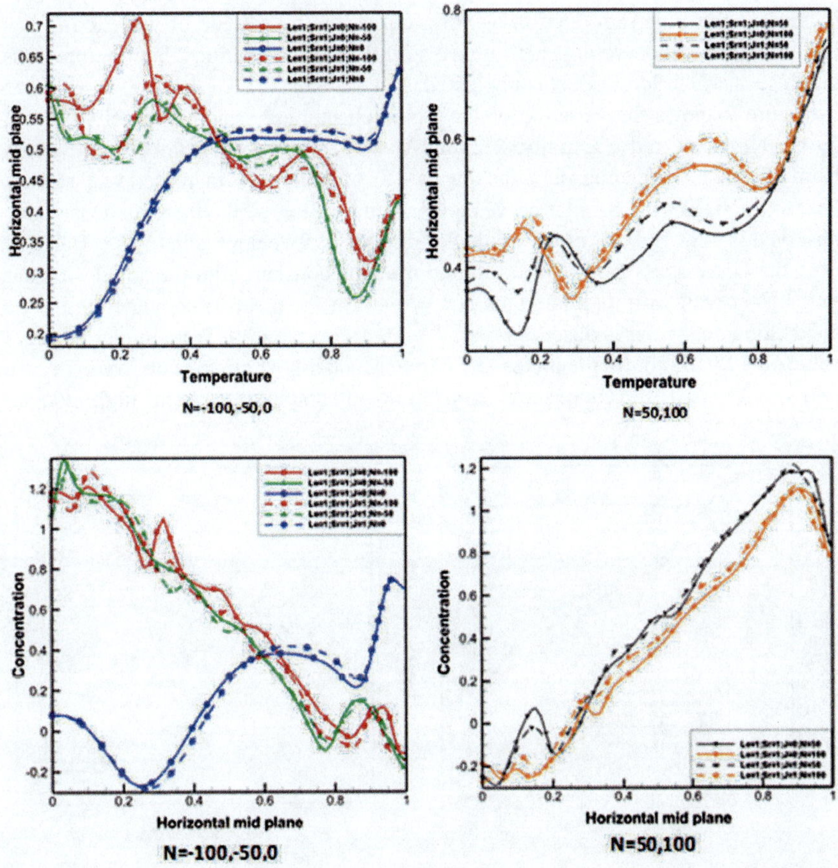

Fig. 6 Mid-plane temperature and concentration variations with Joules heating and Soret effect

that the nonlinearity arises on introduction of Soret effect only. For the negative value of buoyancy ratio ($N = -100$ and $N = -50$), the concentration decreases on moving from left to right in a downward curve (0.85 –.2) while for the positive buoyancy values, i.e. ($N = 50,100$) the concentration increases on moving right to left (0–0.8). The graph for $N = 0$ with Joules heating is overlapping with $N = -50$ curve which also satisfy the argument that the Joules effect is working against the buoyancy effect and it is seen by curve for $N = 0$; $J = 1$ which completely overlaps with each other. When the Joules effect is induced in the medium there is not a major change in the negative buoyancy values as they overlap with the previous values (without Joules heating).

A minor change is observed in the value of $N = 50$ where the final concentration decreases near the wall as compared to the case when the absence of Joules heating is considered. When the Soret effect is also induced in the graph then we observe that the curves become zig-zag from linear shape previously seen. Now, this movement of fluid particles is very high for $N = -100$ and decreases as the buoyancy ratio is increased. With the introduction of Soret effect, the magnitude of vibration is just increased and the curve appears to have slightly more deviations than the previous graph seen when Soret effect is neglected.

Figure 7 shows the variation of local Nusselt number which is largely affected by the buoyancy ratio. Initially, keeping $N = -100$, Nusselt number value starts from -3 and starts decreasing to as low as -38 and then gets increased very rapidly near to the right wall. Similar curve is observed for $N = -50$, where the least value observed is -28 and this effect gradually diminish as we approach to $N = 0$. For $N = 0$, the curve starts from Nu $= -6$ and remains constant with the small dip near to the right wall with its final value Nu $= -4$. For the positive buoyancy ratio, the contrasting behaviour is observed. For $N = 50$, the curve starts from higher negative value of -28 and shows the increasing curve. Much higher effect is observed in case of $N = 100$, where the curve starts from Nu $= -36$ and increase to the highest value

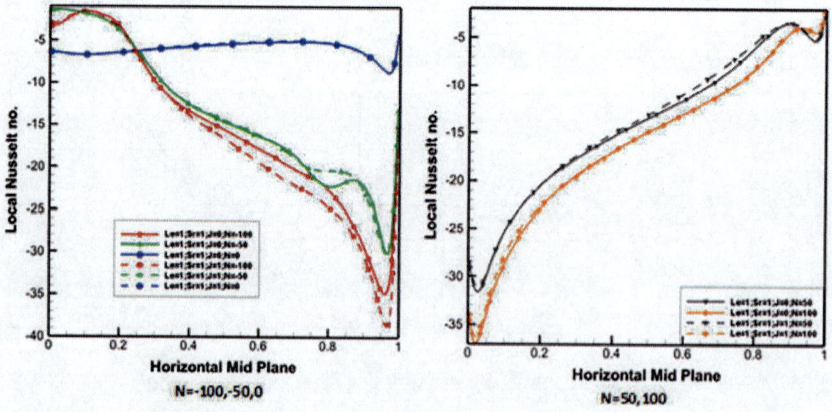

Fig. 7 Local Nusselt numbers for at different buoyancy ratio

Table 2 Comparison of average Nusselt number for present numerical results with Al-Amiri et al. [20] at Le = 1; Ri = 0.01 and Re = 100

Buoyancy ratio (N)	Nu_{avg}		
	Present results	Al-Amiri et al. [20] results	Percentage of error (%)
−100	1.4974	1.5022	0.32
−50	1.7241	1.7350	0.63
0	2.2703	2.2764	0.27
1	2.2825	2.2859	0.15
50	2.7894	2.7998	0.37
100	3.0667	3.0732	0.21

of −3. For both the graph, Joules heating has the negligible effect as it can be seen with the overlapping graphs. On the introduction of Soret effect, a sudden bump is observed near to the left wall in the case of negative buoyancy ratio and near the right wall in the case of positive buoyancy ratio. The curve for $N = 0$ has seen no change irrespective of either Joules heating or Soret effect, thus concluding that the Nusselt number depends only on the buoyancy ratio and is independent of Joules and Soret effect.

The average Sherwood number in this case comes to 11.8449 as observed from the tabular data obtained. For $N = -100$, the average value is 29.5294, whereas, for $N = -50$, the average value is 24.694. On checking the positive buoyancy ratio values, i.e. for $N = 50$ and $N = 100$, contrary effect is observed as the curve starts moving from left to right. On inducing Soret effect, there is a drastic change seen in the behaviour of Sherwood number value with change in distances from the horizontal mid-plane. The curve usually shows a high value for average Sherwood number when the distance from horizontal mid-plane is less, whereas the value drops as distance from the horizontal mid-plane gradually increases as can be observed from the graphs.

The average Nusselt and Sherwood number are obtained for two conditions with and without Soret effect and it was found that there was a significant effect of Joules heating only in the presence of Soret effect (listed in Table 2). Without Soret effect, the maximum deviation obtained was 0.85% for $N = 0$ which is insignificant in comparison with 6.9% deviation in presence of Soret effect. Same is the case in average Sherwood number. The deviation obtained in without Soret effect case was insignificant with the deviation of 0.85% as compared with 12.20% deviation in presence of Soret effect (Table 3).

5 Conclusion

A double-diffusive mixed convection flow in a square cavity is considered in the present study to investigate the heat and fluid flow behaviour with the effects of

Table 3 Variation of average Nu and Sh for different N with and without considering Sr and J

Buoyancy ratio (N)	Average Nu				Average Sh			
	Sr = 0		Sr = 1.0		Sr = 0		Sr = 1.0	
	$J = 0$	$J = 1$	$J = 0$	$J = 1$	$J = 0$	$J = 1$	$J = 0$	$J = 1$
−100	14.7647	14.7649	15.0130	16.0515	29.5294	29.5298	8.9316	7.8425
−50	12.3472	12.3453	13.5594	13.3895	24.6944	24.6907	6.3295	6.1643
0	5.9224	5.8720	5.9225	5.8720	11.8449	11.7440	−0.3240	-0.4300
50	12.8320	12.8308	14.0738	13.7270	25.6640	25.6616	11.1081	11.3861
100	15.2746	15.2752	15.9246	15.7775	30.5492	30.5503	16.3675	16.2827

Soret number (Sr) and Joule's heating parameter (J) using finite volume method. The present results are compared with the reported literatures and found to be in good agreement. The following results were obtained from the present work,

In the absence of Soret number with the negative buoyancy there exist two types of motion, aiding and non-aiding.

The fluid velocity near the top wall remains higher for negative buoyancy with both particle motions in clockwise and anticlockwise direction. While for positive buoyancy, there exists just clockwise motion with the high velocity fluid in the centre of cavity.

As the Joules effect is insignificant without Soret effect, it has a marginal variation on introduction of Soret effect. In the absence of Joules effect, the bulk fluid is moving in upward direction and propel towards right side with very high frequency, while this magnitude decrease by 20% on introduction of Joules effect. The temperature is linearly varying in absence of Soret effect. Once Soret effect is induced in the cavity, diffusion of particles takes place as it is also seen by the zig-zag curve obtained. But the diffusion is absent when buoyancy is kept zero. For local Sherwood number, along with the diffusion of the particles, Soret effect is seen to create an additional effect across the horizontal mid-plane. The local Sherwood number begins with negative value for absence of Soret effect and remains constant in mid-cavity with a dip near the right wall, while on instigating the Soret effect totally contrasting effect is observed. The average Sh is nearly double the Nusselt number for Sr = 0. However, for Sr = 1, the average Nu is found to be increased with marginal value. But Sherwood is decreased by half to one-third.

References

1. Hyun JM, Lee JW (1990) Double-diffusive convection in a rectangle with cooperating horizontal gradients of temperature and concentration. Int J Heat Mass Transfer 33:1605–1617
2. Kafoussias NG, Williams EW (1995) Thermal-diffusion and diffusion-thermo effect on mixed-forced convective and mass transfer boundary layer flow with temperature dependent viscosity. Int J Eng Sci 33(9):1369–1384

3. Teamah MA, El-Maghlany WM (2010) Numerical simulation of double-diffusive mixed convective flow in rectangular enclosure with insulated moving lid. Int J Thermal Sci 49:1625–1638
4. Pal D, Mondal H (2012) Influence of chemical reaction and thermal radiation on mixed convection heat and mass transfer over a stretching sheet in Darcian porous medium with Soret and Dufour effects. Energy Convers Manage 62:102–108
5. Wang J, Mo Y, Zhang Y (2014) Onset of double-diffusive convection in horizontal cavity with Soret and Dufour effects. Int J Heat Mass Transfer 78:1023–1031
6. Wang J, Yang M, Zhang HLY (2016) Oscillatory double-diffusive convection in a horizontal cavity with Soret and Dufour effects. Int J Thermal Sci 106:57–69
7. Reddy S, Chamkha AJ, Soret and Dufour effects on MHD convective flow of Al2O3–water and TiO2–water nanofluids past a stretching sheet in porous media with heat generation/absorption. Adv Powder Technol 27:1207–1218
8. Ren Q, Chan CL (2015) Transient double-diffusive convection in a vertical cavity with Soret and Dufour effects by Lattice Boltzmann Method on CUDA platform. Heat Transfer Thermal Eng V08AT10A025
9. Mohan CG, Satheesh A (2017) Computational investigation of double diffusive mixed convective flow in an enclosed square cavity with Soret effect. Front Heat Mass Transfer 8:1–13
10. Mansour A, Amahmid M, Hasnaoui M (2004) Bourich, Soret effect on double-diffusive multiple solutions in a square porous cavity subject to cross gradients of temperature and concentration. Int Comm Heat Mass Transfer 31:413–440
11. Nithyadevi N, Yang RJ (2009) Double diffusive natural convection in a partially heated enclosure with Soret and Dufour effects. Int J Heat Fluid Flow 30:902–910
12. Khadiri A, Amahmid M, Hasnaoui A (2010) Rtibi, Soret effect on double-diffusive convection in a square porous cavity heated and salted from below. Numer. Heat Transfer Part A 57:848–868
13. Bhuvaneswari M, Sivasankaran S, Kim YJ (2011) Numerical study on double diffusive mixed convection with a Soret effect in a two-sided lid-driven cavity. Numer Heat Transfer Part a 59:543–560
14. Hayat T, Abbasi FM, Al-Yami M, Monaquel S (2014) Slip and Joule heating effects in mixed convection peristaltic transport of nano-fluid with Soret and Dufour effects. J Molecular Liquids 194:93–99
15. Roy K, Murthy PVSN (2015) Soret effect on the double diffusive convection instability due to viscous dissipation in a horizontal porous channel. Int J Heat Mass Transfer 91:700–710
16. Jha AK, Choudhari K, Sharma A (2014) Influence of Soret effect on MHD mixed convection flow of visco-elastic fluid past a vertical surface with Hall effect. Int J Appl Mech Eng 19:79–95
17. Hayat T, Iqbal R, Tanveer A, Alsaedi A (2016) Soret and Dufour effects in MHD peristalsis of pseudo-plastic nano-fluid with chemical reaction. J Molecular Liquids 220:693–706
18. Rahman MM, Alim MA, Sarker MMA (2010) Numerical study on the conjugate effect of joule heating and magneto-hydrodynamics mixed convection in an obstructed lid-driven square cavity. Int Commun Heat Mass Transfer 37:524–534
19. Rahman MM, Saidur R, Rahim NA (2011) Conjugate effect of joule heating and magneto-hydrodynamic on double-diffusive mixed convection in a horizontal channel with an open cavity. Int J Heat Mass Transfer 54:3201–3213
20. Al-Amiri AM, Khanafer I (2007) Pop, Numerical simulation of combined thermal and mass transport in a square lid-driven cavity. Int J Thermal Sci 46:662–671

Dimethyl Adipate-Based Microencapsulated Phase Change Material with Silica Shell for Cool Thermal Energy Storage

Vedanth Narayan Kuchibhotla, G. V. N. Trivedi, and R. Parameshwaran

Abstract Phase change materials (PCM) have the ability to store and release thermal energy. Encapsulation of these energy storage materials overcomes the difficulties that can enable them for a broad range of applications. In the present study, microencapsulation of dimethyl adipate into silica shell was carried through interfacial hydrolysis and polycondensation method. The prepared microencapsulated phase change materials (MPCM) were characterised using a field emission scanning electron microscope, have shown good sphericity with an average particle size of 596 nm. The chemical structure of MPCM obtained using Fourier transform infrared spectroscopy has exhibited good chemical stability between shell and core materials. Latent heat of enthalpy measured using differential scanning calorimetry was around 24 kJ/kg with onset melting and end set melting as 7.33 °C and 11.97 °C, respectively. Furthermore, thermo-gravimetric analysis studies have shown that MPCM exhibited end set temperatures as 180 °C. Due to the inorganic shell coating over the PCM droplets, MPCM has shown an increase in thermal stability. These properties make MPCM as a viable candidate for cool thermal energy storage applications.

Keywords Thermal energy storage · Phase change material · Microencapsulation

1 Introduction

In recent years, global energy demand has increased rapidly due to economic, population growth, and industrial developments. With the incessant depletion of fossil fuel reserves, renewable energy sources have attracted significant research attention from the past decade. Besides, there is also a need in the development of energy storage systems and technologies to improve energy utilisation and reduce energy wastage by maintaining a balance between energy supply and demand.

PCMs are latent thermal energy storage materials that can absorb and release the thermal energy by virtue of its phase transition from solid-to-liquid or vice versa.

V. N. Kuchibhotla · G. V. N. Trivedi · R. Parameshwaran (✉)
Department of Mechanical Engineering, BITS Pilani, Hyderabad 500078, India
e-mail: parameshwaranr@hyderabad.bits-pilani.ac.in

S. Revankar et al. (eds.), *Proceedings of International Conference on Thermofluids*, Lecture Notes in Mechanical Engineering,
https://doi.org/10.1007/978-981-15-7831-1_21

Depending on the chemical nature, these materials are classified as organic- and inorganic-based PCMs. The organic-based PCMs show merits such as concurrent phase change behaviour, high-energy storage density, no phase segregation, non-toxic, and non-corrosive, chemically stable and excellent reliability.

However, utilisation of these materials was hindered by limitations such as incompatibility with surroundings, leakage during the phase transitions. Therefore, to overcome these drawbacks, encapsulation of PCM into organic or inorganic shell materials is a viable approach. This approach would increase the heat transfer rate, prevent degradation during phase transitions, control volume changes during phase change, and reduce chemical reactivity with surroundings [1].

Many researchers have developed various methods for the microencapsulation of the organic-based PCMs into polymer materials, such as urea–formaldehyde [2], melamine–formaldehyde [3], PMMA [4], polystyrene [5]. Furthermore, there were also studies in the utilisation of inorganic-based shell materials as silica [6], $CaCO_3$ [7], TiO_2 [8] because of their excellent thermal conductivity, superior thermal stability compared with polymeric shell materials.

Among the available organic-based PCMs, fatty acid esters have attracted considerable attention in their application for thermal energy storage at low temperatures. These ester-based compounds due to the presence of carbon chains of fatty acids and carboxylic groups are chemically stable, show high phase change enthalpy per unit mass and exhibit small volume changes during phase transition [9]. However, in order to overcome the limitations of these materials during applications, encapsulation of PCM was an intended approach.

Therefore, in this current study, microencapsulation of the dimethyl adipate an organic ester with silica as a shell was carried using the interfacial polycondensation method. The prepared microcapsules were characterised using a field emission scanning electron microscope (FESEM) to understand the surface morphology. The X-ray photoelectron spectroscopy (XPS) was employed to analyse elemental composition microcapsules. Chemical structure and chemical stability studies of the microcapsules were studied using Fourier transform infrared (FTIR) spectroscopy. Phase change characteristics and thermal stability of prepared microcapsules were investigated using differential scanning calorimetry (DSC) and thermo-gravimetric analysis (TGA) techniques, respectively. The obtained results are presented and discussed in detail.

2 Experimental

2.1 *Materials*

Dimethyl adipate (DMA), tetraethyl orthosilicate (TEOS) having a purity more than 99% were purchased from Alfa Aesar and were used as PCM and shell precursor, respectively. The cetrimonium bromide obtained from Sigma Aldrich was used as an emulsifier. Ammonium hydroxide solution (25%) obtained from Finar Limited was utilised as a catalyst. The deionised distilled water and ethanol were used as solvents. All the obtained chemicals were utilised without any further purification as per material safety data sheet provided by the supplier.

2.2 Synthesis of MPCM

Microencapsulation of the DMA with silica as a shell was carried using the interfacial polycondensation method as illustrated in Fig. 1. DMA of 10 ml mixed with 50 ml of deionised distilled water, the mixture was agitated for 45 min at a temperature of 60 °C using a magnetic stirrer. To this mixture, 0.1 g of cetrimonium bromide was added and stirring continued for another 30 min for the formation of stable oil in water (O/W) emulsion.

Simultaneously in another beaker, 10 ml TEOS and 80 ml ethanol are mixed and gently stirred at a temperature of 50 °C for 45 min for the preparation in shell precursor solution. The obtained shell precursor solution was added gradually into the O/W emulsion. Under continuous stirring, 5 ml ammonium hydroxide solution was added to the mixture to initiate the condensation reaction. The mixture was further stirred continuously for 2 h for complete condensation. The resultant mixture turns into a milky white precipitate. Further, the precipitate was oven-dried at the temperature of 70 °C for 8 h to remove the solvents. The resultant microcapsules in the form of white powder were collected for further characterisation.

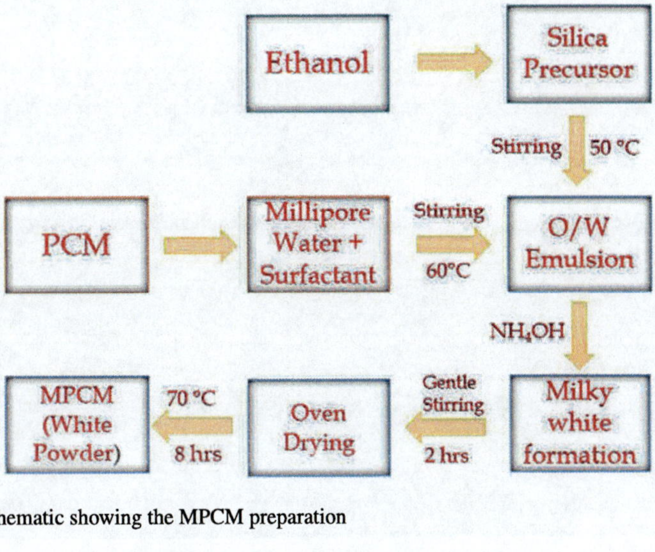

Fig. 1 Schematic showing the MPCM preparation

2.3 *Characterisation Methods*

Surface morphology of the prepared microcapsules was observed using FESEM (APERO, S) at 20 kV-accelerated voltage under high vacuum mode. The elemental composition of MPCM was analysed using X-ray photoelectron spectrometer (VG ESCALAB 250). The chemical structure of microcapsules was studied using FTIR spectrometer (JASCO, FT/IR-4200) recording the IR frequency over a range of 4000–400 cm^{-1}. Phase change properties are investigated using DSC apparatus (DSC-60, SHIMADZU) carried under N_2 atmosphere at 5 °C/min in a temperature range of −20 to 20 °C. Thermal stability of the MPCM was studied using TGA apparatus (DTG-60 SHIMADZU), under N_2 atmosphere, ramping at 10 °C/min from 30 to 600 °C.

3 Results and Discussions

3.1 *Surface Morphology and Particle Size*

The surface morphology of the MPCM observed using FESEM was presented in Fig. 2. The images reveal that morphology of microcapsules was near-spherical with

Fig. 2 Surface morphology observed from FESEM

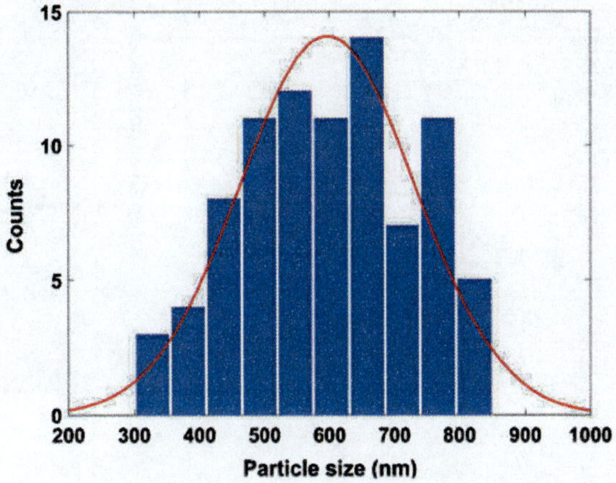

Fig. 3 Average particle size distribution of MPCM

good sphericity. However, the surfaces of the microcapsules are little coarse and are adhered to each other similar to the findings observed in [10]. Due to the mismatch of the condensation rate with the diffusion of TEOS, the interfusion of silica oligomers into the aqueous solution occurs. The aqueous interfused silica oligomers do not contribute to the shell formation over the PCM droplet, and thus, poly condensate themselves results in the formation of silica nanoparticles. These silica nanoparticles get deposited on the polycondensed shell surfaces resulting in the formation of the little coarse surface [10].

The particle size of the microcapsules plays a crucial role in functionality. The smaller the particle increases the surface to volume ratio and structural stability of the microcapsules. The particle size distribution of the prepared microcapsules is obtained from the FESEM images using ImageJ software [11]. The obtained results are illustrated in Fig. 3. The average particle size obtained through controlled reaction conditions was around 596 nm.

3.2 *Elemental Analysis*

The elemental composition of the MPCM was analysed using XPS, and obtained spectra were presented in Fig. 4.

The prepared MPCM has shown around 56 atomic wt% of oxygen, 26 atomic wt% of silica, and 16 atomic wt% of carbon. The presence of Si2p peak at 103.5 eV and O1s peak at 531.5 eV has confirmed the polycondensation of the silica precursors and successful formation of SiO_2 shell over the PCM droplets [12].

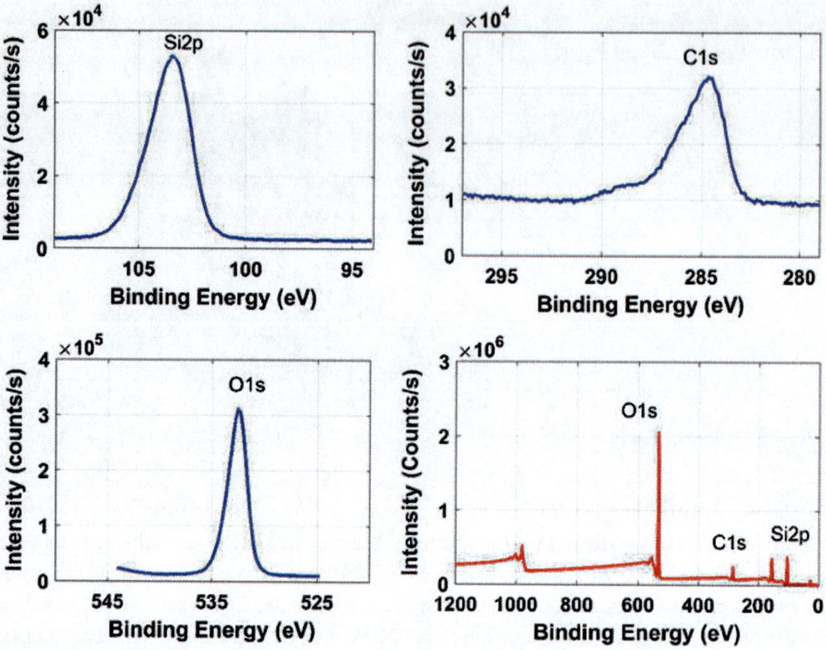

Fig. 4 XPS spectra of the MPCM

3.3 Chemical Structure of MPCM

The chemical structure of the PCM, shell material, and MPCM was studied by FTIR spectrometer, and obtained transmittance spectra were presented in Fig. 5. The DMA has shown an absorption peak near 2954 cm^{-1} attributed to the asymmetric stretching vibration of the methyl group. The absorption band near 1739 cm^{-1} attributed to the C=O stretching vibration of fatty acid methyl ester compound. The C–O stretching vibration absorption bands of ester compound occur near 1178 and 1075 cm^{-1}.

The prepared shell material has exhibited broad and strong absorption near 3425 cm^{-1} corresponds to the Si–OH or O–H functional groups. The broad and weak absorption occurring near 1072 and 456 cm^{-1} attributed to Si–O–Si stretching vibrations [13]. The prepared MPCM have exhibited the absorption spectra of the ester functional group (1739 and 1178 cm^{-1}) and stretching vibrations of silica groups (3425, 1072, and 456 cm^{-1}) without any additional peaks. Therefore, the results signify that the prepared MPCM has shown no chemical interaction between shell and core material and confirms the formation of the shell and core was through physical interaction only. Thus, the shell coating over the PCM helps in retaining their properties.

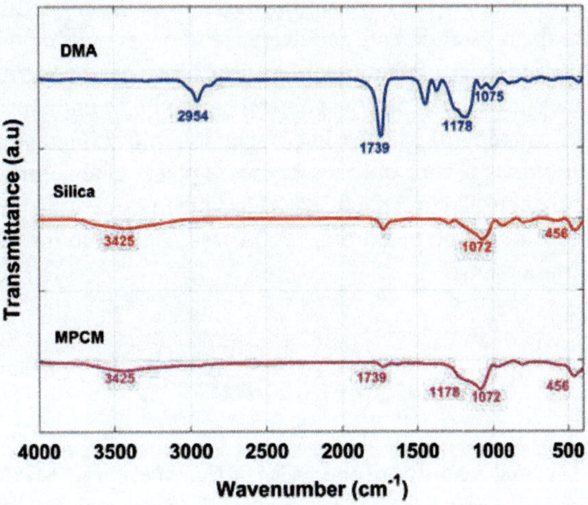

Fig. 5 FTIR spectra of MPCM

3.4 Phase Change Properties of MPCM

Phase change properties are measured using DSC technique, and the results obtained are presented in Fig. 6. The results illustrate that prepared microcapsules have exhibited a single downward peak corresponds to the endothermic process. The corresponding onset and end set temperatures are 7.33 °C and 11.97 °C, respectively. The melting behaviour of the microcapsules was quite similar to the pure PCM with onset melting as 9.24 °C [14]. However, it is evident to note there is a slight decrease in the onset melting temperature for MPCM compared to PCM. This decrease is due to the fact the silica shell acts as confinement around the PCM droplet thereby restricting

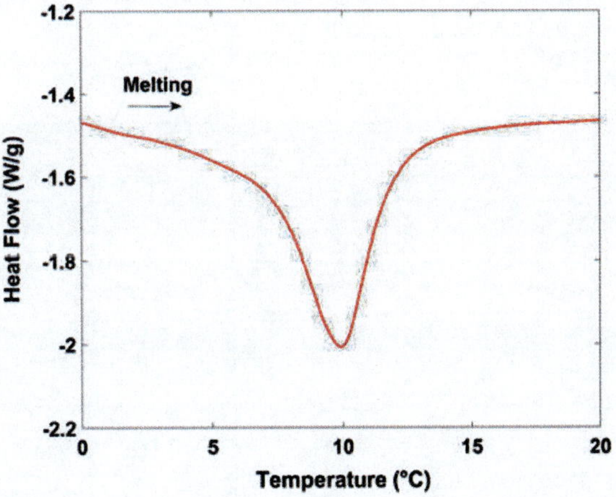

Fig. 6 DSC thermograms of MPCM

the motion of the DMA molecules during the crystallisation process and thus leading to the crystal defects and decrease in onset melting temperature [10].

The area of the endothermic peak represents enthalpy of latent heat that was measured as 23.52 J/g against the pure PCM enthalpy 153.01 J/g [14]. The decrease in enthalpy of MPCM in comparison with dimethyl adipate was because the PCM acting as a core only undergoes a phase change and while the silica shell acts as barrier protecting the PCM. Thus, due to its melting point and thermal energy storage capability, the prepared MPCM was applicable for the cool thermal energy storage applications.

3.5 Thermal Stability of MPCM

Thermal stability of the PCM, silica shell, and MPCM investigated using TGA is presented in Fig. 7. The results illustrate that the DMA has shown a single step weight loss with onset temperature at 130 °C and end set temperatures at 157 °C. The prepared MPCM has experienced two-step weight loss at 120–180 °C and 220–285 °C. The first step of decomposition about 16% was due to the evaporation of hydroxyl compounds absorbed on the shell surface. This step was followed by the decomposition of the PCM present inside the shell because of the breakdown of the ester compounds with an increase in temperature.

The second step decomposition around 12% corresponds to the dehydration and polycondensation of silanols that has not completely formed at low-temperature interfacial polycondensation [10]. The test results illustrate that there is about 20 °C enhancement in the end set decomposition temperature of the MPCM compared to pure PCM. Therefore, this signifies utilisation of the silica as a shell material for microencapsulation has resulted in improved thermal stability.

Fig. 7 TGA thermograms of MPCM

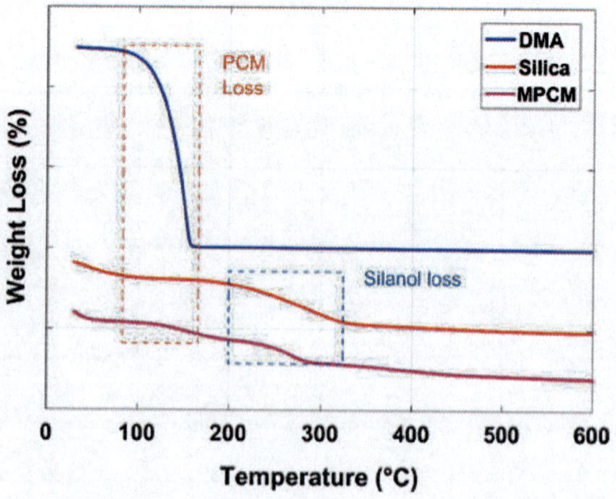

4 Conclusions

The microencapsulation with dimethyl adipate as core and silica as shell material has been synthesised through interfacial hydrolysis and polycondensation method. The following conclusions are obtained.

- Surface morphology obtained from field emission scanning electron microscope has shown good sphericity with an average particle size of 596 nm.
- The elemental analysis carried using X-ray photoelectron spectroscopy has shown the microencapsulation of PCM into inorganic silica shell.
- Chemical structure studies confirmed that the formation of MPCM was through physical interaction.
- Latent heat of enthalpy of the MPCM was around 24 kJ/kg, with onset melting temperature as 7.33 °C.
- Thermal stability studies have signified MPCM shown good thermal stability with a 20 °C increase in the end set temperatures of MPCM compared to PCM.

These findings suggest the prepared MPCM can be considered as a suitable candidate for cool thermal energy storage applications.

Acknowledgements The authors thankfully acknowledge DST-SERB, New Delhi, for providing financial support to carry out this research under DST-SERB ECR scheme (Sanction Order No. ECR/2017/001146). Authors are grateful to research scholar Mr. R. Naresh and Central Analytical Laboratory, BITS Pilani, Hyderabad Campus, for their technical assistance and support in performing various characterisations.

References

1. Shchukina EM, Graham M, Zheng Z, Shchukin DG (2018) Nanoencapsulation of phase change materials for advanced thermal energy storage systems. Chem Soc Rev 47:4156–4175
2. Sarier N, Onder E, Ukuser G (2015) Silver incorporated microencapsulation of n-hexadecane and n-octadecane appropriate for dynamic thermal management in textiles. Thermochim Acta 613:17–27
3. Huang R, Li W, Wang J, Zhang X (2017) Effects of oil-soluble etherified melamine-formaldehyde prepolymers on in situ microencapsulation and macroencapsulation of n-dodecanol. New J Chem 41:9424–9437
4. Wang H, Luo J, Yang Y, Zhao L, Song G, Tang G (2016) Fabrication and characterization of microcapsulated phase change materials with an additional function of thermochromic performance. Sol Energy 139:591–598
5. Sami S, Sadrameli SM, Etesami N (2018) Thermal properties optimization of microencapsulated a renewable and non-toxic phase change material with a polystyrene shell for thermal energy storage systems. Appl Therm Eng 130:1416–1424
6. Alva G, Huang X, Liu L, Fang G (2017) Synthesis and characterization of microencapsulated myristic acid–palmitic acid eutectic mixture as phase change material for thermal energy storage. Appl Energy 203:677–685

7. Yu S, Wang X, Wu D (2014) Microencapsulation of n-octadecane phase change material with calcium carbonate shell for enhancement of thermal conductivity and serving durability: synthesis, microstructure, and performance evaluation. Appl Energy 114:632–643
8. Chai L, Wang X, Wu D (2015) Development of bifunctional microencapsulated phase change materials with crystalline titanium dioxide shell for latent-heat storage and photocatalytic effectiveness. Appl Energy 138:661–674
9. Liston LC, Farnam Y, Krafcik M, Weiss J, Erk K, Tao BY (2016) Binary mixtures of fatty acid methyl esters as phase change materials for low temperature applications. Appl Therm Eng 96:501–507
10. Zhang H, Sun S, Wang X, Wu D (2011) Fabrication of microencapsulated phase change materials based on n-octadecane core and silica shell through interfacial polycondensation. Colloids Surf A 389:104–117
11. Imran HS, Ameelia RA, Kalaiselvam S (2018) Bifunctional nanoencapsulated eutectic phase change material core with SiO_2/SnO_2 nanosphere shell for thermal and electrical energy storage. Mater Des154:291–301
12. Zhu Y, Chi Y, Liang S, Luo X, Chen K, Tian C, Wang J, Zhang L (2018) Novel metal coated nanoencapsulated phase change materials with high thermal conductivity for thermal energy storage. Sol Energy Mater Sol Cells 176:212–221
13. Zheng J, Wei Q (2017) Synthesis of a novel nanoencapsulated n-eicosane phase change material with inorganic silica shell material for enhanced thermal properties through sol-gel route. J Text Sci Eng 7:1–8
14. Moinuddin O, Trivedi GVN, Parameshwaran R, Deshmukh SS (2019) Study on thermal storage properties of microencapsulated organic ester as phase change material for cooling application. Int J Env Anal Chem, 1–10

Estimation of Parameter in Non-Newtonian Third-Grade Fluid Problem by Artificial Neural Network Under Noisy Data

Vijay Kumar Mishra, Sumanta Chaudhuri, Jitendra K. Patel, and Arnab Sengupta

Abstract Estimation of a parameter of third-grade fluid problem by using artificial neural network is reported. The fluid is allowed to flow through two parallel plates and supplied with a constant and uniform heat flux. Inverse analysis is employed by using artificial neural network (ANN) to estimate third-grade fluid parameter. The direct problem is solved by semi-analytical method and least square method (LSM) to compute the temperature profile. The temperature profile and the corresponding parameter of third-grade fluid are perturbed to mimic error in measurement and then used in ANN for training the neurons with error in data. An unknown temperature profile is fed into the trained ANN, and ANN gives the corresponding parameter as output. In the ANN, Levenberg–Marquardt algorithm (LMA) is used to train the ANN. Five different levels of error in measurement are analyzed in the estimation of the parameter.

Keywords Inverse analysis · Third-grade fluid · Artificial neural network · Levenberg–Marquardt algorithm

Nomenclature

A	Third-grade fluid parameter
Ac	Cross-sectional area
A_1, A_2, A_3, A_n	Kinematic tensor
Br	Brinkman number
C_p	Specific heat at constant pressure (kJ/kg K)
D	Differential operator
D/Dt	Material derivative
f	Body force per unit volume

V. K. Mishra (✉) · S. Chaudhuri · J. K. Patel · A. Sengupta
School of Mechanical Engineering, Kalinga Institute of Industrial Technology (KIIT) Deemed to be University, Bhubaneswar, Odisha 751024, India
e-mail: mishra.vdm@gmail.com

S. Revankar et al. (eds.), *Proceedings of International Conference on Thermofluids*, Lecture Notes in Mechanical Engineering,
https://doi.org/10.1007/978-981-15-7831-1_22

g	Function
h	Half depth of channel (m)
k_{th}	Thermal conductivity of the fluid (kJ/m K)
L	Length of the channel (m)
N	Non-dimensional pressure gradient
Nu	Nusselt number
p^*	Dimensional pressure (Pa)
Q	Dimensional flow rate (m^3/s)
q	Heat flux ratio
q_1, q_2	Heat fluxes at lower and upper walls (kJ/m^2)
R	Residual
S	Sum of square of residual
T^*	Dimensional temperature (K)
T^*_m	Bulk mean temperature (K)
$T^*_{w_l}$	Temperature of the lower wall (K)
u	Non-dimensional velocity along axial direction
u_N	Non-dimensional velocity for Newtonian fluid
u^*	Dimensional velocity along axial direction (m/s)
u_0	Average velocity (m/s)
V^*	Velocity vector (m/s)
$v, \tilde{v}$	Velocity functions
x, y, z	Non-dimensional coordinates
x^*, y^*, z^*	Dimensional coordinates (m)
w_i	ith weight function

Greek symbols

ρ	Density of the fluid (kg/m^3)
μ	Dynamic viscosity of the fluid (N s/m^2)
θ	Non-dimensional temperature
θ_N	Non-dimensional temperature for Newtonian fluid
Φ_i	Base function
τ	Stress (N/m^2)

1 Introduction

Analysis of any thermal engineering problem involves computation of any one or a combination of dependent variables such as: velocity, temperature and heat flux, by solving the relevant governing equations. In these type of problems, independent properties (such as: geometric details, fluid and flow properties, and boundary and initial conditions) are known. Such types of problems are categorized as direct problems. Whereas, the computation of independent properties from a prior knowledge of dependent properties comes under inverse types of problems [1, 2]. Inverse problem is useful in performance optimization and designing of any thermal device.

Various industries such as food processing, polymer processing and biological instruments involve non-Newtonian third-grade fluid. Problems involving these fluids are difficult to analyze due to high level of nonlinearity, and hence involve specially developed semi-analytical method to solve such problems [3–5]. One of the widely used method, least square method (LSM) is employed in the present study to solve the direct part of the problem [6–8].

In the inverse problems, various tools such as genetic algorithm, simulated annealing and particle swarm method are extensively employed by various researchers. But, the use of ANN in these types of problems is new and lead to various new researches. ANN has advantages like: not requiring knowledge of the relationship of the parameter involved, can handle nonlinear relationships, capable of handling noisy data, etc. [9, 10].

In the present study, a non-Newtonian third-grade fluid is considered to flow through two parallel plates, both subjected to a constant and uniform heat flux. The governing equations are solved by using semi-analytical method LSM. The temperature profile computed by LSM is used to train ANN. In the ANN, LMA is used for training of neurons. AN unknown temperature profile is fed into the trained ANN to get corresponding third-grade parameter as output.

2 Problem Formulation

Two parallel plates are supplied with a constant and uniform heat flux. A third-grade fluid is flowing through the two parallel plates and as shown schematically in Fig. 1. The assumptions made in the analysis are: flow is thermally and hydrodynamically fully developed, laminar, incompressible, steady-state and fluid properties are constant. Body forces are also neglected in the present study. The length

Fig. 1 Schematic of the third-grade fluid flow through two parallel plates subjected to constant and uniform heat flux

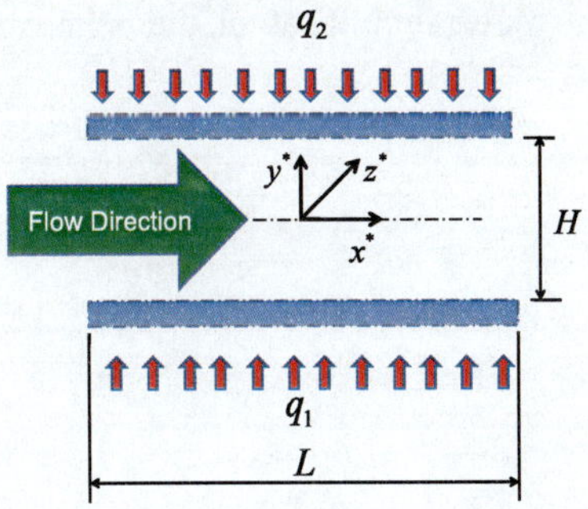

(L) is very large as compared to other two dimensions such as: width (W) and height (H). The relevant governing equations used in the analysis are:

Continuity equation

$$\nabla \cdot V^* = 0 \tag{1}$$

where the velocity vector is represented by V^*.

Momentum equation

$$\rho \frac{DV^*}{Dt} = \nabla \cdot \tau + f \tag{2}$$

It is to be noted here, the linear dependence of stress tensor on strain rate tensor makes analysis of Newtonian fluid comparatively easy. But in case on non-Newtonian fluid, the analysis is difficult as the stress tensor depends on strain rate tensor nonlinearly. Dependence of stress tensor on the strain rate tensor for a third-grade fluid is expressed by

$$\tau = -pI + \mu A_1 + \alpha_1 A_2 + \alpha_2 A_1^2 + \beta_1 A_3 + \beta_2 (A_1 A_2 + A_2 A_1) + \beta_3 (\mathrm{tr} A_1^2) A_1 \tag{3}$$

$$A_1 = (\mathrm{grad}\, V^*) + (\mathrm{grad}\, V^*)^{\mathrm{Transpose}} \tag{4}$$

$$A_n = \frac{\mathrm{d}A_{n-1}}{\mathrm{d}t} + A_{n-1}(\mathrm{grad}\, V^*) + (\mathrm{grad}\, V^*)^{\mathrm{Transpose}} A_{n-1}, \quad n = 1, 2, 3 \tag{5}$$

Energy conservation equation:

$$\rho c_p \frac{\mathrm{d}T^*}{\mathrm{d}t} = \tau : \mathrm{grad}(V^*) - \nabla \cdot (-k_{th} \nabla T^*) \tag{6}$$

Hydro-dynamically fully flow assumption gives rise to

$$v^* = [u^*(y^*),\ 0,\ 0] \tag{7}$$

After simplification, the momentum equations can be written as:

$$\frac{\partial p^*}{\partial x^*} = \mu \frac{\mathrm{d}^2 u^*}{\mathrm{d}y^{*2}} + 2(\beta_2 + \beta_3) \frac{\mathrm{d}}{\mathrm{d}y^*} \left(\frac{\mathrm{d}u^*}{\mathrm{d}y^*} \right)^3 \tag{8}$$

$$\frac{\partial p^*}{\partial y^*} = (2\alpha_1 + \alpha_2) \frac{\mathrm{d}}{\mathrm{d}y^*} \left(\frac{\mathrm{d}u^*}{\mathrm{d}y^*} \right)^2 \tag{9}$$

$$\frac{\partial p^*}{\partial z^*} = 0 \tag{10}$$

Substitution of velocity field from Eqs. (7) to (6) leads to simplification of energy equation as:

$$\rho c_p u^* \frac{\partial T^*}{\partial x^*} = k_{\text{th}}\left(\frac{\partial^2 T^*}{\partial x^{*2}} + \frac{\partial^2 T^*}{\partial y^{*2}}\right) + \mu\left(\frac{\mathrm{d}u^*}{\mathrm{d}y^*}\right)^2 + 2(\beta_2 + \beta_3)\left(\frac{\mathrm{d}u^*}{\mathrm{d}y^*}\right)^4 \quad (11)$$

The second term on the right side of Eq. (11) takes care of viscous dissipation. The thermal conductivity of the fluid is represented by k_{th}. The non-Newtonian parameters lead to addition of the third term in the above equation. Whereas, the third term accounts for the effect of non-Newtonian parameters. The corresponding situation for a Newtonian fluid can be obtained by substituting zero values for β_2 and β_3 in Eq. (11). The temperature profile is a function of only y^*, due to thermally fully developed assumption, and can be expressed as:

$$\theta = \frac{T^* - T^*_{wl}}{T^*_m - T^*_{wl}}$$

Here, bulk mean temperature is represented by T^*_m and temperature of lower plate is represented by T^*_{w1}. The assumption of constant and uniform heat flux gives rise to the following form for non-dimensional temperature along the axial direction:

$$\frac{\partial T^*}{\partial x^*} = \frac{\mathrm{d}T^*_m}{\mathrm{d}x^*} = \text{const.}, \quad \frac{\partial^2 T^*}{\partial x^{*2}} = 0 \quad (12)$$

After substituting the values from Eq. (12) into Eq. (11), we get the following form of energy equation

$$\rho c_p u^* \frac{\partial T^*}{\partial x^*} = k_{\text{th}}\left(\frac{\partial^2 T^*}{\partial y^{*2}}\right) + \mu\left(\frac{\mathrm{d}u^*}{\mathrm{d}y^*}\right)^2 + 2(\beta_2 + \beta_3)\left(\frac{\mathrm{d}u^*}{\mathrm{d}y^*}\right)^4 \quad (13)$$

Boundary conditions: At surface of both the parallel plates, no-slip conditions prevail

$$u * (-h) = 0, \quad u * (h) = 0 \quad (14)$$

Application of a constant and uniform heat flux on both the plates can be written as

$$k_{\text{th}}\left(\frac{\partial T^*}{\partial y^*}\right)_{-h} = \mp q_1, \quad k_{\text{th}}\left(\frac{\partial T^*}{\partial y^*}\right)_{h} = \pm q_2 \quad (15)$$

The temperature can be non-dimensionalized to make the analysis more generalized:

$$\theta = \frac{T^* - T^*_{w_l}}{q_1 h / k_{\text{th}}} \tag{16}$$

Here, q_1 represents the magnitude of the constant and uniform heat flux at the lower plate. The flow velocity and the distance along y-direction can be non-dimensionalized as:

$$u = \frac{u^*}{u_0}, \quad y = \frac{y^*}{h} \tag{17}$$

After non-dimensionalization, the momentum and the energy equation become:

$$\frac{\mathrm{d}^2 u}{\mathrm{d}y^2} + 6A\left(\frac{\mathrm{d}u}{\mathrm{d}y}\right)^2 \frac{\mathrm{d}^2 u}{\mathrm{d}y^2} = N \tag{18}$$

$$\beta u = \frac{\mathrm{d}^2 \theta}{\mathrm{d}y^2} + Br\left(\frac{\mathrm{d}u}{\mathrm{d}y}\right)^2 + 2ABr\left(\frac{\mathrm{d}u}{\mathrm{d}y}\right)^4 \tag{19}$$

$$A = \frac{\beta_2 + \beta_3}{\mu}\left(\frac{u_0}{h}\right)^2, \quad Br = \frac{\mu u_0^2}{q_1 h}, \quad u_0 = \frac{1}{N}\frac{\mathrm{d}p^*}{\mathrm{d}x^*}\frac{h^2}{\mu}, \quad \beta = \frac{\rho c_p u_0}{q_1}\frac{\mathrm{d}T^*}{\mathrm{d}x^*} \tag{20}$$

In Eq. (18), N is pressure gradient along x-direction in non-dimensional form.

For momentum equation, the following boundary conditions in non-dimensional form are used:

$$u(-1) = 0, \quad u(1) = 0 \tag{21}$$

For energy equation, the following boundary conditions in non-dimensional form are used:

$$\theta(-1) = 0 \tag{22}$$

$$\frac{\mathrm{d}\theta}{\mathrm{d}x}(-1) = \mp 1 \tag{23}$$

$$\frac{\mathrm{d}\theta}{\mathrm{d}x}(1) = \pm\frac{q_2}{q_1} = q \tag{24}$$

The governing equations are solved by semi-analytical method LSM.

3 Results and Discussion

By employing LSM, the governing equations are solved with relevant boundary conditions. This constitutes the direct part of the problem. The direct part is validated with the results of Danish et al. [11] and presented in Fig. 2a. Effect of magnetic field is removed by setting $Ha = 0$. Results corresponding to two different values of third-grade fluid parameter A and is considered here for validation. The results from LSM are also validated for Newtonian fluid and as shown in Fig. 2b.

After the completion of direct part, the inverse part of the problem starts. In the inverse problem, the temperature profile obtained from LSM, along with the corresponding third-grade fluid parameter, is fed into the ANN. In the ANN model, LMA is used to train 10 number of neurons and 14 number of cases are fed as input for the training purpose. The ANN model is shown in Fig. 3. The input temperature profile and the corresponding third-grade fluid parameter are perturbed by different levels of error (0, 0.1, 0.5, 1 and 2%) to mimic error in the measurement of data. The ANN is trained for different levels of noise and tested for their capability for estimation of parameter in different noise conditions. An unknown temperature profile is fed into the trained ANN, and the ANN returns the corresponding third-grade fluid parameter as output. During the ANN training, the model is tested with the help of performance curves, histogram and regression analysis. Graphs pertaining to zero noise in the

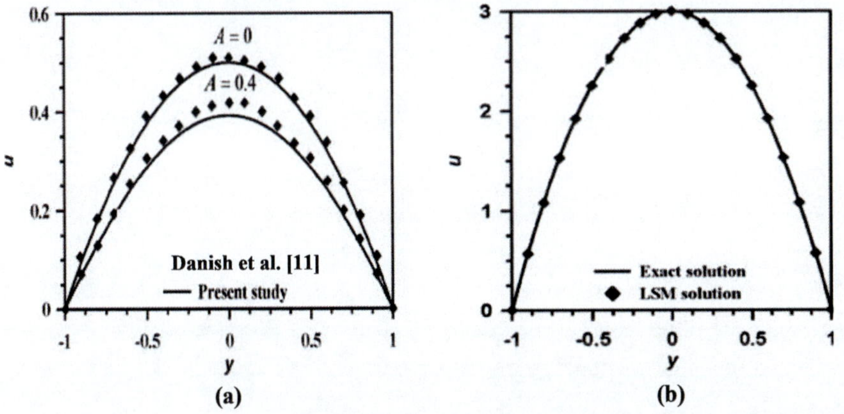

Fig. 2 Validation of results from LSM with exact solution. **a** Velocity profile comparison for third-grade fluid. **b** Velocity profile comparison for Newtonian fluid

Fig. 3 ANN model used for estimation of parameter in inverse analysis

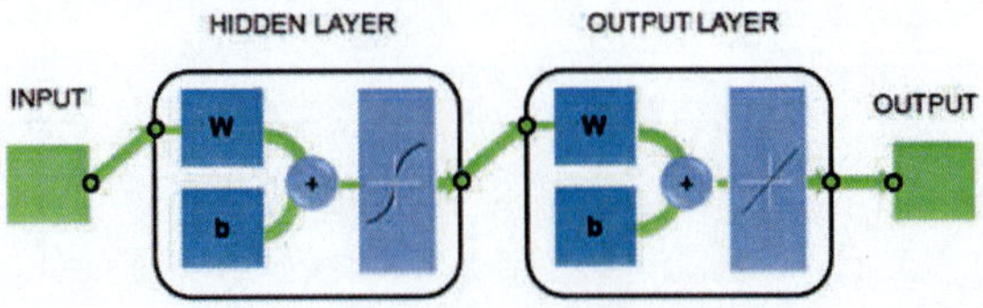

training data is presented here. Figure 4a shows the performance curve during the three different stages of training of ANN. The convergence criterion is met in third iteration of training. In Fig. 4b, histogram of the data used in the ANN training is shown. It can be viewed that a lot of data falls near the zero error line, indicating that the model can handle different kinds of situations. Also, since the maximum height of the data does not falls on zero error line, the model may not give highly accurate result with 14 numbers of input cases. More cases for input and/or higher number of neurons can make the model more robust.

In the regression analysis of the ANN model, the regression coefficient is found to be near one and is shown in Fig. 5a. This shows that LMA as a training algorithm in ANN for these types of problem is quite useful [9, 10].

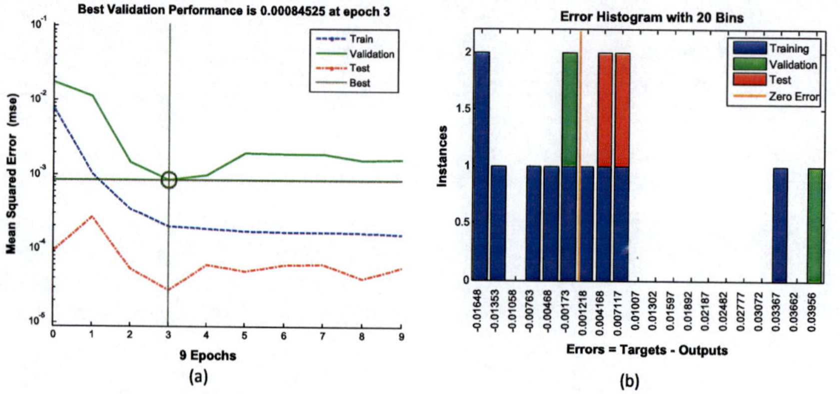

Fig. 4 **a** Performance curve of training in ANN with LMA. **b** Histogram of data used in ANN training

Fig. 5 Regression analysis of ANN model with LMA

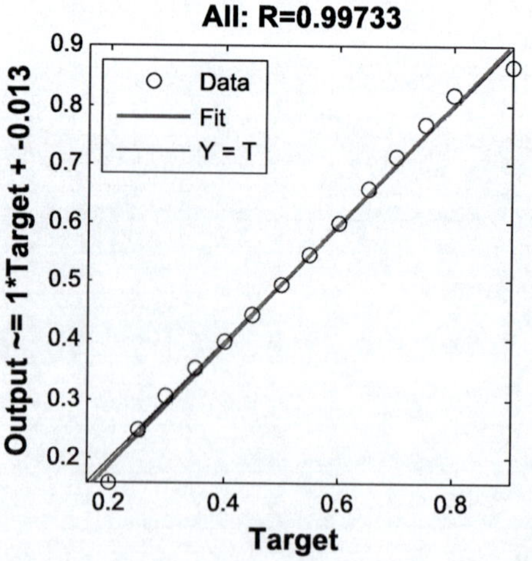

Table 1 Comparison of estimated third-grade fluid parameter with different levels of error in input data

S. No.	% Error in input	Exact value AACT	Estimated value AEST	% Error in estimation
1	0.0	0.85	0.866	1.8824
2	0.1	0.85	0.8733	2.7412
3	0.5	0.85	0.8288	2.4941
4	1.0	0.85	0.8065	5.1176
5	2.0	0.85	0.8707	2.4353

Once the ANN is successfully used in the estimation of the third-grade fluid parameter, attempt is made to use the ANN model under noisy data condition. The temperature profile obtained from LSM and corresponding third-grade fluid parameter is perturbed by different levels of error (0.1, 0.5, 1 and 2%) and fed into ANN for training purpose. Once the ANN model is trained, an unknown temperature profile in fed into ANN, and the ANN gives the corresponding third-grade fluid parameter as output. The results of estimation for data used in the training with different levels of error are listed in Table 1.

It is observed that the accuracy of estimation for all the data with different levels of error is within 5.2%. For data with no error, the ANN does the estimation with high accuracy. Apart from the case when training data has 1% error and 5.12% accuracy in estimation, for all other case, the ANN is able to estimate the parameter with good accuracy. Other training algorithms in ANN like scaled conjugate gradient (SCG) and Bayesian regularization (BR) can also be explored to look for better accuracy in estimation of parameter for these types of problems.

4 Conclusions

Artificial neural network (ANN) is explored in inverse problem involving non-Newtonian third-grade fluid. The temperature profile of third-grade fluid flowing through two parallel plates, subjected to a constant and uniform heat flux, is computed by using least square method (LSM) and a semi-analytical approach. The temperature profile along with corresponding third-grade fluid parameter is then used in the training of artificial neural network (ANN). The data fed into artificial neural network (ANN) contains different levels of noise to mimic error in measurement. The artificial neural network (ANN) is trained by using Levenberg–Marquardt algorithm (LMA). An unknown temperature profile is fed into artificial neural network (ANN), and artificial neural network (ANN) gives back the corresponding third-grade fluid parameter. The estimation of parameter under different levels of noise is under 5.2%. Different ways to improve the accuracy of estimation may be: the use of scaled conjugate gradient (SCG) and Bayesian regularization (BR) as training algorithm, more number of neurons, more data for the training the neurons.

References

1. Dumek V, Druckmuller M, Raudensk, Woodbury KA (1993) Novel approaches to the ICHP: neural networks and expert systems, inverse problems in engineering: theory and practice. In: Proceedings of the first international conference on inverse problems in engineering. ASME no. I00357, pp 275–282
2. Mishra VK, Mishra SC, Basu DN (2017) Simultaneous estimation of parameters in analyzing porous medium combustion—assessment of seven optimization tools. Numer Heat Transf Part A 71(6):666–676
3. Al Mukahal FHH, Wilson SK, Duffy BR (2015) A rivulet of a power-law fluid with constant width draining down a slowly varying substrate. J Non-Newton Fluid Mech 224:30–39
4. Jalil M, Asghar S (2013) Flow of power-law fluid over a stretching surface: a Lie group analysis. Int J Non-Linear Mech 48:65–71
5. Tso CP, Sheela FJ, Hung YM (2010) Viscous dissipation effects of power-law fluid flow within parallel plates with constant heat fluxes. J Non-Newton Fluid Mech 165:625–630
6. Hatami M, Ganji DD (2013) Heat transfer and flow analysis for SA-TiO_2 non-Newtonian nano fluid passing through the porous media between two co-axial cylinders. J Mol Liq 188:155–161
7. Pourmehran O, Rahimi-Gorji M, Gorji-Bandpy M, Ganji DD (2015) Analytical investigation of squeezing unsteady nanofluid flow between parallel plates by LSM and CM. Alex Eng J 54:17–26
8. Fakour M, Vahabzadhe A, Ganji DD, Hatami M (2015) Analytical study of micro polar fluid flow and heat transfer in a channel with permeable walls. J Mole Liq 204:198–204
9. Jambunathan K, Hartle S, Ashforth-Frost S, Fontama VN (1996) Evaluating convective heat transfer coefficients using neural networks. Int J Heat Mass Transf 39:2329–2332
10. Sablani SS (2001) A neural network approach for non-iterative calculation of heat transfer coefficient in fluid-particle systems. Chem Eng Process 40:363–369
11. Danish M, Kumar S, Kumar S (2012) Exact analytical solutions for the Poiseuille and Couette-Poiseuille flow of third grade fluid between parallel plates. Commun Non Linear Sci 17:1089–1097

Role of PCM in Solar Photovoltaic Cooling: An Overview

Pragati Priyadarshini Sahu, Abhilas Swain, and Radha Kanta Sarangi

Abstract The present article put forth a comprehensive review of the latest research works carried out on the cooling techniques for maintaining the required temperature of photovoltaic modules for achieving greater efficiency. The conventional cooling technologies such as water cooling, air cooling and water spray are little discussed in this article as these are not preferred in the present scenario due to their power requirement and complicated arrangement. The phase change materials (PCM) have evolved as a better alternative for the cooling of photovoltaic modules due to their advantageous thermo-physical properties. The recent research investigations on using phase change materials as heat absorbing material are critically reviewed in the present article. The investigations include experimental and numerical studies on single PCM, combined PCM (PCM with nanoparticles, graphite powders), PCM as secondary thermal energy storage. The critical review suggests that further research works are necessary for developing a passive design (not using external power) for obtaining optimum temperature of the solar photovoltaic modules.

Keywords Phase change material · Solar photovoltaic · Efficiency of PV · Photovoltaic thermal

1 Introduction

In the era of growing civilization and industrialization, the requirement of energy is growing. Due to this, the demand for the renewable energy and consequently the solar photovoltaics has played a greater role in the energy generation. However, till date, the technological development of photovoltaics is able to achieve a highest efficiency of about 15–20%. Moreover, the part of the irradiation not converted to electricity increases the temperature of the photovoltaics and the efficiency decreases with increase in the temperature. Thus, it is essential to remove heat from the photovoltaics

P. P. Sahu (✉) · A. Swain · R. K. Sarangi
School of Mechanical Engineering, Kalinga Institute of Industrial Technology (KIIT) Deemed to be University, Bhubaneswar, Odisha 751024, India
e-mail: pragatipsahu009@gmail.com

S. Revankar et al. (eds.), *Proceedings of International Conference on Thermofluids*, Lecture Notes in Mechanical Engineering,
https://doi.org/10.1007/978-981-15-7831-1_23

to optimize the electricity generation. For this objective, researchers are continuously investigating and developing technologies for controlling the temperature of the photovoltaic panels. Some of these are heat exchangers, water sprinklers, phase change materials, heat sinks, transparent coating, etc.

The present article presents a concise review of the latest developments made in this field of research for obtaining the best suitable cooling technology. Along with this a greater emphasis is given to the application of the phase change material (PCM) for cooling of PV cell. The comparison of different ways to use PCM for this application is presented.

Sunlight is one of the perpetual renewable energy sources which is converted into electrical energy through the photovoltaic (PV) system. These photovoltaic systems are gaining popularity for electricity generation because of their anti-pollution technology, no effect on global warming, low cost of operation and low maintenance. As already described, a big problem in case of the photovoltaic is the reduction in efficiency with increase in temperature. For each 1 °C temperature rise of the photovoltaic module surface, there is an efficiency reduction of 0.5% [1].

The above discussion leads to the conclusion that there is a strong need for a suitably designed cooling technology with least power consumption for the optimum temperature of the photovoltaic module and best possible efficiency. Thus, the present article presents an discussion on the current trends of the cooling technologies being investigated for cooling of PV modules and more emphasis is given on the use of phase change materials for this purpose.

2 Review of Other Techniques for Cooling of Photovoltaic

Floating-Tracking-Cooling and Concentrating (FTCC) is a concept applied to get the maximum performance of PV modules. In this arrangement, the modules are arranged over the structures made out up of PVC pipes in a water pool. For cooling purpose, the water sprinklers are used to spray water on the panels for cooling purpose. Due to the water spray, the power production increased by 40%. However, the water sprinkling is capable to decrease the temperature of the certain areas of the photovoltaic modules. Another method adopted for cooling of the PV modules is the hybrid solar PV and thermal methodology in which the cooling fluid is usually air or water. The heat recovered by the air or water is used for the domestic purpose. Akbarzadeh and Wadowski [2] have adopted a similar method and observed an increase in the power output by almost 50%. It is also observed that the maximum temperature could be maintained at 46 °C when allowed for power generation for a time span of 4 hour [2]. Tonui and Tripanagnostopoulos [3] have investigated the performance of PV modules with air as the cooling fluid under natural and forced convection conditions. They observed that the arrangement with fins on the panel has performed best considering air as the working fluid [3].

3 Cooling of Photovoltaics Using PCM

By considering these problems facing in case of solar photovoltaic cell efficiency, so many works were carried out in order to solve the problems stated and observed that cooling is needed for solar photovoltaic cell as like heat exchanger, to minimize wasted solar radiation and high system temperature. Less numbers of authors have published an extensive literature review on cooling solar PV modules by using various techniques which can increase the overall efficiency of the solar conversion system by implementing these techniques, we can reduce the unwanted temperature rise. Some cooling techniques are Floating-Tracking-Concentrating-Cooling system (FTCC), heat sink, cooling by spraying water, forced water circulation, forced air circulation, water immersion cooling technique, use of phase change materials (PCM), cooling by transparent coating (photonic crystal cooling) and thermoelectric cooling.

Phase change material can be used as a latent heat storage in the temperature range 0–120 °C for low temperature applications like solar cooling, etc. [4]. Duffie and William [5] have suggested to select such a phase change material which do not have a tendency to supercool. In a suitable temperature range, material undergoes in a phase change and absorbing a large amount of latent heat is known as phase change materials (PCMs) which can used for a passive heat storage [5]. Ismail and Goncalves [6] have stated PCM as a latent heat storage unit by changing its states form solid to liquid and regaining it back without any external sources [6]. Wirtz et al. [7] have formulated the performance of a dry PCM and resulted positive use of PCM for more heat storage and temperature control of electronics. Without showing adequate change in temperature, the thermal energy can be stored into latent energy by heating and cooling of phase change material. The stored energy can be repaired when the process is overturned. Due to the high latent heat of phase change material during phase change process, it is popularly used in such thermal energy storing process.

Sarı and Kaygusuz [8] have used lauric acid with 95% of purity as a phase change material which has a small temperature change and no subcooling during the solidification and they studied temperature distribution and thermal characteristics of phase change material [8]. Khodadadi et al. [9] and Kibria et al. [10] have reviewed thermo-physical properties of phase change material with dispersed several nanoparticles. Some reviews has been concise on thermal energy storage capacity, long-term stability, encapsulation, temperature range and system-related issues of phase change materials [11, 12].

Waqas and Jie [13] have investigated numerically phase change materials' effect on cooling of photovoltaics in the hot climatic condition in Pakistan during hottest month of summer. They carried out the computational work using the enthalpy method for the melting of PCM and taking mathematical model for the PV panel. It has been observed that the peak module temperature can be lowered by maximum 30 °C which increases the efficiency of the panel by 9–10%. They concluded that the performance will be best when there is a 10–12 °C temperature difference between the melting temperature of the PCM and the atmospheric temperature. Figure 1 shows

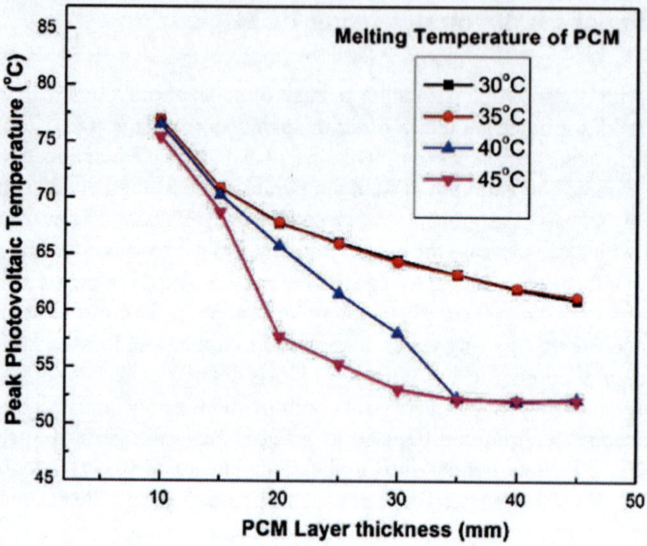

Fig. 1 Melting temperature effect [13]

that with the increase in the melting temperature of PCM, the peak temperature of PV panel decreases. Moreover, higher the thickness of the PCM, the lower will be the peak temperature [13].

Energy balance equation for the solar cell is given in Eq. (1).

$$S * \alpha\tau * \beta_c + S * \alpha\tau * (1 - \beta_c) = U_T * \left(T_c - T_g\right) + U_T * (T_c - T_{bs}) + E \quad (1)$$

In Eq. (1), the first factor represents the solar energy absorbed by the solar cell after transmission, second factor represents the solar energy absorbed after transmission, third factor represents conductive heat transfer between glass and solar cell, fourth section represents rate of energy conducted from solar cell to the back surface of the module, and E suggests the rate of electrical energy available from PV module.

The rate of thermal energy transferred from the conventional PV cell to the back surface of the PV module and then transferred to the ambient can be obtained by following equation

$$U_T * (T_{bs} - T_c) = h_w * (T_{amb} - T_{bs}) \quad (2)$$

where the rate of energy conducted from solar cell to the back surface of PV module is equal to the rate of heat transferred from the back surface of ambient.

Sainthiya and Benewal [14] have carried out an experimental investigation studying effect of front surface cooling of PV panels by flowing water for different flow rate conditions. During their experimentation, a thin layer of water is allowed to flow over the PV panel from top to bottom. The power output and the efficiency

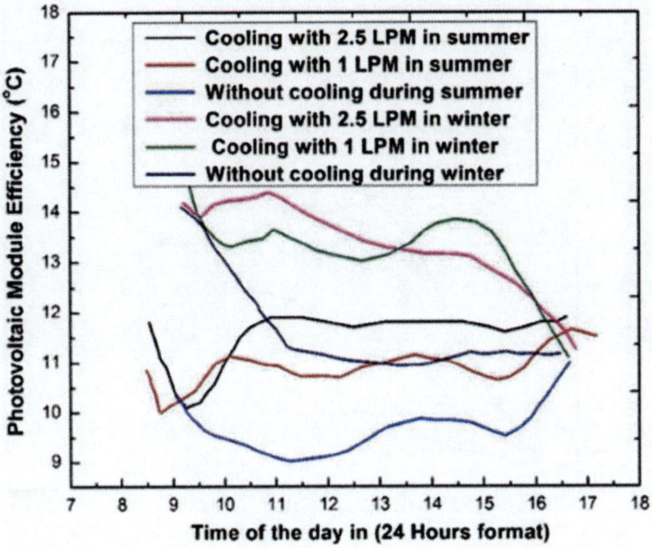

Fig. 2 Effect of water flow cooling or PVT [14]

are observed in both summer and winter weather conditions under four flow rate conditions (1–2.5 LPM). They observed an increase in efficiency of about 20–40%. It was observed that in winter the water cooling is more effective than the summer. Figure 2 shows the variation of the module efficiency during the whole day in both winter and summer for a different water flow rate.

Ahmed et al. [15] have investigated photovoltaic in combination with thermal heat recovery by using water flow at the back surface of the photovoltaic cell in a hot climatic condition of Egypt. The efficiency and the power output showed an improvement because of the water cooling arrangement. They evaluated the overall efficiency of the system by considering the thermal heat gain. One important point needs to be mentioned here that the electrical power consumption for the pump used for the flow of water. The variation of the peak temperature is shown in Fig. 3.

$$\eta_{\text{tot}} = \frac{Q_{\text{th}} + P_{\text{ele}}}{P_{\text{act}}} \tag{3}$$

Abdollahi and Rahimi [16] have investigated a cooling technology for photovoltaics by using water and PCM. The water is allowed to flow at the back side of the PV cell to receive heat. The hot water exits from the cooling path is allowed to pass through a helical coil present in a cylinder for receiving heat by placing phase change material. The PCM consists of 82% coconut oil and 18% sunflower oil leading to a melting temperature of 25–26 °C and latent heat of 308 kJ/Kg. The schematic of the setup is presented in Fig. 4. They compared the performance of the arrangement

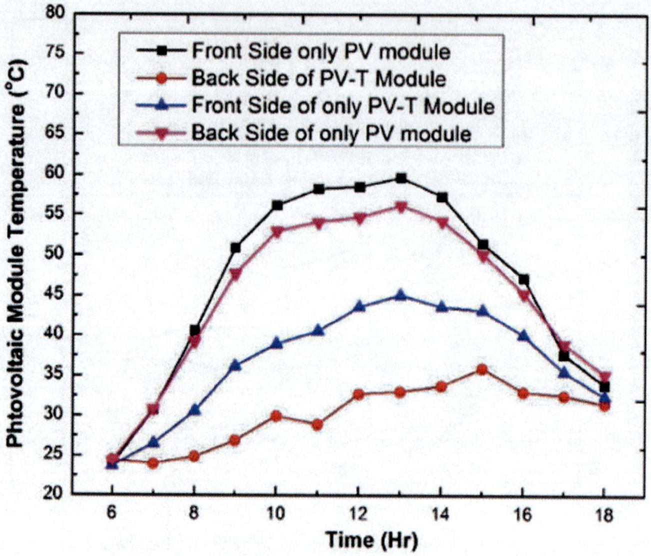

Fig. 3 Comparison of the front and back surface temperature of the photovoltaic cell for (1) only PV cell and (2) PV cell with water cooling (PV/T) [15]

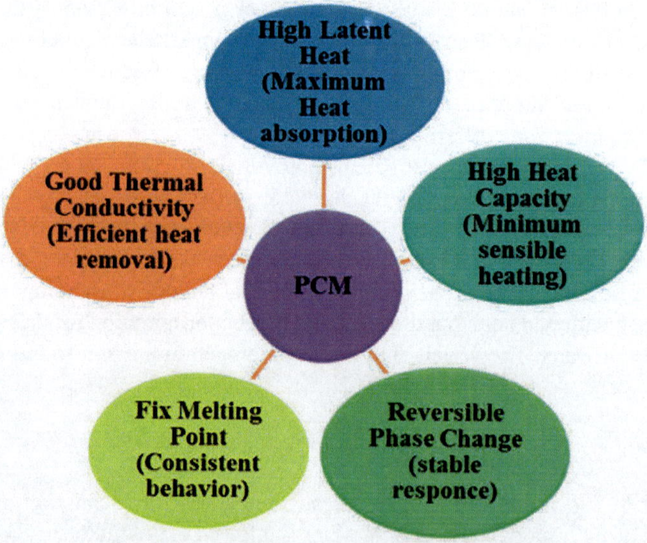

Fig. 4 Properties of a PCM desired for photovoltaic thermal regulation

Table 1 Thermal properties of pure and combined PCM

Properties	Pure PCM	Combined PCM
Thermal conductivity (W/m K)	0.18	92.1
Density (kg/m^3)	900	957
Specific heat (J/kg K)	2500	1900
Melting temperature (°C)	36–60	36–60

with that of only PV by using a term COE which is the efficiency considering the power consumption by pump.

As per the studies, 25–28 °C is the suitable operating temperature of PV cell which provides 80–90% of efficiency. By violating this range and constant solar radiations, it results decrease in conversion rate and efficiency by 0.5% per each degree Celsius rise of surface temperature [17].

Heng et al. [18] have decreased the operating temperature and increase in efficiency of solar panels by using phase change materials (PCM) and developed a 2D finite volume heat transfer model for framing integrated photovoltaic cell with the use of PCM.

Hachem et al. [19] have experimented and stated that pure PCM enhance electrical performance of PV cell by an average of 3%, where transient energy balance presented to analyse thermal behaviour and an average of 5.8% improvement of efficiency while using combined PCM. The thermal conductivity of phase change material which is 0.18 W/m K is enhanced by combining mass ratio of 20% copper powder and 10% of graphite powder with 70% of pure PCM and the enhanced thermal conductivity will be 92.1 W/m K (Table 1).

A PCM (paraffin-based) with 38–43 °C of melting range is integrated at the backside of the solar PV panel and its cooling effect is monitored. The increased PV power output due to cooling produced by PCM is quantified and PV annual electrical energy enhanced by 5.9% in the hot climatic condition [20].

At an adequately constant temperature, PV panel is maintained with the help of high latent heat capacity of PCM. The stored heat later can be used for water heating, space heating and many more. It is viewed that forced air and water cooling techniques are widely used to cooling PV panels as compared to natural ventilation-based cooling as an inadequate method. Without any additional electricity consumption, PCM has the advantages to delaying the temperature rise of PV panels.

Japs et al. [21] have experimented by considering PV with and without PCM and resulted that the generated energy by the panel with PCM is higher than the panel without PCM for 5 out of 25 days while with PCM+ graphite-PV. They got the results that the average energy and economic yields were positive at peak temperature means at the afternoon while it is negative for the rest. Later, Japs et al. [22] have concluded that PCM rigged with graphite, accomplished better than that have not with it. As graphite improves the conductivity of the phase change material which causes expeditious dissipation of heat results better output.

Thermal management performance and the heat transfer characteristics of dry-packed MEPCM were investigated by Tanuwijava et al. [23] via CFD simulations.

And stated that by consolidating the appropriate MEPCM layer in PV module gives improvement in thermal and electrical performance as the output. Ho et al. [24] have concluded that the layer thickness with 5 cm PCM has the best PV cooling performance with a melting temperature of 30 °C. Though, PCM could not fully solidify at night after certain thickness causes poor cooling performance. Experiment was carried out by installing microencapsulated phase change material (MEPCM) which is water insulated, attached to the back of PV panel and floats on a water surface (Table 2).

A noticeable drop in temperature results efficiency improvement while cooling the solar PV module using PCM technique. PCM being restrained by many researchers due to its higher cost for PV module cooling application. Pork fat, a cheaper phase change material have been numerically used by Nižetić et al. [30] and they have developed a numerical model by showing comparison study between conventional PCM and pork fat for cooling PV models. However, not much convincing results have been found, but concluded that the pork fat can be used as a potential PCM. 10–26 °C temperature drop and efficiency increases up to 3.73% have been resulted by Arıcı et al. [31] by developing a numerical model and analysing the performance of PV module cooling using phase change material technique. Many researchers have also used PCM as a cooling agent to improve the efficiency of hybrid PVT. Preet [32] has used water-based PCM cooling technique for cooling of PVT system (Table 3).

Some research articles are mentioned in Table 4 by the use of different types of PCM. They are petroleum jelly, coconut oil and palm oil and RT35.

The above discussion expounds the use of phase change materials in different ways for the cooling of photovoltaic modules such as directly on the back side, combined PCM, PCM with fins, PCM as secondary heat recovery material, etc. It is also observed that different types of PCMs are also used for different ambient conditions and application methodologies. Still, further research is needed to design and develop a passive cooling system for the photovoltaic modules which is possible by using phase change material.

Table 2 Various techniques used for solar PV cooling

Technique used for cooling PV cell	Highlights	References
Water and air-based cooling	• Conventional cooling techniques were reviewed • With limited temperature, degradation of PV cells was studied • Performance of low PV cell was studied which create difficulties • PV cooling using micro-channel, automotive radiator and thermoelectric modules need extensive research which are in developing stage	[25]
Natural/forced, hydraulic, heat pipe and PCM cooling	• To maintain cell temperature, surface area must be raised to maintain the concentration ratio • In ventilated facade system, large temperature fluctuation of PV cell was found • The heat removed from PV cell by PCM-based cooling system was not used properly	[26]
Phase change material	• Large capital investment needed during its low service life and low heat transfer, high maintenances cost which may affect overall efficiency of PV cell • Need to find out its benefit and reliability in a real environmental condition to find market potential	[27]
Single phase fluid	• Less than 10% of efficiency has been achieved by most of the cooling techniques • Air-based cooling is less effective than water-based cooling • Due to irregularity of different conditions tested it is difficult to state the comparison	[28]
Jet, micro-channel, PCM, heat sink and heat pipe	• PCM deteriorates from toxicity, corrosiveness and inflammability • Dumping of PCM after their uses is a problem • For attaining uniform lower temperature in PV panel, micro-channels can be used • Minimum cell temperature can be carried out by hybrid micro-channel jet impingement technique	[29]

Table 3 Summary of current research trends

Important observations	Results	References
• The combined effect of water and phase change material on the PV panel performance is investigated • Effect of Al_2O_3 nanoparticles loading in the phase change material on PV cell performance is conducted • Behind the PV panel, PCM/water implemented and occupation ratio is presented • Phase change material can be an excellent solution for the PV cooling	Compound technique; Al_2O_3 ($\varphi = 1\%$)/PCM mixture ($\lambda_{PCM} = 25\%$) + 75% water (5.31 kg/s m^2) obtains the highest PV performance	[33]
• Optimum option can be PVT and PVT-PCM systems for both heat and electricity • Experimentally, the PV cell temperature reduced to 12 °C • 13.98% and 13.87% are the efficiency of PVT-PCM numerically and experimentally founded, respectively • PVT-PCM systems increases electrical efficiency around 7% with compared to PVT system • Thermal collector with aluminium is used by presenting a design to get better heat transfer performance, which is in PVT-PCM and PVT systems	The module temperature dropped 12.6 and 10.3 °C For PV, the numerical and experimental electrical efficiency are 13.72 and 13.56% and in case of PVT system is 13.85 and 13.74%	[34]
• Sugarcane wax and Al_2O_3 composite have been used as phase change material • Gelatin-gum Arabic used as the polymer shell material	Enhancing the composite, PCM layer thickness by 7 mm from 4 mm could lower the cell's front-facing surface temperature by 4% resulting in raised the PV cell power generation by 12% at the peak time, due to the temperature storage capacity of the composite phase change material	[35]

(continued)

Table 3 (continued)

Important observations	Results	References
• Mathematical model and ANN model have been presented for nanofluid and nano-PCM cooling PVT • The expected linear models are consistent with ANN model and experimental results • The linear predicting models diverse from simple and accurate results	The expected predicting models gained an excellent R^2 result of 0.99 and MSE value of 0.006 RSME of 0.009 for both P-M1, P-M2 models The expected linear prediction models help to decrease the error in furcating future results and determine the best conditions for any solar system in an easier and faster way	[36]
• For a concentrated PV cooling, a combined PCM and water cooling system are developed • Efficiency and thermal power output raised while water flow rate is low • Cell temperature decreases by 60% and the system efficiency is raised by 224% • By using nanofluid, power output enhanced by 2.5% and PCM melting time lowers by 12%	Average temperature reduced up to 60% in case of CPV while comparing with PCM-PV and water cooling system Concentration ratio (CR) at 10 and HTF velocity at 0.01 m/s, the panel temperature did not exceed 78 °C Nanofluid has been used as HTF enhancer which helps to enhances the CPV efficiency by 2.7%	[37]
• Efficiency of conventional PV and the water-based PVT-PCM have been compared • Thermal energy, electrical energy and equivalent thermal exergy have been analysed throughout the year • Efficiency of embodied energy, energy payback time and lifecycle conversion have been analysed • Investigation have been carried out for DC electrical energy production cost and annual cost	Water-based PVT-PCM has 27% longer life cycle transformation efficiency as compared with conventional PV module	[38]

(continued)

Table 3 (continued)

Important observations	Results	References
• Efficiency increased by implementing PCM as PV cooling • PV performance further improved by adding nanoparticle (Al_2O_3) to PCM • Efficiency enhanced by 5.7 and 13.2% in PV-PCM and PV-PCM/nanoparticles, respectively	Efficiency improved with the use of PCM along with Al_2O_3 nanoparticles. Temperature drop measured in case of PV-PCM and PV-PCM/nanoparticle is 8.1 and 10.6 °C, respectively. Similarly, the efficiency gained by 5.7% and 13.2%, respectively	[39]

Table 4 Types of PCM used in some research studies

Details of research work/findings	PCM used	References
Up to 21.6% efficiency improvement has been shown due to heat removal by PCM in Indonesian climatic condition	Petroleum jelly	Indartono et al. [40]
For an ambient temperature range 27–30 °C, palm oil considered as the better PCM which can reduce the cell temperature by 9.6 °C having 102 mm of thickness. And gives an enhanced power output of 23%	Coconut oil and palm oil	Indartono et al. [41]
Experimenting for around 4 h in a peak temperature of 53 °C it gives reduction of cell temperature by 10 °C	RT35	Mahamudul et al. [42]

4 Conclusion

Low heat dissipation rate is the major affecting parameter which increases the temperature of solar PV panel and decreases the system efficiency. Various cooling techniques have been investigated by many researches resulted in the enhanced system efficiency. Latest researchers stated some advantages and disadvantages which is listed in the conclusion section.

- PCM can store large amount of heat at a tiny temperature change
- Changing of phase can occur at stable temperature
- Further, the utilization of absorbed heat can be done
- Maintenance free
- During off sunshine also PCM can works
- High cooling capacity
- Absorption capability of the material degrades over time
- May not attain the same output during hot and cold climate

- Disposal problem after life cycle
- Due to toxic nature of some PCMs cause fire safety issue
- Segregation reduces active volume for heat storage.

Future research must focused on to find suitable techniques for cooling a different solar PV module. Finding of suitable phase change materials will also be a challenging work, and of course, the further utility of harvested heat from PCMs in a broader way.

References

1. Kumar K et al (2007) Standalone photovoltaic (PV) module outdoor testing facility for UAE climate. Submitted to CSEM-UAE Innovation Centre LLC
2. Akbarzadeh A, Wadowski T (1996) Heat pipe-based cooling systems for photovoltaic cells under concentrated solar radiation. Appl Therm Eng 16(1):81–87
3. Tonui JK, Tripanagnostopoulos Y (2007) Improved PV/T solar collectors with heat extraction by forced or natural air circulation. Renew Energy 32(4):623–637
4. Abhat A (1983) Low temperature latent heat thermal energy storage: heat storage materials. Sol Energy 30(4):313–332
5. Duffie JA, Beckman WA (1991) Solar engineering of thermal processes. Wiley, New York
6. Ismail KAR, Goncalves MM (1999) Thermal performance of a PCM storage unit. Energy Convers Manage 40(2):115–138
7. Wirtz RA, Zheng N, Chandra D (1999) Thermal management using "dry" phase change material. In: Fifteenth annual IEEE semiconductor thermal measurement and management symposium (Cat. No. 99CH36306). IEEE
8. Sarı A, Kaygusuz K (2002) Thermal and heat transfer characteristics in a latent heat storage system using lauric acid. Energy Convers Manag 43(18):2493–2507
9. Khodadadi JM, Fan L, Babaei H (2013) Thermal conductivity enhancement of nanostructure-based colloidal suspensions utilized as phase change materials for thermal energy storage: a review. Renew Sustain Energy Rev 24:418–444
10. Kibria MA et al (2015) A review on thermophysical properties of nanoparticle dispersed phase change materials. Energy Convers Manage 95:69–89
11. Zalba B et al (2003) Review on thermal energy storage with phase change: materials, heat transfer analysis and applications. Appl Therm Eng 23(3):251–283
12. Fan L, Khodadadi JM (2011) Thermal conductivity enhancement of phase change materials for thermal energy storage: a review. Renew Sustain Energy Rev 15(1):24–46
13. Waqas A, Jie J (2018) Effectiveness of phase change material for cooling of photovoltaic panel for hot climate. J Sol Energy Eng 140(4):041006
14. Sainthiya H, Beniwal NS (2019) Efficiency enhancement of photovoltaic/thermal module using front surface cooling technique in winter and summer seasons: an experimental investigation. J Energy Res Technol 141(9):091201
15. Salem Ahmed M, Mohamed ASA, Maghrabie HM (2019) Performance evaluation of combined photovoltaic thermal water cooling system for hot climate regions. ASME J Sol Energy Eng 141:041010-1–10
16. Abdollahi N, Rahimi M (2020) Using a novel phase change material-based cooling tower for a photovoltaic module cooling. J Sol Energy Eng 142(2):1–20
17. Ozakin AN, Yakut K, Khalaji MN (2020) Performance analysis of photovoltaic-heat pump (PV/T) combined systems: a comparative numerical study. J Sol Energy Eng 142: 021010-1–10
18. Zhang H, Liu H, Chen H, Guo X, Liang K, Yao P (2018) Research on the performance of flat-box photovoltaic/thermal collector with cooling channels, ASME J Sol Energy Eng 140: 021002-1–10

19. Hachem F et al (2017) Improving the performance of photovoltaic cells using pure and combined phase change materials—experiments and transient energy balance. Renew Energy 107:567–575
20. Hasan A et al (2017) Yearly energy performance of a photovoltaic-phase change material (PV-PCM) system in hot climate. Sol Energy 146:417–429
21. Japs E et al (2014) Energy-economic comparison of photovoltaic modules equipped with a layer of conventional and improved phase-change material. In: 2014 IEEE 40th photovoltaic specialist conference (PVSC). IEEE
22. Japs E et al (2013) Technical investigation of a photovoltaic module with integrated improved phase change material. In: Proceedings of the 28th European photovoltaic solar energy conference and exhibition, vol 28. Paris, Frankreich
23. Tanuwijava A et al (2013) Numerical investigation of the thermal management performance of MEPCM modules for PV applications. Energies 6(8):3922–3936
24. Ho C-J, Chou W-L, Lai C-M (2015) Thermal and electrical performance of a water-surface floating PV integrated with a water-saturated MEPCM layer. Energy Convers Manage 89:862–872
25. Chandrasekar M, Rajkumar S, Valavan D (2015) A review on the thermal regulation techniques for non-integrated flat PV modules mounted on building top. Energy Build 86:692–697
26. Du D, Darkwa Jo, Kokogiannakis G (2013) Thermal management systems for photovoltaics (PV) installations: a critical review. Sol Energy 97:238–254
27. Ma T et al (2015) Using phase change materials in photovoltaic systems for thermal regulation and electrical efficiency improvement: a review and outlook. Renew Sustain Energy Rev 43:1273–1284
28. Reddy SR, Ebadian MA, Lin C-X (2015) A review of PV–T systems: thermal management and efficiency with single phase cooling. Int J Heat Mass Transf 91:861–871
29. Bahaidarah HMS, Baloch AAB, Gandhidasan P (2016) Uniform cooling of photovoltaic panels: a review. Renew Sustain Energy Rev 57:1520–1544
30. Nižetić S et al (2018) Investigation of pork fat as potential novel phase change material for passive cooling applications in photovoltaics. J Clean Prod 170:1006–1016
31. Arıcı M et al (2018) Phase change material based cooling of photovoltaic panel: a simplified numerical model for the optimization of the phase change material layer and general economic evaluation. J Clean Prod 189:738–745
32. Preet S (2018) Water and phase change material based photovoltaic thermal management systems: a review. Renew Sustain Energy Rev 82:791–807
33. Salem MR et al (2019) Performance enhancement of the photovoltaic cells using Al_2O_3/PCM mixture and/or water cooling-techniques. Renew Energy 138:876–890
34. Fayaz H et al (2019) Numerical and outdoor real time experimental investigation of performance of PCM based PVT system. Sol Energy 179:135–150
35. Tangsiriratana E et al (2019) Thermal properties and behavior of microencapsulated sugarcane wax phase change material. Heliyon 5(8): e02184
36. Al-Waeli AHA et al () Mathematical and neural network modeling for predicting and analyzing of nanofluid-nano PCM photovoltaic thermal systems performance. Renew Energy 145:963–980
37. Nasef HA, Nada SA, Hassan H (2019) Integrative passive and active cooling system using PCM and nanofluid for thermal regulation of concentrated photovoltaic solar cells. Energy Convers Manag 199:112065
38. Maatallah T, Zachariah R, Al-Amri FG (2019) Exergo-economic analysis of a serpentine flow type water based photovoltaic thermal system with phase change material (PVT-PCM/water). Sol Energy 193:195–204
39. Nada SA, El-Nagar DH, Hussein HMS (2018) Improving the thermal regulation and efficiency enhancement of PCM-Integrated PV modules using nano particles. Energy Convers Manage 166:735–743
40. Indartono YS, Suwono A, Pratama FY (2016) Improving photovoltaics performance by using yellow petroleum jelly as phase change material. Int J Low-Carbon Technol 11(3):333–337

41. Indartono YS et al (2015) Simulation and experimental study on effect of phase change material thickness to reduce temperature of photovoltaic panel. In: IOP conference series: materials science and engineering, vol 88(1). IOP Publishing
42. Hasan M et al (2014) Development of a temperature regulated photovoltaic module using phase change material for Malaysian weather condition. J Optoelectron Adv Mater Rapid Commun 8:1243–1245

Second Law and Cycle Analysis

Enhancement of Boiling Heat Transfer Using Surfactant Over Surface with Mini-Channels

Abhilas Swain, Radha Kanta Sarangi, Satya Prakash Kar, P. C. Sekhar, and Sandeep Swain

Abstract The present article focuses on experimentally investigating the saturated pool boiling performance of isopropanol and its solution with surfactant sodium lauryl sulphate (SLS) over modified surfaces intended towards application in small heat transfer equipment. A comprehensive review on enhancement of pool boiling heat transfer using different surfactants is also presented. Three types of surfaces such as plain surface, a surface having macro-channels, surface with mini-channels are tested for pool boiling. The best performing surface, the surfaces with mini-channel is tested with three solutions of isopropanol with the above-mentioned surfactant with 200, 600 and 1000 PPM concentration. The highest heat transfer coefficient is observed for the 600 PPM solution with the surface with mini-channels.

Keywords Pool boiling · Heat transfer coefficient · Viscosity · Critical micelle concentration

1 Introduction

Surfactant is a chemical compound which when mixes with the solvents changes its property. In case of pool boiling, when it is added, it decreases the surface tension of the solution [1]. There are four types of the surfactant. These are (1) Non-ionic (2) Anionic, (3) Cationic, (4) Amphoteric. When the surfactant is added to the solution, the surfactant decreases surface tension until the CMC. After CMC is achieved, the extra surfactant is added, it goes to the CMC [1]. When the surface tension decreases, it enhances the vapour formation from the nucleation cavity which leads to the increase in the rate of the heat transfer. Kumar et al. [2] performed the experiment and suggested on different forces acting on the bubble in pool boiling. They discussed the implication of all these forces on the departure frequency, bubble departure diameter and heat transfer on upward, downward and vertical facing heater. It is observed

A. Swain (✉) · R. K. Sarangi · S. P. Kar · P. C. Sekhar · S. Swain
School of Mechanical Engineering, Kalinga Institute of Industrial Technology (KIIT) Deemed to be University, Bhubaneswar, Odisha 751024, India
e-mail: abhilas.swain@gmail.com

S. Revankar et al. (eds.), *Proceedings of International Conference on Thermofluids*, Lecture Notes in Mechanical Engineering,
https://doi.org/10.1007/978-981-15-7831-1_24

that on the downward facing of the heater, there was no bubble departure, but at the same time, bubble departure was present with all orientation of heater [2]. When the surfactant is adsorbed in the liquid–vapour interface, the force of repulsion between the bubbles is seen to be critical in avoiding bubble coalescence and influences the departure away from the heater surface [2]. This is the reason for which departure frequency is more in upward facing heater and less in downward facing heater [2].

Hetsroni et al. [3] conducted an experiment by taking environmentally accepted surfactant alkyl glycosides. They compared the boiling behaviour of the surfactant added solution with water and found that the boiling behaviour of the water is very lower than the surfactant added solution [4]. Bubble generation rates become very faster in case of the surfactant added solution, but in case of pure water, immense coalescence is seen. Adding of surfactant solution initiates activation of nucleation site in the cluster mode. Dikici and Sukaini [5] conducted the experiment by taking three surfactants sodium laurel sulphate (SLS), ECOSURF™ (EH-14) and ECOSURF™ (SA-9). It has been observed that the boiling heat transfer coefficient is much higher in SLS than EH-14 and SA-9. It is observed that boiling heat transfer enhancement is higher 46% for SLS, 30% for EH-14 and 21% for SA-9 as compared to water [6]. In order to extend the research, we have to search for some environmentally safe surfactant like SA-9 and EH-14.

Wang et al. [7] used the two surfactant solutions ethanol and silicone oil which has lower surfactant solution than water and choose three surfactants. These surfactants are cationic surfactant CTAC, anionic surfactant dodecyl benzene sulphonate and non-ionic surfactant of alkyl poly glycoside. Besides the reduction of surface tension, bubble jet and bubble explosion are the phenomena which enhance the heat transfer rate. They also described the bubble jet phenomena in the paper. Due to Marangoni convection, the surfactant molecule transfers from the bottom of the bubble to the top of the bubble which divides the larger bubble into two smaller bubbles with the change of surface tension. It was found that the surfactant solution and ethanol had the best heat transfer performance [7]. Elghanam et al. [8] conducted the experiment of boiling heat transfer by using three surfactants. These are anionic sodium dodecyl sulphate, anionic sodium laurel sulphate and non-ionic Triton-X-100. By adding surfactant to the cooling water improves the heat transfer with the amount of 241% in case of SDS, 185% in case of SLES and 133% in case of Triton-X-100. In a given aqueous solution concentration, when the temperature increases, it increases the pool boiling heat transfer coefficient and active nucleation site density. Heat transfer increases for the SDS and SLS with increasing the concentration, but for the Triton-X-100, it increases up to 500 ppm of concentration and beyond that value, insignificant heat transfer observed [8].

Zhang and Manglik [9] experimented by taking the surfactants sodium dodecyl sulphate, cetyltrimethyl ammonium bromide (CTAB) and octyl phenoxy olyethoxy ethanol (Triton-X-305). Molecular mobility of these surfactants at interface shows the dynamic surface tension, surface wetting which leads to the formation of vapour bubble from a heated surface and changes the heat transfer significantly. Heat transfer coefficient increases when the surfactant concentration less than equal to CMC, and the heat transfer coefficient decreases when the concentration greater than CMC.

An optimum heat transfer enhancement is obtained at the CMC of the surfactant. Gajghate et al. [4] performed the boiling heat transfer experiment by using the NH_4CL (ammonium chloride) surfactant. By the addition of the surfactant, the thermophysical properties of the solution are changed. It is observed that up to 2600 ppm, there is significant increase in heat transfer rate, and beyond that limit, no significant heat transfer is observed [10]. Decrement of surface tension and wettability are two phenomena which play an important role in nucleate boiling heat transfer in case of surfactant ammonium chloride [4].

Wasekar and Manglik [6] conducted his experiment by taking the anionic surfactant like (SDS, SLS) and compared the boiling performance of pure water with aqueous solution of anionic surfactant. Heat transfer is enhanced by the presence of SDS with the early onset of the nucleate boiling. Optimum level of enhancement is observed at critical micelle concentration (CMC). It has been observed that kinetics of the surfactant molecule is different for the room temperature and for the boiling temperature. This has impact on the formation of vapour bubbles along with increments of departure frequency, and it also decreases the tendency of coalescence which causes foaming. Hetsroni et al. [11] studied the bubble growth in saturated pool boiling in water and surfactant solution. They have done the analysis by taking two different heat fluxes $q = 10$ kW/m^2 and $q = 50$ kW/m^2. At lower heat flux, the lifetime, volume and shape of the bubble did not differ much from water. But at higher heat flux, boiling is more vigorous than the normal water. The lifetime of the bubble in the cluster is lesser than the single water bubble. They observed that for the water with increasing heat flux, detachment diameter of water bubbles increases, but in the case of surfactant solution, the detachment diameter of bubble decreases with increasing heat flux. Hetsroni et al. [12] conducted the experiment and stated that not only surface tension but also kinematic viscosity plays a major role in changing the boiling behaviour of a liquid in which surfactant is added. Literature survey [12] showed that by adding surfactant to sea water, it can enhance the boiling process. So, it can become cost effective to an acceptable level [12]. Hetsroni et al. [11] and Hetsroni et al. [3] took the cationic surfactant HABON-G and conducted the experiment and found that the bubbles found in the HABON-G surfactant are very much smaller than vapour formed in water and the heated surface remains covered with it.

Suryanarayan et al. [13] choose the surfactant ammonium dodecyl sulphate (ADS) for performing the experiment as it is human friendly and three times best soluble than SDS. With the addition of ADS, the molecule of ADS disturbs bonding due to cohesion and adhesion, and so the bubble formed at low temperature difference. This leads to formation of wetted bubbles which contain high vapour inside. Due to low adhesive force between liquid and solid surface, it detaches from solid surface so quickly, and bubble flow occurs in upward direction due to buoyancy effect. ADS can be used as best alternatives for SDS. Zicheng et al. [10] conducted the experiment by taking two surfactant SDS, Triton-X-114. The result showed that boiling is more vigorous with smaller size and fast departure frequency. Best heat transfer is achieved at CMC of surfactant. Optimum heat transfer is more in SDS than Triton-X-114 at the same heat flux compared to water. Zicheng et al. revealed that heat transfer enhancement is due to the reduction of the surface tension and change of contact angle. The contact angle of SDS solution is independent on the change of concentration, but for the Triton-X-100, contact angle is inversely proportional to the concentration [10]. Wasekar and Mangklik [14] performed the experiment by taking surfactant Triton-X-100 and the nichrome wire. Throughout the experiment, they tested the heat transfer effect on different concentration of the Triton-X-100. They concluded that up to 500 ppm concentration of Triton-X-100, heat transfer increases, but beyond that, there will be no significant heat transfer [14].

Thus, it can be observed from the literature that the effect of surfactant on boiling heat transfer having organic liquids is not studied. The characteristic of the organic liquids is similar to the liquids used for the electronics cooling. Thus, the present article focuses on the performance of the isopropanol with the concentration of surfactant sodium dodecyl sulphate.

2 Experimental Set-up and Procedure

An experimental set-up is designed and developed for investigating the boiling heat transfer over small sample surfaces. The major components of the experimental facility are the test chamber, variable resistor (variac), direct current power supply, digital multi-metre and temperature indicator. The test chamber is a rectangular chamber made up of the base plate, top cover and the glass walls, and the test chamber is composed of PTFE sheet. The four side walls are made up of glass for visualization purpose. The glass and PTFE sheet can be approximated as insulators for heat loss from the saturated liquid. A copper tube carrying tap water is placed near the top wall which will act as condenser by which the vapour produced can be again be converted to liquid.

In base plate of the test chamber, the test sample made up of copper cylinder of diameter 30 mm and length 40 mm is inserted within another nylon cylinder of 42 mm diameter and 70 mm length. One 6 mm hole is drilled in the copper cylinder in parallel alignment through which the heater passes. A small hole of 2 mm size diameter is drilled at the centre of the copper through which the thermocouple passes. The tip of the T-type thermocouple is just touching the top surface of the copper block whose reading is recorded as the surface temperature. The temperature of the liquid in the test chambered is ensured to be at saturated condition before recording the temperature. The thermocouples are calibrated with a dry block type calibrator, and the error associated is within ±1 °C (Fig. 1).

At the top and the bottom of the glass window, Teflon cover is used because it is both thermal and liquid proof. The glass and Teflon layer are attached together by resin bond that prevents liquid leakage. The boiling phenomenon happening inside the chamber can be visualized through the glass chamber. The AC current supply is provided to the DC power supply through a variable resistor to control the input voltage supply. Thus, by controlling the input to the DC power supply, the required output voltage is achieved. The experiments are conducted for 6, 7, 8, 9 and 10, 11 V,

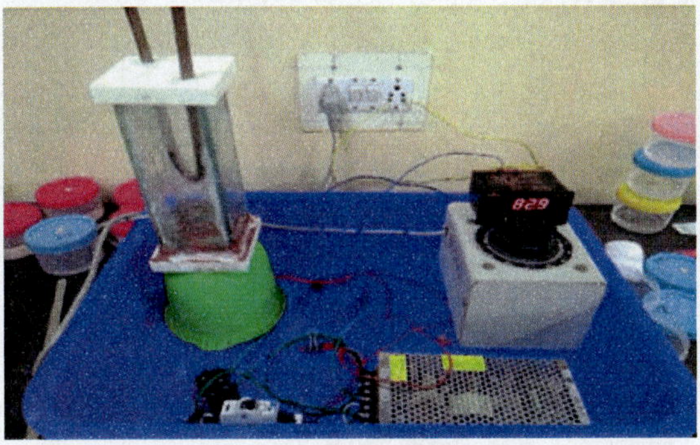

Fig. 1 Photograph of the test chamber

measured by digital multi-metre. Thus, from the voltage and current, the power input to the heater is calculated. By dividing the heater power with the surface area of the sample copper surface, the heat flux is determined. A 12 A magnetic circuit breaker (MCB) is also used to prevent any damage to the electric circuit.

First setting up the variac at a particular voltage level is done to fix the heat flux. Then, the liquid inside the chamber starts heating up, and after reaching to saturation temperature, the boiling occurs. After that, the readings of temperature and voltage are noted down only when the steady state is achieved. Figure 2 shows the different surfaces produced for the experimental study, and the characteristics are mentioned in Table 1. The smooth surface is produced by rubbing the copper surface with a 1200 grade emery paper. The roughness average of smooth surface is 0.5 μm. The surface having macro-channels is produced by scratching the surface by a hacksaw blade

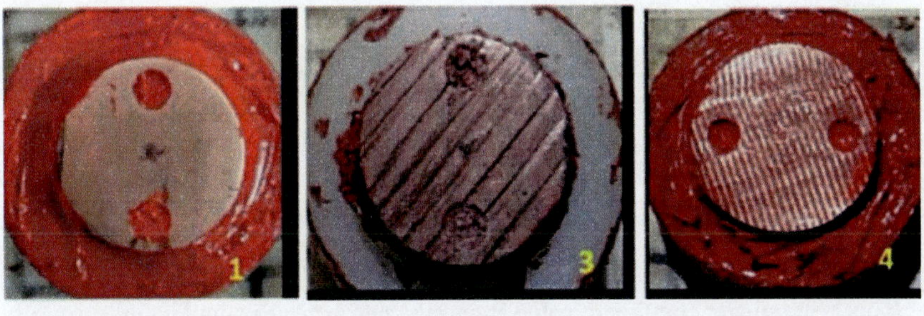

Fig. 2 Different surfaces used to study pool boiling heat transfer. 1 Plain or smooth, 2 Surface having macro-channels. 3 Mini-channels

Table 1 Surface characteristics

Types of surfaces	Surface characteristics
Smooth surfaces	0.5 μm roughness
Macro-channel surfaces	0.5 mm deep and 1 mm width channel
Mini-channel surfaces	0.5 mm width and height 0.8 mm

Table 2 Heat flux values obtained from experiment

Voltage	Power	Heat flux (unit)
7	13.61	19,251.92
8	17.78	25,145.371
9	22.50	31,824.61
10	27.77	39,289.64
11	33.61	47,540.46
11.5	36.73	51,960.55

several times from many directions producing a roughness average of 9–10 μm. The surface with macro-channels is produced by drawing channels through hacksaw blade. The deep channels on the surface are 0.5 mm deep and 1 mm width. The surface with very fine mini-channels is produced by using the knurling tool on the flat face of the copper cylinder. The channels are 0.5 mm width, and the height of walls of mini-channels are 0.8 mm.

The set-up is assembled and checked for any leakage. Initially, the glass chamber is cleaned with water and then with acetone. It is then kept sometime for acetone to be evaporated. Then, the test liquid is poured into the chamber. Then, the electrical power is supplied to the autotransformer. The controlling knob of the autotransformer is rotated to get the required voltage corresponding to the highest heat flux value. The voltage is checked through a digital multi-metre. Then, the liquid is heated until the steady state is achieved. Then, the final temperature readings are recorded. The heat flux values corresponding to the voltage applied are presented in Table 2.

3 Results and Discussion

The heat transfer coefficient is determined from the heat flux applied to the surface divided by the excess wall superheat, i.e. the difference between the surface temperature and the liquid temperature.

$$\mathrm{HTC} = q/(T_s - T_l) \qquad (1)$$

This heat transfer coefficient value shows the performance level of any surface in case of boiling heat transfer. Therefore, the HTC values are presented, compared and discussed here. Figure 3 illustrates the trend of the HTC for saturated pool boiling of isopropanol over all the three surfaces under atmospheric pressure.

Figure 3 shows the variation of the PBHTC with heat flux for all the three surfaces taking isopropanol as the working liquid. Moreover, the variation in PBHTC for isopropanol is similar to that of the acetone, i.e. it is observed to be increasing with heat flux for all the three surfaces. Again, the PBHTC are higher for the surface with mini-channels than the other two types of surfaces.

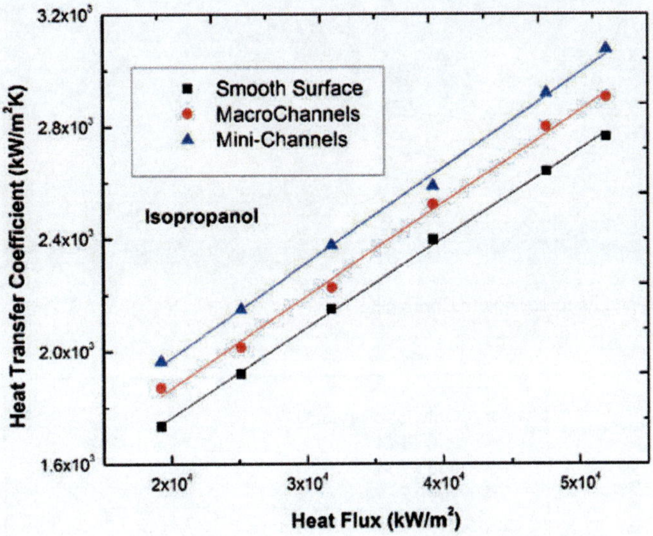

Fig. 3 Variation of HTC with heat flux for isopropanol taken on all surfaces

The following points can be inferred from Fig. 3. (a) The heat transfer coefficients increase with raising the heat flux level applied to the heater surface. (b) The heat transfer coefficients are highest for the mini-channel surface at all heat flux levels and then follows the order as channelled > rough > smooth.

The reason for the first observation is very much well established from previous investigations. The number of activated nucleation sites increases due to increase in the heat input level to the surface. With the higher heat flux level, the surface temperature increases which activates more number of cavities on the surfaces leading to more number of incipience of bubble formation. Thus, due to the latent heat of evaporation dissipation from the surface, the heat transfer coefficients increase. There are two reasons for the second observation, i.e. the mini-channel surface is having highest heat transfer coefficient values. The mini-channel surface is having higher surface area due to which the rate of heat dissipation is higher. The small channels on the surface help in vapour trapping in nucleation sites along the channels like re-entrant cavities. The vapour trapping of the nucleation sites is necessary for bubble formation. It means that the nucleation sites must have little amount of vapour covered with pool of liquid to be able to incipience bubble formation.

To study the effect of surfactant, sodium lauryl sulphate (SDS) is added with isopropanol to get solutions of different concentrations. The solutions prepared are 200, 600 and 1000 PPM. Figure 4 shows the compound and the three solutions.

Figure 5 depicts the comparison between the pool boiling performances of isopropanol over the mini-channel surfaces for all the heat flux values. It has been observed that the heat transfer coefficient values are higher for isopropanol with surfactant than the pure isopropanol. The surfactant mixed with the isopropanol

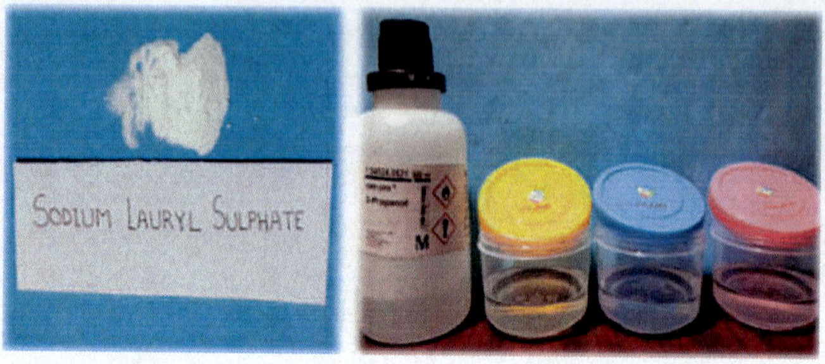

Fig. 4 SLS and its solution with isopropanol

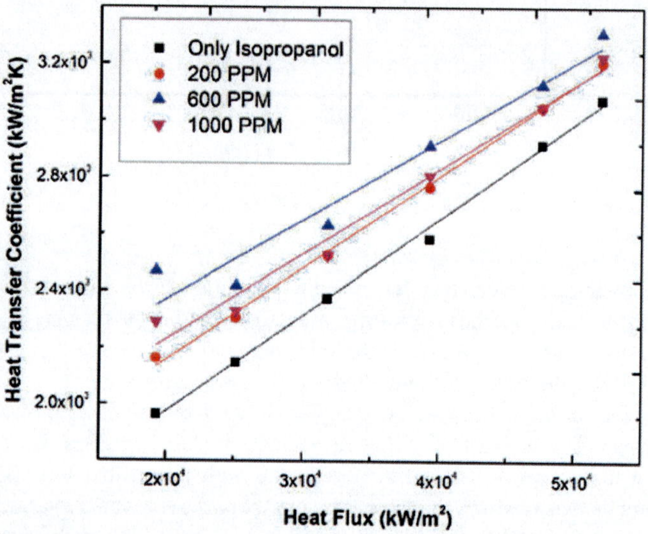

Fig. 5 HTC versus heat flux for surfactant solutions of isopropanol

decreases the surface tension value which leads to more number of active nucleation sites. This increases the heat transfer rate.

It can be seen that the HTC rises with rise in heat flux value for all the three solutions of isopropanol with 200, 600 and 1000 PPM. This is due to the fact that with the increment in heat flux value, the number of active nucleation sites increases from which the bubbles generate. Thus, more number of bubble formation leads to more extraction of latent heat of vapourization from the surface. Therefore, the boiling HTC and heat flux are directly proportional for all the solutions.

However, it is also observed that the heat transfer coefficient is lower for the solution with 1000 PPM than that corresponding to 600 PPM. This happens because of the fact that as the surface tension decreased further, the bubble growth is also hindered.

As the surface tension has two counteracting effects on the bubble dynamics, an optimized amount of surfactant can be the best solution. In this present case, the solution with 600 PPM performs better than the 200 and 1000 PPM solution. Figure 6 shows the variation of wall superheats obtained during the experiments.

The values of pool boiling heat transfer coefficient obtained from the experiment are correlated in terms of heat flux value and concentration of surfactant through the regression modelling approach. The correlation obtained is presented in Eq. (2). The predicted values lie within $\pm 15\%$ of the experimental value as observed in Fig. 7.

$$\mathrm{PBHTC} = 50.09q^{0.37}s^{0.013} \tag{2}$$

Thus, the above investigation showed that the pool boiling heat transfer coefficient of isopropanol on the surface having mini-channels is enhanced by addition of the surfactant. However, as the concentration reaches the critical micelles concentration, the PBHTC starts declining. This is 600 PPM for SLS with isopropanol over the surface with mini-channels. The condition may vary depending on the characteristics of the surface.

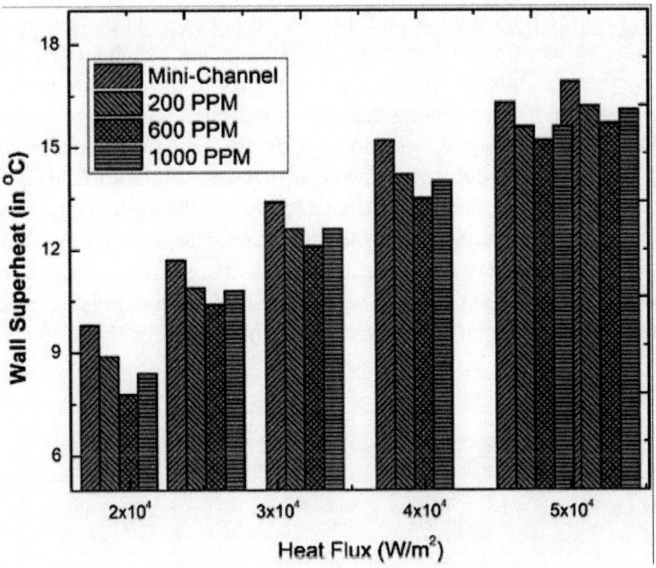

Fig. 6 Variation of wall superheat with heat flux for surfactant solutions of isopropanol over mini-channels surface

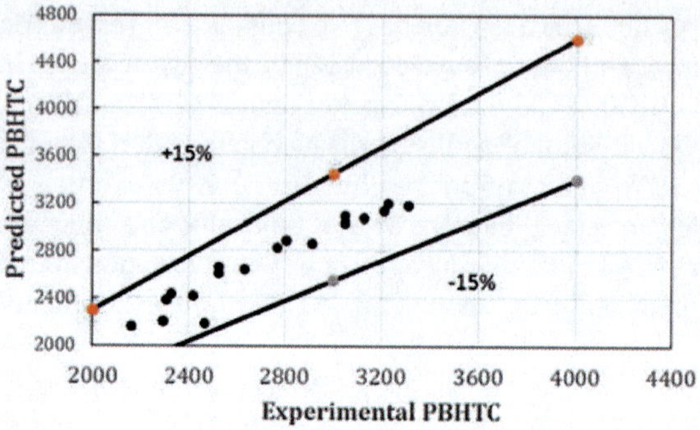

Fig. 7 Comparison of predicted and experimental pool boiling heat transfer coefficient

4 Conclusion

The saturated pool boiling performance of pure isopropanol is studied on three kinds of surfaces such as smooth, surface with macro-channels and surface with mini-channels. The surface with mini-channels outperformed than the smooth and macro-channel surfaces. Then, this surface is chosen for the further study on the effect of surfactant on the pool boiling performance. It is observed that the heat transfer coefficient increases with increase in concentration of surfactant up to a certain point and then decreases. Thus, in the present investigation, the solution with 600 PPM solution is found to be best among the three solutions. Addition of surfactant also decreases the surface tension which enhances the formation of vapour bubbles in the nucleation cavity. These bubbles take away heat content of the surface by the vapour and increase the heat transfer rate. After addition of the surfactant, the departure diameter of the bubbles becomes very less as compared to the normal water due to the decrease of the surface tension. It leads to the increase in heat transfer coefficient. If the PPM is less than 200 ppm, then the heat transfer coefficient is less as compared to the solution with 600 ppm concentration. Again if the concentration is more than 1000 ppm, then also, the HTC decreases due to the formation of critical micelles concentration.

References

1. Ajinath K, Kulkarni K (2019) Review on natural and synthetic surfactant for pool boiling. In: Proceedings of second Shri Chhatrapati Shivaji Maharaj QIP conference on engineering innovations, pp 393–395
2. Kumar N, Raza Q, Raj R (2018) Surfactant aided bubble departure during pool boiling. Int J Therm Sci 131:105–113. https://doi.org/10.1016/j.ijthermalsci.2018.05.025
3. Hetsroni G, Gurevich M, Mosyak A, Rozenblit R, Segal Z (2004) Boiling enhancement with environmentally acceptable surfactants. Int J Heat Fluid Flow 25(5):841–848. https://doi.org/10.1016/j.ijheatfluidflow.2004.05.005

4. Gajghate S, Acharya AR, Pise AT (2014) Experimental study of aqueous ammonium chloride in pool boiling heat transfer. Exp Heat Transfer J Therm Energy Gen Transp Storage Conver 27(2):113–123. https://doi.org/10.1080/08916152.2012.757673
5. Dikici B, Al-Sukaini BQA (2017) Comparison of aqueous surfactant solutions. In: ASME power conference, 2016, POWER2016-59351, pp 1–7. https://doi.org/10.1115/POWER2016-59351
6. Wasekar VM, Manglik RM (2002) The influence of additive molecular weight and ionic nature on the pool boiling performance of aqueous surfactant solutions. Int J Heat Mass Transfer 45(3):483–493. https://doi.org/10.1016/S0017-9310(01)00174-0
7. Wang J, Li FC, Li XB (2016) On the mechanism of boiling heat transfer enhancement by surfactant addition. Int J Heat Mass Transfer 101:800–806. https://doi.org/10.1016/j.ijheatmasstransfer.2016.05.121
8. Elghanam RI, Fawal MMEL, Aziz RA, Skr MH, Khalifa AH (2011) Experimental study of nucleate boiling heat transfer enhancement by using surfactant. Ain Shams Eng J 2(3–4):195–209. https://doi.org/10.1016/j.asej.2011.09.001
9. Zhang J, Manglik RM (2016) Additive adsorption and interfacial characteristics of nucleate pool boiling in aqueous surfactant solutions. J Heat Transfer 127(7):684–691. https://doi.org/10.1115/1.1924626
10. Zicheng H, Jiaqiang G, Xinnan S, Qian W (2011) Pool boiling heat transfer of aqueous surfactant solutions. In: Proceedings—4th international conference on intelligent computation technology and automation. ICICTA 2011, vol 2, pp 841–844. https://doi.org/10.1109/ICICTA.2011.497
11. Hetsroni G, Mosyak A, Pogrebnyak E, Sher I, Segal Z (2006) Bubble growth in saturated pool boiling in water and surfactant solution. Int J Multiph Flow 32(2):159–182. https://doi.org/10.1016/j.ijmultiphaseflow.2005.10.002
12. Hetsroni G, Zakin J, Lin Z, Mosyak A, Pancallo E, Rozenblit R (2001) The effect of surfactants on bubble growth, wall thermal patterns and heat transfer in pool boiling. Int J Heat Mass Transf 44(2):485–497. https://doi.org/10.1016/s0017-9310(00)00099-5
13. Suryanarayana G, Rao G, Balakrishna N (2015) Experimental investigation on pool boiling heat transfer with sodium dodecyl sulfate. Int J Mech Eng Comput Appl 3(2):032–037
14. Wasekar VM, Manglik RM (2000) Pool boiling heat transfer in aqueous solutions of an anionic surfactant. J Heat Mass Transfer 122(4):708–715

An Analytical Investigation of Pressure-Driven Transport and Heat Transfer of Non-Newtonian Third-Grade Fluid Flowing Through Parallel Plates

Sumanta Chaudhuri, Paromita Chakraborty, Rajen Das, Amitesh Ranjan, and Vijay Kumar Mishra

Abstract Flow of a third-grade fluid induced by the difference in pressure, thorough rectangular parallel plates, having different wall temperatures, is revisited considering the effect of viscous dissipation. The governing equations, describing the physical phenomenon, are nonlinear. The non-Newtonian third-grade fluid parameter is considered to be small, and the governing equations are reduced to weakly nonlinear equations. The nonlinear momentum and energy conservation equations, thus obtained, are solved employing perturbation method. Analytical solutions for the velocity and temperature distributions are obtained. Results generated by the perturbation method are compared with that of Danish et al. (Commun Nonlinear Sci Numer Simul 17:1089–1097, 2012 [1]) and an excellent agreement is exhibited within the small range of perturbation parameter (third-grade fluid parameter). Effects of various parameters such as Brinkmann number, non-Newtonian third-grade fluid parameter, temperature ratio of the upper plate and lower plate on the variation of velocity and temperature are discussed. Results indicate that an increase the third-grade fluid parameter results in a decrease in the velocity. Temperature of the fluid decreases with an increase in third-grade fluid parameter and displays an increasing trend with an increase in Brinkman number. The peak temperature is observed to occur not at the upper plate, at a region which is near the upper plate. Results are useful for designing thermal systems applied in the fields of polymer melt flows, food processing, flow of slurry, etc.

Keywords Third-grade fluid · Pressure-driven flow · Homotopy perturbation method · Viscous dissipation

S. Chaudhuri · P. Chakraborty · R. Das · A. Ranjan · V. K. Mishra (✉)
Kalinga Institute of Industrial Technology (KIIT) Deemed to be University, Bhubaneswar, Odisha 751024, India
e-mail: vijay.mishrafme@kiit.ac.in

S. Revankar et al. (eds.), *Proceedings of International Conference on Thermofluids*, Lecture Notes in Mechanical Engineering,
https://doi.org/10.1007/978-981-15-7831-1_25

1 Introduction

Transport phenomenon of non-Newtonian fluids has drawn the attention of numerous researchers because of its diverse engineering applications in food processing, flow of polymer melts, slurry, etc. Researchers, engaged in this filed for describing the transport characteristics of non-Newtonian fluids, have proposed several non-Newtonian fluid models. Power law model, Sisko fluid model, viscoelastic model, third-grade fluid model, Bingham plastic model, etc., are some of the non-Newtonian fluid models frequently employed for modeling the behavior of these kinds of fluids. Power law model is frequently used which is a two-parameter model, and it can describe the fluid behavior within a short interval of shear stress and strain rate. On the other hand, the Sisko fluid model can be regarded as in improvement over the power law model as it has three parameters in the constitutive equation. Consequently, the Sisko model can describe the fluid behavior in a larger range of shear stress and strain rate data. Viscoelastic model attempts to capture the flow behavior considering the fluid having both the characteristics of elastic solid and viscous fluids. Bingham plastic is one which models the fluid behavior which exhibits fluid characteristics only after a certain stress limit is crossed. Third-grade fluid model, also termed as Rivlin–Ericson fluid of grade three, has been employed by several researchers for modeling polymer melts, food items in diverse engineering applications.

Flow of a third-grade fluid, through a set of parallel plates, has been studied by Siddiqui et al. [2], and analytical solution for the velocity is obtained by employing Adomian decomposition method (ADM) [2, 3]. Bhatti et al. [4] investigated flow of and heat transfer of third-grade fluid through parallel plates employing homotopy perturbation method (HPM) [4–6]. Danish et al. [1] presented exact analytical solutions for the velocity and flow rate for pressure difference-driven (Poiseuille) and combined pressure difference and shear-driven (Couette–Poiseuille) flow of a third-grade fluid through a set of parallel plates. Researchers have also examined the effect of externally imposed electrical and magnetic field on the transport characteristics of third-grade fluids. Li et al. [7] investigated the effect of rotating electroosmosis on the flow of third-grade fluid through a set of two micro-parallel plates. Thermal characteristics of heat transfer in wire coating analysis of third-grade fluid under the influence of externally imposed magnetic field have been studied by Nayak et al. [8]. In the governing equations, temperature-dependent viscosity is considered; homotopy analysis method (HAM) is employed to solve the nonlinear equations. Wang et al. [9] studied electromagnetohydrodynamic (EMHD) forced convection of a third-grade fluid through parallel plates, employing traditional perturbation method and the effect of Brinkmann number, non-Newtonian third-grade fluid parameter, on the dimensionless temperature, velocity, and Nusselt number has been analyzed. Chaudhuri and Sahoo [10] examined effect of the aspect ratio on the characteristics of a third-grade fluid in magnetohydrodynamic (MHD) flow through a rectangular channel.

The preceding discussion clearly highlights the importance of exploring flow and heat transfer of third-grade fluid and different aspects such as their behavior under the influence of externally imposed electrical and magnetic field. However, many of the issues are yet to be explored in detail and need further investigation. Further, few studies can be revisited for improvement in the reported results. For example, Danish et al. [1] presented exact analytical solution for the velocity fields of pressure difference-driven flow and combined pressure difference and shear-driven flow of a third-grade fluid through parallel plates. Temperature distribution, however, was not considered in the study. Siddiqui et al. [4] studied temperature distribution applying HPM. In the present study, transport characteristics of a third-grade fluid flowing through a set of parallel plates are revisited and results are generated employing traditional perturbation method. Construction of a proper homotopy is required in HPM and choosing a guess solution is also essential. If the homotopy is not constructed properly and guess solution is not chosen judiciously, results generated may not be accurate. Compared to this, implementation of perturbation method is easier. Additional results on temperature distribution are presented, which can serve useful for designing of thermal systems handling polymer melts, slurry, food items in the industries.

2 Problem Formulation

The schematic of the physical problem, addressed in the present study, is pictorially shown in Fig. 1. Flow of a third-grade fluid, induced by pressure difference, through a set of parallel plates has been considered. The upper and lower walls are maintained at constant temperatures T_l and T_u, respectively. Flow considered in the current study is assumed to be steady, laminar, hydrodynamically fully developed, and incompressible. Further, properties are assumed to be independent of temperature. The gap between the pates $2H$ is considered to be very small. Representative velocity and temperature profile are displayed in Fig. 1.

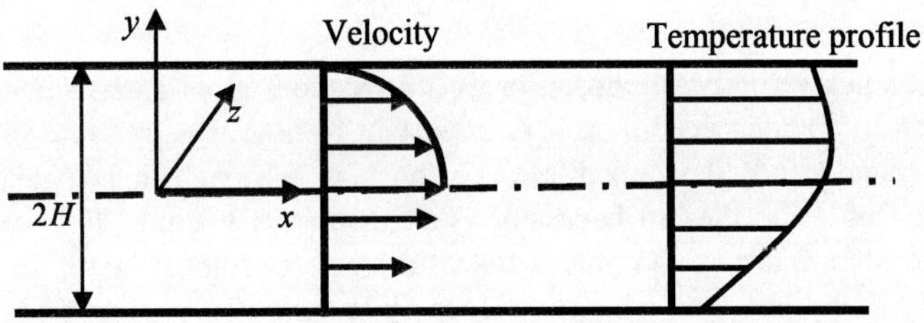

Fig. 1 Schematic of the flow of a third-grade fluid through a set of parallel plates maintained at constant temperature

2.1 Mathematical Formulation

Governing equations, describing transport phenomena of flow as well as heat transfer are obtained from the conservation of momentum and energy. The equations, in their non-dimensional forms, are presented in the present study. Conservation of momentum and energy equations, in non-dimensional forms, is presented as follows:

$$\frac{\mathrm{d}^2u}{\mathrm{d}y^2} + 6A\frac{\mathrm{d}^2u}{\mathrm{d}y^2}\left(\frac{\mathrm{d}u}{\mathrm{d}y}\right)^2 = N \tag{1}$$

$$\frac{\mathrm{d}^2T}{\mathrm{d}y^2} + Br\left(\frac{\mathrm{d}u}{\mathrm{d}y}\right)^2 + 2ABr\left(\frac{\mathrm{d}u}{\mathrm{d}y}\right)^4 = 0 \tag{2}$$

where *u*, *T*, *y*, *A*, *Br*, *N* are non-dimensional velocity, temperature, non-dimensional coordinate normal to the direction of flow (axial direction), non-Newtonian third-grade fluid parameter, Brinkman number, and dimensionless pressure gradient, respectively. The dimensionless variables, and parameters are defined as follows:

$$u = \frac{u^*}{u_0},\quad y = \frac{y^*}{H},\quad T = \frac{T^*}{T_l},\quad A = \frac{\beta u_0^2}{\mu H^2},\quad Br = \frac{\mu u_0^2}{KT_l},\quad N = \left(\frac{\mathrm{d}p}{\mathrm{d}x}\right)\frac{H^2}{\mu u_0} \tag{3}$$

where u^*, u_0, y^*, H, T^*, *TL*, K, $\mathrm{d}p/\mathrm{d}x$ are dimensional velocity, reference velocity, dimensional coordinate along the direction perpendicular to the axis, half of the gap between the parallel plates, dimensional temperature, temperature of the lower plate, thermal conductivity of the fluid, and the pressure gradient along the axial direction, respectively. In the present study, no inherent reference velocity exists and the same has to be defined. Reference velocity can be chosen as follows:

$$u_0 = \left(\frac{\mathrm{d}p}{\mathrm{d}x}\right)\frac{H^2}{\mu} \tag{4}$$

The reference velocity so chosen in Eq. (4) reduces $N = 1$; the same value of N is chosen by Danish et al. [1]. It is important to note that pressure gradient in hydrodynamically fully developed flow is a constant. (The right-hand side of Eq. (1) is a function of x and the left-hand side is a function of y alone. The condition is satisfied for all x and y, if and only if both sides are constant.) Therefore, pressure varies as a straight line with a negative slope up to the outlet starting from the fully developed region. It is to be noted that the convective term $u\frac{\partial T}{\partial x}$ approximated to be zero as a consequence of the assumption of narrow gap between the parallel plates. Therefore, the right-hand side of the non-dimensional energy equation [Eq. (2)] is set to zero. It is important to note that the assumption of narrow gap between the plates, finally discards the effect of the convective term, and reduces the problem to that of a conduction heat transfer including the effect of viscous heat generation.

This is a valid assumption as the flow is considered through large parallel plates, which is equivalent to the flow through a channel which is very large along the transverse direction compared to the gap between the plates. Equation (1) is an ordinary differential equation (ODE), nonlinear in nature, of order two, which requires two boundary conditions to obtain an unique solution. The boundary conditions, in the non-dimensional forms, needed for solving Eq. (1) are presented as follows (no-slip boundary condition at the upper and lower walls):

$$u(-1) = u(+1) = 0 \tag{5}$$

Equation (2), too, is a second-order equation requiring two boundary conditions which are as follows:

$$T(-1) = 1, \quad T(+1) = \frac{T_u}{T_l} = T_r \tag{6}$$

where T_{u} and T_{l} are temperatures of upper and lower plates, respectively. For solving Eqs. (1) and (2), perturbation method is employed, which is an effective analytical technique for tackling nonlinear partial differential equation (PDE) and ordinary differential equations (ODE). Perturbation technique can generate accurate results for weakly nonlinear equations requiring the presence of a small parameter in the governing equation. In the present study, it is assumed that the non-Newtonian third-grade fluid parameter A is small, implying weak non-Newtonian effect. Considering A to be a very small quantity ε ($A = \varepsilon$), the governing equations are reduced to the following:

$$\frac{\mathrm{d}^2 u}{\mathrm{d}y^2} + 6\varepsilon \frac{\mathrm{d}^2 u}{\mathrm{d}y^2}\left(\frac{\mathrm{d}u}{\mathrm{d}y}\right)^2 = 1 \tag{7}$$

$$\frac{\mathrm{d}^2 T}{\mathrm{d}y^2} + Br\left(\frac{\mathrm{d}u}{\mathrm{d}y}\right)^2 + 2\varepsilon Br\left(\frac{\mathrm{d}u}{\mathrm{d}y}\right)^4 = 0 \tag{8}$$

For solving Eqs. (7) and (8), u and T are expanded with respect to ε as follows:

$$u = u_0 + \varepsilon u_1 + \varepsilon^2 u_2 + \cdots, \quad T = T_0 + \varepsilon T_1 + \varepsilon^2 T_2 + \cdots \tag{9}$$

Substituting u and T, from Eq. (9) into Eqs. (7) and (8) and equating the coefficients of ε_0, ε_1, and ε_2 to zero, the leading-order (zeroth order), first-order, and second-order equations are obtained as follows:

2.2 0th-order equations and boundary conditions:

$$\frac{d^2 u_0}{dy^2} = 1 \tag{10}$$

$$\frac{d^2 T_0}{dy^2} + Br\left(\frac{du_0}{dy}\right)^2 = 0 \tag{11}$$

Boundary conditions (no-slip and no temperature jump boundary condition):

$$\text{At } y = -1, \quad u_0 = 0, \quad T_0 = 1 \tag{12}$$

$$\text{At } y = -1, \quad u_0 = 0, \quad T = T_r \tag{13}$$

First-order equation are as follows (no-slip and no temperature jump boundary condition):

$$\frac{d^2 u_1}{dy^2} + 6\frac{d^2 u_0}{dy^2}\left(\frac{du_0}{dy}\right)^2 = 0 \tag{14}$$

$$\frac{d^2 T_1}{dy^2} + 2Br\frac{du_0}{dy}\frac{du_1}{dy} + 2Br\left(\frac{du_0}{dy}\right)^4 = 0 \tag{15}$$

Boundary conditions for the first-order equation:

$$\text{At } y = -1, \quad u_1 = 0, \quad T_1 = 0 \tag{16}$$

$$\text{At } y = 1, \quad u_1 = 0, \quad T_1 = 0 \tag{17}$$

Second-order equations and boundary conditions are given as below:

$$\frac{d^2 u_2}{dy^2} + 6\frac{d^2 u_1}{dy^2}\left(\frac{du_0}{dy}\right)^2 + 12\frac{d^2 u_0}{dy^2}\frac{du_0}{dy}\frac{du_1}{dy} = 0 \tag{18}$$

$$\frac{d^2 T_2}{dy^2} + Br\left(\frac{du_1}{dy}\right)^2 + 2Br\frac{du_0}{dy}\frac{du_2}{dy} + 8Br\left(\frac{du_0}{dy}\right)^3\frac{du_1}{dy} = 0 \tag{19}$$

Boundary conditions:

$$\text{At } y = -1, \quad u_2 = 0, \quad T_2 = 0 \tag{20}$$

$$\text{At } y = 1, \quad u_2 = 0, \quad T_2 = 0 \tag{21}$$

Solution for zeroth-order equations.

Equation (10) is solved and the boundary conditions presented by Eqs. (12) and (13) are used to find the leading-order or zeroth-order solution as follows:

$$u_0 = 0.5(y^2 - 1) \tag{22}$$

u_0 obtained from Eq. (22) is substituted in Eq. (11) and using the boundary conditions given by Eqs. (12) and (13), solution for the leading-order temperature distribution is obtained as given below:

$$T_0 = 0.5T_r(1 + y) + 0.5(1 - y) + \frac{Br}{12}(1 - y^4) \tag{23}$$

Solution to first-order equations.

Substituting Eq. (22) in Eq. (14) and using the boundary conditions presented by Eqs. (13.1) and (13.2), the solution for the second-order velocity is obtained as follows:

$$u_1 = \frac{1}{2}(1 - y^4) \tag{24}$$

Using Eq. (24) in Eq. (15), and applying the boundary conditions given by Eqs. (16) and (17), the solution for the temperature distribution of first-order equation is obtained as follows:

$$T_1 = \frac{Br}{30}(y^6 - 1) \tag{25}$$

Solution of the second-order equations.

Substituting the velocity distributions for the zeroth and first order from Eqs. (22) and (24) in Eq. (18) and utilizing boundary conditions given by Eqs. (20) and (21), solution for the velocity distribution of second-order equation is obtained as below:

$$u_2 = 2(y^6 - 1) \tag{26}$$

Substituting velocity distributions for the zeroth, first, and second-order velocity distributions from Eqs. (22), (24) and (26) in Eq. (19) and utilizing boundary conditions given by Eqs. (20) and (21), solution for the temperature distribution of second-order equation is obtained as follows:

$$T_2 = \frac{3}{14}Br(1 - y^8) \tag{27}$$

Substituting the solutions for the zeroth-order, first-order, and second-order velocity distributions and temperature distributions in Eq. (9) velocity and temperature distributions are obtained as below:

$$u = \frac{1}{2}(y^2 - 1) + \varepsilon\frac{1}{2}(1 - y^4) + \varepsilon^2 2(y^6 - 1) \tag{28}$$

$$T = \frac{T_r}{2}(1 + y) + \frac{1}{2}(1 - y) + \frac{Br}{12}(1 - y^4) + \varepsilon\frac{Br}{30}(y^6 - 1) + \varepsilon^2\frac{3}{14}Br(1 - y^8) \tag{29}$$

3 Results and Discussions

Figure 2 exhibits the comparison of the results obtained from the study of Danish et al. [1] and that of the current study. As the results of the current study are generated by employing traditional perturbation method, perturbation parameter ε (third-grade fluid parameter) is varied up to 0.1. From Fig. 2, is evident that the results generated by perturbation method are in close agreement with that of the exact results of Danish et al. [1]. Figure 2 clearly reveals that the velocity decreases significantly when the non-Newtonian third-grade fluid parameter decreases. The resistance towards the flow increases with an increase in the non-Newtonian third-grade fluid parameter. This factor leads to a decrease in the velocity. It is to be noted that velocity is zero at one wall and then increases with increase in y. Velocity is maximum at the center and then again decreases gradually. Finally reaching zero at the other wall. Effect of the temperature ratio of the upper and lower plate is displayed in Fig. 2. It is evident that the effect of the temperature ratio on the temperature distribution is significant. If the temperature of the upper wall is higher compared to the lower wall

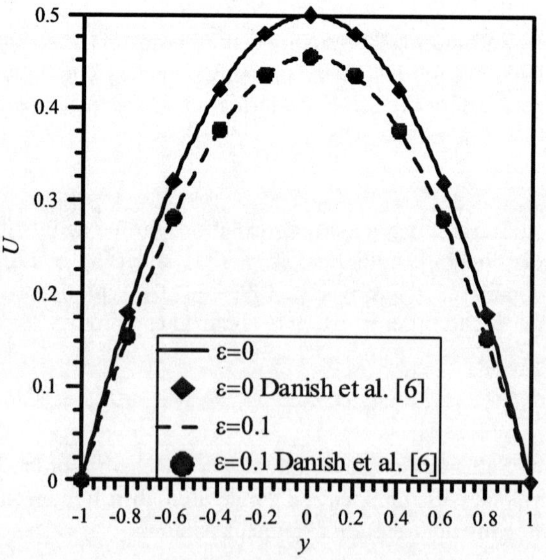

Fig. 2 Comparison of dimensionless velocity profile

temperature, magnitude of heat transfer from the upper plate to the lower one will increase resulting in an increase in the fluid temperature (Fig. 3).

Effect of the viscous dissipation, captured in Brinkman number *Br*, is displayed in Fig. 4. It is clearly indicated by Fig. 4 that temperature of the fluid increases when Br increases. Higher values of *Br* indicate higher viscous heat generation which causes

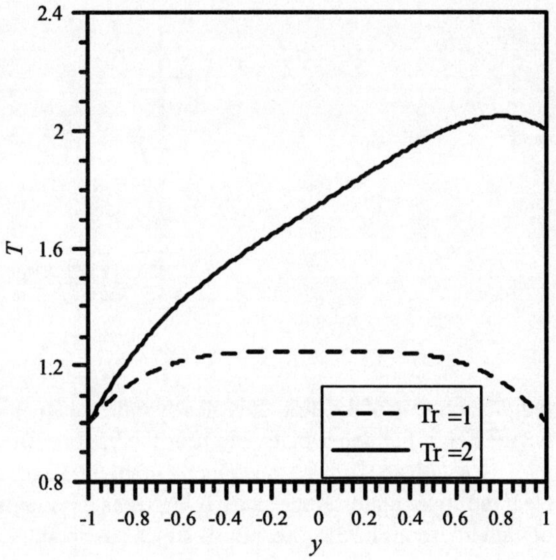

Fig. 3 Temperature distribution for different temperature ratio

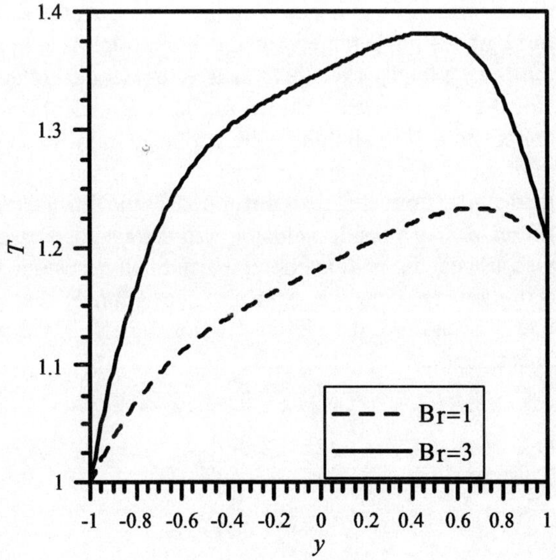

Fig. 4 Dimensionless temperature distribution when *Br* varies

Fig. 5 When ε varies

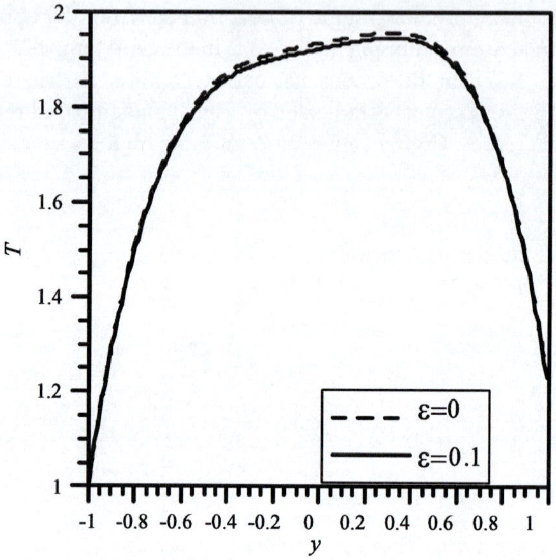

temperature rise of the fluid. It is clearly indicated by Fig. 4, that the peak temperature occurs not in the upper plate, but near the upper plate. As viscous dissipation effect increases with an increase the velocity gradient, near the plate its effect is significant. Near the upper plate, temperature is higher as it is maintained at higher level. Because of higher viscous heating near the plate, the temperature increases further and reaches its maximum. At higher values of the *Br*, for maintaining the plate temperatures at constant level, higher heat has to be removed from the plates, which requires cooling of the plate to be enhanced. Figure 5 presents the influence of the non-Newtonian third-grade fluid parameter on the dimensionless temperature distribution. When the third-grade fluid parameter increases, temperature of the fluid is observed to decrease. However, this decrease is marginal as it is evident from the figure. This decrease in temperature of the fluid may be attributed to the decrease in fluid velocity when the third-grade fluid parameter increases. As discussed earlier, for values of the third-grade fluid parameter, resistance towards the flow increases. Because of this, velocity decreases. As a result, velocity gradient also decreases, consequently reducing the viscous heat dissipation effect, which finally causes a decrease in the temperature. It is important to note that in the absence of convective term, the temperature profile is linear. Because of the viscous dissipation effect, the temperature deviates from the linear trend.

4 Conclusion

Flow of non-Newtonian third-grade fluids, induced by pressure, through a set of parallel plates, maintained at different temperatures, is revisited by employing traditional perturbation method. Perturbation technique, a widely used analytical tool is employed to revisit the problem for examining the effect of Brinkman number, third-grade fluid parameter, and temperature ratio on the temperature distribution of the fluid. The following conclusions are drawn from the study. Velocity decreases with an increase in the third-grade fluid parameter due to higher resistance offered to the flow. Velocity is significantly affected by the change in third-grade fluid parameter. Temperature of the fluid decreases with an increase in the non-Newtonian third-grade fluid parameter. Compared to the effect of the third-grade fluid parameter on velocity, effect on temperature is marginal. Brinkman number affects the temperature significantly presenting a substantial increase in temperature with an increase in Brinkman number. Temperature of the fluid increases significantly with an increase in the temperature ratio. In the current study, properties such as thermal conductivity, third-grade fluid parameter are assumed to be independent of temperature which is reasonable only for marginal temperature difference between the upper plate and lower one. Numerical techniques may be implemented for solving the problems in case of the properties depending on temperature. The results of the current study can serve useful for validation of the numerical results.

References

1. Danish M, Kumar S, Kumar S (2012) Exact analytical solutions for the Poiseuille and Couette-Poiseuille flow of a third grade fluid between parallel plates. Commun Nonlinear Sci Numer Simul 17:1089–1097
2. Siddiqui AM, Hameed M, Siddiqui BM, Ghori QK (2010) Use of Adomian decomposition method in the study of parallel plate flow of a third grade fluid. Commun Nonlinear Sci Numer Simul 15:2388–2399
3. Babolian E, Vahidi AR, Cordshooli GA (2005) Solving differential equations by decomposition method. Appl Math Comput 167(1):1150–1155
4. Bhatti MM, Zeeshan A, Ellahi R (2016) Endoscope analysis on peristaltic blood flow of Sisko fluids with Titanium magneto-nano particles. Comput Biol Med 78(1):29–41
5. Zeeshan A, Ali N, Ahmed R, Waqus M, Khan WA (2019) A mathematical frame work for peristaltic flow analysis of non-Newtonian Sisko fluid in an undulating porous curved channel with heat and mass transfer effect. Comput Methods Programs Biomed 182:105040
6. He JH (2006) Homotopy perturbation method for solving boundary value problems. Phys Lett A 350:87–88
7. Li SX, Jian YJ, Xie ZY, Liu QS, Li FQ (2015) Rotating electro-osmotic flow of third grade fluids between two microparallel plates. Colloids Surf A 470:240–247
8. Nayak MK, Dash GC, Singh LP (2014) Steady MHD flow and heat transfer of a third grade fluid in wire coating analysis with temperature dependent viscosity. Int J Heat Mass Transf 79:1087–1095
9. Wang L, Jian Y, Liu Q, Li F, Chang L (2016) Electromagnetohydrodynamic flow and heat transfer of third grade fluids between two micro-parallel plates. Colloids Surf A 494:87–94

10. Chaudhuri S, Sahoo S (2018) Effect of the aspect ratio on flow characteristics of magnetohydrodynamic (MHD) third grade fluid flow through a rectangular channel. Sadhana 43:106

Design and Analysis of a Naturally Ventilated Fog Cooled Greenhouse Integrated with Solar Desalination System

Md Iftikar Ahmed, Sk Arafat Zaman, and Sudip Ghosh

Abstract This paper presents the design and performance of a naturally ventilated fog cooled greenhouse coupled with solar still. The greenhouse is East–West oriented, and its microclimate is maintained using evaporative cooling through foggers. The canopy-mounted solar still produces freshwater from brackish or saline water. The analysis is performed on a representative day for the month of May and December for Kolkata located in the state of West Bengal and representing hot and humid climatic conditions. Computer codes are developed in-house to develop thermodynamic model for analysing the performance of the integrated system. The study reveals that the solar still can produce maximum 26.39 kg of fresh water on a typical summer day. It is also found that the greenhouse temperature can be maintained at a level 6 °C below the ambient during peak sunshine hours on a typical summer day.

Keywords Greenhouse · Solar still · Desalination · Fog cooling · Natural ventilation

1 Introduction

India receives abundant solar radiation all through the year. The hot and arid areas of India find it difficult to have open field cultivation due to scarcity of irrigation water and very high ambient temperature, whereas coastal areas find difficulty in cultivating

M. I. Ahmed (✉) · S. A. Zaman · S. Ghosh
Department of Mechanical Engineering, Indian Institute of Engineering Science and Technology, Shibpur, Howrah, West Bengal 711103, India
e-mail: mdiahmed49@gmail.com

S. A. Zaman
e-mail: skarz1995@gmail.com

S. Ghosh
e-mail: sudipghosh.becollege@gmail.com

S. Revankar et al. (eds.), *Proceedings of International Conference on Thermofluids*, Lecture Notes in Mechanical Engineering,
https://doi.org/10.1007/978-981-15-7831-1_26

crops due to availability of saline water. The unwanted excessive radiation is detrimental for the growth of the plants. To overcome from these challenges, an idea of a greenhouse cum solar still is presented in this paper [1]. Solar desalination is one of the simplest and economical techniques of purifying contaminated water. The greenhouse provides a suitable microclimate for the growth of the plants using evaporative cooling, whereas the solar still on the roof of the greenhouse produces freshwater from the saline water, and the distillate can meet the irrigation requirements [1].

In view of this, many researchers have experimentally and theoretically studied the performance of greenhouse cum solar desalination system. Sathish et al. [2] experimentally examined that the performance of solar still is highly enhanced by using metal matrix structure as a heat storage media. Singh et al. [3] developed a microclimate model of a greenhouse by considering heat and mass transfer process which predicted the temperature of inside air, plants and plastic cover. Mari et al. [4] found that the solar intensity and PAR in greenhouse reduce by 52% when the solar still is installed on roof of greenhouse. Salah et al. [1] found that transparent roof solar still is more effective than transparent roof photo-voltaic in heating the greenhouse in winter. Chaibi [5] observed that selective glasses have a higher solar radiation absorption capacity and freshwater production compared to standard glass and plastic material. Ghosal et al. [6] studied greenhouse combined with a solar desalination system using flowing brackish water and found that with the increase of mass flow rate of saline water, the distillate output as well as greenhouse temperature reduces.

In this paper, a thermodynamic model for a greenhouse cum solar desalination system is developed, and numerical computations are performed for a representative day in May and December for Kolkata region. Computer codes were developed in-house for analysing the performance of the integrated system.

2 System Description

A schematic diagram of a greenhouse integrated with solar distillation system on the south roof is shown in Fig. 1. The greenhouse, with a floor area of 35 square metre, is naturally ventilated using continuous roof vent and side vents and is oriented in east–west direction. It is also equipped with fog cooling system to cool the inside air by evaporative cooling. Solar energy enters the solar still units through glass roof and is retained by basin liner. A part of solar intensity is retained by saline water, and the rest is transferred to the greenhouse through conductive, convective and radiative heat transfer. The water gets heated up, and the water vapour formed ascends towards the lower temperature glass panel and distills into liquid leaving behind the salts in the basin which is removed from time to time through a drain pipe. The condensed water is collected in a trough and finally stored in a storage tank, which is used for irrigating the plants inside the greenhouse.

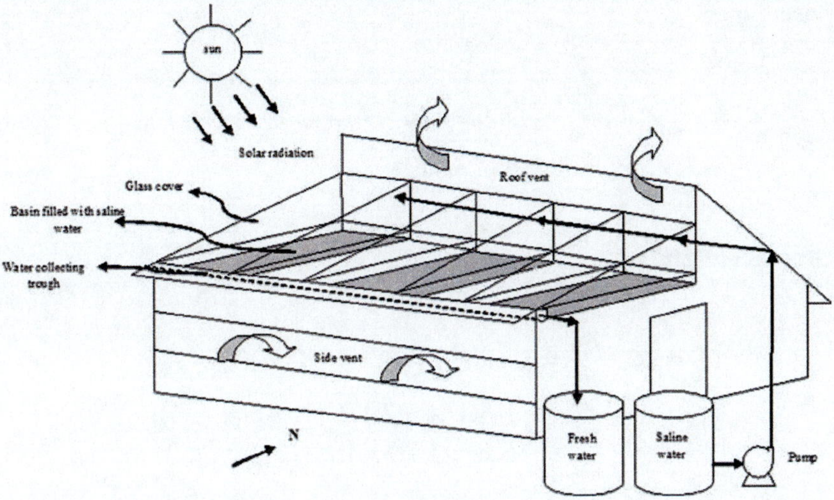

Fig. 1 Schematic diagram of the greenhouse integrated with solar distillation system

3 Mathematical Model

Equation (1) describes the total solar intensity incident on any inclined surface [7],

$$I_t = I_b r_b + I_d r_d + (I_b + I_d) r_r \tag{1}$$

where $\boldsymbol{I_b}$ and $\boldsymbol{I_d}$ are the beam and diffuse components of solar intensity. $\boldsymbol{r_b}$, $\boldsymbol{r_d}$ and $\boldsymbol{r_r}$ are tilt coefficients for beam, diffuse and reflected radiation, respectively [7].

The energy equations of the various parts of the solar still coupled greenhouse are described in Eqs. (2)–(5) [8, 9].

Glass Cover

$$\begin{aligned} m_g C_{pg} \frac{dT_g}{dt} &= I_t A_g \alpha_g + h_{con\_wg} A_w (T_w - T_g) \\ &+ h_{evp\_wg} A_w (T_w - T_g) + h_{rad\_wg} A_w (T_w - T_g) \\ &- h_{con\_ga} A_g (T_g - T_a) - h_{rad\_gsky} A_g (T_g - T_{sky}) \end{aligned} \tag{2}$$

Brackish Water

$$\begin{aligned} m_w C_{pw} \frac{dT_w}{dt} &= I_t A_w \alpha_w \tau_g + h_{con\_bw} A_b (T_b - T_w) \\ &- h_{rad\_wg} A_w (T_w - T_g) - h_{con\_wg} A_w (T_w - T_g) \\ &- h_{evp\_wg} A_w (T_w - T_g) \end{aligned} \tag{3}$$

Basin Liner

$$m_b C_{pb}\frac{dT_b}{dt} = I_t A_b \alpha_b \tau_g \tau_w - h_{con\_bw} A_b (T_b - T_w) - U_b A_b (T_b - T_{gh}) \tag{4}$$

Greenhouse Air

$$m_{gh} C_p \frac{dT_{gh}}{dt} = \tau_c \alpha_c I_t (1 - SF) A_{pa} + U_b A_b (T_b - T_{gh}) - U A_c (T_{gh} - T_a) - m_{va} C_p (T_{gh} - T_a) - h_{fg} \gamma m_{fog} - h_{fg} A_f LAI \kappa (T_p - T_{gh}) \tag{5}$$

Hourly distillate output (kg/h) and instantaneous efficiency of a solar still are described in Eqs. (6) and (7) [8].

$$m_{ew} = \frac{h_{evp\_wg} A_w (T_w - T_g)}{h_{fg}} \times 3600 \tag{6}$$

$$\eta = \frac{h_{evp\_wg} (T_w - T_g)}{I_t} \times 100 \tag{7}$$

4 Numerical Solution Procedure

The total solar intensity incident on an inclined surface is calculated using the hourly data of diffuse and beam radiations for the Kolkata region. Computer codes are developed for solving the energy equations of different elements of solar still cum greenhouse. Initially, the temperature of glass cover, basin liner, saline water and greenhouse air is taken equal to the ambient. Energy equations of the solar still are solved simultaneously, and the result is used in solving the energy equation of the greenhouse air. The design parameters are indicated in Table 1.

The hourly variation of solar irradiance I_t on a typical summer and winter day for Kolkata region, which is used as input for the numerical calculations, is shown in Fig. 2.

Table 1 Design parameters

Parameters	Value
Absorptivity of basin liner (α_b)	0.95
Transmissivity of glass panel (τ_g)	0.9
Absorptivity of water (α_w)	0.3
Greenhouse covering area	102 m^2
Free wind velocity	1 m/s
Fog evaporation factor	0.55
Shading factor	0.75
Coefficient of discharge	0.64

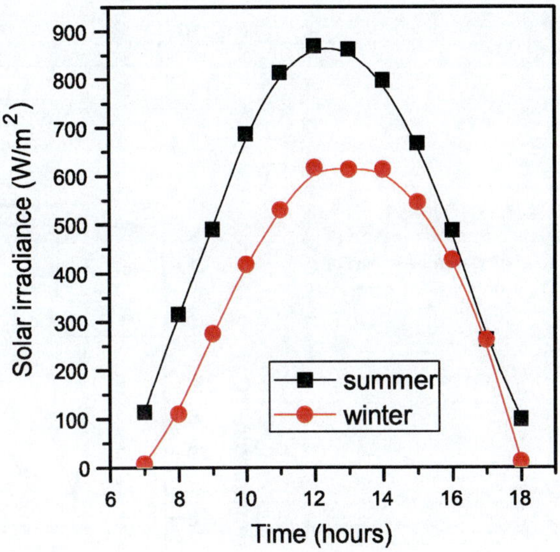

Fig. 2 Hourly variation of solar irradiance (I_t) on a typical summer and winter day

5 Results and Discussion

Figures 3 and 4 show hourly variation of glass cover temperature and basin water temperature with the ambient on a typical summer and winter day. It indicates that variation in the glass temperature and water temperature is in accordance with the variation in the solar intensity. A marginal temperature difference is maintained between the glass cover and water throughout the day, which results in the production of the distillate output. There is a maximum temperature difference of 7.88 °C and 9.84 °C between the glass cover and water at peak sunshine hour on a typical summer and winter day, respectively, which gives maximum distillate at that time.

Figure 5 shows hourly variation of distillate mass from solar still with time on a typical summer and winter day. A maximum output of 3.91 kg/h and 2 kg/h is obtained at peak sunshine hour on a typical summer and winter day when the solar

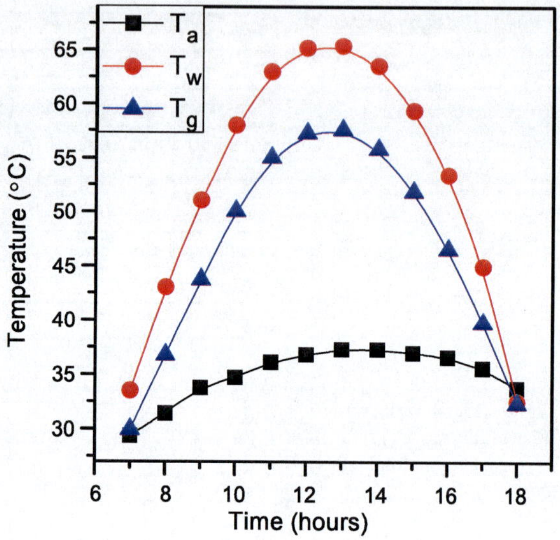

Fig. 3 Variation of glass cover temperature (T_g) and basin water temperature (T_w) on a typical summer day

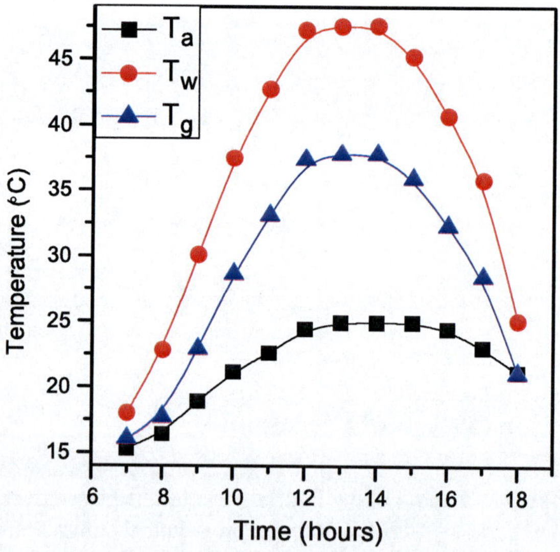

Fig. 4 Variation of glass cover temperature (T_g) and basin water temperature (T_w) on a typical winter day

intensities are 869.9 W/m^2 and 616.2 W/m^2, respectively. This result is in good agreement with the result of literature [10] mentioned in the reference.

Figure 6 shows hourly variation of the instantaneous efficiency of solar still on a typical summer and winter day. It is noticed that the instantaneous efficiency during the peak sunshine hours is 43.6% and 28.7% for a basin water depth of 10 mm on a summer and winter day, respectively. This result nearly matches with the result of the literature [11] mentioned in the reference.

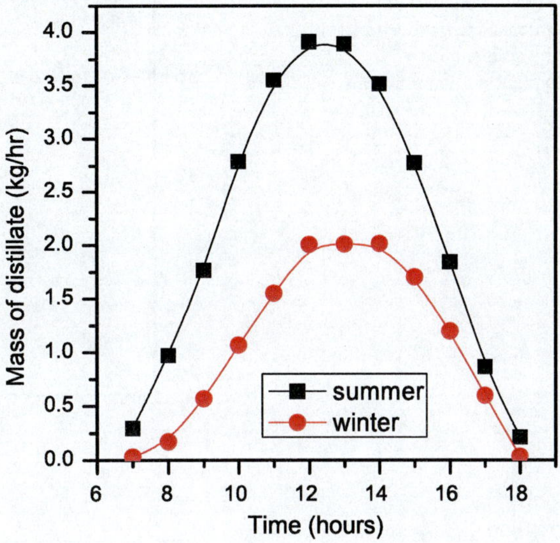

Fig. 5 Variation of mass of distillate from solar still with time on a typical summer and winter day

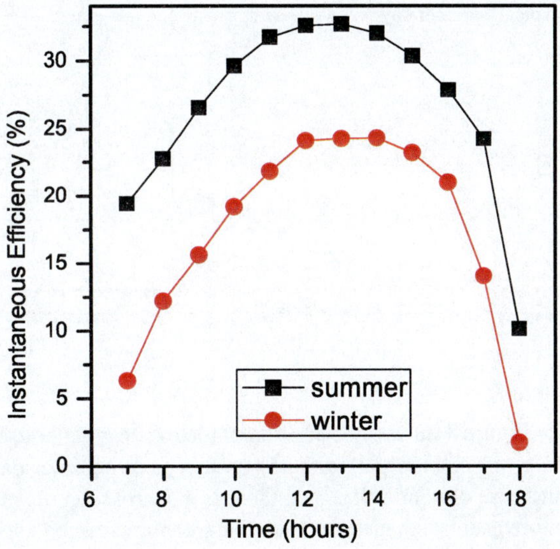

Fig. 6 Variation of instantaneous efficiency of solar still on a typical summer and winter day

Figure 7 depicts hourly variation of the inside greenhouse temperature with the ambient on a typical summer day. It is noticed that using 75% shading factor, 1.2 ACM ventilation rate and assuming 50% relative humidity, the inside greenhouse temperature is maintained 6 °C below the ambient during the peak sunshine hours. The peak inside temperature of the greenhouse is noticed as 29 °C when the peak ambient temperature is 36.7 °C. This result is in good agreement with the result of literature [12].

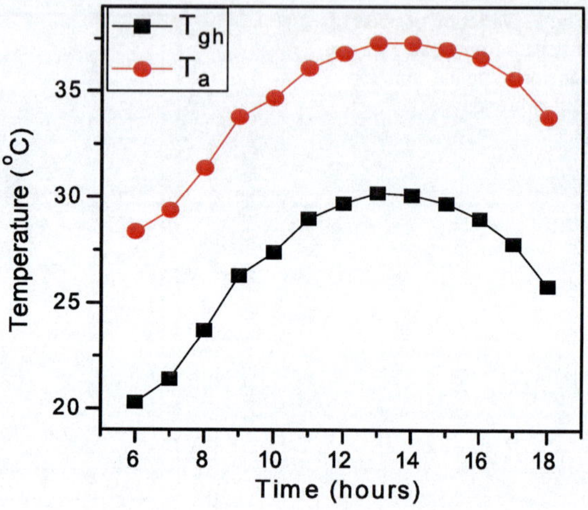

Fig. 7 Variation of greenhouse temperature with the ambient on a typical summer day

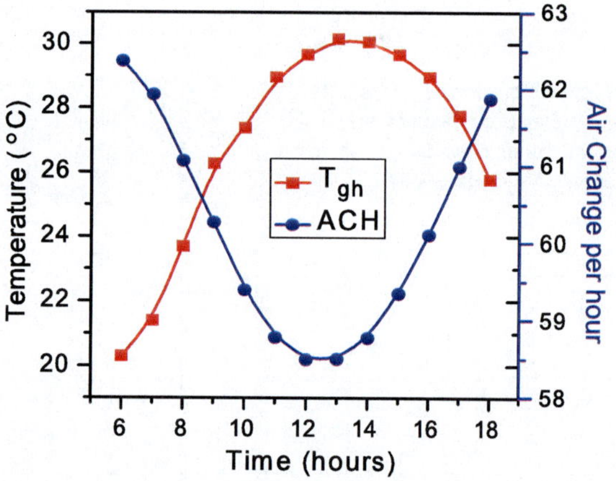

Fig. 8 Variation of greenhouse temperature with ventilation rate on a typical summer day

Figure 8 shows the variation of the inside greenhouse temperature with ventilation rate (air change per minute) on a typical summer day. It is noticed that with the increase of ventilation rate, there is a decrease in the greenhouse air temperature, as more heat is exhausted ouside the greenhouse due to ventilation. This result matches with the result of the literature [13].

6 Conclusions

The study reveals the following conclusions:

- The maximum distillate output during peak sunshine hours varies from 2 kg/h to 3.91 kg/h in winter and summer, respectively.
- The overall mass of freshwater output during the daytime varies from 12.95 kg to 26.39 kg in winter and summer, respectively.
- The instantaneous efficiency of the solar still during the peak sunshine hours varies from 28.7% to 43.6% in winter and summer, respectively.
- The inside greenhouse temperature is maintained 6 °C below the ambient during the peak sunshine hours on a typical summer day.
- With the increase in ventilation rate, the greenhouse air temperature drops.

References

1. Salah AH, Hassan GE, Fath H, Elhelw M, Elsherbiny S (2017) Analytical investigation of different operational scenarios of a novel greenhouse combined with solar stills. Appl Therm Eng 122:297–310
2. Sathish D, Veeramanikandan M, Tamilselvan R (2019) Design and fabrication of single slope solar still using metal matrix structure as energy storage. Mater Today Proc 27:1–5
3. Singh MC, Singh JP, Singh KG (2018) Development of a microclimate model for prediction of temperatures inside a naturally ventilated greenhouse under cucumber crop in soilless media. Comput Electron Agric 154:227–238
4. Mari EG, Colomer RPG, Blaise-Ombrecht CA (2007) Performance analysis of a solar still integrated in a greenhouse. Desalination 203:435–443
5. Chaibi MT (2000) Analysis by simulation of a solar still integrated in a greenhouse roof. Desalination 128:123–138
6. Ghosal MK, Tiwari GN, Srivastava NSL (2003) Thermal modeling of a controlled environment greenhouse cum solar distillation for composite and warm humid climates of India. Desalination 151:293–308
7. Sukhatme SP, Nayak JK (2009) Solar energy: principles of thermal collection and storage, 3rd edn. McGraw-Hill Education, India
8. Velmurugan V, Gopalakrishnan M, Raghu R, Srithar K (2008) Single basin solar still with fin for enhancing productivity. Energy Convers Manag 49:2602–2608
9. Abdel-Ghany AM, Kozai T (2006) Dynamic modeling of the environment in a naturally ventilated, fog-cooled greenhouse. Renew Energy 31:1521–1539
10. Srivastava NSL, Din M, Tiwari GN (2000) Performance evaluation of distillation-cum-greenhouse for a warm and humid climate. Desalination 128:67–80
11. Varun Raj S, Muthu Manokar A (2017) Design and analysis of solar still. Mater Today Proc 4:9179–9185
12. Misra D, Ghosh S (2017) Microclimatic modeling and analysis of a Fog-cooled naturally ventilated greenhouse. Int J Environ Agric Biotechnol 2:997–1001
13. Ganguly A, Ghosh S (2007) Modeling and analysis of a fan–pad ventilated floricultural greenhouse. Energy Build 39:1092–1097

Performance Assessment of a Steam Gasification-Based Hybrid Cogeneration System

Sk Arafat Zaman, Dibyendu Roy, and Sudip Ghosh

Abstract In this article, a hybrid cogeneration scheme based on a solid oxide fuel cell (SOFC)—gas turbine system and utilizing syngas derived from municipal solid wastes (MSW)—has been investigated thermodynamically. Steam gasification of municipal solid wastes produces hydrogen-rich gas that is consumed by the SOFC stack as fuel. An externally fired air turbine (EFAT) draws the heat of combustion of the SOFC exhaust via a heat exchanger, while the clean turbine exhaust air itself is fed to the cathode of the SOFC. A bottoming heat recovery steam generator (HRSG) further recovers waste heat to produce process steam required for the gasifier as well as for process heater. Results of the study for base case parameters show that the integrated system can have a maximum electrical efficiency of about 50%, while substantial fuel saving (more than 51% when compared with separate power and steam plants) because of cogeneration of steam required for gasification system.

Keywords SOFC · Steam gasification · Gas turbine · MSW

1 Introduction

Municipal solid waste (MSW) is considered as one of the most favorable forms of renewable sources, and currently, it has emerged as a preferable choice for fossil fuel alternative. Alarming situation of environmental condition across the globe is the reason for growing interest of renewable power plant with higher efficiency with lower environmental impact. Solid oxide fuel cell (SOFC) which operates at very high-temperature (600–1000 °C) [1] is an ideal alternative to conventional power generation plant. Lesser CO_2 emission and higher efficiency make high-temperature fuel cells more appealing as a power generating unit [2]. Different SOFC-based systems were modelled and investigated by previous researchers. Roy et al. [2] investigated a biomass-based SOFC-externally fired gas turbine-organic Rankine

S. A. Zaman · D. Roy · S. Ghosh (✉)
Department of Mechanical Engineering, Indian Institute of Engineering Science and Technology, Shibpur, Howrah, West Bengal 711103, India
e-mail: sudipghosh.becollege@gmail.com

S. Revankar et al. (eds.), *Proceedings of International Conference on Thermofluids*, Lecture Notes in Mechanical Engineering,
https://doi.org/10.1007/978-981-15-7831-1_27

cycle (ORC) integrated system and it was observed that the highest efficiency of the system near 50%. Mishra et al. [3] investigated a reversible solid oxide fuel cell (RSOFC) plant integrated with solar PV/T and they reported highest efficiency to be 77%. Ghosh et al. [4] investigated proton-conducting RSOFC integrated solar thermal power plant and they found that maximum efficiency around 65% at cell operation temperature 873 K and current density of 500 A/m^2. Wang et al. [5] investigated integration of SOFC with process utility plant and they reported the improvement in efficiency and reduction in emission as fuel cell is added to industries. Biomass gasification-based modelling and analyzing have been done before by previous researchers. Pilatau et al. [6] investigated performance of an integrated gasification combined cycle (IGCC) plant coupled with diesel engine and they found the possibility of CO_2 exhaust reduction to atmosphere by 1.5 times in combined cycle system with biomass gasifier. Matelli [7] investigated performance assessment of a biomass-based cogeneration system in which he used a shell for quick prototyping. Pande et al. [8] performed experimental and numerical analyses for designing two-pot biomass cookstove and they found that combined mathematical and computational helps designing cookstove. Keche et al. [9] simulated air gasification process employing ASPEN PLUS software and they reported higher conversion efficiency with babul wood. Cebrucean et al. [10] investigated the performance assessment of coal-based power systems co-fired with biomass and they reported negative values of CO_2 exhaust capturing carbon from the plant. Rivera-Tinoco and Bouallou [11] investigated biomass incineration process related to CO_2 emissions and reduction of 135 Mt of carbon was observed to produce hydrogen using steam methane reforming.

In this paper, preliminary thermodynamic assessment of a steam gasification-based SOFC-externally fired air turbine (EFAT) integrated cogeneration plant has been carried out. Syngas is produced from municipal solid wastes and power is generated utilizing a SOFC and EFAT module. Required steam for steam gasification is obtained by utilizing the heat content of outlet gas stream using a HRSG, from where we also get additional process heat.

2 Model Description

The simplified flow diagram of the proposed system is shown in Fig. 1.The integrated system mainly consists of two subunits: steam gasification unit and power generation unit.

Municipal waste management (MSWM) system is used to pre-process the MSW before feeding it to the gasifier. MSWM encompasses production, storage, gathering, transfer, transport, processing, and disposal sequentially. A detailed discussion of MSWM system can be found in the work of Das and Bhattacharyya [12].

MSW is fed to the gasifier unit after being processed by the MSWM system. H_2-rich syngas is obtained from the steam gasification unit, which is then fed to the anode of the SOFC after being processed and conditioned at the high-temperature gas cleaning unit. Syngas is heated to the required inlet temperature of SOFC by a

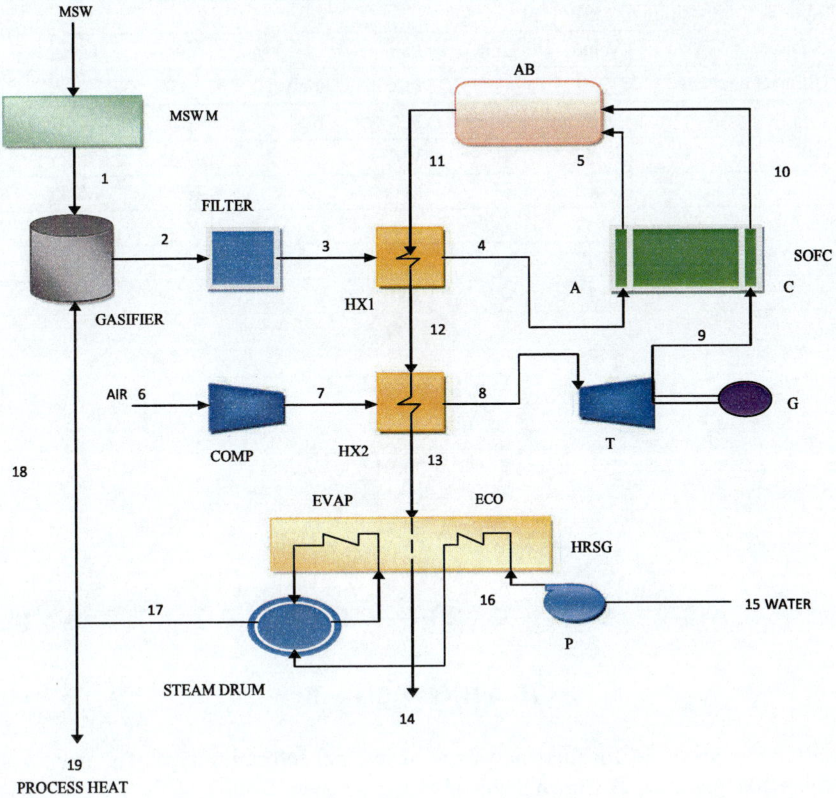

Fig. 1 Schematic of MSW-based SOFC-EFAT cogeneration plant

heat exchanger (HX1). Air is compressed and is preheated through heat exchanger (HX2), which then drives an air turbine and then is fed to the cathode of SOFC. The combustible outlet gases from the anode channel and the unutilized air from cathode outlet are fed to an afterburner (AB). The gas stream departing the after burner is passed through both heat exchangers HX1 and HX2, respectively. And finally, a HRSG utilizes the heat content of the gas stream from where steam for gasification and process heat is generated.

3 Thermodynamic Modeling and Analysis

The composition of biomass used (municipal solid wastes) is shown in Table 1

Thermodynamic equilibrium model is used in steam gasification. The major chemical reactions considered in the steam gasification process are as shown below [13]

Table 1 Composition of MSW [13]

Parameters	Value	Unit	Parameters	Value	Unit
Ultimate analysis			Proximate analysis		
C	30.77	%	FC	7.70	%
O	17.30	%	VM	46.15	%
H	4.62	%	LHV	13.31	MJ/kg
S	0.39	%	–	–	–
N	0.77	%	–	–	–
ASH	46.15	%	ASH	46.15	%

$$C + CO_2 \rightarrow 2CO \tag{1}$$

$$C + H_2O \rightarrow CO + H_2 \tag{2}$$

$$CO + H_2O \rightarrow CO_2 + H_2 \tag{3}$$

$$C + 2H_2 \rightarrow CH_4 \tag{4}$$

$$CH_4 + H_2O \rightarrow CO + 3H_2 \tag{5}$$

SOFC considered for this analysis is of internal reforming type. The reactions which take place are as shown below [14]

$$CH_4 + H_2O \rightarrow CO + 3H_2 \tag{6}$$

$$CO + H_2O \rightarrow H_2 + CO_2 \tag{7}$$

$$H_2 + 0.5O_2 \rightarrow H_2O \tag{8}$$

The reversible cell voltage is calculated employing Nernst equation as in Eq. (9).

$$V_N = -\frac{\Delta g}{2F} - \frac{RT}{2F} \ln \frac{P_{H_2O}}{P_{H_2} P_{O_2}^{0.5}} \tag{9}$$

The actual cell voltage (V_{act}) is obtained by deducting the total polarization losses from the reversible voltage as shown in Eq. (10)

$$V_{act} = V_N - V_{loss} \tag{10}$$

The equations to calculate different voltage losses can be found in [14]

Total power generated at the SOFC unit (or array) is estimated as Eq. (11)

$$P_{\mathrm{SOFC}} = N_{\mathrm{cell}} * V_{\mathrm{act}} * i_{\mathrm{cell}} * N_{\mathrm{stack}} \tag{11}$$

where

N_{Stack} Total number of stacks.
N_{cell} Total number of cells.

Auxiliary power required by the pump is calculated as shown in Eq. (12).

$$P_{\mathrm{pump}} = m_{\mathrm{water}} * (h_{16} - h_{15}) \tag{12}$$

Power generation from the air turbine is computed using Eq. (13).

$$P_{\mathrm{T}} = m_{\mathrm{air}} * (h_8 - h_9) \tag{13}$$

Auxiliary power required by the air compressor is given by Eq. (14) as shown below.

$$P_{\mathrm{COMP}} = m_{\mathrm{air}} * (h_7 - h_6) \tag{14}$$

The total power output from plant generated is computed by Eq. (15).

$$P_{\mathrm{net}} = P_{\mathrm{SOFC}} + P_{\mathrm{T}} - P_{\mathrm{COMP}} - P_{\mathrm{pump}} \tag{15}$$

System electrical efficiency can be obtained as shown below employing Eq. (16).

$$\eta_{\mathrm{elec}} = \frac{P_{\mathrm{net}}}{(m_{\mathrm{b}} * \mathrm{LHV_b}) + (m_{\mathrm{steam}} * \mathrm{LHV_{steam}})} \tag{16}$$

where m_{b} and m_{steam} are the mass flow rate of biomass and steam, respectively; $\mathrm{LHV_b}$, $\mathrm{LHV_{steam}}$ are the lower heating values of biomass and steam, respectively

Total process heat is computed employing Eq. (17).

$$Q_{\mathrm{heat}} = m_{19} * h_{19} \tag{17}$$

The fuel energy saving ratio (FESR) when compared with separate power and steam plants is estimated as shown below [15] using Eq. (18).

$$\mathrm{FESR} = \frac{\Delta F}{\frac{P_{\mathrm{net}}}{\eta_{\mathrm{e}}} + \frac{Q_{\mathrm{heat}}}{\eta_{\mathrm{b}}}} \tag{18}$$

where

$$\Delta F = \frac{P_{\text{net}}}{\eta_e} + \frac{Q_{\text{heat}}}{\eta_b} - (m_b * \text{LHV}_b) \tag{19}$$

where η_e and η_b are the efficiencies of the standard power system and boiler plant which are assumed to be 40% and 90%, respectively [15].

4 Results and Discussion

The volume concentration on dry basis of syngas produced after steam gasification is shown in Table 2

The base case input parameters employed are given in Table 3

In this segment, the effect of main operating and designing plant parameter, viz. current density of solid oxide fuel cell and pressure ratio of compressor on plant performance has been discussed.

Obtained results from the system under base case configurations are given in Table 4

Figure 2 shows the variation of current density on voltage and power density of SOFC. It can be observed that power density of SOFC first increases and then decreases after 7000 A/m^2 with the increase in current density and voltage output

Table 2 Composition of syngas components

Components	Volume concentration (%)	Components	Volume concentration (%)
CO	23.92	N_2	0.24
CO_2	21.07	CH_4	0.03
H_2	54.74	–	–

Table 3 Base case input parameters for the system [14]

Parameters	Value	Parameters	Value
Current density of (A/m^2)	2000	Gasification temperature (°C)	800
Cell area (A_{cell}) (m^2)	0.01	S/B	1
SOFC operating temperature (°C)	850	Number of cells (N_{cell})	500
Fuel utilization ratio of SOFC	0.80	Number of stacks (N_{stack})	20
Oxygen utilization ratio of SOFC	0.17	Pressure ratio of compressor	1.6
		–	–

Table 4 Base case performance indicators

Parameters	Value	Parameters	Value
P_{net} (kW)	176.62	Electrical efficiency (%)	47.70
Q_{heat} (kW)	162.64	FESR (%)	50.27

from SOFC decreases with the rise in current density. Operating voltage of SOFC decreases with the rise in current density as the voltage polarization losses also with the rise in current density. And the power density increases with the rise in current density because more syngas is drawn by the anode of SOFC as current density increases, which results in increased SOFC power output.

Figure 3 shows the effect of current density on process heat output and net power output. The process heat output of the system increases linearly with the rise in current density of SOFC from 1000 to 9000 A/m^2 and the net power output increases in a nonlinear fashion [2, 16] as shown in Fig. 3. The compressor power also increases as current density increases. But the combined effect of SOFC and GT power increase is much more than compressor power increase. The range of net power output obtained is 93.5–406.47 kW and the range of process heat output obtained is 80.62–729.88 kW.

Figure 4 shows the variation of electrical efficiency and fuel energy saving ratio of the system with varying current density of SOFC for a constant pressure ratio of compressor. It is observed that both electrical efficiency and FESR decreases in a similar manner with the increase in current density. This is because though the net

Fig. 2 Effect of current density of SOFC on voltage and power density of SOFC

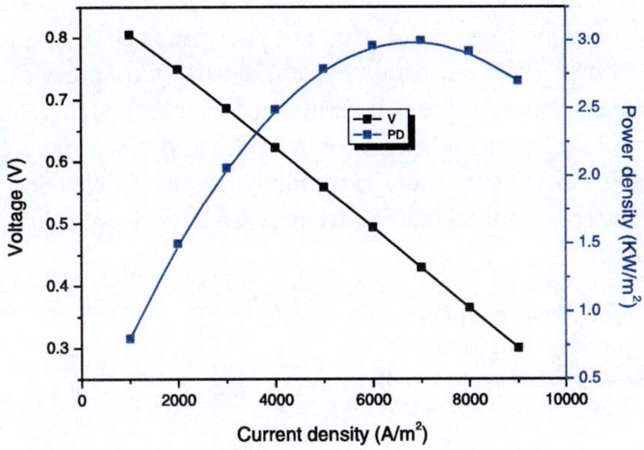

Fig. 3 Effect of current density of SOFC on net power of plant and process heat

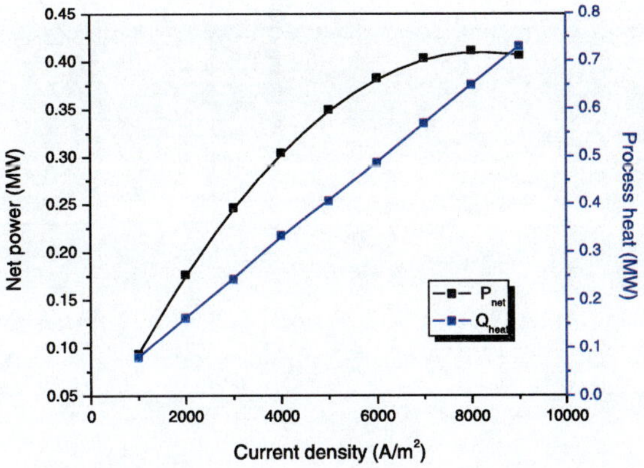

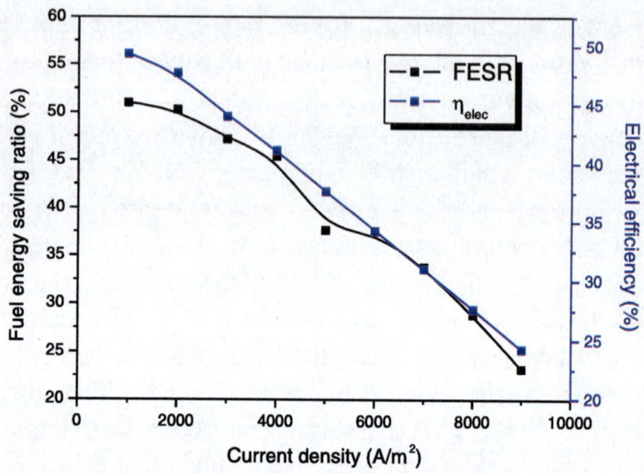

Fig. 4 Impact of SOFC current density on FESR and electrical efficiency of plant

output power increases with the increase in current density, the energy input to the system in the form of biomass and steam increases as their mass flow rate increases with the rise in current density. The electrical efficiency obtained for the system is in the range of 24.22–49.34% and FESR obtained is in the range of 23.22–51.01%.

Figure 5 shows the effect of pressure ratio of compressor on FESR and electrical efficiency for a constant current density. Compression ratio beyond 4.4 has not been taken into account as it results in lower cathode channel temperature of air [17]. As pressure ratio increases, both FESR and electrical efficiency increases. The reason for this is as pressure ratio increases, net power and process heat output from the system increase, leading to increase in both electrical efficiency and FESR.

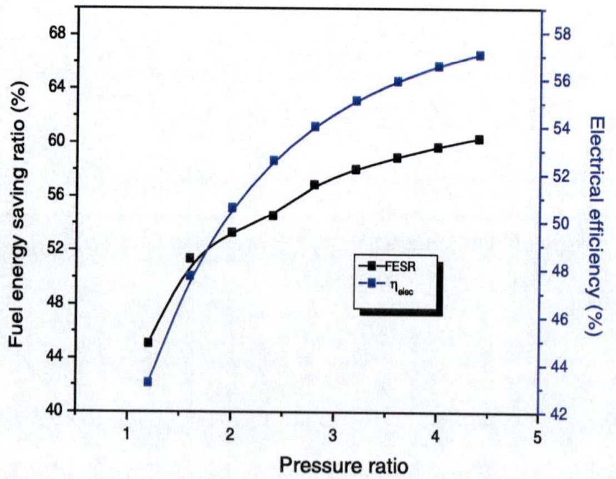

Fig. 5 Influence of pressure ratio of compressor on FESR and electrical efficiency of plant

5 Conclusion

Thermodynamic study of a hybrid cogeneration system based on a solid oxide fuel cell (SOFC)—gas turbine system and utilizing syngas produced from municipal solid wastes (MSW)—has been carried out. The influence of major operating parameters is examined on performance parameters of the plant. The main findings are as shown below

- At higher current density, the values of both process heat output and net power output increase and their maximum values are 729.88 kW and 406.47 kW, respectively.
- At elevated levels of current density, the values of electrical efficiency and fuel energy saving ratio decrease and their maximum values are found to be 49.34% and 51.01%, respectively.
- As current density of SOFC increases, power density of SOFC increases for a while and then decreases and voltage output from SOFC decreases linearly and their maximum values are 2.7 and 0.8045, respectively.
- With the rise of pressure ratio, both electrical efficiency and FESR increase in a similar manner for a constant current density and their maximum values are 57.07% and 60.35%, respectively.

References

1. Rangel-Hernández VH, Fang Q, Blum L, Ramírez-Minguela JJ, Zaleta-Aguilar A (2019) Impact of operational and design variables on the thermodynamic behavior of a simulated 500 kw ng-fueled solid oxide fuel cell stack. Energy Convers Manag 204:112283. https://doi.org/10.1016/j.enconman.2019.112283
2. Roy D, Samanta S, Ghosh S (2019) Techno-economic and environmental analyses of a biomass based system employing solid oxide fuel cell, externally fired gas turbine and organic Rankine cycle. J Clean Prod. https://doi.org/10.1016/j.jclepro.2019.03.261
3. Mishra AK, Roy D, Ghosh S (2018) Reversible solid oxide fuel cell connected to solar PV/T system: cell electrochemical modelling and analysis. IOP Conf Ser Mater Sci Eng 377:0–6 (2018). https://doi.org/10.1088/1757-899X/377/1/012077
4. Ghosh A, Roy D, Ghosh S (2019) Proton conducting reversible SOFC integrated in a solar thermal power generation system. J Phys Conf Ser 1240. https://doi.org/10.1088/1742-6596/1240/1/012112
5. Wang B, Zhang N, Hwang S, Kim J (2013) Process integration of solid oxide fuel cells with process utility systems. Clean Techn Environ Policy 15(5):801–815. https://doi.org/10.1007/s10098-013-0643-1.
6. Pilatau AY, Viarshyna HA, Gorbunov AV, Nozhenko OS, Maciel HS, Baranov VY, Mucha OV, Maurao R, Lacava PT, Liapeshko I, Filho GP, Matus A (2014) Analysis of syngas formation and ecological efficiency for the system of treating biomass waste and other solid fuels with CO_2 recuperation based on integrated gasification combined cycle with diesel engine. J Braz Soc Mech Sci Eng 36(4):673–679. https://doi.org/10.1007/s40430-014-0166-7.
7. Matelli JA (2016) Conceptual design of biomass-fired cogeneration plant through a knowledge-based system. J Braz Soc Mech Sci Eng 38:535–549. https://doi.org/10.1007/s40430-015-0326-4

8. Pande RR, Sharma SK, Kalamkar VR (2019) Experimental and numerical analyses for designing two-pot biomass cookstove. J Braz Soc Mech Sci Eng 41:1–18. https://doi.org/10.1007/s40430-019-1839-z
9. Keche AJ, Prasad A, Gaddale R (2015) Simulation of biomass gasification in downdraft gasifier for different biomass fuels using ASPEN PLUS. Clean Technol Envion Policy 17:465–473. https://doi.org/10.1007/s10098-014-0804-x
10. Cebrucean D, Cebrucean V, Ionel I (2019) Modeling and performance analysis of subcritical and supercritical coal-fired power plants with biomass co-firing and CO_2 capture. Clean Technol Environ Policy. https://doi.org/10.1007/s10098-019-01774-1
11. Rivera-Tinoco R, Bouallou C (2010) Using biomass as an energy source with low CO_2 emissions. Clean Technol Environ Policy 12:171–175. https://doi.org/10.1007/s10098-009-0241-4
12. Das S, Bhattacharyya BK (2015) Performance evaluation of the proposed and existing waste management system: economic analysis. In: Proceedings of the 2015 international conference on Operations Excellence and Service Engineering.Orlando, Florida, USA, pp 267-277.
13. Pala LPR, Wang Q, Kolb G, Hessel V (2017) Steam gasification of biomass with subsequent syngas adjustment using shift reaction for syngas production: An Aspen plus model. Renew Energy 101:484–492. https://doi.org/10.1016/j.renene.2016.08.069
14. Roy D, Samanta S, Ghosh S (2019) Energetic and exergetic analyses of a solid oxide fuel cell (SOFC) module coupled with an organic rankine cycle. In: Saha P, Subbarao P, Sikarwar B (eds) Advances in fluid and thermal engineering. Lecture notes in mechanical engineering. Springer, Singapore
15. Ghosh S, De S (2006) Energy analysis of a cogeneration plant using coal gasification and solid oxide fuel cell. Int J Energy Res 31:345–363. https://doi.org/10.1016/j.energy.2005.01.011
16. Lin Y, Beale SB (2006) Performance predictions in solid oxide fuel cells. Appl Math Model 30:1485–1496. https://doi.org/10.1016/j.apm.2006.03.009
17. Roy D, Samanta S, Ghosh S (2018) Thermodynamic analysis of a biomass based solid oxide fuel cell integrated advanced power generation system. IOP Conf Ser Mater Sci Eng 377. https://doi.org/10.1088/1757-899X/377/1/012210.

Heat and Mass Transfer

Experimental Study of Thermal Contact Conductance for Selected Interstitial Materials

Mohammad Asif and Alok Kumar

Abstract Enhancement of thermal contact conductance (TCC) has found numerous scopes in a variety of thermal systems such as nuclear power reactor, microelectronics, heat exchangers, dry sliding contacts, etc. Better heat dissipation is the main objective in all such applications where a higher value of TCC is the prime concern at the contact of heat source and sink. In the present investigation, axial heat flow experiments have been carried out on bare metallic joints as well as for the contacts with interface materials for the improvement of TCC. Steady state methodology is employed for the estimation of TCC at the interface of two materials under atmospheric conditions. Stainless steel (304) has been chosen as the specimens' material due to its low thermal contact conductance value and extensive range of applications. Silicone-based thermal grease and graphene pastes are employed as thermal interface materials between the specimens. The thickness of paste is an important factor since large thickness will further reduce the thermal contact conductance. Therefore, TCC is estimated and studied with the varying thickness of thermal paste. Moreover, the thermal pastes have also been tested for varying contact pressures (0–5 MPa) and interface temperatures (30–100 °C). Eventually, the results of thermal contact conductance with thermal interface materials have been compared with those for bare metallic contacts. Results demonstrate the applicability and suitability of two types of thermal pastes for their application in the present range of operating parameters.

Keywords Thermal contact conductance · Interface material · Silicon paste · Graphene paste

1 Introduction

Interfacial heat transfer has significant role for the design and performance of any thermal system especially metal casting, metal forming, heat exchangers, electronic devices, IC engines, nuclear reactor and space applications. The efficient design

M. Asif (✉) · A. Kumar
Department of Mechanical Engineering, A.M.U., Aligarh, India
e-mail: masif@zhcet.ac.in

S. Revankar et al. (eds.), *Proceedings of International Conference on Thermofluids*, Lecture Notes in Mechanical Engineering,
https://doi.org/10.1007/978-981-15-7831-1_28

and performance of a thermal system sometimes demand the minimization of heat flow to decrease heat losses or sometimes, and its maximization may be desired to avoid excessively high temperature gradients for a given heat flux. It leads to the enhancement of system performance and safety and results in the reduction of its manufacturing cost. Thermal contact conductance (TCC) is an important parameter to relate the interfacial heat transfer between two bodies. Therefore, in many applications, higher TCC values are the prime concern such as in microelectronics, heat exchangers, in dry sliding contacts, while there are numerous thermal systems where a higher values of thermal contact resistance are the ultimate goal, for example, low conductivity structural supports for storage of cryogenic fluids, thermal isolation of spacecraft systems, etc. [1]. Thermal contact conductance for bare metallic contact is found to be very low especially for low thermal conductivity materials and/or for low contact pressures [2]. The enhancement of thermal contact conductance is generally carried out by applying thermal pastes, metallic foils and metallic coatings at the interface. Essentially, the purpose of these thermal interface materials (TIM) is to fill the microscopic voids and cavities at the interface with a higher thermal conductivity material thereby to enhance the thermal contact conductance at the interface [3].

Antonetti and Yovanovich [4] presented a work to enhance TCC for nominally flat surfaces by using metallic coatings. The model was well supported with the experiments on nickel specimens. The specimens were prepared with silver coating and glass-bead-blasted. They concluded that the smoother bare contacting surface presented greater enhancement [4]. Peterson and Fletcher [5] performed experiments on chemically polished aluminum with anodized coatings having thickness 60.9–163.8 micron, kept in contact with an unanodized aluminum sample [5]. Salerno et al. [6] experimentally estimated the thermal contact conductance using indium foil and Apiezon-N grease in the low temperature range (1.6–6.0 K). The applied forces were in the range of 22–670 N. The specimens were made of OFHC copper, aluminum, brass and stainless steel. They found that the application of indium foil and Apiezon grease made an enhancement of TCC over uncoated surfaces.[6]. Marotta et al. [7] studied the sintered copper coatings on a sintered ferrous substrate to improve the TCC. It was found that the improvement was greater for higher thickness of coatings [7]. Wolff and Schneider [8] studied the effects of various parameters on the TCC of steel contacts using different types of interface materials. They concluded that silicone grease and Ag paint presented better improvement in TCC than metallic foils [8]. Kshirsagar et al. [9] worked with thin layers of silicon nitride coating on OFHC copper. They reported a significant reduction in TCC with silicon nitride coatings [9]. Further, Kshirsagar et al. [10] carried out experiments on gold-coated OFHC copper. It has been resulted an improvement in TCC while using gold coatings and further increased with contact pressure [10]. Jeng et al. [11] conducted the investigation to study the effects of diamond film coatings on TCC. It has been noted an increase in TCC by varying the thickness of diamond film [11]. Misra and Nagaraju [12] carried out experiments on gold-plated OFHC copper and brass in different environments [12]. Merrill and Garimella [3] presented experimental and numerical study on coated metallic contacts. The experiments had been carried out on Cu, Br and Al of varying roughness and with Ag, Ni and Sn coatings of varying thicknesses. They found a strong connection between the coating material and substrate material [3]. Zheng et al. [13] experimentally studied the thermal contact resistance for super-alloy (GH600) material and 3-D carbon–carbon composite material in contact. The carbon fiber sheet has been utilized as an interface material [13].

Thermal pastes have wide application to improve the TCC at different interfaces for the heat transfers from the power semiconductor module to heat sink in electronic cooling applications. Thermal pastes basically consist of thermally conductive particles generally metallic oxides suspended in a carrier medium, mostly silicone. In the present work, two types of thermal pastes have been studied, viz. silicon-based thermal grease and graphene powder-based thermal paste. Graphene oxide has excellent thermal, chemical and mechanical properties [14]. Hence, these two types of interstitial materials, silicon paste and graphene paste have been tested for varying thicknesses between stainless steel contacts under varying loading and temperature conditions.

2 Experimentation

The experimental apparatus consists of loading plates, heating system, cooling system, insulating material, loading system and temperature measurement system. Two cylindrical samples having 3.0 cm height and 2.5 cm diameter are kept in contact. The axial heat flow has been established in the specimens in contact by supplying heat from the top side and cooling from the bottom side in the heat flow column. Heating system consists of a heating block made of copper in which two cartridge-heating rods each of 150 W have been inserted for the generation of high density heat flux. The applied heat flux is controlled manually with the help of dimmerstat, connected with voltmeter and ammeter. To extract the heat in downward direction axially, a proper cooling system has been arranged. A cylindrical cooling block made of copper is used in which chilled water is supplied by a chiller. The heating block and cooling block are insulated from the other ends by insulating blocks. In addition, radial insulation of glass wool is provided to minimize convection losses. Eight Alumel-Chromel (K-Type) thermocouples have been utilized to measure the axial temperatures in both the samples. To perform experiment for various loading conditions, a proper loading system comprises a screw jack, load cell and load indicator.

The test samples are made of stainless steel and placed in nominally flat contact with or without applying thermal pastes. The system is started at a particular load and keeps on running until the attainment of steady state condition. Subsequently, the loading is changed to new level, and again, the experiment is run in the similar way, and reading is taken up to the steady state. The methodology is repeated again in the similar manner for all the experiments with and without interstitial materials. A block diagram of the experimental unit and a pictorial representation of experimental assembly have been presented in Fig. 1.

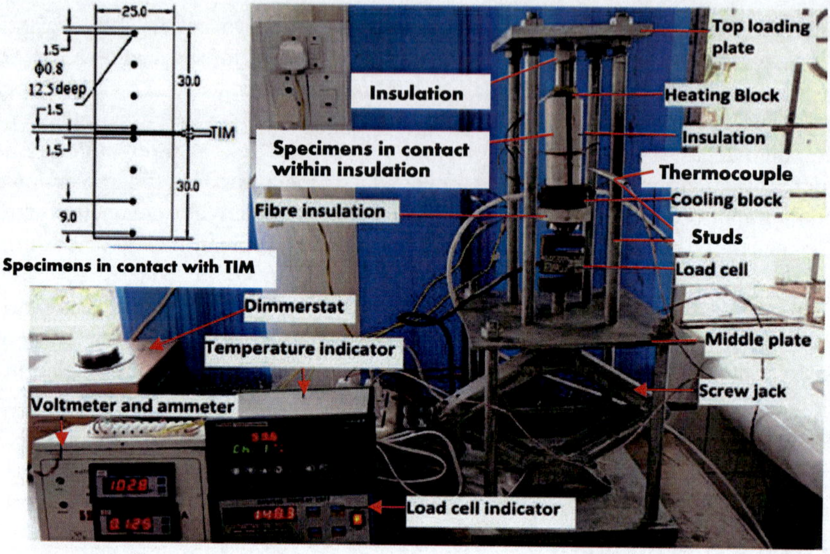

Fig. 1 Schematic of specimens in contact and pictorial representation of experimental setup

3 Data Analysis

First of all, the experiments have been performed on the stainless steel 304 specimens with no interface material for different loading and temperature conditions under the atmospheric environment. The applied loads are varied as 50, 100, 150 and 200 kg which are equivalent to contact pressures 1, 2, 3 and 4 MPa. Further, mean interface temperature has been varied in the range of 30–80 °C by changing the input heat flux. Subsequently, experiments have been carried out with interface materials in three varying thicknesses between the contacting specimens for varying loading and temperature conditions in the similar ranges as stated above.

Thermal contact conductance (TCC) is estimated by using a well-known and reliable steady state methodology utilizing the expression: $\text{TCC} = \frac{q}{\Delta T}$.

Here, 'q' is the average heat flux across the interface with: $q = (q_1 + q_2)/2$, where q_1 and q_2 are the heat fluxes in the upper and lower specimens, respectively. Heat fluxes q_1 and q_2 are calculated by the Fourier's law of conduction in both the specimens utilizing their axial temperature data of steady state. Further temperature drop 'ΔT' has been estimated by the extrapolation of temperatures at the boundaries of the samples assuming linear temperature profile in both the specimens.

4 Results and Discussion

The results of TCC are evaluated for bare metallic specimens of SS 304 in contact and with interface materials in different thicknesses under varying loading and temperature conditions. Three thicknesses of each silicon grease and graphene paste have been examined under four loading conditions and for four set of heat inputs.

4.1 Bare SS Contacts

The results of TCC for SS contacts have been shown in Fig. 2 for different loading conditions from 1 to 4 MPa but keeping constant mean interface temperature as 46 °C. From Fig. 2, it has been noted that TCC increases with contact pressure. This may be attributed to the deformation of asperities with the pressure and therefore increasing heat transfers across the real contact asperities, thereby increasing TCC. Moreover, it has been noted from Fig. 2 that percentage increment in TCC was 44.64% with the four times of contact pressure.

Further, results of TCC with mean interface temperatures for bare SS contact have been presented in Fig. 3 for constant loading condition of 4 MPa. From Fig. 3, it has been observed that TCC increases with mean interface temperature. This may be due to change in thermo-mechanical properties of specimens such as thermal conductivity, micro hardness, elasticity as well as thermal properties like thermal conductivity, thermal diffusivity of interstitial air with temperature. It is noted from Fig. 3 that TCC variation is 21% for 39 °C rise in temperature from 37 to 76 °C. Some researcher has reported low variation of TCC with mean interface temperature [15], but their experiments and predictions are for solid spot conductance. However, the present results of TCC are estimated for joint contact conductance which is the

Fig. 2 TCC with interface pressure

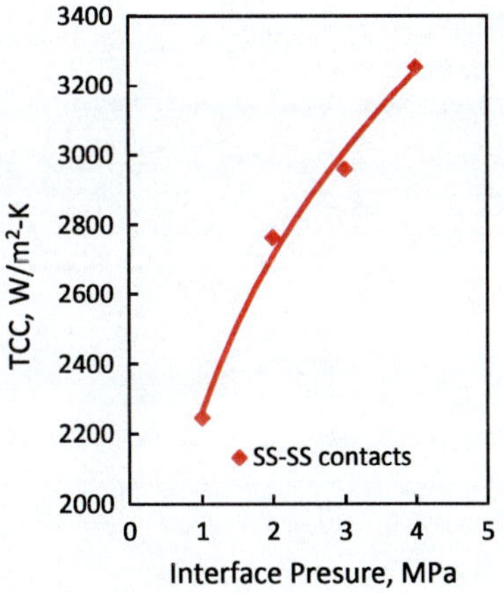

Fig. 3 TCC with mean interface temperature at 4 MPa

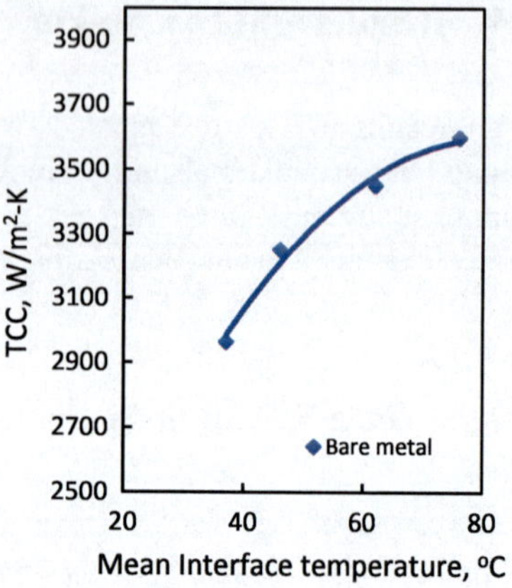

combination of solid spot conductance and gap conductance. Hence, TCC under atmospheric conditions is influenced by the change in temperature.

4.2 *Silicon Grease as Interstitial Material*

Results of TCC for flat SS contact with silicon grease as interstitial material have been presented with contact pressure for various thicknesses of silicon grease in Fig. 4. Results of TCC with silicon grease have been compared with the bare metals and found that TCC is improved from 148 to 166% at lowest pressure with the increase of silicon paste thickness from 22 to 30.5 micron. Further, percentage improvement with the thickness is 186 to 379% at highest contact pressure. Furthermore, from Fig. 4, it has been observed that percentage improvement is more for higher thickness of

Fig. 4 TCC with interface pressure for different thickness of silicon grease

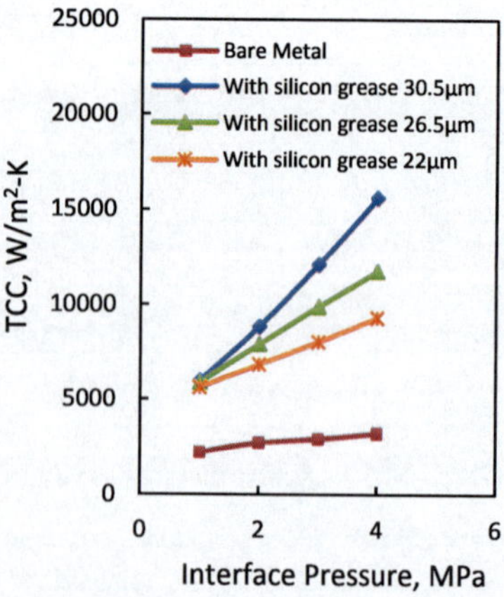

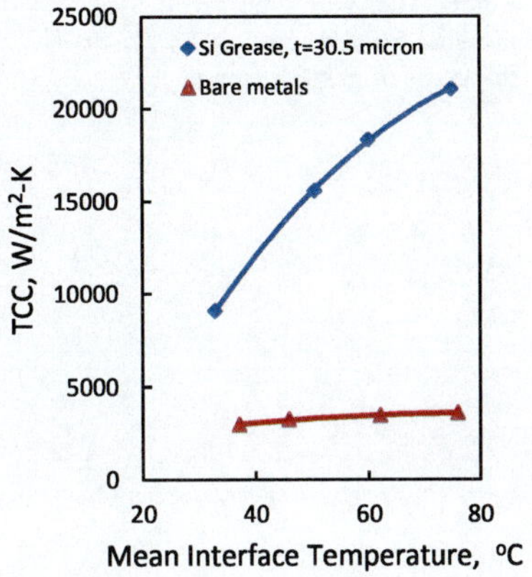

Fig. 5 TCC with mean interface temperature at 4 MPa with silicon grease

silicon paste. In addition, it was noted that TCC of SS pairs using silicon grease is in the range of 5664–15,583 W/m^2-°C for the contact pressure range of 1–4 MPa. Yovanovich et al. [16] experimentally estimated the TCC of an aluminum heat sink-ceramic package assembly with silicon grease as interface material in the range of 29,850 to 46,948 W/m^2-K for interface pressures of 7.0–350 kPa [16].

Further, Fig. 5 shows the variation of TCC with mean interface temperatures at 4 MPa for stainless steel contacts applying 30.5 micron thickness of silicon paste. From Fig. 5, it has been found that TCC keeps on increasing with mean interface temperature. This may be attributed to the change in thermo-mechanical properties of specimens and interstitial material with temperature. It has been noted that percentage change of TCC is 131% with 42 °C change of temperature from 33 to 75 °C. Therefore, change in TCC is higher with temperature with Si grease as compared to bare metal.

4.3 Graphene Paste as Interstitial Material

Results of TCC with graphene paste as interstitial material between the SS contact have been plotted with contact pressure in Fig. 6 for various thicknesses of graphene paste. It has been found that TCC improvement was from 461 to 676% at lowest pressure compared with the bare metal for graphene paste thickness from 27.6 to 39.8 micron. Further, percentage improvement in TCC with the thickness is noted to be 590 to 884% at the highest contact pressure. Moreover, it may be observed from Fig. 6 that percentage improvement is more for higher thickness of graphene paste.

Figure 7 shows the variation of TCC with mean interface temperatures for 27.6 micron thickness of graphene paste and 1 MPa contact pressure. From Fig. 7, it is noted that TCC has increasing trend with mean interface temperature. Further, it has been observed that percentage change of TCC is 71% with the change of

Fig. 6 TCC with interface pressure for different thickness of graphene paste

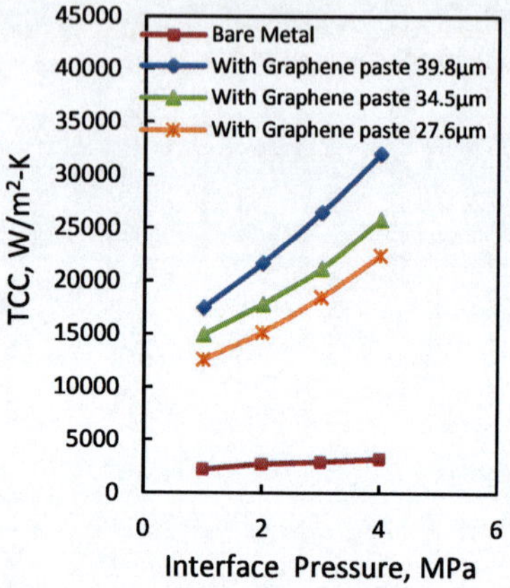

Fig. 7 TCC with mean interface temperature at 1 MPa for graphene paste

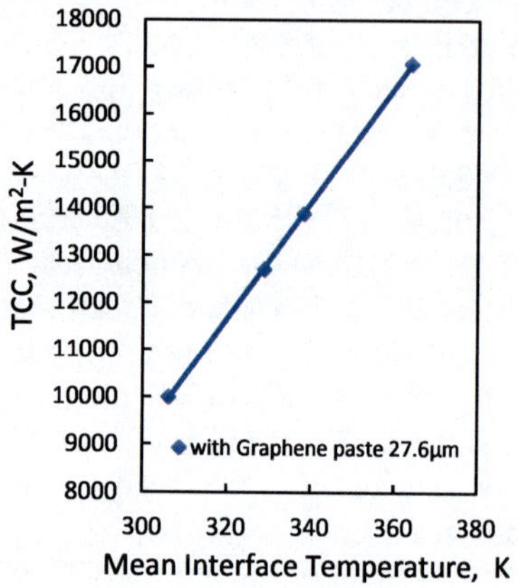

temperature from 33 to 90 °C which is less than the percentage change for silicon grease.

From Figs. 4 and 6, it may be noted that for almost same thickness of paste, the range of TCC is 7912 to 11,720 W/m^2-K for silicon paste while 12,607 to 22,421 W/m^2-K for graphene paste under 1–4 MPa contact pressure. Therefore, from the above discussion, graphene paste was found to be better interstitial material for the enhancement in TCC for the present range of parameters. However, it has been found from the calculation for the present specimen size that the silicon grease (SN752: Rs. 200/- per 100 g ~ Rs. 1.00 per mm) is 5 times cheaper than graphene paste (Rs. 1144/- per 100 g ~ Rs. 5.60 per mm).

5 Conclusions

The TCC of stainless materials (SS 304) in nominally flat contact has been investigated employing two important thermal pastes, viz. silicon grease and graphene paste as interstitial materials. The TCC has been evaluated for three thicknesses of thermal interface materials under varying loading and temperature conditions. From the investigation, the following are the important outcomes:

- Overall TCC was found to increase with contact pressure with and without interstitial materials.
- TCC with thermal pastes at the interface has been compared with bare SS contact and found that graphene paste shows higher improvement in TCC as compared to silicon paste for all the pressure range.
- Further, percentage improvement in TCC is found to be faster for higher thickness value of graphene paste and silicon paste.
- Moreover, it has been found that TCC keeps on increasing with mean interface temperature for SS contact with and without interface material.
- Further, TCC with silicon paste is found to increase faster with the change of temperature as compared to graphene paste.
- Furthermore, it has been found that cost wise silicon grease is cheaper, while performance wise graphene paste is better for the same thickness of two pastes.

Findings from the present investigation can be used for the better enhancement of TCC for thermal management applications for the present range of parameters.

References

1. Madhusudana CV (1996) Thermal contact conductance. Springer, New York
2. Lewis DV, Perkins HC (1968) Heat transfer at the interface of stainless steel and aluminum—the conditions on the directional effect. Int J Heat Mass Transf 11:1371–1383
3. Merrill CT, Garimella SV (2011) Measurement and prediction of thermal contact resistance across coated joints. Exp Heat Transf 24(2):179–200
4. Antonetti VW, Yovanovich MM (1984) Enhancement of thermal contact conductance by metallic coatings: theory and experiment. J Heat Transf 107:513–519
5. Peterson GP, Fletcher LS (1990) Measurement of the thermal contact conductance and thermal conductivity of anodized aluminum coatings. J Heat Transf 112:579–585
6. Salerno LJ, Kittel P, Spivak AL (1994) Thermal contact conductance of pressed metallic contacts augmented with indium foil or Apiezon grease at liquid helium temperatures. Cryogenics 37(8):649–654
7. Marotta E, Fletcher LS, Aikawa T, Maki K, Aoki Y (1999) Thermal contact conductance of sintered copper coatings on ferro-alloy. J Heat Transf 121:177–182
8. Wolff EG, Schneider DA (1998) Prediction of thermal contact resistance between polished surfaces. Int J Heat Mass Transf 41:3469–3482
9. Kshirsagar B, Nagaraju J, Murthy MVK (2003) Thermal contact conductance of silicon nitride-coated OFHC copper contacts. Exp Heat Transf 16:273–279
10. Kshirsagar B, Misra P, Jampana N, Murthy MVK (2005) Thermal contact conductance across gold-coated OFHC copper contacts in different media. J Heat Transf 127(6):657–659

11. Jeng YR, Chen JT, Cheng CY (2006) Thermal contact conductance of coated surfaces. Wear 260:159–167
12. Misra P, Nagaraju J (2010) Thermal gap conductance at low contact pressures (<1 MPa): effect of gold plating and plating thickness. Int J Heat Mass Transf 53:5373–5379
13. Zheng X, Liu D, Wei D, Shang X (2011) Experimental investigation of high temperature thermal contact resistance with interface material. Theoret Appl Mech Lett 1:051009-1–4
14. Graphene Products Technical Detail (2019) https://www.graphene-bond.com
15. Wahid SMS, Madhusudana CV, Leonardi E (2003) Solid spot conductance at low contact pressure. Exp Therm Fluid Sci 28(6):489–494
16. Yovanovich MM, Culham JR, Teertstra P (1997) Calculating interface resistance. Electron Cool 54:1–9

Boiling Front and Boiling Temperature in Microchannels Under Non-uniform Heat Flux Distribution

Radha Kanta Sarangi, Abhilas Swain, Satya Prakash Kar, and P. C. Sekhar

Abstract Boiling front, onset boiling temperature, and pressure drop for microchannel in non-uniform heat flux condition have been investigated in this paper. Homogeneous flow model has been assumed with constant heat flux boundary condition. Model, results have been shown for water as working fluid. Unlike most of the models from literature, fixed outlet pressure and inlet temperature have been assumed making the model analysis more realistic. The governing equations have been solved to predict the effects of boiling front, onset boiling temperature, and pressure drop on various parameters like outlet pressure, mass flow rate, and inlet temperature.

Keywords Boiling front · Pressure drop · Homogeneous flow model · Two-phase flow

1 Introduction

Boiling heat transfer in microchannel is based on heat transfer of liquid–vapor phase change phenomena. There are various models for predicting the heat transfer and pressure drop in two-phase flow. But in non-uniform heat flux condition, for microchannels, the modeling studies are limited. So there is a need to analyze the boiling studies for non-uniform heat flux in microchannels due to its various applications starting from electronic cooling to biological systems. In these type of applications, heat dissipation density should be very high for reducing the overall size. This leads to decrease in coolant inventory and weight.

The advantages of microchannels include high heat transfer rate, lower mass flow rate, smaller volume, and weight. The pressure drop is significantly higher for these channels. The benefits of high heat transfer coefficients and high heat carrying capacity through latent heat transport come at a penalty of high pressure drop [1]. Lower and higher limits of diameter in case of microchannel are not very clear

R. K. Sarangi (✉) · A. Swain · S. P. Kar · P. C. Sekhar
School of Mechanical Engineering, Kalinga Institute of Industrial Technology (KIIT) Deemed to be University, Bhubaneswar, Odisha 751024, India
e-mail: sarangi_radhakanta@yahoo.co.in

S. Revankar et al. (eds.), *Proceedings of International Conference on Thermofluids*, Lecture Notes in Mechanical Engineering,
https://doi.org/10.1007/978-981-15-7831-1_29

from literature. In the field of microelectronics applications, the range of hydraulic diameter below 200 μm is becoming important [2]. Considering this, the normal range of hydraulic diameter attributed to microchannel is 20–200 μm.

Two most important parameters for analyzing heat transfer in microchannels are pressure drop and heat transfer. These parameters depend strongly on boiling front. The boiling front needs to be determined first, for predicting the pressure drop. As the liquid flows through a microchannel, nucleation occurs over cavities within a certain size range. The size of cavities depends on the flow conditions. Depending on these cavities, there is always a degree of wall super heat for boiling to start. Super heat necessary for nucleation may be expressed by Hsu and Graham [3]. As channel size decreases, wall super heat to initiate boiling increases. Wall super heat of 20 °C is required for 200 μm channel and wall super heat for 50 μm channel is 30 °C with water [4]. In the simulation, wall super heat is not considered and it is assumed that boiling starts at saturation temperature corresponding to local pressure.

Literature indicates the similarities between major two-phase flow regimes between microchannel and macrochannels. Triplett et al. [5], Damianides and Westwater [6], Mandhane et al. [7], Mishima and Ishii [8] observed the different flow regimes with the help of high speed photography and substantiated it. In the present research, modeling has been done for homogeneous flow boiling. Pressure drop characteristics of microchannels have been investigated by various researchers Mala and Li [9], Brutin et al. [10], Li et al. [11] and Xu et al. [12]. According to them, there is an early transition and magnitude of pressure drop is higher than the pressure drop predicted by Poiseuille law. If the surface is rough and channel hydraulic diameter is less than 100 μm, this deviation is more pronounced. In the present work, Reynolds number falls in the range of 100–200. Hence, laminar flow is assumed and traditional Poiseuille law is used for calculating the pressure drop. Studying single and two-phase flow in microchannels, Watel [13] observed that the single-phase flow and boiling characteristics that have been established for large channels (diameter greater than 5 mm) may not be applicable to microchannel flows. This is due to the presence of laminar boundary layer and perhaps a vastly bubble generation mechanism within a microchannel. In the present model, it is assumed that, the flow is fully developed from the beginning.

The effect of non-uniform heat flux in microchannel has been investigated experimentally by Ritchey et al. [14] and Alam et al. [15]. They concluded that, in non-uniform heat flux, input pressure drop and heat transfer characteristics are different from uniform heat flux input. In the research work of Sarangi et al. [16], boiling front, wall temperature, and film temperature have been investigated for homogeneous model flow model in uniform heat flux input. This paper investigates, boiling heat transfer in microchannels for non-uniform heat flux in case of, homogeneous flow.

2 Setup Description and Modelling

For analyzing the non-uniform heat flux channels, four case studies are considered. The width, depth, and length are taken as 50 μm, 75 μm, and 12 mm, respectively. Total heat input 2.5 W was distributed non-uniformly along the length. Heat input of 2.25 W (90%) was concentrated along one-third of the channel (4 mm) while the remaining 10% (0.25 W) was distributed over 8 mm in various combinations. These combinations consist of four different cases which are given in Fig. 1.

Two-phase flow in microchannel is shown in Fig. 2. The flow is along z-direction. At $z_{c,}$ boiling starts and after that, it is a mixture of liquid and vapor. In this model, like the model of Sarangi et al. [16], exit pressure and inlet temperature are known. So, it is required to start from outlet to inlet for finding the boiling front ($z = z_c$) and subsequent analysis. Therefore, single-phase analysis is carried out from 0 to Z_c and two-phase analysis is carried out from Z_c to l.

2.1 Pressure Drop

As the mass flow rate is small and tube hydraulic diameter is also very small, Reynolds number is always less than 200 and flow is laminar. So, pressure drop in single phase

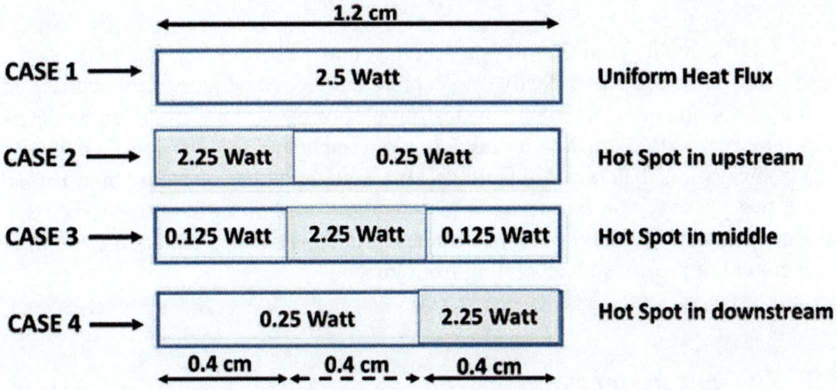

Fig. 1 Heat flux arrangement in the microchannels

Fig. 2 Schematic of flow in microchannel

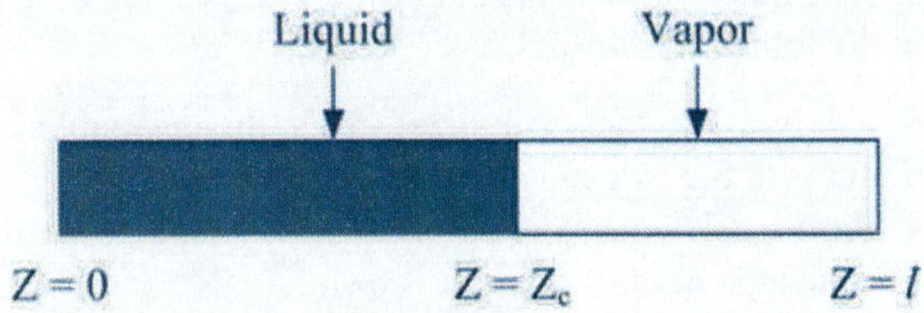

is determined by Haygen-Poiseuille law. Two-phase pressure drop is determined by following equation in homogeneous flow like Sarangi et al. [16].

$$\left(\frac{\mathrm{d}p}{\mathrm{d}z}\right)_{2\phi} = \frac{f_{2\phi} G^2}{2\rho_{2\phi} d_h} + \frac{1}{A_{\mathrm{CH}}} \frac{\mathrm{d}}{\mathrm{d}z} \left(\frac{G^2 A_{\mathrm{CH}}}{\rho_{2\phi}}\right) \tag{1}$$

where $f_{2\phi}$ is the two-phase friction factor, G is the mass flux, A_{CH} is the channel cross-sectional area, and d_h denotes the hydraulic diameter of the channel. The first and second terms on the RHS denote the frictional pressure drop and acceleration pressure drops, respectively. The variations in fluid properties like due to very high pressure drops across the microchannel, the thermo-physical properties along the length of the microchannel cannot be neglected because of large pressure drop in microchannel. Two-phase friction factor, $f_{20\phi}$ and two-phase density, $p_{2\phi}$ is determined by following Eq. 2 and 3. Two-phase viscosity is calculated by the viscosity model proposed by Duckler et al. [17] as given in Eq. 4.

$$f_{2\phi} = \frac{\mu_{2\phi}}{\mu_l} \tag{2}$$

$$\frac{1}{\rho_{2\phi}} = \frac{x}{\rho_v} + \frac{1-x}{\rho_l} \tag{3}$$

$$\mu_{2\phi} = \frac{x v_g \mu_g + (1-x) v_f \mu_f}{x v_g + (1-x) v_f} \tag{4}$$

where x is the quality of the liquid vapor mixture and the subscripts f and g denote the liquid and vapor phases, respectively. As pressure and temperature vary at different locations continuously, thermos-physical properties of working fluid can be determined by curve fitted equation as a function of temperature or pressure. Exit quality (x_e) is always less than one in this model. If the exit quality (x_e) is less than 1, then entire two-phase region corresponds to saturation condition. Fluid temperature in single phase can be found out by considering energy balance equation from inlet temperature of previous node starting from inlet.

2.2 *Computer Algorithm*

Homogeneous model equations are straight forward and those equations can be applied iteratively for each node after discretization. The algorithm can be described in the following steps.

- The total length is discretized into sufficient number of nodes and then the analysis starts at the exit and ends at the inlet.
- At the exit node, p_{out} is given and all the properties at the node next to it are evaluated at this pressure.

- Then, using Eq. (1), dp/dz is calculated and added in to the pressure at the previous node to get the local pressure at this node.
- At this local pressure, liquid saturation temperature and other thermos-physical properties are calculated.
- Then, this method is adopted for the next upstream nodes and the code is repeated till inlet.

3 Results and Discussions

The simulation was first carried out for validating the experimental results of Zhang et al. [18] for uniform heat flux condition. In their experimental work, boiling front and pressure drop have been investigated for 50 × 70 μm channel of 2 cm length and water as working fluid. The simulation result under predicted the boiling front by 30% and over predicted the two-phase pressure drop by 26%. Then model is used to study the effects of non-uniform heat input along the flow direction. The heat flux was made to vary spatially along z. The boiling front for non-uniform heating can be higher or lower than the uniform heating case depending on the location of the hot spot. In a given hot spot condition, boiling front can be adjusted by appropriate mass flow rate (m), outlet pressure (p_{out}), and inlet temperature (T_{in}).

Figure 3 shows the variation of boiling front with the mass flow rate for the four cases of the heat input arrangement. Figure 3 shows that, as mass flow increases, boiling front also increases due to higher liquid inventory. Higher liquid inventory

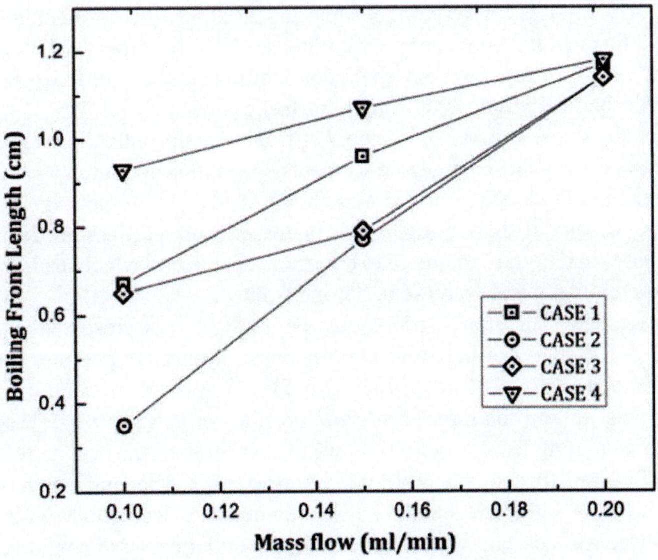

Fig. 3 Boiling front variation with mass flow rate for different cases (T_{in} 20 °C and p_{out} 1.01325 bar)

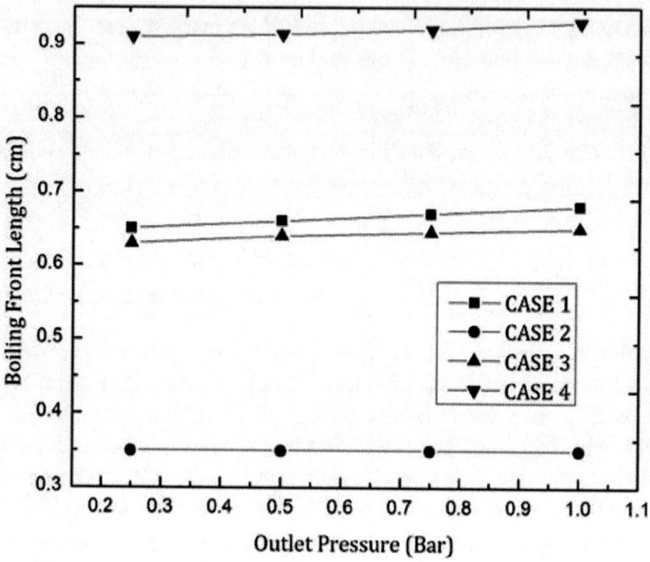

Fig. 4 Boiling front variation with outlet pressure (m 0.1 ml/min and $T_{in} = 20$ °C)

requires more amount of heat to boil and as a result, boiling front also increases. It can be seen that, boiling front variation with mass flow rate is linear in uniform heat flux zone. Heat flux is uniform all along for case 1. So, corresponding boiling front variation is also linear. Similarly, for case 4, boiling starts from 0.93 cm and varies linearly. But it is not a continuous line and the slopes are different in two zones (lower heat flux and hot spot zone). As boiling starts at two different heat flux zones that is low heat flux zone and hot spot zone, boiling front is not linear throughout and slope of the line will change from one zone to other.

Figure 4 shows the variation of the boiling front with the outlet pressure condition for the four cases of heat input arrangements. As outlet pressure (p_{out}) increases, boiling front also increases but the magnitude of increase is not significant. The increase is because of the fact that, with increase in outlet pressure, local boiling point increases and liquid is required to take more heat to boil which leads to increase in liquid travel length and increase in boiling front.

Variation of boiling front with inlet temperature (T_{in}) is presented in Fig. 5. It shows the decrease in boiling front with increase in inlet temperature for a given mass flow rate (m) and outlet pressure (p_{out}) condition. As inlet temperature increases, amount of heat to boil the liquid decreases for a given mass flow rate which leads to decrease in boiling front. Figure 6 shows the variation of the boiling temperature with mass flow rate for pressure outlet condition of 1.01325 bar and inlet temperature of 20 °C. It can be observed that the boiling temperature decreases with increase in mass flow rate for a certain inlet temperature and outlet pressure condition. This is

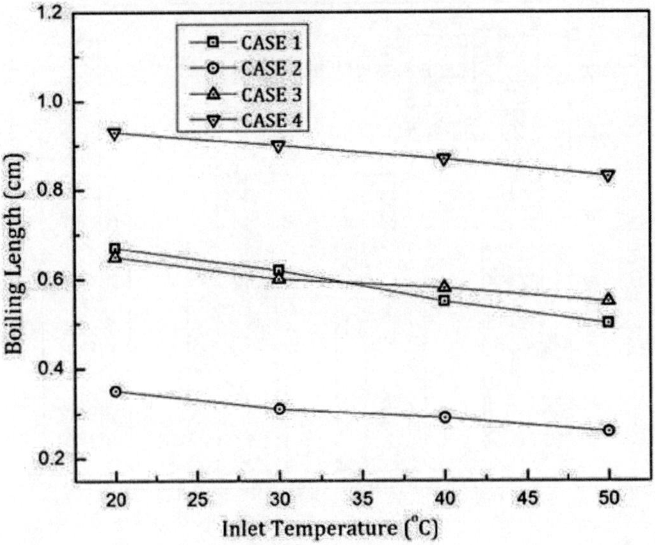

Fig. 5 Boiling front variation with inlet temperature (m 0.1 ml/min and p_{out} 1.01325 Bar)

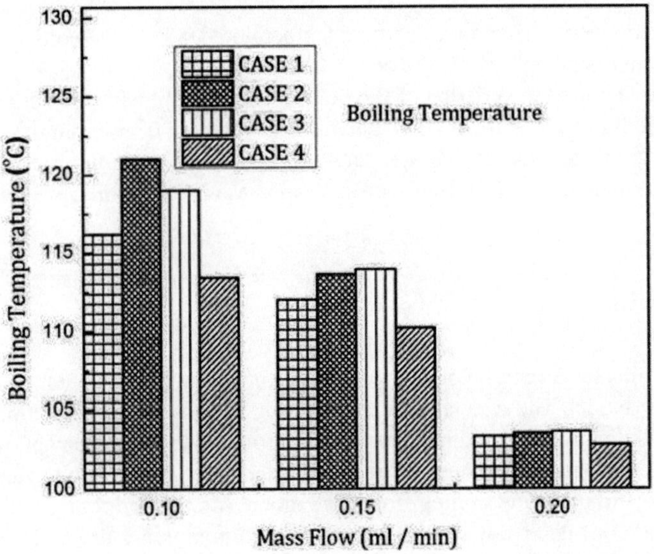

Fig. 6 Boiling temperature variation with mass flow rate (T_{in} = 20 °C and p_{out} 1.01325 bar)

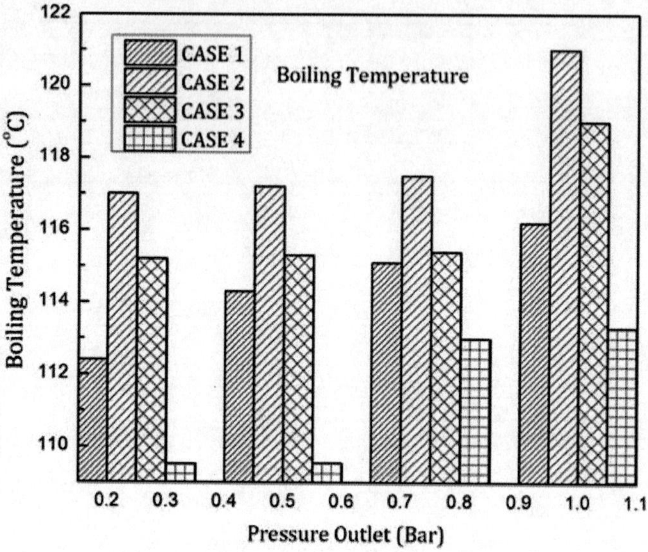

Fig. 7 Boiling temperature variation with outlet pressure ($m = 0.3$ ml/min and $T_{\text{in}} = 20$ °C)

because reduction in single-phase liquid temperature leads to late boiling and lower boiling temperature.

Figure 7 shows the variation of the boiling temperature with the outlet pressure for inlet liquid temperature of 20 °C and mass flow rate of 0.1 ml/min. It is observed from the graph that the boiling temperature increases with outlet pressure due to increase in local pressure and subsequently, increase in local boiling temperature.

4 Conclusions

Boiling fronts have been predicted for non-uniform heat flux channels by homogeneous flow model. The results show that, boiling front is a strong function of mass flow rate, outlet pressure and inlet temperature for a given microchannel. It also can be concluded that, wall temperature and film temperature are strong functions of heat flux distribution and boiling front. The model can be used for optimum design of channels for effective heat transfer and minimum pressure drop. Hot spots can be located by analyzing the wall temperature variations along the channel. Other boiling models like slug flow models, annular flow models, or different models in combination according to quality may be incorporated for accurate results.

References

1. Reay DA (2002) Compact heat exchangers enhancement and heat pumps. Int J Refrig 25:460–470
2. Kandlikar SG (2002) Fundamental issues related to flow boiling in microchannels. Exp Therm Fluid Sch 26:389–407
3. Hsu YY, Graham RW (1964) An analytical and experimental study of thermal boundary layer and Ebulition cycle in nucleate boiling. NASA TN-D-594
4. Fukano T, Kariyasaki A (1993) Characteristics of gas-liquid two phase flow in a capillary. Nucl Eng Des 141:59–68
5. Triplett KA, Ghiaasiaan SM, Abdel-Khalik ST, Sadowski DL (1999) Gas liquid two phase flow in microchannel. Int J Multiph Flow 25(3):377–394
6. Damianides CA, Westwater JW (1998) Two phase flow pattern in a compact heat exchanger and in small tubes. In: Proceedings of the 2nd U.K. national conference on heat transfer, pp 1257–1268
7. Mandhane JM, Gregory GA, Aziz KA (1974) A flow pattern for gas-liquid for horizontal pipes. Int J Multiph Flow 1:537–553
8. Mishima K, Ishii M (1984) Flow regime transition criteria for two phase flow in vertical tubes. Int J Heat Mass Transfer 27:723–737
9. Mala GM, Li DQ (1999) Flow characteristics in micro tubes. Int J Heat Fluid Flow 20:142–148
10. Brutin B, Topin F, Tardist L (2002) An efficient method to calculate friction factor of laminar flow in capillary tubes. In: Proceedings of the international conference on mini channels and microchannels, paper ICMM-1037. Rochester, Apr 2002, pp 24–25
11. Li ZX, Du D, Guo ZY (2000) Experimental study on flow characteristics of liquid in circular micro tubes. In: Proceedings of international conferences on heat transfer and transport phenomenon in micro scales. Banff, Canada, 15–20 Oct 2000
12. Xu B, Ooi KT, Wong NT, Liu CY, Choi WK (1999) Liquid flow in microchannels. In: Proceedings of the ASME/JSME, joint thermal engineering conference. San Diego, CA, 15–19 Mar 1999
13. Watel B (2003) Review of saturated flow boiling in small passage of compact heat exchangers. Int J Therm Sci 42:107–140
14. Ritchey SN, Weibel JA, Garimella SV (2014) Local measurement of flow boiling heat transfer in an array of non-uniformly heated microchannels. Int J Heat Mass Transf 71:206–216
15. Alam T, Lee PS, Yap CR, Jin L (2013) A comparative study of flow boiling heat transfer and pressure drop characteristics in microgap and microchannel heat sink and an evaluation of microgap heat sink for hotspot mitigation. Int J Heat Mass Transf 58:35–347
16. Sarangi RK, Bhattacharya A, Prasher RS (2009) Numerical modeling of boiling heat transfer in micro channel. Appl Therm Eng 29:300–309
17. Duckler AE, Wicks M, Cleveland RG (1964) Pressure drop and hold-up in two phase flow. AIChE J 10:38–51
18. Zhang L, Koo JM, Jiang L, Asheghi M, Goodson KE, Santiago JG, Kenny TW (2002) Measurements and modeling of two-phase flow in microchannels with nearly constant heat flux boundary conditions. J MEMS 11:12–19

Natural Convection Heat Transfer on the Strip Heaters Flushed on the Vertical Flat Plate: A Numerical Study

Akurathi Rupendra Siva Prasad, Medisetti Venkata Krishna, and Pradeep S. Jakkareddy

Abstract In the present study, a CFD analysis of a vertical plate with heat generating sources subjected to natural convection phenomenon is performed. The effect of each individual strip heater over the flat plate assembly which mimics the cooling of the electronic systems is studied. The geometry, the grid generation and the simulations with appropriate boundary conditions are carried out using COMSOL Multiphysics. Mesh independence test has been carried out. Prior to the actual simulations, the numerical code is validated. A test case of constant temperature vertical plate is simulated and the obtained results, i.e., the average heat transfer coefficient is compared with the standard empirical correlation results. Influence of the strip heaters on the vertical plate is elucidated in terms of temperature, velocity and heat transfer coefficient variations.

Keywords Free convection · CFD analysis · Vertical plate · Heat sources

1 Introduction

For many new age electronics and mechanical components, lifespan is decreasing due to heat generation and transfer through them. Convection is the mode of energy transfer between a solid and fluid that is in motion [1]. Many researchers have performed analytical, numerical and experimental studies on vertical plate with different modes of heat transfer and placing heaters at different location on vertical plate and by varying the heat generation values of heaters. Few pertinent articles are relooked here. Chen et al. [2] conducted an experimental study on the effect on cooling due to spacing between electronic devices. Authors concluded that equispacing arrangement in the electronic devices is not preferable. Gururaja et al. [3] investigated on the effect of mixed convection and surface emissivity on the vertical parallel plate channel containing discrete heat sources. Authors observed that the

A. Rupendra Siva Prasad · M. Venkata Krishna · P. S. Jakkareddy (✉)
Department of Mechanical Engineering, Amrita School of Engineering, Amrita Vishwa Vidyapeetham, Bengaluru 560035, India
e-mail: js_pradeep@blr.amrita.edu

S. Revankar et al. (eds.), *Proceedings of International Conference on Thermofluids*, Lecture Notes in Mechanical Engineering,
https://doi.org/10.1007/978-981-15-7831-1_30

local, average, maximum temperature decreases with surface emissivity. Sarper et al. [4] reported the effect of natural convection in a vertical channel with two flush mounted heat sources numerically and experimentally. From the study, they found that the temperature of the heated walls decreases from the leading edge to the trailing edge, and if the distance between the heat sources increases, the convective heat transfer rate also increases. In the cooling of electronic systems, where natural convection induced due to discrete heating finds a scope for a better understanding of the heat transfer characteristics. In the present study, numerical investigation of steady state, laminar, conjugate heat transfer on vertical plate with three heaters mounted at different locations is carried out by varying the heat generation inputs.

2 Numerical Methodology

Steady state, two dimensional, laminar, natural convection heat transfer simulations are carried out on the vertical plate embedded with three discrete heaters and the fluid domain using a software package COMSOL [5]. A flat plate is of width 0.015 m and height 0.155 m, and the heat generating sources each with width of 0.01 m and height 0.025 m located at distance of 0.0122 and 0.0622 and 0.1122, respectively, from the leading edge is shown in Fig. 1. A fluid domain of dimensions 0.175 × 0.15 m is considered in the current study. The materials used for the heat sources and the plate are Aluminum and Bakelite. Air is the working fluid under different conditions. A computational domain is shown in Fig. 2.

The assumptions and the governing equations of the present study are as follows:

I. The Boussinesq approximation which relates density change to temperature change is used in formulating buoyancy force in the momentum equation.
II. Dissipation effect is neglected in the energy equation.

The continuity equation

$$\frac{\partial u}{\partial x} + \frac{\partial v}{\partial y} = 0 \tag{1}$$

The momentum equation

$$u\frac{\partial v}{\partial x} + v\frac{\partial v}{\partial y} = \beta g(T - T_\infty) + v\frac{\partial^2 v}{\partial x^2} \tag{2}$$

The energy equation

$$u\frac{\partial T}{\partial x} + v\frac{\partial T}{\partial y} = \alpha\left(\frac{\partial^2 T}{\partial x^2} + \frac{\partial^2 T}{\partial y^2}\right) + q_v \tag{3}$$

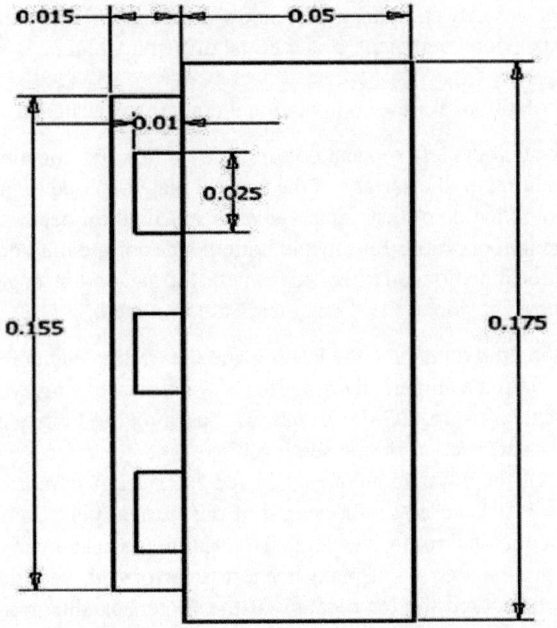

Fig. 1 Geometry of the three heaters flushed against the vertical plate and air domain (all dimensions are in m)

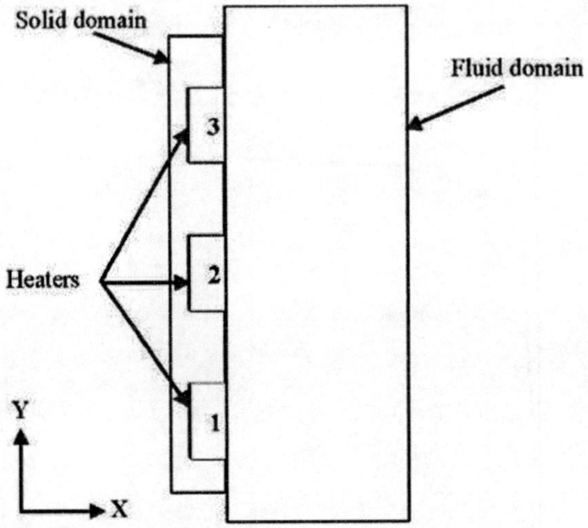

Fig. 2 Computational domain

where u, v is the velocity component in x- and y-directions, T is the temperature, β is the thermal expansion coefficient, α is thermal diffusivity and q_v is volumetric heat generation.

Boundary conditions for the computational domain are assigned as follows:

- Adiabatic wall at the left, top and bottom side of the solid domain.
- No slip condition on the surface of the vertical plate exposed to quiescent air.
- Different combination of heat input values is specified for heaters.
- Open boundary conditions for top and bottom sides of the fluid domain.
- Left side of the fluid domain extended from the top and bottom edge of the vertical plate and the right side of the fluid domain are insulated.

Length of the fluid domain at the leading and the trailing edge of a vertical plate is extended to avoid the numerical error. This is a sort of tweaking, which is required in this numerical software, COMSOL 5.5 [5] (based on the finite element method) to attain the convergence in the present problem. The convergence criteria are set to 1×10^{-6}. All the simulations are valid for a Rayleigh number range of 1.51×10^6 to 1.81×10^6 (based on the length of the vertical plate). The unstructured triangular computational mesh is generated to capture the heat transfer phenomenon of the present problem. Grid independence test is performed. Variation between the number of elements and the temperature at the heater1 is shown in Fig. 3. From Fig. 3, we can observe that the value of temperature of heater1 remains unaltered after 35,374 elements. This indicates 35,374 elements are sufficient to carry out the further simulations.

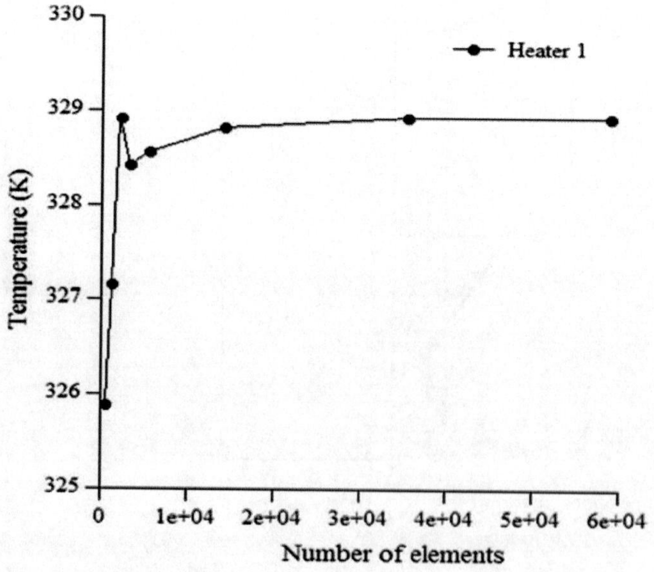

Fig. 3 Grid independence test

3 Results and Discussions

Before conducting the actual simulations of three discrete heaters embedded in the vertical plate under natural convection, the numerical code employed in the present study had to be validated. For this, an exercise was taken up, where a vertical plate with a constant temperature of 92 °C is simulated. The average heat transfer coefficient obtained from the numerical simulations is compared with the empirical correlation $Nu = 0.59Ra^{0.25}$ [6]. The average percentage difference of heat transfer coefficient between the two is found to be reasonable (~2.9%).

In the present study, numerical simulations are carried out for different combination of heat sources ($Q1 > Q2 > Q3$, $Q2 > Q1 > Q3$, $Q3 > Q1 > Q2$ and $Q1 = Q2 = Q3$) flush against the front surface of the vertical bakelite plate exposed to ambient still air (Prandtl number = 0.7). The two-dimensional contour plots of temperature and velocity distribution are shown in Figs. 4 and 5. Temperature variations in the solid domain are in the range of 300–340 K. Velocity variations are from 0 to 0.25 m/s in the fluid domain. A maximum temperature of 329, 330, 331 and 340 K is attained for the heat inputs of case (a) = 10, 5 and 2.5 W (at heater 1), case (b) = 5, 10 and 2.5 (at heater 2), case (c) = 5, 2.5 and 10 (at heater 3) and case (d) = 10, 10 and 10 W (at heater 2 and 3).

Determining the temperature distributions of wall for different discrete heating inputs is of importance to thermal engineers for designing the electronic equipment, thus augments the rate of heat transfer. This piece of information would guide us to select the proper sequencing of electronic chips to operate under prescribed temperature safely. The unevenly distribution of the temperature at the front surface of the vertical plate embedded with three discrete heaters of varying intensity is clearly seen in Fig. 6. At higher heat input in all the four cases considered in the present problem, the temperature gradient at the corresponding heater is high. Thus, the heat transfer from the heaters having high intensity (at heater 3, 1, 2 and 3) to the working fluid (air) is more, when, compared to the remaining regions of the solid domain. We can also observe that for the case of heat input $Q1 = Q2 = Q3 = 10$ W, the plate temperature in most part of the region is high. The reason for this is more concentrated heat input in the heated region. Potted in Fig. 7a–c are the temperature variations in the axial directions from the heaters flush mounted on the vertical plate at Rayleigh number of order 10^6 for different combination of heat inputs. A maximum temperature of 340 K is found at heater 3 (see Fig. 7c).

Figure 8 depicts the temperature profile in the fluid domain in the longitudinal direction at an offset distance of 1 mm from the solid domain. One can observe that temperature is varying in the increasing and decreasing fashion in the thermal boundary layer. As the fluid comes in contact with the heated region, there will be energy exchange between the two, driven by the temperature difference and the convection current will setup due to the buoyancy effect. For the given heat input ($Q1, Q2, Q3$) to the three discrete heaters, one can see that there is a heat conduction and convection from the heaters to the unheated zone of the plate and to the fluid. Temperature drop and rise is evident in Figs. 6 and 8. The maximum temperature of

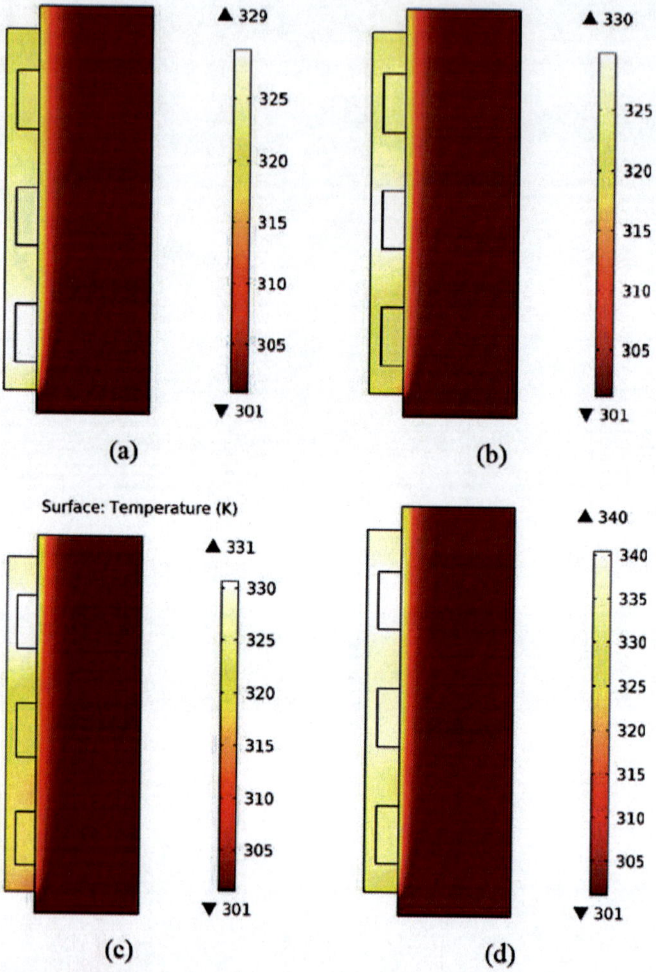

Fig. 4 Temperature contours of the computational domain with different combination of heat sources. **a** 10, 5 and 2.5 W, **b** 5, 10 and 2.5, **c** 5, 2.5 and 10 and **d** 10, 10 and 10 W

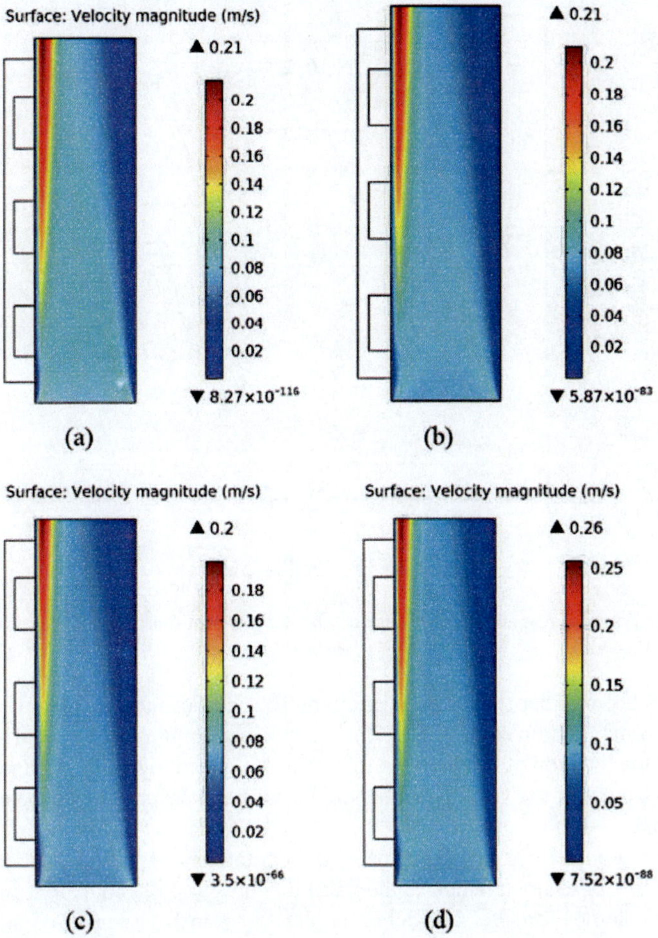

Fig. 5 Velocity contours of the computational domain with different combination of heat sources. **a** 10, 5 and 2.5 W, **b** 5, 10 and 2.5, **c** 5, 2.5 and 10 and **d** 10, 10 and 10 W

328 K is attained at heater 1 and the corresponding fluid temperature of 322 K at a distance of 1 mm from the heater. It is worth noting that, in some of the combination of heat inputs, the heat lost by the heater will rise the temperature of the fluid, which can be higher than the temperature of the unheated surface. Thus, the heat transfer is induced from the fluid to unheated regions of the vertical plate. The variation of velocity in the longitudinal direction at an offset distance of 1 mm from the wall is shown in Fig. 9.

Shown in Fig. 10a–c are the development of velocity profile in the axial direction. The velocity of the air is accelerated near the heaters having high heat intensity. From Fig. 10a–c, we can observe that for any combination of heat input in the current study, the air velocity at a fast speed attains the fully developed condition for

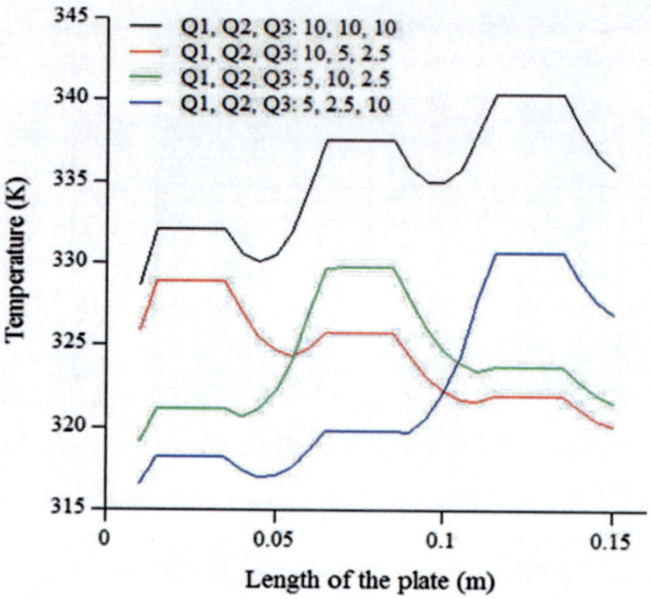

Fig. 6 Variation of temperature for different combination of heat sources on the vertical plate

lower Grashof number (based on heater length). Grashof number is in the order of 10^4–10^6. A maximum velocity of 0.22 m/s is obtained for the case of equal heat input (10 W) at the location of heater 3. An influence of the heating values is significant in the development of the velocity profiles, which affects the cooling of the electronic components.

The heat transfer coefficient is one of the parameters required in understanding the heat transfer characteristics of the natural convection phenomenon. In Fig. 11, space variable heat transfer coefficient is plotted against the length of the vertical plate assembly. Heat transfer coefficient is found to be larger for the case $Q1 = Q2 = Q3 = 10$ W. This indicates average heat transfer is better from the heated wall. For the case of $Q1$, $Q2$ and $Q3$ of 5, 2.5 and 10 W, there is a sharp rise of the local heat transfer coefficient at the location of third heater in the wall. In general, the variation of the local heat transfer coefficient is smooth in the case of a vertical plate with constant temperature. Whereas in the case of different combination of heaters flush mounted on a vertical plate, the local heat transfer coefficient is increasing and decreasing depending on the intensity of the heaters (see Fig. 11).

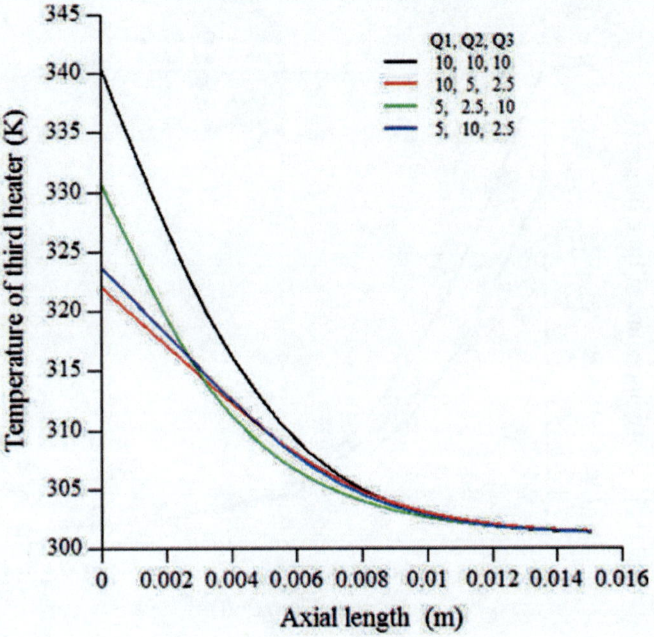

(a)

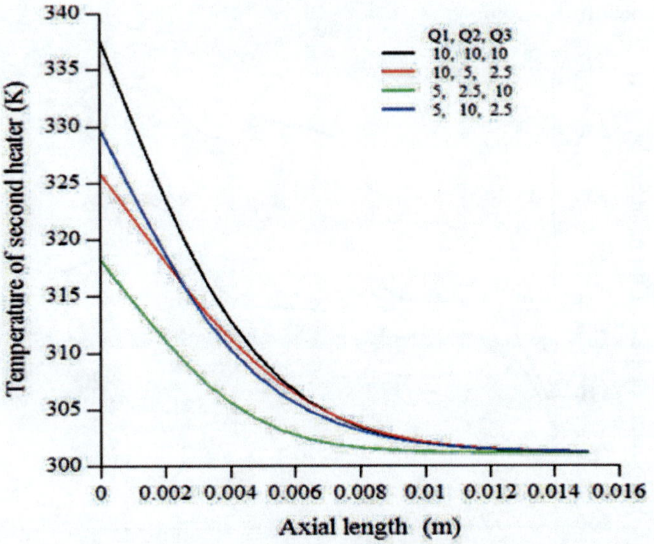

(b)

Fig. 7 **a–c** Development of temperature profiles in the axial direction

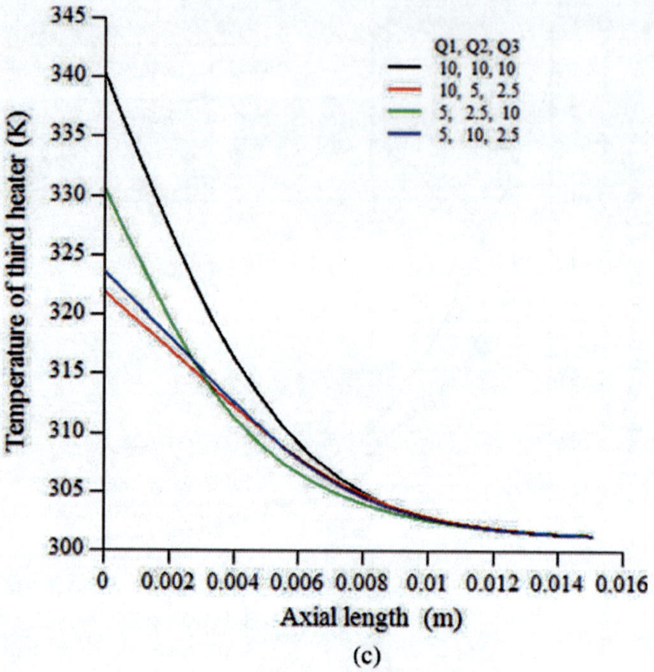

Fig. 7 (continued)

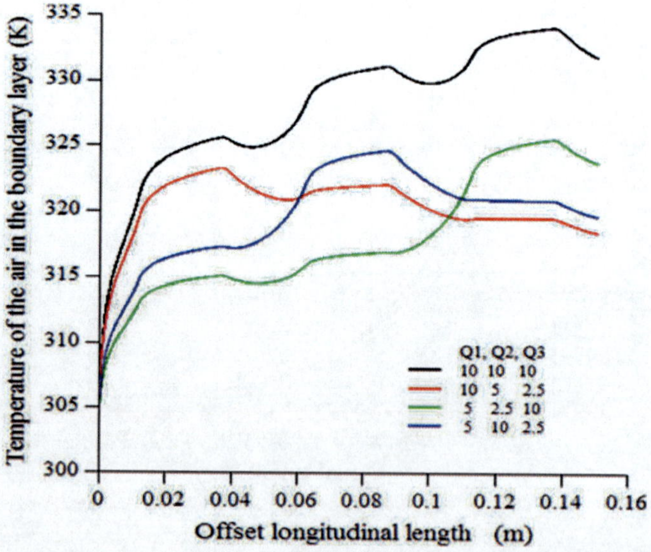

Fig. 8 Temperature distribution of the air in the longitudinal direction at a distance of 1 mm offset from the vertical plate

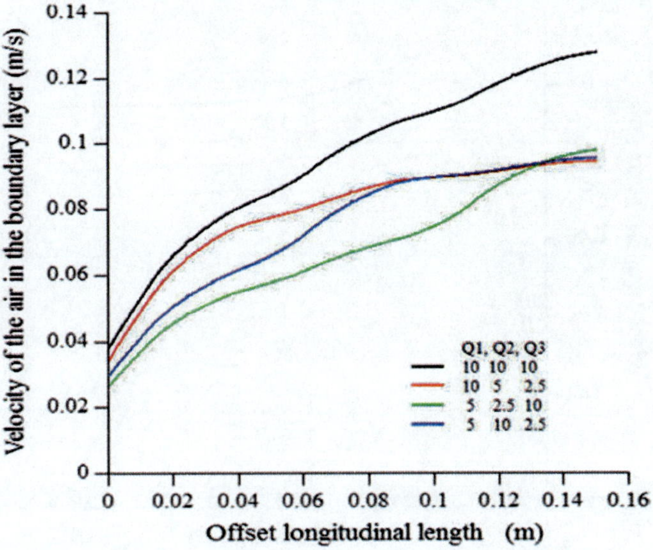

Fig. 9 Velocity distribution of air in the boundary layer at a distance of 1 mm

4 Conclusions

Numerical investigation on a vertical plate with three heaters has been studied in detail. The simulations are carried out for the steady, laminar, 2D, natural convection for different values of heat sources using COSMOL [5]. In order to validate the numerical code, first we have taken a case of vertical plate with constant wall temperature condition. We found that the percentage difference of average heat transfer coefficient between simulation and empirical correlation was 2.9%. This exercise verifies the adequacy of the numerical scheme.

The CFD simulations carried out on three heaters flush mounted in a vertical plate reveal that the temperature variation is not uniform due to spreading of heat to the unheated surface and heat convection from the heated surface. Heater strength influences the local heat transfer coefficient of the plate. These results are useful for the optimal thermal design of electronic components.

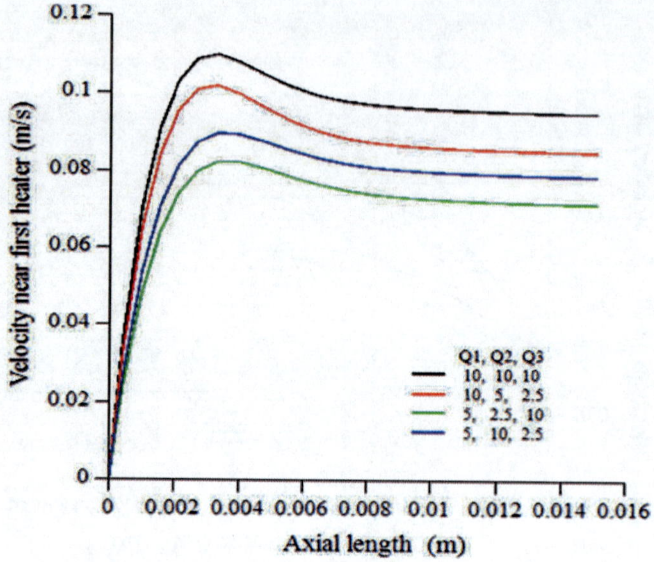

(a)

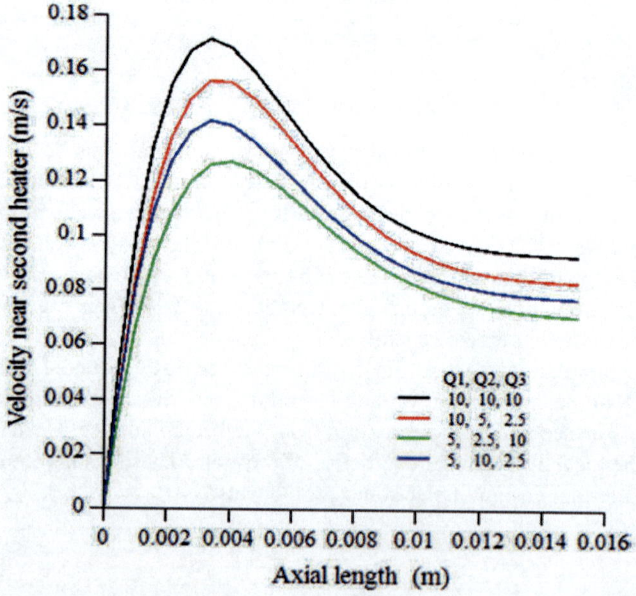

(b)

Fig. 10 Axial variations of velocity from the heaters

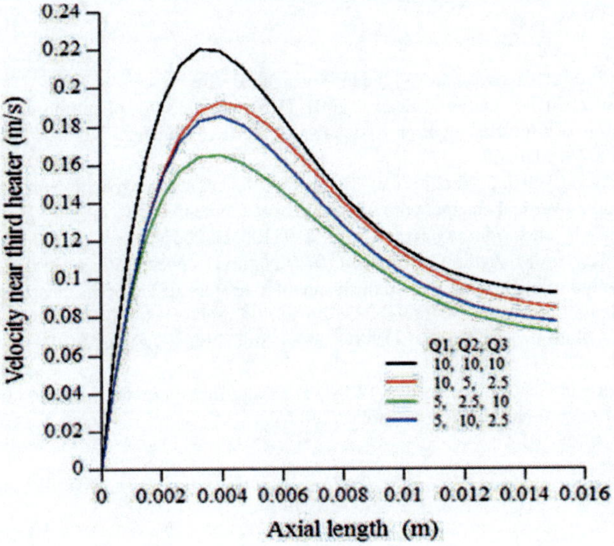

(c)

Fig. 10 (continued)

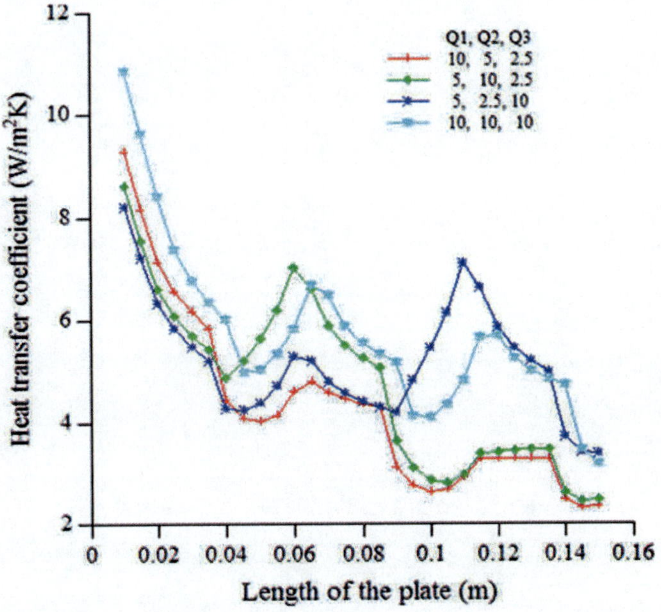

Fig. 11 Distributions of local heat transfer coefficient along the vertical length of the plate. There are three discrete heaters embedded in the wall

References

1. Latif MJ (2006) Heat convection, 1st edn. Springer, Berlin
2. Chen S, Liu Y, Chan S, Leung C, Chan T (2001) Experimental study of optimum spacing problem in the cooling of simulated electronic package. Heat Mass Transfer 37:251–257. https://doi.org/10.1007/s002310000168
3. Gururaja Rao C, Balaji C, Venkateshan SP (2002) Effect of surface radiation on conjugate mixed convection in a vertical channel with a discrete heat source in each wall. Int J Heat Mass Transf 45:3331–3347. https://doi.org/10.1016/s0017-9310(02)00061-3
4. Sarper B, Saglam M, Aydin O, Avci M (2018) Natural convection in a parallel-plate vertical channel with discrete heating by two flush-mounted heaters: effect of the clearance between the heaters. Heat Mass Transf 54:1069–1083. https://doi.org/10.1007/s00231-017-2203-4
5. COMSOL, Multiphysics V.5.3., COMSOL AB., Stockholm, Sweden, 2017. https://www.comsol.com
6. Kothandaraman CP, Subramanyan S (2014) Heat and mass transfer data hand book, 8th edn. New Age International Publishers, India

Pool Boiling Heat Transfer Using Isopropyl Alcohol and Ammonium Chloride Surfactant

Sandeep Swain, Abhilas Swain, and Satya Prakash Kar

Abstract This research article mainly focuses on the enhancement of pool boiling heat transfer using isopropyl alcohol and its solution with the surfactant ammonium chloride on a heating surface, which can be used in heat transfer device. A review on the effect surfactant (ammonium chloride) with different solvent on the heat transfer process also given. The solution is used as the working fluid to measure the heat transfer performance. After conducting the experiment, the result shows that in case of 200 PPM concentration, heat transfer coefficient is more than in case of 400 and 600 PPM, and it is due to the high viscosity and critical micelle concentration (CMC). The experimental data are modeled through the radial basis neural network (RBN) for the prediction of wall superheat, and heat transfer coefficient was given. The RBN model is able to predict the experimental data with a good accuracy level in comparison with the empirical and semiempirical correlations.

Keywords Enhancement · Pool boiling · Surfactant · Heat transfer · Heat transfer coefficient · Viscosity · Critical micelle concentration · Radial basis network

1 Introduction

Boiling is a rapid vaporization of the liquid, which occurs when a liquid is heated to its boiling point. Boiling is a liquid–vapor phase change process, and it uses the latent heat of vaporization to remove the heat from a surface by small temperature difference [1]. Boiling is divided into two types (i) Pool boiling (ii) Flow boiling. Pool boiling is used in many industries like nuclear power plant, aircraft, and space technology, refrigeration air conditioning, electronics cooling, and waste heat recovery [1]. Pool boiling heat transfer can be enhanced by addition of different surfactant in the solvent and by modifying the surface structure. Addition of surfactant compels the fast boiling and increases the heat transfer process. By adding the surfactant into

S. Swain · A. Swain · S. P. Kar (✉)
School of Mechanical Engineering, Kalinga Institute of Industrial Technology (KIIT) Deemed to be University, Bhubaneswar, Odisha 751024, India
e-mail: satyapkar@gmail.com

S. Revankar et al. (eds.), *Proceedings of International Conference on Thermofluids*, Lecture Notes in Mechanical Engineering,
https://doi.org/10.1007/978-981-15-7831-1_31

the working fluid, it affects the properties of the fluid like surface tension, thermal conductivity which ultimately leads to the increase of heat transfer coefficient in the pool boiling process [2]. Surfactants are the organic amphiphilic compounds. It has both hydrophilic head and hydrophobic tail [2]. After the addition of the surfactant to the working fluid, it get absorbed by liquid–vapor interface, and its head remains toward the liquid, and tail remains toward the vapor [2]. Addition of surfactant in the working fluid decreases the surface tension of the liquid which increases the rate of formation of bubbles, and this bubble takes away latent heat from the surface to the upper part of the liquid inside the pool. Surfactant with lower molecular weight decreases the surface tension quickly as compared to surfactant with high molecular weight. This type of surfactant diffuses very rapidly and enhances the bubble formation in a rapid rate, which enhances the heat transfer coefficient. Heat transfer rate increases by increasing the concentration of the surfactant. However, for the some of the surfactant, heat transfer increases up to a certain concentration, and beyond that concentration, it decreases [3]. Water bubble detached from the heated surface when buoyancy force becomes more than surface tension force.

Gajghate et al. [4] experimented the effect of ammonium chloride as a surfactant on the pool boiling heat transfer on a nichrome wire as a heating surface. Thermophysical properties of the solution change after addition of the NH4CL [4]. It has been found from the experiment that heat transfer rate increases up to the 2600-ppm concentration of the NH4CL and then significant heat transfer rate is not observed [4]. Depression of the wettability and surface tension are the two important phenomena, which play a vital role in the case of NH4Cl surfactant when it is added to the distilled water [4]. Kumar and Mathew [5] experimented by taking the different concentration of ammonium chloride with water. They observed that heat transfer observed up to a certain range and beyond that range, no heat transfer observed [5]. The trend of the pool boiling curve moves toward low excess temperature side in the presence of NH4CL in the water [5].

Acharya et al. [6] investigated the ammonium chloride as surfactant for heat transfer enhancement in pool boiling. Ammonium chloride was chosen as surfactant and mixed with water at variable concentration. It was found that heat transfer coefficient of surfactant solution was normal as equal to the HTC of normal water without surfactant up to the 800-PPM because of its solubility limit [6]. After that concentration, heat transfer rate increases up to 2800 PPM, and beyond that, concentration HTC coefficient found to be constant or decreasing slightly [6]. This is because the surfactant shows the heat transfer coefficient up to its solubility limit [6]. Bubble form in the surfactant solution is smaller than the bubble formed in the pure water, and it remains covered on the heating surface [6]. Due to the presence of surfactant in the liquid, average bubble velocity increases [6].

From the above discussed extensive literature survey, it can be understood that addition of surfactant decreases the surface tension of the working fluid and affect the different thermo-physical properties of the fluid which ultimately gives rise to the enhancement of pool boiling heat transfer. The vapor bubble, which occurs during the pool boiling, takes away heat from the heated surface, and by adding different surfactant, we can enhance the pool boiling heat transfer. For the ammonium chloride surfactant, it has certain value above which by the excess addition of ammonium chloride, it does not affect much to the heat transfer process. However, the effect of ammonium chloride as surfactant with any organic liquid has not been studied.

2 Experimental Setup and Procedure

A test arrangement is made for doing the experiment on the above said goals. The parts of the experimental setup are the test chamber, direct current power supply (DC power supply), variable resistor (variac), computerized multi-meter, and temperature pointer. The base and top front of the test chamber was made out of PTFE sheet. The four side dividers are comprised of glass for boiling visualization reason. The glass and PTFE sheet can be approximated as protectors for heat loss from the fluid during experiment. A U shape copper tube was inserted in the top PTFE sheet in which the water will be inserted for the condensation process due to which vapor produced can be converted into water again (Fig. 1).

A copper cylinder of 30 mm diameter and length 40 mm is inserted inside the nylon cylinder, which is having diameter 42 mm and length 70 mm. This nylon cylinder was inserted into the lower part of the Teflon block sheet by making a hole in the Teflon block. One hole of 6 mm was made in copper cylinder in which heater will be inserted for the heating the surface as shown in Fig. 2. For the thermocouple, a 2 mm hole is made in the copper cylinder, which will measure the surface temperature, and

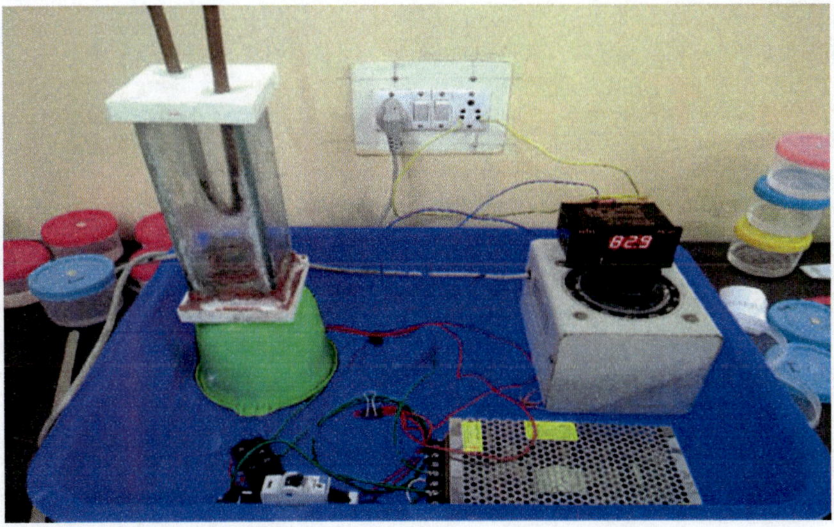

Fig. 1 Complete experimental setup

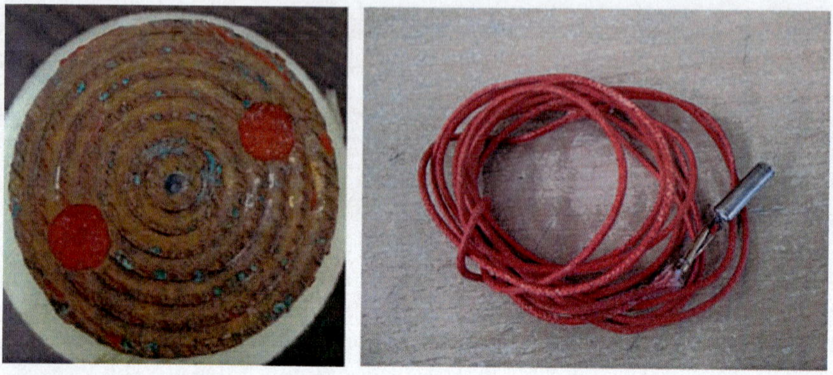

Fig. 2 Copper surface and cartridge heater

it will be reflected on the temperature. The upper part of the thermocouple touched the upper part of the copper cylinder.

Teflon sheet is used for the purpose of thermal proof and liquid proof at both the ends of the chamber. Resin bond is used for the attaching both glass chamber and Teflon which also work as the liquid proof between these two. The AC current supply is given to the DC control supply through a variable resistor to control the input voltage supply. A miniature circuit breaker (MCB) is used in between heater and DC supplier in order to prevent the short circuit. To measure the temperature of the surface and liquid, *T* type thermocouple is used. A temperature indicator is connected to the thermocouple for indication of the temperature. The temperature indicator with thermocouple is calibrated with a dry block type calibrator and was found that the error associated was within ± 1 °C.

Figure 2 shows the copper surface which is used in the experiment. The mini channels on the surface were prepared by machining process and are made as shown in figure. The solution is made by taking three different concentration of the ammonium chloride, and it was mixed with the isopropyl alcohol of 100 ml each and kept in three different containers. The concentration of the solution is taken as 200, 400 and 600 PPM. Figure 3 shows the bubble formation on heater surface which takes away the heat from the surface to the upper part of the solution.

3 Results and Discussion

Figure 4 shows the comparison of pool boiling performance of isopropyl alcohol and ammonium chloride as surfactant on a copper surfaces for all the heat flux value. On 200-PPM concentration, the heat transfer coefficient is more than 400 and 600 PPM. Therefore, by increasing the concentration of surfactant, the heat transfer coefficient decreases. The important reason for that is due to change in viscosity and due to the critical micelles concentration. There might be chances that due to high viscosity, the

Fig. 3 Vapor bubbles of pool boiling during experiment

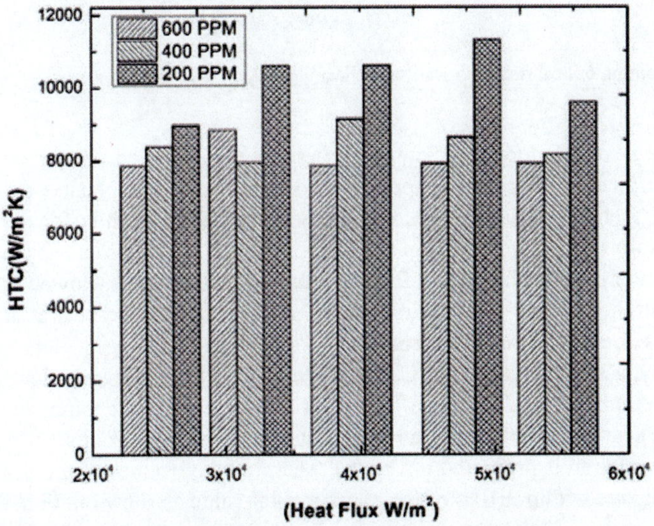

Fig. 4 HTC comparison of pool boiling with isopropyl alcohol and NH_4CL

convection over heated surface becomes slow. After critical micelle concentration, the extra surfactant added to the solution directly goes to the micelles.

It has seen that the heat transfer coefficient for 200-PPM concentration increases for all heat flux value and suddenly it decreases at higher heat flux values. In the case of 400 PPM, the heat transfer coefficient initially increases slightly and then decreases for all heat flux values. For the 600 PPM of concentration first increases then with the increasing heat flux value it decreases and again it rises with increase in heat flux value. The temperature of the surface depends on the heat transfer values.

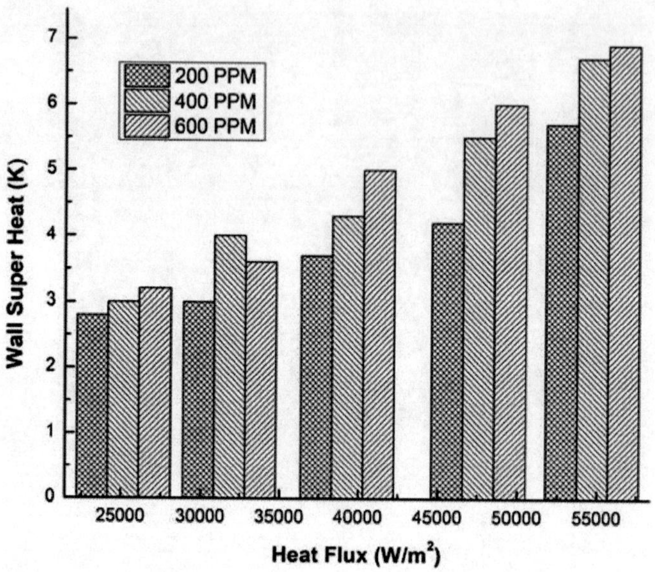

Fig. 5 Variation of heat flux with wall super heat

When the heat transfer coefficient is higher, then it extracted more heat from the boiling surface. It is very much applicable to electronics components cooling.

Figure 5 shows variation of wall superheat with the heat flux for the different concentrations of the surfactant in the isopropanol. It can be observed that the wall superheat increases with heat flux for all the three concentrations. However, the lower concentration, i.e., 200 PPM solution has the lowest wall superheat at all the heat flux levels among the three concentrations.

Initially, a semiempirical correlation is developed through regression modeling. The semiempirical model is able to predict the heat transfer coefficient as output taking the heat flux and the concentration of the surfactant as the output. The semiempirical correlation is presented in Eq. (1). The predicted values of the heat transfer coefficients are within ±10% of the experimental value as shown in Fig. 6.

$$h = 20660.175C^{-0.24}q^{0.049} \tag{1}$$

Then, the experimental data are modeled through the radial basis neural network (RBN) for the prediction of wall superheat and heat transfer coefficient. The RBN models are single layer neural network having capacity to predict the experimental data with sufficient accuracy.

Radial basis function network is almost like a usual multilayer network capable of approximating functions and carrying out pattern recognition [7]. The radial basis functions applied to the inputs and network weights are associated to form a linear

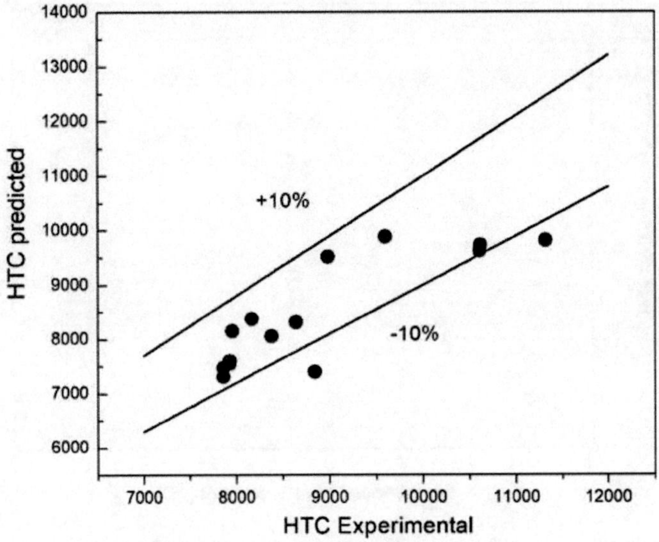

Fig. 6 Comparison between experimental HTC and predicted HTC

combination. The RBN serves as a model but works on the basis of interpolation function for a set of inputs and outputs [7].

The interpolation function is given by

$$S(x) = \sum_{i=1}^{n} Wiv_i(x) \quad \text{Where } v_i(x) = \varnothing(\text{norm}(x - x_i) \tag{2}$$

In terms of Eucledian distance, the radial basics function is of different types. In the following, the radial basis function is given.

1. Multiquadratic function $\varnothing(r) = (r^2 + c)^{1/2}$
2. $\varnothing(r) = r^n, \quad n = 1, 2, 3, \ldots$ (3)
3. $\varnothing(r) = \exp(-r^2)$

Thc RBN model is able to predict the experimental data with a good accuracy level in comparison with the empirical and semiempirical correlations as can be visualized in Figs. 7 and 8. Figure 7 shows the comparison of the predicted wall superheat with that obtained from experiment, and similarly, Fig. 8 shows for the heat transfer coefficient. It can be observed from both the figures that the predicted values are within ±5% of the experimental data. Thus, the radial basis function models are more accurate than the semiempirical models.

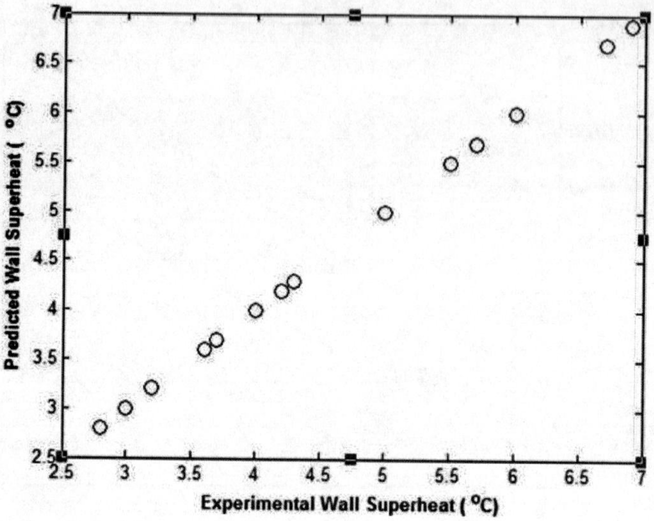

Fig.7 Comparison of the experimental and predicted wall superheat from RBN model

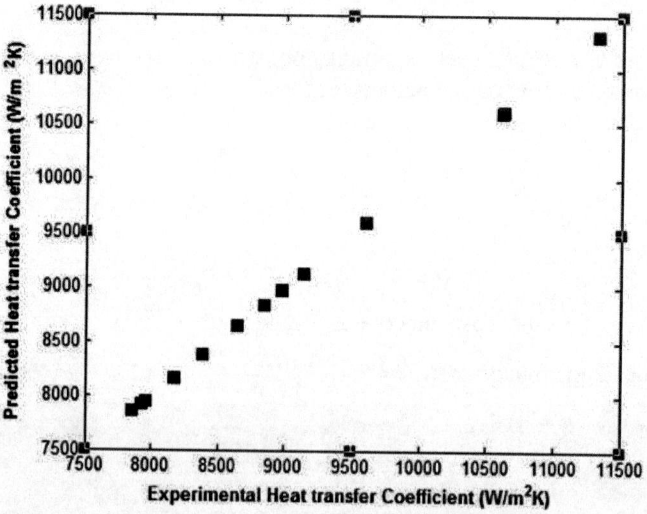

Fig. 8 Comparison of the experimental and predicted heat transfer coefficient from RBN model

4 Conclusion

The pool boiling experiment was conducted over a machined surface by taking the working fluid as isopropyl alcohol and adding aluminum chloride surfactant in it. A literature survey was done for the surfactant aluminum chloride as it decreases the surface tension when added to the working fluid which enhances the heat transfer in pool boiling. From the current experiment, it is confirmed that for the 200 PPM concentration of the ammonium chloride, the working fluid shows the best heat transfer rate, but beyond that concentration of surfactant, the heat transfer does not change appreciably.

As the pool boiling is one of the promising media of heat transfer, it has a broad application in different industries. So, different working fluid with different surfactant should be tested to measure which has the best and effective combination to be used in industries application, and different available surfactant should be added to the isopropyl alcohol and tested which has the highest heat transfer coefficient.

References

1. Kumar N, Raza MQ, Raj R (2018) Surfactant aided bubble departure during pool boiling. Int J Therm Sci 131:105–113
2. Liang G, Mudawar I (2018) Review of pool boiling enhancement with additives and nanofluids. Int J Heat Mass Transf 124:423–453
3. Hetsroni G, Gurevich M, Mosyak A, Rozenblit R, Segal Z (2004) Boiling enhancement with environmentally acceptable surfactants. Int J Heat Fluid Flow 25:841–848
4. Gajghate S, Acharya AR, Pise AT (2015) Experimental study of aqueous ammonium chloride in pool boiling heat transfer. Exp Heat Transf J Therm Energy Gen Transp Storage Conver 27:37–41
5. Kumar R, Mathew R (2016) Effects of surfactants (NH_4Cl) behaviors on nucleate pool boiling over Nichrome wire. Int J Eng Sci Res Technol 5(9):251–255
6. Acharya AR, Pise AT, Momin II (2011) Ammonium chloride as surfactant for heat transfer enhancement in pool boiling. Int J Eng Technol 3(3):323–326
7. Swain A, Das MK (2014) Artificial intelligence approach for the prediction of heat transfer coefficient in boiling over tube bundles. Proc Inst Mech Eng Part C J Mech Eng Sci 228(10):1680–1688

Numerical Simulation on Impact of a Liquid Droplet on a Deep Liquid Pool for Low Impact Velocities

Vineet Kumar Tiwari, Tanmoy Mondal, and Akshoy Ranjan Paul

Abstract Droplet impact finds application in paint industries, spray coating, and aeration and hence becomes problem of engineering value. Velocity of impact, geometry, as well as the medium through which the droplet travels before it impacts the liquid surface plays a key role for the occurrence of droplet coalescence and droplet bouncing. In the present work attention has been given to low impact velocities ranging from 0.2 to 0.6 m/s for a droplet of diameter of 3 mm. Transition from coalescence to bouncing is observed at a velocity of 2 m/s. During the bouncing a secondary droplet forms and detaches from the parent droplet before it coalesces. Volume of fluids (VOF) method has been used to carry out the numerical simulation. VOF model is used for two or more immiscible fluids by solving a single set of momentum equation and it tracks the volume fraction of all the phases throughout the flow domain. Interface calculation has been done using Geometric Reconstruction Scheme using a Commercial software package.

Keywords Pinch off · Volume of fluids · ANSYS FLUENT · Secondary droplet

1 Introduction

Droplet impact in liquid pool is a common phenomenon in our day to day life like rain drops, drops from shower but it has also good applicability in spray coating, agriculture (the impact of a raindrop onto the surface of a puddle enhances soil erosion and produces secondary droplets that can act as carriers of spores and bacteria) [1] and many more things. Thus it is a problem of engineering significance.

When a droplet falls in a liquid pool interface dynamics depends on several parameters like velocity of impact on the liquid pool, geometry, the medium through which it travels before it impacts the liquid pool, angle of impact and also on the viscosity and density of the liquid and drop as well [1–7]. Many literatures have studied these

V. K. Tiwari (✉) · T. Mondal · A. R. Paul
Motilal Nehru National Institute of Technology Allahabad, Prayagraj, Uttar Pradesh 211004, India
e-mail: vineettiwari522.vt@gmail.com

S. Revankar et al. (eds.), *Proceedings of International Conference on Thermofluids*, Lecture Notes in Mechanical Engineering,
https://doi.org/10.1007/978-981-15-7831-1_32

parameters but only few of them address the effect of droplet behaviour by changing the impact velocity.

Hasan et al. [2] at impact velocity of 2.4 m/s, show the phenomena of underwater sound during rain as an outcome of air bubble entrainment in the liquid. Morton et al. [3] in their study have shown the mechanism of formation of bubble due to crater collapse from droplet impact at various velocities of impact. Murat and Donald [4] performed computational analysis of the droplet impact velocity and at solar and lunar surface values of gravity using VOF method in FLUENT. Ray et al. [5] utilized level set formulation to carry forward the work by Morton et al. [3] in order to establish the conditions under which bubble entrapment and jet formation occurs. Orozco and Eduardo in their experimental study followed by computation at an impact velocity of 1.8 m/s suggested that for a jet and its breakup require a balance between surface tension, capillary and viscous forces [1]. Inertia of the pool liquid and drop viscosity are key factors to determine the maximum crater depth is quoted by Jain et al. [6].

Among several available literatures Morton et al. [3] and Ray et al. [5] considered liquid as pure water and studied the impact of water droplet in liquid pool by varying the impact velocity. Morton et al. [3] used high frame rate camera to investigate the falling droplet. The study was conducted at droplet diameter of 3 mm and impact velocity ranging from 0.8 to 2.64 m/s. Ray et al. [5] revised the work on the same droplet diameter using combined level set and volume of fluid method (CLSVOF) in the impact velocity range of 1–4 m/s. Their work reveals a state of transition from coalescence to bouncing with increasing velocity of impact. But still there is void in the study of low impact velocity range as the lowest velocity of impact that has been studied is 0.8 m/s [3].

One such attempt has been made in the current study using commercially available software ANSYS FLUENT. Volume of fluid method (VOF) has been used to study the impact phenomena in the low impact velocity range from 0.2 to 0.6 m/s for a droplet of 3 mm diameter.

Figure 1 schematically shows a situation when a water droplet begins to impact on a deep water pool. The figure shows H as the initial height of the pool, D as droplet

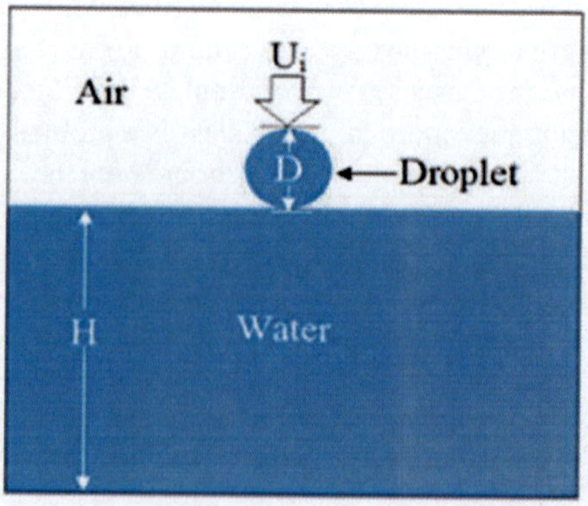

Fig. 1 Schematic representation of a water droplet as it begins to impact on a deep water pool

diameter and U_i as the initial impact velocity. For the present study, air is considered as the surrounding fluid and water as the fluid for the droplet as well as pool.

2 Governing Equation and Numerical Methodology

The flow is considered to be two-dimensional (2D) and incompressible. The volume of fluids (VOF) method in FLUENT platform has been used to simulate the present flow problem. In VOF formulation, coupled Navier–Stokes equations (continuity and momentum equations) and VOF equations are solved simultaneously. Properties of each control volume is determined by the presence of component phases in the cell. VOF method solves single momentum equation in the whole domain and depends on the volume fraction of all phases. The dimensional form of the governing equations are as follows.

Defining the volume fraction

$$\alpha = \begin{cases} 0 & \text{Gas} \\ 0 < \alpha < 1 & \text{Gas-Liquid Interface} \\ 1 & \text{Liquid} \end{cases} \tag{1}$$

where α is the volume fraction of liquid.

Density (ρ) and viscosity (μ) are calculated as:

Density:

$$\rho = \alpha\rho_l + (1-\alpha)\rho_g \tag{2}$$

Viscosity:

$$\mu = \alpha\mu_l + (1-\alpha)\mu_g \tag{3}$$

Continuity equation:

$$\nabla \cdot (\vec{u}) = 0 \tag{4}$$

Momentum equation:

$$\frac{\partial(\rho\vec{u})}{\partial t} + \nabla \cdot (\rho\vec{u}\vec{u}) = -\nabla p + \nabla \cdot \left[\mu\left(\nabla\vec{u} + \nabla^T\vec{u}\right)\right] + \sigma\kappa_f n_f \delta\left(x - x_f\right) + \rho\vec{g} \tag{5}$$

Interface is traced with the following VOF equation:

$$\frac{\partial\alpha}{\partial t} + \nabla \cdot (\alpha\vec{u}) = \alpha(\nabla \cdot \vec{u}) \tag{6}$$

where l is the subscript to denote liquid phase and g is to denote gas phase respectively, κ_f is interface curvature, x is position vector, x_f is interface position, n_f and is interface normal and δ is the delta function. Values of density and viscosity are taken from Eqs. (2) and (3) and put in transport equation.

Piecewise linear interface calculation scheme (PLIC) which is available in FLUENT [8] as Geometric Reconstruction Scheme has been used to model the interface. Volume fraction of a phase is calculated with the help of explicit scheme applied to interface tracking equation i.e. Eq. (6) and integrated over control volume. For Pressure–Velocity coupling PISO (Pressure implicit with splitting of operator) algorithm has been used. As the length scale and time scales are small they have been non dimensionalized with respect to the droplet diameter D and the ratio of droplet diameter D to that with impact velocity U_i as $t_{\text{ref}} = D/U_i$. Second order upwinding is used to discretize advection terms.

3 Boundary Conditions

The computational domain is shown schematically in Fig. 2. It is an axisymmetric domain. On the right side is a free slip boundary condition and an axisymmetric boundary condition is set on the left hand side of the computational domain. Lower boundary is a solid wall and no slip condition is considered. Drop of diameter 3 mm has been made using the patch option in fluent solver by assigning its center coordinates and radius. The drop has been patched with an initial velocity of impact U_i. Outflow Boundary condition has been assigned to the top of domain. Pictorial representation of the above mentioned boundary conditions are visible in Fig. 2

Fig. 2 Boundary conditions

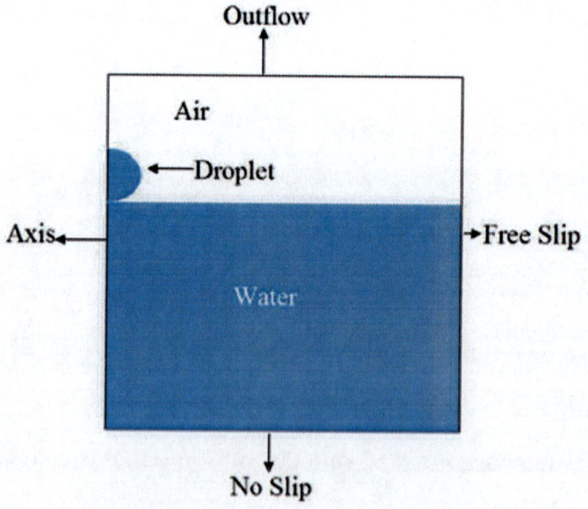

4 Results and Discussion

4.1 Validation

To ensure the accuracy of the present computational work validation of the numerical method has been done. Non dimensionalized crater depth (H/D) has been plotted against the non dimensionalized time and the results have been compared with the work by Morton et al. [3].The drop diameter D has been chosen to be 3 mm and the impact velocity U_i has been chosen to be 1.55 m/s in consistence with parameters of Morton et al. [3] (Fig. 3). Also qualitative validation has been shown at different time steps in Fig. 4. The non dimensionalized time steps at which the results of interface dynamics are compared qualitatively are 6.2 and 8.6.

Initial deviation is due to the drop being completely spherical (in computational work) as it strikes the pool surface because the drop falling from a certain height in the work by Morton et al. [3] is not completely spherical as it deforms as it travels through a certain height before striking the pool.

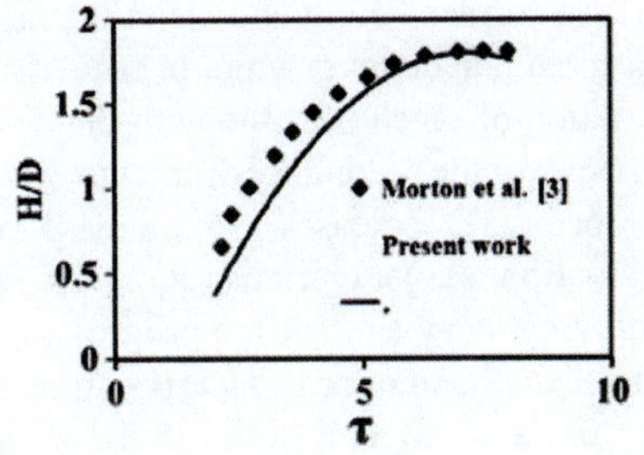

Fig. 3 Validation of variation of crater depth with time (non-dimensional) from the work of Morton et al. [3]

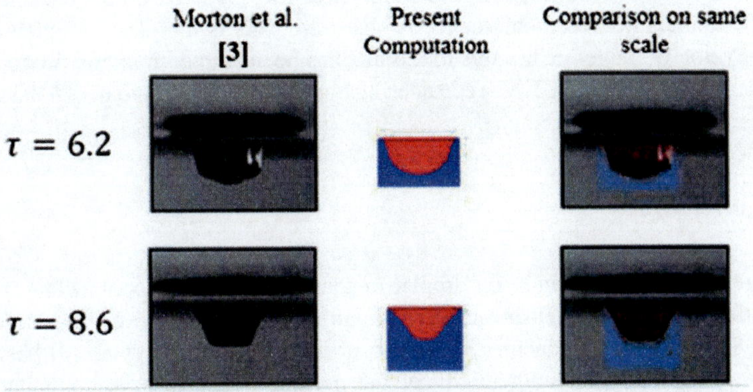

Fig. 4 Air-water interface comparison with the work by Morton et al. [3] at two time instants

4.2 Results of Droplet Impact at Different Impact Velocities

The dynamics of air–water interface at various time instants for a given impact velocities are shown in Fig. 5. As seen, a clear pinch off of secondary droplet is visible at a time instant $\tau = 0.8$ for the impact velocity of 0.2 m/s. In the range of $\tau = 1.006$ to 1.173 for impact velocity of 0.4 m/s secondary droplet can be observed to reach the limiting case of pinch off but still attached to the water in the pool, i.e. disintegration of secondary droplet is not observed while in case of impact velocity of 0.6 m/s full coalescence is observed and for sure there is no pinch off at any time instant. This is in agreement with the work of Morton et al. [3] which reports no pinch off up to impact velocity of 0.8 m/s. A deep crater as compared to other cases (impact velocity of 0.2 and 0.4 m/s) is observed and a column of water is seen rising after drop coalesces for the time range of $\tau = 0.32$ to 8.64 for impact velocity of 0.6 m/s. Figure 6a shows the variation of the crater depth with the impact velocities from which it is evident that there is a sharp decrease of height of free surface as soon as the pinch off takes place (for $U_i = 0.2$ and 0.4 m/s) and again the free surface height begins to rise when pinched off droplet coalesces. It also indicates the formation of shallow crater for the pinch off case. However, in case of a complete coalescence there is gradual drop in free surface height and also the crater thus formed is deep. From Fig. 5 it is also observed that crater is wider in case of a complete coalescence. Figure 6b shows the variation of pinch off time with the different values of impact velocities. As indicated, time for the secondary droplet to pinch off (or about to pinch off for complete coalescence case) increases with increase in impact velocity.

Other important observation is a jet circumscribing the droplet (initiated at $\tau = 0.32$ at $U_i = 0.6$ m/s) moving away as the time progresses can be seen for impact velocity of 0.6 m/s but no such phenomena is occurring for smaller impact velocities i.e. at impact velocity of 0.4 and 0.2 m/s. In the initial stages of impact droplet is seen to be spreading which leads to formation of circumscribing plateau around the region of impact of droplet in case of impact at 0.4 and 0.2 m/s.

In extension to the work by Ray et al. [5], which addresses bubble entrapment and formation of jet to a minimum impact velocity of 1 m/s whereas in the present work droplet bouncing before coalescence is observed. This is unique as a much lower impact velocity range i.e. 0.2 m/s to 0.6 m/s has been studied numerically (using a commercial software FLUENT) considering both droplet and liquid pool to be water.

5 Conclusion

The present numerical study on droplet impact in deep liquid pool at low impact velocities has been carried out in 2D domain in commercially available software ANSYS FLUENT. Satisfactory validation of work by Morton et al. [3] has been carried out initially for authenticity of the present work. To save computational effort only half domain has been used as the problem is symmetric about the axis.

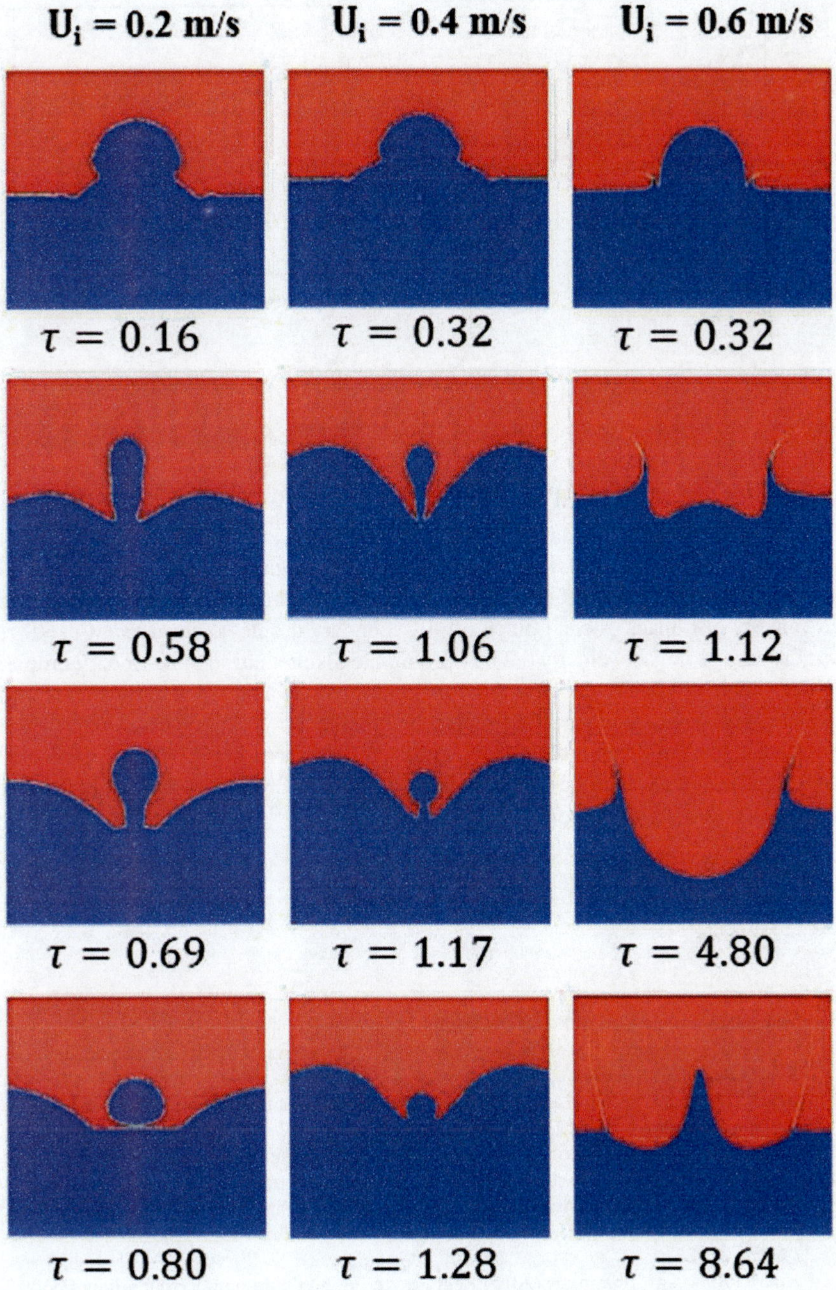

Fig. 5 Instantaneous air–water interface variation with time for different impact velocity

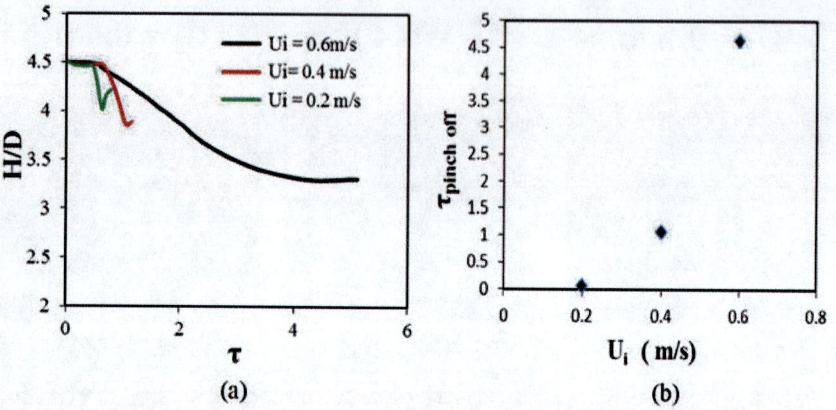

Fig. 6 **a** Variation of interface height of free surface with time. **b** Time of pinch off at different impact velocities

Present study reveals that as the impact velocity is lowered there is transition from complete coalescence of droplet to formation of secondary droplet followed by its pinch off. Also the pinch off time and crater depth are found to increase with increase in velocity of impact. A clear pinch off of secondary droplet is observed at $\tau = 0.2$, for the case of impact velocity 0.2 m/s. Complete disintegration of secondary droplet is not visible at impact velocity of 0.4 m/s but it can be concluded to be a threshold velocity below which clear pinch off can be obtained.

Plateau formation around the droplet impact region is reported for impact velocity of 0.2 and 0.4 m/s velocity of impact whereas a circumscribing jet region has formed for impact velocity of 0.6 m/s.

References

1. Castillo-Orozco E et al (2015) Droplet impact on deep liquid pools: Rayleigh jet to formation of secondary droplets. Phys Rev E 92(5):053022
2. Oguz HN, Prosperetti A (1990) Bubble entrainment by the impact of drops on liquid surfaces. J Fluid Mech 219:143–179
3. David M, Rudman M, Jong-Leng L (2000) An investigation of the flow regimes resulting from splashing drops. Phys Fluids 12(4):747–763
4. Dinc M, Gray DD (2012) Drop impact on a wet surface: computational investigation of gravity and drop shape. advances in fluid mechanics, heat and mass transfer, proceedings of the 10th world scientific and engineering academy and society (wseas) international conference on fluid mechanics and aerodynamics (FMA'12), Istanbul, Turkey, ISBN: 978-1-61804-114-2, pp 374–379, Aug 21–23
5. Bahni R, Biswas G, Sharma A (2015) Regimes during liquid drop impact on a liquid pool. J Fluid Mech 768:492–523
6. Jain U et al. (2019) Deep pool water-impacts of viscous oil droplets. Soft Matter 15(23):4629–4638

7. Kaarle VH, Malalasekera W (2007) An introduction to computational fluid dynamics: the finite volume method. Pearson Education, pp 193–196
8. ANSYS FLUENT user guide, release 14.5, Nov 2012 and ANSYS FLUENT theory guide 12.0, Apr 2009

Enhancement of Thermal Performance of Parabolic Trough Collector Using Cavity Receiver

Bilkan John Hemrom, Uttam Rana, and Aritra Ganguly

Abstract In the present work, the use of rectangular perforated fins in a *V* cavity absorber tube has been studied for the enhancement of the thermal performance of the solar parabolic trough collector system. Four different cases of fin geometries have been considered, and numerical analysis has been done to study the laminar flow and heat transfer characteristics of the absorber tube. The continuity, momentum, and energy equations have been solved by using commercial software package. The fins with the perforation diameter of 4 mm and pitch ratio of 22.1 show the maximum heat transfer. The optimal case shows the maximum value of thermo-hydraulic performance as 1.049, and the increase in Nusselt number is 1.3 times as compared to a smooth absorber tube.

Keywords *V*-shaped cavity · Perforated fins · Heat transfer · Thermo-hydraulic performance

1 Introduction

The parabolic trough collector is a linear concentrating collector having a wide range of applications varying from water heating, space heating, desalination, power production, refrigeration, air conditioning, and many more [1]. One of the significant components of PTC is its receiver tube. Presently, evacuated tube receiver is more popular because of its high thermal efficiency, but its high cost and loss of vacuum after extended use are some of its drawbacks. Hence, a black body cavity receiver with a good insulation covering and a small aperture for solar radiation inlet can be considered as the prominent alternative. Many researchers have investigated the use of a blackbody cavity receiver in linear concentrating solar concentrators. Singh et al. [2] examined the different design parameters of the cavity receiver of trapezoidal shape used with the linear Fresnel reflector. Bader et al. [3] studied the *v* corrugated

B. J. Hemrom (✉) · U. Rana · A. Ganguly
Department of Mechanical Engineering, Indian Institute of Engineering Science and Technology, Shibpur, Howrah, West Bengal 711103, India
e-mail: bjhemrom.bjh@gmail.com

S. Revankar et al. (eds.), *Proceedings of International Conference on Thermofluids*, Lecture Notes in Mechanical Engineering,
https://doi.org/10.1007/978-981-15-7831-1_33

tube with single and double glazed aperture windows. A novel cavity receiver design with a center tube and fins inclined at some angle was proposed by Liang et al. [4]. Li et al. [5] studied the various design parameters for designing a new arc-shaped linear cavity receiver with the lunate channel. Zhai et al. [6] investigated different shapes of cavity receiver, i.e., circle, semicircle, square, and triangle, and found the triangle cavity receiver to have the highest thermal and optical efficiency.

Along with various cavity receiver shapes, researchers also studied the different techniques of improving the thermal performance of the parabolic trough receiver. One of the most accessible and effective methods is by using passive inserts. Different inserts like twisted tape, wavy tape, vortex generator, longitudinal fins, toroidal rings, etc., have been studied with parabolic trough collectors to enhance the thermal performance of the receiver tube. Bellos et al. [7] used perforated plate, twisted tape, and internal fins with both type of receiver tube, i.e., evacuated and non-evacuated to study the heat transfer enhancement. In the present study, a numerical simulation has been carried out to explore the flow and heat transfer characteristic of *V* cavity receiver tube with the perforated rectangular fins.

2 Model Description

2.1 Physical Modeling

In the present study, a 3D model of the receiver tube has been drawn based on the design parameters provided by Xiao et al. [8]. The computational domain is the segment area of circular pipe having an outer radius of 30 mm (Fig. 1). The rectangular

Fig. 1 3D diagram of the cavity receiver

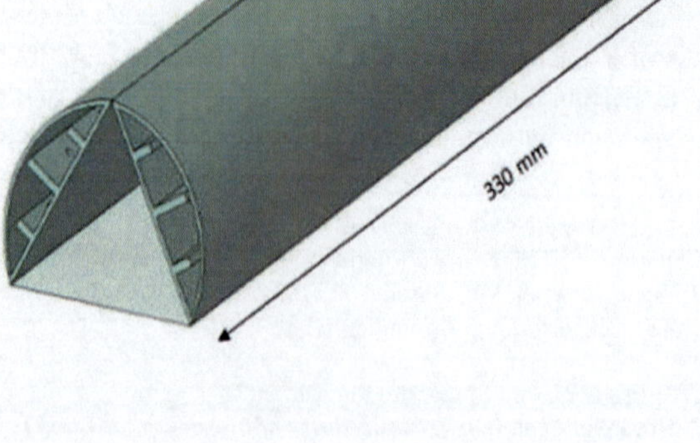

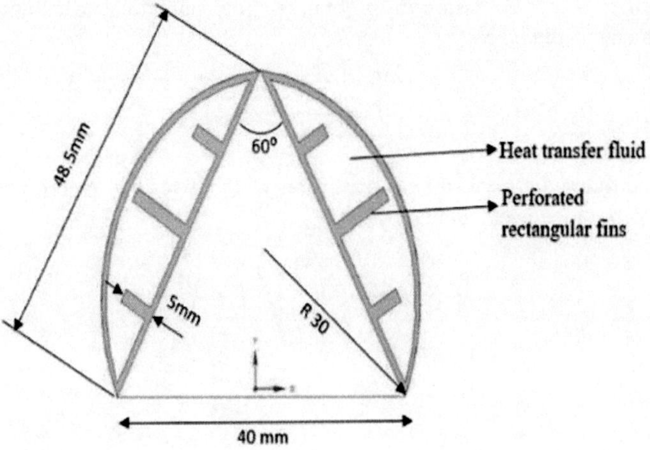

Fig. 2 Schematic diagram of the cavity receiver

perforated fins are used inside the tube to increase the heat transfer (Fig. 2). The perforated fins have been studied for two different perforation diameters. Table 2 shows the various configurations of perforated rectangular fins used in the present CFD investigation. The flow inside the tube has been considered laminar hence the analysis has been done for Reynolds number ranging from 200 to 1000. The heat transfer fluid used is water, and the material for the tube has been considered as aluminum. Table 3 shows the thermo-physical properties of the cavity receiver and water at the temperature considered for the present analysis.

2.2 *Mathematical Modeling*

The following equations are used for the calculation of heat transfer performance parameters:

Nusselt number is calculated by

$$N_u = \frac{\overline{h}D}{k} \tag{1}$$

Average heat transfer coefficient between the heating surface and heat transfer fluid

$$\overline{h} = \frac{Q_u}{A_s\left(T_w - T_{fm}\right)} = \frac{\dot{m}c_p(T_o - T_i)}{A_s\left[T_w - \left(\frac{T_o+T_i}{2}\right)\right]} \tag{2}$$

Using the value of pressure drop obtained from numerical calculation friction factor can be calculated by

$$f = \frac{\Delta P.2D}{\rho L V^2} \tag{3}$$

Thermal hydraulic performance parameter which gives the overall performance of the absorber tube is calculated by

$$\text{THPP} = \left(\frac{Nu}{Nu_s}\right) \Big/ \left(\frac{f}{f_s}\right)^{\frac{1}{3}} \tag{4}$$

2.3 Optical Modeling

Irradiance profile on the V-shaped surface of the cavity receiver estimated by Xiao et al. [8], using Monte Carlo ray tracing (MCRT) technique, has been used in the present analysis. The heat flux profile has been calculated considering the direct normal irradiance of 1000 W/m^2. Gaussian functions were generated based on the heat flux data using curve fitting function in MATLAB. These Gaussian functions were developed into a subroutine program and were treated as a heat flux wall boundary condition on the V-shaped surface of the cavity receiver tube.

2.4 Numerical Modeling

The conservation equations namely continuity, momentum, and energy equation are solved by the finite volume method using commercial software ANSYS FLUENT [9]. The discretization of the momentum and energy equations has been done by second-order upwind scheme to achieve the higher-order accuracy at cell faces. The pressure–velocity coupling has been established by using SIMPLE algorithm. The convergence criteria for the residuals values obtained by solving the continuity and momentum equations is 10^{-04}, whereas for energy equation, the value of residuals has been kept below 10^{-07}. Along with the residuals value, convergence has also been ensured by the constant value of the area-weighted average of pressure and temperature at the outlet.

2.4.1 Governing Equations

The fluid flow through the parabolic trough collector receiver equipped with perforated rectangular fins has been considered laminar. The flow is steady state and incompressible. The conservation equations can be written as follows.

Continuity equation:

$$\frac{\partial}{\partial x_j}(\rho u_j) = 0 \tag{5}$$

Momentum equation:

$$\frac{\partial}{\partial x_j}(\rho u_i u_j) = -\frac{\partial P}{\partial x_i} + \frac{\partial}{\partial x_j}\left[\mu\left(\frac{\partial u_i}{\partial x_j} + \frac{\partial u_j}{\partial x_i}\right)\right] \tag{6}$$

Energy equation:

$$\frac{\partial}{\partial x_j}(u_j T) = \frac{\partial}{\partial x_j}\left[\left(\frac{k}{\rho c_p}\right)\frac{\partial T}{\partial x_j}\right] \tag{7}$$

where T is the static temperature, P is the static pressure, k, ρ, and c_p are thermal conductivity, density, and specific heat capacity of the heat transfer fluid, respectively.

2.4.2 Boundary Conditions

The conservation Eqs. (5)–(7) are solved in a segregated manner subject to the boundary conditions as given in Tables 1 and 2.

Table 3 represents the important Thermo-physical properties of cavity receiver and water at temperature considered for present CFD analysis.

2.4.3 Validation of the Model

The reliability of the present model has been confirmed by validating its results with the existing experimental findings of Xiao et al. [8]. Figure 6 portrays the variation of axial temperature distribution as predicted by the present numerical simulation and that obtained by the experiment. From the figure, it can be observed that the present numerical simulation very closely approximates the experimental findings under identical operating conditions.

Table 1 Boundary conditions applied at different surfaces or walls

Inlet	$u_i = 0, u_j = 0, u_k = \dot{m} \big/ \rho_f A_{in}, T = T_{in}, P = P_o$
Outlet	The fully developed condition has been assumed $\frac{\partial u}{\partial z} = \frac{\partial v}{\partial z} = \frac{\partial w}{\partial z} = \frac{\partial T}{\partial z} = 0$
Heating surface	Heat flux boundary condition
Fins surface	Coupled boundary condition
Absorber tube inner surface	No slip boundary condition
Absorber back surface	Convective heat transfer coefficient is applied $h_{cv} = \frac{0.48k_a(G_r.P_r)^{0.25}}{d_s}, G_r = \frac{g\beta\Delta T d_s^3}{v_a^2}$ d_s represents the outer diameter of the tube, v_a and k_a are the viscosity and thermal conductivity of air, respectively, at the mean temperature T_m. $T_m = 0.5(T_s + T_a)$, T_s is absorber back surface temperature and T_a represents temperature of air. β represents volume expansion coefficient, $\beta = \frac{1}{T_m}$, $\Delta T = T_s - T_a$. The equivalent hydraulic diameter for the cavity receiver is calculated by $D = 4A_c/P$

Table 2 Four different configurations of rectangular perforated fins

Fin configurations	Hydraulic diameter (mm)	Perforation dia (mm)	Perforation pitch (mm)
1		2	33.1
2	14.06	2	22.1
3		4	33.1
4		4	22.1

Table 3 Thermo-physical properties of cavity receiver and water at temperature

Properties	Water	Receiver pipe
Density (kg/m^3)	997.13	2719
Specific heat 'C_p' (J/kg K)	4180	871
Thermal conductivity 'k' (W/mK)	0.606	202.4
Viscosity	0.000891	NA
Prandtl number	6.145	NA

3 Results and Discussions

The graphs of Nusselt number and friction factor against Reynolds number have been plotted to discuss the heat transfer and friction characteristics of the V cavity receiver equipped with perforated rectangular fins.

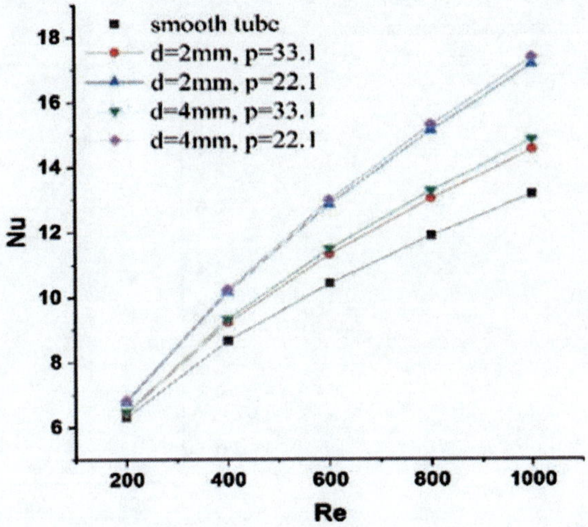

Fig. 3 Effect of Reynolds number on Nusselt number

3.1 Effect of Reynolds Number on Heat Transfer Performance Parameters

Figure 3 compares the plot of Nusselt number against Reynolds number, for a tube fitted with different fin geometries and smooth tube for the perforation pitch of 22.1 and 33.1. It is found that perforated fins have a higher heat transfer coefficient compared to that of the plain tube. Moreover, there is a significant rise of Nusselt number values between fins with perforation of 4 mm diameter compared to that with perforation of 2 mm. Increase in number of perforations also increases the heat transfer. Figures 7 and 8 show the temperature contour plot for a smooth tube and tube having fins, respectively. The blue colored area gets reduced when fins are used, which signifies the more amount of heat transfer between heating surface and fluid.

3.2 Effect of Reynolds Number on Average Friction Factor

From Darcy's friction factor formula, we can see that the friction factor for the smooth tube is the function of the Reynolds number only. When perforated rectangular fins are introduced in the tube, the friction factor will become a function of both Reynolds number and roughness parameters (*d* and *p*). Figure 4 shows the comparison of friction factor variation with Reynolds number for a smooth tube and tube inserted with rectangular perforated fins. It can be observed from the plot that the average friction factor for the tube with fins is relatively higher than the smooth tube. As the fluid flows along the tube, the fins increase the surface area and hence increases

Fig. 4 Effect of Reynolds number on friction factor

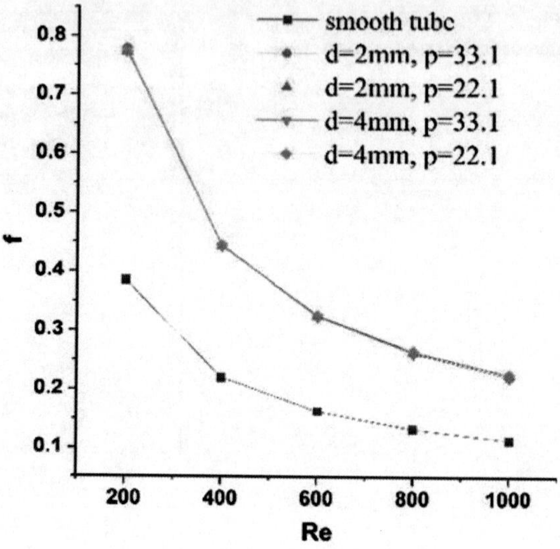

the friction factor. The plot also shows that, with increase in Reynolds number the average friction factor decreases.

3.3 *Effect of Reynolds Number on Thermo-Hydraulic Performance*

The thermo-hydraulic performance parameter (THPP) gives the idea of the overall performance of the parabolic trough collector cavity receiver. Figure 5 shows the variation of THPP of rectangular perforated fins for different pitch values at Reynolds number varying from 200 to 1000 at perforation diameter of 2 and 4 mm. The plot shows that there is an increase in THPP values with the increment in Reynolds number for each case. Most probable reason could be, the pressure drop does not increase significantly when perforated fins are used, and hence there is not much increment in pumping work. The maximum value of THPP was found to be 1.049 for the diameter of 4 mm and pitch of 22.1 at the Reynolds no. 1000 (Figs. 6, 7 and 8).

4 Conclusions

In the present work, the performance of a *V* cavity receiver using perforated rectangular fins has been studied. Similar types of works are not available at least in the open literature. The prime aim of the study is to observe the thermal performance

Fig. 5 Variation of THPP with Re number

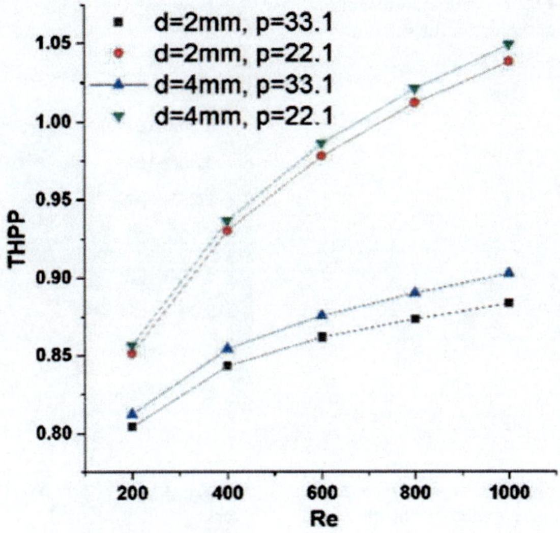

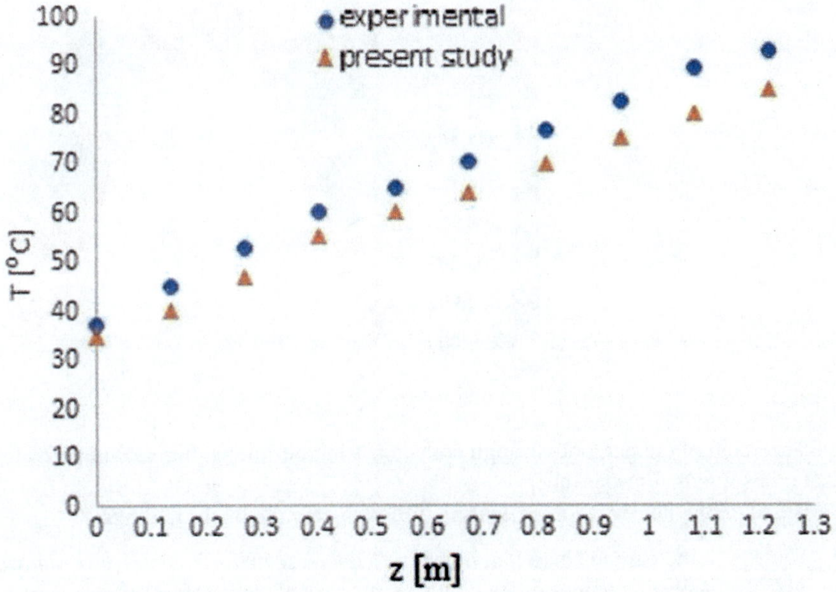

Fig. 6 Variation of temperature along axial direction

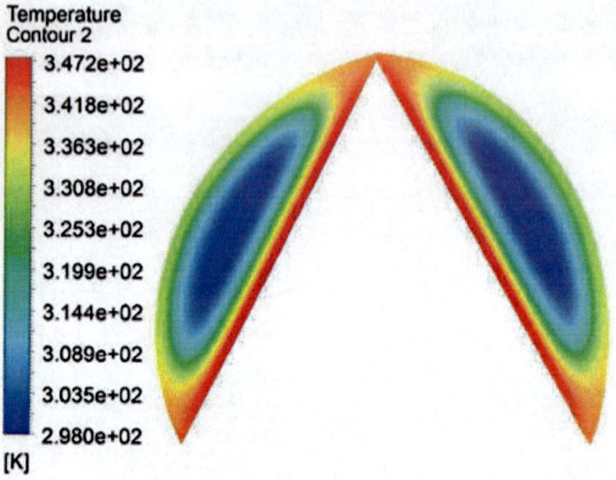

Fig. 7 Temperature contour at the outlet for smooth absorber tube

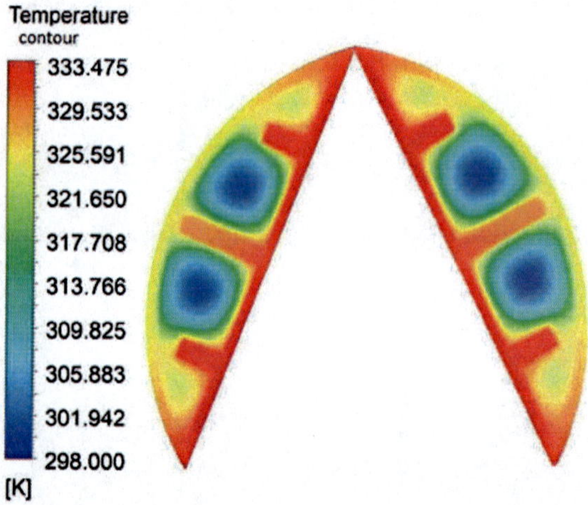

Fig. 8 Temperature contour at the outlet for absorber tube with fins

enhancement of the parabolic trough collector using perforated fins as compared to that of a smooth absorber tube.

Based on the numerical investigation, following are the major findings:

- There is an increase in Nusselt number and a decrease in the friction factor values with the increasing values of Reynolds number for all cases of fin geometries.
- The decrease of pitch ratio results in an increase in both the Nusselt number and friction factor.
- The fin with perforation diameter of 4 mm and pitch ratio 22.1 shows the maximum heat transfer and hence is the thermally efficient case.
- The maximum THPP value is found to be 1.049 and 1.038 for the perforation diameter of 4 mm and 2 mm, respectively, at the pitch ratio of 22.1.

- The increase in the Nusselt number for thermally efficient case is found to be 1.3 times higher than the smooth tube.

References

1. Kalogirou SA (1999) Applications of artificial neural networks in energy systems. Energy Convers Manag 40:1073–1087
2. Singh PL, Sarviya RM, Bhagoria JL (2010) Heat loss study of trapezoidal cavity absorbers for linear solar concentrating collector. Energy Convers Manag 51:329–337
3. Bader R, Pedretti A, Steinfeld A (2012) Experimental and numerical heat transfer analysis of an air-based cavity-receiver for solar trough concentrators. J Sol Energy Eng 134
4. Liang H, Zhu C, Fan M, You S, Zhang H, Xia J (2018) Study on the thermal performance of a novel cavity receiver for parabolic trough solar collectors. Appl Energy 222:790–798
5. Li X, Chang H, Duan C, Zheng Y, Shu S (2019) Thermal performance analysis of a novel linear cavity receiver for parabolic trough solar collectors. Appl Energy 237:431–439
6. Zhai H, Dai Y, Wu J, Wang R (2007) Study on trough receiver for linear concentrating solar collector. ISES Sol World Congr 1:711–715
7. Bellos E, Tzivanidis C (2018) Enhancing the performance of evacuated and non-evacuated parabolic trough collectors using twisted tape inserts, perforated plate inserts and internally finned absorber. Energies 11:1129–1157
8. Xiao X, Zhang P, Shao DD, Li M (2014) Experimental and numerical heat transfer analysis of a *V*-cavity absorber for linear parabolic trough solar collector. Energy Conver Manag 86:49–59
9. ANSYS Inc (2015) CFD flow modeling software and solutions from FLUENT. https://www.fluent.com/

Laminar Mixed Convection Over a Rotating Vertical Hollow Cylinder Exposed in the Air Medium

Basanta Kumar Rana and Jnana Ranjan Senapati

Abstract The present numerical analysis focuses on the thermo-fluid characteristics of laminar mixed convection from a rotating hollow tube placed vertically with a variation of Rayleigh number ($10^4 \leq \text{Ra} \leq 10^8$), Reynolds number ($\text{Re}_D < 2100$) and cylinder aspect ratio ($2.5 \leq L/D \leq 20$). A comprehensive comparative study has been performed to estimate the augmentation in heat loss rate from the pipe wall due to rotational effect. A series of numerical simulations are conducted to understand the detailed flow behavior around the cylinder for various Reynolds numbers. Thermal plumes and velocity vectors are pictorially presented to have clear understanding of physics of problem.

Keywords Rotating cylinder · Mixed convection · Vortex · And heat transfer

Nomenclature

D	Tube diameter, m
g	Gravitational acceleration, m/s^2
h	Heat transfer coefficient, W/m^2 K
L	Length of the cylinder, m
L/D	Tube aspect ratio
Nu	Nusselt number
Pr	Prandtl number
Q	The heat loss rate, W
r	Radial coordinate

B. K. Rana (✉)
School of Mechanical Engineering, Kalinga Institute of Industrial Technology (KIIT) Deemed to be University, Bhubaneswar, Odisha 751024, India
e-mail: basanta.uce@gmail.com

J. R. Senapati
Department of Mechanical Engineering, National Institute of Technology Rourkela, Rourkela, Odisha 769008, India

S. Revankar et al. (eds.), *Proceedings of International Conference on Thermofluids*, Lecture Notes in Mechanical Engineering,
https://doi.org/10.1007/978-981-15-7831-1_34

Ra_L Rayleigh number
Re_D Reynolds number
Ri_D Richardson number
T Fluid temperature, K
T_w Tube wall temperature, K
T_∞ Atmospheric temperature, K
v_r Velocity in radial direction, m/s
v_z Velocity in axial direction, m/s
z Axial coordinate

Greek Symbols

α Thermal diffusivity, m^2/s
β Coefficient of thermal expansion, 1/K
μ Dynamic viscosity, Pa s
ν Kinematic viscosity, m^2/s
ρ Fluid density, kg/m^3

1 Introduction

Convection heat transfer (natural, forced, and mixed convection) is considered as one of the most effective and convenient ways of cooling for many engineering components. The convection heat transfer takes part a very crucial role in making heat removal from electrical and electronics equipment, heat exchangers, components of air conditioning, nuclear reactors, dry cooling towers, during casting and forging process, components of solar system and many more due to which this kind of heat transfer is very popular and inexpensive among all modes of heat transfer. A wide range of applications are found in case of mixed convection around the rotating cylinders like centrifugal casting, heat loss from turbo machineries, heat transfer from turbine rotors or electrical motor shaft, cooling of rotating bearings, heat transfer from rotating condenser pipes in seawater distillation process, drying of papers on rollers in the paper industry, cooling of ship funnel etc. A plethora of studies are performed on mixed convection around the solid vertical cylinder. Few early efforts [1–5] have been made related to heat transfer study around the cylinder in order to propose Nusselt number correlation for the pipe wall. Heckel et al. [6] have presented the mixed convection effect on cylinders placed vertically for various working fluids with different Prandtl number (0.1–100) to determine the Nusselt number locally for entire mixed convection regime. Ishak [7] investigated on steady mixed convection over cylinder surface by using similarity solution and found out the dependency of mixed convection on curvature factor and Pr. Kumari and Nath [8] analysed the influence of mixed convection from a thin tubes placed vertically with locally heating/cooling and injection/suction effects in order to find out the increase/decrease pattern of Nusselt number and skin friction coefficient with variation of flow parameters. Lee et al. [9, 10] investigated the natural convection from slender cylinder placed vertically with employment of varying temperature on wall and heat flux in order to find out the rate of heat loss which is a function of different parameters like Prandtl number, curvature, etc. Jarall and Campo [11] experimentally investigated steady laminar free convection around electrically heated cylinder placed vertically and proposed suitable relationship between local Nusselt numbers with local Rayleigh number under the constant heat flux condition. Senapati et al. [12] have performed numerical simulations for vertical cylinders with annular fins for both laminar and turbulent regimes and reported that heat transfer rate first increases and then decreases after some maximum point with addition to the fins. Acharya and Dash [13] have performed numerical simulations to study the turbulent natural convection heat flow from solid or hollow tube palced vertically and developed correction for Nu and air flow rate in terms of Ra and *L/D*. Goodrich and Marcum [14] have carried out experiments to analyze the free convection from heated vertical cylinder and developed Nu correlation for laminar, transition and turbulent regimes.

No study has been reported till now, where laminar mixed convection from a hollow rotating cylinder hanged vertically is being studied. A comparison of the present study with natural convention heat transfer from hollow vertical cylinder (non-rotating) is also presented here. In this study, various pertinent input parameters such as aspect ratio (*L/D*), Reynolds number, and Rayleigh number are varied to observe the thermo-fluid characteristics extensively. Static temperature contours and velocity vector plots are illustrated to have complete understanding of the flow physics around the cylinder.

2 Problem Description

Figure 1 depicts a schematic model in order to represent the rotating vertical hollow cylinder having a negligible wall thickness, which is suspended in the air medium. Here, L is denoted as pipe length, D is denoted as tube diameter, and ω is symbolized the angular speed of pipe around the centerline axis. Figure 2 shows the details of the computational domain and mesh structure. The present problem has been modelled as two-dimensional axisymmetric which is very clear from the schematic sketch of computational domain (Fig. 2a). The isothermal condition is imposed on both inner and outer wall of the hollow pipe ($T_w = 326$ K). Computation domain boundaries (total four edges) are subjected to pressure outlet (three edges) and axis (one edge) as depicted in Fig. 2a. Figure 2b represents the mesh generated in the domain. The heat loss from the tube would be felt near the cylinder wall. Hence, the finer grids are formed close to the wall, and the mesh is relatively coarser away from the cylinder wall.

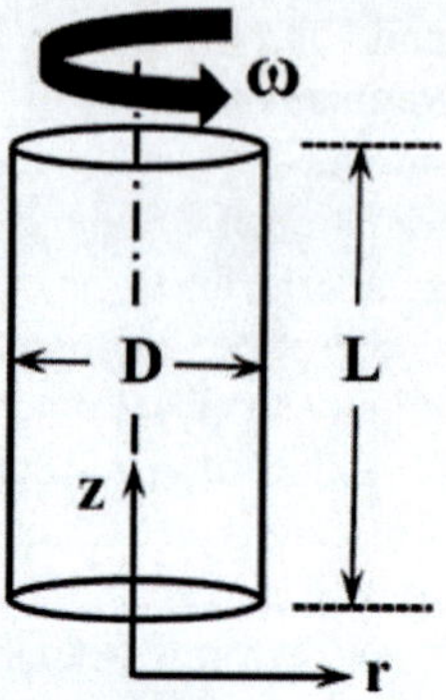

Fig. 1 The 3D view of physical model of hollow cylinder exposed to air medium

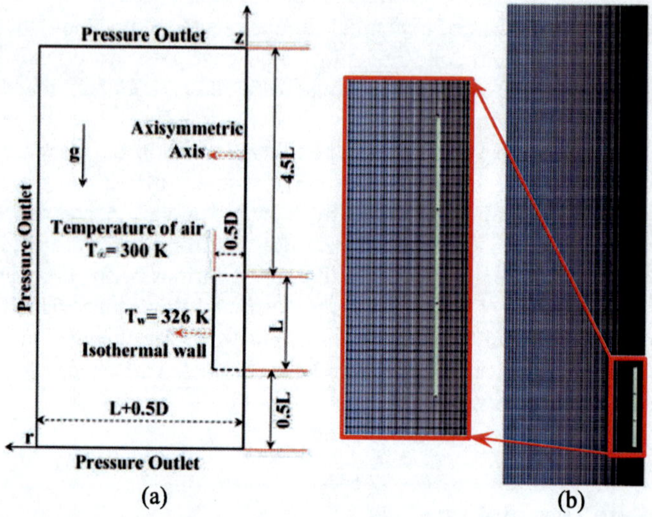

Fig. 2 Computational domain. **a** Schematic sketch. **b** Mesh layout

3 Numerical Methodology

The flow is assumed to be two dimensional, steady and laminar. An incompressible ideal gas used as working fluid inside the computational domain. The following differential equations need to be solved.

Mass balance equation:

$$\nabla \cdot \vec{U} = 0 \tag{1}$$

Momentum balance equation:

Table 1 Boundary conditions

At the cylinder wall	$T = T_{\mathrm{w}},\ v_r = v_z = 0$ and $v_\theta = \omega r$
Axisymmetric condition on the z-axis (right side of the domain)	$\frac{\partial Q}{\partial \theta} = 0$
On the outer edge of the computational domain (left, top, and bottom)	$(p = p_\infty,\ T = T_\infty)$

$$\rho\left(\vec{U}\cdot\nabla\right)\vec{U} = -\nabla p + \mu\nabla^2\vec{U} + \rho\vec{g} \tag{2}$$

Energy balance equation:

$$\left(\vec{U}\cdot\nabla\right)T = \alpha\nabla^2\vec{U} \tag{3}$$

where is $\vec{U} = \hat{e}_r v_r + \hat{e}_\theta v_\theta + \hat{e}_z v_z$ the velocity vector in the flow field.

Boussinesq approximation has been employed in the momentum equation in order to estimate the natural convection phenomenon in detail. All the simulations have been carried out with constant cylinder wall temperature, i.e. 326 K and ambient temperature is 300 K.

Boundary conditions need to be mentioned to solve Eqs. (1)–(3). The present problem is based on axisymmetric conditions. Hence, the z-axis (right side of the domain Fig. 2a) is considered as axis. Table 1 contains all the boundary conditions for the present problem.

The following pertinent non-dimensional heat transfer parameters are defined to describe the thermo-fluid characteristic on a rotating vertical tube.

Rayleigh number: $\mathrm{Ra}_L = \frac{g\beta(T_w - T_\infty)L^3}{\nu\alpha}$.

Reynolds number: $\mathrm{Re}_D = \frac{(\omega D/2)D}{\nu} = \frac{\omega D^2}{2\nu}$.

Richardson number: $\mathrm{Ri}_D = \frac{\mathrm{Ra}_L}{\mathrm{Re}_D^2\,\mathrm{Pr}}$.

Average Nusselt number: $\mathrm{Nu} = \frac{hL}{k}$ where $h = \frac{Q}{A_{\mathrm{wall}}(T_w - T_\infty)}$.

Numerical results have been validated with the existing literature data [15] and found the excellent matching which is shown in Fig. 3.

4 Results and Discussion

4.1 Impact of Re_D on Thermal Plume

Figure 4 shows contours of temperature around the rotating/non-rotating cylinder placed vertically in the air for different Reynolds numbers (Re_D) for a specific Rayleigh number and aspect ratio. It is evident from figure that the thermal plumes cover a larger area as Re_D rises. With the rise in Re_D, the boundary layer thickness

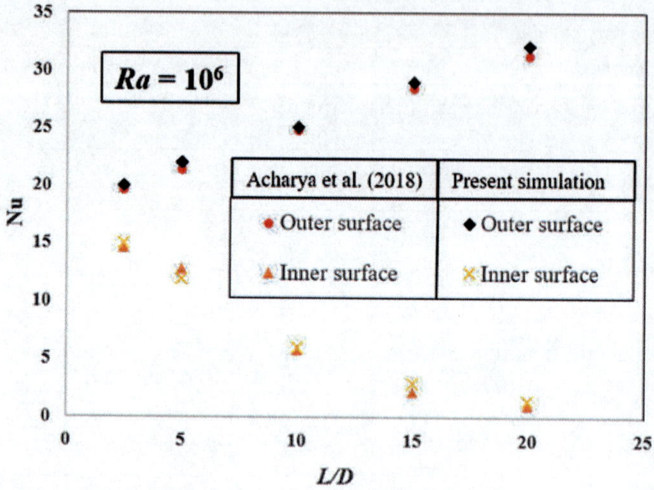

Fig. 3 Comparison between Acharya et al. [15] and this study of average Nu of outer and inner wall of the tube for Ra = 10^6

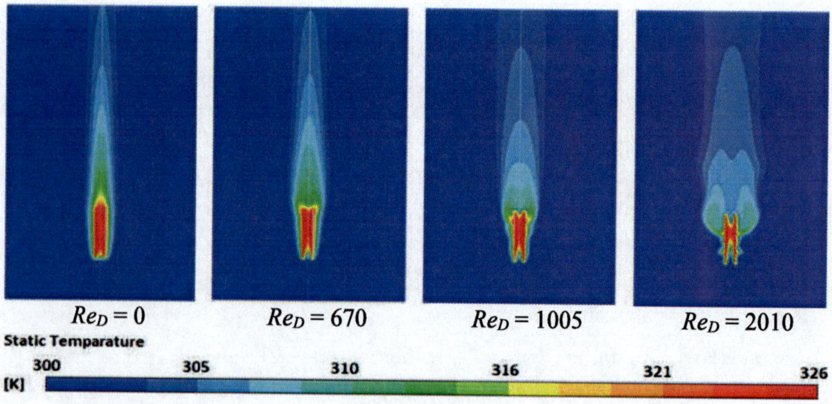

Fig. 4 Variation of temperature contour for different Reynolds numbers with Ra = 10^4 and L/D = 5

decreases due to which heat loss rate from the tube wall gets enhanced. Variation of the plume structure from non-rotating natural convection to rotating mixed convection is clearly observed from Fig. 4.

4.2 *Impact of* L/D *on Plumes*

The variation of structure of thermal plume for different aspect ratio of the pipe (*L*/*D*) with constant Rayleigh number (Ra $= 10^5$) and Reynolds number ($\mathrm{Re}_D =$ 1500) is shown in Fig. 5. It has been observed from the temperature contours that for a larger *L*/*D* ratio (*D* is varied, while *L* is fixed), the shape of thermal plume is thinner in comparison to low *L*/*D*. As *L/D* goes on decreasing from 15 to 2.5, the diameter of the cylinder increases due to surface area of cylinder wall increases monotonically which is depicted in the isotherms. As in the natural convection, the flow is buoyancy-driven, and plumes are directed axially upward. The thermal buoyant plume encounters larger hindrance at higher *L/D* inside the hollow cylinder as the flow has to move up in a smaller diameter cylinder. Hence, the convective flow through the hollow pipe is remarkably much higher at the low *L*/*D* in comparison to high *L*/*D*. Hence, the plume achieves a comparatively low temperature inside the hollow tube at low aspect ratio compared to high aspect ratio because the flow experiences less resistance through the tube. It is quite obvious that the heat loss from wall surfaces of vertical hollow cylinders at low aspect ratio is higher than that of high aspect ratio for both stationary and rotating cases. It has been noticed that the percentage of rise of the heat loss rate from non-rotating ($\mathrm{Re}_D = 0$) to rotating ($\mathrm{Re}_D \neq 0$) is much higher at high *L/D* ratio for a given Rayleigh number.

Table 2 presents the percentage rise in heat transfer rate for different *L/D* ratios from non-rotating to rotating for Ra $= 10^5$. When the stationary hollow cylinder is subjected to rotation about its axis, the motion of the cylinder adds supplementary heat transfer along with a natural convection effect, which acts as a mixed convection situation. So, the heat loss rises with rotating speed of the cylinder. It can be predicted with reference to the table that the heat loss (*Q*) is almost 11 times higher in case

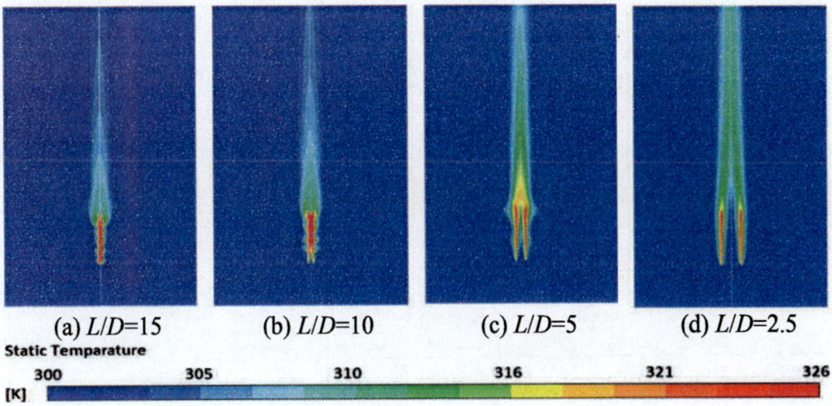

Fig. 5 Variation of temperature contour for different aspect ratios (*L*/*D*) with a fixed Ra $= 10^5$ and $\mathrm{Re}_D = 1500$

Table 2 Percentage rate of increase of heat transfer for different *L/D* ratio from $Re_D = 0$ to $Re_D = 1500$ for with fixed $Ra = 10^5$

L/D	*Q* (W) at $Re_D = 0$	*Q* (W) at $Re_D = 1500$	Increase of *Q* (%)
15	0.61	0.95	56.47
10	0.81	1.20	48.05
5	1.70	2.12	24.58
2.5	3.52	3.72	5.70

of $L/D = 15$ in comparison to $L/D = 2.5$. It is also clear from the simulated snaps given in Fig. 5.

4.3 Impact of Rotation on the Velocity Vector

Figure 6 presents the detailed velocity vectors to understand the fluid flow behavior around the cylinder with and without rotation for a given $Ra = 10^4$ and $L/D = 5$ with varying Re_D. It has been observed only natural convection current moving in upward direction in case of stationary vertical hollow cylinder. However, the effect of rotation influences the flow behavior around the cylinder. The velocity of flow is maximum near the cylinder wall for rotating cylinder cases, whereas for stationary cylinder the flow velocity is very small near the wall due to boundary layer effect. Also, when the speed of rotation increases, a larger region around the cylinder wall

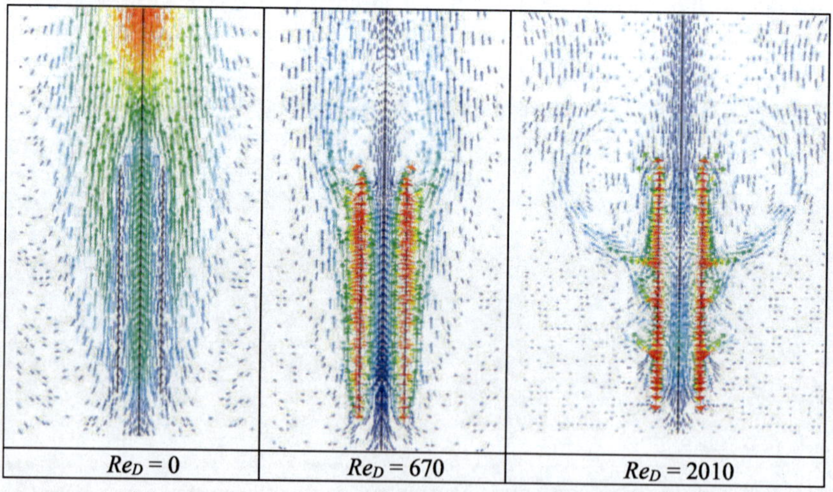

Fig. 6 Behavior of velocity vectors of fluid nearby the tube wall with and without rotation with a fixed aspect ratio

with high velocity is observed. It is clearly seen when the vector plots for $Re_D = 670$ and $Re_D = 2010$ are compared.

4.4 Impact of Re_D and Ra on the Heat Loss Rate

Figure 7 shows the influence of *L*/*D* on the heat loss rate for different Re_D for a given Rayleigh number. Here, two cases of Ra are presented in order to understand the pattern of increase of rate of heat transfer with wide range of *L/D* ratio with variation of Re_D. First of all, it has been noticed that *L/D* ratio decreases gradually; the rate of heat flow from the cylinder surface increases for both rotating and stationary cases due to rise in surface area. But it is quite interesting to notice that the percentage in rise of rate of heat transfer in comparison to stationary cylinder increases with Re_D for particular *L/D* ratio and Ra due to larger rotational velocity of the cylinder. Also, it has been observed that the percentage rise in heat loss rises with *L/D* for a given Re_D. As the Reynolds number (Re_D) is based on hollow pipe diameter and with the rise of aspect ratio (*L/D*), basically diameter decreases due to which angular velocity is higher for greater *L/D* for a fixed Re_D leads to significant amount of percentage increase in heat transfer at high *L/D* ratio for particular Ra. Further, the discussion has been extended for different Ra where one can find the heat loss rises with Ra because of higher amount air flow rate, so, the heat loss from the hollow pipe wall is marginally higher. But the percentage increase of heat transfer rate from stationary to rotating cylinder decreases from low Ra to high Ra because higher effect of natural convection compared to forced convection, which is clearly shown in Fig. 7a, b.

The influence of area-weighted average Nusselt number (Nu) due to Richardson number (Ri) for different *L/D* ratio with a fixed $Re_D = 500$ has been presented in Fig. 8. One can notice the monotonic increase of overall Nu with Ri for the particular *L/D* ratio due to higher amount of flow passes through the cylinder surface which helps for the higher amount of heat removal from the hollow tube walls. Variation of *L/D* on Nu_{avg} is very important aspect that should be elaborated in detail for this kind

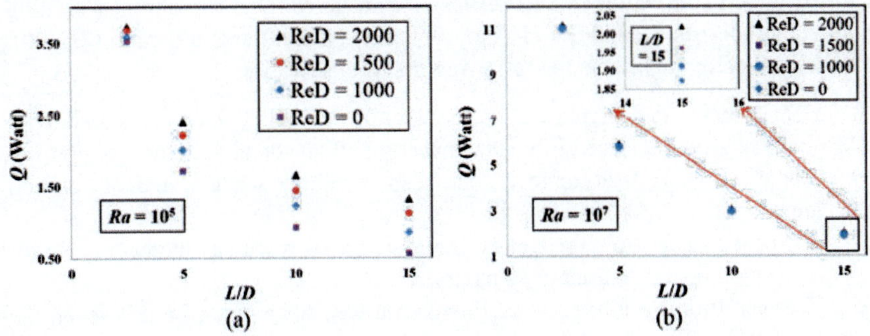

Fig. 7 Variation of Q (W) with *L/D* for **a** Ra $= 10^5$ and **b** Ra $= 10^7$

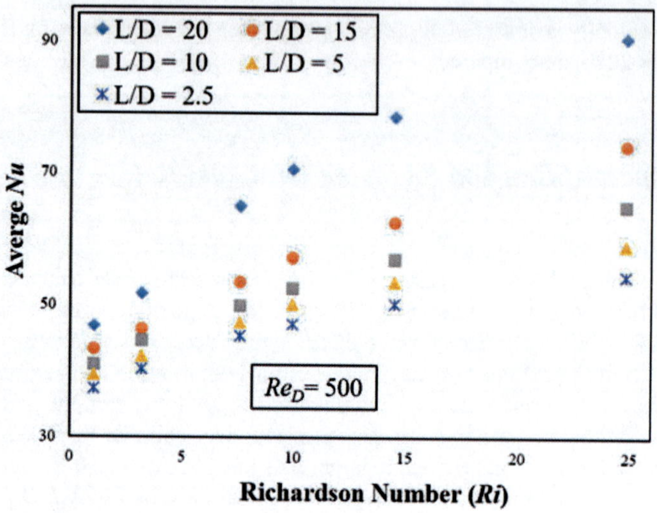

Fig. 8 Plot of Nu_{avg} verses Ri for different *L*/*D* ratio with $Re_D = 500$

of study. It has already been mentioned that with decrease in aspect ratio, the diameter of the hollow pipe increase continuously which leads to increase in surface area of the tube wall. But the expression for the area-weighted average Nusselt number is presented by $\mathrm{Nu} = (hL)/k = (QL)/\{A(T_w - T_\infty)k\}$ where $Q = hA(T_w - T_\infty)$. It is clearly understood from the expression that as cylinder wall surface area (*A*) goes on increasing, the heat removal (*Q*) from the walls of the tube also rises with decrease in *L/D* ratio.

5 Conclusion

A detailed numerical investigation has been explored in order to understand the laminar mixed convection characteristics around the vertical rotating/non-rotating hollow cylinder for various Ra (10^4 to 10^8), Re_D (<2100) and *L/D* ratio (2.5–20). The present study highlights the following major observation:

I. Heat transfer rate increases for the rotating cylinder when compared with stationary cylinder due to forced convection effect added with the free convection effect. The heat removal rate also increases when speed of rotation increases.
II. The heat transfer from the walls of hollow pipe rises with an increase in Ra for both rotating and stationary cylinders.
III. The heat loss is observed to be rising with Re_D for a given Ra. However, the percentage of increase heat transfer rate happens to be less while moving from non-rotating to rotating with an increase in Ra with a fixed aspect ratio.

IV. Velocity vectors for both rotating and non-rotating cylinders are shown where the maximum velocity is observed near the cylinder wall, which results in heat transfer enhancement.

References

1. Eigenson LS (1940) Les lois gouvernant la transmission de la chaleur aux gaz biatomique par les parois des cylindres verticaux dans le cas de convection naturelle. Dokl Akad Nauk SSSR 26:440–444
2. Kyte JR, Madden AJ, Piret EL (1953) Natural-convection heat transfer at reduced pressure-spheres and cylinders. Chem Eng Prog 49(12):653–662
3. Mueller AC (1942) Heat transfer from wires to air in parallel flow. Trans Amer Inst Chem Eng 38:613–627
4. Touloukian YS, Hawkins GA, and Jakob M (1948) Heat transfer free convection from heated vertical surfaces to liquids. Trans ASME 70:13–18
5. Fujii T (1959) Experimental studies of free convection heat transfer. Bull JSME 2(8):555–558
6. Heckel JJ, Chen TS, Armaly BF (1989) Mixed convection along slender vertical cylinders with variable surface temperature. Int J Heat Mass Transf 32(8):1431–1442
7. Ishak A (2009) Mixed convection boundary layer flow over a vertical cylinder with prescribed surface heat flux. J Phys A Math Theor 42(19):195501
8. Kumari M, Nath G (2004) Mixed convection boundary layer flow over a thin vertical cylinder with localized injection/suction and cooling/heating. Int J Heat Mass Transf 47(5):969–976
9. Lee HR, Chen TS, Armaly BF (1988) Natural convection along slender vertical cylinders with variable surface temperature. J Heat Trans 110(1):103–108
10. Lee SL, Chen TS, Armaly BF (1986) Mixed convection along isothermal vertical cylinders and needles. In: International heat transfer conference digital library. Begel House Inc
11. Jarall S, Campo A (2005) Experimental study of natural convection from electrically heated vertical cylinders immersed in air. Exp Heat Transf 18(3):127–134
12. Senapati JR, Dash SK, Roy S (2017) Numerical investigation of natural convection heat transfer from vertical cylinder with annular fins. Int J Therm Sci 111:146–159
13. Acharya S, Dash SK (2020) Turbulent natural convection heat transfer from a vertical hollow cylinder suspended in air: A numerical approach. Therm Sci Eng Progr 15:100449
14. Goodrich SS, Marcum WR (2019) Natural convection heat transfer and boundary layer transition for vertical heated cylinders. Exp Thermal Fluid Sci 105:367–380
15. Acharya S, Agrawal S, Dash SK (2018) Numerical analysis of natural convection heat transfer from a vertical hollow cylinder suspended in air. J Heat Transf 140(5):052501

Computational Analysis of Static Flow Instabilities in Supercritical Natural Circulation Loop

Santosh Kumar Rai, Pardeep Kumar, and Vinay Panwar

Abstract The present paper is to study and explore the possibility of static instability, Ledinegg flow excursion, in a rectangular supercritical natural circulation loop (RSCNCL). Due to drastic change of thermo-physical properties of fluid near the pseudo-critical point, various static and dynamic instabilities might occur in the loop. The instability may be either Ledinegg excursion or density wave oscillations (DWOs). Many previous researchers have analyzed dynamic instability, DWO, in SCNCL, whereas the Ledinegg excursive instability is not studied in depth. In the present research work, a mathematical model will be implemented to predict the Ledinegg instability in SCNCL. The numerical investigation is successful to find whether there would be any Ledinegg instability in the case of SCNCL or not.

Keywords Ledinegg excursive instability · Supercritical water reactor · Rectangular natural circulation loop

Nomenclature

A	Cross sectional area (m^2)
D	Hydraulic diameter (m)
f	Friction factor
G	Mass flux density (kg/m^2)
H	Loop height (m)
h	Specific enthalpy
K	Loss coefficient
L	Length of the section (m)

S. K. Rai · P. Kumar
Meerut Institute of Engineering and Technology, Meerut, Uttar Pradesh 250005, India

S. K. Rai · V. Panwar (✉)
SRM Institute of Science and Technology, Delhi-NCR Campus, Delhi-Meerut Road, Modinagar, Ghaziabad, Uttar Pradesh, India
e-mail: vinaypanwar86@gmail.com

S. Revankar et al. (eds.), *Proceedings of International Conference on Thermofluids*, Lecture Notes in Mechanical Engineering,
https://doi.org/10.1007/978-981-15-7831-1_35

P	Pressure (N/m^2)
q''	Heat flux (W/m^2)
Q	Total power (W)
Re	Reynolds number
W	Mass flow rate (kg/s)
x	Axial distance (m)

Greek Symbols

ρ	Density (kg/m^3)
μ	Dynamic viscosity of fluid (Pa s)
θ	Angle

Subscripts

h	Hot
crit	Critical
l	Liqued

1 Introduction

The super critical water nuclear reactors (SCWNR) is one of the highly feature reactor of the GIF–IV program for safety and cost effective power generation, and that why SCWNR currently being examined rigorously worldwide by many researcher [1]. The SCWNRs are also known as light water reactors. Its working temperature (674 K) and pressure (22.064 MPa) are very high, which is above the critical point of thermodynamic. The expected thermal efficiency of SCW is around 45–47%, which is significantly higher thermal performance than the sub-critical water reactors. The light water reactor is very compact in size compare to the sub-critical based plant due to absence of the coolant drum, pressurize and steam generator etc. SCWNR is designed by taking the concept of two superior technologies namely LWRs and the supercritical boiler, which is generating the high power at low cost [2, 3].

Stability is an indispensable point of concern for the SCNCLs. In case, the velocity oscillates, but the amplitude of the velocity is not growing and remains the same in time, the state of the system is said to be "neutrally stable". If the system further oscillations and whose amplitudes increase in time and may lead to flow reversal, such kind of the system is known as "unstable" and if the oscillation amplitudes decrease with time and the system regains their original position later on, this system is known as "stable". The flow excursion at supercritical condition is due to the sharp and significant change of density of the coolant close to the pseudo-critical point. This flow instability must be avoided as the mechanical vibration of the components can initiate and if sustain for long can cause failure of those components due to thermal fatigue. Such kind of flow instability can create the problem to control the system in SCWNRs. It is also affect the heat transfer behaviour. The literature review indicates that the flow instability of NC in boiling systems has already been analyzed extensively by many researchers, while that at the supercritical conditions is the subject of interest at present and needs more studies specifically for SCNCLs [4].

A large variety of thermal–hydraulic (TH) instabilities may occur in a nuclear reactor. The TH instabilities are recognized as periodic time oscillations of diverse quantities in a machine such as mass flow rate, inlet velocity or heated wall surface temperature.

There are mainly two groups of TH instabilities, static and dynamic ones. The static instabilities are independent of time, whereas the dynamic instabilities are theoretically studied by using the time dependent conservation equations. Among the dynamic instabilities, the pressure drop oscillations and the DWOs are the most studied ones and received a large attention in the nuclear fraternity. The excitation of standing waves in a single phase or liquid and vapour (two phase) compressible system with a frequency in the acoustic range may cause resonance of pressure waves leading to pressure drop oscillations—which are also called as "acoustic instabilities". The perturbation in the mass flow flowing through the heated channel at some conducive condition may trigger DWOs. The DWOs result from the multiple feedback relationships among mass flow rate, density, pressure drop and other TH parameters. Among the static instabilities, Ledinegg excursion is one of them which may exist in a NCL under supercritical condition. The present work will focus mainly on the Ledinegg instability of RSCNCL where no pump is used. For a force circulation system, pump provides the driving pressure head and whereas, gravity provides that in the case of a natural circulation system. The relationships among the mass flux or mass flow rate and input power governs the mechanisms of the Ledinegg instability in a RSCNCL and how it does so will be studied in present work through numerical investigation.

2 Mathematical Modeling

The Ledinegg excursive instability in a rectangular-SCNCL, with fixed radius and horizontally placed heat source in bottom of the loop and sink placed at the top of the loop, which is shown in Fig. 1a, was analyzed by Chatoorgoon [5] using 1-D steady state flow equation and it was opined that the instability will occur when $\partial \boldsymbol{G}/\partial \boldsymbol{Q} < 0$ (where $\boldsymbol{G} = \boldsymbol{W}/\boldsymbol{A}$) and it is shown in Fig. 1b.

The present investigation is carried out using similar 1-D steady state conservation equations [4, 7–12] but with different methodology so as to find whether the criterion

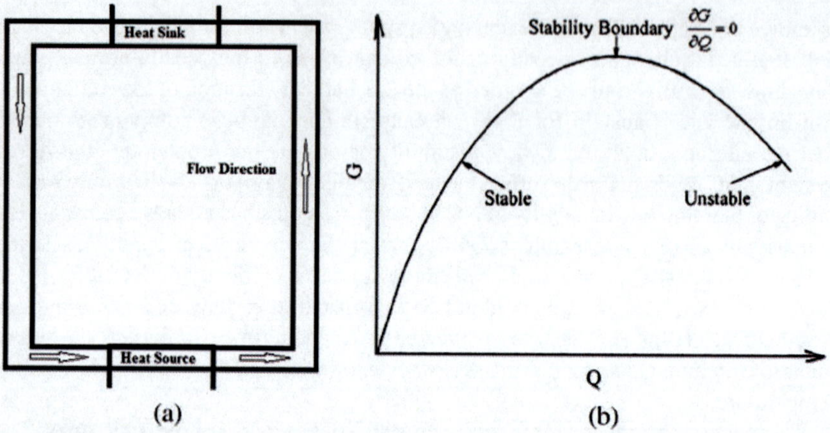

Fig. 1 **a** Schematic diagram of loop **b** Stability boundary of SCNCL at supercritical pressure [6]

given by Chatoorgoon [6] satisfies or not and thus we can determine the possibility of Ledinegg instability for the case study under consideration.

Continuity equations:

$$\frac{\mathrm{d}w}{\mathrm{d}x} = 0 \tag{1}$$

Axial momentum equation:

$$\frac{1}{A^2}\frac{\mathrm{d}}{\mathrm{d}x}\left(\frac{w^2}{\rho}\right) + \rho g \cos\theta + \frac{w^2}{2\rho A^2}\left(\frac{f}{D} + K\right) + \frac{\mathrm{d}p}{\mathrm{d}x} = 0 \tag{2}$$

Energy equation:

$$\frac{1}{A}\frac{\mathrm{d}(wh)}{\mathrm{d}x} = \begin{cases} q_h'' or - q_c'' \\ 0 \end{cases} \tag{3}$$

The following thermodynamic equation of state is used for the calculation of density:

$$\rho = f(p, h) \tag{4}$$

The friction factor is calculated based on the local Reynolds number using the following correlations, Eq. (5) and Eq. (6):

$$f_{\text{laminar}} = \frac{64}{\text{Re}} \tag{5}$$

$$f_{\text{turbulant}} = \frac{0.316}{\text{Re}^{0.25}} \tag{6}$$

To avoid the discontinuity during transition from laminar to turbulent flow, f_{laminar} is used when $\text{Re} < \text{Re}_{\text{crit}}$ and $f_{\text{turbulent}}$ is used when $\text{Re} > \text{Re}_{\text{crit}}$. The Re_{crit} is obtained by equating Eq. (5) with Eq. (6).

3 Result and Discussion

The SCNCL considered in the present study is shown in Fig. 2. Its geometrical dimensions and operating conditions, with which the simulations are carried out, are listed in Table 1.

The variation of mass flow rate (W) with increasing inlet power for given value of heater inlet enthalpy 1000 kJ/kg, is shown in Fig. 3a. It is found that on increasing

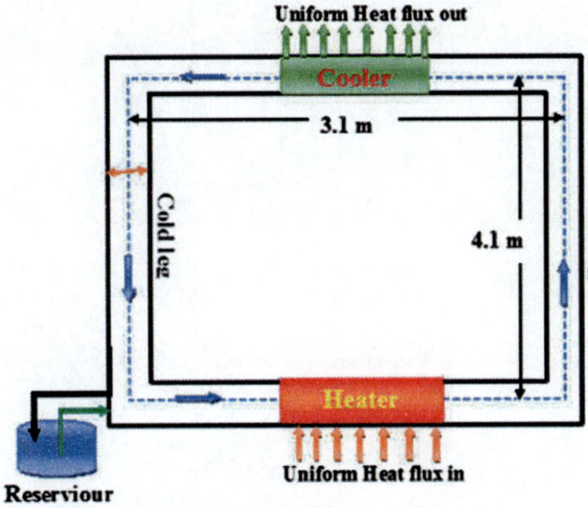

Fig. 2 Schematic diagram of SCNCL

Table 1 Geometrical dimensions and operating parameters

Name	Parameters
Loop radius	0.0072 m
Heater section length	1.3 m
Cooler section length	1.2 m
Riser height (hot leg)	4.1 m
Length of down comer (cold leg)	4.1 m
Pressure	25 MPa
Inlet enthalpy	1000–1500 kJ/kg
Power	0–300 kW

Fig. 3 **a** Mass flow rate (W) versus inlet power (Q) at 1000 kJ/kg inlet enthalpy. **b** Mass flow rate (W) versus inlet power (Q) at 1200 kJ/kg inlet enthalpy. **c** Mass flow rate (W) versus inlet power (Q) at 1400 kJ/kg inlet enthalpy. **d** Mass flow rate (W) versus inlet power (Q) at 1500 kJ/kg inlet enthalpy

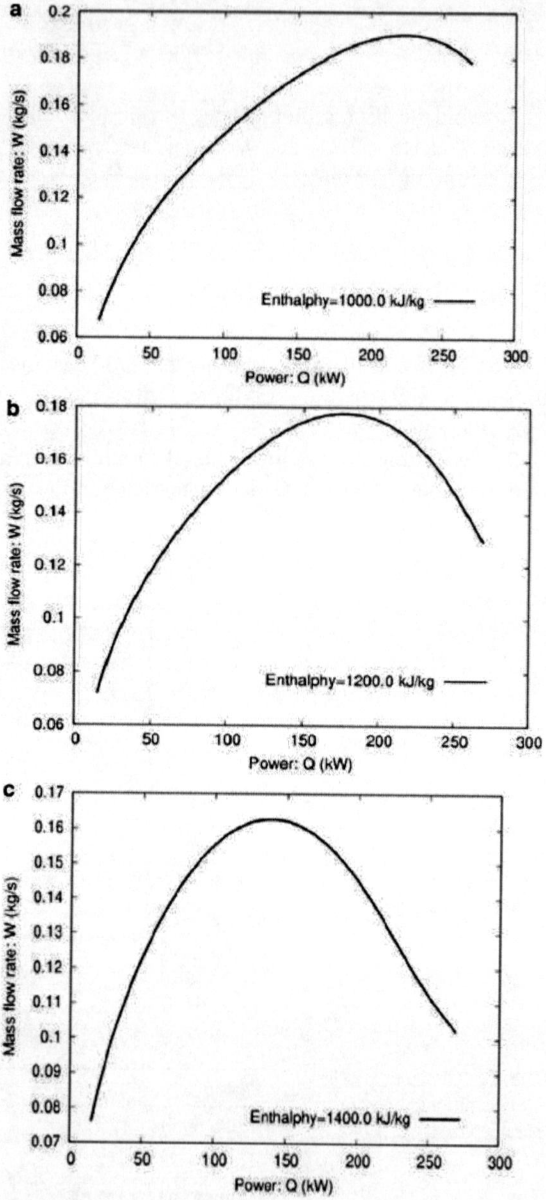

Fig. 3 (continued)

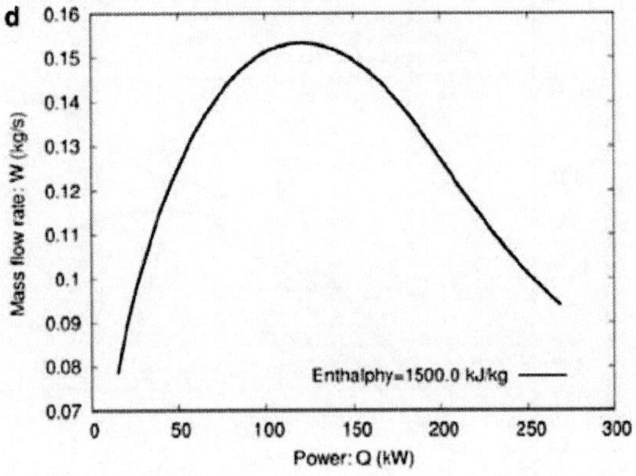

heater inlet power, the mass flow rate initially increases and then decreases. The similar simulations are conducted for different heater inlet enthalpies keeping the geometry of the loop and other operating conditions the same. The variation of mass flow rate (W) with the increase in heater inlet power for different heater inlet enthalpies, 1200, 1400, 1500 kJ/kg, are shown in Fig. 3b–d respectively and in all these figures, the similar trend is observed, i.e., the mass flow rate initially increases and then decreases with the increase in power. To determine and compare the effect of different heater inlet enthalpies on the mass flux versus power profile, the results of Fig. 3a–d are superimposed and shown in Fig. 4. It is found that at lesser power range, when the power is kept constant, the mass flow rate increases with the increase in heater inlet enthalpy. The trend gets reversed at higher power range, i.e., when the power is high and if kept constant, the mass flow rate decreases with the increase in heater inlet enthalpy. For the given inlet enthalpies 1000, 1200,1400 and 1500 kJ/kg, the respective maximum mass flow rates, as shown in Fig. 4, are observed to be 0.192, 0.178, 0.163 and 0.155 kg/s and those occur at a power 230,180, 150 and 122 kW respectively. It is also observed that beyond the peak mass flow rate, the slope turns out to be negative, i.e., $\left(\frac{\partial G}{\partial Q} < 0\right)$, indicating the possibility of Ledinegg excursion in the SCNCL.

The results demonstrate that there will be Ledinegg instability for the given RSCNCL and operating range under consideration as per the stability criterion provided by Chatoorgoon [6].

4 Conclusion

A mathematical model is developed to resolve the steady state equation for mass, momentum (axial) and energy conservation equations for a RSCNCL. Various parametric studies are carried out based on the mathematical model. The simulation results for different heater and cooler powers and heater inlet enthalpies are analyzed

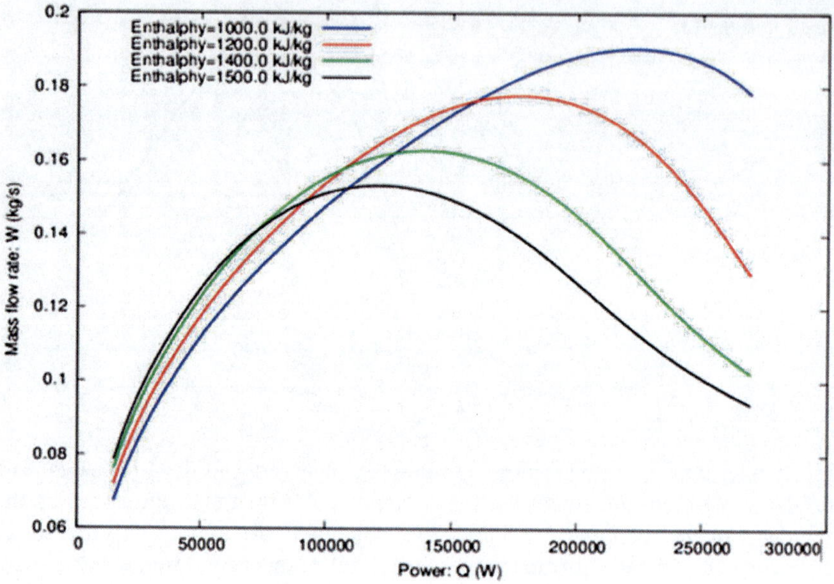

Fig. 4 Variation of mass flow rate (*W*) versus inlet power (*Q*) for different inlet enthalpy

to find and predict the Ledinegg instability in a RSCNCL. This method is simple and reliable.

References

1. Buongiorno J, MacDonald PE (2003) Supercritical water reactor (SCWR) progress report for the FY-03 generation-IV R&D activities for the development of the SCWR in the U.S. 2003, INEEL/EXT-03-01210
2. Jiyang Y, Shuwei C, Ran L, Bingxue Q (2011) Analysis of Ledinegg flow instability in natural circulation at supercritical pressure. Progr Nucl Energy 53(6):775–779
3. Sarkar Milan KS, Tilak A, Basu DN (2014) A state-of-the-art review of recent advances in supercritical natural circulation loops for nuclear applications. Ann Nucl Energy 73:250–263
4. Vijayan PK, Nayak AK (2010) Introduction to instabilities in natural circulation systems. Lecture Notes for T-06: In: Joint ICTP-IAEA course on natural circulation phenomena and passive safety systems in advanced water cooled reactors Trieste, Italy: 2152–2122, http://indico.ictp.it/event/a09155/session/13/contribution/11/material/0/0.pdf
5. Chatoorgoon V (2001) Stability of supercritical fluid flow in a single channel natural convection loop. Int J Heat Mass Transfer 44(10):1963–1972
6. Chatoorgoon V, Voodi A, Fraser D (2005) The stability boundary for super critical flow in natural convection loops. Part I: H_2O studies. Nucl Eng Des 235(24):2570–2580
7. Sharma M, Pilkhwal DS, Vijayan PK, Saha D, Sinha RK (2010) Steady state and linear stability analysis of a supercritical water natural circulation loop. Nucl Eng Des 240(3):588–597
8. Rai SK, Dutta G, Sheorey T, Stability analysis of supercritical water natural circulation loop with vertical heater and cooler. In: Proceedings of the 24th national and 2nd international

ISHMT-ASTFE heat and mass transfer conference (IHMTC-2017). BITS Pilani, Hyderabad, India, 27–30 Dec 2017

9. Upadhyay R, Rai SK, Dutta D (2018) Numerical analysis of density wave instability and heat transfer deterioration in a supercritical water reactor. J Mech Sci Technol 32(3):1063–1070
10. Swapnalee BT, Vijayan PK, Sharma M, Pilkhwal DS (2012) Steady state flow and static instability of supercritical natural circulation loops. Nucl Eng Des 245:99–112
11. Jain PK, Uddin R (2008) Numerical analysis of supercritical flow instabilities in a natural circulation loop. Thesis
12. Sharma M, Vijayan PK, Asako Y (2013) Steady state and stability characteristics of natural circulation loops operating with carbon dioxide at supercritical pressures for open and closed loop boundary conditions. Nucl Eng Des 265:737–754

Effect of Loop Geometry on the Flow Dynamics of a Single-Phase Natural Circulation Loop

Ritabrata Saha, Koushik Ghosh, Achintya Mukhopadhyay, and Swarnendu Sen

Abstract In thermal reactors, like PWR and PHWR, single-phase natural circulation loop (NCL) is used to remove heat from the reactors in the absence of forced circulation. Different types of flow patterns (like steady, oscillatory and chaotic flow) are observed in an NCL depending on the geometrical configuration and operating conditions. In the present study, a square geometry is considered as the basic design of the loop. The changes in flow pattern are observed due to a change in the aspect ratio of the loop geometry and the tube diameter of the loop. Numerical simulations are carried out using MATLAB-based Simulink model, and a comparative assessment is made between various loop configurations for different heater powers. The initial transient part of the time-series data is neglected for the proper investigation of the flow dynamics. In this work, the phase change of the working fluid is not considered. This limits the maximum heater power to 800 W. From the results, we found that with the change in the aspect ratio or tube diameter flow behaviour changes. Accordingly, its effect the heat transfer from the source to sink.

Keyword Natural circulation loop · Natural convection · Flow reversal · Instability

Abbreviations

A	Area of cross section (m^2)
A_{in}	Internal cross-sectional area of loop (m^2)
A_0	External cross-sectional area of loop (m^2)
C_p	Specific heat of fluid at constant pressure (J/kg K)
C_{pw}	Specific heat of wall at constant pressure (J/kg K)
C_{pex}	Specific heat of coolant at constant pressure (J/kg K)
d_{in}	Internal loop diameter (mm)
Dz	Change in length (m)

R. Saha · K. Ghosh · A. Mukhopadhyay · S. Sen (✉)
Department of Mechanical Engineering, Jadavpur University, Kolkata 700032, India
e-mail: sen.swarnendu@gmail.com

S. Revankar et al. (eds.), *Proceedings of International Conference on Thermofluids*, Lecture Notes in Mechanical Engineering,
https://doi.org/10.1007/978-981-15-7831-1_36

F	Friction factor
T	Time (s)
T_f	Temperature of fluid (°C)
T_w	Loop wall temperature (°C)
T_{ex}	Coolant temperature (°C)
T_{ref}	Reference temperature (°C)
T_a	Ambient temperature (°C)
L_{loop}	Total loop length (m)
k_f	Thermal conductivity of fluid (W/m K)
k_w	Thermal conductivity of wall (W/m K)
k_{ex}	Thermal conductivity of heat exchanger (W/m K)
v_f	Fluid velocity (m/s)
V_w	Wall volume (m^3)
V_{ex}	Heat exchanger fluid volume (m^3)
h_i	Heat transfer coefficient between fluid & wall (W/m^2 K)
h_o	Heat transfer coefficient between wall & ambient (W/m^2 K)
h_{ex}	Heat transfer coefficient between wall & coolant (W/m^2 K)
$\dot{Q}$	Volumetric heat generation (W/m^3)
β_{av}	Thermal volumetric expansion coefficient (K^{-1})
μ_f	Dynamic viscosity of fluid (kg/m s)
υ	Kinematic viscosity of fluid (m^2/s)
ρ_f	Fluid density (kg/m^3)
ρ_w	Density of wall (kg/m^3)
ρ_{ex}	Coolant density (kg/m^3)

1 Introduction

Natural circulation refers to the fluid flow inside a closed loop due to density difference caused, for example, by the temperature difference between different parts of the loop. Therefore, natural circulation or natural circulation loop (NCL) provides a very significant option for an effective way of fluid transport inside a closed-loop without employing external devices like pumps, fans, or blowers.

In the absence of electromechanical devices which may fail to perform, NCL adds more reliability to the systems. Heating at a location related to the relatively lower arm and cooling at a place associated with the relatively higher arm produce a favourable density gradient which causes the warmer lighter fluid to move upward due to the buoyancy force, and cooler denser fluid to come down due to the gravitational force. Therefore, from a high-temperature source, thermal energy is transported to a low-temperature sink without any direct contact and any rotary device.

For this reason, as a cooling system, NCL is widely used in nuclear power plants [1, 2]. In PWR and PHWR, single-phase NCL is used to remove heat from the reactors during the absence of forced circulation. In BWR and AHWR, two-phase NCL is used to remove heat from the reactors in the absence of forced circulation. All the advanced reactors are tending to use NCL for the cooling purpose during normal operating condition along with the accidental situation due to increased reliability. Some other areas like electrical machine rotor cooling, solar water heater systems, geothermal energy systems also use NCL for transferring heat or cooling of the system.

The most typical characteristic of NCLs is their very high sensitivity to the operating state and susceptibility to unstable behaviour, mainly due to the fully coupled nature of fluid flow velocity and liquid temperature. This unstable behaviour is a significant problem in the field of NCL. Due to this unstable phenomenon temperature difference between two vertical arms is affected and might be reversed. This may cause the periodic inversion of the flow direction, which can affect the heat eviction from the thermal source and may cause a system failure.

This typical characteristic motivated the researchers to explore the flow characteristics of NCL in the past few decades. Misale et al. [3] observe the flow dynamics of a single-phase NCL experimentally and numerically. They used CATHARE and RELAP codes for the numerical simulations. From the comparative study, they found good agreement between numerical results and experimental results. Jiang et al. [4] experimentally and numerically observed the flow dynamics for a copper toroidal thermosyphon. They observed that usual glass torus is less stable than a copper torus.

The group of Lee and Pan [5, 6] observed nonlinear flow dynamics of a nuclear coupled two-phase NCL and perform the stability analysis using stability mapping. They found that two instability regions are present there. One is a type-I instability region, and another one is a type-II instability region. Durgaprasad et al. [7] observed the flow instabilities produced due to the parallel channel configuration in a boiling natural circulation system. They used lumped parameter mathematical model and RELAP5/MOD3.3 for the numerical simulation and compare the results. They found that different types of flow oscillations are produced due to the different operating conditions. They also performed parametric studies for a single power condition and found that results are changing significantly. Kozmenkov et al. [8] numerically and experimentally observed flashing-induced instability in natural circulation systems. They used RELAP5/MOD3.3 for the numerical simulation. They found that numerical results are showing good agreement with experimental results. They found that oscillation frequency increases with the increase in core inlet temperature. Hwang et al. [9] observed flow dynamics of an advanced PWR from the experimental data. They performed stability analysis using frequency-domain linear stability analysis code. For this purpose, they developed a code called ALFS_F. They also studied the parametric effect on the stability and found that onset of flow instability changes with the parametric changes.

Rabiee et al. [10] perform an experimental study on Shiraz University natural circulation loop (SHUNCL). They also carried out numerical simulations using RELAP5/MOD 3.2 code and compared the results. They found that depending upon the operating condition, type-I dynamic instability grows-up. Misale [11] performs experiments to study the effect of the power steps on the flow dynamics of a single-phase NCL. He found that with the increase in power steps frequency and amplitude of the flow oscillations increases. Pini et al. [12] perform numerical simulations to observe the effect of internal heat generation on the flow dynamics stability using RELAP5 and CFD code. They found that flow dynamics stability can be improved with the help of internal heat generation.

Mishra and Singh [13] perform a linear stability analysis using the stability map of a two-phase natural circulation loop. They found that in the flow dynamics, both supercritical and subcritical bifurcations are present. Cong et al. [14] perform numerical simulations to observe the flow dynamics of a single-phase laminar natural circulation under swing conditions. They found that startup performances and behaviour of the steady-state operations are significantly changing with the change in swing conditions like swing period, maximum swing angle, swing centre and swing phase. Lei et al. [15] experimentally and numerically observe the effect of flow resistance on NCL. They found that depending upon the heater power flow resistance affects the flow dynamics of the NCL. Krishnani and Basu [16] numerically observed the effect of the inclination on the flow-dynamic stability. For this purpose, they used a 3D computational model. They found that depending on the inclination angle onset of the flow dynamics changes.

Cammi et al. [17] carried out numerical simulations based on information entropy methodology. After that, they perform the stability analysis of a single-phase NCL using stability mapping. They found that there are different regimes in the flow dynamics. Cammi et al. [18] carried out numerical simulations to observe the effect of the pipe wall material on the flow dynamics of the NCL. For the numerical simulations, they used object-oriented (O–O) one-dimensional model and RELAP5. From the results, they found that both the models show similar behaviour and pipe wall material has a critical part in the stability of the system. Luzzi et al. [19] performed a stability analysis using the stability map approach to study the flow dynamics of a single-phase NCL. This study is done by the semi-analytical and numerical models on time-series data sets which are obtained from the experiments. From the study, it is found that different regimes are present in the flow dynamics of the NCL.

Borreani et al. [20] perform numerical simulations and experiments for a single-phase NCL. Numerical studies are done using ANSYS-FLUENT 14. They found an excellent agreement between numerical results and experimental results. Goudarzi and Talebi [21] perform numerical investigations on single-phase natural circulation loop. They observed flow instability associated with the NCL using nonlinear stability analysis method. They found that with the change in the parameters like loop diameter, cooler length affects the flow dynamics significantly. They also performed a comparative study for different orientations of the loop and found that flow dynamics change significantly.

Compared to two-phase NCLs, single-phase NCLs are less prone to the unstable phenomenon. Accordingly, single-phase NCLs are used more in the field of nuclear engineering and many other engineering sectors who are using NCLs to transport heat from one place to another. The aim of the present study is to observe the effect of the aspect ratio (Horizontal: Vertical) of the loop geometry and the tube diameter on the flow dynamics. There is significant work is present in NCL with other geometry in single phase. But with the square geometry, there is no work is present. However, change in geometry has significant role in flow dynamics. For this purpose, we consider a square geometry for the initial investigation and consider as a base configuration. Numerical simulations are done using MATLAB-based Simulink model. With the steady increase in the operating power, flow pattern of the NCL is changing from steady-state behaviour to flow reversal through the oscillatory behaviour. With the change in aspect ratio, flow pattern is changing. From the detailed investigation of the flow dynamics, it is found that oscillatory flow is purely periodic, and flow reversal is chaotic in nature [22]. Now, as the chaotic flow pattern is not suitable for a cooling system, we must choose the proper aspect ratio of the loop so that the stability of the loop increases.

2 Numerical Modelling

NCL loop geometry which is schematically shown in Fig. 1 and numerical solving procedure for the generation of time-series data in the present work is discussed in detail in our previous work [23]. Therefore, we have not discussed it here for the compactness of the paper.

Governing equations like mass conservation equation, momentum equation and energy equation for the loop fluid, coolant and wall are similar to our earlier works [23–25] and are mentioned briefly below.

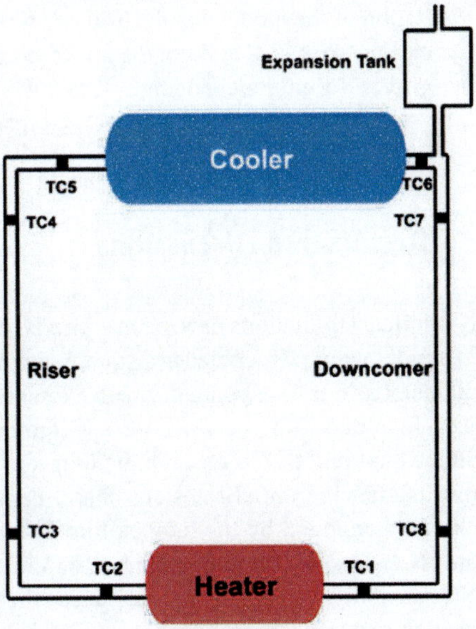

Fig. 1 Schematic diagram of the loop geometry of a square natural circulation loop

$$\frac{\partial \rho_{\mathrm{f}}}{\partial \mathrm{t}} + \frac{\partial}{\partial \mathrm{z}}((\rho v)_{\mathrm{f}}) = 0 \tag{1}$$

Loop fluid mass flux (G_{f}) = (ρv)

$$\frac{\mathrm{d}G_{\mathrm{f}}}{\mathrm{d}t} L_{\mathrm{loop}} = -\frac{f G_{\mathrm{f}}^2}{2\rho_f d_{\mathrm{in}}} L_{\mathrm{loop}} + \oint \rho_{\mathrm{f}} g \beta_{\mathrm{av}} (T_{\mathrm{f}} - T_{\mathrm{ref}}) \mathrm{d}z \tag{2}$$

$$\frac{\partial}{\partial t}(\rho_{\mathrm{f}} C_{\mathrm{p}} T_{\mathrm{f}}) + \frac{\partial}{\partial z}(G_{\mathrm{f}} C_{\mathrm{p}} T_{\mathrm{f}}) = \frac{\partial}{\partial z}\left(k_{\mathrm{f}} \frac{\partial T_{\mathrm{f}}}{\partial z}\right) - \frac{4h_i (T_{\mathrm{f}} - T_{\mathrm{w}})}{d_{\mathrm{in}}} \tag{3}$$

$$\frac{\partial}{\partial t}(\rho_{\mathrm{ex}} C_{\mathrm{pex}} T_{\mathrm{ex}}) + \frac{\partial}{\partial z}(G_{\mathrm{ex}} C_{\mathrm{pex}} T_{\mathrm{ex}}) = \frac{\partial}{\partial z}\left(k_{\mathrm{ex}} \frac{\partial T_{\mathrm{ex}}}{\partial z}\right) + \frac{h_{\mathrm{ex}} A_0 (T_{\mathrm{w}} - T_{\mathrm{ex}})}{V_{\mathrm{ex}}} \tag{4}$$

The heat conduction equation for tube wall is

$$\frac{\partial}{\partial t}(\rho_{\mathrm{w}} C_{\mathrm{pw}} T_{\mathrm{w}}) = \frac{\partial}{\partial z}\left(k_{\mathrm{w}} \frac{\partial T_{\mathrm{w}}}{\partial z}\right) + \frac{h_i A_i (T_{\mathrm{f}} - T_{\mathrm{w}})}{V_{\mathrm{w}}} - \frac{h_0 A_0 (T_{\mathrm{w}} - T_{\mathrm{a}})}{V_{\mathrm{w}}} + \frac{\dot{Q}}{V_{\mathrm{w}}} \tag{5}$$

These governing equations are numerically solved, using MATLAB-based Simulink model, as we consider this problem as a one-dimensional transient conjugate heat transfer problem. For this purpose, the governing equations are discretized and solved using finite volume technique. Several closure relations are used in the numerical model to solve the governing equations which were furnished in detail in our earlier works [23–25] and not discussed here for the compactness of the paper. Validation of the model and the grid and time independence study were performed in our earlier work [23] and not discussed here for the compactness of the paper. From the grid and time independence study, we found that 5 mm grid size with 0.005 s time step is sufficient for the converged results.

3 Results and Discussions

Numerical simulations have been carried out using MATLAB-based Simulink model in single phase. We consider a square geometry (aspect ratio 1:1) with 30 mm tube diameter as our base geometry with a total loop length 4 m. In the cooler, we provide 0.04 kg/s mass flow rate with 20 °C temperature, and this is constant throughout all the simulations. To observe the long-time flow dynamics, we are showing all the results after the initial transient phase. Driving force for the fluid circulation in an NCL is generated by the temperature difference between the two vertical limbs of the NCL. Due to this temperature difference, a density difference is produced in the NCL, and this density difference causes the fluid to flow by balancing the buoyancy, friction and gravitational forces.

3.1 Effect of Variation of Aspect Ratio of the Loop

NCL (Fig. 1) in single phase which is consider as the base case for the simulations. Here, we are doing the simulation for the different aspect ratio of the loop so that we can find out the effects of the aspect ratio on the heat transfer and loop fluid flow dynamics. Therefore, for the base case aspect ratio, Horizontal: Vertical is 1:1. In this condition, we found that with the increase in heater power flow pattern changes from steady to flow reversal through the oscillatory regime, as shown in Fig. 2.

Various power plants are using NCLs as a heat transferring device for transfer of heat from one place to another in the plant. Now in the power plant, flow-reversal condition affects the heat transfer. Therefore, this is an undesirable phenomenon for the plants. So, in view of the safety of the plants, we need to suppress the flow-reversal conditions. Therefore, we are varying the aspect ratio of the geometry to see its effect on the flow dynamics. When we are changing the aspect ratio, we kept total loop length and tube diameter the same as the base condition, which is 4 m and 30 mm, respectively. Hence, to attain different aspect ratios, we need to change both the horizontal and vertical sections of the loop. From Fig. 2, we can observe that for aspect ratio (H: V) 1:1, which is the base case, flow-reversal characteristics are

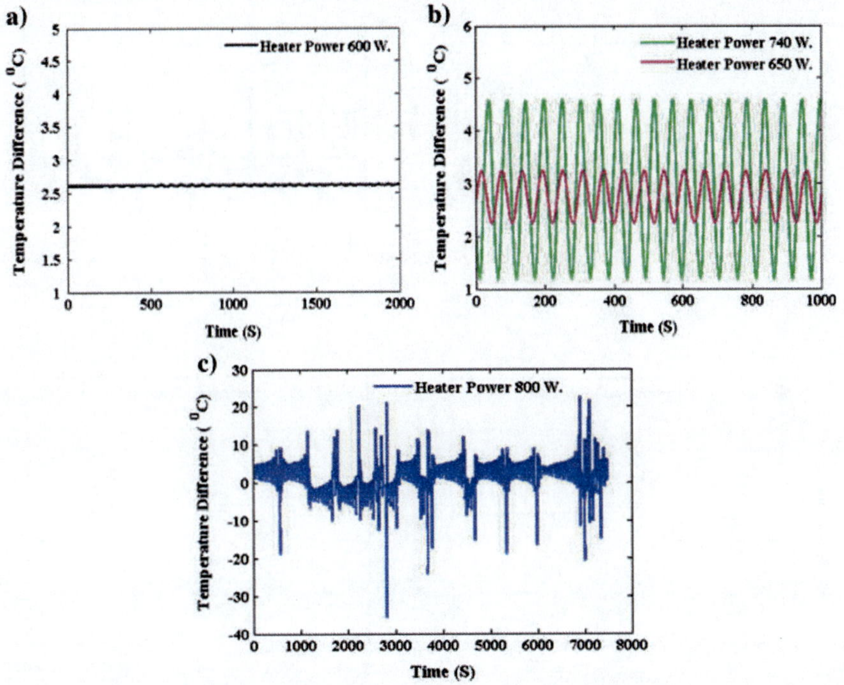

Fig. 2 Variation of temperature difference with time for **a** steady state, **b** oscillatory, **c** flow reversal condition

observed at 800 W heater power. At this heater power, when we change the aspect ratio to 1:2 or 1:3 variation of temperature difference shows similar flow-reversal characteristics, as shown in Fig. 3. But the frequency of flow reversal increases steadily. Hence, with the decrease in aspect ratio, instability of the system increases. Also, from the detailed investigation, we found that for aspect ratio 1:2 and 1:3 for any heater power in the single-phase regime, we found only flow-reversal characteristics. To attain the aspect ratio, 1:2 or 1:3 horizontal limb decreases in length and vertical limb increases in length. Due to the increase in vertical limb acceleration due to the buoyancy force decreases. Hence, the tendency of the flow-reversibility increases and the system becomes unstable for any heater power condition.

At 800 W heater power, when we change the aspect ratio to 2:1 or 3:1 variation of temperature difference shows steady-state flow, as shown in Fig. 3. With the increase in aspect ratio from 1:1 to 2:1 system became stable, and accordingly, heat transfer became better. Further, an increase in aspect ratio to 3:1, we found further increase in heat transfer. To attain the aspect ratio, 2:1 or 3:1 horizontal limb increases in length and vertical limb decreases in length, which causes an increase in acceleration due to the buoyancy force. Hence, the fluid flow became one-directional, and the system becomes stable. However, due to the increase in horizontal length flow velocity produced by the buoyancy force at relatively low heater power level is not sufficient

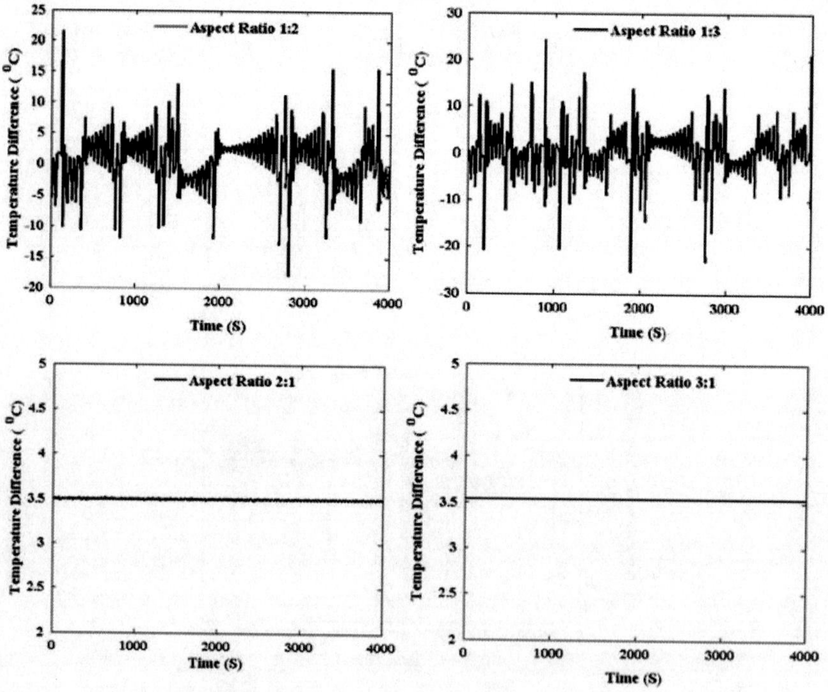

Fig. 3 Variation of temperature difference with time for different aspect ratio at 800 W

to overcome friction force. Therefore at these situations, we found no flow in the tube. From the detailed investigation, we found that after getting sufficient acceleration caused by the buoyancy force to overcome friction force flow in the NCL for aspect ratio 2:1 and 3:1 starts. Therefore for the relatively low heater powers, fluid flow is not being established and after fluid flow starts, we found only steady-state flow dynamics in these two conditions for any heater power.

3.2 *Effect of Variation of the Tube Diameter*

Here, we are doing the simulation for the different tube diameter so that we can find out the effects of the tube diameter on the heat transfer and loop fluid flow dynamics for our square NCL (Fig. 1) in single phase which is the base case for the study. For the base case, tube diameter is 30 mm. When we are changing the tube diameter, we kept total loop length same as the base condition, which is 4 m.

Variation of temperature difference with time for different tube diameter at 700 W heater power is shown in Fig. 4. From the figure, we can observe that with the increase in tube diameter flow instability increases. For 30 mm tube diameter (base case), we found that flow dynamics shows oscillatory nature. However, when we move to 40-mm-tube diameter flow dynamics changes to flow reversal. Due to the increase in tube diameter wall and liquid volume increase, which decrease the fluid temperature. Accordingly, the acceleration produced by the buoyancy force decreases, which accelerate the tendency of the flow-reversibility. Hence, instability in the system increases.

Further, an increase in tube diameter, (i.e. at 50 mm diameter) flow reversal is observed, and the occurrence of flow reversal increases in comparison with the 40-mm-tube diameter case. Due to the change in flow dynamics, heat transfer is also affected. From a detailed investigation, we also find that for the base case three

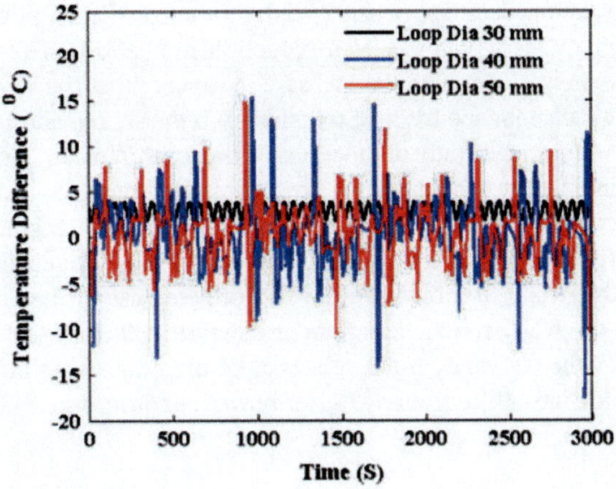

Fig. 4 Variation of temperature difference with time for different tube diameter at 700 W

types of flow dynamics (steady, oscillatory, flow reversal) can be observed with the increase in heater power (Fig. 2) but for the 40 and 50 mm tube diameter only flow-reversal phenomenon can be observed. If we decrease the tube diameter from the base condition (30 mm), we find that flow is not being established due to the high friction force.

Further, from a detailed investigation, we found that for a square geometry total loop length to diameter ratio (*L*/*D*) has a significant role in flow dynamics. We found that if the *L*/*D* ratio is equal or greater than 200, then due to the strong friction force flow is not being established at any heater power condition. Also, we found that if the *L*/*D* ratio is less than or equal to 100 due to the weak acceleration produced by the buoyancy force, tendency of the flow-reversibility increases which produce flow instability for any heater power condition.

4 Conclusion

Numerical simulations are being carried out using MATLAB-based Simulink model in single phase to study the effect of the aspect ratio of the loop, and tube diameter on the heat transfer and the flow dynamics of the loop. For this purpose, a square geometry (aspect ratio 1:1) with 30 mm tube diameter has been considered as a base geometry and condition as the base case with whom we compare the results. From the flow dynamics of the base geometry, it is found that three types of flow dynamics occurred with the increase in heater power. For relatively low heater power, we found steady-state flow dynamics, and then with the increase in power, we found an oscillatory flow, and for further increase in heater power, we found flow reversal.

From the study of the effect of the aspect ratio of the loop, it can be concluded that with the decrease in aspect ratio, instability of the system increases and with the increase in aspect ratio, the stability of the system increases. But for the high aspect ratio flow velocity produced by the buoyancy force at relatively low heater power level is not sufficient to overcome friction force, which causes no flow in the tube. With the decrease in aspect ratio, for a fixed heater power frequency of flow reversal increases which causes a reduction in heater transfer from the source and affects the plant. But with the increase in aspect ratio, the stability of the system increases, which enhance the heat transfer from the source and benefit the plant.

From the study of the effect of the tube diameter, it can be concluded that with the increase in tube diameter flow instability increases. Further, from the study, we can conclude that for a square geometry, total loop length to diameter ratio (L/D) has a significant role in flow dynamics. If L/D ratio is equal or greater than 200, then due to the strong friction force flow is not being established at any heater power condition and if *L*/*D* ratio is less than or equal to 100 due to the weak acceleration produced by the buoyancy force, tendency of the flow-reversibility increases which produces flow instability for any heater power condition.

Acknowledgements The first author owes to the UGC for issuing the fellowship under UPE - Phase-II Program.

References

1. Sinha RK, Kakodkar A (2006) Design and development of the AHWR—the Indian thorium fuelled innovative nuclear reactor. Nucl Eng Des 236:683–700
2. Basu DN, Bhattacharyya S, Das PK (2014) A review of modern advances in analyses and applications of single-phase natural circulation loop in nuclear thermal hydraulics. Nucl Eng Des 280:326–348
3. Misale M, Frogheri M, D'auria F, Fontani E, Garcia A (1999) Analysis of single-phase natural circulation experiments by system codes. Int J Thermal Sci 38:977–983
4. Jiang YY, Shoji M, Naruse M (2002) Boundary condition effects on the flow stability in a toroidal thermosyphon. Int J Heat Fluid Flow 23:81–91
5. Lee JD, Pan C (2005) Nonlinear analysis for a nuclear-coupled two-phase natural circulation loop. Nucl Eng Des 235:613–626
6. Lee JD, Pan C, Chen SW (2015) Nonlinear dynamic analysis of a two-phase natural circulation loop with multiple nuclear-coupled boiling channels. Ann Nucl Energy 80:77–94
7. DurgaPrasad GV, Pandey M, Pradhan SK, Gupta SK (2008) Study of flow instabilities in double-channel natural circulation boiling systems. Nucl Eng Des 238:1750–1761
8. Kozmenkov Y, Rohde U, Manera A (2012) Validation of the RELAP5 code for the modelling of flashing-induced instabilities under natural-circulation conditions using experimental data from the CIRCUS test facility. Nucl Eng Des 243:168–175
9. Hwang DH, Kwon H, Kim SJ (2015) Investigation of two-phase flow instabilities under advanced PWR conditions. Ann Nucl Energy 80:135–143
10. Rabiee A, Mirzaee MM, Nematollahi MR, Atf A (2016) Experimental and numerical investigation of natural circulation stability of the SHUNCL thermal-hydraulic loop. Prog Nucl Energy 93:386–396
11. Misale M (2016) Experimental study on the influence of power steps on the thermohydraulic behaviour of a natural circulation loop. Int J Heat Mass Transf 2016(99):782–791
12. Pini A, Cammi A, Luzzi L (2016) Analytical and numerical investigation of the heat exchange effect on the dynamic behaviour of natural circulation with internally heated fluids. Chem Eng Sci 145:108–125
13. Mishra AM, Singh S (2016) Non-linear stability analysis of parallel channels with natural circulation. Nucl Eng Des 309:136–150
14. Cong T, Peng M, Zhang Z (2017) Preliminary application of CFD methodology on analysis of the single-phase laminar natural circulation under swing conditions. Prog Nucl Energy 97:1–10
15. Lei W, Yang L, Hai-Jun J, Jun W (2017) Innovative flow-resistance performance in the single-phase natural circulation loop and relevant experiment verification. Int J Heat Mass Transf 107:66–73
16. Krishnani M, Basu DN (2017) Computational stability appraisal of rectangular natural circulation loop : effect of loop inclination. Ann Nucl Energy 107:17–30
17. Cammi A, Misale M, Devia F, Cauzzi MT, Pini A, Luzzi L (2017) Stability analysis by means of information entropy : assessment of a novel method against natural circulation experimental data. Chem Eng Sci 166:220–234
18. Cammi A, Luzzi L, Pini A (2016) The influence of the wall thermal inertia over a single-phase natural convection loop with internally heated fluids. Chem Eng Sci 153:411–433
19. Luzzi L, Misale M, Devia F, Pini A, Cauzzi MT, Fanale F, Cammi A (2017) Assessment of analytical and numerical models on experimental data for the study of single-phase natural circulation dynamics in a vertical loop. Chem Eng Sci 162:262–283

20. Borreani W, Chersola D, Lomonaco G, Misale M (2017) Assessment of a 2D CFD Model for a single-phase natural circulation loop. Int J Heat Technol 35:S300–S306
21. Goudarzi N, Talebi S (2018) Heat removal ability for different orientations of single-phase natural circulation loops using the entransy method. Ann Nucl Energy 111:509–522
22. Saha R, Ghosh K, Mukhopadhyay A, Sen S (2018) Dynamic characterization of a single-phase square natural circulation loop. Appl Therm Eng 128:1126–1138
23. Saha R, Sen S, Mookherjee S, Ghosh K, Mukhopadhyay A, Sanyal D (2015) Experimental and numerical investigation of a single-phase square natural circulation loop. J Heat Transf 137:121010-1–121010-8
24. Saha R (2013) Experimental and numerical study of a single-phase square natural circulation loop. Jadavpur University, Kolkata
25. Saha R, Ghosh K, Mukhopadhyay A, Sen S (2017) Flow dynamics study of a single-phase square NCL using recurrence plot and recurrence quantification. Indian Acad Sci Conf Ser 1:67–75

Refrigeration and Air Conditioning

A Comparative Analysis Between Indoor and Outdoor Thermal Comfort Parameters of Railway Pantry Car

Md. Sarfaraz Alam, Arunachalam Muthiah, and Urmi Salve

Abstract Indian railway pantry car kitchen plays a vital contribution to the catering system, which provides the meal to every onboard passenger. It is an integral part of every "long and short distance train." The objective of this study was to compare the indoor and outdoor thermal comfort parameters of the Indian railway pantry car kitchen. Data were collected on 14 railway pantry car coaches for field measurements during summer and winter season. Descriptive analysis and Mann–Whitney test were performed in this study. A descriptive analysis result of this study revealed that the measured indoor thermal comfort parameters found more as compare to outdoor conditions except for air velocity cases. The statistical analysis result shows that there is a significant difference between air temperature ($p < 0.05$) and globe temperature ($p < 0.05$) values in indoor and outdoor conditions. There is no significant difference found in the case of relative humidity ($p > 0.05$) in both conditions. However, in the case of air velocity ($p > 0.05$), outdoor condition exceeds than indoor condition. Further, the design intervention of workplace would be suggested to improve the thermal comfort of the chef.

Keywords Thermal comfort · Railway pantry car · Mann–Whitney test

Nomenclature

T_a (°C)	Air temperature
T_g (°C)	Globe temperature
RH (%)	Relative humidity
v_a (m/s)	Air velocity
Z	Standard normal distribution

Md. . Sarfaraz Alam (✉) · A. Muthiah · U. Salve
Department of Design, Indian Institute of Technology Guwahati, Guwahati 781039, India
e-mail: sarfaraz8665@gmail.com

S. Revankar et al. (eds.), *Proceedings of International Conference on Thermofluids*, Lecture Notes in Mechanical Engineering,
https://doi.org/10.1007/978-981-15-7831-1_37

1 Introduction

Thermal comfort refers to the "thermal balance" between the "human body" and the environment and also the appropriate balance between "body heat production and heat dissipation" [1, 2], and it contains a critical impact on the human body. Six major factors that influence human thermal comfort, including four environmental factors, viz., "air temperature, globe temperature, relative humidity, and air velocity." Whereas two individual factors, viz., clothing level and metabolic rate [3, 4]. Thermal comfort investigation commonly has done in the field of the (i) commercial building environment, viz., residential, hospital, school, and office, etc.; and (ii) passengers comfort in the vehicle, viz., car, bus, train, and aeroplane [5–7]. However, some kinds of studies have established in the commercial kitchen environment, such as restaurant kitchen, hotel kitchen, and rural house kitchen in both developing and developed countries [8, 9]. In which, evaluated thermal environmental factors, personal factors, and thermal perception of workers during work periods. But no researchers have yet focused on the "thermal environment of the railway pantry kitchen." However, Alam et al. [10] have setup an initial research on the "thermal comfort of chefs in pantry car kitchens," but they investigated with few sample sizes. The "Indian railway pantry car" is a particular type of kitchen that prepares the food of the onboard passengers [11]. It has an essential role in the Indian railway catering system, which is an integrated part of every long and short distance trains running across the country. Currently, 338 trains are running with pantry car coaches. A pantry car customized railway coach is the working space for 3–5 chefs, 40–50 waiters, and 2–3 [10, 12]. In the present scenario, there are two types of the railway pantry coach model available, namely (i) ICF "Integral Coach Factory" and (ii) LHB "Linke Hofmann Busch." LHB pantry car coach model has air conditioning facilities. Whereas ICF pantry car coach models are non-air-conditioned [12, 13]. Although according to Alam et al. [10] in terms of the thermal environment at the time of cooking, there is no significant difference between "air-conditioned and non-air-conditioned" pantry car coaches based on the physical measurement evaluation. Chefs use the same type of appliances for cooking in both pantry car coaches like "gas stove," "heater," "oven," "kettle," "deep fryers," "soup warmer," etc. Such types of appliances lead to generated heat and humidity at period of "food preparation inside the pantry car due to this indoor environment of the pantry car" most of the time hot and humid [14]. These types of hot and humid environments create an uncomfortable working situation for pantry car chefs and workers, which will directly affect work efficiency and productivity.

From the above literature, it is apparent that thermal comfort is required for railway pantry car chefs. The present study aims to evaluate thermal comfort parameters in the Indian railway pantry car. Therefore, the objective of this study was to compare the indoor and outdoor thermal comfort parameters of the Indian railway pantry car kitchen.

2 Method

2.1 Sample and Period

In this study, 14 Indian railway pantry car coaches were taken for field investigation, in which six pantry car coaches taken “during the summer season,” and eight pantry car coaches taken “during the winter season.” While “two climatic zones, viz., humid and subtropical, and tropical wet and dry” considered according to the “Koppen climate classification system.” The work activity level of the “chefs in the pantry car kitchen” is very high. Therefore, this study was selected for 15 days period for both seasons. The data collected for “the summer season in August,” while December month considered for “the winter season.” In this research, “both seasons of data collected in 2018.”

2.2 Field Measurement

In this study, all four environmental factors of thermal comfort parameters in both indoor and outdoor conditions have measured. The air temperature (T_a), relative humidity (RH), and air velocity (v_a) were measured with the help of Kestrel anemometer 4500, while globe temperature (T_g) was “measured with the help of 6-in. black-globe thermometer.” All thermal comfort parameters have been “measured according to ASHRAE 55 Standard” [15], in which T_a, RH, v_a “measured at 1 ft (0.3 m) near the workstation (cooking zone)” and “43-in. (1.1 m) above the workstation floor in the pantry car kitchen.” While “T_g” was measured in the “center position of the cooking zone.” Indoor and outdoor “Ta” measured with the instrument range “−29.0 to + 70.0 °C and 1 °C accuracy.” Similarly, indoor and outdoor “RH” measured with the instrument range 5–95% and ±3% accuracy. While indoor and outdoor “v_a” measured with the instrument range 0.4–60 m/s and ±0.1 m/s accuracy. However, “T_g” measured with the instrument range “−5° to + 95 °C” in both “indoor and outdoor conditions.” All indoor and outdoor data of “thermal comfort parameters” were recorded at the time of cooking period; “breakfast—7:00 am, lunch—11:30 am, snack—4:00 pm, and dinner—6:30 pm.”

3 Result and Discussion

In this study, for comparison between indoor and outdoor thermal comfort parameters based on the descriptive analysis and the Mann–Whitney test, Eq. (1) was used. However, for all statistical analysis, SPS2016 version software was incorporated.

$$r = 1 - \frac{2U_1}{n_1 n_2} \tag{1}$$

where
r is rank-biserial; *u* is unfavorable proportion; *n* is number of pairs.

3.1 Indoor and Outdoor Environment

All "thermal comfort" environmental parameters carried out in this research throughout the summer and winter seasons in "humid and subtropical and tropical wet and dry" climatic zones. During the field measurement, the average value of indoor thermal comfort parameters was calculated, such as; T_a (30.42 °C), T_g (28.68 °C), RH (68.98%), and v_a (0.03 m/s). Similarly, the average value of outdoor thermal comfort parameters was calculated, such as T_a (25.78 °C), T_g (25.94 °C), RH (66.84%), and v_a (1.49 m/s). From the descriptive analysis, it shows that the measured indoor thermal comfort parameters found more as compare to the outdoor thermal comfort parameters except for the air velocity case. Both conditions measured thermal comfort parameters that do not comply the ASHRAE 55 standards [15]. According to the ASHRAE, comfort temperature range for subjects from 21 to 26 °C. While "comfortable range of RH between 30 and 60%, optimum range of RH between 40 and 60% and considered ideal RH for 50%" [16]. Similarly, "air velocity should be for summer (<0.25 m/s) and winter (<0.15 m/s)" according to ASHRAE 55 standard [17, 18].

Significant differences between indoor and outdoor thermal comfort parameters have been investigated in this study using the Mann–Whitney test. The result of Mann–Whitney test is depicted in Tables 1 and 2.

For air temperature (T_a) case, test indicated, on average, that the "T_a" values of the indoor conditions (*Mean Rank* = 71.63, *n* = 56) significantly exceed of the outdoor conditions (Mean Rank = 41.37, *n* = 56), $U = 720.500$, $z = -4.938$, $p = 0.000$, two tailed. Similarly, for globe temperature (T_g) case, test indicated, on average, that

Table 1 Mann–Whitney test ranks for both indoor and outdoor thermal comfort parameters

Thermal comfort variables	Condition	*n*	Mean rank	Sum of ranks
T_a(°C)	Indoor	56	71.63	4011.50
	Outdoor	56	41.37	2316.50
T_g(°C)	Indoor	56	66.42	3719.50
	Outdoor	56	46.58	2608.50
RH%	Indoor	56	57.31	3209.50
	Outdoor	56	55.69	3118.50
v_a (m/s)	Indoor	56	32.96	1845.50
	Outdoor	56	80.04	4482.50

Table 2 Mann–Whitney test statistics for thermal comfort parameters

	T_a(°C)	T_g(°C)	RH%	v_a (m/s)
Mann–Whitney U	720.500	1012.500	1522.500	249.500
Wilcoxon W	2316.500	2608.500	3118.500	1845.500
Z	−4.938	−3.239	−0.265	−8.093
Asymp. Sig. (2-tailed)	0.000	0.001	0.791	0.000

the "T_g" values of the indoor conditions (*Mean Rank* = 66.42, $n = 56$) significantly exceed of the outdoor conditions (*Mean Rank* = 46.58, $n = 56$), $U = 1012.500$, $z = -3.239$, $p = 0.001$, two tailed. While for relative humidity (RH) case, test indicated that there no difference between indoor condition (*Mean Rank* = 57.31, $n = 56$) and outdoor condition (*Mean Rank* = 55.69, $n = 56$) significantly, $U = 1522.500$, $z = -0.265$, $p = 0.791$, two tailed. However, for air velocity (v_a) case, test indicated, on average, that the "v_a" values of the outdoor conditions (*Mean Rank* = 80.04, $n = 56$) significantly exceed of the indoor conditions (*Mean Rank* = 32.96, $n = 56$), $U = 249.500$, $z = -8.093$, $p = 0.001$, two tailed.

The statistical test result revealed that there was a significant difference in air temperature, globe temperature between both "indoor and outdoor environmental conditions" of the pantry cars. Whereas, there was no considerable difference in "relative humidity in both" conditions. Although outdoor condition air velocity was significantly found more as compare to the indoor condition air velocity of the pantry car. As distinguished in the previous study, most of the cooks closed the party car window and door during the cooking period [10, 18]. Therefore, very little air movement is found inside the pantry as it creates a problem for food preparation. Due to this the "indoor environment becomes hot and humid" of the pantry cars.

4 Conclusion

This study attempt to understand the current situation of the "indoor and outdoor" thermal environmental condition of "Indian railway pantry car kitchen" during the working period. The field measurement of the present research was carried out throughout the winter and summer seasons. The analysis of the survey results is as follows:

- The measured thermal comfort parameter in all indoor and outdoor environmental conditions does not comply with the recommended limits of the ASHRAE 55 standard.
- Based on statistical analysis, there is difference in the air temperature and globe temperature values of "indoor and outdoor conditions." However, in the case of relative humidity, there is no difference between "indoor and outdoor conditions."

While in the case of air velocity, outdoor values significantly found more as compare to indoor values.

Only physical parameters were considered in this study; further, subjective perception needs to be evaluated. These findings may have an important role in the future to establish comfort zones for improved performance of pantry car chefs.

References

1. Song G (2009) Thermal insulation properties of textiles and clothing. Text Cold Weather Apparel, pp 19–32
2. Rahmillah FI, Tumanggor AHU, Sari AD (2017) The analysis of thermal comfort in kitchen. IOP Conf Ser Mater Sci Eng 215(1):012033
3. Pinto NM, Xavier AAP, Hatakeyama K (2015) Thermal comfort in industrial environment: conditions and parameters. Procedia Manuf 3:4999–5006
4. Simion M, Socaciu L, Unguresan P (2016) Factors which influence the thermal comfort inside of vehicles. Energy Procedia 85:472–480
5. Teli D, James PA, Jentsch MF (2013) Thermal comfort in naturally ventilated primary school classrooms. Build Res Inf 41(3):301–316
6. Spentzou E, Cook MJ, Emmitt S (2018) Natural ventilation strategies for indoor thermal comfort in Mediterranean apartments. Build Simul 11(1):175–191
7. Neacsu C, Tabacu I, Ivanescu M, Vieru I (2017) The evaluation of the overall thermal comfort inside a vehicle. IOP Conf Ser Mater Sci Eng 252(1):012031
8. Ravindra K, Agarwal N, Kaur-Sidhu M, Mor S (2019) Appraisal of thermal comfort in rural household kitchens of Punjab, India and adaptation strategies for better health. Environ Int 124:431–440
9. Simone A, Olesen BW, Stoops JL, Watkins AW (2013) Thermal comfort in commercial kitchens (RP-1469): procedure and physical measurements (Part 1). Hvac&r Res 19(8):1001–1015
10. Alam MS, Arunachalam M, Salve UR (2019) A pilot study on thermal comfort in Indian Railway pantry car chefs. J Phys Conf Ser 1240(1):012033
11. Patil P, Mukul S, Mathur S (2012) How to improve catering services in Indian Railways. Int Union Railw, 7–52
12. Pethkar G, Jadhav S, Dhumal P, Bhosale P, Ahir PR (2015) Automated pantry car system in India using RF module and GSM technology. Int J Adv Res in Electron Commun Eng 4(4):794–797
13. Alam MS, Arunachalam M, Salve U (2017) A comparative study of thermal environment between two varieties of pantry car available in Indian Railway. In: 15th international conference on humanizing work and work environment. Department of Mechanical Engineering, Aligarh Muslim University, Aligarh, India, pp 21–27
14. Matsuzuki H, Ito A, Ayabe M, Haruyama Y, Tomita S, Katamoto S, Muto T (2011) The effects of work environments on thermal strain on workers in commercial kitchens. Ind Health 49(5):605–613
15. American Society of Heating, Refrigerating and Air-Conditioning Engineers (2010) Thermal environmental conditions for human occupancy. ASHRAE Standard-55, Atlanta
16. American Society of Heating, Refrigerating and Air-Conditioning Engineers (2016) Ventilation for acceptable indoor air quality. ASHRAE Standard-55, Atlanta
17. American Society of Heating, Refrigerating and Air-Conditioning Engineers. Thermal environmental conditions for human occupancy, Atlanta: ASHRAE Standard-55 (2004)
18. Alam MS, Muthiah A, Salve UR (2019) Thermal comfort of the kitchen in pantry cars on Indian railways. Instrum Mesure Métrol 18(5):465–477

Performance Optimization of an Air Conditioning System at Different Mass Flow Rates and Temperatures

Taliv Hussain, Adnan Hafiz, Sameer Raza, Abdur Rahman, Tabrez Ahmed, and Pragati Agarwal

Abstract The objective of this paper is to find out the optimum flow rate to achieve optimum COP. In order to achieve the objective, the duct was fabricated, and a blower was attached at inlet of the duct. The vapour compression refrigeration system and duct were combined together. Different flow rate were measured for different ambient temperatures. The flow rates (cfm) were selected as 173.97, 212.45, 260.67, 287.09 and 337.52 for ambient temperatures 30, 32, 34, 36 and 38 °C. The outlet temperature and specific humidity were also monitored. The optimum COP was found to be 2.83 at flow rate (cfm) 287.09 and at 36 °C.

Keywords Air conditioning duct · Ambient temperature · Flow rate and COP

1 Introduction

Refrigeration, cooling and heating processes are crucial in a number of everyday situations including the air conditioning and heating of buildings, hotels, restaurants, hospitals, operation theatres, automobiles and transportation. Refrigeration also includes large-scale industrial implementation, particularly in the produce of ice, dehydration of gases, commercial and domestic refrigerators, large-scale warehouses for preservation and storage of foods and a host of other industrial and commercial services [1]. Usage of refrigerants in the petroleum industry include separation of volatile hydrocarbons, lubricating-oil purification and low-temperature reactions. Condensing and evaporation processes in refrigeration structure are due to the heat transfer occurring by way of phase change in refrigerants. Therefore, the model of a cooling system mainly based on the properties of the refrigerants. As the standard of living is improving, more and more people search for a comfortable living. Since from the last two decades air conditioner has become an effective comfort providing device [2].

T. Hussain (✉) · A. Hafiz · S. Raza · A. Rahman · T. Ahmed · P. Agarwal
Department of Mechanical Engineering, A.M.U, Aligarh, India
e-mail: hussaintaliv@gmail.com

S. Revankar et al. (eds.), *Proceedings of International Conference on Thermofluids*, Lecture Notes in Mechanical Engineering,
https://doi.org/10.1007/978-981-15-7831-1_38

Foremost objective of an air conditioning system is to provide the highest comfort to the entire air conditioning zone. To obtain this objective, an analysis of performance of the cooling machine will be eventually inadequate. Scope should be broadening to an analysis of airflow over the whole air conditioning apparatus. Especially in hot climate, air conditioning consumes a huge amount of electrical energy. Working of an air conditioner at raised ambient temperatures results in a lesser coefficient of performance (COP). This closure comes directly from surveying the carnot cycle. The COP relation, $COP = T_{evap}/(T_{cond} - T_{evap})$, shows that there is decrement in COP when there is increment in the condenser temperature at a constant evaporation temperature. For the refrigerants operating in the vapour compression cycle, the COP degradation is higher than that for the carnot cycle, and it differs among fluids. The two most convincing fundamental thermodynamic properties which affect this COP degradation are a refrigerant's critical temperature and molar heat capacity. For a given application, a fluid tend to have a lower COP at a lower critical temperature [3]. The humidity and temperature of the air in buildings need to be controlled to prevent mould growth, maintain human comfort and ensure building durability. This requires the removal and addition of latent and sensible energy, which means that the air in buildings requires cooling, heating, dehumidifying and humidifying to varying degrees depending on the position of the building. Recent research in the heating, ventilation and air conditioning (HVAC) field has refined the techniques for conditioning this air, but it is typically done in the same way as it was done earlier: vapour compression heat pumps, electric resistance heaters, heating with direct combustion and cooling and dehumidifying with vapour compression air conditioners [4].

There are few problems about these current practices, concerns that must be communicated in any sustainable energy future. In USA, conditioning air accounts for 48% of the main energy used in buildings [5]. Cooling and dehumidification mostly from vapour compression systems account for a notable part of the peak electric requirement in hot climates. Controlling humidity with vapour compression systems is also becoming much hard as energy improvements often lower the building's sensible load but do not affect the latent load [6, 7]. Researchers are pursuing substitutes to these previous exercises especially for cooling and dehumidification [8, 9]. In this area, advancements in artificial membranes enable new possibilities. While for industrial separations traditionally used, such as reverse osmosis and gas separation, membranes provide a mode to selectively transfer water vapour from one fluid to another fluid [10, 11]. The coupling between sensible and latent energy enables a variety of potentially energy-efficient HVAC processes. Extensive works have already been performed, when some concentrate on improving efficiencies of the components including heat exchanger and compressor while others focus by optimizing the cycle mode on ameliorating the overall cycle efficiency. Nikolaidis and Probert [12] measured analytically the effect of change in condenser and evaporator temperatures using R22 of two-stage vapour compression refrigeration plant. Devotta et al. [13] evaluated the suitability of few selected fluids as alternatives to R22 for air conditioners. The refrigerants studied are HC290, R407c, R134a, R410a and three blends of R134a, R32 and R125. Evaporator and condenser temperatures were fixed, neglecting the pressure drops. Pressure ratio of R407c was more than that of R22 for the range of evaporator temperatures due to higher discharge pressure. They also have described that discharge temperature of R407c was considerably lower than that of R22. Specific compressor displacement of R407c was same as compare to R22. Halimic et al. [14] had compared the performance of R290, R410a and R134a with R12 by using in vapour compression refrigeration system which was originally designed for R12. Due to the similar performance of R134a in comparison with R12, R134a can be substituted in the same system without any addition in the system components. But in reference to global warming potential, R410a gave out better results [15]. Spatz and Motta [16] had mainly concentrated on replacement of R12 with R410a through experimental investigation of medium-temperature vapour compression refrigeration cycles. In terms of thermodynamic analysis, comparison of pressure drop and heat transfer characteristics, R410a provided best performance as that of R12, R404a and R290a. Cabello et al. [17] had studied about the effect of operating parameters on first law efficiency (COP), cooling capacity and work input of single-stage vapour compression refrigeration system. There was great influence on energetic parameters due to change in suction pressure, evaporating and condensing temperatures. Venkataramanamurthy et al. [18] conducted an experimental test for the investigation of the comparisons of energy flow and second law efficiency of R22 and its substitutes R-436b in vapour compression refrigeration system.

2 Experimental Section

Figure 1 shows the experimental set-up. It consists of a compressor, condenser, expansion device and evaporator. The refrigerant used in the set-up is R-410a. Figure 2 shows blower attachment with the duct. Figure 3 represents the different measuring devices used in the set-up. Figure 4 shows the condenser fan. Figure 3 shows the duct system model. A duct system is a path for air flow, and so duct delivers cool air to all areas in the home.

The various formulae which are used for the calculation are as given through Eqs. 1, 2 and 3:

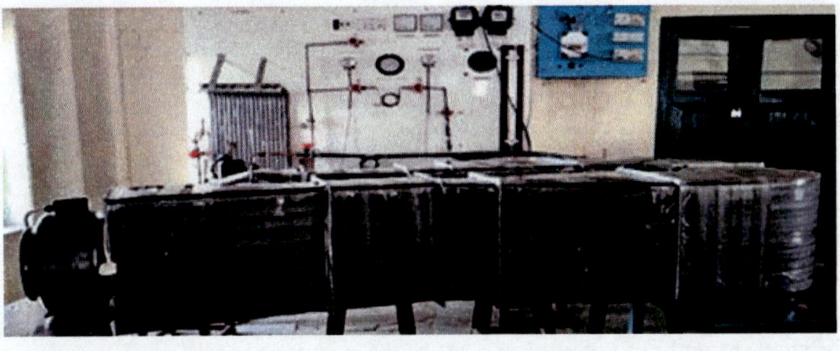

Fig. 1 Experimental set-up

Fig. 2 Blower attached to a duct

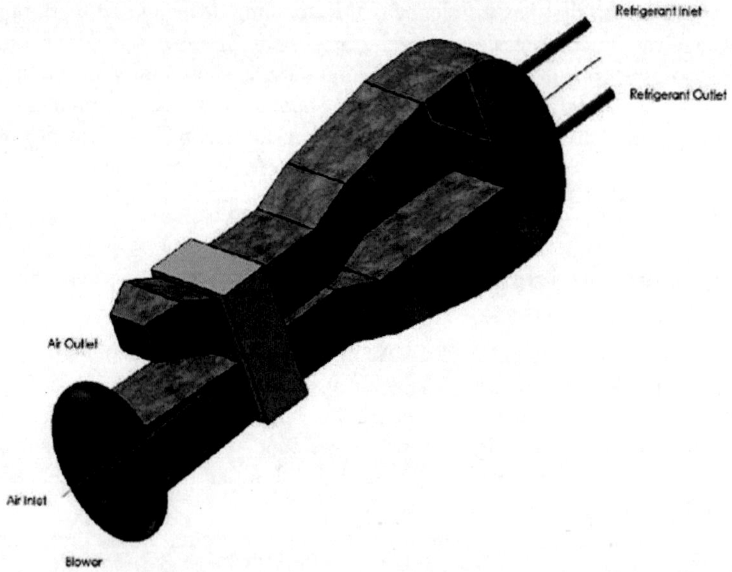

Fig. 3 Duct system model

$$\text{Compressor Work, } W = (h_2 - h_1) \tag{1}$$

$$\text{Cooling effect produced, } Q = (h_1 - h_4) \tag{2}$$

$$\text{COP} = Q/W \tag{3}$$

where

h_1Refrigerant enthalpy at entrance of the compressor in kJ/kg;

Table 1 Instrumentation details

Energy meter	Used for measuring the power consumption of the heater and compressor
Ammeter and voltmeter	Used for monitoring the power input to the compressor
Pressure gauges	Used for measuring the discharge and suction pressure
Rotameter	Used for the measurement of volume flow rate of the refrigerant
Thermometers	Used for measuring the temperatures at different state points

h_2Refrigerant enthalpy at end of the compressor in kJ/kg;
h_3Refrigerant enthalpy at end of the condenser kJ/kg;
h_4Refrigerant enthalpy at entrance of the evaporator in kJ/kg (Table 1).

3 Results and Discussion

In order to improve the COP of our air conditioning unit using the refrigerant R410a, we have carried out the experiment at various air flow rates at different ambient temperatures. We have compared the COP of the system at different flow rates and ambient temperature. We have also monitored the outlet temperatures and specific humidity of the duct at different flow rates.

3.1 Effect of Flow Rate on the COP

The above graph Fig. 4 shows the variation of coefficient of performance with the flow rate. As the flow rate increases, the coefficient of performance increases for each ambient temperature. The optimum coefficient of performance is obtained at 287.9 CFM. With increase in flow rate of air passing through evaporator, the more amount of heat is given to evaporator, so coefficient of performance increases, but the ratio

Fig. 4 Variation of COP with flow rate

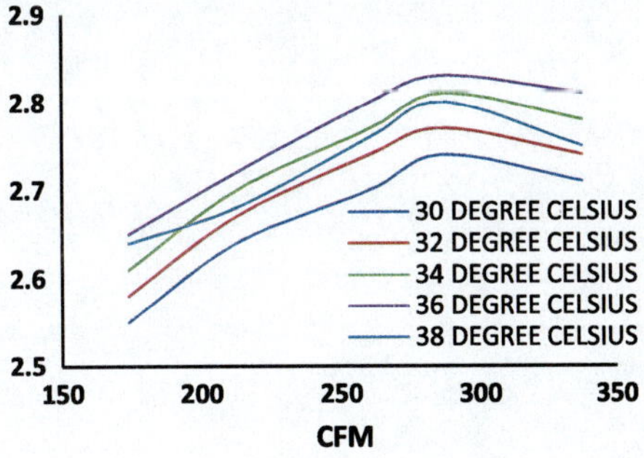

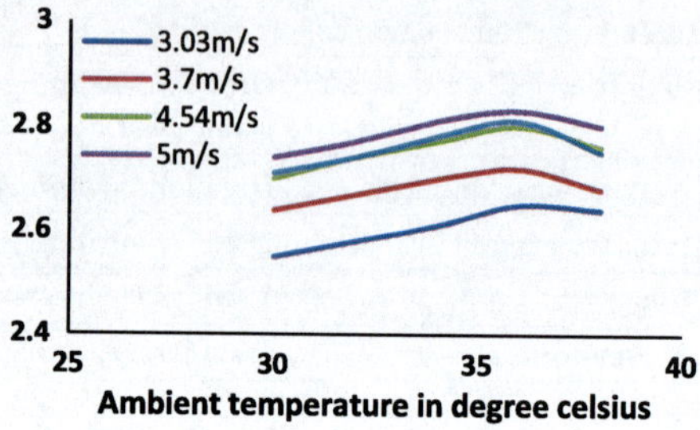

Fig. 5 Variation of COP with ambient temperature

of amount of air in contact with the cooling coil to the total amount decreases, so an optimum flow rate is obtained.

3.2 *Effect of Ambient Temperature on the COP*

Figure 5 shows the graph between coefficient of performance and ambient temperature. With increase in ambient temperature, the temperature difference between the cooling coil and ambient temperature increases, there will be more effective heat exchange, and hence, coefficient of performance increases.

3.3 *Effect of Flow Rate on Outlet Temperature of Air*

Figure 6 shows the variation of outlet temperature of air with volume flow rate of air. With increase in volume flow rate, conditioned air temperature increases because bypass factor increases with increases in flow rate.

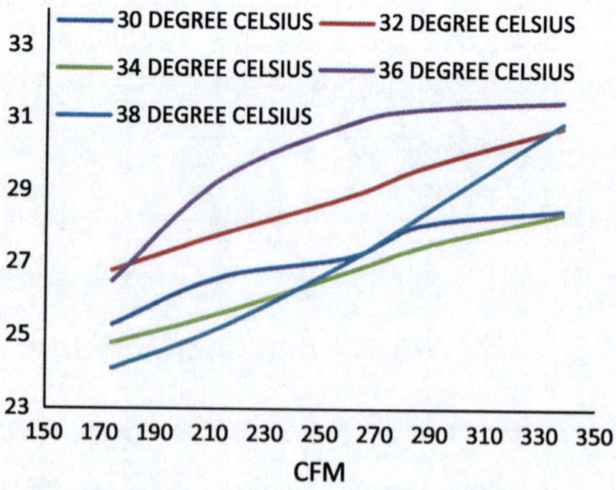

Fig. 6 Variation of outlet temperature of air with flow rate

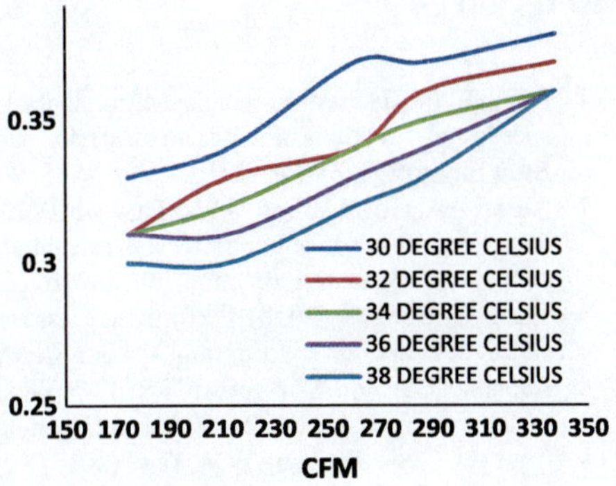

Fig. 7 Variation of bypass factor with flow rate

3.4 *Effect of Flow Rate on Bypass Factor*

Figure 7 shows the plot of bypass factor with flow rate. With the increase of flow rate of air, the more amount of air passes un-touching the coil.

With increase in flow rate of air passing through evaporator, the more amount of heat is given to evaporator, so coefficient of performance increases, but the ratio of amount of air in contact with the cooling coil to the total amount of decreases, so an optimum flow rate is obtained. With increase in ambient temperature, the temperature difference between the cooling coil and ambient temperature increases, there will be more effective heat exchange, and hence, coefficient of performance increases.

4 Conclusions

Air conditioning units are used in residential as well as industrials buildings. More than 25% of the building energy costs in HVAC systems. So there is a need of more efficient HVAC systems. On the basic of results obtained by the experiments, the following conclusions can be drawn:

- As the flow rate increases, the coefficient of performance increases for each ambient temperature. The optimum coefficient of performance is obtained at 287.90 CFM.
- With increase in volume flow rate, conditioned air temperature increases because bypass factor increases with increases in flow rate. However, with the increase of flow rate of air, the more amount of air passes without touching the coil.

References

1. Doiphode P, Jadhav A, Kumar M, Samanta I (2019) Investigating the influence of different speeds and directions of winds on the performance of split air conditioning systems. Sci Technol Built Environ 25(3):356–371
2. Ansari AA, Goyal V, Yahya SM, Hussain T (2018) Experimental investigation for performance enhancement of a vapor compression refrigeration system by employing several types of water-cooled condenser. Sci Technol Built Environ 24(7):793–802
3. Wang G, Song L (2018) Performance assessment of variable frequency drives in heating, ventilation, and air-conditioning systems. Sci Technol Built Environ 24(10):1075–1083
4. Kapadia RG, Jain S, Agarwal RS (2009) Transient characteristics of split air-conditioning systems using R-22 and R-410A as refrigerants. HVAC R Res 15(3):617–649
5. Chua KJ, Chou SK, Yang WM, Yan J (2013) Achieving better energy-efficient air conditioning–a review of technologies and strategies. Appl Energy 104:87–104
6. Ansari AA, Khan F, Chaturvedi P, Goyal V, Raj A, Hussain T (2018) Theoretical and experimental performance evaluation of vapour compression refrigeration system using organic and inorganic pads in evaporative cooled condenser. Adv Sci Eng Med 10(3–4):374–378
7. Doiphode P, Jadhav A, Kumar M, Samanta I (2018) Thermo-flow performance analysis of split air conditioner under high-speed winds around buildings. Adv Build Energy Res, 1–24
8. Hussain T, Khan F, Ansari AA, Chaturvedi P, Yahya SM (2018) Performance improvement of vapour compression refrigeration system using Al2O3 nanofluid. IOP Conf Ser Mater Sci Eng 377(1):012155
9. Mairaj M, Siddiqui SA, Hafiz A (2015) Energetic and exergetic analysis of some models of vapour absorption chillers using lithium bromide and water. J Basic Appl Eng Res 2(4):326–329
10. Hussain T (2019) Experimental investigation of an air conditioner unit by varying different air flow rates using air and evaporatively cooled condenser. IOP Conf Ser Mater Sci Eng 691(1):012027
11. Ansari MA, Hussain T, Khan SA, Mustafa S, Goyal V (2019) Experimental investigation on VCRS by using R-134a and R-410a of air and evaporative cooled condenser. IOP Conf Ser Mater Sci Eng 691(1):012071
12. Nikolaidis C, Probert D (1998) Exergy-method analysis of a two-stage vapour-compression refrigeration-plants performance. Appl Energy 60(4):241–256
13. Devotta S, Waghmare AV, Sawant NN, Domkundwar BM (2001) Alternatives to HCFC-22 for air conditioners. Appl Therm Eng 21(6):703–715
14. Halimic E, Ross D, Agnew B, Anderson A, Potts I (2003) A comparison of the operating performance of alternative refrigerants. Appl Therm Eng 23(12):1441–1451
15. Alok P, Sahu D (2019) Numerical simulation of capillary tube for selected refrigerants using homogeneous equilibrium model. Int J Air-Condition Refrig 26(4):1950001
16. Spatz MW, Motta SFY (2004) An evaluation of options for replacing HCFC-22 in medium temperature refrigeration systems. Int J Refrig 27(5):475–483
17. Cabello R, Navarro-Esbrí J, Llopis R, Torrella E (2007) Analysis of the variation mechanism in the main energetic parameters in a single-stage vapour compression plant. Appl Therm Eng 27(1):167–176
18. Venkataramanamurthy VP, Kumar Senthil P (2010) Experimental comparative energy, exergy flow and second law efficiency analysis of R22, R436b vapour compression refrigeration cycles. Int J Sci Technol 2(5):1399–1412

Thermo-physical Investigation of Vegetable Oil-Based Nano-lubricant

Om Prakash, Ashwani Kumar, and Subrata Kumar Ghosh

Abstract This work aims to study thermo-physical properties, including thermal conductivity, viscosity, viscosity index, and rheology of vegetable oil-based nano-lubricant. Synthesis of nano-lubricant was done using vegetable seed oil (castor oil) and copper oxide (CuO) nano-particles using a magnetic stirrer and probe-type sonicator for breaking agglomerate. The thermal conductivity of samples was measured using transient hot-wire liquid thermal conductivity meter, and the effect of nano-particles was analyzed. Viscosity and viscosity index (VI) of the nano-lubricants thus prepared have been analyzed by Stabinger viscometer. The variation of viscosity of nano-lubricant with the shear rate was reported using an advanced air bearing rheometric system. The viscosity variation with respect to shear rate showed a decreasing trend, whereas the viscosity variation with respect to concentration showed an increasing trend. The results show that it is possible to increase the stability of samples in terms of viscosity variation with temperature and shear rate using nano-particles, hence makes it suitable for lubricant applications.

Keywords Nano-lubricant · Viscosity · Thermal conductivity · Rheology

1 Introduction

The surge in energy consumption and demand for economically restricted-energy sources have made us move toward efficient use of energy and energy savings [1–3].

O. Prakash · S. K. Ghosh
Department of Mechanical Engineering, Indian School of Mines, Indian Institute of Technology, Dhanbad, Jharkhand 826004, India
e-mail: ompr.18mt0149@mech.iitism.ac.in

S. K. Ghosh
e-mail: subrata@iitism.ac.in

A. Kumar (✉)
Kalinga Institute of Industrial Technology, Deemed To Be University, Bhubaneswar, Odisha 751024, India
e-mail: kumarashwani004@gmail.com

S. Revankar et al. (eds.), *Proceedings of International Conference on Thermofluids*, Lecture Notes in Mechanical Engineering,
https://doi.org/10.1007/978-981-15-7831-1_39

Environmental challenges posed by lubricant use, such as toxicity, have made us pursue the development of biodegradable and environmentally friendly industrial fluids [4]. Lubrication is an essential element for any known system involving the relative motion of parts in contact with each other. Lubricant development for any mechanical system is known to be a complicated process. It involves some crucial steps such as selection of base oil (stock), selection of additives to serve individual needs like corrosion inhibition, oxidation resistance, or extreme pressure lubrication, etc., and evaluating combination of additives that can work along to serve the needful along with optimization between performance and cost ending with full-scale test [5]. Environmental degradation and toxicity of mineral and synthetic lubricants reflected in the near past have stepped-up the need for environmental-friendly lubricant and industrial oil many-folds. Vegetable oil owing to their biodegradable nature comes up as a viable alternative. Most of the vegetable oil is an ester-based containing triglycerides. Triglycerides being more polar over petroleum-oil have high adherence to the metal surface, but it also has a short-coming of low thermo-oxidation stability and corrosion inhibition [5]. Study of thermo-physical properties of the lubricants in development has seen importance in previous works. The effect of concentration on viscosity for CuO nano-fluid was analyzed as engine oil and reported along with the tribological study of the same [6]. Effect of temperature and concentration on rheological property was analyzed as thermo-physical properties for various hybrid nano-lubricants out of MWCNT/SiO_2 in SAE 40 oil and reported, and an empirical relation was also established for proper rheological study different form normal relations [7]. Work has also been executed on the study of rheological properties experimentally and then the development of the model on neural networks and its sensitivity analysis for validation of the same [8]. Carbon nanostructures have been a center of attraction for nano-scientist in their works and proceedings; the rheological study was also done using graphene solute in organic media glycerol followed by characterizations and rheological studies revealing the shear function of nano-fluid being a function of mass function and temperature [9].

Thermal conductivity enhancement of engine oil using SAE50 motor oil and ZnO nano-oxide was reported by the preparation and characterization of nano-lubricant and then thermal conductivity measurement using KD2 pro thermal analyzer device over a temperature range of 25–55 °C, and finally a correlation for results were developed [10]. An experimental study to reveal the effect of MWCNT on thermal properties like viscosity, flash-point, and thermal conductivity as a parameter affecting the quality application of engine oil was presented in earlier works [11]. CuO microparticle suspension stability was analyzed by FTIR and thermal stability by the TGA method; the suspension was made in API GL-5 gear oil with the purpose of enhancement of thermal and physical properties. DSC thermographs were used to study the thermal effects, and enthalpy interaction prediction for the sample micro-suspension was done and presented [12].

There have been limited studies presented over the enhancement of bio-stocks as industrial fluids. Thus, the presented study highlights the investigation of various thermo-physical properties of a vegetable seed oil-based nano-lubricant. Castor oil

Table 1 Sample nomenclature

Sample	Designation	Detail
Plain oil	PO	Plain oil without additive
Nano-lubricant (0.05%)	No.05	Plain oil with 0.05 wt% additive concentration
Nano-lubricant (0.10%)	No.10	Plain oil with 0.10 wt% additive concentration

is selected as base stock, and CuO nano-particles were dispersed by mixing and sonication processes to form a stable particle-oil suspension. Viscosity and its variation with the temperature being the most crucial property of concern for the base stock (principal constituent for any lubricant) as it is the last line of defense for lubricant is reported for different concentrations along with a temperature range of 20–100 °C. Also, a variation of viscosity w.r.t shear rate being other important properties for industrial fluids was also studied using an advance air bearing rheological setup over a shear strain range of 0–40 s^{-1} at 40 °C. Finally, the thermal conductivity of the lubricant samples crucial for heat conduction, which in turn affects the viscosity of lubricant was studied to understand the effect of metal oxide nano-particle additive.

2 Materials and Methods

2.1 Sample Preparation

Samples of nano-lubricant were prepared by adding dehydrated CuO nano-particles (40–70 nm in size) in castor oil. The details of the samples are presented in Table 1. Nano-particles have been mixed by stirring over magnetic stirrer at 500 rpm for 3 h, followed by ultra-sonication using probe-type sonicator (ultrasonic processor; piezo-U-sonic) for 1 h. Obtained samples have been kept under inspection for 2 h to observe stability following which no sedimentation was observed.

2.2 Viscosity and Viscosity Index

Lubricants serve an essential purpose of frictional loss optimization and minimization of wear of component applied to cooling or heat dissipation and contaminant suspension [13]. The evaluation of oil viscosity is a critical parameter for both fresh and used oil (condition monitoring). For every particular application, there is a prescribed ASTM standard. For condition monitoring, change in viscosity can reveal important conclusions about oil, such as oxidation of base stock, build-up, shear breakdown of additives or dilution.[14]. Viscosity index is a measure of variation of viscosity with the temperature.

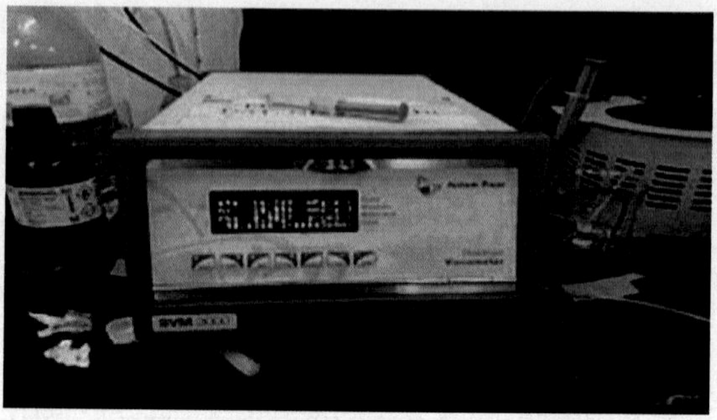

Fig. 1 Photographic view of Stabinger viscometer

A higher VI value signifies that the fluid undergoes a lower change in viscosity when subjected to different temperatures [15]. Viscosity and viscosity index of samples have been measured using Stabinger viscometer (SVM 3000; Anton Paar), the photographic view of viscometer is shown in Fig. 1. A sample of 2 ml is taken for the measurement of viscosity at a pre-fixed temperature. In this work, viscosity was measured at an increment of 10 °C in the range of 20–100 °C.

2.3 *Rheology*

Industrial fluid often undergoes a situation where it faces a variety of shear zone in the same machinery, as in the case of rotating machines like compressors, transmission fluids, and gear oils. The same oil passes over different shear zones; hence, the evaluation of the shear response of lubricant is essential; in rheology, the lubricant is subjected to shear strain rate and variation of viscosity is observed which tell us about the shear response of oil. This test helps predict shear thinning and thickening of the oil, which can disrupt the desired lubrication system. Advanced air bearing rheometer (Bohlin Gemini 2; Malvern Instruments, UK) was used in this study for measurement of viscosity within shear strain rate range: 0–40 s^{-1} at a temperature of 40 °C which is depicted in Fig. 2. Rheometer works on air pressure generated from an attached compressor at 6 bars. A cup and bowl arrangement is used to generate the shear rate, and the gap between the cup and bowl is maintained at 150μ.

2.4 *Thermal Conductivity*

Industrial fluids (lubricants) not only reduces friction and wear but are also needs to be efficient in heat transfer as in the case of engine oil, for this purpose, thermal conductivity is a very crucial property to be measured as it directly affects the efficiency of heat transfer. Thermal conductivity is also a crucial parameter of performance and design [16]. Thermal conductivity also affects the samples viscosity indirectly

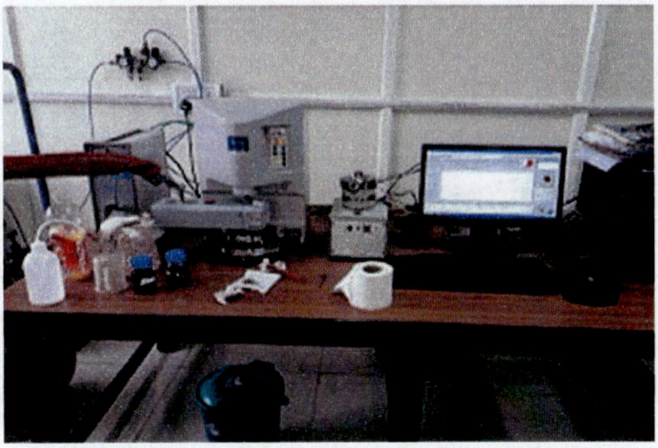

Fig. 2 Advance air bearing rheological system

as it controls the heat transfer capacity of lubricant. The thermal conductivity of the samples has been measured using transient hot-wire liquid thermal conductivity meter (THW-L2; thermotest thermo-physical instruments). Thermal conductivity meters measure the thermal conductivity value by the technique outlined in (ASTM D7896-14) [17]. A sample of 20 ml in isothermal condition has been taken in a slit-cylindrical container at 20 °C, and then the hot-wire sensor was immersed into it to measure the thermal conductivity at set temperature. Thermal conductivity meter is shown in Fig. 3

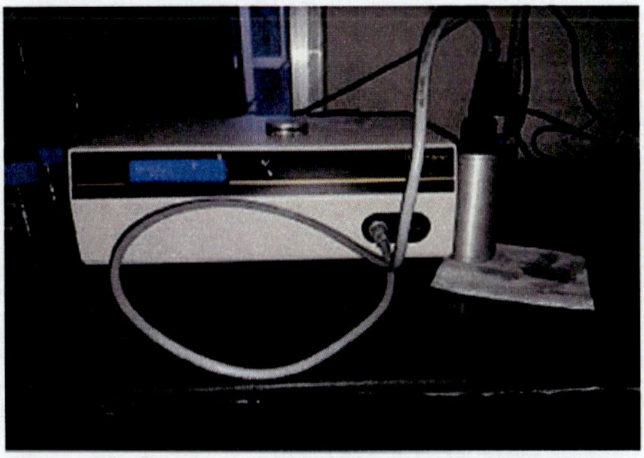

Fig. 3 Transient hot-wire thermal conductivity meter

3 Results and Discussion

3.1 Viscosity

Variation of viscosity w.r.t change in temperature is depicted in Figs. 4 and 5. The viscosity of the nano-lubricant decreases with an increase in temperature, as is the case of the base oil. The decrease in viscosity of the fluid with temperature can be attributed to the weakling of the intermolecular forces due to temperature. It was also seen that the viscosity slightly increase with the increase in concentration. The increase in viscosity can be attributed to nano-particle between the fluid layer of lubricants, causing ease in movement of fluid layer decreasing resistance to flow, i.e., viscosity but as concentration increases agglomeration of a particle causes formation asymmetric and larger particles, thereby hindering the ease of flow thus increasing the viscosity [6].

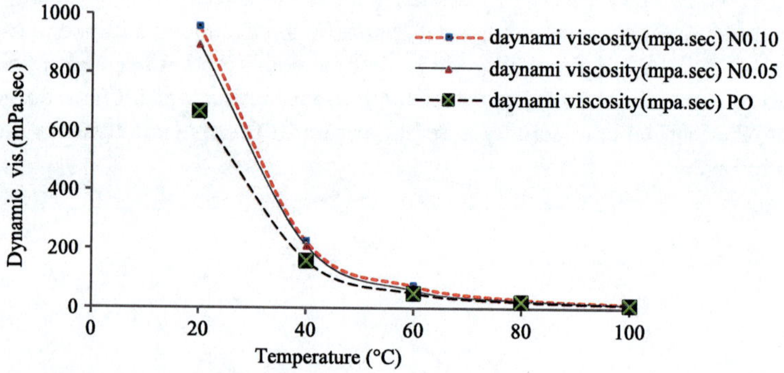

Fig. 4 Variation of dynamic viscosity (m Pa s) versus temperature

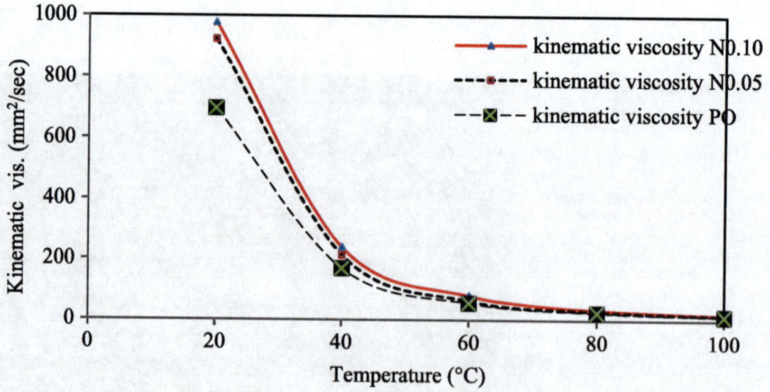

Fig. 5 Variation of kinematic viscosity (mm^2/s) versus temperature

Table 2 Viscosity index of samples

Sample	VI
PO	85.8
N0.05	86.6
No.10	87.2

The VI of the sample has been reported in Table 2. It can be seen that the VI of the samples increase with the increase in concentration this is due to the fact that the thermal conductivity increases with concentration enhancing the heat transfer efficiency, thereby a lesser change in viscosity can be seen at higher concentration, and thereby a high VI is obtained.

3.2 Rheology

The variation of viscosity with shear rate is shown in Fig. 6. It can be observed that the oil sample undergoes a shear thinning at a higher shear rate, and viscosity decreases with a decrease in concentration. The same trend is observed for all samples. The reason of shear thinning at a higher shear rate is due to the introduction of high shear and the particle aggregates are broken in much smaller size, thereby enhancing the symmetry of particle and these smaller aggregates being present between oil layers enhance the ease of flow, and thereby causes a reduction in the viscosity of sample [18]. It can also be deduced from the curve that at all shear rates, the viscosity increases with the increase in the concentration of nano-particle, i.e., the volume fraction of the suspension, which is in accordance with the conventional Einstein model for infinitely dilute suspension of the non-interacting hard sphere [19].

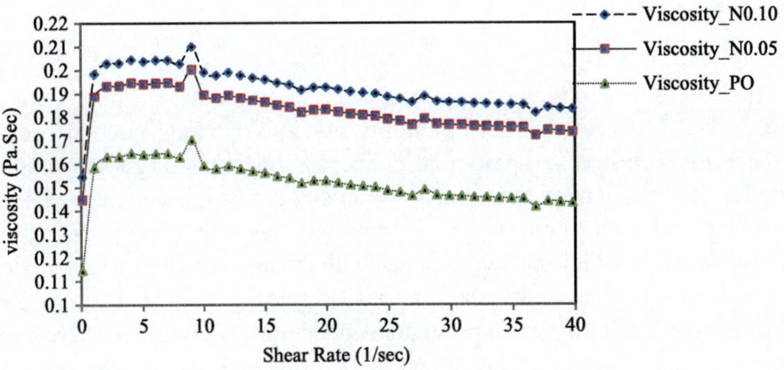

Fig. 6 Variation of viscosity (Pa s) versus shear rate at 40 °C

Fig. 7 Variation of thermal conductivity with concentration of nano-lubricants

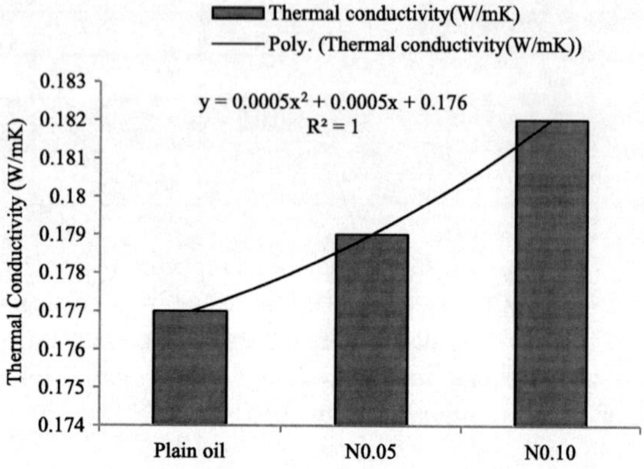

3.3 *Thermal Conductivity*

Variation of thermal conductivity can be seen with respect to concentration at a constant temperature of 20 °C for the castor oil-based nano-lubricants. Clearly, the enhancement of thermal conductivity can be seen above in Fig. 7. The variation of thermal conductivity was increasing in nature and found to be nonlinear w.r.t increase in concentration.

The presence of metal oxide nano-particles causes an increment in the surface interface for heat exchange inside the fluid, thereby increasing fluid-particle interfacial heat exchange, and thus an increment in heat transfer properties can be observed [20]. Enhancement of 2.82% in thermal conductivity was obtained by the suspension of copper metal oxide nano-particle in base stock over plain base oil.

4 Conclusions

Viscosity decreases with temperature, and the same trend has been seen for all the samples. It was observed that both the kinematic and dynamic viscosity decreased as temperature increases at a particular concentration of nano-lubricants and plain oil samples. It has also been seen that the viscosity across the sample is observed to be increasing with concentration, i.e., the viscosity has a direct proportionality with the volume fraction, which is in accordance with the theoretical Einstein model. The result shows that viscosity with shear rate is decreasing, i.e., the nano-lubricant is found to be shear thinning in experimental behavior. The nano-particle owing to its geometry was expected to create an ease of flow and decrease the viscosity of the sample at a higher shear rate, which would cause the breakdown of aggregates. A nonlinear increment in thermal conductivity can be observed w.r.t concentration in the case of nano-lubricants due to the enhancement of fluid-particle interfacial heat exchange. The obtained results show that the thermo-physical properties of vegetable oil can be enhanced with the addition of nano-particles. This shows that the nano-lubricant has the potential to replace mineral oil for lubrication applications.

References

1. Huminic A, Huminic G, Fleaca C, Dumitrache F, Morjan I (2015) Thermal conductivity, viscosity and surface tension of nanofluids based on FeC nano-particles. Powder Technol 284:78–84
2. Zhai Y, Li L, Wang J, Li Z (2019) Evaluation of surfactant on stability and thermal performance of Al_2O_3-ethylene glycol (EG) nanofluids. Powder Technol 343:215–224
3. Esfe MH, Raki HR, Emami MRS, Afrand M (2019) Viscosity and rheological properties of antifreeze based nanofluid containing hybrid nano-powders of MWCNTs and TiO_2 under different temperature conditions. Powder Technol 342:808–816
4. Hsien WLY (2015) Towards green lubrication in machining. Springer, Singapore
5. Perez JM (2000) Oxidative properties of lubricants using thermal analysis. Thermochim Acta 357:47–56
6. Jatti VS, Singh TP (2015) Copper oxide nano-particles as friction-reduction and anti-wear additives in lubricating oil. J Mech Sci Technol 29(2):793–798
7. Esfe MH, Afrand M, Yan WM, Yarmand H, Toghraie D, Dahari M (2016) Effects of temperature and concentration on rheological behavior of MWCNTs/SiO_2 (20–80)-SAE40 hybrid nano-lubricant. Int Commun Heat Mass Transf 76:133–138
8. Esfe MH, Saedodin S, Rejvani M, Shahram J (2017) Experimental investigation, model development and sensitivity analysis of rheological behavior of ZnO/10W40 nano-lubricants for automotive applications. Phys E 90:194–203
9. Moghaddam MB, Goharshadi EK, Entezari MH, Nancarrow P (2013) Preparation, characterization, and rheological properties of graphene–glycerol nanofluids. Chem Eng J 231:365–372
10. Yang L, Mao M, Huang JN, Ji W (2019) Enhancing the thermal conductivity of SAE 50 engine oil by adding zinc oxide nano-powder: an experimental study. Powder Technol 356:335–341
11. Ahmadi H, Rashidi A, Nouralishahi A, Mohtasebi SS (2013) Preparation and thermal properties of oil-based nanofluid from multi-walled carbon nanotubes and engine oil as nano-lubricant. Int Commun Heat Mass Transf 46:142–147
12. Maheswaran R, Sunil J, Vettumperumal R, Subhash SV (2018) Stability analysis of CuO suspended API GL-5 gear lubricant sol. J Mol Liq 249:617–622
13. Caines AJ, Haycock RF, Hillier JE (2004) Automotive lubricants reference book, vol 354. Wiley
14. Jakoby B, Buskies M, Scherer M, Henzler S, Eisenschmid H, Schatz O (2001) A novel multifunctional oil condition sensor. In: Advanced microsystems for automotive applications 2001. Springer, Berlin, Heidelberg, pp 157–165
15. Braga JWB, dos Santos Junior AA, Martins IS (2014) Determination of viscosity index in lubricant oils by infrared spectroscopy and PLSR. Fuel 120:171–178
16. Wrenick S, Sutor P, Pangilinan H, Schwarz EE (2005) Heat transfer properties of engine oils. In: World tribology congress III. American Society of Mechanical Engineers Digital Collection, pp 595–596
17. Nwaoha C, Idem R, Supap T, Saiwan C, Tontiwachwuthikul P, Rongwong W, Benamor A (2017) Heat duty, heat of absorption, sensible heat and heat of vaporization of 2–Amino–2–Methyl–1–Propanol (AMP), Piperazine (PZ) and Monoethanolamine (MEA) tri–solvent blend for carbon dioxide (CO_2) capture. Chem Eng Sci 170:26–35
18. Rashin MN, Hemalatha J (2013) Viscosity studies on novel copper oxide–coconut oil nanofluid. Exp Thermal Fluid Sci 48:67–72
19. Einstein A (1906) On the theory of the Brownian movement. AnnPhys 19(4):371–381
20. Karthikeyan NR, Philip J, Raj B (2008) Effect of clustering on the thermal conductivity of nanofluids. Mater Chem Phys 109(1):50–55

Numerical Analysis of Laser-Assisted Cryopreservation of Biological Tissues

Saurabh Das and Satya Prakash Kar

Abstract Cryopreservation of tissue is an important aspect of bioengineering, and it does not involve any ice formation, rather tissue is solidified by intense increase in viscosity. This can only be brought by applying cooling rates in the order of 10^5 K/min. An external heating source followed by a sudden cooling is suggested here. The heat source in this study is assumed to be a Gaussian laser, and the cryogen used to cool down the tissue would be liquid nitrogen. This numerical study tries to understand the viability and feasibility of laser-assisted cryopreservation technique while accounting for convective–radiative heat loss and following a dual phase lag conduction equation. The algebraic equations obtained after discretization of the governing equations by finite volume method (FVM) are then solved by the tri-diagonal matrix method. Enthalpy-porosity technique is used to capture the frozen tissues. Cooling rates at various locations are calculated, and the feasibility of this approach is tested.

Keywords Vitrification · Cryopreservation · Dual phase lag · Radiative transport equation · Convective heat loss

1 Introduction

The viable preservation of tissues beyond the current normal requires an approach different from normal [1], while the current preservation techniques aim at freezing the cell, leading to crystallization of water to ice [2]. Since it is a temperature-dependent phase change, the properties of ice differ to that of water [3]. An increased specific volume leads to rupture of cell membranes, resulting in cell death [4]. The objective of this paper is to study an alternative form of cell preservation, known

S. Das
CSIR- Institute of Minerals and Material Technology, M. Tech. Scholar at KIIT, Bhubaneswar, India

S. P. Kar (✉)
Kalinga Institute of Industrial Technology, Bhubaneswar, Odisha 751024, India
e-mail: satyapkar@gmail.com

S. Revankar et al. (eds.), *Proceedings of International Conference on Thermofluids*, Lecture Notes in Mechanical Engineering,
https://doi.org/10.1007/978-981-15-7831-1_40

Table 1 Properties of laser

Laser power (kW)	Beam radius (mm)	Operational time (μs)
5	1.5	60

as vitrification. Vitrification is the solidification of fluids by increasing the viscosity drastically. While studying heat transfer in biological tissues, the Pennes bioheat equation is applied [5]. However, recent studies have suggested the presence of time lags during conduction because of heat flux and that taken for the temperature gradient to be fully developed [6]. Thus, a dual-phase-lag equation is incorporated to study heat transfer [7]. Jasinski et al. analyzed the heat transfer process in tissue sample subjected to laser irradiation, using the dual-phase-lag equation [8].

Radiative heat loss is incorporated into the model by implementing the Radiative transport Equation [9, 10]. A radiation source term depending on the laser irradiation is calculated.

From the critical literature survey, it is observed that the study of cryopreservation of tissue using vitrification by laser source with application of dual-phase-lag equation is rarely reported.

2 Mathematical Modeling

2.1 *Physical Domain*

A two-dimensional domain has been taken into consideration. The dimensions of this domain are 25 mm and 5 mm, respectively, in the x and y directions (Table 1). The properties of Gaussian laser which is irradiating the surface of the tissue are depicted in Table 1, and the tissue properties are depicted in Table 2 (Fig. 1).

2.2 *Boundary Conditions*

For the heating of the domain, the boundary conditions are given in Table 3.

These boundary conditions are maintained for 60 μs, following which the boundary conditions for cooling the domain are put in effect (Table 4).

2.3 *Numerical Method*

Enthalpy-Porosity Method:

An enthalpy-porosity method is used to model the solidification/melting process. The melt interface is not tracked explicitly, instead the liquid fraction (fraction of

Table 2 Properties of tissue [11]

Parameter	Value	Unit
Conductivity in solid phase (k_s)	0.6	W/m K
Conductivity in liquid phase (k_l)	2.24 + 0.005975(273-T)$^{1.156}$	W/m K
Specific heat in solid phase (C_s)	4200	J/kg K
Specific heat in liquid phase(C_l)	7.16 T + 138	J/kg K
Density of solid phase (ρ_s)	999	kg/m^3
Density of liquid phase (ρ_l)	921	kg/m^3
Latent heat (L)	335	J/kg
Melting point temperature (T_m)	273.47	K
Free stream temperature (T_{inf})	110	K
Initial temperature ($T_{initial}$)	300	K

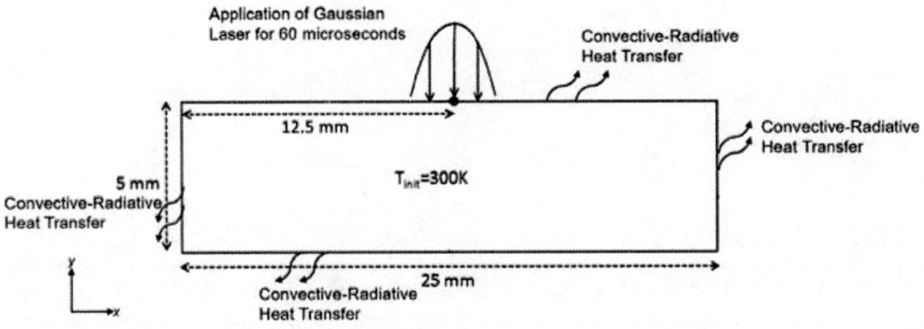

Fig. 1 Schematic diagram of the calculation domain

Table 3 Boundary conditions for heating of domain

Location	Boundary condition
Left wall ($x = 0, 0 \leq y \leq W$) :	$-k\frac{\partial T}{\partial x} = 0$
Right wall ($x = L, 0 \leq y \leq W$):	$-k\frac{\partial T}{\partial x} = 0$
Bottom wall ($y = 0, 0 \leq x \leq L$):	$-k\frac{\partial T}{\partial x} = 0$
Top wall ($y = W, 0 \leq x \leq L$):	$-k\frac{\partial T}{\partial x} = I_o\alpha * \exp\left(-\frac{r^2}{R^2}\right) + h(T(x, W, t) - T_\infty)$

Table 4 Boundary conditions for cooling of domain

Location	Boundary condition
Left wall ($x = 0, 0 \leq y \leq W$)	$-k\frac{\partial T}{\partial x} = h(T(0, y, t) - T_\infty) + \sigma T^4(0, y, t) - T^4$
Right wall ($x = L, 0 \leq y < W$):	$-k\frac{\partial T}{\partial x} = h(T(L, y, t) - T_\infty) + \sigma T^4(L, y, t) - T^4$
Bottom wall ($y = 0, 0 \leq x \leq L$):	$-k\frac{\partial T}{\partial x} = h(T(x, 0, t) - T_\infty) + \sigma T^4(x, 0, t) - T^4$
Top wall ($y = W, 0 \leq x \leq L$):	$-k\frac{\partial T}{\partial x} = h(T(x, W, t) - T_\infty) + \sigma T^4(x, W, t) - T^4$

cell volume in liquid form) is associated with each cell in the domain. Since the freezing process takes place over a region of temperature, there will be a mushy or transitional zone where the liquid fraction will vary from 0 to 1. This 'mushy' zone is treated as a pseudoporous media in which porosity decreases from 1 to 0 as the material solidifies.

The enthalpy of the material is defined by the following equation, as a sum of sensible enthalpy h and the latent heat ΔH:

$$H = h + \Delta H \tag{1}$$

where

$$h = h_{\text{ref}} + \int_{T_{\text{ref}}}^{T} c_{\text{p}} \text{d}T \tag{2}$$

The liquid fraction β can be defined as 0, if $T < T_{\text{solidus}}$; as 1, if $T < T_{\text{liquidus}}$ and as

$$\beta = \frac{T - T_{\text{solidus}}}{T_{\text{liquidus}} - T_{\text{solidus}}} \quad \text{if} \quad T_{\text{solidus}} < T < T_{\text{liquidus}} \tag{3}$$

The latent heat content ΔH can be written as

$$\Delta H = \beta L \tag{4}$$

For solidification/melting problems, the energy equation is written as

$$\frac{\partial}{\partial t}(\rho H) + \nabla \cdot (\rho \vec{v} H) = \nabla \cdot (k \nabla T) + S \tag{5}$$

The solution for temperature is a continuous iteration between the energy equation (Eq. 5) and liquid fraction equation. (Eq. 3) The temperature obtained from solving energy equation and this value is used to update the liquid fraction as well as sensible enthalpy, and the new enthalpy value is then used in the energy equation.

Calculation of radiation source term

To obtain the heat produced due to irradiation, we need to first calculate the value of G, the incident radiation using the following transport equation (Eq. 6).

$$\nabla \cdot (\Gamma \nabla G) - \kappa_a G + 4\kappa_a n^2 \sigma T^4 = S \quad (6)$$

where κ_a is the absorption coefficient, G is the incident radiation, n is the refractive index, σ and is the Stefan–Boltzmann constant [6–10, 12].

Γ is a parameter defined by the following equation:

$$\Gamma = \frac{1}{3(\kappa_a + \kappa_s) - c\kappa_s} \quad (7)$$

Here, κ_a and κ_s are absorption and scattering coefficients, respectively, and c is the anisotropy factor. Once the values of G and Γ are calculated, radiative heat flux can be calculated as,

$$q_r = -\Gamma \nabla G \quad (8)$$

This will be the radiation source term for our equation.

3 Results and Discussions

3.1 *Grid Independence Test*

The domain was tested to finalize the most optimum mesh size. Thus, a grid independence test was conducted. It is found that results seem to stabilize at a mesh size of 25 × 125, and it was taken as the mesh size for this study (Fig. 2).

The most optimal time step was found to be 10^{-9} s for the heating phase and 10^{-3} s for the cooling phase.

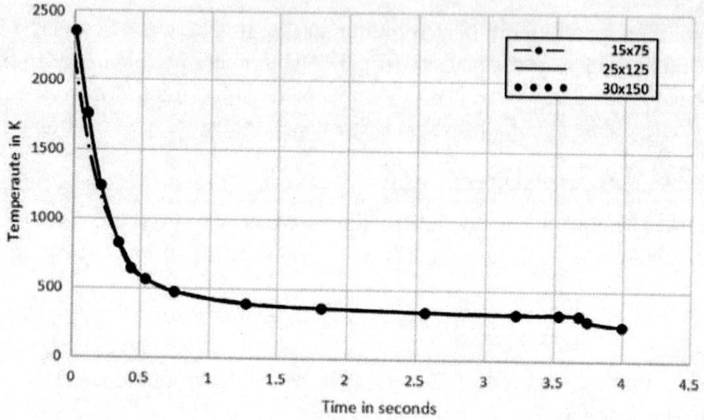

Fig. 2 Grid independence study

3.2 *Validation*

The radiation model was benchmarked against the results obtained by Liu et al. [13]. The thawing fronts at different temperatures were calculated and compared and were found to be in agreement (Figs. 3 and 4).

To validate for the radiation model, the optical parameters of the tissue were changed and the results were observed to be in agreement with that obtained in the literature (Figs. 5 and 6).

Fig. 3 −1 °C thawing front

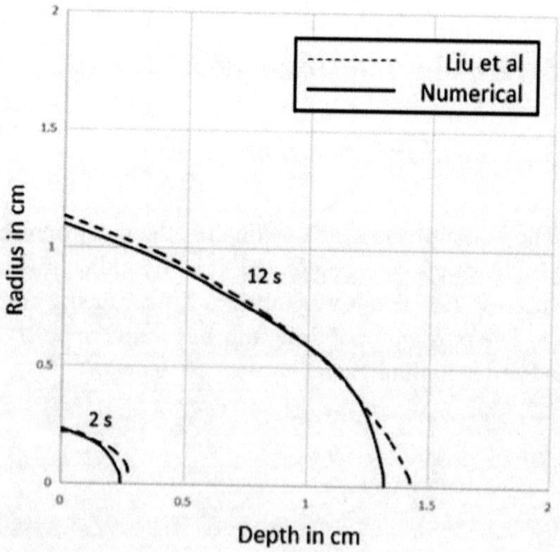

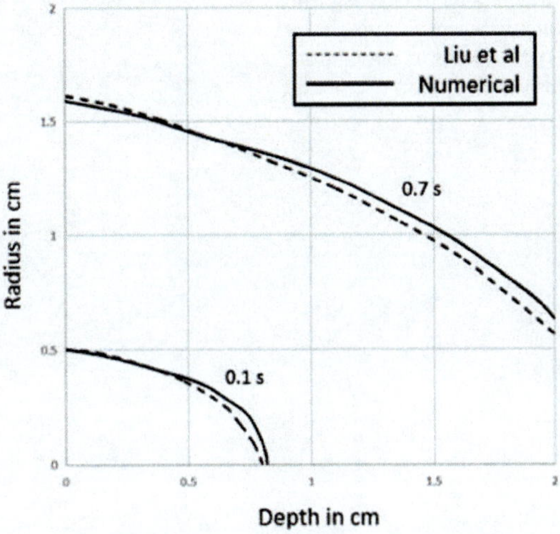

Fig. 4 −8 °C thawing front

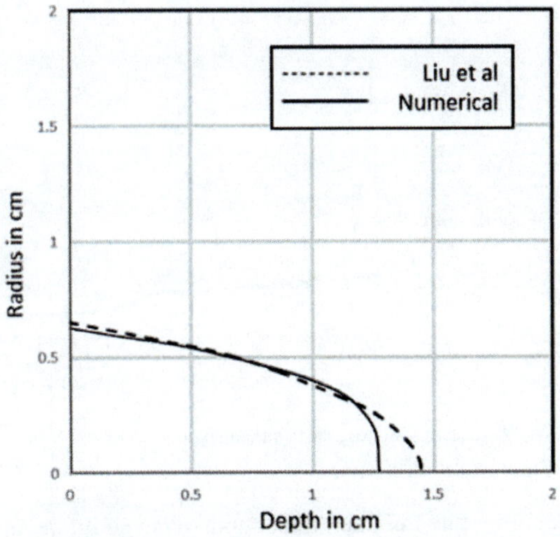

Fig. 5 −1 °C Thawing front at 10 s when anisotropy factor is increased from 0.8 to 0.95

3.3 Temperature Variations and Discussion

A significant increase in the cooling rate is observed when radiation and convective heat losses are considered. The average cooling rate was found to be 13,200 K/min. According to literature dealing with cryopreservation through vitrification, this is enough for vitrification of tissues without the aid of CPAs (Figs. 7 and 8).

Figure 9 shows a declining cooling rate at different points in the tissue. The cooling rate dips to about 2000 K/min at a depth of 1.875 mm. A trend is observed that the

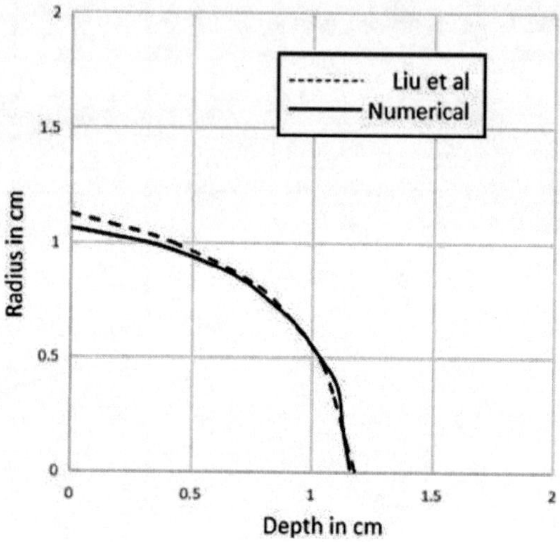

Fig. 6 −1 °C Thawing front at 10 s when scattering coefficient is increased from 100 to 200 cm^{-1}

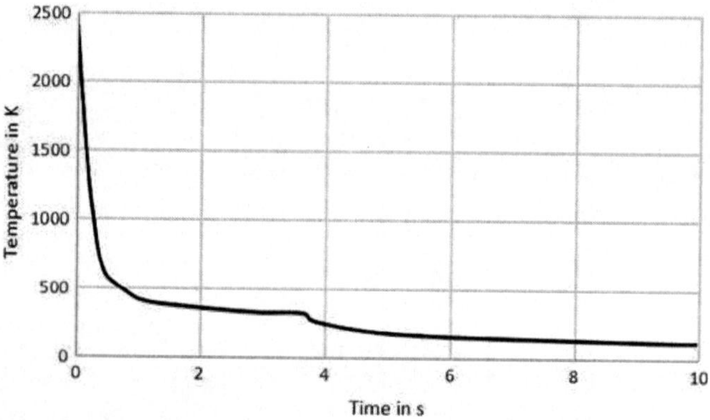

Fig. 7 Thermal history of tissue surface

cooling rate declines with depth. This could be an indication that a surface heating method may not be the most optimal solution for vitrifying larger tissue samples.

It is observed that the use of cryoprotective agents for tissue cryopreservation and with them, their toxic effects can be bypassed entirely by vitrification at an extremely high cooling rate. A significant increase in cooling rate by the inclusion of free convective heat loss opens up the possibility of exploring the utility of forced convective heat transfer in this model. It is observed that due to the thermophysical properties of tissues, heating effect cannot penetrate deep enough to make this method viable for larger tissues. Thus, this method might be only suitable for smaller tissues like blood vessels and cartilage. Surface irradiation is found to not be viable for larger tissues samples. This method can be further enhanced by the addition of cryoprotective agents, which may reduce the required cooling rate. However, mass

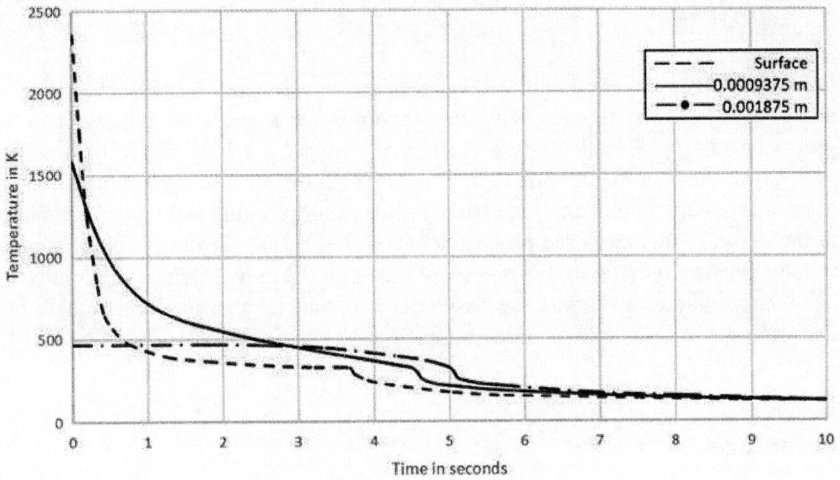

Fig. 8 Thermal history at different locations in the tissue sample

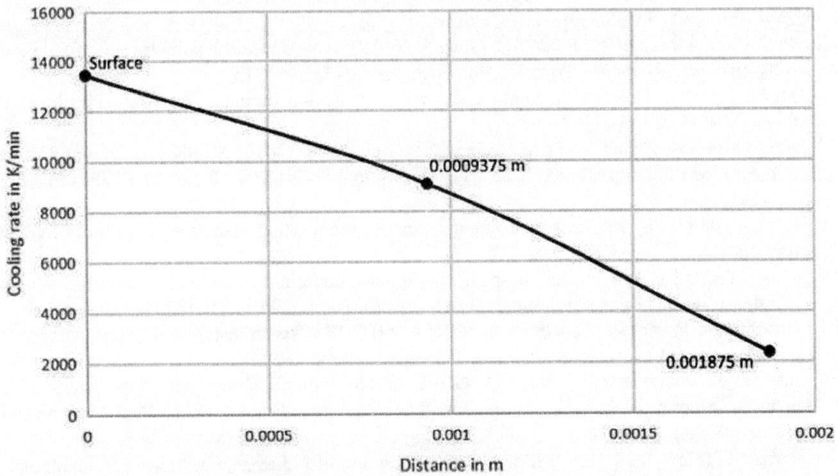

Fig. 9 Cooling rates at different positions in the tissue

transfer and its associated aspects should be introduced. This is beyond the scope of current study and warrants further research.

4 Conclusion

In modern bioengineering and medical scenario, the demand for preserved tissues which have retained their biophysical properties far exceeds the supply. This is primarily due to the difficulties in storage of tissues for a long period of time. In such cases, vitrification appears to be an interesting prospect in cryopreservation of tissues. It is numerically analyzed here. The declining cooling rate across the depth of the tissue in this proposed process suggests that surface heating is not the most suitable heating method in this process. Volumetric heating provides uniformity in temperature variations across the tissue domain, and as such is more suitable for uniform vitrification across a tissue sample.

References

1. Deepak K, Charalampopoulos T, Devireddy R (2004) Numerical investigation of a novel method to vitrify biological tissues using pulsed lasers and cryogenic temperatures. In: ASME 2004 heat transfer/fluids engineering summer conference. American Society of Mechanical Engineers Digital Collection, pp 701–706
2. Taylor MJ, Weegam BP, Baicu SC, Giwa SE (2019) New approaches to cryopreservation of cells, tissues, and organs. Transfus Med Hemother 46(3):197–215
3. Mazur P (1984) Freezing of living cells: mechanisms and implications. Am J Physiol-Cell Physiol 247(3):C125–C142
4. Xu F, Moon S, Zhang X, Shao L, Song YS, Demirci U (2010) Multi-scale heat and mass transfer modelling of cell and tissue cryopreservation. Philos Trans R Soc Math Phys Eng Sci 368(1912):561–583
5. Pennes HH (1948) Analysis of tissue and arterial blood temperatures in the resting human forearm. J Appl Physiol 1(2):93–122
6. Zhang Y (2009) Generalized dual-phase lag bioheat equations based on nonequilibrium heat transfer in living biological tissues. Int J Heat Mass Transf 52(21–22):4829–4834
7. Hooshmand P, Moradi A, Khezry B (2015) (2015): Bioheat transfer analysis of biological tissues induced by laser irradiation. Int J Therm Sci 90:214–223
8. Jasiński M, Majchrzak E, Turchan L (2016) (2016): Numerical analysis of the interactions between laser and soft tissues using generalized dual-phase lag equation. Appl Math Model 40(2):750–762
9. Farmer JT, Howell JR (1994) Monte Carlo prediction of radiative heat transfer in inhomogeneous, anisotropic, nongray media. J Thermophys Heat Transf 8(1):133–139
10. Hunter B, Guo Z (2015) Numerical smearing, ray effect, and angular false scattering in radiation transfer computation. Int J Heat Mass Transf 81:63–74
11. Kandra D (2004) Tissue interactions with lasers and liquid nitrogen: a novel cryopreservation method
12. Siegel R, Howell JR (2001) Thermal radiation heat transfer, 4th edn
13. Zhao JM, Liu LH (2008) Spectral element method with adaptive artificial diffusion for solving the radiative transfer equation. Numer Heat Transf Part B Fundam 53(6):536–554

Combustion Analysis

Experimental and Numerical Study of Temperature Distribution on Float Glass Along the Wall

Raj Kumar Mishra, Ankit Dasgotra, Mahesh Kumar Tiwari, Akhilesh Gupta, Ravi Kumar, and Pavan K. Sharma

Abstract Present work represents the experimental and numerical study of temperature distribution over float glass under different fire conditions. A total of 9 experiments have been carried out, on fully exposed float glass of $300 \times 300 \times 4$ mm^3 pin supported along the wall with diesel line burner of 1020×150 mm^2 with continuous fuel supply to investigate the effect of thermal load and temperature distribution on glass. Pins were fully insulated from glass to avoid heat transfer between the two. Four K-type thermocouples were placed on the exposed surface while one was placed at the center of glass rear surface. Another three K-type thermocouples were mounted at 0.5 mm, 1 mm, and 1.5 mm from the wall surface, respectively, to measure the surrounding air temperature between the glass and fire source. Heat flux sensor was mounted at a parallel distance of 25 mm from the glass to register the incident heat per unit area upon the exposed surface. Parameters measured during the experiments were heat flux, temperature distribution over float glass, i.e., exposed surface temperature, rear surface temperature, air temperature between the glass and fire source. Maximum temperature difference between exposed and rear surface is noted at the time of glass breakage.

Keywords Heat flux · Temperature distribution · Line burner · Exposed float glass

1 Introduction

Float glass is most commonly used glass that uses raw materials, typically consisting of sand, soda ash (sodium carbonate), dolomite, limestone, and salt cake (sodium sulfate) [1]. Safety from fire hazard, i.e., presence of flame or the risk of an uncontrolled fire, is one of the major issues in structural building or compartment, and

R. K. Mishra (✉) · A. Dasgotra · M. K. Tiwari · A. Gupta · R. Kumar
Department of Mechanical and Industrial Engineering, IIT Roorkee, Roorkee, Uttarakhand 247667, India
e-mail: rajiit2015@gmail.com

P. K. Sharma
Reactor Safety Division, Bhabha Atomic Research Center, Mumbai, Maharashtra 400085, India

S. Revankar et al. (eds.), *Proceedings of International Conference on Thermofluids*, Lecture Notes in Mechanical Engineering,
https://doi.org/10.1007/978-981-15-7831-1_41

so need to be maintained, monitored, and modified time to time [2]. In most of the cases, fire initiation and propagation largely depend on availability of fresh air and its continuous supply [3]. Glass acts as the most fragile part of building and increases the probability of fire hazard abruptly due to fallout, as cavity formed due to fallout allows more and more fresh air for fire to develop and become uncontrolled [4]. Very limited work has been carried out on temperature distribution and breakage behavior of fully exposed glass under fire condition along the wall [5]. The present experimental work therefore tries to fill the gap by examining the temperature distribution on fully exposed glass along the wall subjected to movable fire source.

2 Methodology

Figure 1a shows the experimental facility at laboratory and Fig. 1b shows computational domain used to carry out temperature distribution analysis over the float glass. A continuous diesel fuel supply system with line burner is being used. To change the heat flux incident on glass surface, burner is moved horizontally as well as vertically with respect to float glass. Nine sets of different locations of fire source are taken from lower glass edge to vary the heat flux by first keeping vertical distance fixed as 15 cm and varying horizontal distance as 10 cm, 15 cm, and 20 cm for each experiment 1, 2, and 3 (result shown in Figs. 2, 3, and 4 respectively). Similarly, vertical distance is again fixed as 20 cm and 25 cm, respectively, and same set of horizontal distance is varied, i.e., 10 cm, 15 cm, and 20 cm for experiment no. 4, 5, 6, 7, 8, and 9 (result shown in Figs. 5, 6, 7, 8, 9, and 10 respectively). A water cooled heat flux sensor (SBG01) is placed parallel to the glass at a distance of 25 mm with a measuring range of 0–50 kW/m^2 is shown in Fig. 1a and marked as HF (corresponding result shown

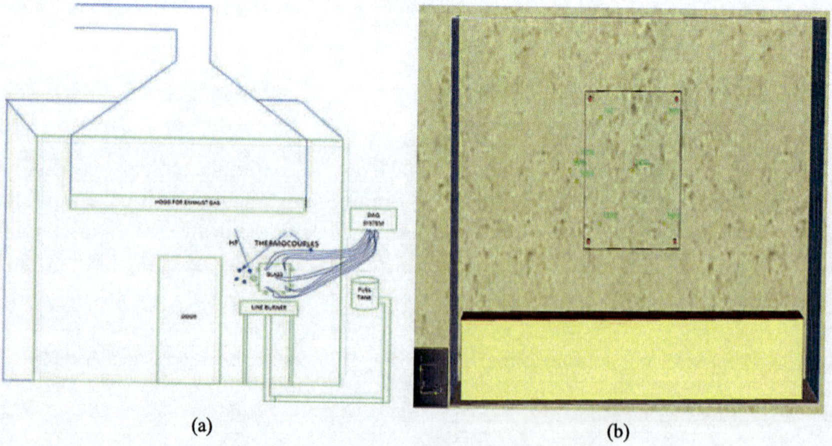

Fig. 1 Schematic diagram of **a** experiment setup, **b** computational domain

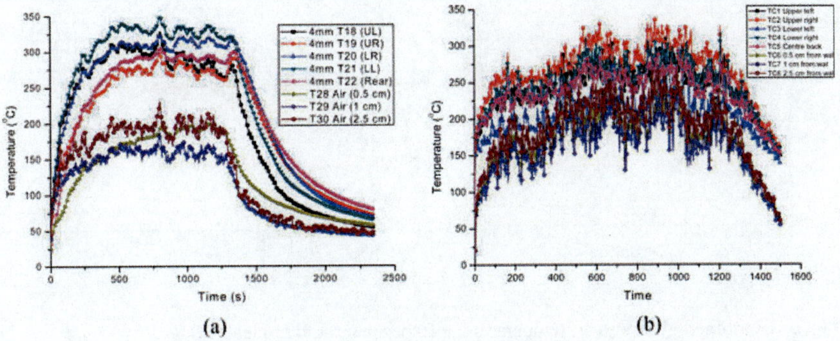

Fig. 2 Combined glass and air temperature. **a** Experimental, **b** model-FDS

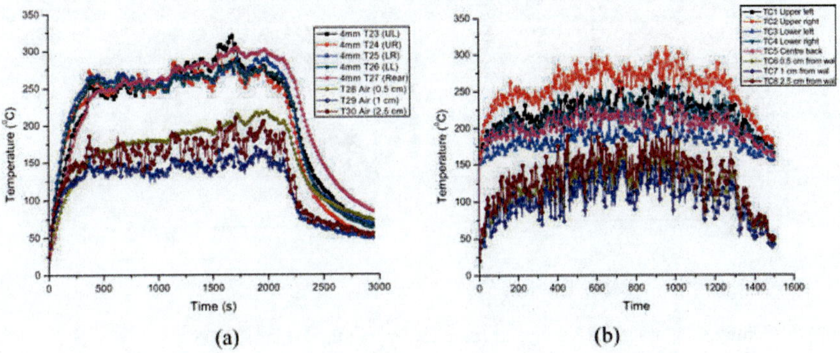

Fig. 3 Combined glass and air temperature. **a** Experimental, **b** model-FDS

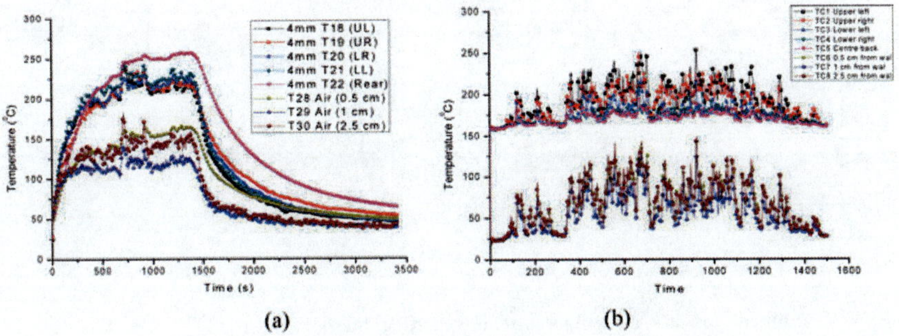

Fig. 4 Combined glass and air temperature. **a** Experimental, **b** model-FDS

in Fig. 11a). Also, the numerical simulation related to experimental work is carried out in Fire Research Laboratory, IIT Roorkee, using Fire Dynamics Simulator (FDS

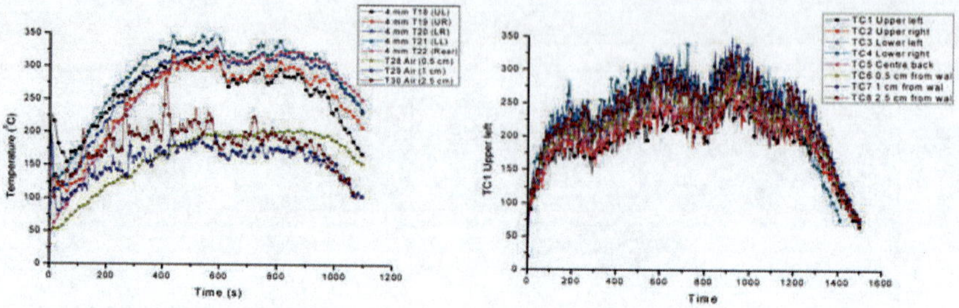

(a) (b)

Fig. 5 Combined glass and air temperature. **a** Experimental, **b** model-FDS

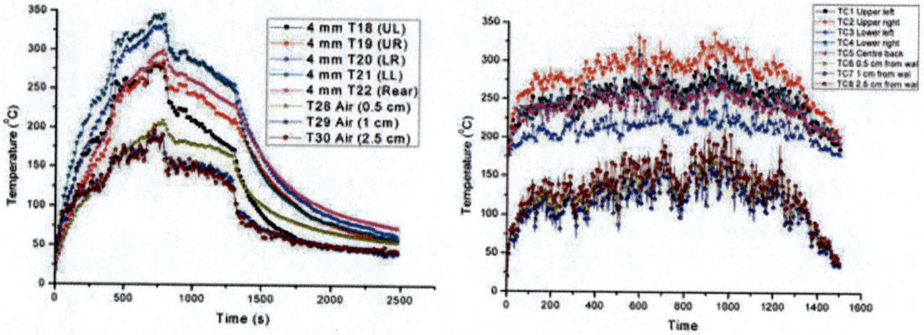

(a) (b)

Fig. 6 Combined glass and air temperature. **a** Experimental, **b** model-FDS

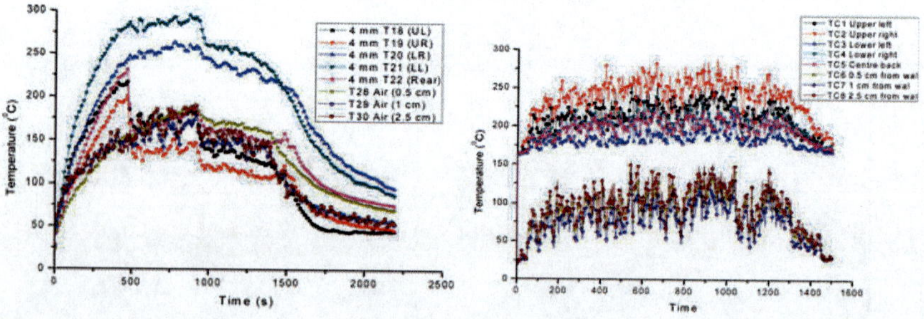

(a) (b)

Fig. 7 Combined glass and air temperature. **a** Experimental, **b** model-FDS

is developped by NIST USA) to investgate the heat flux and temperature distribution on the float glass.

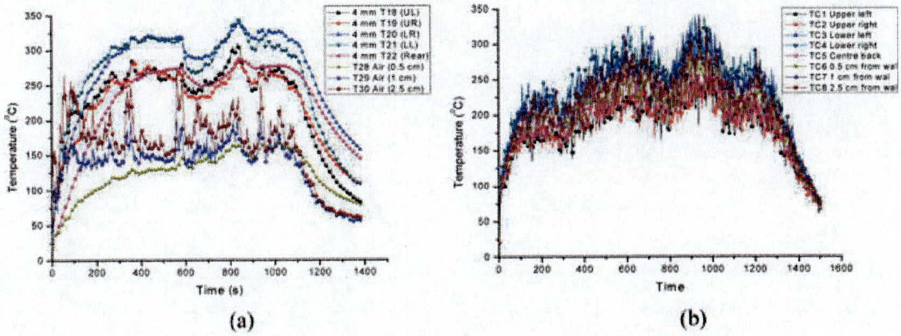

Fig. 8 Combined glass and air temperature. **a** Experimental, **b** model-FDS

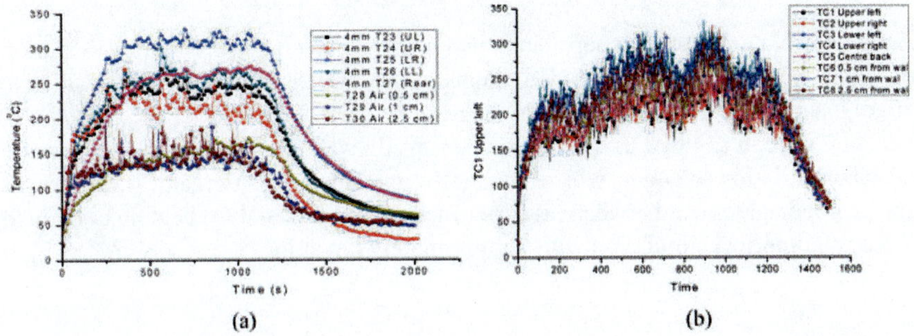

Fig. 9 Combined glass and air temperature. **a** Experimental, **b** model-FDS

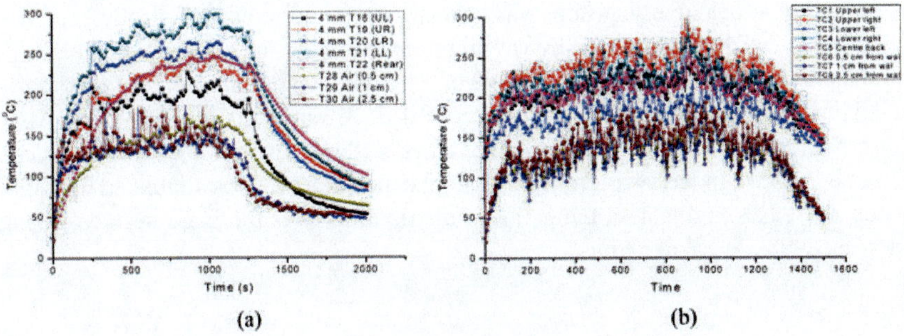

Fig. 10 Combined glass and air temperature. **a** Experimental, **b** model-FDS

3 Results

Figure 2a, b depicts the experimental as well as simulated results for temperature distribution on 4 mm float glass and air when fire source kept 15 cm vertical and 10 cm horizontal from the lower glass edge. It is observed that simulated result

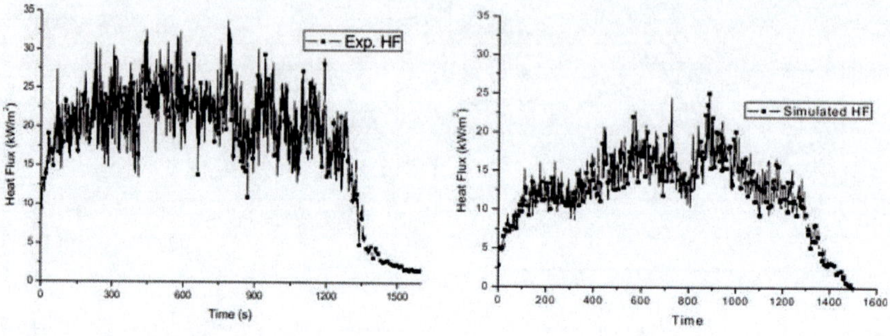

(a) (b)

Fig. 11 Heat flux. **a** Experimental, **b** model-FDS

satisfactorily agrees with experimental result with an average temperature deviation of 20–25 °C due to some experimental uncertainty and unaccountability of radiative effect in simulation as well [6]. From the results below (Figs. 2, 3, 4, 5, 6, 7, 8, 9, 10, and 11), we can say that as fire source is moved away from the glass, temperature distribution is lower which will result more time for glass to break [7]. Figure 11 shows good agreement between (a) experimental heat flux and (b) heat flux obtained through numerical simulation with approximately deviation of 5–6 kW/m^2.

4 Discussion

A continuous fuel supply system was arranged for continuous and steady-state fire condition. Thermal load varied by varying the position of fire source along the horizontal and vertical directions [8]. Range of maximum and minimum heat flux registered during glass breakage was in between 10–15 kW/m^2 and 20–25 kW/m^2. Also, for other fire source positions, experimental as well as numerical heat flux has been obtained. As we move away from glass, experimental as well as simulated heat flux value decreases giving less temperature distribution over the glass surface giving increased glass breakage time.

5 Conclusion

Temperature distribution and heat flux over the float glass have been investigated experimentally as well as numerically with good agreement. Peak experimental temperature for Exp. 1 to 3 is found to be 350, 300, and 250 °C, respectively, while simulated value found to be 325, 275, and 225 °C, respectively. Similarly, for Exp. 4 to 6 experimental value found to be 350, 330, and 300 °C and simulated value found to be 320, 310, and 275 °C, respectively. For Exp. 7 to 9 experimental value found to be 380, 325, and 300 °C while simulated value found to be 330, 310, and 280 °C, respectively. Average experimental heat flux registered for exp. 1 was 25 kW/m^2 while simulated heat flux value found to be 19 kW/m^2. With the help of above parameters, glass breakage time could be predicted well with further extended analysis.

References

1. https://en.wikipedia.org/wiki/Float_glass
2. Gardon R (1961) A review of radiant heat transfer in glass. J Am Ceram Soc 44:305–313
3. Shields TJ, Silcock GWH, Flood M (2002) Performance of a single glazing assembly exposed to a fire in the centre of an enclosure. Fire Mater 26(2):51–75
4. Skelly MJ (1990) An experimental investigation of glass breakage in compartment fires. M. S. Thesis, Virginia Polytechnic Institute
5. Emmons HW (1988) Window glass breakage by fire. Home Fire Project Technical Report No. 77, Harvard University, Cambridge
6. Wang Q, Zhang Y, Sun J, Wen J, Dembele S (2011) Temperature and thermal stress Simulation of window glass exposed to fire. Procedia Eng 11:452–460
7. Shields TJ, Silcock GWH, Hassani SKS (1997–1998) The behavior of glazing in an enclosure fire. J Appl Fire Sci 7(2):145–163
8. Zhang J, Lu S, Li Q, Yuen R, Chen B, Yuan M, Li C (2012) Smoke filling in closed compartments with elevated fire sources. Fire Saf J 54:14–23

Numerical Study on Crude Oil Pool Fire Behavior in an Enclosure

Avinash Chaudhary, Mahesh Kumar Tiwari, Akhilesh Gupta, and Surendra Kumar

Abstract This paper analyzes the experimental results on crude oil pool fire obtained in an enclosure fire experiment, by using FDS software. Experiment is conducted in an enclosure of volume 64 m^3. A cylinder pan of diameter 0.5 m is used to burn jatropha oil fuel. Fire dynamics simulator (FDS) is used to perform the numerical simulations in which heat release rate of jatropha oil measured using the mass loss method was inputted. Simulation results show that calculated temperatures are in good agreement with the experimental findings. Overall, simulation using FDS provides an effective way to study the burning behavior of crude oil pool fire.

Keywords Jatropha oil · Enclosure fire · Heat release rate · Fire dynamics simulator

1 Introduction

With the widespread use of biodiesel, it becomes necessary to consider the various safety aspects regarding its storage and transportation. Experimental investigation and numerical analysis are the two methods to study the behavior of enclosure fire. In a full-scale fire investigation, focus is on the reproduction of the scenario of real fire, and this requires lot of budget and space for carrying out the experiments. Also, as the experiments are limited, it is difficult to extend the tests or to point out the main conclusions from the limited number of experiments. However, data of the

A. Chaudhary (✉)
School of Mechanical Engineering, Kalinga Institute of Industrial Technology (KIIT) Deemed to be University, Bhubaneswar, Odisha, India
e-mail: avinashmoond005@gmail.com

M. K. Tiwari · A. Gupta
Department of Mechanical and Industrial Engineering, Indian Institute of Technology Roorkee, Roorkee, Uttarakhand 247667, India

S. Kumar
Department of Chemical Engineering, Indian Institute of Technology Roorkee, Roorkee, Uttarakhand 247667, India

S. Revankar et al. (eds.), *Proceedings of International Conference on Thermofluids*, Lecture Notes in Mechanical Engineering,
https://doi.org/10.1007/978-981-15-7831-1_42

full-scale experiments can be compared with the available correlation and numerical simulations results to validate the model.

A number of experimental and numerical investigations in fire dynamics have been carried on petroleum fuels by many researchers using CFD tool packages such as JASMINE, FLUENT, SOFIE, and FDS [1–8]. Among the various software available, fire dynamics simulator (FDS) is widely used for fire modeling studies. FDS solves numerically a large eddy simulation form of Navier–Stokes equation for low speed flows, with an emphasis on smoke and heat transport from fires. Many researchers have used FDS for simulating pool fire experiments. LES simulations of buoyant flows with a Smagorinsky turbulence model showed that three-dimensional computational with large grid volumes perform better than the two-dimensional counterparts [1]. Simulation of large-scale pool fire using FDS shows that the results match well with experimental findings of Ryder et al. [2]. A review of modeling challenges done by Trouve and Wang [6] for LES treatment of enclosure fires showed that simulations of the enclosure fire require many physical sub-models to describe a range of multi-physics phenomenon occurring in fire. FireFOAM has been used to model purely buoyant fire plumes with different heat release rates. The simulations results were compared with McCaffrey's experiments, and it showed good agreements with experimental results of Wang et al. [8]. Recently, Paudel and Hostikka [9] studied the model uncertainty propagation in compartment fire and reported that the uncertainty in the heat flux measurement is a measure component of overall uncertainty.

The present paper reports validation of experimental results of Chaudhary et al. [10] by using FDS software. Additional results are presented in terms of smoke layer height and hot gas layer temperature at a depth of 0.5 m below ceiling. The heat release rate measured using mass loss method was inputted into FDS for performing simulation. Simulation results also shows the effectiveness of simulation technique in real fire scenario.

2 Experimental Study

Chaudhary et al. [10] conducted experiments in an enclosure of dimension 4 m × 4 m × 4 m, as shown in Fig. 1. Fire is produced by burning of jatropha oil in pan diameter of 0.5 m. Natural ventilation is provided by an opening on front wall of enclosure. Initial fuel quantity is 10.095 kg, and fuel is filled in the pan up to a height of 0.045 m. The fuel is heated by using an oil heating assembly. Fuel pan assembly is kept on weighing platform to record the fuel consumption rate. Heat release rate is determined using mass loss method. Temperature is measured at location of ceiling, corner and centerline of door opening. Figure 2 shows the positions of the thermocouples at 0.5 m below ceiling.

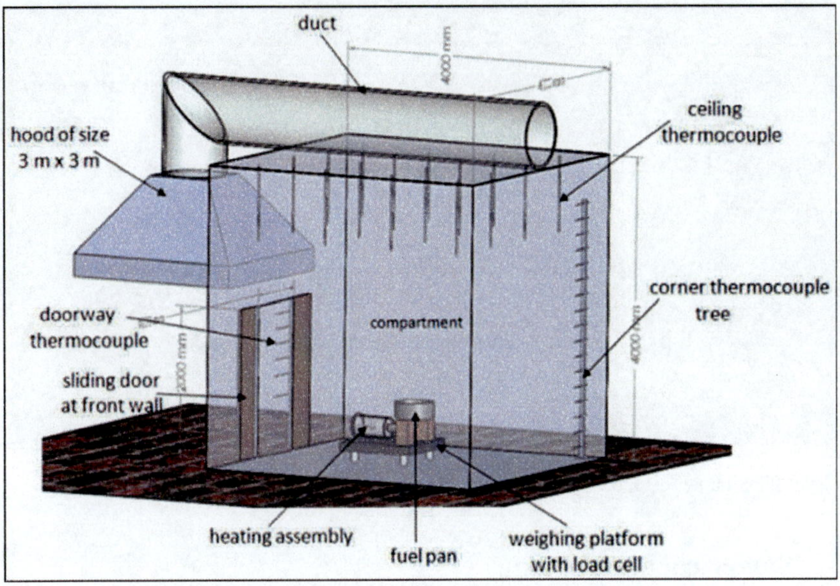

Fig. 1 Details of experimental test enclosure

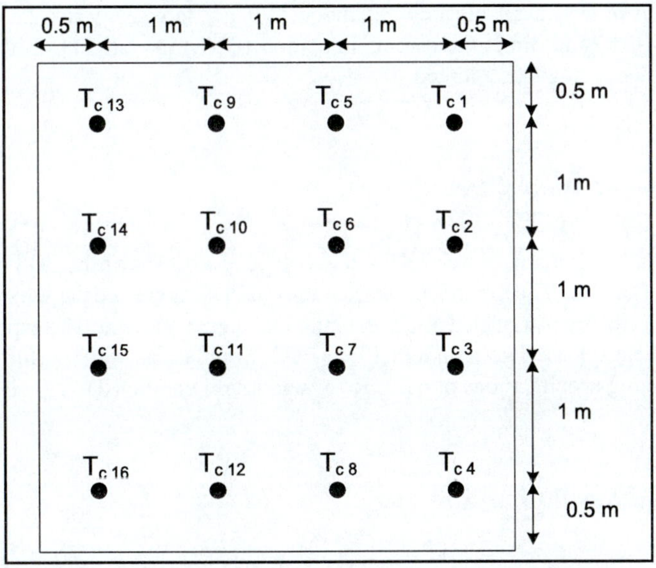

Fig. 2 Schematic showing positions of thermocouples at 0.5 m below the ceiling

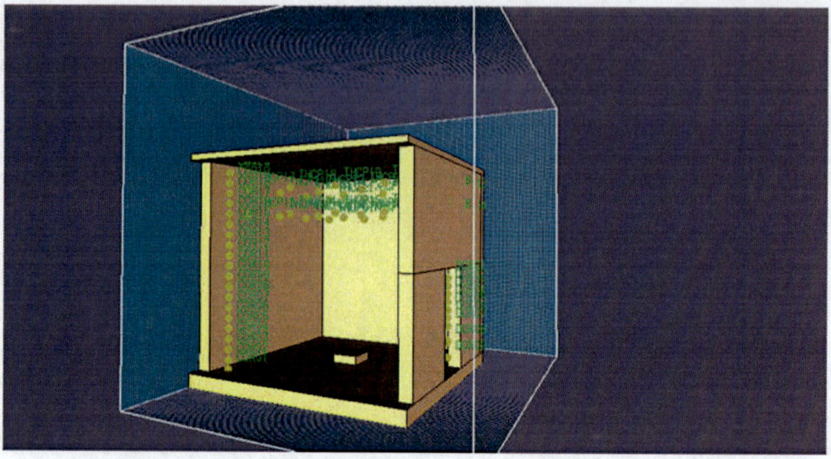

Fig. 3 FDS numerical model with 0.05 m grid size

3 Numerical Simulation

Fire dynamics simulator (FDS) version 6.2.0 is used for performing the simulation. The various models available are $k - \varepsilon$ model [11–13], large eddy simulation (LES) model [14–16] and direct numerical simulation (DNS) model [17]. In this paper, LES turbulence model is adopted.

3.1 Computational Details

The calculation domain set was $7.46 \times 6 \times 6\ \text{m}^3$. Figure 3 shows the FDS model for grid size of 0.05 m. The thermocouple positions in FDS model are the same as in test enclosure. The fire is modeled as a horizontal rectangular surface. The experimental obtained heat release rate is inputted into FDS. Simulations were performed on a system having specifications of 6 GB RAM and Intel Pentium (R) 2.90 GHz.

3.2 Grid Resolution Analysis

Simulation results depend upon the grid size. Axially symmetric flames using FDS were simulated, and a characteristic length of around 0.05 was reported for best results by Ma and Quintiere [18]. A ratio D^*/Δ between 4 and 16 has been recommended to accurately resolve the fire events [19]. Equation (1) is used to determine the grid size, for the present study.

Table 1 Simulated domain with total number of cells for different mesh sizes

Mesh	Mesh size Δ (m)	Simulated domain $x \times y \times z$ (m^3)	Total number of cells
Coarse	0.12	7.46 × 6 × 6	138,240
Medium	0.05	7.46 × 6 × 6	2,343,750

$$D^* = \left(\frac{\dot{Q}}{\rho_\infty C_P T_\infty \sqrt{g}} \right)^{2/5} \tag{1}$$

Here D^* is characteristics fire size in m; $\dot{Q}$ is fire power (heat release rate) in kW;C_P is air specific heat in kJ/kgK; ρ_∞ is ambient air density in kg/m^3; T_∞ is ambient temperature in K; g is gravitational acceleration in m/s^2. The average heat release rate obtained in the jatropha oil pool fire is estimated to be 198.2 kW. So D^* is computed to be 0.495 for this fire size. Table 1 gives the value of mesh size considered in the simulation.

Two grid sizes of 0.12 and 0.05 m are chosen to verify the independence of FDS simulation results on the grid size. Comparison of FDS predication with experimental results is presented at some selected locations. Among simulation parameters, ambient temperature was set as 30 °C, ambient pressure 10^5 Pa, relative humidity 54%, simulation time 2200 s, simulation type LES.

4 Results and Discussion

4.1 Heat Release Rate

Figure 4 represents the variation of mass loss rate and corresponding heat release rate. The heat release rate is calculated using Eq. (2).

$$\dot{Q} = \dot{m} \Delta H_c \chi \tag{2}$$

where $\dot{m}$ is mass loss rate (kg/s), ΔH_c is heat of combustion (38,800 kJ/kg), combustion efficiency χ is assumed to be 0.85 [10].

After the ignition of jatropha oil fuel, heat release rate rises sharply to 180 kW at 500 s; this time period shows the growth phase. After that, fire behavior shows steady profile up to 1710s. Later heat release rate starts to decrease, pointing out the decay phase. The highest heat release rate achieved is 225.0 kW corresponding to 1480 s. The curve can be divided into four parts including (I) ignition; (II) growth; (III) steady profile; and (IV) decay. The experimentally obtained heat release rate is used as input to FDS for performing simulation.

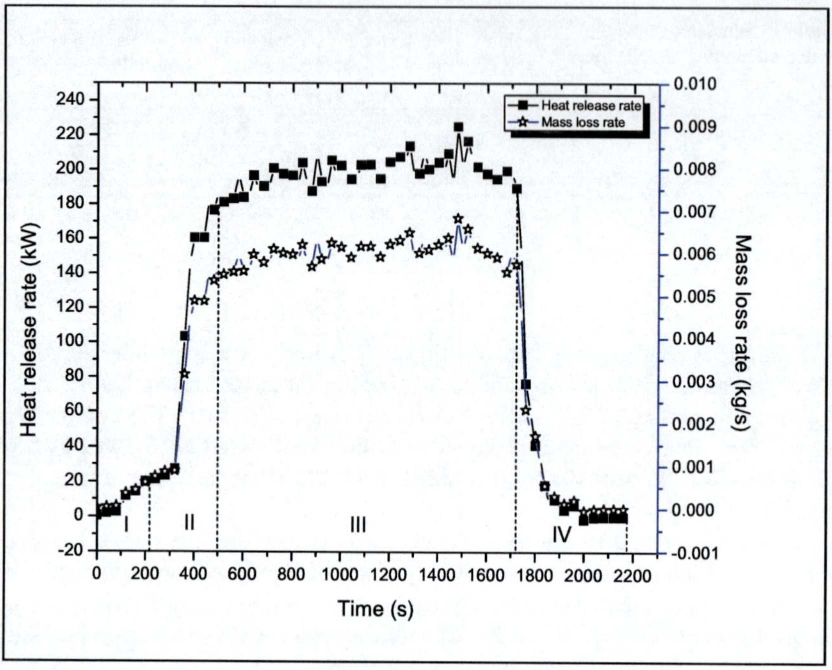

Fig. 4 Mass loss rate and heat release rate variations with time for pool size of 0.5 m

4.2 Hot Gas Layer Temperature

Figure 5 shows the comparison between the temperature under the ceiling (hot gas layer temperature) for T_{C1} and T_{C3} from the experimental and FDS results. During the phases (I) and (II), FDS results are very close to measured value. Hot gas layer temperature during experiment was 180–185 °C. Experimental results are higher than the FDS results just before the start of decay phase, maximum percentage deviation 20%. During the decay phase, FDS results drops down more than experimental results. The results verify that the hot layer temperature in the fire enclosure can be simulated very well by FDS.

4.3 Doorway Temperature

Figure 6 shows the temperature at center of the door for experimental as well as simulation. Both types of flows exist at the door: outward and inward. Above the neutral plane height, hot gases generated in the enclosure leave the enclosure, while to sustain the combustion, fresh air enters from below the neutral plane. From Fig. 6, it could be seen that simulation results matched well with experimental results.

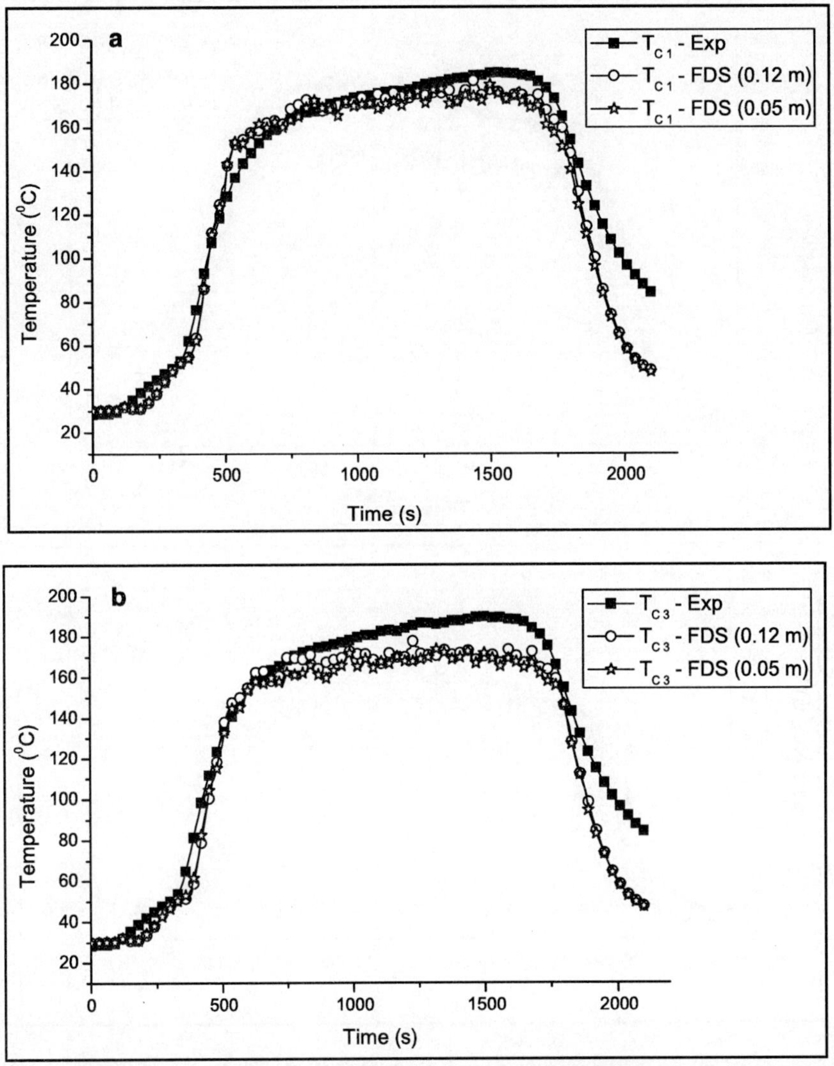

Fig. 5 **a** Comparison of hot gas layer temperature T_{c1} for experimental and numerical study. **b** Comparison of hot gas layer temperature T_{c3} for experimental and numerical study

4.4 Smoke Layer Height

Two methods have been used to calculate the smoke layer height, i.e., by analyzing the vertical temperature profile and by visual analysis. Visual analysis refers to the analysis of video recording which was done with Fujifilm HS50EXR digital camera during the whole duration of the experiment.

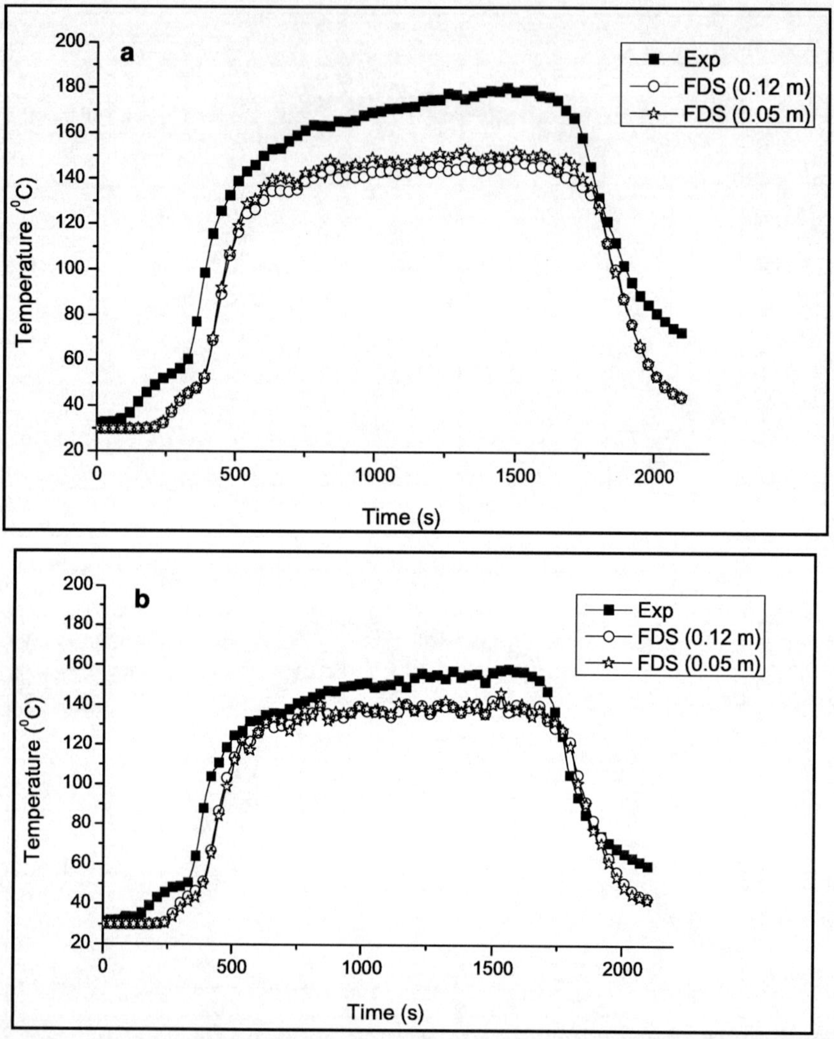

Fig. 6 **a** Comparison of temperature for door height 1.8 m. **b** Comparison of temperature for door height 1.5 m

Determination of smoke layer height is based on the studies presented in the work [20, 21]. The smoke layer height (shown in Fig. 7) is not same during experiment and in simulation results. Experimentally determined smoke layer height is approximately at 1.1–1.2 m while FDS predicted at grid size of 0.05 m lies at 1 m; but at grid size of 0.12 m, smoke layer height lies at 0.7–0.8 m. After height of 1.2 m, predicted temperature at both grid sizes does not change much and matched with the experimental findings (Fig. 5). Thus, in the upper zone of enclosure, changing the

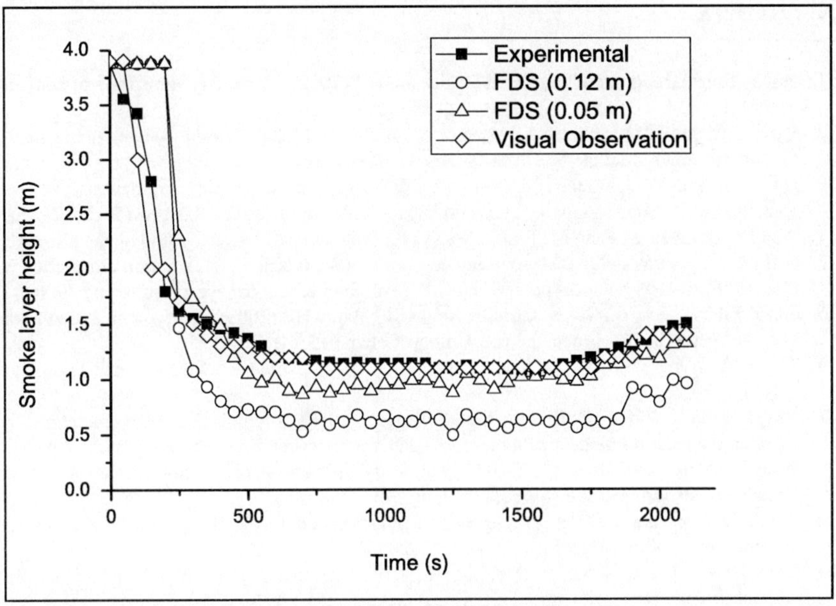

Fig. 7 Smoke layer height for the experiment and simulation results

grid size does not show much effect; but in the lower zone, better agreement is found between the experimental values and FDS results at medium grid size of 0.05 m.

5 Conclusions

In this paper, an attempt has been made to validate the experimental results of 0.5 m jatropha oil pool fire with simulation results using FDS. Simulation results matched well with experimental findings in upper region of the enclosure. Changing the grid size from 0.12 to 0.05 m does not have any significant impact on simulation results in upper region, while in lower region of the enclosure, better simulation results are obtained at grid size of 0.05 m.

Acknowledgements The present work is supported by Bhabha Atomic Research Centre, Mumbai, under the Grant No: DAE 507 MID.

References

1. Xin Y, Gore J, Mcgrattan KB, Rehm RG, Baum HR (2002) Large eddy simulation of buoyant turbulent pool fires. Proc Combust Inst 29:259–266
2. Ryder NL, Sutula JA, Schemel CF, Hamer AJ, Brunt VV (2004) Consequence modeling using the fire dynamics simulator. J Hazard Mater 115:149–154
3. Hu Z, Utiskul Y, Quintiere JG, Trouve A (2007) Towards large eddy simulations of flame extinction and carbon monoxide emission in enclosure fires. Proc Combust Inst 31:2537–2545
4. Wen JX, Dembele S, Yang M, Tam V, Wang J (2008) Numerical investigation on the effectiveness of water spray deluge in providing cooling. Smoke dilution and radiation attenuation in fires. In: Fire safety science–proceedings of the ninth international symposium, pp 639–650
5. Xin Y, Filatyev SA, Biswas K, Gore JP, Rehm RG, Baum HR (2008) Fire dynamics simulation of a one-meter diameter methane fire. Combust Flame 153(4):499–509
6. Trouve A, Wang Y (2010) Large eddy simulation of enclosure fires. Int J Comput Fluid Dyn 24(10):449–466
7. Yang D, Hu LH, Jiang YQ, Huo R, Zhu S, Zhao XY (2010) Comparison of FDS predictions by different combustion models with measured data for enclosure fires. Fire Saf J 45(5):298–313
8. Wang Y, Chatterjee P, de Ris JL (2011) Large eddy simulation of fire plumes. In: Proceedings of the Combustion Institute, vol 33, No. 2, pp 2473–2480
9. Paudel D, Hostikka S (2019) Propagation of model uncertainty in the stochastic simulations of a compartment fire. Fire Technol 55:2027–2054
10. Chaudhary A, Gupta A, Kumar S, Kumar R (2017) Thermal environment induced by jatropha oil pool fire in a compartment. J Therm Anal Calorim 127:2397–2415
11. Yan Z, Holmstedt G (1999) A two equation turbulent model and its application to a buoyant diffusion flame. Int J Heat Mass Transf 42(7):1305–1315
12. Wang HY, Joulain P, Most JM (1999) Modeling on burning of large scale vertical parallel surfaces with fire induced flow. Fire Saf J 32(3):241–271
13. Liu F, Wen JX (2002) The effect of turbulence modeling on the CFD simulation of buoyant diffusion flames. Fire Saf J 37(2):125–150
14. Hostikka S, Mcgrattan KB, Hamins A (2002) Numerical modeling of pool fires using LES and finite volume method for radiation. in: fire safety science—proceedings of the seventh international symposium, pp. 383–394
15. Ferraris S, Wen JX, Dembele S (2005) Large eddy simulation of a large scale methane pool fire. In: Fire safety science—proceedings of the eighth international symposium, pp. 963–974
16. McGrattan KB, Baum HR, Rehm RG (1998) Large eddy simulations of smoke movement. Fire Saf J 30:161–178
17. Clement, J.M. and Fleischmann, C.M. Experimental verification of the fire dynamics simulator hydrodynamic model. In: Fire safety science—proceedings of the seventh international symposium, pp. 839–851
18. Ma TG, Quintiere JG (2003) Numerical simulation of axi-symmetric fire plumes: accuracy and limitations. Fire Saf J 38:467–492
19. NRC (2007) Verification and validation of selected fire models for nuclear power plant applications. In: NUREG-1824, 2007. U.S. Nuclear Regulatory Commission Washington D.C.
20. He Y (1997) On experimental data reduction for zone model validation. J Fire Sci 15:144–161
21. He Y, Fernando A, Luo M (1998) Determination of interface height from measured parameter profile in enclosure fire experiment. Fire Saf J 31:19–38

Thermal and Emissions Characteristics of Pressurized Kerosene Stoves with Selected Commercial Burners

Gyan Sagar Sinha, Lav K. Kaushik, and P. Muthukumar

Abstract Kerosene is still an alternative for cleaner cooking than other conventional biomass cooking sources mostly used in rural areas of the developing countries. Kerosene stoves are cheap and easily available. The commercial burners for pressurized kerosene cooking stoves that are available in the Indian market have been the main focus in the present article. This paper also highlights the performances of various kerosene cook-stoves. From available kerosene cook-stoves, venus burner assisted cook-stoves has the highest thermal efficiency of ~58%, whereas cook-stoves with roarer and silencer burner results in thermal efficiency of ~55% and ~54%, respectively. For the input power range of 1.5–3 kW, the CO and NO_x emission for venus burner was found in the range of 610–915 ppm and 19–35 ppm, respectively.

Keywords Pressurized cook-stove · Commercial Kerosene burners · Thermal efficiency · Emissions

1 Introduction

India and other developing countries are facing the energy crisis, which affects not only human health but also creates an environmental imbalance. In several village areas, people are still relying on kerosene and biomass for their cooking activities [1]. According to the 68th round of the NSSO, upto two-thirds of households in rural India still rely on firewood and cow dung for their primary cooking fuel needs. The Government of India has made efforts to enhance access to clean cooking energy by promoting kerosene, biogas, improved cook-stoves (ICS) and LPG through various policies and programmes [2].

G. S. Sinha
Kalinga Institute of Industrial Technology, Bhubaneswar, Odisha, India

L. K. Kaushik (✉) · P. Muthukumar
Indian Institute of Technology Guwahati, Guwahati, Assam, India
e-mail: love.kaushik90@gmail.com

S. Revankar et al. (eds.), *Proceedings of International Conference on Thermofluids*, Lecture Notes in Mechanical Engineering,
https://doi.org/10.1007/978-981-15-7831-1_43

Kerosene contains hydrocarbon ranging from C9 to C16. The stoichiometric combustion equation of kerosene can be similar to decane as given in Eq. 1.

$$C_{10}H_{22} + 15.5(O_2 + 3.76N_2) \rightarrow 10CO_2 + 11H_2O + 58.28N_2 \quad (1)$$

However, incomplete combustion results in emission of carbon monoxide (CO) and oxides of Nitrogen (NO_x). Sulphur impurities may result in emission of Sulphur dioxide (SO_2) but it is in negligible quantity hence not under consideration here. The flashpoint of standard kerosene is 37–65 °C, its auto-ignition temperature is 220 °C, and it has a higher calorific value of ~43 MJ/kg.

In the conventional burner, combustion happens in an open environment (gaseous environment), also known as free flame combustion (FFC). The temperature gradient is large during FFC, which leads to more thermal NO_x formation [3]. The thermal efficiency of biomass stoves are (5–10%) far less than the kerosene stoves [4]. Two types of kerosene stoves are available in market viz., pressurized kerosene cook-stove and wick cook-stove. The thermal efficiency of wick cook-stoves is very less as compared to the pressurized kerosene cook-stoves. The average thermal efficiency of pressurized kerosene cook-stove varies between 40 to 45% [5, 6]. The combustion in LPG cook-stoves is clean as compared to kerosene pressurized cook-stoves. Renewable energy sources are forward steps to reduce the energy crisis and pollutants emissions. But due to the high cost and technical difficulties, LPG and renewable energy are not feasible for the low-income peoples in the villages as well as at cities. Use of LPG in village areas is more complex. It is because of cost limitation and poor distribution network of LPG. The Government of India is giving a huge subsidy for kerosene. Figure 1 shows the subsidy, which has been disbursed by the Government of India for the past five years [7].

Reduction of kerosene consumption or utilization of the maximum energy is the only possible way forward to reduce the huge subsidies which are given by

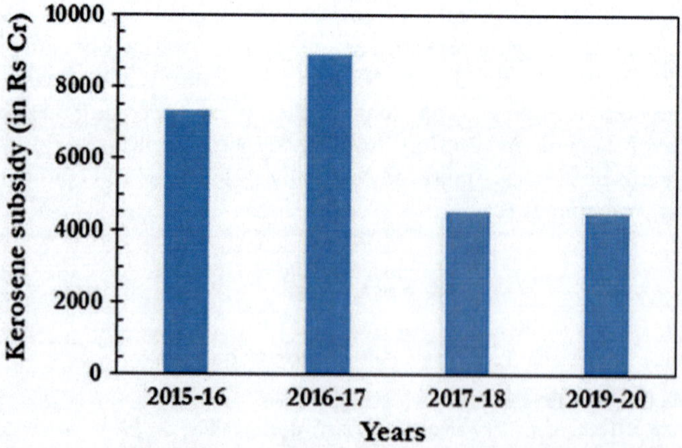

Fig. 1 Trend of expenditure on subsidies by the Government of India on kerosene [7]

the Government of India. For the purpose described above, a benchmark study is mandatory. Performance analysis of pressurized kerosene cook-stove with different commercial burners, which are available in the Indian market has been done in present work that includes thermal efficiency and emissions of NO_x and CO.

2 Construction of Pressurized Kerosene Cook-Stove

There are two types of kerosene stove viz., wick stove and pressurized stove used for the cooking application. Due to better combustion, pressurized kerosene stove is preferred. However, the emissions of CO and NO_x in kerosene pressurized stoves are higher than the WHO (World Health Organization) prescribed limits [8]. The thermal efficiency of pressurized kerosene stove is measured according to BIS (Bureau of Indian standard) guidelines (IS 10109:2002) [9]. Kerosene cook-stove burner consists of a pair of ascending and descending tubes, which is shown in Fig. 2.

Both ascending and descending tube meet to the burner head. Mid of the descending tubes consist of a spray nozzle. A rising tube that is connected to a fuel tank consists of an ascending tube. Stable combustion in pressurized kerosene cooking stove takes place at the surface of the metallic burner, which is shown in Fig. 3. Kerosene stove with pressurized type has been designed in such a way that the combustion that occurs in kerosene stove is always a fuel-rich type. The kerosene is typically burned as a suspension of droplets generated by atomization process. The vapour mix with surrounding air and ignite.

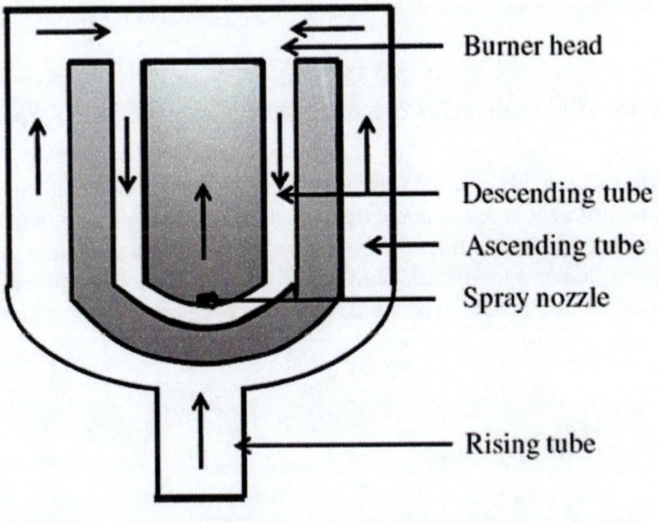

Fig. 2 Schematic of the stove burner [10]

Fig. 3 Illustrative interpretation of the free flame in a traditional cooking kerosene stove

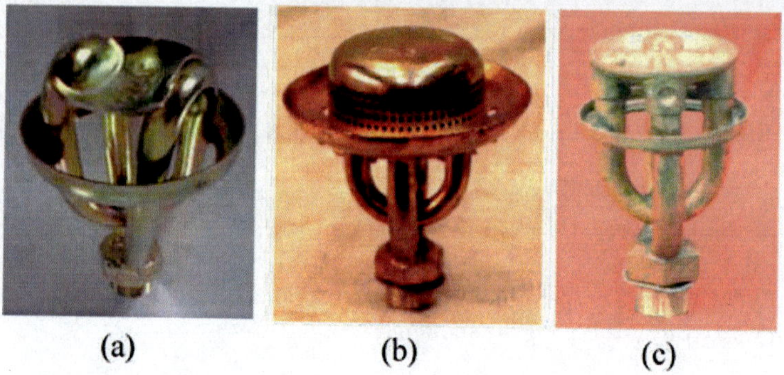

Fig. 4 Conventional burners available in the Indian market: **a** Venus burner, **b** Silencer burner, **c** Roarer burner

2.1 Types of Commercial Burner for Kerosene Cook-Stoves

For performing the analysis of kerosene pressurized stoves, a comprehensive survey was done on different types of the burner which are available in the Indian market. There are three types of burners which are available in Indian market viz., venus burner, roarer burner and silencer burner. Figure 4 presents the different types of burners. Due to the unique type of design of the venus burner, it has a better performance [11].

3 Experimental Method

In the present work, the thermal efficiency is measured according to BIS guidelines, which is called the water boiling test (WBT). For this purpose, a small amount of kerosene is poured to the spirit cup for ignition (as shown in Fig. 5). Burner head receives heat from the flame, which helps converts liquid kerosene to vapour. Kerosene vapour mixes with air and combustion takes place over the surface of the burner head, which is shown in Fig. 3. Burner head continuously receives the heat from the combustion flame and convert liquid kerosene to vapour and cycle continues. For measuring thermal efficiency, a known quantity of water is placed in aluminium pan of known diameter. Weight of the water with the pan and lid is measured in weighing machine (least count: 1 g). For the initial weight of the cook-stove, burning cook-stove is kept in the platform (dimension: 400 mm × 300 mm) of weighing machine and is then kept down and at same time pan with water is kept above the cook-stove. Water is heated up to 90 ± 0.5 °C (T_2) from room temperature (T_1). Temperature is measured by the glass in mercury thermometer (accuracy ±0.5 °C). An aluminium stirrer is used to maintain a uniform temperature of the water. After reaching the 90 ± 0.5°C, the cook-stove is quenched and the final weight is measured. The difference between the initial and final weight of the cook-stove gives fuel consumption. The time duration for heating process is also noted. Thermal efficiency is the ratio of heat consumed by water and pan to the heat given by fuel. In BIS standard formula for measuring thermal efficiency has given, which is given in Eq. 2.

$$\eta_{th} = \frac{\left(m_p \times C_p + m_w \times C_w\right) \times (T_2 - T_1)}{m_f \times \text{CV}} \tag{2}$$

where, C_p, C_w, m_w, m_p, m_f and CV is the specific heat of the pan, the specific heat of water, the mass of water, the mass of vessel including lid and stirrer, the mass of kerosene consumed and lower calorific value of kerosene, respectively. All experiment were repeated three times, and the average has been taken as a final result. The uncertainty in thermal efficiency was found as ±1.4%.

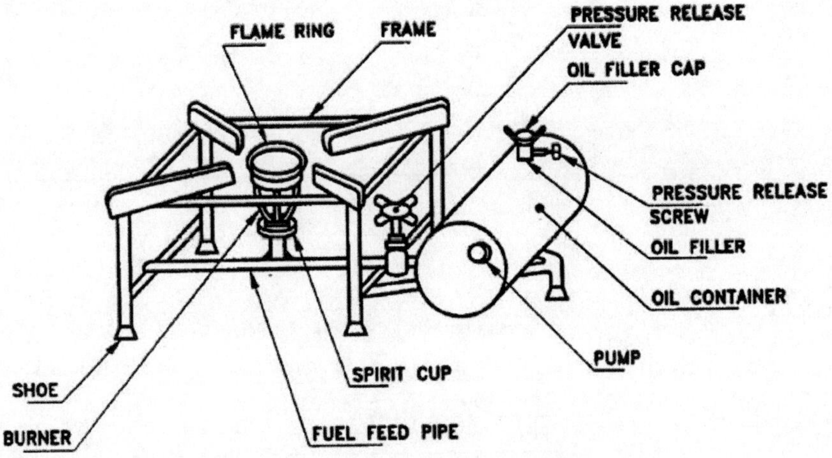

Fig. 5 Schematic of a BIS specified kerosene pressurized stove [8]

4 Results and Discussion

In this section, thermal efficiency and emission (CO and NO_x) results have been presented. For this purpose, three burners viz., venus burner, roarer burner and silencer burner have been selected. For different input power, the thermal efficiency of pressurized kerosene cook-stoves is shown in Fig. 6.

From the analysis of the experiments, it was observed that cook-stove with venus burner resulted in the maximum thermal efficiency of 58%. However, other experimented burners viz., roarer and silencer show maximum thermal efficiency as 55% and 54%, respectively. Table 1 shows the range of thermal efficiencies of the experimented burners. From Fig. 6, it was observed that for all the burners, the thermal efficiency decreases with increasing input power. The decrease in thermal efficiency in the present study when compared to previous works followed a similar trend [12, 13]. The reason behind thermal efficiency reduction was increasing heat loss with increasing input power. With an increase in input power, the height of the flame increases, which results in more convective heat loss [14, 15].

For measuring pollutant emissions, the guidelines prescribed in IS 10109:2002 was used [8]. A hood was fabricated, which was used for isolating the flue gases. By using extraction hood, flue gas sampling was done by a portable flue gas analyzer (Make: Testo-340). Since venus burner has the highest thermal efficiency, emission

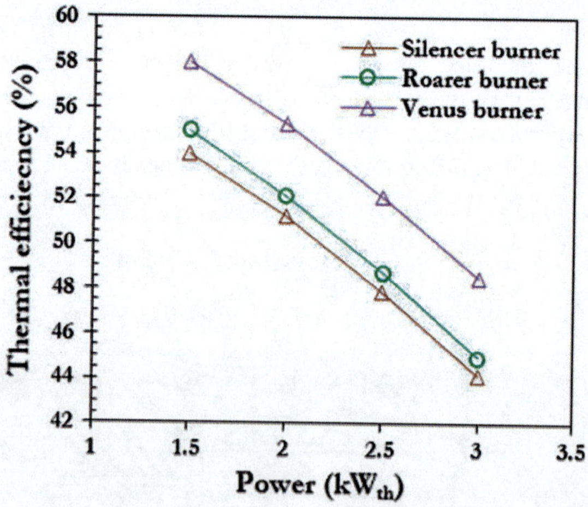

Fig. 6 Thermal efficiency of different conventional burners

Table 1 Thermal efficiency of different types of burners

Types of burners	Thermal efficiency (%)	
	Maximum	Minimum
Venus burner	58	48.5
Roarer burner	55	45.0
Silencer burner	54	44.1

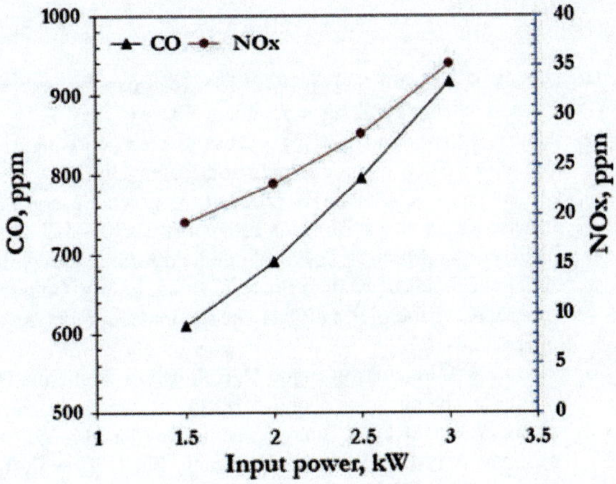

Fig. 7 Emission characteristics of a pressurized kerosene cooking stove with venus burner

from the same was measured. Figure 7 portrays the CO and NO_x emissions for different input power.

For input power range of 1.5–3 kW, the CO and NO_x were found in the range of 610–915 ppm and 19–35 ppm, respectively. Unlike thermal efficiency, pollutants emission increase with input power. Increasing input power leads to higher fuel-rich conditions resulting in more CO emissions. The local temperature of the flame increases with increasing input power and results in increased NO_x emission. The behaviour of emissions is found to follow a similar pattern, as reported in the earlier investigations [12–15]. However, pollutants emission of kerosene stove was over the safe limits prescribed by WHO [16]. So there is an immediate need for modification of existing burner so that pollutants emission can be reduced further and use the fuel energy to the maximum extent.

5 Conclusions

The present research work analyzes different commercial pressurized kerosene burners available in the Indian market. The impact of varying input power on thermal efficiency and emission characteristics have been investigated experimentally. The maximum thermal efficiency was found with venus-type burner. Venus burner resulted in a maximum of 58% thermal efficiency at input power 1.5 kW. For the same operating condition, 55% and 54% were the maximum thermal efficiency of roarer and silencer type burners, respectively. Thus in the venus-type burner, an ample amount of kerosene can be saved.

References

1. Patnaik S, Tripathi S, Jain A (2019) Roadmap for access to clean cooking energy in India. Council on Energy, Environment and Water (CEEW)
2. Patnaik S, Tripathi S (2017) Access to clean cooking energy in India. State of the Sector, Council on Energy, Environment and Water (CEEW)
3. Bhattacharya SC, Salam PA (2002) Low greenhouse gas biomass options for cooking in the developing countries. Biomass Bioenergy 22:305–317
4. Dendukuri G, Mittal JP (1993) Some field experiences with improved chulhas in introduced in rural households of Andhra Pradesh, India. Energy Convers Manag **34**, 457–464
5. Thukral K, Bhandari PM (1994) The rationale for reducing the subsidy on LPG in India. Energy Policy 22:81–87
6. Makonese T, Pemberton-Pigott C, Robinson J, Kimemia D, Annegarn H (2012) Performance evaluation and emission characterization of three kerosene stoves using a heterogeneous stove testing protocol (HTP). Energy Sustain Dev 16:344–351
7. Vaishnav A (2019) Demand for Grants 2019–20 analysis, petroleum and natural gas PRS legislative research. Institute for Policy Research Studies
8. Kandpal B, Maheshwari RC, Kandpal TC (1995) Indoor air pollution from domestic cooking stoves using coal kerosene and LPG. Energy Convers Manage 36:1067–1072
9. Bureau of Indian Standard (BIS) (2002) Burners for oil pressure stoves and oil pressure heaters, specification (2nd Revision): IS 10109 (2002)
10. Sharma M, Mishra SC, Mahanta P (2011) An experimental investigation on efficiency improvement of a conventional kerosene pressure stove. Int J Energy Clean Environ 12:79–93
11. Sinha GS (2017) Development and performance analysis of self-aspirated porous radiant burners for kerosene pressure stove. Ph.D. thesis. IIT Guwahati, Guwahati. https://gyan.iitg.ernet.in/handle/123456789/902
12. Kaushik LK, Muthukumar P (2019) Performance assessment of a porous radiant cook stove fueled with blend of waste vegetable oil (WVO) and kerosene. Energy Procedia 158:2391–2396
13. Kaushik LK, Muthukumar P (2018) Experimental analysis of a porous radiant pressurized cook stove by using a blend of waste cooking oil (WCO) and kerosene. In: 2018 international conference and utility exhibition on green energy for sustainable development (ICUE). IEEE (2018)
14. Sharma M, Mahanta P, Mishra SC (2016) Usability of porous burner in kerosene pressure stove: an experimental investigation aided by energy and exergy analyses. Energy 103:251–260
15. Sharma M, Mishra SC, Mahanta P (2016) Effect of burner configuration and operating parameters on the performance of kerosene pressure stove with submerged porous medium combustion. Appl Therm Eng 107:516–523
16. World Health Organization (2010) WHO guidelines for indoor air quality: selected pollutants. Switzerland World Health Organization, Geneva (2010).

Experimental Study on Elevated Methanol Pool Fires in a Compartment

Mahesh Kumar Tiwari, Akhilesh Gupta, Ravi Kumar, Raj Kumar Mishra, Avinash Chaudhary, and Pavan K. Sharma

Abstract The present paper reports an experimental investigation to explore the effect of fuel pan elevation on fire scenario due to methanol pool fires in a fixed ventilated compartment of size 4 m (height) × 4 m (length) × 4 m (width) having a door opening of 2 m (H_d) × 1 m (W). A reasonably large pool size of diameter 0.8 m is used in experiments so that the results are useful for actual plant fire hazard analysis (FHA) in future. The combustion characteristic of methanol is investigated in 1/32 ratio of fuel surface area to compartment floor area. Fuel pan was kept far toward rear wall center with initial fuel depth of 0.045 m and lip height of 0.055 m, respectively, instead of center (conventionally popular). The three fire source heights '*h*' for a circular pool fire experiments are 0.3, 0.6, and 0.9 m from the floor. Various parameters like heat release rate (HRR), flame centerline temperature, hot gas temperature, and heat flux are measured. It is noted that no one case flame is impinged to hot gas region. The difference between hot gas layer temperature and upper heat flux is found to be higher in Exp. 3 as comparison of other two experiments (Exp. 1 and Exp. 2). The HRR is also found to be higher for Exp. 3 due to availability of less height between hot region and fire source, which resulted in more evaporation of fuel.

Keywords Elevated fire · Compartment fire · Large-scale heat release calorimeter

M. K. Tiwari (✉) · A. Gupta · R. Kumar · R. K. Mishra
Department of Mechanical & Industrial Engineering, IIT Roorkee, Uttarakhand 247667, India
e-mail: mahesh.tiwari035@gmail.com

A. Chaudhary
Department of Mechanical Engineering, KIIT Deemed To Be University, Bhubaneswar, Odisha 751024, India

P. K. Sharma
Reactor Safety Division, Bhabha Atomic Research Center, Mumbai, Maharashtra 400085, India

S. Revankar et al. (eds.), *Proceedings of International Conference on Thermofluids*, Lecture Notes in Mechanical Engineering,
https://doi.org/10.1007/978-981-15-7831-1_44

1 Introduction

The information about development of fire within the compartment is essential for fire safety evaluation. The pool fire deals with heat transfer, fluid mechanics, and combustion kinetics [1]. Most of earlier researches have been focused on a fire source in a geometrically centered location inside a compartment [2, 3]. Tran [4] carried out experiments of bench-scale and full-scale compartment fires with different burner locations to validate and develop a wall fire model in over ventilated fire conditions. He found that there were significant changes in heat release rate (HRR) and time to flashover when the burner was located at either the centerline of the rear wall or the rear corner. Zhang et al. [5] describe the fire source level in an enclosure may be at above the floor surface, such as several oil pipelines in chemical, nuclear, thermal power plant, equipment, platforms in ship cabin, structure beneath the surface of ground, and other industries. This situation also represents any elevated fire of 0.8 m dia size. Different authors have performed experiments on elevated fires in compartments. Most of the works, however, are related to ceiling vented or closed compartment with standard fuel [6, 7]. An investigation of smoke filling behavior on elevated pool fires in closed compartment concluded that smoke filling time for the fire position at the floor level was less than elevated fire conditions. One gaseous burner was elevated inside the vertically ventilated compartment and attached to 6.1 m hallway with the forced ventilation at the end of hallway [8]. It has reported an increase in rate of CO generation within the exhaust gases moving out through the compartment with increase in height of fire source. Another work has used forced-ventilation condition to investigate temperature environment within the enclosure, when the air inlet was located at higher elevation and its subsequent effects on the fuel burning [9]. While addressing fire safety, burning behavior of the fuel provides one of the best input for defining the fire. The fuel burning behavior of pool fires with reference to increase in pool fire height provides useful data for designing safer buildings. To summarize, not many comprehensive and efficient description of burning behavior of methanol fuel in an enclosed space with elevated fire sources has been reported earlier in sufficiently naturally ventilated situation.

The present experimental work tries to fill the gap by examining thermal environment inside a large compartment subjected to methanol fire. The HRR, flame temperature, hot region gases temperature, incident heat flux at the fire source adjacent wall of the compartment, etc., were measured. This study also gives information about how much energy has targeted a fixed point of adjacent wall with respect to elevation of fire source which is an input to FHA in nuclear power plants (NPPs).

2 Methodology

Figure 1 shows the experimental facility developed in a compartment of size 4 m × 4 m × 4 m having door opening 2 m × 1 m ($H_d \times W$) with wall thickness 0.27 m made

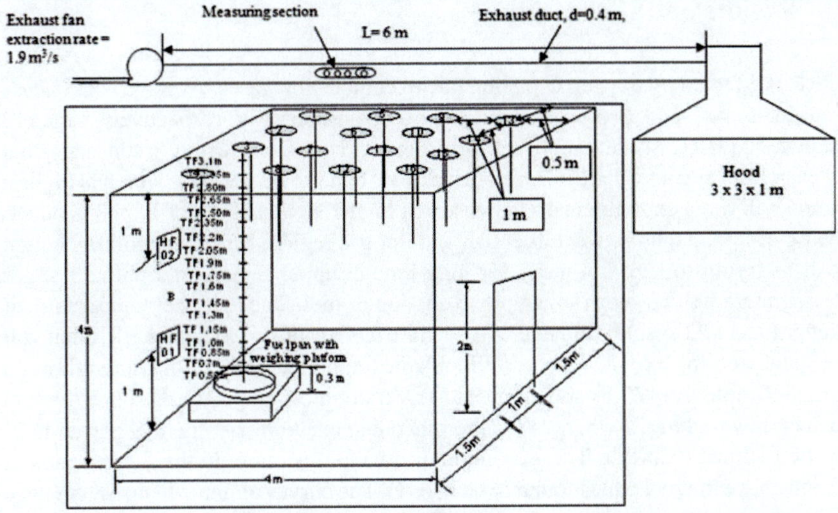

Fig. 1 Schematic of fire test compartment with fuel pan at 0.3 m above the floor (Exp. 1)

of normal bricks plastered on both side to investigate the full-scale fire experiments. The large-scale cone calorimeter has given heat release rate reading based on oxygen consumption theory with uncertainty of ±14% [10].

Three methanol pool fire experiments are used in the current study. A cylindrical mild steel fuel pan of diameter 0.8 m and lip height 0.10 m with initial fuel level of 0.045 m from the bottom surface is used. The fuel pan in compartment was elevated by 15, 30, and 45% with respect to fixed door height (H_d) of 2 m. The door was fully open during the all experiments. To change the fuel pan height, a flexible stand was prepared using two small diameter pipes of different diameters such that inner pipe slides into the outer pipe and the stand height is fixed by fixing bolts and nuts at specified height. The first experiment did not require the stand as the height of top surface of the weighing platform itself was 0.3 m from floor level. The weighing platform mounted on adjacent to back rear wall having accuracy of ±1 g. Water-cooled heat flux sensor-type SBG01 is used in these experiments. Two heat flux sensors with measuring range of 0 to 50 kW/m^2 are as shown in Fig. 1, i.e., HF01 and HF02. The HF01 heat flux sensor was placed at a distance of 1 m above from the floor, and HF02 was placed 1 m below from the ceiling at center of rear wall of the compartment. One K-type thermocouple is mounted over each heat flux sensor for measuring temperature at the surface of it.

3 Results and Discussion

HRR is affected by oxygen depletion due to combustion gases. Average heat release rate for Exp. 1 and Exp. 2 is found to be 260 and 265 kW, respectively. Figure 2 depicts the HRR of methanol fuel for different cases along with O_2 concentration. Meanwhile, average HRR in case of Exp. 3 is found to be 320 kW, which is highest among all three experiments. The reason of higher HRR in case of Exp. 3 is due to being of less height between fuel pan and hot gas region. Since less column height will be trying to kept hot gases for long time at upper region, and plume hot gas temperature will be keep rising upto extinction of fuel. Due to higher temperature of hot gas, the fuel evaporation rate will be more and resulting the higher HRR that will be generated for Exp. 3. The centerline flame temperatures were measured through thermocouple tree '*B*' as shown in Fig. 1. Variation of temperatures above the fuel pan is shown in Fig. 3 where '*Y*' is thermocouple location over the fuel pan and '*D*' is the fuel pan diameter. The maximum average temperature in the flame region at 0.3 m above the fuel pan is found to be 730 °C. The curves of flame temperature have been shown that flame temperature for Exp. 1 is found to be lower than the Exp. 2 and Exp. 3, and it may be due to column height that is higher for lower position of pan, so that more flame length will be covered by fresh air entrained through the door. Further increasing the fuel pan height will be affected to the cooling of flame length. Figure 4 illustrates that peak heat flux is achieved 23 kW/m^2 at HF02 for Exp. 3 due to minimum distance between HF02 and flame zone. The heat flux values are much affected by changing distance between continuous flame zone and heat flux sensors. The temperatures over the heat flux are being gradually increasing upto the extinction period (Fig. 5; Table 1).

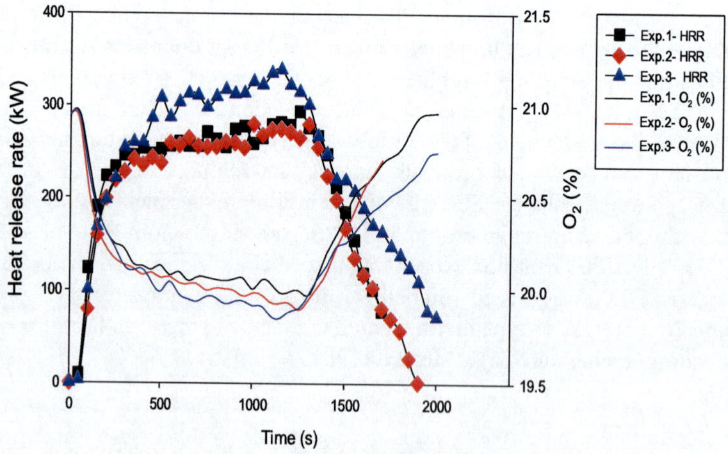

Fig. 2 HRR and O_2 concentration measured by LSHR calorimeter

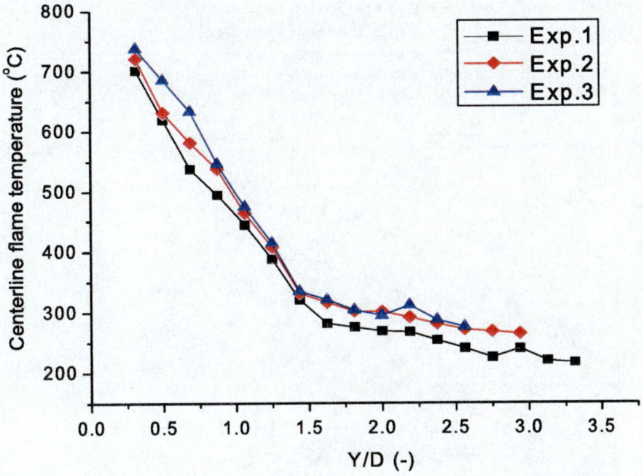

Fig. 3 Variation of flame temperature for different experiments

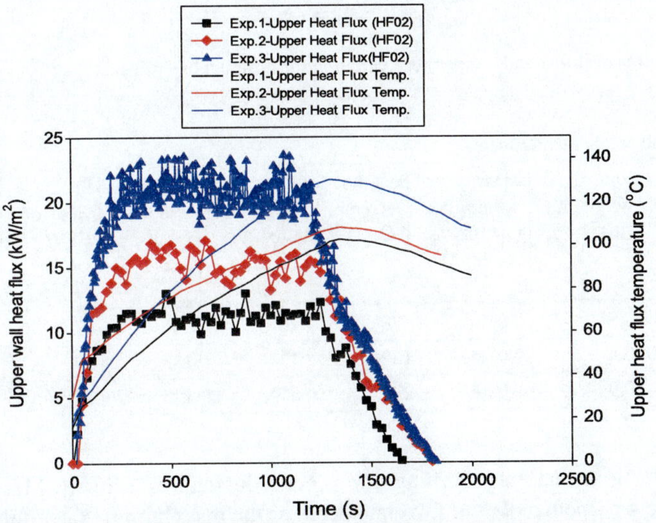

Fig. 4 Upper heat flux and temperature variation for different experiment

Above obtained experimental results show that increased fire source height does not relatively change to mass burning rate due to availability of sufficient oxygen at the base of the fire through ventilation. Previous literature [8] has been given that while elevating fire source height the flame interacts with stratified layer at upper hot region of the compartment. Therefore, oxygen availability gets reduced at the base of the fire due to occurring of less buoyant effect around the luminous flame zone. The buoyant force depends upon the temperature difference between flame and near the flame surrounding. In current study, flame merging with stratified layer is

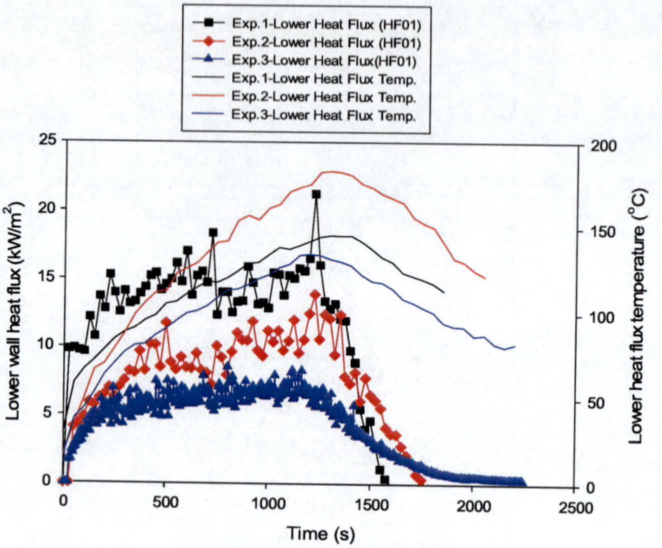

Fig. 5 Lower heat flux and temperature variation for different experiment

Table 1 Results of experiments

Exp. no.	Pan height, h, (m)	Mass burning rate, m″ (kg/m^2 s)	Peak hot gas temp. (°C)	Peak value of flux HF01, (kW/m^2)	Peak temp. at HF01, (°C)	Peak value of HF02, kW/m^2	Peak temp. at HF02, (°C)
1	0.3	0.0186	170	16	140	12	100
2	0.6	0.0188	180	12	185	17	105
3	0.9	0.020	220	6	125	23	130

not occurring so that mass burning rate is not reducing, respectively. Mass burning rate of the 4 m methanol pool fire considered as the mass burning rate of the infinite diameter of pool fires $\dot{m}''_{\infty}$ is taken as 0.017 kg/m^2s [11]. The mass burning rate is found to be 0.0158 kg/m^2s for fuel pan diameter of 0.6 m [12]. The maximum mass burning rate has been found 0.020 kg/m^2s in present study, which is close to mass burning rate of the infinite diameter of pool fires obtained from literature, therefore no need to do further experiment with increasing size of fuel pan in this compartment.

4 Conclusions

Naturally ventilated compartment fire experiments with methanol fuel as fuel source have been conducted three elevating height of fuel pan. HRR is the most important in fire hazard assessment, and it is interrelated with mass burning rate of fuel. Present results of mass burning rate have shown complete combustion of pool fire for all three fire source heights. It reveals that oxygen availability is sufficient for fuel burning in all taken fire source height. In present study, upper layer temperature is in the range of 170–220 °C with increasing fuel pan height. The higher temperature environment will expose higher energy to upper half wall region. The data generated can be used for validation purpose and to design passive fire protection system.

References

1. Karlsson B, Quintiere JG (2002) enclosure fire dynamics. CRC Press, Newyork
2. Quintiere JG, Rinkinen WJ, Jones WW (1988) The effect of room openings on fire plume entrainment. Combust Sci Technol 26:193–201
3. Merci B, Vandevelde P (2007) Experimental study of natural roof ventilation in full-scale enclosure fire tests in a small compartment. Fire Saf J 46:523–535
4. Tran, H. C.: Experimental aspects of validating a compartment wall fire model. In: Proceedings of 5th international fire conference, Interflam 90, London, England, 3–6 Sept, pp 13–24
5. Zhang J, Lu S, Li R, Yuen R, Yuan M, Li C (2013) Impacts of elevation on pool fire behavior in a closed compartment: a study based upon a distinct stratification phenomenon. J Fire Sci 31:178–193
6. Zhang J, Lu S, Li Q, Yuen R, Chen B, Yuan M, Li C (2012) Smoke filling in closed compartments with elevated fire sources. Fire Saf J 54:14–23
7. Zhang J, Lu S, Li Q, Li C, Yuan M, Yuen R (2013) Experimental study on elevated fires in a ceiling vented compartment. J Therm Sci 22:377–382
8. Mounaud LG (2004) A parametric study of the effect of fire source elevation in a compartment. Master Thesis, Virginia Polytechnic Institute and State University, USA
9. Backovsky J, Foote K, Alvares NJ (1989) Temperature profiles in forced-ventilation enclosure fires. Fire Safety Science, Hemisphere Publishing Corporation, Tokyo, Japan, pp 315–324
10. Huggett C (1980) Estimation of rate of heat release by means of oxygen consumption measurements. Fire Mater 4:61–65
11. Babrauskas V (1983) Estimating large pool fire burning rates. Fire Technol 251–26
12. Sahu D, Jain S, Shashi, Gupta A.: Experimental study on methanol pool fires under low ventilated compartment. Procedia Earth Planet Sci 1:507–515
13. Tiwari MK, Gupta A, Kumar R, Sharma PK (2019) Full scale experimental studies on liquid fuel fires in a compartment. J Energy Environ Sustain 8:24–27

Soot Formation Characteristic of Impinging Diesel and Biodiesel Blended Sprays at Diesel Engine-Like Conditions

Sanaur Rehman and Shah Shahood Alam

Abstract In the present experimental work, soot formation characteristics of impinging sprays of two fuels are analyzed at diesel engine-like operating conditions in the constant volume combustion chamber. Comparative study is conducted between diesel fuel and biodiesel blend (50% diesel and 50% biodiesel by v/v) under similar operating conditions in terms of soot formation. Hot surface ignition approach is employed for the combustion of sprays. Biodiesel developed through conventional transesterification process from waste vegetable oil is used. Hot surface temperature and cylinder air pressure are varied from 623 to 723 K and 20 bar to 40 bar respectively. Photo transducer is used for indicating and recording the whole combustion process on-screen of four-channel scopemeters. It is found that soot formation during combustion of impinging sprays of both fuels significantly reduced with increase in hot surface temperature and cylinder air pressure. It is also noted that soot formation during combustion of biodiesel blend sprays (BD50) is appreciably lower than soot formation of diesel fuel sprays at all operating conditions in present study. Biodiesel blend appears to be a better choice with respect to soot formation emissions in high-speed, high-injection pressure, small size, direct injection diesel engines having significant fuel impingement.

Keywords Soot formation · Waste vegetable oil · Biodiesel · Impinging spray

1 Introduction

Since its invention, a great deal of research has been focused on the diesel engine design and on the findings of suitable fuel for the engine for improving its performance. Many researches have been carried out on the vegetable oils as alternative fuels and reported vegetable oils as potential fuels for the engine [1, 2]. However,

S. Rehman (✉) · S. S. Alam
Department of Mechanical Engineering, ZHCET, Aligarh Muslim University, Aligarh 202002, India
e-mail: sanaur.rehman@zhcet.ac.in

S. Revankar et al. (eds.), *Proceedings of International Conference on Thermofluids*, Lecture Notes in Mechanical Engineering,
https://doi.org/10.1007/978-981-15-7831-1_45

using pure vegetable oil in diesel engine can cause numerous engine-related problems due to higher viscosity and lower volatility of vegetable oil in comparison to diesel fuel [3]. Over the years, biodiesel has been emerged as a possible alternative fuel for diesel engine due to insignificant changes in engine hardware [4]. Biodiesel is gaining importance because it is sulphur-free, biodegradable, highly oxygenated, renewable and non-toxic fuel. In the U.S. and Europe, biodiesel is produced from soyabean oil and rapeseed oil respectively using methanol.

Biodiesel blended diesel fuel sprays affect engine performance significantly [5, 6]. Further research is required for understanding the effects of impinging biodiesel blended sprays on emission formation in the engines. Impingement of fuel sprays occur on piston/surfaces in high-speed small-sized direct injection engines employing high-pressure common rail injection systems [7, 8]. Therefore, the current study is focused on the soot formation (SF) characteristic of vegetable oil-based biodiesel blend and also its SF comparison with the petroleum-based diesel fuel at typical diesel engine-like operating conditions in constant volume chamber (CVCC).

2 Experimental Setup

Experimental setup and its block diagram used in the present study are shown in Figs. 1 and 2, respectively. Commercial diesel fuel (DF) and waste vegetable oil based biodiesel blended fuel (50% diesel + 50% biodiesel by v/v) are used for comparative study at diesel engine like conditions in CVCC. DF properties are shown in the study [9]. Biodiesel used in present work is prepared from waste vegetable oil through well-known transesterification process. The diagrams and specifications of components employed in experimental setup are given in study [9]. Various components of the setup are given in Table 1. Experimental setup has a cylindrical combustion chamber with mechanical fuel injector and hot surface plate and both are mounted on opposite sides of the chamber. Heating element placed behind hot surface plate heated it to desired temperature. Fuel injector has single-hole pintle nozzle for producing hollow conical sprays, which impinge on hot surface/plate. Bosch fuel injection pump is mounted on mechanical system and connected with fuel injector via fuel line. Plunger of injection pump is actuated by lever and lever is operated by hand manually. Fuel injection pressure (IP) is obtained with pressure gauge placed on fuel injection pump. Cylinder air pressure (CP) or pre-firing pressure is obtained with pressure gauge placed on combustion chamber. Photo transducer (sensor) mounted on one side of combustion chamber opposite to hot surface plate to detect the combustion process. K-type thermocouples indicate hot surface temperature (HST) and cylinder air temperature on different indicators. Inlet valve and exhaust valve are mounted on same side of combustion chamber. Reciprocating air compressor delivered compressed air at various desired pressures.

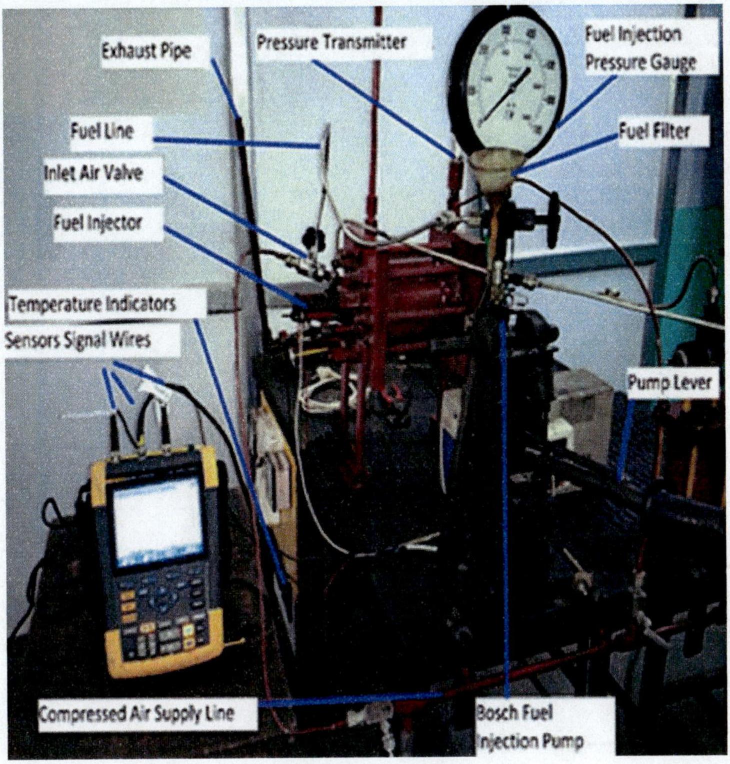

Fig. 1 Experimental setup

2.1 Experimental Procedure

For carrying out experiments on setup, first of all, temperature of hot surface plate is maintained at desired temperature, i.e. HST of 623 K or 673 K or 723 K. After maintaining HST, fresh air is supplied to the combustion chamber at desired CP such as 20, 30 and 40 bar. Once HST and CP are maintained in the chamber, fuel is injected into the chamber through injector and nozzle by the injection pump at IP of 400 bar. As soon as the injection and subsequent combustion events occur in the chamber, image on screen of the scopemeter is held and saved for estimating SF. Photosensor indicates whole combustion process on the screen of scopemeter as shown in Figs. 3 and 4. After saving the image of scopemeter at one operating condition, another reading is taken in the same manner as discussed above at other operating condition. At each operating condition, four readings are taken and average of four readings is noted as final value to ensure accuracy in measurement.

Experimental conditions or operating conditions considered in present study are diesel engine-like conditions and is given in Table 2.

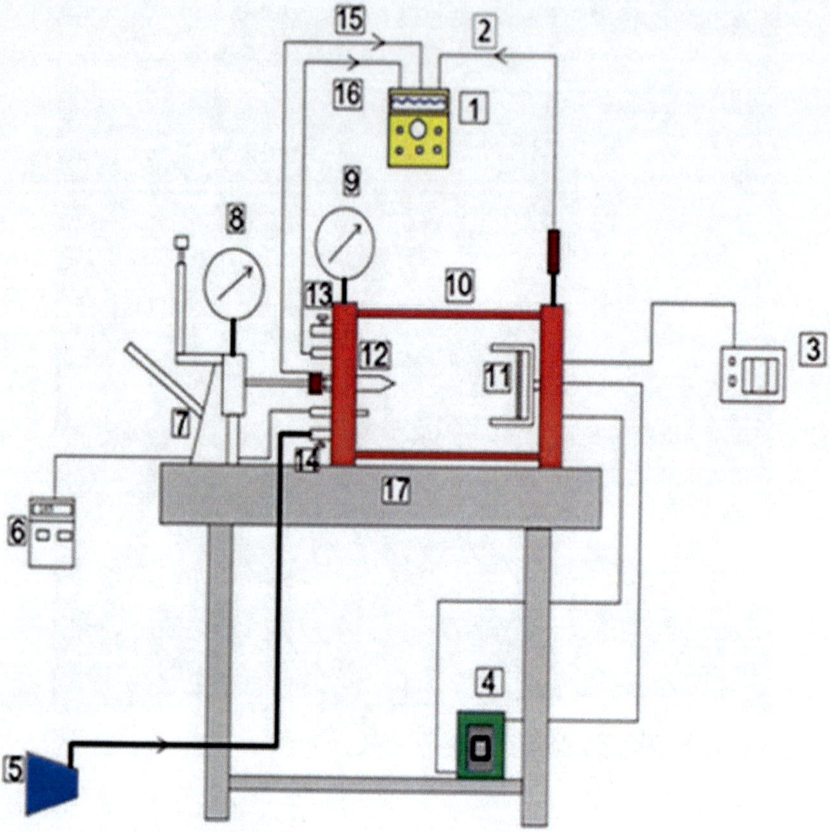

Fig. 2 Block diagram of experimental setup

3 Results and Discussion

3.1 Soot Formation Characteristic

The entire combustion process can also be characterized by the flame or natural luminosity [10]. The flame luminosity mainly consists of soot incandescence and chemiluminescence. The soot incandescence is the major source of flame luminosity [10, 11]. The chemiluminescence comes from OH, CH, and C2 radicals mainly and its intensity is much weaker than the intensity of soot incandescence [12]. Flame or natural luminosity also indicates the soot emissions because soot incandescence depends on soot concentration and temperature [13] and soot incandescence is the main part of flame luminosity. Flame or natural luminosity is the indicator of soot formation [11].

Table 1 Components of experimental setup

S. No	Components
2	Four-channel scopemeter
2	Signal from piezoresistive pressure sensor
3	Temperature controller and indicator
4	Transformer
5	Reciprocating air compressor
6	Air temperature indicator
7	Fuel pump assembly
8	Pressure gauge for IP
9	Pressure gauge for CP
10	Control volume combustion chamber
11	Hot surface plate for ignition
12	Single-hole pintle nozzle
13	Exhaust valve
14	Inlet valve
15	Signal from needle lift sensor
16	Signal from photosensor
17	Table for holding experimental setup

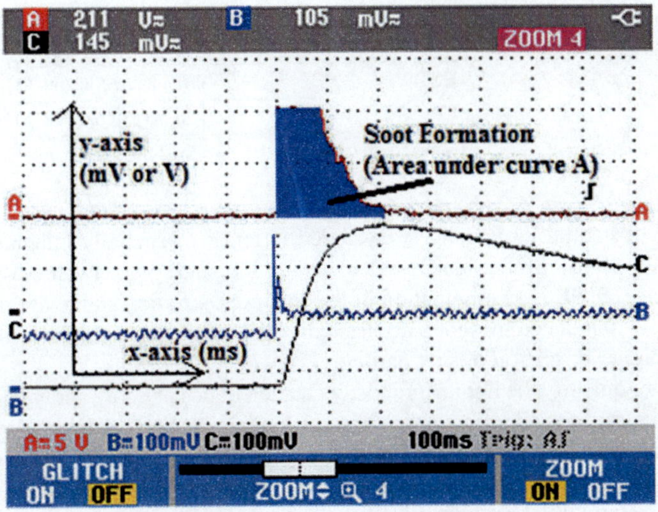

Fig. 3 Zoom image depicting area under the curve of channel B for the estimation of soot formation during diesel fuel spray combustion at operating condition (HST = 623 K, IP = 400 bar and CP = 20 bar)

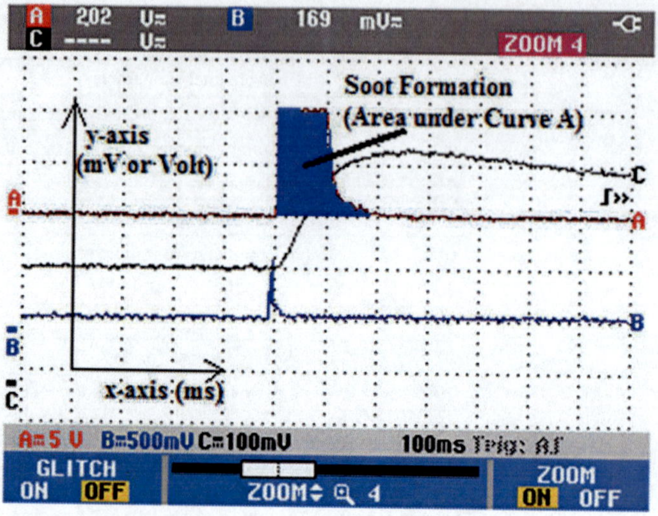

Fig. 4 Zoom image depicting area under the curve of channel A for estimation of soot formation during BD50 blend spray combustion at operating condition (HST = 673 K, IP = 400 bar and CP = 30 bar)

Table 2 Experimental conditions

Operating parameter	Typical values
IP	400 bar
HST	623 K, 673 K and 723 K
CP	20 bar, 30 bar and 40 bar
Injected quantity	0.12 ml@400 bar

Flame or natural luminosity is measured in terms of spatially integrated flame luminosity (SINL). SINL value is calculated by integrating pixel values over entire flame image. SINL is a combination of the soot concentration and temperature [14, 15] and it is dominated by soot particles radiation. Therefore higher values of SINL indicate higher formation of soot particles [14, 16].

To better quantify the flame luminosity characteristics of entire spray combustion duration or process, time-integrated natural luminosity (TINL) is employed [11]. The TINL is an indicator of soot formation (SF) level during whole combustion duration [11, 14]. TINL values are calculated by integrating SINL with time [11]. In the present study, TINL value is calculated as area under the flame curve A as shown in Figs. 3 and 4. The flame luminosity curve A is obtained as a function of time over the complete combustion process or DOC. Therefore, area under curve A indicates the level of SF during whole combustion process and it is denoted as TINL. TINL values are indicated in auxiliary units (a.u.).

3.2 Effects of HST and CP on SF

As shown in Fig. 5 and 6, SF during fuel spray combustion due to impingement of both fuels, i.e. diesel and BD50 sprays are significantly affected by HST and CP values. It is found that SF of both diesel and BD50 sprays substantially reduce with increase in HST from 623 to 723 K and CP from 20 to 40 bar.

Moreover, it can be noticed from Fig. 6 that with increase in HST, SF level decreases gradually in impingement assisted spray combustion. The decrease in SF with increase in HST may due to the reason that increase in HST results in corresponding increase in temperature of compressed cylinder air and therefore oxidation of fuel-rich regions increases with rising compressed cylinder air temperature. The reduction in SF with increase in CP as shown in Fig. 5 and it is may be due to the fact that increasing CP increases density of cylinder air and this causes enhancement in diesel and BD50 blend spray-air mixing process. Also, air entrainment into the fuel spray increases with the rise in CP. These factors may contribute to appreciable reduction in SF at higher CP.

3.3 Comparison of SF of DF and BD50 Sprays

It is found in Figs. 5 and 6 that SF (TINL values) in case of BD50 blend spray combustion is significantly smaller than SF during diesel fuel spray combustion at all diesel engine-like operating conditions in the present study. Vegetable oil based biodiesel blend used in present study (i.e. BD50) has smaller SF than diesel fuel due to different properties of biodiesel blend as compared to diesel fuel. Biodiesel blend (BD50) has higher cetane number, higher oxygen content, almost no sulphur and no aromatics and these properties cause significant reduction in exhaust emissions such as CO, HC and soot formation (SF) of biodiesel blends [17, 18]. Biodiesel blend (BD50) has relatively higher oxygen content than diesel fuel, therefore, biodiesel blend fuel–air mixture is more uniform and homogeneous and also biodiesel blend fuel–air mixture is more near to stoichiometric conditions. Therefore all these factors cause more complete, faster and clean combustion of biodiesel blend sprays than diesel fuel sprays.

It is observed that lowest percentage reduction in SF is achieved at HST of 673 K at all CP values. It is deduced in the present study that biodiesel blends appear to be a better choice in terms of soot formation emissions for modern high-speed, high-injection pressure, small-sized direct injection diesel engines having significant fuel impingement on piston/surfaces.

Fig. 5 Effect of CP on SF of DF and BD50 Sprays at **a** HST of 623 K, **b** HST of 673 K and **c** HST of 723 K

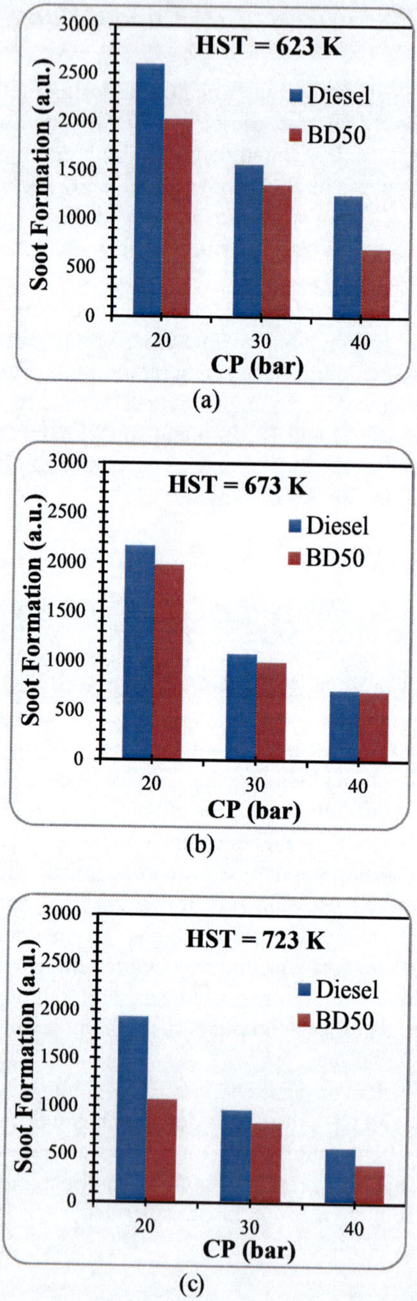

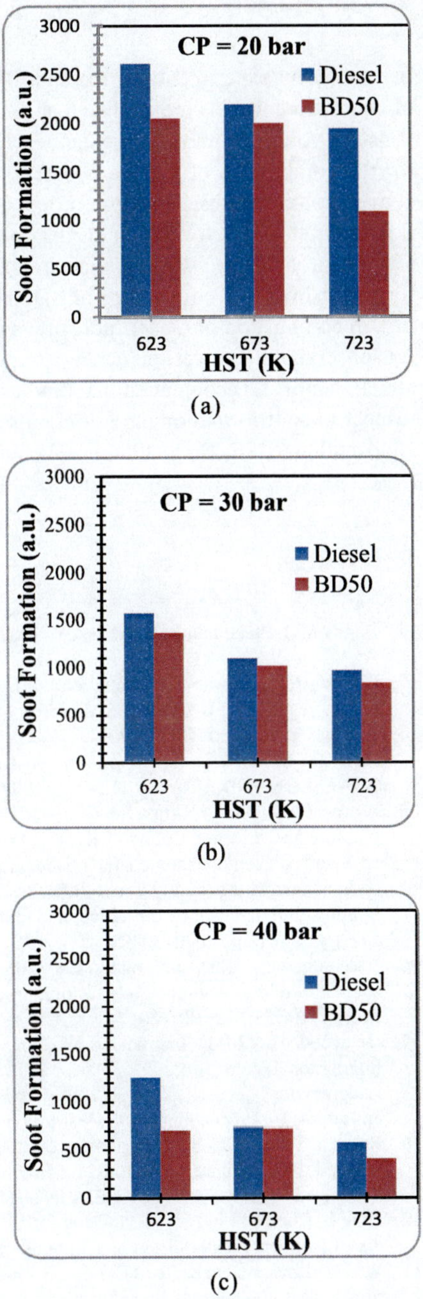

Fig. 6 Effect of CP on SF of DF and BD50 Sprays at **a** CP of 20 bar, **b** at CP of 30 bar and **c** at CP of 40 bar

4 Conclusions

In the present experimental work, soot formation characteristic of impinging sprays of two different fuels are analyzed at diesel engine-like operating conditions in the constant volume combustion chamber. Hot surface ignition approach is employed for the combustion of sprays. Biodiesel produced from waste vegetable oil using convention-al transesterification process is considered for making biodiesel blend. It is found that soot formation during combustion of impinging sprays of both fuels significantly reduced with increase in HST and CP values. It is also noted that soot formation in case of combustion of BD50 blend is appreciably lower than soot formation in combustion of diesel fuel sprays at all operating condition in present study. It is observed that lowest percentage reduction in SF is achieved at HST of 673 K at all CP values. It is concluded that biodiesel blend appears to be a better choice with respect to soot formation emissions in modern high speed, high-injection pressure and small-size direct injection diesel engines having significant fuel impingement.

References

1. Bagby MO, Freedman B, Schwab AW (1987) Seed oils for diesel fuels sources and properties. ASAE paper 87-1583
2. Goering CE, Schwab AW, Dangherty MJ, Pryde EH, Heakin AJ (1982) Fuel properties of eleven vegetable oils. Trans. ASAE 25(6):1472–1477
3. Korus RA, Mousetis TL, Lloyd L (1982) Polymerization of vegetable oils in vegetable oils fuels. In: Proceedings of the international conference on plant and vegetable oils as fuels. ASAE, Fargo, ND, USA, pp 218–223 (1982)
4. Agarwal AK, Dhar A, Gupta JG, Kim W, Choi K, Lee CS, Park S (2015) Effect of fuel injection pressure and injection timing of Karanja biodiesel blends on fuel spray, engine performance, emissions and combustion characteristics. Energy Convers Manag 91:302–314
5. Aldhaidhaw M, Miron L, Chiriac R, Badescu V (2018) Autoignition process in compression ignition engine fueled by diesel fuel and biodiesel with 20% rapeseed biofuel in diesel fuel. J. Energy Eng. 144(5):04018049
6. Aldhaidhaw M, Chiriac R, Badescu V (2017) Ignition delay, combustion and emission characteristics of diesel engine fueled with rapeseed biodiesel—a literature review. Renew Sustain Energy Rev 73:178–186 (2017).
7. Heywood JB (2011) Internal combustion engines fundamentals, 2nd edn. McGraw Hill International, New York
8. Ladommatos N, Xiao Z, Zhao H (2005) The effect of piston bowl temperature on diesel exhaust emissions. In: Proceedings of the Institution of Mechanical Engineers 219(3), 371–388
9. Rehman S, Zaidi K (2014) Experimental study of hot surface temperature effect on diesel and blended diesel exhaust emissions. In: International Conference on energy and Environment, Organized by IST and JNTUHCEH, JNTUH, Hyderabad, 15–17 Dec 2014
10. Feng L, Chen B, Liu H, Yao M, Geng C (2017) Combustion characteristics of wall-impinging diesel fuel spray under different wall temperatures. SAE Technical Paper 2017–01–2251
11. Wu H, Nithyanandan K, Lee TH, Lee CF, Zhang C (2014) Spray and combustion characteristics of neat acetone-butanol-ethanol, n-butanol, and diesel in a constant volume chamber. Energy Fuels 28:6380–6391
12. Dec J, Espey C (1998) Chemiluminescence imaging of autoignition in a DI diesel engine. SAE Paper 982685

13. Zhao H, Ladommatos N (1996) Optical diagnostics for soot and temperature measurement in diesel engines. Prog. Energy Combust. Sci. 24(3):221–255, 1998.FLEXChip Signal Processor (MC68175/D), Motorola
14. Du W, Zhang Q, Zhang Z, Lou J, Bao W (2018) Effects of injection pressure on ignition and combustion characteristics of impinging diesel spray. Appl. Energy 226:1163–1168
15. Zhang J, Jing W, Fang T (2012) High speed imaging of OH* chemiluminescence and natural luminosity of low temperature diesel spray combustion. Fuel 99:226–234
16. Liu Y, Li J, Jin C (2015) Fuel spray and combustion characteristics of butanol blends in a constant volume combustion chamber. Energy Convers Manag 105:1059–1069
17. Monyem A, Van Gerpen JH, Canakci M (2001) The effect of timing and oxidation on emissions from biodiesel-fueled engines. Trans ASAE 44(1):35–42
18. Graboski MS, McCormick RL (1998) Combustion of fat and vegetable oil derived fuels in diesel engines. Prog Energy Combust Sci 24:125–164

Effect of Inlet Swirl and Turbulence Levels on Combustion Performance in a Model Kerosene Spray Gas Turbine Combustor

Prakash Ghose and A. Datta

Abstract A computational simulation has been performed for Kerosene ($C_{12}H_{23}$) fuel through a 3D gas turbine combustor model. This numerical work is conducted to study the effect of temperature distribution pattern with different swirl and initial turbulent quantities. The turbulence effect is solved by the realizable two-equation model, discrete phase model (DPM). Eulerian–Lagrangian approach is used to simulate the interaction between continuous and discrete phase. The pressure-swirl atomizer model is also used here and the reaction was performed by constrained equilibrium chemical reaction model. Turbulent-chemistry interaction has been incorporated with the help of presumed beta probability density function (PDF). The effect of radiation on the temperature field is also incorporated by using discrete ordinate (DO) radiation model. A noticeable change in temperature pattern due to swirl can be proven by this simulation. Results also hint that the inlet turbulent intensity and length scale variation might not affect temperature.

Keywords Combustor · Discrete phase model · Turbulence effect · Discrete ordinate · Constrained equilibrium

Nomenclature

φ	Scalar quantity
D_s	Swirler diameter
D	Mass diffusivity
D_h	Hub diameter
Re	Reynolds Number

P. Ghose (✉)
School of Mechanical Engineering, Kalinga Institute of Industrial Technology Deemed To Be University, Bhubaneswar, Odisha 751024, India
e-mail: pghosefme@kiit.ac.in

A. Datta
Jadavpur University, Kolkata 700098, India

S. Revankar et al. (eds.), *Proceedings of International Conference on Thermofluids*, Lecture Notes in Mechanical Engineering,
https://doi.org/10.1007/978-981-15-7831-1_46

G_k Turbulent KE formation due to mean velocity gradient
ε Dissipation rate of turbulent KE
σ Prandtl number
l Length scale
SN Swirl number
τ Stress tensor
ρ Density

Super script

$'$ Fluctuation quantity
$''$ Fluctuation quantity including density fluctuation
P Pressure
ξ Mixture fraction
u Velocity vector
I Turbulent intensity
C Model constant
(~) Favre average
(_) Mean
$\Gamma\xi$ Turbulent Schmidt number
μ Viscosity
k Turbulent kinetic energy

Subscript

avg Average
ii Coordinate
jj Coordinate

1 Introduction

In gas turbine combustor, combustion of liquid jet fuel takes place to generate required heat. The hot burnt gas expands through turbine or nozzle to get desired work. In combustor, the air entered through compressor in three passages. The major part of the air (primary air) is used for combustion, secondary air completes the combustion by burning the remaining fuel, and carbon monoxide and the tertiary air reduce the temperature of the product through dilution [1–3], by which the turbine blade life can be extended. The primary air moves into the combustor through a swirler having various vane angle. Due to swirl, the pressure near the inlet region along the axis falls and creates an adverse pressure gradient. As a result, a central recirculation zone (CRZ) is formed. Swirling motion has certain advances, e.g., it restricts the flame stretching and thus reduces the flame blowout probability. Effects of inlet swirl levels on isothermal flow pattern in can combustors have been studied by different researchers [1, 2, 4, 5]. On the other hand, Kenbar et al. [6] investigated the effect of swirl level on both cold and reacting flow experimentally of a gas-fired burner with peripheral fuel injection.

Turbulence level in flow entering the combustor may also have an effect over the mean flow parameters such as velocity, pressure, and temperature distributions. Many researchers studied the effects of turbulence on the flow field, spray droplets residence time, drag and lift, etc.[5, 7, 8]. The variation of turbulent intensity and turbulent kinetic energy inside the flow field of the combustor has also received much attention [9–12]. Kushari et al. [13] have conducted experiments to study the effect of swirl and Reynolds number on turbulent intensity, dissipation rate, and turbulent kinetic energy. They found that with higher swirl number and Reynolds number, the turbulent parameters near the inlet increases, but turbulence levels subsequently decreases toward the exit of the combustor.

The turbulent combustion modeling in gas turbine combustors has either considered the eddy dissipation model [14–17] or uses an assumed probability density function (PDF) [18–21] to describe the interaction between turbulence and chemistry. Obieglo et al. [22] compared different output parameters of a non-premixed combustion, predicted by three different combustion models, with experiments and found that all the models predict almost equal results. However, the computational cost in PDF approach is much less, and therefore, this is often preferred over the other methods, particularly when complexities like pollutant formation and radiation have also taken into consideration.

In the present work, the effects of inlet swirl and turbulence levels variation on temperature distributions at outlet and wall have been studied in a model combustor as shown in Fig. 1. The combustor has two coaxial air entries such as; primary and secondary entry. The primary air is supplied through a swirler, having provision to vary the vane angle in order to impart the different swirl level. The turbulence intensity at inlet varies by varying the turbulent intensity and length scale. The fuel is injected from pressure-swirl injector placed at the center of the combustor. Experiment has also been performed in a test rig having same geometry as the geometry used for computational work attached with a 60° vane angle swirler for validation purpose.

2 Computational Model

Computational simulations has been performed in three 3D combustor geometry, where all the dimensions are kept same except the swirler part. Three different swirlers are used having vane angle of 40°, 50°, and 60°. The air inlets for primary

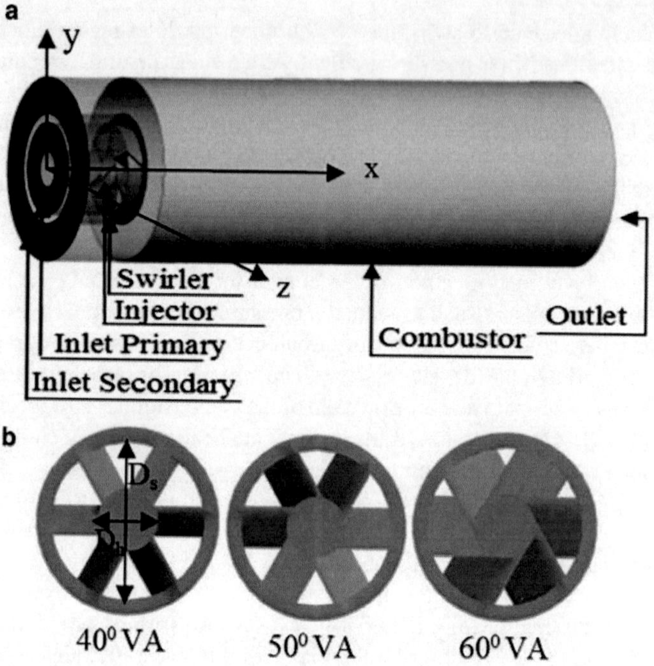

Fig. 1 **a** Computational model swirler, **b** swirler with different vane angle (VA)

and secondary air are kept at 40 mm behind to the swirler position. Various Favre-averaged (density-weighted average) governing equations are solved in Eulerian frame. The turbulence quantities are computed using realizable k-ε turbulent model. A particular cone angle is used for injection of liquid fuel. Linearized instability sheet atomization model (LISA) of Schmidt et al. [23] is considered to compute the primary breakup. The mean droplet diameter is obtained from LISA. Rosin–Rammler distribution function is used to evaluate the droplet class by considering a dispersion parameter of 3.5 in the spray. The Lagrangian approach has been adopted to track the movement of the evaporating droplets stochastically. The gas-phase combustion is assumed to be infinitely fast. An presumed β-PDF is considered to incorporate the turbulent effect on gas-phase reaction. Discrete (DO) ordinate model [24, 25] is used to incorporate the radiative heat transfer effect in gas phase only.

2.1 *Governing Equations*

ANSYS Fluent [26], CFD software package is used to compute the conservation of mass, momentum, energy, and scalar quantities. The universal forms of the Favre-averaged (density-weighted) governing equations for Eulerian phase are as follows:

Conservation of mass equation

$$\frac{\partial \overline{\rho}}{\partial t} + \frac{\partial}{\partial x_i} \overline{\rho}\,\tilde{u}_i = 0 \tag{1}$$

Conservation of momentum equation

$$\frac{\partial}{\partial t}(\overline{\rho}\,\tilde{u}_i) + \frac{\partial}{\partial x_j}(\overline{\rho}\,\tilde{u}_i\,\tilde{u}_j) = -\frac{\partial \overline{p}}{\partial x_i} + \frac{\partial}{\partial x_j}\left(\overline{\tau}_{ij} - \overline{\rho\, u_i''\, u_j''}\right) \tag{2}$$

Conservation of mixture fraction

$$\frac{\partial}{\partial t}\left(\overline{\rho}\tilde{\xi}\right) + \frac{\partial}{\partial x_j}\left(\overline{\rho}\,\tilde{u}_j\tilde{\xi}\right) = \frac{\partial}{\partial x_j}\left(\Gamma_\xi \frac{\partial \tilde{\xi}}{\partial x_i}\right) \tag{3}$$

Conservation of scalar quantities

$$\frac{\partial}{\partial t}\overline{\rho}\tilde{\phi} + \frac{\partial}{\partial x_j}\overline{\rho}\,\tilde{u}_j\tilde{\phi} = \frac{\partial}{\partial x_j}\overline{\rho}D\frac{\partial \tilde{\phi}}{\partial x_j} + \overline{S}_\phi \tag{4}$$

Realizable k–ε model [27] is used to simulate turbulent quantities. The governing equations for k and ε equations are expressed as:

$$\frac{\partial}{\partial x_i}(\overline{\rho}k\tilde{u}_i) = \frac{\partial}{\partial x_j}\left[\left(\mu + \frac{\mu_t}{\sigma_k}\right)\frac{\partial k}{\partial x_j}\right] + G_k - \overline{\rho}\varepsilon \tag{5}$$

$$\frac{\partial}{\partial x_i}(\overline{\rho}\varepsilon\,\tilde{u}_i) = \frac{\partial}{\partial x_j}\left[\left(\mu + \frac{\mu_t}{\sigma_\varepsilon}\right)\frac{\partial \varepsilon}{\partial x_j}\right] + C_{1\varepsilon}\frac{\varepsilon}{k}G_k - C_{2\varepsilon}\overline{\rho}\frac{\varepsilon^2}{k} \tag{6}$$

$$G_k = \mu_t\left(\frac{\partial \tilde{u}_i}{\partial x_j} + \frac{\partial \tilde{u}_j}{\partial x_i}\right)\frac{\partial \tilde{u}_i}{\partial x_j} \tag{7}$$

$$\mu_t = C_\mu \overline{\rho}\frac{k^2}{\varepsilon} \tag{8}$$

The combustion of kerosene fuel has been simulated by using an equivalent hydrocarbon $C_{12}H_{23}$ [28]. Constrained equilibrium modeling approach for combustion has been used by assuming that gas-phase chemical reactions are infinitely fast. The rich flammability limit is set as 0.095, where equivalence ratio is kept within the RFL. In order to obtain appropriate mean scalar quantities from instantaneous mean mixture fraction through turbulent-chemistry interaction, a presumed β-PDF model is used.

In liquid phase, 40 numbers of droplets classes of different size are injected from injector. The velocity of droplets are computed from the pressure difference between injector inside and ambient. A particular cone angle for droplet dispersion has been taken. Each droplet movements are tracked within the computational domain. The momentum interaction between two phases are taken into account by considering the drag on droplet by gas phase. Spherical drag law is used to incorporate the drag effect. Heating of droplet takes place through both convection and radiation. The convective heat transfer coefficient is evaluated by using the Ranz–Marshall correlation [29, 30]. The evaporation of droplet is tracked till complete evaporation, and the evaporated fuel has been added as source terms in gas-phase equations.

2.2 Operating Conditions

Total eight numbers of K-type thermocouples can be incorporated with Ag-paste in pots provided on the combustor wall. The space in between corresponding thermocouples are kept at an interval of 50 mm along the axis. There is a provision to allow the last thermocouple moves in traversed direction to measure the exit temperature in radial direction.

The simulation is performed with various operating conditions that are listed in Table 1. Average velocities at both inlets are given. The turbulent intensity and the length scale at inlet conditions are set by using a proposed empirical relations given in literature [26]. At the outlet plane, atmospheric pressure condition is given. Heat loss to the surrounding is considered from the combustor wall both due to convection and radiation. In DPM, the reflect condition is given on the solid walls and escape condition at the inlet and outlet planes for liquid droplets.

Table 1 Operating conditions

Flow conditions	Primary air	Secondary air	Others
Bulk velocity (m/s)	6.147	2.94	–
Temperature (K)	300	300	–
Swirl number	2.142	0.0	–
Air-to-fuel ratio	–	–	110:1
Operating pressure (atm)	–	–	1.0
Turbulent intensity	4%	4%	–
Length scale	0.07D	0.07D	–
Injected fuel temperature (K)	–	–	300
Spray cone angle (°)	–	–	36
Injector pressure (bar)	–	–	5
Droplet class	–	–	20

3 Results and Discussion

The analysis is able to capture the temperature distribution. Figure 2a, b depicts the comparison of the simulated and measured temperatures along the wall of the combustor and across the exit plane, respectively. From Fig. 2a, it has been observed that there is a well agreement between predicted wall temperature measured value, except within the region close to the combustor inlet. However, certain deviation from measured has been observed near central position at exit plane. Near-wall region outlet temperature prediction is much better.

The over prediction at outlet is more near the axis, and the maximum difference is about 15%. This over prediction can be attributed to the combustion model that has been employed. As we have assumed the reactions to be infinitely fast, a higher temperature is predicted in the flame due to slower rate of scalar transport. As a result, the predicted gas temperatures at the core of the combustor are found to be higher than the actual measured values.

3.1 *Effect of Swirl on Temperature Distribution*

Afterwards, three different inlet swirl numbers of the primary air flow have been considered in the simulation by varying the vane angle of the swirler. The swirl number (SN) is expressed as,

$$\mathrm{SN} = \frac{2}{3}\frac{\left(D_\mathrm{s}^3 - D_\mathrm{h}^3\right)}{\left(D_\mathrm{s}^2 - D_\mathrm{h}^2\right)\left(D_\mathrm{s} - D_\mathrm{h}\right)}\tan\theta \quad (9)$$

where D_S and D_h are the tip and hub diameters of the swirler and θ is the vane angle. Three different vane angles of 40°, 50°, and 60° are considered for which the swirl numbers are 1.038, 1.474, and 2.142, respectively.

Streamline plots in Fig. 3a–c indicate the flow patterns within the combustor for the three different swirl conditions. It is clearly observed from the figures that a

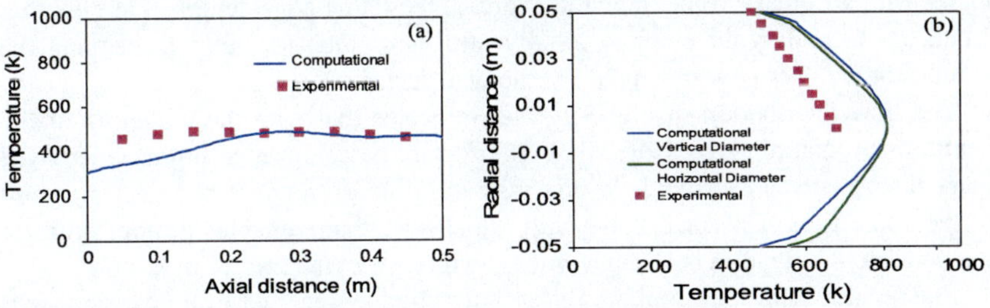

Fig. 2 **a** Wall temperature distribution, **b** outlet temperature distribution

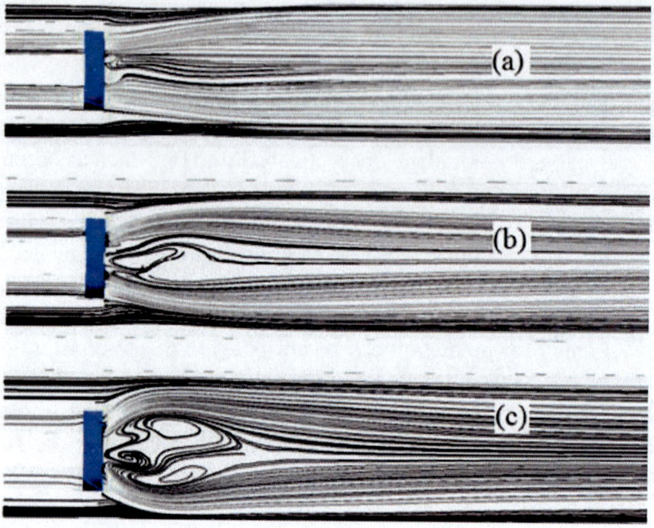

Fig. 3 Streamline contours on x–y plane **a** 1.038, **b** 1.474 (c) 2.142

much stronger central recirculation zone is observed as the swirl number increases. Moreover, the flow does not remain axisymmetric particularly in the recirculating zone. This is because of the influence by the swirler over which the primary air enters into the combustor. At a much downstream region, the flow becomes nearly parallel as indicated by the streamline distributions. The central recirculation zone close to the injector which is placed at the hub of the swirler helps to stabilize the spray flame.

Figure 4a–c show the temperature distributions in a central plane of the combustor for the three different swirl numbers. The highest temperature regions in the figures indicate the flame region, where the evaporated liquid fuel reacts with oxygen from primary air following the combustion model employed. Subsequently, the product gas mixes with the co-flowing secondary air to bring down the temperature. It is evident from the figures that at higher swirl number, the flame length decreases considerably. The strong reverse flow offers much higher drag on the fuel droplets as a result of which the spray cannot penetrate deeper into the combustor. The evaporated vapor mixes with air closer to the inlet and shortens the flame considerably. On the other hand, the secondary air entering axially into the combustor helps in keeping the combustor wall cool by grazing along the wall surface.

The flow distribution in Fig. 3 further indicates that with the widening of the central recirculation zone, better flow mixing can be achieved at higher inlet swirl level. The secondary air tends to mix with the primary product at an earlier point in the combustor. This fact along with the shorter flame enables a more uniform temperature distribution at the exit when the inlet swirl number is high.

Figure 5a presents the combined effects of radiation, secondary air flow, and swirl on the combustor wall temperature for three different swirled condition. At immediate entry to the combustion chamber, the wall temperature for each case of swirl is closed

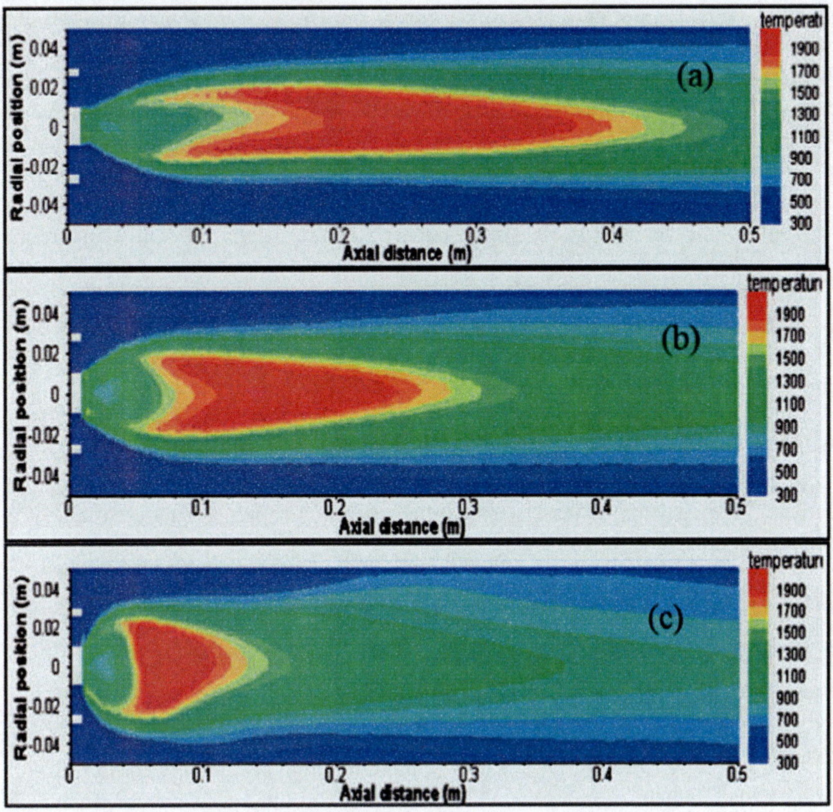

Fig. 4 Temperature contours on *x*–*y* plane **a** 1.038, **b** 1.474, **c** 2.142

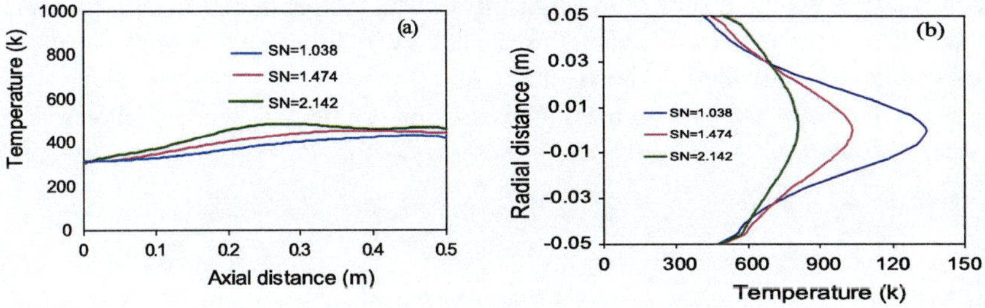

Fig. 5. Temperature distribution at different swirl number **a** wall, **b** outlet

to 300 K (inlet temperature of secondary air). The temperature shows an increasing trend toward the downstream for each swirl number except for SN = 2.142. First of all due to mild swirl with SN = 1.038, the height of flame is less. Therefore, in the inlet region, the distance in between higher temperature region of flame and wall increases which leads to reduce the heat transfer rate due to radiation because air is considered as a participating media. The cooling effect on the wall by the secondary air flow has an significant impact on wall temperature distribution. With high swirl, mixing of primary and secondary air stream is being better. Therefore, wall temperature increases toward the exit. That seems the enhancement of uniformity of temperature towards the downstream. But with the highest swirl, it is used in this simulation (SN = 2.142), maximum wall temperature occurred at near the half of the combustor length and remains nearly constant toward outlet. This is because proper diffusion due to strong swirling effect.

With considering upper half of Fig. 5b, it is very much clear that with strong swirl temperature at the exit become more uniform. With low swirl due to the poor mixing of product gases and secondary air, distribution of temperature is not much uniform, and its value is least at the wall. Again due to lengthy flame, the central outlet temperature is maximum.

3.2 *Effect of Inlet Turbulent Intensity and Length Scale on Temperature Distribution*

From Fig. 6, it can be observed that even though length scale is increased from 0.7 to 20% of combustor diameter, by keeping inlet turbulent intensity constant, any significance variation is not at all found on wall temperature distribution. But by using an increased value of turbulent intensity up to 20%, a little change on wall temperature is found. It shows the uniformity of wall temperature. But still the effect of turbulent intensity as well as length scale has a very minute effect on flow pattern and temperature distribution shown in Fig. 6.

From the result analysis, an overall observation can be presented as follows:

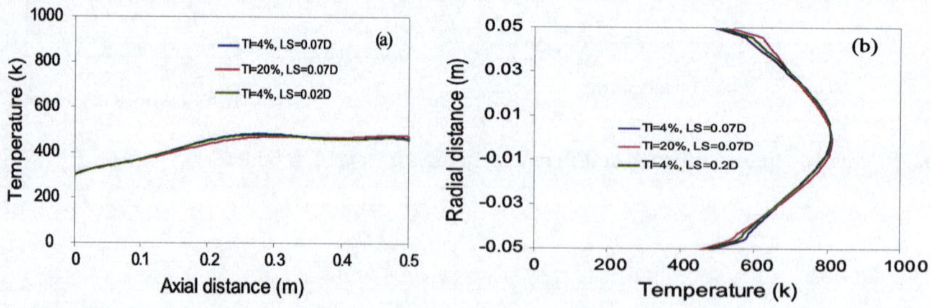

Fig. 6. Temperature distribution at different inlet turbulent intensity and length scale **a** wall, **b** outlet

- A strong CTRZ is formed with SN = 2.142, but flow is not exactly axisymmetric.
- Wall temperature is more with higher swirl number, but with the same condition outlet temperature pattern is uniform.
- There is a negligible effect of turbulent intensity and length scale at inlet on the temperature profile of the combustor.
- It is found that even though the problem appears axisymmetric, but due to the vane thickness and position, it is not exactly axisymmetric.

4 Conclusion

Computational modeling has been performed for a model can combustor to investigate the effect of swirl level and inlet turbulent parameters on flow and temperature field. Three different inlet swirl numbers of the primary air flow have been considered in the simulation by varying the vane angle of the swirler. The swirl number for different vane angles are evaluated as 1.038, 1.474, and 2.142, respectively. Three combination of turbulent intensity and turbulent length scale are used such as TI = 4% & $l = 0.007D$, TI = 20% & $l = 0.007D$ and TI = 4% & $l = 0.002$, respectively. The results are validated with the experiments conducted in laboratory. A considerable change in temperature pattern caused by swirl is found, but interestingly, the inlet turbulent intensity and length scale variation have least impact on it.

References

1. Mondal S, Datta A, Sarkar A (2006) Aerodynamics of film cooling jet interaction in a dump combustor. Numer Heat Transf Part 50:361–379
2. Mondal S, Datta A, Sarkar A (2004) Influence of side wall expansion angle and swirl generator on flow pattern in a model combustor calculated with k–ε model. Int J Therm Sci 43:901–914
3. Li L, Liu T, Peng XF (2005) Flow characteristics in an annular burner with fully film cooling. Appl Therm Eng 25(17–18):3013–3024
4. Zhang M, Hu Z, He G, Liu P (2010) Large eddy simulation of kerosene spray combustion in a model scramjet chamber. Proc IMechE Part G J Aerosp Eng 224(9)
5. Guo B, Langrish TAG, Fletcher DF (2002) CFD simulation of precision in sudden expansion pipe flows in low inlet swirl. Appl Math Model 26(1):1–15
6. Kenbar AMA, Beltagui SA, Maccallum NRL (1995) Combustion aerodynamics of a gas-fired furnace with peripheral fuel injection. Exp Thermal Fluid Sci 10(3):335–346
7. Holger H, Bec J, Grauer R (2013) Effect of turbulent fluctuations on the drag and lift forces on a towed sphere and its boundary layer. J Fluid Mech 721:155–179
8. Chu C, Chiu YH, Chen YJ (2009) Turbulence effects on the discharge coefficient and mean flow rate of wind-driven cross-ventilation. Build Environ 44(10):2064–2072
9. Rahim A, Veeravalli SV, Singh SN (2002) Effect of inlet swirl and dump gap on the wall pressure distribution of a model can combustor. Indian J Eng Mater Sci 9:472–479
10. Rahim A, Singh SN, Veeravalli SV (2007) Liner dome shape effect on the annulus flow characteristics with and without swirl for a can-combustor model. Proc IMechE Part A J Power Energy 221(3):359–369

11. Reddy A, Panduranga SRI, Chakravarthy SR (2006) Swirler flow field characteristics in a sudden expansion combustor geometry. J Propul Power 22(4):800–808
12. Baoyu G, Tim L, Fletcher AG, David F (2001) Numerical simulation of unsteady turbulent flow in axi-symmetric sudden expansion. J Fluids Eng 123(3):574–586
13. Yadav NP, Kushari A (2005) Effect of swirl on the turbulent behaviour of a dump combustor flow. Proc IMechE Vol. 224 Part G J Aerosp Eng 224(6):705–717
14. Magnussen BF, Hjertager BH (1976) On mathematical models of turbulent combustion with special emphasis on soot formation and combustion. In: 16th symposium (International) on combustion. The Combustion Institute
15. Taoufik G, Kamel G, Rachid S (2009) Numerical study of the swirl effect on a coaxial jet combustor flame including radiative heat transfer. Numer Heat Transf Part A 56(11):897–913
16. Seang-wock L, Yasuhiro D (2005) Simulation for high efficiency liquefied petroleum gas engine development. Proc Inst Mech Eng Part D J Automob Eng 218(10): 1201–1208
17. Kim HJ, Heo NI, Kim YM, Kang SM, Ahn JH (2002) Numerical modeling of combustion processes and pollutants formation in direct injection diesel engine. In: ASME pressure vessels and piping conference, pp 121–131
18. Wen Z, Yun S, Thomson MJ, Lightstone MF (2003) Modeling soot formation in turbulent kerosene/air jet diffusion flames. Combust Flame 135(3):323–340
19. Epaminondas M (2009) Ignition of turbulent non-premixed flames. Progr Energy Combust Sci 35(1):57–97
20. Jiang Q, Zhang C, Jiang J (2003) A sensitivity study of NOx emission to the change in the input variables of a FGR industrial furnace. In: ASME heat transfer summer conference, pp 95–102
21. Malalasekera W, Ranga-Dinesh KKJ, Ibrahim SS, Masri AR (2008) LES of Recirculation and vortex breakdown in swirling flames. Combust Sci Tech 180(5):809–832
22. Obieglo A, Gass J, Poulikakos D (2000) Comparative study of modeling a hydrogen non-premixed turbulent flame. Combust Flame 122:176–194
23. Schmidt DP, Nouar IP, Senecal K, Rutland CJ, Martin JK, Reitz RD (1999) Pressure-swirl atomization in the near field. SAE Paper 122(1–2), 176–194
24. Raithby GD, Chui EH (1990) A finite-volume method for predicting a radiant heat transfer in enclosures with participating media. J Heat Transf 112:415–423
25. Chui E. H., Raithby. G. D.: Computation of Radiant Heat Transfer on a Non- Orthogonal Mesh Using the Finite-Volume Method. Numerical Heat Transfer, Part B, 23(3), 269–288 (1993).
26. ANSYS FLUENT theory guide, Release 13.0, November (2010).
27. Zhu J, Shih TH (1994) Computation of confined co-flow jets with three turbulence models. Int J Numer Meth Fluids 19(10):939–956
28. Pandey V., De A., Kushari, A.: Reacting Flow and Emission Characteristics in a Liquid. International Journal of Chem. Tech. Research 5(2,) 912–917 (2005).
29. Ranz WE, Marshall WR (1952a) Evaporation from Drops, Part I. Chem Eng Prog 48(3):141–146
30. Ranz WE, Marshall WR (1952b) Evaporation from Drops, Part II. Chem Eng Prog 48(4):173–180

Performance and Emission Analysis of CI Engine Fueled with Waste Cooking Biodiesel Blends at Different Compression Ratios

Rahul Mohanty and Vinod Kotebavi

Abstract Biofuels can be considered as one of the alternative to replace the conventional liquid fuels partially as they can be used as blends in IC engines. Due to continuous use of petroleum sources in automobiles, it causes the depletion of world petroleum reserves, and it is expected to last for few years. Many active researches are in progress in finding the alternative for this conventional fuel. One of the alternative fuels whose use is rapidly growing is biodiesel. In the present work, biodiesel is produced from waste cooking oil which is obtained from local hotels and restaurants by using transesterification process. Various properties of diesel, biodiesel, and its blends are found. Experiments are carried out at various loads to study the performance of a single-cylinder, multi-fuel engine at two different compression ratios. The compression ratios used are 17.5 and 19.0. Different performance and emission parameters are evaluated for different blends and are compared with each other. Among all the results, B20 biodiesel blends show the best performance as compared to other biodiesel blends. And with 30% waste cooking oil biodiesel blend, lowest emissions of oxides of nitrogen (NO_x) were observed. For B20 blend, CO was reduced by 35%, CO_2 by 29.5%, and fuel consumption was reduced by 10%. Same time, 8.5% more in efficiency was observed. When B30 was used increased by 15.1%, fuel consumption reduced by 10%, and HC reduced by 20% and CO_2 was reduced by 16%.

Keywords Biodiesel · Transesterification · Blends · Waste cooking oil · Performance and emission characteristics

1 Introduction

The consumption of petroleum-based diesel can be reduced by different alternate fuels. One of the alternate fuels is vegetable oil which is a renewable. This vegetable

R. Mohanty · V. Kotebavi (✉)
Department of Mechanical Engineering, Amrita School of Engineering, Amrita Vishwa Vidyapeetham, Bengaluru, India
e-mail: k_vinod@blr.amrita.edu

S. Revankar et al. (eds.), *Proceedings of International Conference on Thermofluids*, Lecture Notes in Mechanical Engineering,
https://doi.org/10.1007/978-981-15-7831-1_47

oil can be used in engines in different forms like straight vegetable oil, esterified vegetable oil, or blends of this with diesel. We have made an attempt to run the engine by using blends of esterified waste cooking oil. Many researchers are working on these vegetable oil blends to study the impact of it on the performance the engine. Shetty et al. [1], reported that with 40% blend with waste cooking oil, a pressurized kerosene stove will have more efficiency (at 1.5 bar) than with pure kerosene. Subramanian et al. [2] did study of effect of diethyl ether on direction injection CI engine running on water–diesel emulsion. They found that diethyl ether can significantly reduce smoke density and NO_x levels without affecting the brake thermal efficiency. Tejesh et al. [3] performed test with blends of sesame and palm oil. They blended this dual biodiesel up to 40% in diesel. They found reduction in CO, HC, and without much reduction in efficiency. Kotebevi et al. [4] experimented by blending waste cooking oil biodiesel with diesel. They observed a considerable reduction in emissions like CO, UBHC, and smoke density with a small drop in thermal efficiency. Sunil Kumar et al. [5] did study with dual biodiesel. They used Simarouba and Neem biodiesel blends in diesel engine. They found reduction of HC and CO with percentage increase in blends. Kumar et al. [6] conducted study on CI engine at various compression ratio using ethanol–diesel blends. They found reduced smoke density, increase in NO_x, and increase in thermal efficiency with increase in compression ratio. Krishna Murthy et al. [7] used canola oil blends for their study, and they found better performance with 30% blends. Korus et al. [8] studied the polymerization of vegetable oils. Agarwal et al. [9] investigated the effect of injection pressure and timing on the performance of Diesel engine with karanja biodiesel blends. They found that increase in injection pressure increases the thermal efficiency. Aldhaidhawi et al. [10] investigated the effect of 20% rape seed biodiesel blend on delay period, under full load at different engine speeds. They observed shorter ignition delay for the blend. Most of the literature mainly propose a low blending order upto 20% with pure diesel for optimum performance. In this work, the stress is given on higher blending order, say, 40% biodiesel mixed with diesel and the performance and emission.

2 Methodology

Waste cooking oil is the oil obtained after frying the food items. The vegetable oil will not be suitable for consumption after repeated frying of food items because of high FFA content. This waste cooking oil will be still viscous and cannot be used directly in engines. Viscosity of this oil has to be reduced by some methods. We used transesterification process after filtering the oil. Then, the blends were prepared with different percentages. Blends are designated as B0, B10, B20, B30, and B40 where the number represents the percentage of biodiesel in the blend. Different instruments like bomb calorimeter, Saybolt viscometer, open-cup apparatus, etc., are used to find the different properties of diesel and blends.

2.1 Properties of Biodiesel and Waste Cooking Oil

The different properties like viscosity, flash and fire point, calorific value, specific gravity, iodine value, FFA content, pour point, etc., of diesel and biodiesel are measured and are tabulated in Table 1. The properties like viscosity, calorific value, flash point, and fire point of blends are obtained and are tabulated in Table 2.

The viscosity for waste cooking oil biodiesel is measured at various temperatures, and a graph of viscosity vs. temperature is plotted in Fig. 1.

Table 1 Properties of fuels

Fuel property	Units	Pure diesel	Waste cooking oil	Biodiesel
Density@ 27 °c	kg/m^3	820	918.4	866.0
Flash point	°C	64	305	170
Kinematic viscosiy@40 °C	cst	3.613	45.34	4.027
Cetane number	N/A	47	30	66
Calorific value	MJ/kg	42.5	35.82	38.034
Fire point	°C	82	345	190
Cloud point	°C	−18 to + 20 °F	24	13

Table 2 Properties of biodiesel blends

Sl. no.	Blend	Density (kg/m^3)	Viscosity (CST)	Flash point (°C)	Fire point (°C)	Calorific value (kj/kg)
1	Diesel	823	2.72	52	63	43,018
2	B10	827	3.02	59	64	42,986
3	B20	838	3.67	61	68	43,482
4	B30	846	3.92	67	76	39,887
5	B40	855	4.03	73	82	39,127

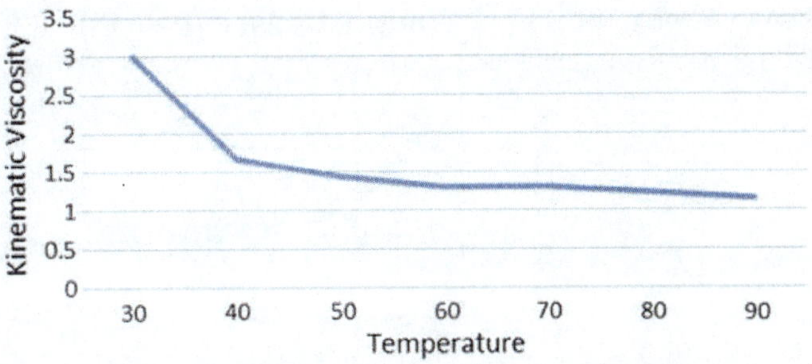

Fig. 1 Temperature versus kinematic viscosity for waste cooking oil biodiesel

2.2 Engine Specifications

Tests are conducted on single-cylinder, multi-fuel VCR engine. It is a four-stroke computer-based diesel engine coupled with eddy current dynamometer which can run on both petrol and diesel. The compression ratios can be varied from 6:1 to 10:1 for petrol and 14:1 to 20:1 for diesel. Provisions are made for mounting both spark plug and diesel injector on cylinder head. A PCB sensor is mounted on the top of the cylinder to measure the pressure inside the cylinder at different crank angles. Performance characteristics like thermal efficiency and specific fuel consumption are calculated, and emission characteristics measured are CO, HC, NOx, CO_2, and smoke density. Tests are conducted at two different compression ratios (17.5 and 19) with pure diesel and blends, varying percentages from 10 to 40 at five different loads.

3 Results and Discussions

3.1 Performance Characteristics

The variation of specific fuel consumption with torque for compression ratio of 17.5 and 19 is shown in Figs. 2 and 3, respectively. It can be seen from the graph that the BSFC is lower for diesel at all compression ratios and increases as the blend amount is increased, and then, the specific fuel consumption is greater while varying the percentages of blend. BSFC of B10, B20, B30, and B40 is more than that of diesel at partial load. Higher compression results in better combustion which reduces the mass of fuel per unit brake power. But BSFC stayed more than diesel in case of waste cooking oil biodiesel diesel blends. Figure 2 shows the comparison of specific fuel consumption of blends with the diesel. The graph shows that B20 has better performance. The reason could be higher combustion efficiency for B20 blend at this compression ratio.

Fig. 2 BSFC versus torque for compression ratio 17.5

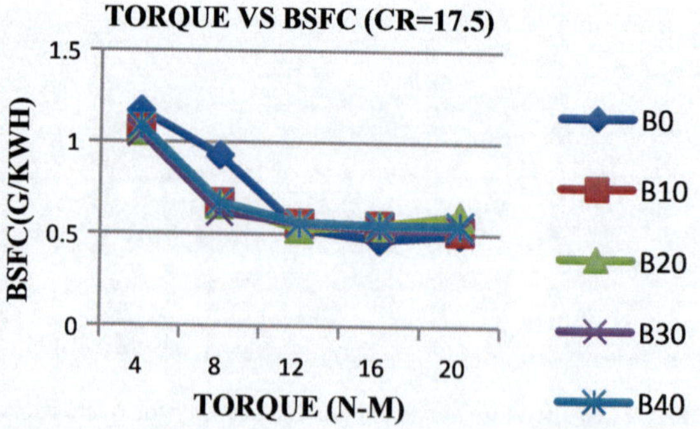

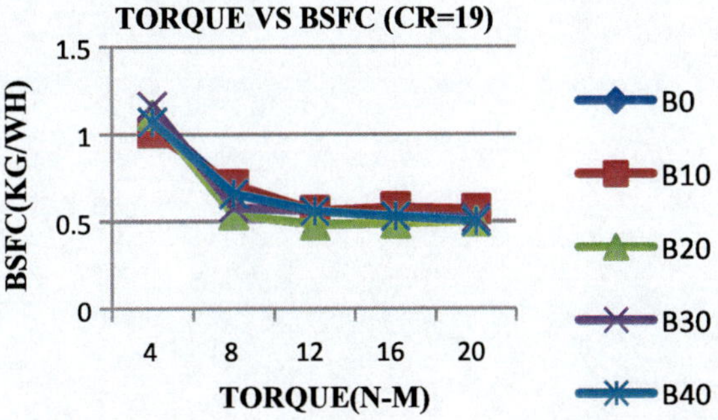

Fig. 3 BSFC versus torque for compression ratio 19

The variation of specific fuel consumption with torque for blends and diesel at compression ratio 19 is shown in Fig. 3. The graph shows that B30 has better performance. The reason could be higher combustion efficiency for B30 blend at this compression ratio.

The impact of torque variation on brake thermal efficiency (BTE) has been analyzed and is plotted in Figs. 4 and 5. A marginal reduction in thermal efficiency is observed with increases in percentages of blends. This is because of greater viscosity and lower calorific value of biodiesel blends. The effect of blending on brake thermal

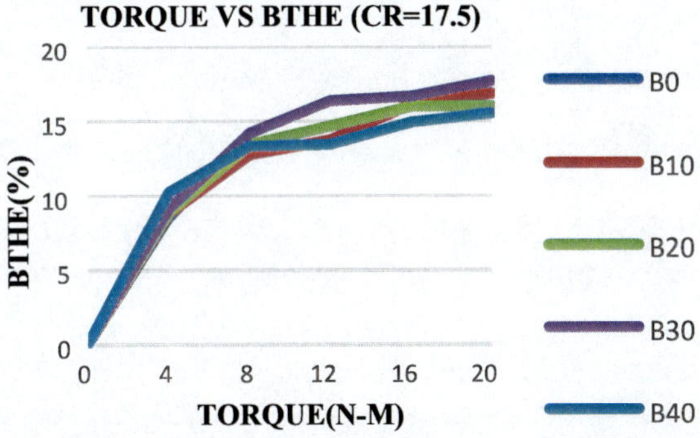

Fig. 4 Variation of brake thermal efficiency with torque for CR = 17.5

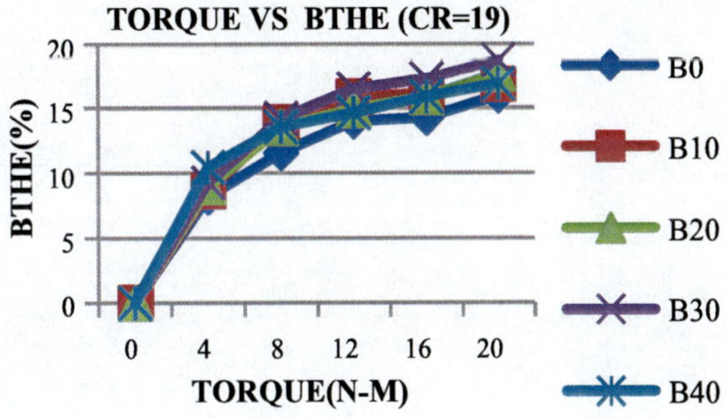

Fig. 5 Variation of brake thermal efficiency with torque for CR = 19

efficiency is shown in Fig. 4. It shows that B30 has the better performance compared to all other combinations.

The effect of blend percentage on thermal efficiency at different load is shown in Fig. 5. The graph shows that B30 has the better performance compared other blends.

3.2 Emission and Combustion Characteristics

Throughout the study, CO HC, CO_2, and NO_x emissions of the engine are obtained at different loads for all blends used. Figures 6, 7, 8, 9, 10, 11, 12, and 13 shows the variations of CO, HC, CO_2, and NO_x emissions for different blends used at different compression ratio. It shows that the CO emissions decreases with increase in compression because the combustion efficiency increases with compression ratio. Same time, increase in CO_2 and NO_x are observed because as the combustion efficiency increases combustion temperature also increases. This is the main reason for higher nitrogen oxides emissions. The CO and HC emissions were little higher for B10 blend.

Variations of cylinder pressure with crank angle for 100% diesel at different load is shown in Fig. 14. Figure (a) is for partial load and (b) for full load. It is observed that the peak pressure is 5° after TDC for this case.

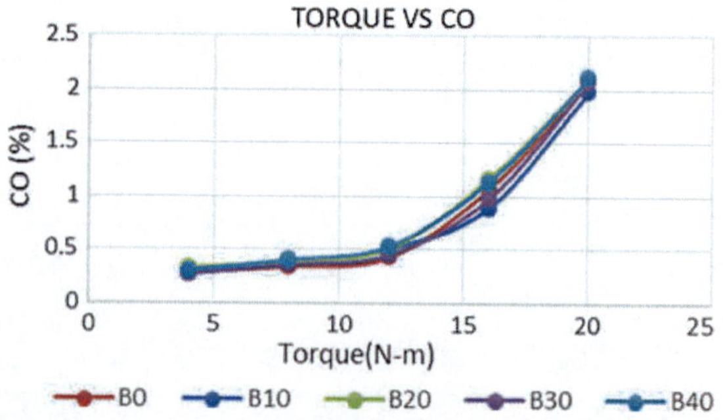

Fig. 6 CO emission results (CR = 17.5)

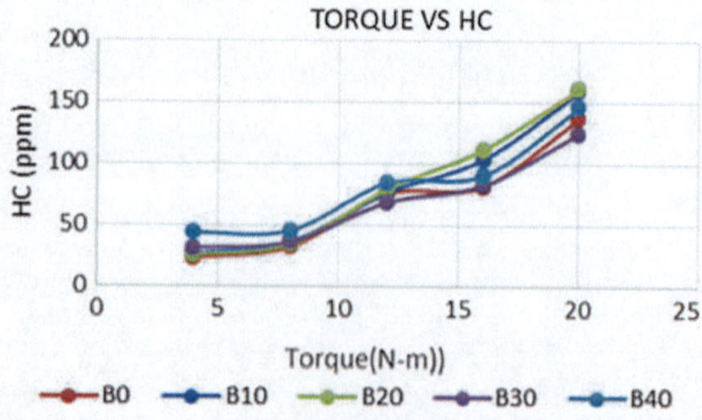

Fig. 7 HC emission results (CR = 17.5)

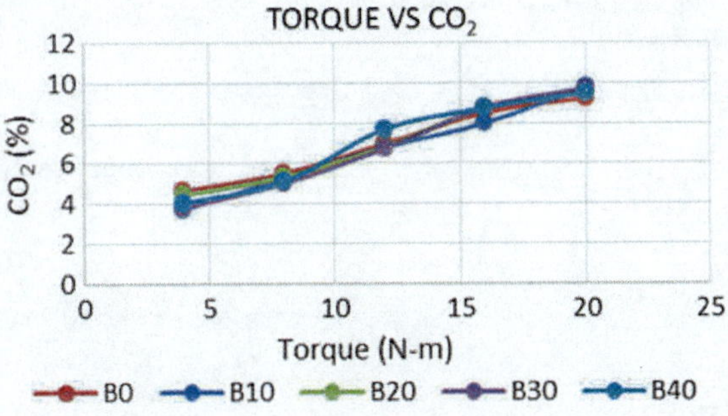

Fig. 8 CO2 emission results (CR = 17.5)

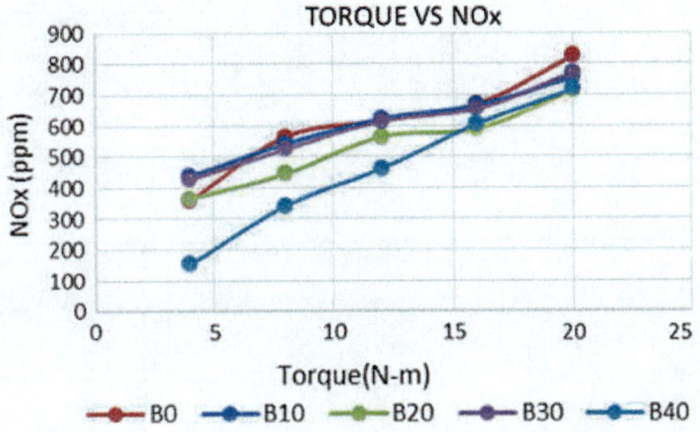

Fig. 9 NO_x emission results (CR = 17.5)

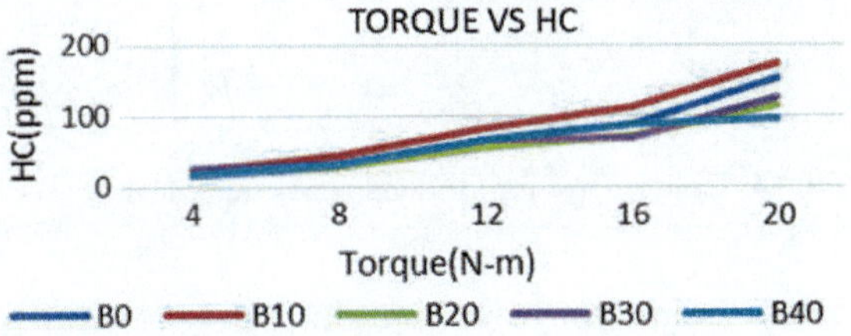

Fig. 10 HC emission results (CR = 19)

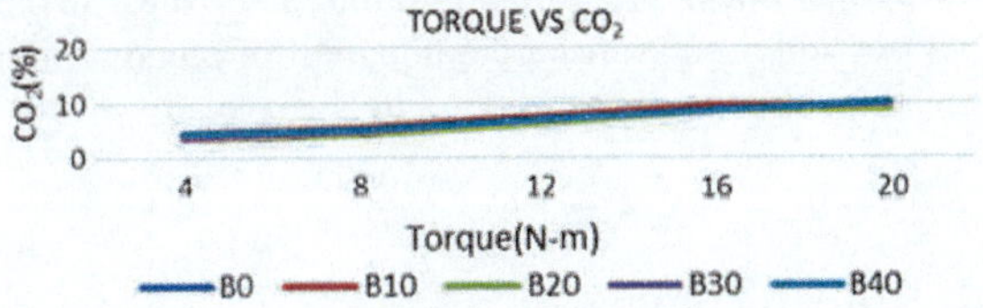

Fig. 11 CO2 emission results (CR = 19)

It is noted that the peak pressure elevation rate at compression ratio 19 was approximately found to be 63 bars for 100% load, and the maximum pressure elevation rate is decreased with decrease in compression ratio. But in compression ratio 17.5, B20 shows the better performance, and the cylinder pressure is found to be 63.5 bars. Improved performance and cylinder pressure is found at compression ratio, and for

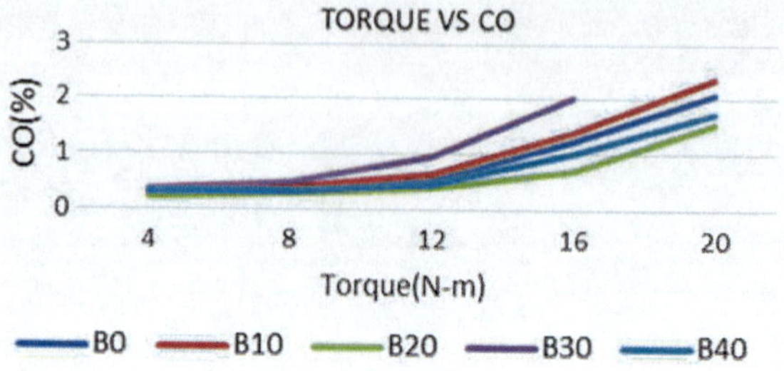

Fig. 12 CO emission results (CR = 19)

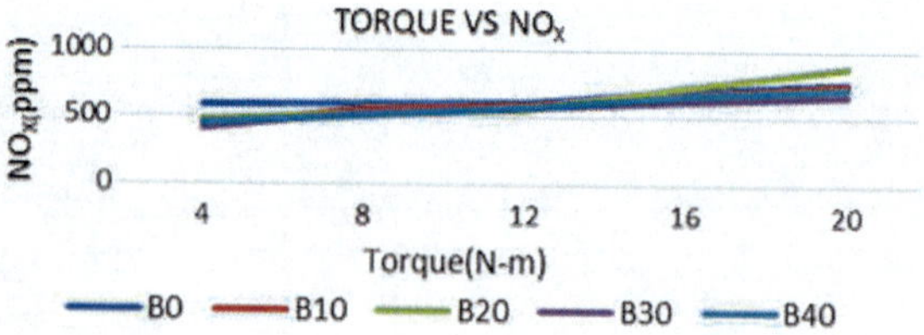

Fig. 13. NO_x emission results (CR=19)

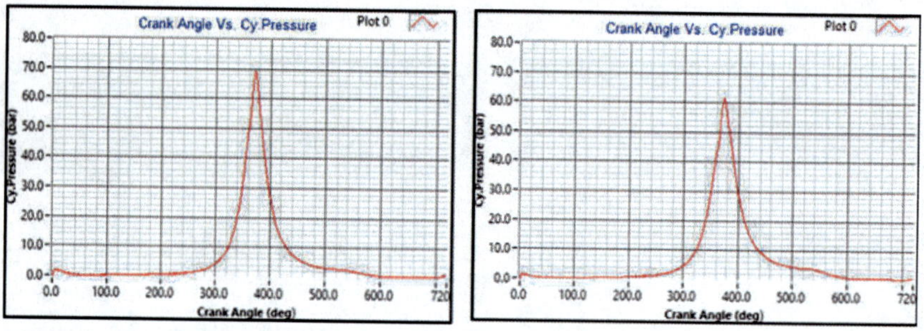

Fig. 14 Cylinder pressure variation versus crank angle

blends the increase in pressure is also more [11, 12]. After getting all the results, we concluded that B40 shows the better performance as compared to standard diesel, as the cylinder/peak pressure are found to be 70 bar in compression ratio 19. When the peak cylinder pressures are compared for blends and pure diesel, blends resulted in higher peak pressure compared to pure diesel.

4 Conclusions

Single-cylinder, multi-fuel, constant-speed CI engine ran successfully with the biodiesel blends. The properties of fuel used are comparable with that of pure diesel. For blends, CO and HC are lower compared to diesel fuel. NO_x and CO_2 emissions are boosted by increasing the proportion of biodiesel in the blends because of higher combustion temperature. Increase in carbon dioxide and nitric oxide emissions are found with increase in compression ratio. When the performance characteristics are compared, B20 with compression ratio 17.5 gave better results compared with other blends. Increase in thermal efficiency with increase in compression ratio is observed, and thermal efficiency is found to be highest for 40% biodiesel blend. Variations of peak pressure with load and compression ratio are observed in p-ɵ diagram.

References

1. Shetty D, Sahu D, Kumar R, Bekal S, Performance and emission characteristics of Pongamia oil-kerosene blend used in commercial kerosene stove. Energy Power 5(1A):19–27
2. Subramanian KA, Ramesh A (2002) Use of diethyl ether along with water-diesel emulsion in a Di diesel engine. SAE Paper number: 2002-01-2720
3. Tejesh P, Kotebavi V, Shyam P, Prasad PSD (2018) Performance and emission characteristics of a CI engine fuelled with palm and sesame oil blended diesel. Int J Veh Struct Syst 10(5):342–346
4. Kotebavi V, Shetty D, Sahu D (2016) Performance and emission characteristics of a CI engine run on waste cooking oil-diesel blends. ARPN J Eng Appl Sci 6(3):55–60
5. Sunil Kumar M, Vinod Kotebavi M, Performance analysis of a diesel engine fuelled with blends of neem and Simarouba oils. Indian J Sci Tech 9(45):1–5. https://doi.org/10.17485/ijst/2016.
6. Kurre SK, Pandey S, Saxena M (2013) Effect of compression ratio on diesel engine performance and emission with diesel-ethanol. Int J Sci Eng Res 4(10)
7. Krishna Murthy KSS, Kotebavi VM (2019) Study on performance and emission characteristics of CI engine fueled with canola oil—diesel blends. In: AIP conference proceedings 2200, 020049
8. Korus RA, Mousetis TL, Lloyd L (1982) Polymerization of vegetable oils in vegetable oils fuels. In: Proceedings of the international conference on Plant and Vegetable Oils as Fuels, ASAE, Fargo, pp 218–223
9. Agarwal AK, Dhar A, Gupta JG, Kim W, Choi K, Lee CS, Park S (2015) Effect of fuel injection pressure and injection timing of Karanja biodiesel blends on fuel spray, engine performance, emissions and combustion characteristics. Energy Convers Manag 91:302–314
10. Aldhaidhaw M, Miron L, Chiriac R, Badescu V (2018) Auto ignition process in compression ignition engine fuelled by diesel fuel and biodiesel with 20% rapeseed biofuel in diesel fuel. J Energy Eng 144(5):04018049
11. Aldhaidhaw M, Chiriac R, Badescu V (2017) Ignition delay, combustion and emission characteristics of diesel engine fueled with rapeseed biodiesel—a literature review, Renew Sustain Energy Rev 73:178–186
12. Graboski MS, McCormick RL (1998) Combustion of fat and vegetable oil derived fuels in diesel engines. Prog Energy Combust Sci 24:125–164

Effect of Compression Ratio on Thermal Efficiency and Cycle-by-Cycle Variation of a Biogas-Fueled SI Engine

Santosh Kumar Hotta, Anil Kumar Rout, and Niranjan Sahoo

Abstract In this study, a gasoline-fueled, single-cylinder, variable compression ratio (VCR), spark ignition (SI) engine of rated indicated power 4.5 kW was successfully retrofitted and tested with raw biogas. Further, this study also examine the consequence of operating compression ratio (CR) on the brake thermal efficiency (BTE) and cycle-by-cycle variations of the raw biogas-operated SI engine working in between CR 10 and CR 14. As observed, the developed cylinder pressure profiles are different and unique for each cycle, even though the operating conditions of the engine remain unaltered. The COV of IMEP increases with increasing CR, and the the maximum COV_{IMEP} was found to be 6.51% at CR 14 and is within the acceptable limit. However, the COV of peak cylinder pressure behaves very differently and reduces with rise in CR of the engine. The maximum BTE of the engine was found 30.3% at CR 12 and 1400 rpm.

Keywords Compression ratio · Cycle-by-cycle variation · Brake thermal efficiency · Biogas engine

1 Introduction

The cycle-by-cyclic variation (CCV) is an extensively accepted phenomenon, explicitly observed in spark ignition (SI) engines. In the presence of cyclic dispersion, the successive cycles are dissimilar as the flame propagation progress differently, resulting in variation of cylinder pressure curve [1]. The difference in combustion

S. K. Hotta (✉)
School of Mechanical Engineering, Kalinga Institute of Industrial Technology, Deemed To Be University, Bhubaneswar, Odisha 751024, India
e-mail: santoshk.hottafme@kiit.ac.in

A. K. Rout
Centre for Energy, Indian Institute of Technology Guwahati, Guwahati, India

N. Sahoo
Department of Mechanical Engineering, Indian Institute of Technology Guwahati, Guwahati, India

S. Revankar et al. (eds.), *Proceedings of International Conference on Thermofluids*, Lecture Notes in Mechanical Engineering,
https://doi.org/10.1007/978-981-15-7831-1_48

process from cycle to cycle is defined as the cycle-by-cycle variation (CCV) [2]. The cylinder pressure curves in internal combustion (IC) engine are uniquely related to the combustion phenomenon. Thus, the fluctuations in combustion process get directly reflected in the in-cylinder pressure [3]. The investigation of CCV is quite essential in a SI engine, because in most of the cases, the optimum ignition timing (IT) or the maximum brake torque timing (MBT) is fixed for the average cycles. Hence, the cycles having faster flame propagation burns much earlier than the average cycles and has an over advanced IT. Similarly, cycles having slower flame propagation speed burn comparatively slower than the average cycles and have a retarded IT. Hence, only few of the cycles are allowed to burn with the MBT and results in loss of power and efficiency that could have been achieved with the correct IT. Further, the intense cyclic variation limits the operating condition of the SI engine. Combustion having much advanced ignition timing initiates fastest burning cycles and are subjected to knock, hence limit the octane requirement and operating CR of the engine. Similarly, the combustion process having retarded ignition timing initiates slower burning cycle and is mostly subjected to incomplete combustion and limit the lean operating limit or may limit the amount of exhaust gas recirculation of the engine [3, 4]. It has been reported that, if these intense cyclic dispersion could have been reduced, there would be a chance to increase the output power approximately by 10% with the same fuel consumption [5]. Similarly, if all the cycles are ignited at the optimum rate and timing, the fuel economy could have been improved by 6% [6]. Further, with reduced combustion instabilities, the limit of practical level of lean fueling and EGR can be improved [7].

The fluctuations in the mass of fuel–air mixture inducted to the engine cylinder, motion of the fuel–air mixture and mixture of the fresh charges and residuals in each cycle in the region of the spark plug are the major factors which intensifies the cyclic dispersion in the SI engine [2, 3]. These fluctuations deeply affect the mixture density in the locality of spark plug, and the magnitude fluctuates randomly in each cycle. Further, this fluctuation initiates random convection of the spark kernel and heat transfer from the burning kernel [8]. The initial stage of flame growth profoundly influences the subsequent stages of combustion. However, the early stage of flame development, growth, and propagation are deeply influenced by the local fuel–air mixture in cylinder movement of the charge and density of exhaust gas in the gap of spark plug region during and after ignition [2]. The overall conclusion is that any phenomenon that may to slow down the process of flame propagation during the stages of flame development increases the cyclic dispersion of the engine [2, 4].

The operating compression ratio (CR) of the engine seems to be an effective tool to reduce the cyclic dispersion of IMEP. The COV of IMEP reduces slightly with higher functioning CR of the engine. The injection, ignition timing, engine load, speed, CR, and configuration of the injector deeply influence the cyclic variation of direct-injection SI engine [9]. Many of the literature enclosed have reported the cyclic variability and methods to reduce the cyclic dispersion mostly for the gasoline, LPG, and hydrogen-operated SI engine. Only few of the literature talk about the effect of CR on the cyclic dispersion. Hence, the purpose of the current research is to analyze the consequence of operating CR on the cyclic dispersion and BTE of the biogas-driven SI engine in lean-burn operational environment.

2 Methodology

In the recent study, a gasoline-fueled, single-cylinder, manifold-injected, variable compression ratio (VCR) SI engine of rated capacity 4.5 kW is used as the experimental test rig and shown in Fig. 1. The base experiential setup is developed by the Apex Innovation Pvt. Ltd. for various octane fuel applications and researches. However, in the current investigation, the engine is retrofitted with a biogas induction (Sl. no. 5 and 14, Fig. 1) and preheating system (Sl. no. 5, Fig. 1) for biogas application. The engine is linked with an eddy current dynamometer for loading, and the applied load is controlled by manually operating the controller of the variable resistor attached to the dynamometer. The load applied on the engine creates an angular displacement of the coupled dynamometer and is detected by a precalibrated piezoelectric load cell. The load cell reflects the reading by a digital load meter attached on the engine dashboard. The experimental setup is facilitated with a tilting block VCR mechanism. The CR of the test bench varies from CR 10 to CR 15. The setup which was initially designed to operate with gasoline was later modified and demonstrated successfully to operate with biogas having 55.6% of CH_4, 42.3% of CO_2, and 2.1% of N2 as the stand-alone fuel [10].

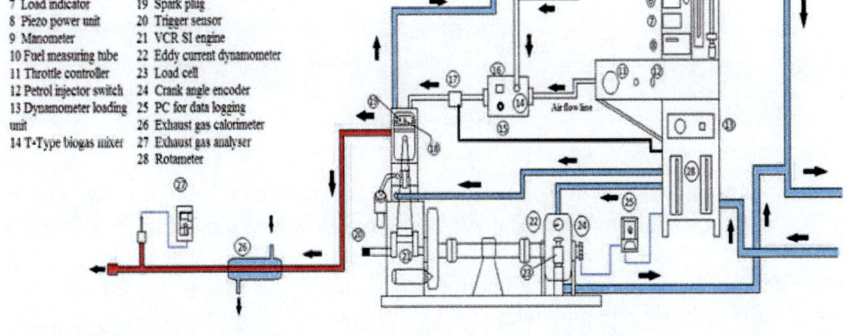

Fig. 1 Experimental test rig

The engine was well instrumented to trace the cylinder pressure, crank angle, and engine speed for each degree rotation of the crank shaft. To measure the pressure of the combustion chamber (in-cylinder pressure), a piezo pressure sensor with inbuilt amplifier (PCB Piezotronics Inc., M111A22) and nose reduction cable is attached on the engine head and exposed to the combustion chamber. Similarly, a Kubler-made optical crank angle encoder is also integrated with the engine shaft to capture the speed and signal for each degree revolution of the crank. The crank angle and pressure signals captured are further transferred to the computer through a piezo powering unit and stored in a data logger (NI USB 6210). Further, post-processing and analysis of the captured signals are done by a LabView-based program called "Enginesoft." The mass flow rate of air is traced by a differential pressure air flow transmitter and an open U-tube manometer connected to the air drum. Similarly, the biogas consumption rate is traced by a variable area rotameter. A reprogrammable, open electronic control unit (ECU) was connected with the pressure sensor, crank angle encoder, trigger sensor, ignition coil, and throttle position sensor and was used to configure the fuel injection time, fuel injection angle, and ignition timing. The detailed experimental setup and procedure for operating the engine in biogas mode are explained in detail in the earlier published research paper [11, 12].

To investigate the cycle-by-cycle variation of the raw biogas-operated SI engine at different operating CRs, the coefficient of variation of indicative mean effective pressure (COV_{IMEP}) and peak cylinder pressure (COV_{PP}) are calculated by using Eqs. 1 and 2, respectively. The COV_{IMEP} is the ratio of standard deviation in the indicated mean effective pressure (σ_{IMEP}) to the mean of indicated mean effective pressure (μ_{IMEP}) [7, 9]. Similarly, the IMEP is the ratio of net work produced W_N (calculated from the P–V plot area) to the displacement volume of the engine cylinder (V_d) and is obtained by using Eqs. 3 and 4.

$$COV_{IMEP} = \frac{\sigma_{IMEP}}{\mu_{IMEP}} \times 100 \tag{1}$$

$$COV_{PP} = \frac{\sigma_{PP}}{\mu_{PP}} \times 100 \tag{2}$$

$$IMEP = \frac{W_N}{V_d} \tag{3}$$

$$W_N = \int P dV \tag{4}$$

In the present investigation, the engine speed is varied from 1400 to 1700 rpm for each operating CR of the biogas-fueled SI engine. However, the CR of the engine is altered in between CR 10 and CR 14. For each operating CR and speed configuration of the engine, 50 conjunctive data points of pressure at each crank angle have been recorded. Hence, for a complete cycle, 36,000 pressure points have been recorded, and the same has been further processed to find out the IMEP, COV_{IMEP}, and COV_{PP} based on the captured pressure data for each and every cycle.

3 Results and Discussion

The cycle-by-cycle variation of cylinder pressure for 50 consecutive cycles at CR 12 and 1600 rpm of the biogas-operated SI engine is depicted in Fig. 2. It was noticed that the cylinder pressure trace evolves differently for each cycle, even though the operating condition remains unaltered. As shown, few cycles are identified as fast burning, average burning, and slow burning at the same operating condition.

Similarly, the average of these 50 consecutive cyclic data of cylinder pressure are plotted with respect to the crank angle for all CRs at 1600 rpm and full throttle operating condition of the biogas engine and shown in Fig. 3. It was noticed that the peak cylinder pressure increases with increasing CR and shifts toward the top dead center (TDC). Despite of having higher peak cylinder pressure (34.60 bar) at CR 14, the maximum brake power output was noticed at CR 12. This is mainly because of the area under *P*–*V* curve, which was found maximum at CR 12.

Additionally, Fig. 4 shows the rapid fluctuation of peak cylinder pressure of 50 conjunctive cycles at 1600 rpm for all CR except CR10. It has been observed that the peak cylinder pressure fluctuates continuously even though the operating condition remains same at a particular CR. As noticed, the window of fluctuating peak pressure becomes narrow with rising CR. At CR 10, 11, 12, 13, and 14, the fluctuating pressure window were found to be 4.667, 8.267, 7.029, 6.632, and 6.166 bar, respectively.

Figure 5 and 6 show the effect of CR on the COV_{IMEP} and COV_{PP} all through the speed array of the biogas-fueled SI engine. The operating condition of the engine remained unaltered for each CR of the engine. As noticed, the COV of IMEP increases with increased CR of the engine. This explains that with increasing CR, the cyclic variation of indicated mean effective pressure becomes severe. Since, the IT was set

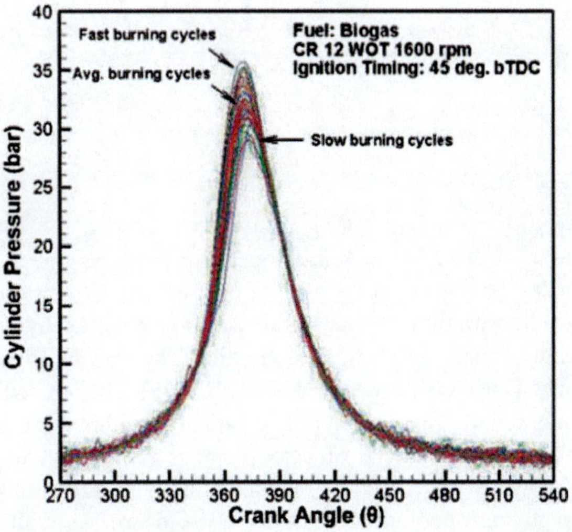

Fig. 2 Cyclic variation of cylinder pressure at CR 12 and 1600 rpm

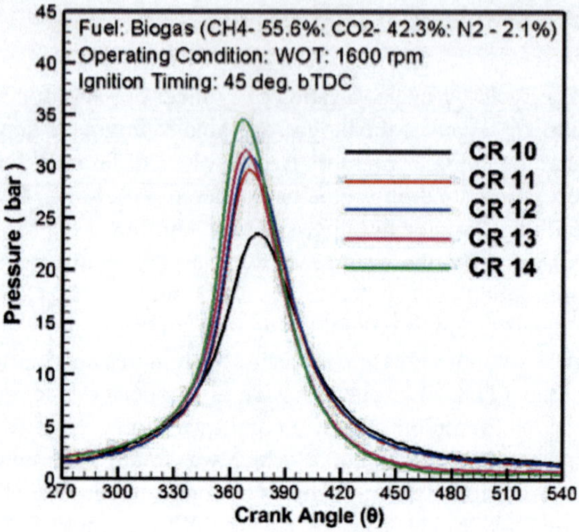

Fig.3 Average cylinder pressure at different CR

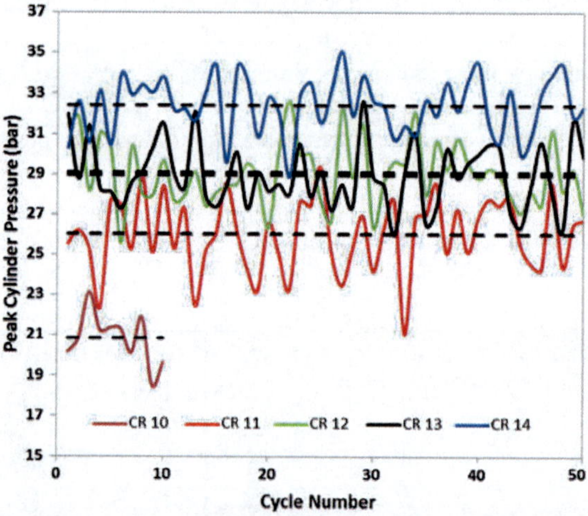

Fig. 4 Variation of peak cylinder pressure of 50 consecutive cycles

to 45° bTDC for the biogas operation in the engine at all operating CRs; there may be chance of knock because of over advanced ignition timing, which intensifies the cyclic variation. However, this could be resolved by retarding the ignition timing at higher operating compression ratios. The vehicle drivability problem usually results when the COV_{IMEP} exceed about 10% [8]. The noticed range of maximum COV_{IMEP} was found to be 6.51% at CR 14 and is within the acceptable limit. Further, it was found that the COV of peak pressure decreases with rise in operating CR of the engine. Since the range of fluctuating peak pressure becomes narrow and reduces at higher operating CRs, the COV_{PP} decreases with increased CR of the engine. However, COV_{PP} increases with rising speed of the engine.

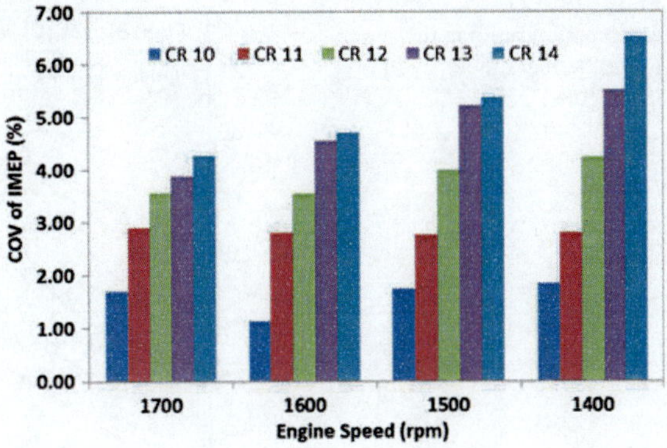

Fig. 5 Effect of compression ratio on the COV of IMEP

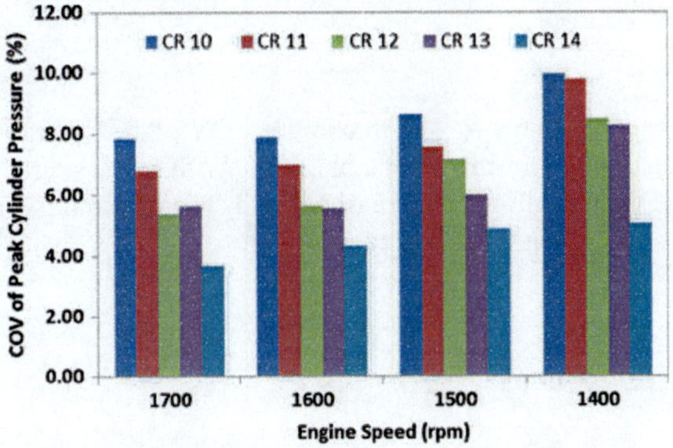

Fig. 6 Effect of compression ratio on the COV of peak cylinder pressure

Figure 7 illustrates the effect of CR on the BTE of the biogas-driven SI engine. As observed, the maximum BTE of the engine was 30.3% at CR 12 and 1400 rpm. However, the BTE was observed maximum only at CR 12, regardless the operating speed of the engine. As it has been already mentioned that the maximum BP output of the engine was observed at CR 12, and the mass flow rate of biogas consumption remains unaltered at all compression ratios of the engine, and hence, it is evident that the maximum BTE will appear at CR 12.

4 Conclusions

The effect of CR on the cyclic variability and thermal efficiency of a biogas-fueled SI engine has been investigated in the current study. The followings are the overall conclusions drawn from the study. The cylinder pressure trace evolves differently for each cycle, even though the operating condition remains unaltered. With increasing CR, the COV of IMEP increases, and the window of fluctuating peak pressure

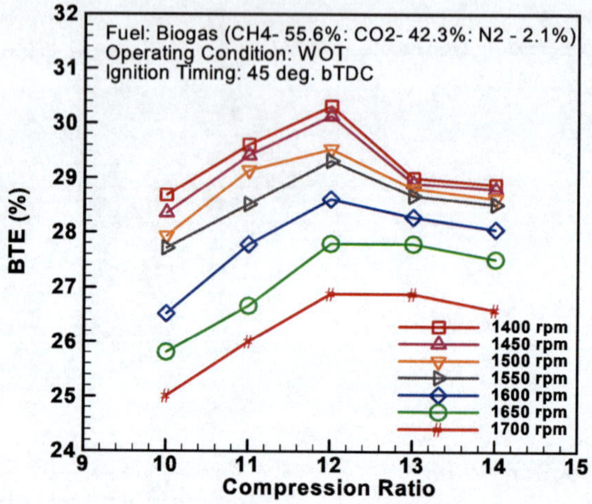

Fig. 7 Effect of compression ratio on the brake thermal efficiency of the engine

becomes narrow. The maximum COV of IMEP was found to be 6.51% at CR 14 and is within the acceptable limit. With increasing CR, the COV of peak cylinder pressure follows a decreasing trend. The maximum BTE of the engine was found to be 30.3% at CR 12 and 1400 rpm.

References

1. Aydın K (2011) Effect of engine parameters on cyclic variations in spark ignition engines. In: 6th international advanced technologies symposium (IATS'11), Elazığ, Turkey
2. Richard S (1999) Introduction to internal combustion engines. Palgrave, New York
3. Heywood JB (1988) Internal combustion engine fundamentals. McGraw-Hill Companies Inc., New York
4. Galloni E (2009) Analyses about parameters that affect cyclic variation in a spark ignition engine. Appl Therm Eng 29(5–6):1131–1137
5. Soltau JP (1960) Cylinder pressure variations in petrol engines. Proc Inst Mech Eng Automobile Div 14(1):99–117
6. Lyon D (1986) Knock and cyclic dispersion in a spark ignition engine, petroleum based fuels and automotive applications. Institution of Mechanical Engineers, London
7. Green Jr JB, Daw CS, Armfield JS, Finney CEA, Wagner RM, Drallmeier JA, Kennel MB, Durbetaki P (1999) Time irreversibility and comparison of cyclic-variability models. SAE Trans, 355–362
8. Ozdor N, Dulger M, Sher E (1994) Cyclic variability in spark ignition engines, a literature survey. SAE Paper no. 940987
9. Gong C, Huang K, Chen Y, Jia J, Su Y, Liu X (2011) Cycle-by-cycle combustion variation in a DISI engine fueled with methanol. Fuel 90(8):2817–2819
10. Hotta SK, Sahoo N, Mohanty K, Mekonen MW (2017) Effect of air-fuel ratio on the performance and emission characteristic of a biogas fuelled spark ignition engine. In: 7th International conference on theoretical, applied, computational and experimental mechanics, IIT Kharagpur, India

11. Hotta SK, Sahoo N, Mohanty K (2019) Comparative assessment of a spark ignition engine fueled with gasoline and raw biogas. Renew Energy 134:1307–1319
12. Hotta SK, Sahoo N, Mohanty K, Kulkarni V (2020) Ignition timing and compression ratio as effective means for the improvement in the operating characteristics of a biogas fueled spark ignition engine. Renew Energy 150:854–867

CFD Simulation of Gasoline and Methanol Combustion in a Twin Spark IC Engine

Praveen Alok and Debjyoti Sahu

Abstract In this current work, the outcome of four sets of spark timings on engine performance parameters is studied. CFD simulations are carried out on a variable compression ratio twin spark SI engine for both methanol and gasoline. Simulations were able to predict performance parameters with high accuracy. To simulate flow physics, combustion in the engine, ANSYS Fluent is employed. A parametric study of the effect of spark timing and compression ratio under methanol operation at 1600 RPM is performed. The results yield a promising strategy in deciding spark advance angle. Both methanol and gasoline show increased mean effective pressure at twin spark of 26°–28° BTDC.

Keywords Combustion simulation · Methanol fuel · DTSi engine · Spark advance angle

1 Introduction

The derived methanol and ethanol have almost alike and superior combustion characteristics in regular gasoline engine [1]. Methanol has good octane number and high flammability limit for combustion in IC engines [2, 3]. Yajia et al. experimentally obtained data regarding the spray characteristics of ethanol, methanol, and gasoline in a direct injection spark ignition (DISI) engine. It was reported that fuel type affects on spray penetration. Ethanol has lower penetration than gasoline [4]. Pefley executed both computational and experimental analysis on a gasoline engine to run on ethanol and methanol fuels [5]. The computational investigation provided insight to the effects of fuel types, compression ratio, rear axle ratio, and reduced cooling loads on performance parameters as well as emission characteristics. The study found that

P. Alok
Simulation Engineer, CADFEM India, Hyderabad 500082, India

D. Sahu (✉)
School of Mechanical Engineering, Kalinga Institute of Industrial Technology, Bhubaneswar 751024, India
e-mail: debjyoti.sahufme@kiit.ac.in

S. Revankar et al. (eds.), *Proceedings of International Conference on Thermofluids*, Lecture Notes in Mechanical Engineering,
https://doi.org/10.1007/978-981-15-7831-1_49

as the spark happening at TDC or delayed (spark happens after compression), then the emissions level often decreases, since mixture is burned at low pressure and temperature generation of pollutant unlikely. On the other hand, as the compression ratio increases, the power and efficiency increase due to higher turbulent condition. Yamin and Dado performed a computational analysis of a variable compression ratio (6.82–10) engine and reported the power output [6]. The performance analysis revealed that a maximum indicated power may increase upto 62% at 10:1 compression ratio and at optimum speed due to higher turbulent condition.

Conventionally for SI engine combustion simulation, a premixed combustion model is preferred [7]. Joshi and Borse used a software package (OpenFoam) to simulate SI engine combustion [8]. The effect of spark advance, number of spark plugs and engine speed variation on engine performance are interesting [8]. Few researchers worked on cold flow compression and combustion simulations vis-a-vis temperature and pressure for various crank angles [9]. Full open cycle simulations can be performed using the commercial code CONVERGE [10]. The simulation of combustion process can be studied with detailed chemistry along with adaptive mesh refinement (AMR) to track the reaction zone during combustion [11]. In-cylinder pressure, heat release, and flame morphology along with actual photographic images are reported by Mobasheri et al. [12]. Variations due to changing fuel type and air–fuel ratio were well captured [12]. Simulation work is also reported by Hepkaya et al. to present fluid flow pattern and combustion characteristics of a single cylinder spark ignition engine [13]. The complete engine cycle was investigated using Reynolds-averaged Navier–Stokes (RANS) and coherent flame model (CFM).

Here, an SI engine was simulated for four different spark advance angles between 32° and 24° to obtain the optimal spark timing for best performance. The greater ignition advance might enhance the cylinder average pressure quickly but increases the work needed to make piston move in the upward direction. If the spark initiates too retort, the combustion process terminates with resulting minimum cylinder pressure [14–17]. Instead of fixing spark plug on an engine at different angle and position, if the work can be imitated in CFD, then it saves lot of resources in the laboratory.

2 Methodology

CFD allows users to easily vary parameters. ANSYS Fluent is utilized here as a tool in investigating the feasibility of operating a twin spark engine on varying fuels. The first step is to develop a three-dimensional solid model, generate a simple mesh of this model, and apply combustion modeling principles to proceed with simulation. Engine test data from earlier published paper will be used to imitate operating conditions [14] (Table 1).

Table 1 Engine specifications [14]

Sl. no.	Particulars	Details
1	Engine type	Four stroke, single cylinder
2	Displacement volume	330 cc
3	Bore and stroke	76 mm, 72.6 mm
4	Rated power	6.02 kW at 1800 rpm
5	Compression ratio	6:1 to 12:1
6	Spark plug dimension	12 mm × 1 mm

2.1 *CFD Simulation*

Once a complete model of the fluid volume was obtained, meshing in ANSYS Workbench was undertaken (Fig. 1). Factors under consideration when building the mesh included optimal accuracy at a manageable number of elements. This ensures that a computational model will produce repeatable results with reasonable computational time. The time step is calculated by the solver based on the engine RPM and number of degrees of rotation (in this case 720°). Transient simulation is performed. In-cylinder dynamic mesh is used to simulate the movement of piston and valves [18, 19]. To tune the movement of inlet and exhaust valve with respect to degree of rotation, profile files are generated using notepad. 20 iterations per time step are used to ensure convergence at each and every time step. Standard k-ε model is considered for turbulence. Boundary conditions are given in Table 2. The convergence criteria for continuity and momentum were 1E−04 and for energy was 1E−06, respectively.

Commercial gasoline is considered as fuel for which calorific value is assumed to be 44 MJ/kg. Reagent grade methanol is considered as fuel for which calorific value is assumed to be 23 MJ/kg [14].

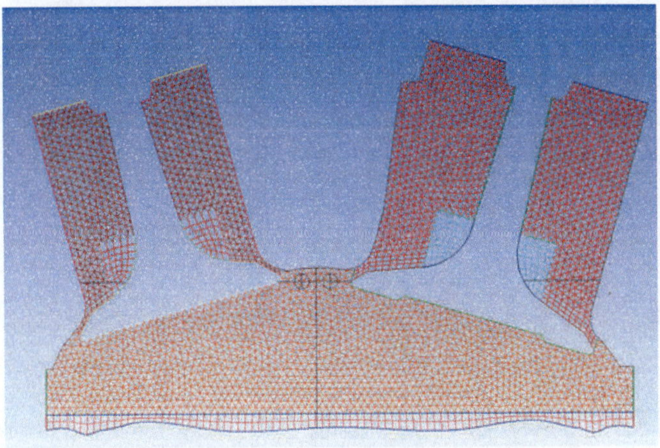

Fig. 1 Meshing of the cylinder head; twin spark plug marked by circle

Table 2 Boundary conditions for simulation

Sl. no.	Particulars	Details
1	Engine speed for simulation	1600 rpm
2	Pressure inlet for simulation	101,418 Pa
3	Fuel temperature for simulation	300 K

3 Results and Discussion

3.1 Flow Visualization

Nowadays, computational fluid dynamics (CFD) is used for optimization of spark advance angle. SI engines are comparatively simpler to simulate in computer than CI engines as physics involved in understanding the spray characteristics is not complicated [8]. 32°–24° before the end of the compression, spark initiates the combustion here. Two spark plugs in twin spark ignition system are located strategically to result efficient flame propagation. Higher the speed of the engine higher should be the spark advance angle to avail optimum combustion duration [20].

Computational fluid dynamics (CFD) is preferred to study combustion in the engine. The in-cylinder combustion can be visualized on a computer screen which is reliable [7]. Variation of temperature in the cylinder head region is captured during four strokes and shown in Figs. 2, 3, 4 and 5. The temperature contour inside the cylinder indicates that the rise in temperature due to combustion is interesting as an approximate temperature at each corner can be predicted. Thus, any portion which is under thermal stress can also be located.

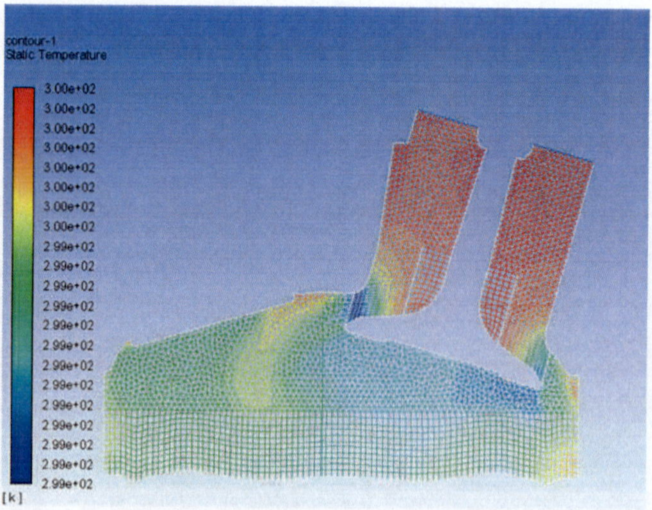

Fig. 2 Variation of temp. when the inlet valve is open

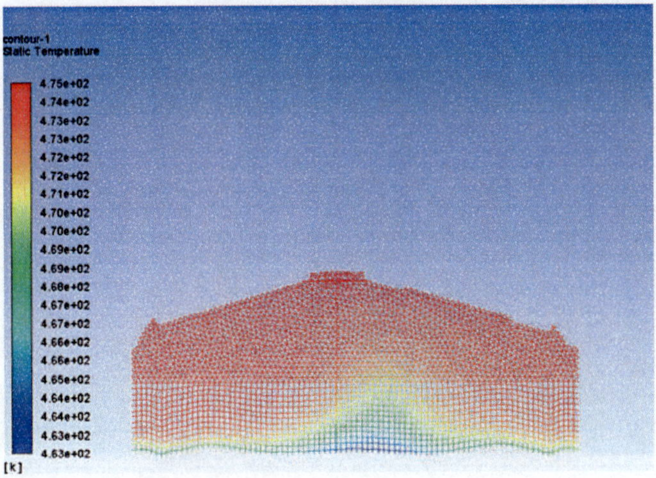

Fig. 3 Variation of temp. during compression

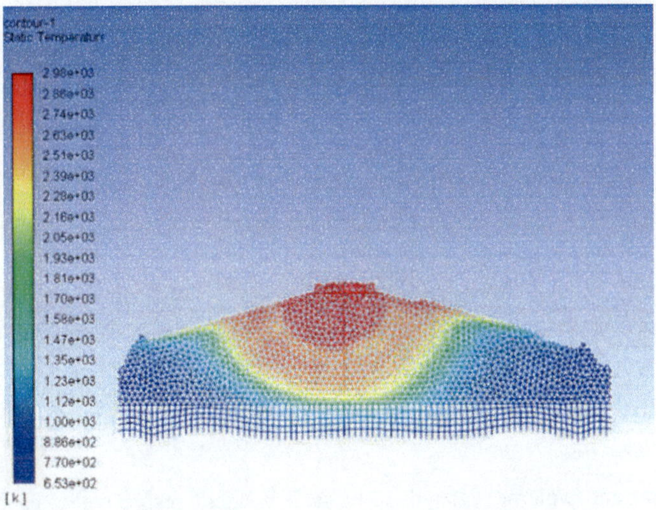

Fig. 4 Variation of temp. during spark

Pressure and volume plotted in y and x axes, respectively, have shown gasoline combustion with single spark at 24° in Fig. 6. It depicts the way work is developed in the engine. Because the temperature rise due to combustion leads to the adiabatic flame temperature, the pressure developed applies the force on the piston so that actual work is generated. Therefore, the more area within the plot area signifies more work (p-V diagram of gasoline combustion with twin spark at 24°–26° BTDC is sohwn in Fig. 7).

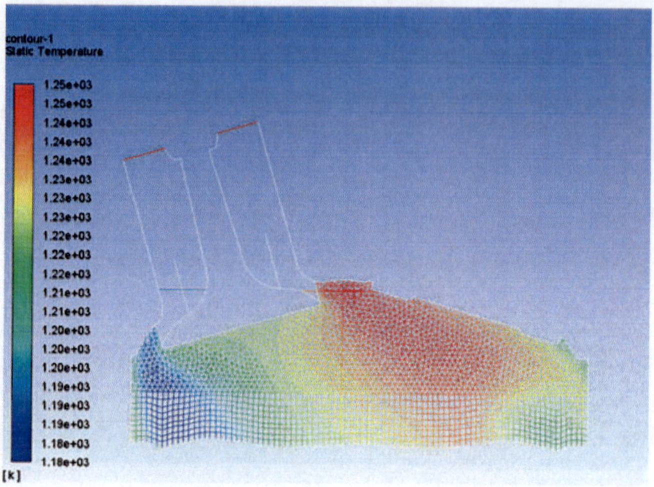

Fig. 5 Variation of temp. during exhaust

Fig. 6. *p–V* diagram of gasoline combustion with singe spark at 24° BTDC

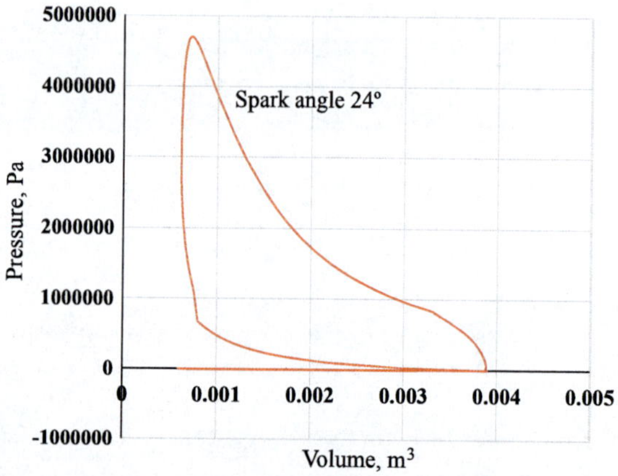

Pressure and volume plotted in *y* and *x* axes, respectively, have shown methanol combustion with twin spark at 24°–26° BTDC in Fig. 8. Comparing with *p–V* diagram of gasoline engine, methanol engine's *p–V* diagram has a sharp peak. But the width of the *p–V* diagram is narrower hence it will produce lesser power when compared to gasoline engine. However, fuel flow rate is similar in all simulations and methanol has lesser calorific value, thus expected to develop lesser power.

Pressure distribution against crank angle is as shown in Figs. 9 and 10 for single spark and twin spark, respectively. Pattern remains similar for all the graphs. Enrichment of oxygen which leads to easier combustion and better laminar flame speed of methanol might be the reasons for maximizing the pressure rise during combustion in this present work. However, compared to gasoline, the change in power output

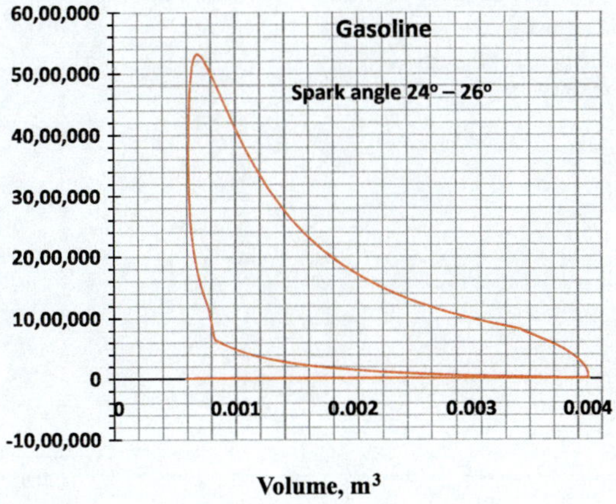

Fig. 7. *p–V* diagram of gasoline combustion with twin spark at 24°–26° BTDC

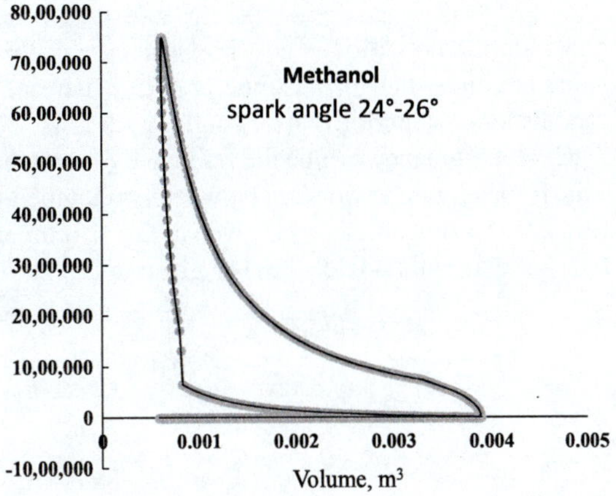

Fig. 8. *p–V* diagram of methanol combustion with twin spark at 24°–26° BTDC

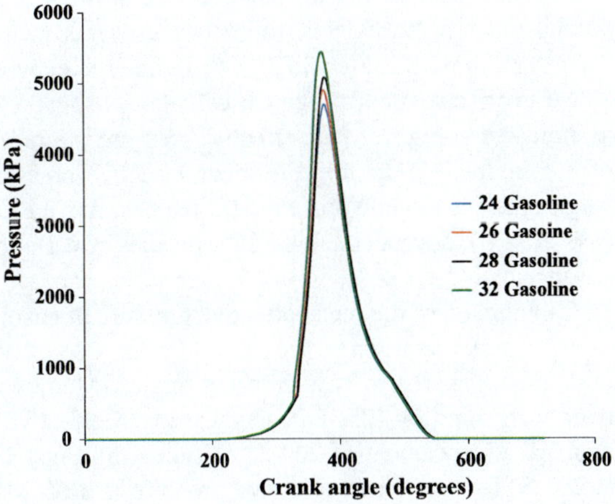

Fig. 9 Pressure distribution for gasoline combustion for single spark

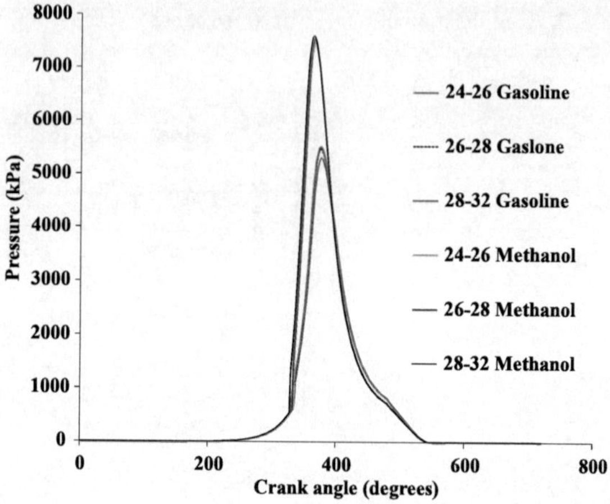

Fig. 10 Pressure distribution for gasoline and methanol combustion for twin spark

under similar condition is not similar in case of methanol fuel because the calorific value is almost half of gasoline [21–23]. Increase in peak pressure in twin spark establishes its superiority over single spark [24].

Carnot efficiency can be the reference for any thermal system performance analysis. If the engine is working between two temperature reservoirs T_H (high) and T_L (low), the Carnot efficiency is given in Eq. 1. Temperature can be approximated from the *p–V* diagram (Boyle's law) as given in Eq. 2.

$$\eta_{carnot} = 1 - \frac{T_L}{T_H} \tag{1}$$

$$T = \frac{pV}{NR} \tag{2}$$

N is the amount of substance and *R* is the gas constant. Mass of air–fuel in combustion is same and *R* is similar because air is the main constituent of the mixture. Therefore, more p value will signify more efficiency if *V* is also similar.

Mean effective pressure can be calculated using the *p–V* diagram by discretization as indicated in Fig. 7. Mean effective pressure is directly indicating the power output of the engine. Comparing the mean effective pressure of gasoline combustion in single spark is as shown in Fig. 11. Highest mean effective pressure is observed in case of spark advancement of 26° and the lowest one is observed in case of spark advance of 24°.

Comparison of the mean effective pressure of gasoline and methanol combustion in twin spark condition is shown in Fig. 12. Use of twin spark plugs in this engine initiates the sparks in two different angles. Many showcased superior results for twin spark ignition [25, 26]. Highest mean effective pressure is observed in case of spark advancement of 26°-28°. Comparing Figs. 11 and 12, it is confirmed that at 26° BTDC the mean effective pressure is 1.65 MPa whereas at 26°–28° mean

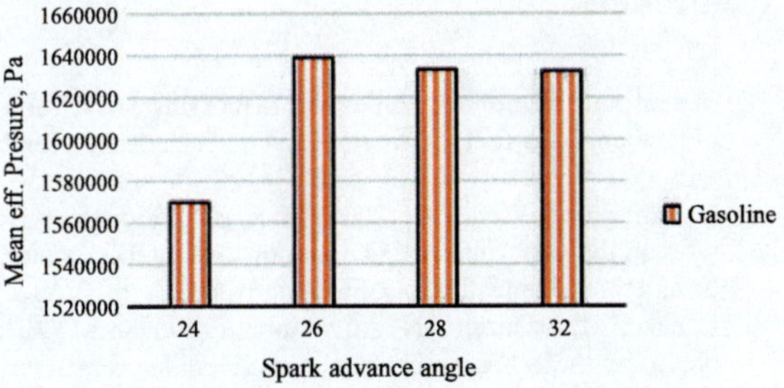

Fig. 11 Mean effective pressure for gasoline combustion for single spark

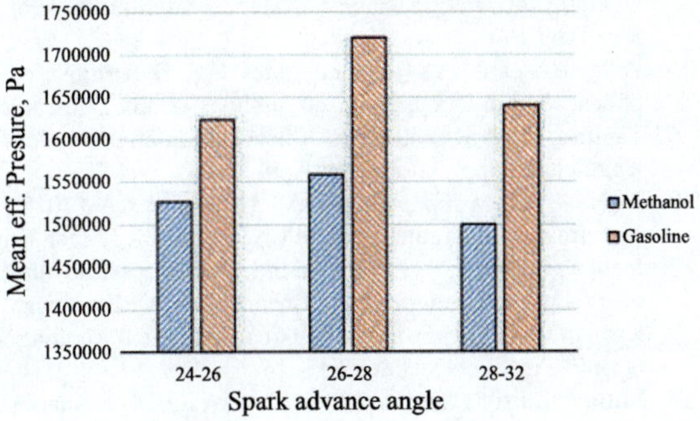

Fig. 12 Mean effective pressure for gasoline and methanol combustion for twin spark

effective pressure is 1.72 MPa for gasoline. Methanol combustion at same spark advance (26°–28°) yields mean effective pressure of 1.56 MPa which is higher than other combinations of 24°–26° and 28°–32°, respectively. The trend matches already published experimental results [14].

4 Conclusion

The spark timings and its effects on power output vis-à-vis mean effective pressure are studied using CFD software package imitating the dimensions of a real engine. Pressure generated due to combustion is observed at several twin spark combinations between 32° and 24° BTDC. Among single spark combinations, near 26° BTDC engine performance is found to be increased to a considerable limit. Comparing the results at 26°–28° BTDC can be considered as optimum spark timing with enhanced mean effective pressure for both methanol and gasoline.

References

1. Vancoillie J, Demuynck J, Sileghem L, de-Ginste MV, Verhelst S, Brabant I (2013) The potential of methanol as a fuel for flex-fuel and dedicated spark ignition engines. Appl Energy 102:140–149
2. Yousufuddin S, Mehdi SN (2008) Effect of ignition timings, equivalence ratio, and compression ratio on the performance and emission characteristics of variable compression ratio SI engine using ethanol-unleaded gasoline blends. IJE Trans B Appl 21:97–106
3. Canakci M, Ozsezen AN, Alptekin E, Eyidogan M (2013) Impact of alcohol-gasoline fuel blends on the exhaust emission of an SI engine. Renew Energy 52:111–117
4. Yajia M, Xu W et al (2009) An experimental and numerical investigation on characteristics of methanol and ethanol sprays from a multi-hole DISI injector. In: The 13th annual conference on liquid atomization and Spra systems. Wuxin: Instiute of Automotive Engineering, Shanghai Jiao Tong University
5. Pefley RK et al (1984) United States. U.S. Department of Energy. Research and development of neat alcohol fuel usage in automobiles. National Technical Information Service, Springfield
6. Yamin J, Dado M (2004) Performance Simulation of a four-stroke engine with variable stroke-length and compression ratio. Appl Energy 77:447–463
7. Joshi A, Borse SL, Alblawi A, Alshqirate AA (2018) Effect of spark plug position on performance of engine using OPENFOAM. Int J Mech Eng Technol 9(11):114–120
8. Joshi A, Borse S (2017) Study of the effect of spark advance engine speed variation and number of spark plugs on engine performance using CFD software. J Ocean Mech Aerosp 48:1–9
9. Kannan BT, Srivathsan PS (2016) Numerical simulation of spark ignition engine using OpenFOAM. Perspect Sci 8:13–15
10. Millo F, Luisi S, Borean F, Stroppiana A (2014) Numerical and experimental investigation on combustion characteristics of a spark ignition engine with an early intake valve closing load control. Fuel 121:298–310
11. Battistoni M, Mariani F, Risi F, Poggiani C (2015) Combustion CFD modeling of a spark ignited optical access engine fueled with gasoline and ethanol. Energy Procedia 82:424–431
12. Mobasheri R, Fotrosy Y, Jalalifar S (2009) Modelling of a spark ignition engine combustion a computational and experimental study of combustion process effects on NOx emission. Asian J Appl Sci 2(4):318–330
13. Hepkaya E, Karaaslan S, Uslu S, Dinler N, Yucel N (2010) A case study of combustion modelling in a spark ignition engine using coherent flame model. J Therm Sci Technol 34(2):111–121
14. Antony AJ, Shetty DAS (2018) Combustion analysis of single cylinder twin spark ignition engine at different gasoline-methanol blends. Pollut Res 37(1):155–159
15. Wu X, Daniel R, Tian G, Xu H, Huang Z, Richardson D (2012) Dual-injection: the flexible, bi-fuel concept for spark-ignition engines fueled with various gasoline and biofuel blends. Appl Energy 88:2305–2314
16. Chen RH, Chiang LB, Wu MH (2010) Gasoline displacement and NOx reduction in an SI engine by aqueous alcohol injection. Fuel 89:604–610
17. Elfasakhany A (2015) Investigations on the effects of ethanol-methanol-gasoline blends in a spark-ignition engine: performance and emission analysis. Eng Sci Technol Int J 18:713–719
18. Montorfano A, Piscaglia F, Onorati A (2015) An extension of the dynamic mesh handling with topological changes for LES of ICE in OpenFOAM. SAE world congress 2015 Detroit Michigan. 10.4271/2015-01-0384
19. Patil VM, Agrawal A (2013) Cylinder cold flow CFD simulation of IC engine using hybrid approach. World J Sci Technol. https://doi.org/10.13140/RG.2.1.2128.2642
20. Wani MM (2018) Thermodynamic evaluation of methanol as an alternative fuel to gasoline in a single cylinder spark ignition engine. Energy Power 8(2):35–45
21. Olah GA, Goeppert A, Prakash GKS (2006) Beyond oil and gas: the methanol economy. Wiley-VCH

22. Ward PF, Teague JM (1996) Fifteen years of fuel methanol distribution. In: XI international symposium alcohol fuels. (www.energy.ca.gov/papers/CEC-999-1996-017.PDF)
23. Richard L, Bechtold MB, Goodman Timbario TA (2007) Use of methanol as a transportation fuel. Prepared for the Methanol Institute. (www.methanol.org/wp-content/uploads/2016/06/Methanol-Use-in-Transportation.pdf)
24. Ramtilak A, Joseph A, Sivakumar G, Bhat S (2005) Digital twin spark ignition for improved fuel economy and emissions on four stroke engines. SAE technical paper, vol. 26. (https://doi.org/10.4271/2005-26-008)
25. Fortea C, Bianchi GM, Corti M, Fantoni S (2015) Evaluation of the effects of a twin spark ignition system on combustion stability of a high performance PFI engine. Energy Procedia 81:897–906
26. Li WY, Li Z, Xie KC (2009) The development of methanol industry and methanol fuel in China. Energy Sourc Part Recovery Utilization Environ Effects 31:1673–1679

Thermofluids in Energy Applications

Performance Study on Flat Plate Solar Water Heater with Copper Nanoparticles

R. Praveen Bharathwaj, M. B. Varun Pradeep, Joe Jones Raju, A. Satheesh, and P. Padmanathan

Abstract Evaluating the improvement in efficiency and heat transfer rate in heat exchangers using additives has been the centre of attraction for many researchers. In this study, a solar flat plate water heater is modelled using ANSYS Fluent, and the simulated results are compared with experimental data. For the simulation, a single-phase model is used with equivalent correlations for the thermo-physical properties of the nanofluid. Using water as the working fluid, the performance was evaluated for a set flow rate and solar insolation. The study is extended by modelling the same in ANSYS Fluent and using water and CuNP mixture as working fluid to analyse the enhancement of heat transfer in thermal absorbers. The pressure drop in the system is analysed and the data is used to justify the usage of nanoparticles in solar thermal absorbers. The efficiency of the system for various volume fractions of nanoparticles is optimized by introducing an efficiency index.

Keywords Solar thermal energy · Nanoparticles · Optimization · Efficiency index · Flat plate collector

Nomenclature

k	Thermal conductivity
Pe	Peclet number
S	Efficiency index
T	Temperature
u	X-velocity
v	Y-velocity
w	Z-velocity
x	Position/length

R. Praveen Bharathwaj · M. B. Varun Pradeep · J. J. Raju · A. Satheesh (✉) · P. Padmanathan
School of Mechanical Engineering, Vellore Institute of Technology, Vellore, Tamil Nadu 632014, India
e-mail: satheesh.a@vit.ac.in

S. Revankar et al. (eds.), *Proceedings of International Conference on Thermofluids*, Lecture Notes in Mechanical Engineering,
https://doi.org/10.1007/978-981-15-7831-1_50

Greek symbols

α	Thermal diffusivity
ϑ	Kinematic viscosity
ΔP	Overall pressure drop in pipe
ΔT	Rise in temperature of nanofluid
ρ	Density
φ	Concentration percentage of nanoparticles

Subscripts

nb	Base fluid
nf	Nanofluid
np	Nanoparticles
i	Dummy variable

1 Introduction

Solar energy is the driver of all other energy systems on this planet (wind, ocean thermal, biomass). Extracting it to its fullest potential is absolutely essential for sustainable growth. Solar energy can be extracted mainly through photovoltaic modules or solar thermal absorbers. According to the global solar thermal energy council, the thermal demand for comfort has levelled the electricity demand in the recent past [1]. Since most of the energy demand being thermal mostly for comfort, establishing effective passive thermal absorbers becomes crucial. To obtain higher efficiency nanoparticles can be implemented, even though many techniques and design adaptations have increased the heat transfer efficiency, nanoparticles have the best improvement factor [2]. Augmenting nanoparticles increases thermal conductivity at the cost of specific heat of the base fluid and pumping power. Optimizing the volume fraction of nanoparticle to obtain the maximum efficiency is one of the main challenges for any thermal system designer. In order to effectively design for any operating condition, there is a need to understand how nanofluids work for wide range of temperature and pressure for various fluids. This detailed study is lacking at present and the existing studies show erratic results and rendering itself worthless. One example is that Jana et al. and Xuan et al. [3, 4] have experimented on the same nanofluid having different volume fraction (0.3, 7.5%) but have arrived at the same thermal enhancement of 78%. This demands a lot of simulated data considering the cost and time of experimental setup.

There are many devices to capture solar radiation like parabolic trough collectors, fresnel reflectors, parabolic dish collector and central tower collectors [5]. Flat plate heaters are selected since they are uncomplicated to install on rooftops of residential houses. Other thermal absorbers are for electricity production and not for direct utilization of thermal energy. Previously, few studies have been conducted in this domain of nanofluids and flat plate heaters and this study is looking to build on this existing knowledge [6–11, 31, 32]. In recent times, nanoparticles like Al_2O_3, TiO_2, Fe, CuO, Cu, SWCNTs and MWCNTs have been researched greatly. Major application in thermal domain has been in HVAC systems, heat pipes and heat exchangers. But application in small-scaled solar heating equipment has been very low. In this study, the properties of nanofluid and improvement in heat transfer by augmenting CuNP (2.5, 4, 5, 7.5) vol % in a solar flat plate water heater are analysed. The study is of two parts experimental and simulation. The simulation is conducted via ANSYS Fluent and the simulation is validated by the experimental data.

2 Selection of Nanoparticles

The thermal conducting abilities of a nanofluid is different for each nanoparticle and the concentration used. Lee et al. have studied the effectiveness of Al_2O_3 with 4.3% concentration [12]. Singh et al. have investigated the use of silicon-carbide water at 4% concentration for heat transfer applications [13]. Ponmani et al. have characterized the thermal and electrical properties of CuO and ZnO nanoparticles at less than 50 nm size [14]. Esfe et al. have shown the properties of 98 nm size Fe at 0.05% concentration [15]. Xuan and Li have used Cu nanoparticles at different concentrations and found high thermal conductivity at 7.5% concentration [16]. The properties observed by these studies are shown in Table 1. From Table 1, it is observed that Cu nanoparticles exhibit a very high percentage of thermal enhancement at 78%. Therefore, Cu nanoparticle is selected for this study in hope for maximum thermal conductivity.

Table 1 Various nanofluids and their enhancement %

Sl. No.	Nanoparticle / Size (nm)	Base fluid	Concentration %	k_{nf}/k_{bf}	Enhancement (%)
1	Al_2O_3	Water	4.3	1.32	32
2	SiCN/100	Water	4	1.22	22
3	ZnO/<50	Water	0.08	1.12	12
4	Fe/98	Water	0.05	1.05	5
5	Cu	Water	7.5	1.78	78

3 Physical Properties of Nanoparticles

3.1 *Effective Thermal Conductivity*

Thermal conductivity of a nanofluid k_{nf} is higher than the base fluid k_{bf}. The enhancement of thermal conductivity in nanofluids is measured as $k_{\mathrm{nf}}/k_{\mathrm{bf}}$ which is the effective thermal conductivity. Some of the studies have given different mathematical expressions for effective thermal conductivity of a nanofluid [17–25]. Gupte et al. have equated the effective thermal conductivity based on the effect of translational motion of nanoparticles in terms of Peclet number [17].

$$k_{\mathrm{nf}}/k_{\mathrm{bf}} = (0.0556\mathrm{Pe} + 0.1649\mathrm{Pe}^2 - 0.0391\mathrm{Pe}^3 + 0.0034\mathrm{Pe}^4) \quad (1)$$

Timofeeva et al. took into account the agglomeration of the nanoparticle in [18]

$$k_{\mathrm{nf}}/k_{\mathrm{bf}} = 1 + 3\varphi \quad (2)$$

Pak and Cho have taken the assumption that the heat transfer enhancement is due to suspended particles in the nanofluid [19].

$$k_{\mathrm{nf}}/k_{\mathrm{bf}} = 1 + 7.47\varphi \quad (3)$$

In this study, the correlation (3) is used to calculate the effective thermal conductivity of the nanofluid containing Cu. This correlation stands good for a wide range of nanoparticles like Al_2O_3, TiO_2, Fe, CuO, Cu, SWCNTs and MWCNTs. Therefore, this correlation is chosen due to its simplicity and minimum deviation that is found between the theoretical and observed values.

3.2 *Density*

There is an increase in the density of the base fluid on addition of nanoparticle. The effective density of the nanofluid ρ_{nf} is equated by Asirvatham et al. in terms of the density of base fluid ρ_{bf} and the density of nanoparticles ρ_{np}[20]

$$\rho_{\mathrm{nf}} = (1 - \varphi)\rho_{\mathrm{bf}} + \varphi\rho_{\mathrm{np}} \quad (4)$$

In their experimental work, Pak and Choi have confirmed the correctness of (4) [19].

3.3 Specific Heat

The specific heat of nanoparticles, $(Cp)_{np}$ is lower than the specific heat of the base fluid $(Cp)_{bf}$. As a result, the effective specific heat of the nanofluid, $(Cp)_{nf}$ is lower than the specific heat of the base fluid. Xuan and Roetzel found the equation for calculation of the effective specific heat [26]

$$(\rho Cp)_{nf} = (1 - \varphi)(\rho Cp)_{bf} + \varphi(Cp)_{np} \tag{5}$$

3.4 Viscosity

Viscosity of nanofluids is the primary deciding factor of pressure drop in a flow. Analytically calculating it to the most accurate value is absolutely essential. Many research works have been conducted in this region and various curve fitted correlation have been formulated. Einstein's equation has been widely used in many research works [20, 27–30],

$$\mu_{nf} = \mu_{bf}(1 + 2.5\varphi) \tag{6}$$

But higher-order term provides us with higher accuracy. Mahbubul et al. have stated improved Einstein with second-order term which is utilized for this study [30].

$$\mu_{nf} = \mu_{bf}(1 + 2.5\varphi + 6.5\varphi^2) \tag{7}$$

4 Experimental Section

4.1 Experimental Setup

The apparatus used is a Venus solar water heater (Model-halo domestic; Fig. 1) with a variable flow setting. The apparatus used is of a non-concentrating type so it can fully utilize the diffused and direct components of the incoming radiation. In the setup, copper tubes of 12 mm diameter and 1900 mm long are mechanically polished and then coated with nickel layer by electrolysis deposition technique. Thermocouples are provided to record the temperature of glass cover, water inlet, water outlet and absorber plate. A rotameter is provided in the flow circuit to control the mass flow rate of water flowing in the pipes. As water flows through the pipes it gets heated up.

Fig. 1 Solar water heater setup

Hot water is stored in the overhead tank above the absorber contraption. Insulation material is stacked below the pipes and on the other side of the absorber setup.

4.2 Experimental Procedure

Experiment was performed with pure water as working fluid for a volume flow rate of 10 LPH which was measured by rotameter. The Reynolds number of the flow was 492.382, inlet temperature of water was 302 K and ambient temperature was 304 K. Solar insolation received was 430 W/m^2. Solar insolation was measured using pyranometer (direct beam radiation + diffused radiation). Data was collected after the setup has reached steady state. While conducting the experiment, the wall temperatures at different locations and the outlet temperature of water were measured by a series of thermocouples adhered to the walls of the fluid carriers. Temperatures were measured using a contact digital thermometer. The wall temperature of absorber was 344.2 K. Many readings were taken to improve the accuracy of the observation. From the obtained data, efficiency was calculated which is to be maximized.

The experiment was mostly conducted during the mid-day of December 10 to December 14, 2019. Location of the solar flat plate collector was 12.9165 N 79.1325 E. The insolation was in appreciable range even though December. The above conditions were replicated into an analysis and it was run for pure water, and the results were in great correlation with the experimental data.

5 Simulation Procedure

5.1 Governing Equations

Continuity equation:

$$\frac{\partial u}{\partial x} + \frac{\partial v}{\partial y} = 0 \tag{8}$$

Navier–Stokes equation:

$$\frac{\partial u}{\partial t} + \overline{u}.\nabla u = -\frac{1}{\rho}\frac{\partial P}{\partial x} + \vartheta.\nabla^2 u \tag{9}$$

Energy equation:

$$\frac{\partial T}{\partial t} + \overline{u}.\nabla T = \alpha\nabla^2 T \tag{10}$$

These are the equations governing any fluid flow. Equation (9) contains density of the working fluid as one of the parameters. Apart from enhancement in thermal conductivity and changes in viscosity, density also affects the thermal properties and flow characteristics of the fluid.

5.2 CAD Modelling and Meshing

The simulation was the exact replication of the experimental conditions and the simulation was carried out using ANSYS Fluent module. Few parameters like fouling, scaling and erosion of the tube interior were neglected as the real setup was fairly new and produces the specified output. Natural and forced convection over the flat plate was neglected since the air was still when the experiment was conducted. There will be small eddies on the leeward side of the flat plate which also carries a part of available energy. The trapping of long wavelength radiations by the glass plate was also neglected since its effect on thermal efficiency is minor when looked at the direct beam radiations. Adaptive meshes and boundary fitted mesh were utilized to reduce the computational cost and improve the accuracy of result. Grid was refined to the best of the processor's capability (Fig. 2).

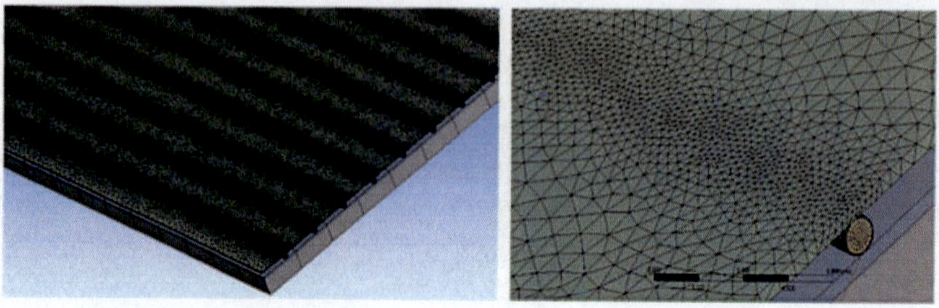

Fig. 2 Meshing of the CAD model

5.3 Boundary Conditions

Fluid inlet, outlet, heat flux, ambient temperature, pipe and absorber temperature form the experiments were used here as boundary conditions (Fig. 3).

6 Result and Discussion

A small deviation of 4.6875% in outlet temperature of experimental setup and simulation of water is obtained for the same boundary conditions mentioned in Fig. 3. This value of error can be justified by small variations in solar flux and wind velocities during the experiment and other irreversibilities in the experimental setup. On stating that the simulation agrees with the experiment, it is extended for different CuNP volumetric percentages (2.5, 4, 5, 7.5%).

Figure 4 shows us the diffusion of thermal energy through the absorber plate to fluid element. The insulation effect can also be visualized through this contour.

Figure 5 clearly presents us with the pressure drop in the pipe. Since in this work, the flow is laminar and the drop in pressure is linear

From Fig. 6, it is observed that the temperature nearly saturates at $x = 0.7$ m. this also shows the steep heat transfer rate at initial stage. From Fig. 8, it is evident that the temperature achieved by water is much less than the nanofluids presented in the

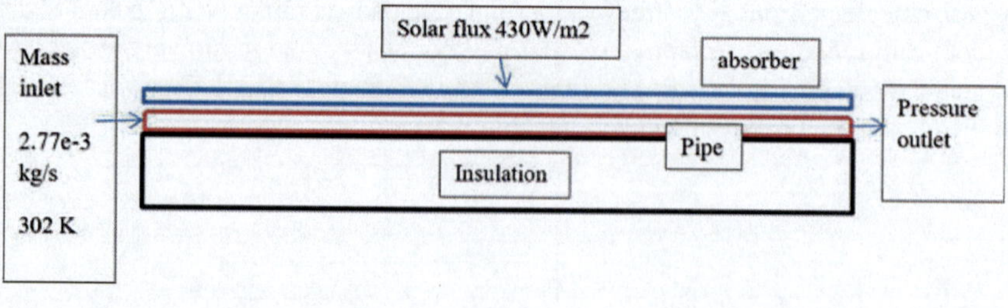

Fig. 3 Boundary conditions used for the simulation

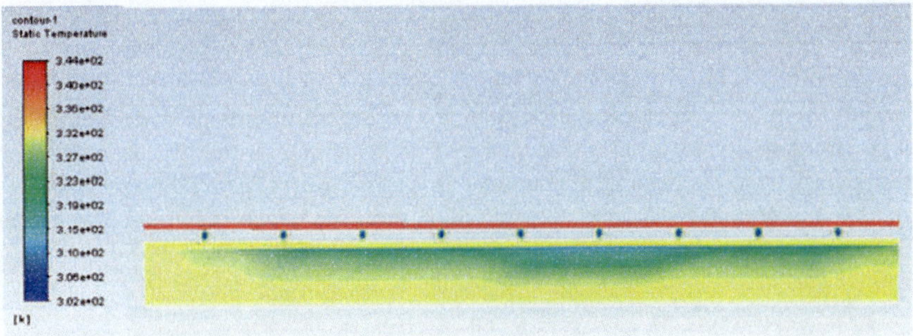

Fig. 4 Temperature contour of a cut section in *y–z* plane

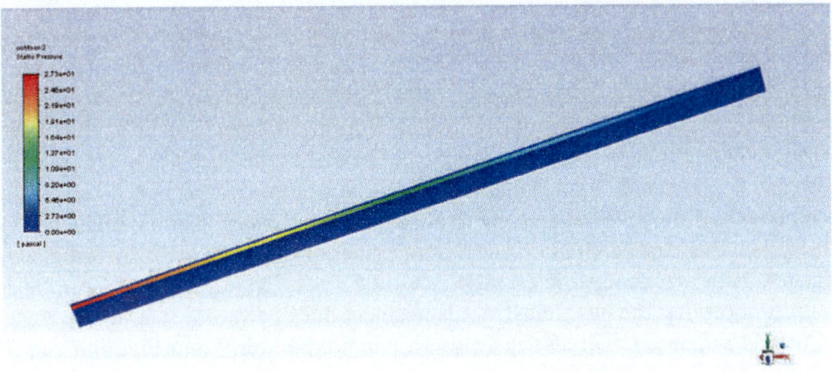

Fig. 5 Pressure contour of a cut section in *x–y* plane for 7.5% CuNP

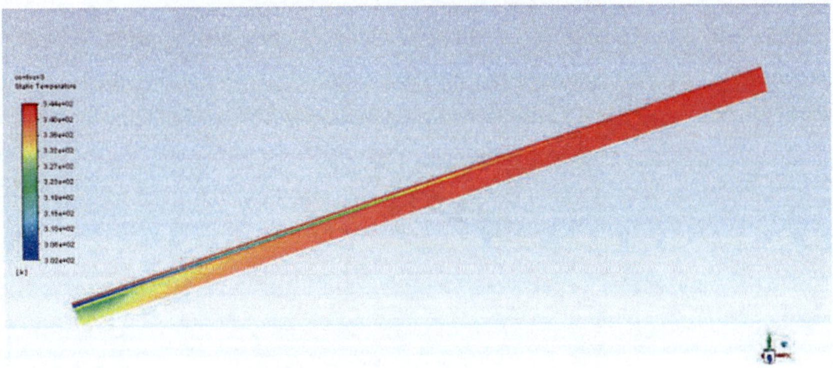

Fig. 6 Temperature contour of a cut section in *x–y* plane 7.5% CuNP

study. For 7.5% CuNP, there is a rise of 1.8215% in outlet temperature and 20.187% increase in heat transfer. Similarly, the other cases present with similar results but the magnitude is lesser than 7.5% CuNP. From Fig. 9, it is seen that the 7.5% CuNP reaches the plate temperature faster than any of the nanofluids used. Therefore, by augmenting the nanoparticles size of the heat exchangers or thermal absorbers can be reduced which is desirable parameter in any industry. The study was restricted to 7.5% CuNP because of the sedimentation problem which is observed at higher CuNP volume fractions.

6.1 Effect of Cu Nanoparticles on Heat Transfer and Overall Efficiency

From literature, it is evident that the rate of heat transfer is enhanced due to augmentation of various nanoparticles, and the present study supports the claim. From Fig. 7, a rapid heat transfer at entry can be observed from the steep slope till $x = 0.6$ m. The rate of heat transfer subdues along the pipe due to the decrease in temperature difference between wall and fluid which can be observed from Fig. 8. When the temperature difference between the plate and water reduces, thermal efficiency of the system is enhanced. Figure 8 clearly shows the enhancement in thermal efficiency due to addition of nanoparticles.

Just observing the heat transfer enhancement does not justify the improvement in overall efficiency as it also depends on pumping power. From Fig. 9, it can be

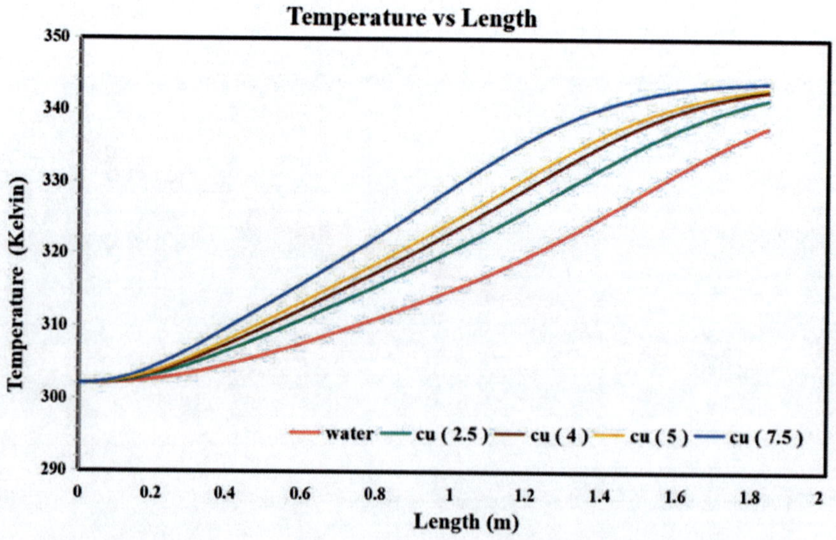

Fig. 7 Temperature of working fluid at different volume fractions of CuNP along the pipe

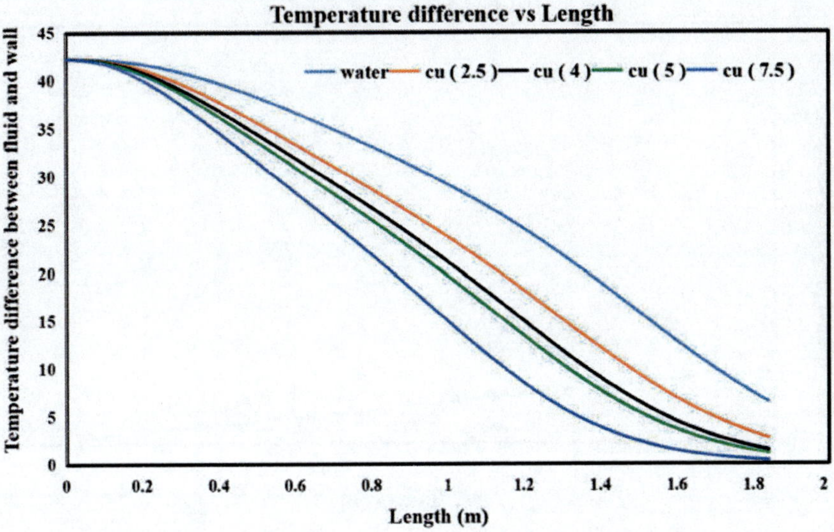

Fig. 8 Temperature difference between pipe wall of the pipe and fluid at different volume fractions of CuNP along the pipe

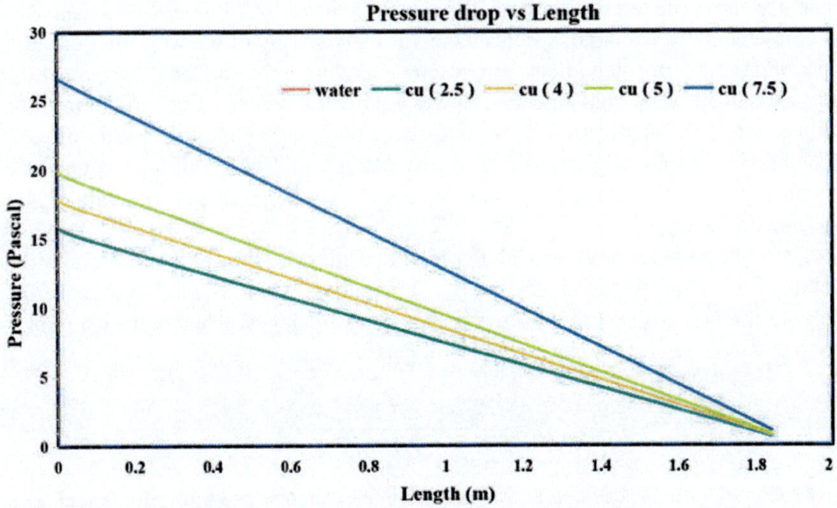

Fig. 9 Pressure drop (Pa) for different volume fractions of CuNP along the pipe

observed that 7.5% CuNP has a drop of 26.0084 Pa, whereas 2.5% CuNP has nearly the same pressure drop but has a mild increase in outlet temperature. Figure 10 shows the pressure difference between water and CuNP of different volumetric percentages.

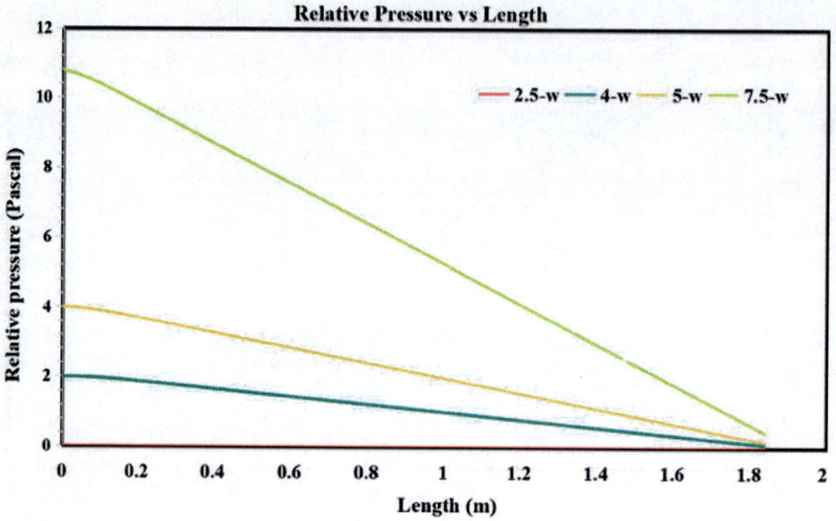

Fig. 10 Pressure drop in nanofluids with respect to water along the pipes

It can be clearly inferred that the pressure difference between water and 2.5% CuNP is nearly zero. The relative drop of 7.5% CuNP is nearly 2.75 times of 4% CuNP.

Implementation of nanoparticles increases the viscosity of the working fluid. But the volumetric percentage must be chosen depending upon the length scales of the heat exchanger or thermal absorber. Lower volumetric percentage of CuNP must be utilized when the application is for long-range pumping where thermal dissipation becomes inevitable. Higher volumetric percentage of CuNP must be augmented in short-range rapid thermal heat exchangers where the motive is to obtain higher thermal efficiency.

The volume percentage of CuNP which provides the best efficiency cannot be defined arbitrarily. An efficiency index (*S*) is defined which is used to identify the volume fraction which provides the best overall efficiency. This expression holds good everywhere since it is a nondimensional number.

$$S = \frac{\rho_{\text{nf}} \times (C_{\text{p}})_{\text{nf}} \times \Delta T}{\Delta P} \tag{11}$$

Efficiency index is defined as the ratio of heat exchange and pumping power. The relative heat exchange must be higher than increase in pumping power due to addition of CuNP for a better performance. Here, higher the value of S, better the overall efficiency. Increase in vol% of nanoparticles decreases the specific heat of the nanofluid and the density of the nanofluid increases with increase in vol%. Efficiency index depends on all the four parameters (ΔT, ΔP, ρ_{nf}, C_{p}). These parameters are a function of vol%; therefore, the efficiency index is directly related to vol%. From Table 2, it is inferred that 4% CuNP provides the highest overall efficiency. Even though

Table 2 Variation of efficiency index (S) with volume fraction of CuNP

Volume % of nanoparticles	ΔT(°C)	ΔP(Pa)	ρ_{nf}(kg/m^3)	$(C_p)_{nf}$(J/kgK)	$S \times 10^6$
2.5	39.514	15.4561	1182.7955	4089.30	12.3653
4	40.761	17.4476	1320.4448	4032.80	12.4404
5	41.119	19.4160	1382.2110	3994.30	11.6922
7.5	41.905	26.0143	1581.6265	3899.4	9.9347

7.5% gives the highest thermal enhancement, 4% CuNP is much more economical and energy efficient. Therefore, the study recommends 4% CuNP for solar flat plate heaters.

$$S = \frac{\rho_{\text{nf}} \times (C_{\text{p}})_{\text{nf}} \times (\Delta T - T_{\text{loss}})}{\Delta P} \quad (12)$$

Equation (12) is adjusted for long-range fluid pumping where thermal losses and pumping power are very high. By calculating the value of efficiency index, the volume fraction can be clearly defined which must be used for that specific application.

7 Conclusions

The heat transfer enhancement capabilities of a heat exchangers are greatly improved in the presence of nanoparticles in the working fluid. Even for a small volume fraction there is considerable increase in the outlet temperature. As the volume fraction of nanoparticle in the working medium increases so does the heat transfer. But with increase in volume fraction comes with a cost of higher pumping power and in some cases sedimentation and erosion. Factors such as particle collision, Brownian motion and other nanoscale energy transport phenomenon contribute to energy exchange. Restricting the calculations to a single-phase study did not affect the results drastically. Even though nanofluids are two phase fluids, it more or less behaves like a pure liquid in actual working conditions. Finding the optimal range for balanced performance is essential. So, to quantify the range, an efficiency index (S) is used and from the calculations, 4% CuNP is found to give best heat transfer enhancement for minimal pumping power.

References

1. Global Solar Thermal Energy Council, https://www.solarthermalworld.org/news/rising-demand-solar-heat-large-buildings-and-industry. Last accessed 2019/12/15
2. Godson L et al (2010) Enhancement of heat transfer using nanofluids—an overview. Renew Sustain Energy Rev 14(2):629–641

3. Soumen J, Salehi-Khojin A, Zhong WH (2007) Enhancement of fluid thermal conductivity by the addition of single and hybrid nano-additives. Thermochim Acta 462(1–2):45–55
4. Yimin X, Li Q (2000) Heat transfer enhancement of nanofluids. Int J Heat Fluid Flow 21(1):58–64
5. Sirajuddin SMR et al (2017) A review paper on electricity generation from solar energy. Int J Res Appl Sci Eng Technol 5(issue IX):1884–1889
6. Rehena N, Alim MA (2014) Thermal performance of nanofluid filled solar flat plate collector. Int J Heat Technol 33(2):17–24
7. Ranjith PV, Karim AA (2016) A comparative study on the experimental and computational analysis of solar flat plate collector using an alternate working fluid. Procedia Technol 24:546–553
8. Chaji H et al (2013) Experimental study on thermal efficiency of flat plate solar collector using TiO_2/water nanofluid. Mod Appl Sci 7(10):60–69
9. Yuvarajan D, Munuswamy DB (2016) Analysis on the influence of nanoparticles of alumina, copper oxide, and zirconium oxide on the performance of a flat-plate solar water heater. Energy Fuels 30(11):9908–9913
10. Joe MJ, Iniyan S (2015) Performance of copper oxide/water nanofluid in a flat plate solar water heater under natural and forced circulations. Energy Convers Manage 95:160–169
11. Heera P, Shanmugam S (2015) Nanoparticle characterization and application: an overview. Int J Curr Microbiol App Sci 4(8):379–386
12. Lee S et al (1999) Measuring thermal conductivity of fluids containing oxide nanoparticles. J Heat Transf 121(2):280–289
13. Singh D et al (2009) An investigation of silicon carbide-water nanofluid for heat transfer applications. J Appl Phys 105(6):064306
14. Swaminathan P et al (2014) Formation and characterization of thermal and electrical properties of CuO and ZnO nanofluids in xanthan gum. Colloids Surf 443:37–43
15. Esfe MH et al (2015) An experimental study on the effect of diameter on thermal conductivity and dynamic viscosity of Fe/water nanofluids. J Thermal Anal Calorim 119(3):1817–1824
16. Yimin X, Li Q (2000) Heat transfer enhancement of nanofluids. Int J Heat Fluid Flow 21(1):58–64
17. Gupte SK, Advani SG, Huq P (1995) Role of micro-convection due to non-affine motion of particles in a mono-disperse suspension. Int J Heat Mass Transf 38(16):2945–2958
18. Timofeeva EV et al (2007) Thermal conductivity and particle agglomeration in alumina nanofluids: experiment and theory. Phys Rev E 76(6):061203
19. Choon PB, Cho YI (1998) Hydrodynamic and heat transfer study of dispersed fluids with submicron metallic oxide particles. Experim Heat Transf Int J 11(2):151–170
20. Godson AL et al (2009) Experimental study on forced convective heat transfer with low volume fraction of CuO/water nanofluid. Energies 2(1):97–119
21. Seok Pil J, Choi SU (2004) Role of Brownian motion in the enhanced thermal conductivity of nanofluids. Appl Phys Lett 84(21): 4316–4318
22. Wei WANG (2017) A comprehensive model for the enhanced thermal conductivity of nanofluids. J Adv Res Phys 3(2)
23. William E, Fish J, Keblinski P (2006) Role of Brownian motion hydrodynamics on nanofluid thermal conductivity. Appl Phys Lett 88(9):093116
24. Bao Y (2008) Thermal conductivity equations based on Brownian motion in suspensions of nanoparticles (nanofluids). J Heat Transf 130(4):042408
25. Maxwell JC (1873) A treatise on electricity and magnetism. Clarendon Press, Oxford
26. Xuan Y, Roetzel W (2000) Conceptions for heat transfer correlation of nanofluids. Int J Heat Mass Transf 43:3701–3707
27. Mahbubul IM et al (2013) Thermophysical properties and heat transfer performance of Al_2O_3/R-134a nano refrigerants. Int J Heat Mass Transf 57(1):100–108
28. Mahbubul IM, Rahman S, Amalina MA (2012) Investigation of viscosity of R123-TiO_2 nanorefrigerant. Int J Mech Mater Eng 7(2):146–151

29. Mahbubul IM, Saidur R, Amalina MA (2013) Heat transfer and pressure drop characteristics of Al_2O_3-R141b nanorefrigerant in horizontal smooth circular tube. Procedia Eng 56:323–329
30. Kedzierski MA (2011) Effect of Al2O3 nanolubricant on R134a pool boiling heat transfer. Int J Refrig 34(2):498–508
31. Karami M, Raisee M, Delfani S (2014) Numerical investigation of nanofluid-based solar collectors. In: IOP conference series: materials science and engineering, vol 64(1). IOP Publishing
32. Sandesh CS, Pise AT, Madane PA (2012) Performance of nanofluid-charged solar water heater by solar tracking system. In: IEEE-international conference on advances in engineering, science and management (ICAESM-2012). IEEE, pp 247–253

Experimentation with a Solar Air Heater Coupled with Evacuated Tubes and Heat Pipes

P. Devasena, D. Dimple, A. Varshith, K. Anvesh Reddy, and Vinod Kotebavi

Abstract The demand for energy is increasing worldwide and the best way to meet this demand is a clean renewable energy source. Among the various options available, solar energy can be considered as the best renewable energy source. This paper deals with design, fabrication, and performance study of a solar air heater coupled with evacuated tubes and heat pipes. Heat pipes are used to absorb the solar energy and evacuated tubes are used to prevent the convective losses. A selective absorber coating is used to coat the inner side of the evacuated tube. Air is forced in to the duct of size 10 cm × 10 cm by a fan, which is driven by a photovoltaic panel. Six heat pipes and six evacuated tubes are used to absorb the solar radiation. A solar flux meter is used in to measure incoming solar radiation. Experiments are conducted with and without evacuated tubes. For both the configurations, the variations of thermal efficiency with respect to time have been reported. The maximum thermal efficiency obtained is 35% for evacuated tube with heat pipe. The average efficiency of evacuated tube with heat pipe is 1.5 times that of heat pipe alone.

Keywords Solar heater · Heat pipe · Evacuated tube · Flux meter

1 Introduction

Air heaters are generally used to heat air for different purposes. There are different types of air heaters such as radiant and space heaters and also forced air products. The principle behind forced air products is that the temperature of the air is controlled by allowing it to circulate across a heat source using a fan or a blower. The heater makes it possible for the air flowing through it to absorb the thermal energy and exit it at a higher temperature. Whereas, the principle behind radiant heaters is that they radiate heat from a point source and in turn heat the different objects which are near to that point source. The air mass in the surroundings is heated due to a free flowing

P. Devasena · D. Dimple · A. Varshith · K. A. Reddy · V. Kotebavi (✉)
Department of Mechanical Engineering, Amrita School of Engineering, Amrita Vishwa Vidyapeetham, Bengaluru, India
e-mail: k_vinod@blr.amrita.edu

S. Revankar et al. (eds.), *Proceedings of International Conference on Thermofluids*, Lecture Notes in Mechanical Engineering,
https://doi.org/10.1007/978-981-15-7831-1_51

convection cycle which is, in turn, formed due to the temperature differential. These systems consume large amount of power since it needs power for heating air and for circulation of air.

In solar air heater, solar energy is captured by using some absorbing medium and then it is transferred to air in the form of heat. This system can be used for process heating, drying of food products, heating buildings, etc. Since we get solar energy at free of cost, we can call this as the cost effective method and also solar energy is clean renewable energy.

A heat pipe is basically a heat exchanger and it transfers heat by conduction and phase transition. It has two sections, an evaporator and a condenser. In evaporator section, it absorbs heat from the source and the liquid inside the heat pipe absorbs this heat and changes it phase. This vapor then moves to condenser section and rejects its heat. Same time, it condenses back to liquid. Since the heat transfer coefficients for boiling and condensation are high, heat pipes are considered as most effective heat exchangers. The evacuated tubes normally consist of two concentric glass tubes which are made from borosilicate. Radiation is allowed to pass through it as the outer tube is transparent in nature. An absorber coating which has high absorptivity but low reflectivity is used to coat the inner tube. The gap between two glass tubes is evacuated. This reduces the convective heat losses.

Many researchers have studied different designs. Jasim et al. performed an experimental study of a solar air heater having longitudinal baffles [1]. They obtained a maximum efficiency of 68% with seven baffles. Mahmood conducted experimental study on double-pass solar air heater [2]. He compared the performance of the system both by using phase change material and without using it. His results show that with use of phase change material, the outlet temperature of air compared to the ambient temperature was increased as well as enhanced. He also concluded that the efficiency of his system with phase change material was 56% and without it was 47%.

Kumar et al. conducted experiments on solar collectors using evacuated tubes [3]. From their results, they came to a conclusion that the efficiency of evacuated tube solar collector using an absorber plate can be increased significantly. Deng et al. conducted experimental study on the concept of deposition of dust on the transparent glass cover of flat plate solar air collector with single pass [4]. They performed the analysis of single-pass flat plate solar air collector with surface having severe dust deposition and a clean surface. It was seen that the efficiency for the case of a surface with severe dust deposition decreased from 10.7 to 2.1% when the range of temperature difference is from 0 to 0.04, in their results. In the same case, the optical efficiency is decreased by 8.39% which is opposite in the case of clean cover surface. Pramod et al. analyzed the performance of a solar still with evacuated tubes and heat pipe [5]. The efficiency achieved with evacuated tube was 40% more than without evacuated tube. Bhushan et al. conducted experimental investigation of double pass solar air heater [6]. They concluded that thermal efficiency greatly depends on time, solar flux, mass flow rate of air. They obtained a maximum efficiency of 58%.

Warrier et al. conducted study on thermal performance of a solar air heater which was fitted with delta winglets [7]. They conducted experimental and numerical study with and without holes on winglets. They found enhancement of thermal performance by 20–150% with punched hole delta winglets. Harikrishnan et al. studied the performance of a solar heat pipe using different working fluids and different fill ratios [8]. They obtained a better performance with acetone when compared to water and methanol. Fill ratio of 25% was found optimum and heat pipe with inclination 60° gave better results when compared to 35°. Ganesh et al. conducted experiments on solar still coupling with evacuated tubes [9]. Their output was doubled when they coupled heat pipe and evacuated tubes with solar still. Vijayasarathy et al. designed a solar distillation system by eliminating the basin with heat pipe and heat exchanger [10]. They obtained an efficiency of 23%. The objective of this paper is to design and build a solar air heater using heat pipe and evacuated tubes and study the performance of the system.

2 Experimental Setup

The prime objective of the study is, heating the air using solar energy. The experimental setup is as shown in Fig. 1. It consists of six evacuated tubes with heat pipes inside, a duct, fan, and a PV cell panel. The length of evacuated tube is 1.5 m and the inner as well as the outer diameters are 3.7 cm and 4.7 cm, respectively. This evacuated tube has two concentric glass tubes and the gap between the tubes is evacuated. The evacuation helps in reducing the convective heat losses in tubes. Evacuated tubes makes an angle of 15° to the ground. Inside the evacuated tube, the evaporator section of the heat pipe is placed and is then exposed to solar radiation. The fluid inside the heat pipe absorbs heat and transfers it to condenser section which is placed inside the duct. The duct is of 1 m length and 10 cm × 10 cm in cross section. The air is forced through the duct by fan inside the duct and this fan is driven by a 12 V DC motor

Fig. 1 Solar air heater with evacuated tube and heat pipe

which in turn is powered by a solar panel. The rated power of the panel is 10 W. No external power supply is used in this system. The heat from the condenser section is absorbed by the air inside of the duct which then flows out. The temperature of air is noted at the entry and exit of the duct using thermocouples. Velocity of air at the outlet is measured by an anemometer. The solar flux is measured by a solar flux meter. The experiments are also conducted with heat pipe alone also. In this mode the evacuated tubes are removed and only heat pipes are used.

3 Results and Discussions

A solar air heater with two different configurations (1) with heat pipe and evacuated tube, (2) heat pipe alone were designed and fabricated. The experiments were conducted in Bangalore (12.97°N, 77.59°E). Readings were taken on two different days from morning to evening for each configuration. Various values recorded were solar flux, inlet and outlet temperatures of air and velocity of air for an interval of 15 min. The variation of solar flux during the day for both days is shown in Fig. 2. We can observe variation of solar flux from 600 to 900 W/m^2 during the day. The average value of solar flux for both the days is almost same.

Using solar flux, temperature difference of air and air flow rate, we calculated the efficiency of the system for both configurations. Efficiency of the system has been calculated using below formulas.

$$\text{Rate of heat absorption by air} = m_{\text{air}} \times \left(C_{\text{p}}\right) \times (\text{temp difference in air}) \tag{1}$$

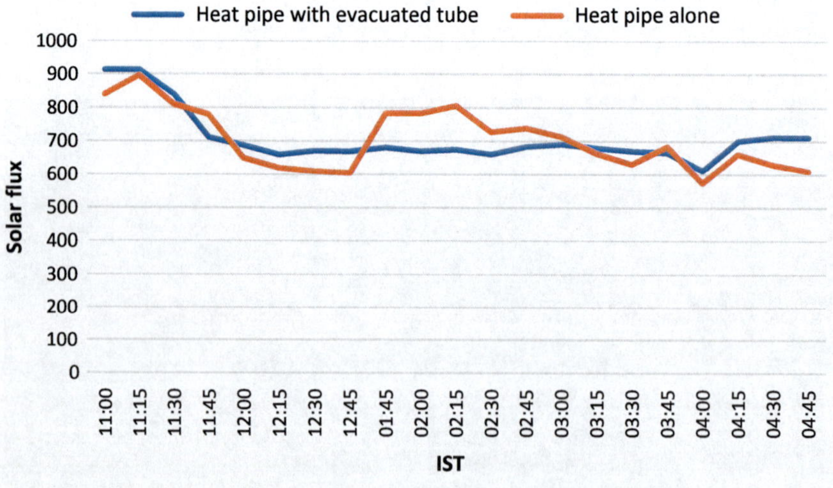

Fig. 2 Time versus solar flux

where "m_{air}" is defined as the mass of air flowing through the duct

$$m_{\text{air}} = \text{density of air} \times (\text{c/s area of the duct}) \times (\text{velocity of flow}) \quad (2)$$

$$\text{Radiation Input} = (\text{Solar flux}) \times (\text{Total absorber area}) \quad (3)$$

$$\text{Efficiency} = (\text{Rate of heat absorption by air}) \div (\text{Radiation Input}) \quad (4)$$

During the experimentation, the velocity of air was almost constant, i.e., 3 m/s and the mass flow rate was about 0.035 kg/s. The variation of efficiency with time is shown in Fig. 3 for both configurations. From Fig. 3, we can observe that the efficiency first increases from morning to afternoon then it decreases in the evening. Efficiency reaches maximum during afternoon. Efficiency for heat pipe with evacuated tube varies from 16 to 35%, whereas for heat pipe alone, it varies from 10 to 27%. The average efficiency of evacuated tube with heat pipe is 27% whereas for heat pipe alone it is 19%. The efficiency of evacuated tube with heat pipe is always more than that of heat pipe alone. The average incident solar flux for both configurations is almost same but the average efficiency of evacuated tube with heat pipe is 50% more than that of heat pipe alone. This is because the evacuated tube prevents the convective loss and also increases the absorptivity of solar radiation.

In various sectors such as agricultural, domestic, commercial as well as industrial sectors, energy in the form of heat is required for various applications. It is necessary

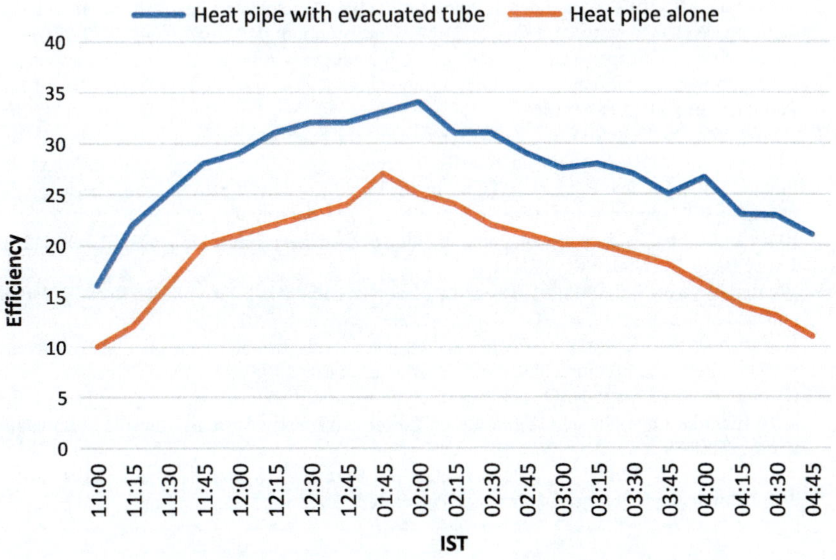

Fig. 3 Time versus efficiency

to design a cost efficient method of producing thermal energy to reduce the gap between the demand and supply of sustainable energy. Here, an attempt was made to design, fabricate, and experimenting on solar air heater using heat pipes and evacuated tubes where the design with evacuated tubes found to be more efficient.

4 Conclusion

Compact solar air heater system reduces the space by replacing glass cover and still with evacuated tubes. Experiments are conducted here with and without evacuated tubes. The average flow rate of air obtained was around 0.035 kg/s. The average efficiency of evacuated tube with heat pipe was about 27% whereas for heat pipe it was 19%. The efficiency can be increased by using more number of evacuated tubes and heat pipes. This result can be used for many other applications like water heating, space heating, agricultural products drying, etc. It has negligible maintenance since it is not using any pump or fan. This clean and renewable energy dependent appliance will have many utilization of energy for different applications.

References

1. Afaq Jasim M (2017) Experimental study of a solar air heater with a new arrangement of transverse longitudinal baffles. J Sol Energy Eng 139(3):031004
2. Mahmood AS (2019) Experimental study on double-pass solar air heater with and without using phase change material. J Eng 25(2):1–17. https://doi.org/10.31026/j.eng.2019.02.01
3. Kumar SBS, Chinnapandian M (2018) Experimental study on the thermal efficiency of the forced convection evacuated tube solar air collectors with and without absorber plate. J Ind Pollut Control 34(1):1839–1846
4. Deng J, Yang X, Yang M, Wang Z (2015) Experimental study of a single-pass flat plate solar air collector with severe dust deposition on the transparent glass cover. Energy Procedia 70:32–40
5. Pramod BVN, Prudhvi Raj J, Hari Krishnan SS, Vinod Kotebavi (2018) Performance analysis of a solar still coupled with evacuated heat pipes. IOP Conf Ser Mater Sci Eng 310:012149
6. Bhushan C, Sing SN. Experimental investigation of double pass solar air heater. ELK Asia Pac J Spec Issue. ISBN: 978-81-930411-8-5
7. Warrier HU, Kotebavi VM (2016) Heat transfer enhancement in solar air heater duct fitted with punched hole delta winglets. IOP Conf Ser Mater Sci Eng 149:012225
8. Harikrishnan SS, Kotebavi V (2016) Performance study of solar heat pipe with different working fluids and fill ratios. IOP Conf Ser Mater Sci Eng 149:012224
9. Ganesh A, Vinod Kotebavi (2016) Pyramid type solar still coupled with evacuated tubes. In: 2016 Biennial International Conference on Power and Energy Systems Towards Sustainable Energy (PESTSE)
10. Vijayasarathy B, Kumar D, Vijayakrishnan K, Kumar A, Vinod Kotebavi (2019) Solar distillation using heat pipes and heat exchanger. IOP Conf Ser Mater Sci Eng 577:012025

Design of Low-Cost Solar Powered E-Rickshaw: A Case Study

Daya Shankar, Manisha Biswas, and Rupa Datta

Abstract Concern has increased over the use of fossil fuels in passenger transport vehicle. In Indian cities, autorickshaws are making a huge quantity of air effluence. There are two types of rickshaw are seen running on the road, first one manual type which is pulled by human being and another one is motorized which is using electricity. Here, the proposed e-Rickshaw is designed with respect to a stable design at optimum ground clearance and its aerodynamics. The functionality, construction and mechanisms of solar power-driven rickshaw are briefly discussed in this paper. This paper shows how this conceptual autorickshaw provides passenger comfortable and safe ride upto a speed of 25–30 km/h.

Keywords E-rickshaw · Aerodynamic drag · Solar energy

1 Introduction

Pollution caused by the adaptation of petrol and diesel into automobiles is already high. In India, roads are crowded with two and three-wheeler which are used for transport [1]. 'Rickshaw' or the 'human-created vehicle', initially used in Bangladesh in 1887 [2]. In transportation, it is the oldest system to transport people in the Indian subcontinent. The size of the rickshaw is small so it is easy to fit in narrow roads. The lesser polluting auto rickshaw may use the conventional solar panels [3]. The total no of 635,698 three-wheel vehicles was wholesaled during the economic year that ended in March, 2017 and 2.5 million autorickshaws were running on the road at the same time [4]. Earlier rickshaws were manual and pulled by men which required lots of physical energy. But now it's motorized hence high speed is achieved with less human effort. Cost of the Rickshaw has gone up. Automobile companies have introduced CNG and LPG autorickshaws to reduce pollution problems [5]. Best option to reduce pollution is by using maximum solar energy because it's less prices. Motorized rickshaw can use solar power more conveniently [6]. We will use solar

D. Shankar (✉) · M. Biswas · R. Datta
Department of Mechanical Engineering, ICFAI University Tripura, Agartala, India
e-mail: iitgdaya@gmail.com

S. Revankar et al. (eds.), *Proceedings of International Conference on Thermofluids*, Lecture Notes in Mechanical Engineering,
https://doi.org/10.1007/978-981-15-7831-1_52

panel so we can use it in cloudy weather also [7]. Additional energy feasting of 6% per kilometre is measured for solar vehicles. Reddy et al. discussed about the problems which arise because of fossil fuel use in vehicles [8]. The electric auto rickshaw fitted with solar panel result less pollution. In future, vehicles will explore using various energy sources [9].

In motorized rickshaw or E-rickshaw, noise and air pollution will be less by ~30% [10]. A solar-powered rickshaw can be a good alternative to CNG [11]. These rickshaws have the ability to move at a competitive speed [12]. Motorized rickshaw helps to integrate applying solar energy. It is more comfortable for the passengers as there is no noise like conventional auto rickshaw [13]. Passenger's space is ~98% open; the e-rickshaw provides excellent heat management of the system. Here this concept e-rickshaw is described with all constructional details including its aerodynamic drag.

2 Model Development

The e-rickshaw is built on a light steel frame that mounts three wheels. Then the next body part is designed by determining the load arising because of the passengers. A typical e-rickshaw can carry up to four passengers at a time. E rickshaw works differently from conventional rickshaw with respect to the DC motor, battery and suspension system. In proposed system, it combines microhybrid systems like electric and thermal energy system. Several hybridization methods are available like series hybridization via solar power, series, parallel, series–parallel (composed in both) and complex parallel hybridization.

2.1 Principle and Process

The principle of the operating planned E-rickshaw is represented in Fig. 1 where the solar PV panel to be kept at the roof. Solar PV panels are expected to charge directly; the regulator is connected to control the power surge. The brush less direct current (BLDC) electric motor can be used to transform the battery power into mechanical drive energy. BLDC motors remain consistent, easy to handle and run with least noise with less frictional damages. The BLDC motor controller carried along with the motor senses the location of the stator and energizes the rotor consequently engaging the Hall Effect sensor. Rotation of the motor is transferred to the wheels through difference gears. The improved vehicle circuit including solar PV panels is shown in Fig. 1.

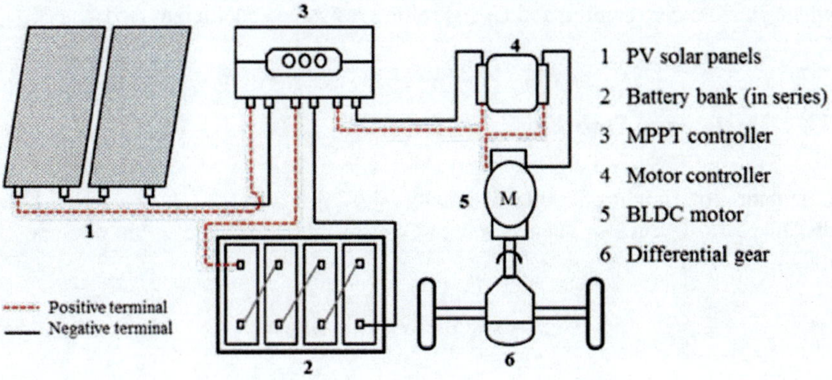

Fig. 1 Line diagram of full operation of solar power E-rickshaw [8]

2.2 Solar Energy Harvesting

The design of any automobile component in India should follow three principles which are—Optimization, safety, comfort. Importance of this vehicle is to accommodate solar panels and accessories for the vehicle existed [14].

The maximum radiation on the panel is given below in Eq. (1).

$$s_{\text{module}} = s_{\text{incident}} \sin((90 - \phi + \delta) + \beta) \tag{1}$$

Here, S_{module} is solar radiation; S_{incident} is solar radiation which is perpendicular to sun; Φ is latitude; δ is declination angle in degree; β it is tilt angle of the module.

2.3 Aerodynamic Drag

Calculation of wind load is defined as the affect on the support structure which exerts stress on the outer side of the structure. The velocity of vehicle varies exponentially.

Wind load calculation is given in Eq. (2).

$$F_{\text{w}} = 0.613(AC_{\text{D}}V^2) \tag{2}$$

Here, C_{D} = Coefficient of drag; A = Area; V = velocity of vehicle; Power necessity as well as choice of motor; Power required to propel is enough to overcome wind drag.

Power required (P_{r}) $\geq$ wind drag + rolling resistance + gravitational force;

$$P_{\text{T}} \geq 0.5C_{\text{D}}AV^2 + W(C_{\text{RR}}\cos\theta + \sin\theta) \tag{3}$$

where P_r is Power required and C_{RR} is rolling resistance coefficient [15].

2.4 Design for Stability

Calculation of Height of center of gravity is given in Eq. (4) (H_{CG} = wheel base distance × difference in vehicle weight/Total weight of vehicle × tan of slope of track).

$$H_{CG} = W_b(W_{FR} - W_F)/m_v \tan\theta \quad (4)$$

where, H_{CG} = height of center of gravity; M_v = mass of vehicle; W_b = distance of wheelbase (1008 mm); θ as 2°; W_{FR} = weight of front tires; W_F = rear wheel; Difference of $W_{FR} - W_F = 9.4$ kg. The final specification is given in Table 1 (Fig. 2).

Table 1 Technically significant dimensions

S. No.	Components	Specification
1	Required speed (v)	20 km/h
2	Net weight	180 kg
3	Driver weight	65 kg
4	Total weight	245 kg
5	Solar panel weight	40 kg
6	Total weight with load	390 kg

Fig. 2 E-rickshaw prototype

2.5 Solar Panel Selection

Here, two solar panels are proposed whose capacity are 40 W, 15 V each. Total area required is 1.6 m^2 and the average panel efficiency ~20% [5]. However the weight of the panel can be reduced to 1/10th of the modern flexible solar panels are used. To develop the required voltage using solar panel, the time duration is 720 Wh/80 W = 9 h. Popularity of solar-based autorickshaw is found in Bangladesh, China, India and coincides with the demand for solar panels in these countries. From the literature, it is found that availability and research on new solar panels is at exponential growth in India [16–19].

2.6 Transmission System

When a vehicle is moving, an inductive force is applied which is equal to the sum of all the opposite forces. The movement of the vehicle causes an additional attractive effort which should be less than the total resistance. The speed of the vehicle gets reduced by the axis of resistance.

3 Result and Discussion

In today's date, autorickshaws plays a vital role in automobile industry. The distance to be travelled by the vehicle per day is considered 50 km (15 trips of 5 km). The battery charging cost was assumed to be INR 10 per kWh [5]. In order to access solar energy, PV panel can be introduced. The maximum velocity is 20 km/h hence regenerative braking system is not applicable here to save energy. CO_2 emissions at the power plant can be reduced by accessing solar energy on spot in place regular grid-dependent charging.

3.1 Case Study of On-Road Performance

Figure 3 depicts the relation between battery discharge rate and vehicle speed. This calculation is considering proposed e-rickshaw not having any load. The graph shows battery discharge rate is increasing with the increase in speed. The maximum speed at each loading condition depends on the pulling ability of the motor. The maximum discharge rate attained was 650 W with a load of 390 kg moving at a speed of 20 km/h. This calculation of the proposed e-rickshaw is carried out considering a straight flat track of 2 km distance for each case and follows the trend reported in literature [5].

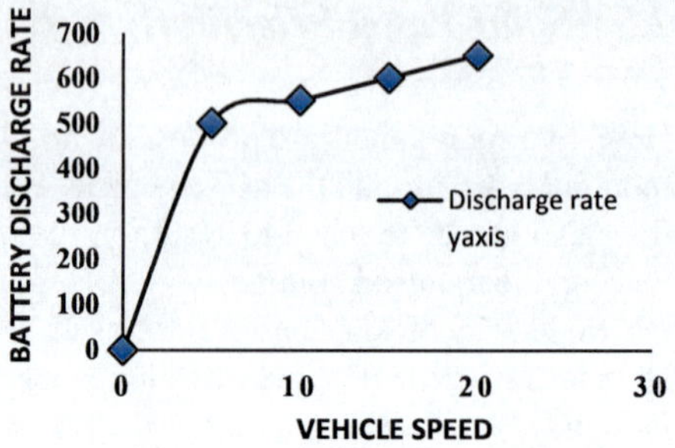

Fig. 3 Impact on battery discharge rate and vehicle speed

3.2 *Benefits of Solar E-rickshaw*

- Noise-Free: Solar E-rickshaws will be noise-free and silent than the typical auto rickshaws and BLDC motor is even more silent proposed in this study.
- Pollution Free: since the soar E-rickshaws use mostly solar power to operate, they will not cause pollution and will reduce carbon dioxide emissions, as opposed to ordinary E-rickshaws [5].
- Efficiency: Solar-powered e-rickshaw will use 40% less power than petrol auto rickshaw and therefore more sustainable.
- Battery Life: After 70–80% charge, the solar-powered rickshaw can run at least 60–70 km and it saves energy by 25–30% than normal e-rickshaw. Because of continuous charging during day time battery will never drain out fully as happens in case of regular e-rickshaw which increases battery life with electronic charge controller.
- Heat Management: Since the passenger space is 98% open, no extra cooling or ventilation system required for rural transport as of now.
- Aerodynamics: Since the frontal portion is covered like a typical automobile and its streamline design makes the best case scenario in reducing the aerodynamic drag under 25–30 km/h speed condition.

4 Conclusions

This concept of e-rickshaw will play a very important role in the development of sustainable rural transport. Also, it will help achieve carbon-neutral growth and less dependency on grid power. Using the solar panel to save fossil fuel can bring in energy security. Efforts should be partially focused on enhancing the efficiency of electrical systems and all mechanical components. Current design takes care of heat management in passenger space and aerodynamic drag at front under low and medium speed condition.

Acknowledgements Authors are thankful to the contribution of Dr. Debjyoti Sahu, Mr. Utkarsh Singh and Mr. Subhrajit Sen from KIIT Deemed to be University, Bhubaneswar India for the literature survey.

References

1. Lukic S, Mulhall P, Emadi A (2008) Energy autonomous solar/battery auto rickshaw. J Asian Electr Veh 6(2):1135–1143
2. Chien C, Huang T, Liao T, Kuo T (2014) Design and development of solar power-assisted manual/electric wheelchair. J Rehabil Res Dev 51(9):1411–1426
3. Mallik A, Arman MA, Pal SK (2017) Solar based micro hybridized auto-rickshaw and its feasibility analysis for Bangladesh. In: Proceedings of the international conference on mechanical engineering and renewable energy 2017, Chittagong, pp 201–206. https://www.cuet.ac.bd/icmere/files2017f/ICMERE2017-PI-366.pdf
4. Sehgal L, Prasath RA, Rehalia A (2015) Proposal of integral mounted solar charging and external solar charging station for an electric rickshaw in Delhi. Am J Renew Sustain Energy 1(3):86–89
5. Malik S (2015) Design of solar assisted electric rickshaw: a step towards green transport system. International seminar on renewable energy and sustainable development. Royal University of Bhutan, Thimphu
6. Saleh N, Mushtaq K, Zaidi AA, Abbasoglu S, Ahmed SF (2016) Design and performance analysis of a solar powered hybrid rickshaw for commercial use in Pakistan. J Environ Sci Technol 9(6):472–480
7. Ranjan C, Akhtar SJ, Tiwari AK, Choubey NK, Kumar R, Rajwar L, Kumar R (2017) Design and fabrication of tri rickshaw with solar charging station. Int J Mech Eng Technol 8(7):358–366
8. Reddy KS, Aravindhan S, Mallick TK (2017) Techno-economic investigation of solar powered electric auto-rickshaw for a sustainable transport system. Energies 10:754
9. Mulhall P, Naviwala M, Lukic SM, Braband J, Emadi A (2007) Entrepreneurial projects program at Illinois Institute of Technology: solar/battery hybrid three-wheel auto rickshaw for India. In: Proceedings of the 2007 IEEE vehicle power and propulsion conference (VPPC), Arlington, pp 682–689
10. Vezzini A, Sharan H, Umanand L (2005) Low-pollution threewheeler autorickshaw with power-assist series hybrid and novel variable DC-link voltage system. J Indian Inst Sci 1–14
11. Rahman S (2008) Future mass rapid transit in Dhaka city: options, issues and realities. Jahangirnagar Plan Rev 6:69–81
12. Lukic S, Mulhall P, Emadi A (2008) Energy autonomous solar/battery auto rickshaw. J Asian Electr Veh 6:1135–1143
13. Phutane PS, Neware A, Bhivagade K, Mugale A, Paul A (2019) An improvement in performance in E-rickshaw. Int J Adv Res Electr Electron Instrum Eng. https://doi.org/10.15662/IJAREEIE.2019.0810009
14. Sharma AK, Kothari DP (2016) Potential in India for use of solar PV in charging of electric 2 & 3 wheelers. J Res 02(04):10–13
15. Anderson LG (2015) Rolling resistance modelling from functional data analysis to asset management system. PhD thesis, Department of Science, Systems and Models, IMFUFA, Roskilde University. https://milne.ruc.dk/imfufatekster/pdf/503.pdf
16. Raina G, Sinha S (2019) Outlook on the Indian scenario of solar energy strategies: policies and challenges. Energy Strat Rev 24:331–341
17. Sharma A (2011) A comprehensive study of solar power in India and world. Renew Sustain Energy Rev 15(4):1767–1776

18. Saxena T, Kumar M, Kesari JP (2017) Commercial and technological feasibility study of using solar e-rickshaw for semi-urban areas. Int J Dev Res 7(12):17432–17439
19. Sulepeth V, Kalashetty S (2019) A review of feasibility study on solar operated auto rickshaw as sustainable transport system for urban and rural areas. Int J Adv Acad Stud 1(1):34–37

Integrated MSW to Energy and Hot Water Generation Plant for Indian Cities: Thermal Performance Prediction

Pradip Mondal, Shambhunath Barman, and Samiran Samanta

Abstract The present study reports the model development and thermodynamic analyses of an integrated municipal solid waste (MSW) fuelled externally fired air turbine (AT) plant and hot water generation plant for Indian cities. Waste generated from a typical Indian municipal town (Kandi) is considered to be the fuel input to the plant. Dry and segregated MSW is burnt in the combustion chamber of dual combustor-air heating (DCAH) unit which then heat-up the working fluid (air) of the topping AT cycle. Exhaust heat from the DCAH unit is utilized to produce hot water to meet the demand of town. Effect of topping compressor's pressure ratio ($r_p = 4$–16), air turbine inlet temperature (ATIT = 900, 1000 and 1100 °C) and hot end temperature difference (HETD = 70–150 °C) on the energetic, exergetic and environmental (3-E) performance of the plant is analyzed and reported here. It is observed that the plant can deliver 1100 kWe of electricity and about 2450 L of hot water (hourly basis) at the base case scenario. Maximum exergy is destroyed at the DCAH unit and at the WHU, respectively. Sustainability index (environmental parameter) value is found to be 1.56 at the base case. Furthermore, it is observed that these plant operational parameters can be influential for evaluation of 3-E performance of the plant.

Keywords Waste management · Waste to energy · Hot water generation · Exergy

1 Introduction

Globally as well as in every Indian urban sector, management and disposal of municipal solid waste (MSW) and reuse of wastewater are being the primary concern for the governing authority as well as for the living beings. Both these issues are creating

P. Mondal · S. Barman
Department of Mechanical Engineering, National Institute of Technology Sikkim, Ravangla 737139, India

S. Samanta (✉)
School of Mechanical Engineering, Kalinga Institute of Industrial Technology (KIIT) Deemed to be University, Bhubaneswar 751024, India
e-mail: samirankgec@gmail.com

S. Revankar et al. (eds.), *Proceedings of International Conference on Thermofluids*, Lecture Notes in Mechanical Engineering,
https://doi.org/10.1007/978-981-15-7831-1_53

severe threats for the environment as well as for the humankind and other animals. Furthermore, to cater the ever-increasing energy demand and deployment of clean energy technologies, energy researchers are paying attention towards the generation of power from renewable/waste through efficient ways [1]. Conventional MSW management is executed via collection, transportation and dumping to lower land areas [2]. But it is experienced that improper technique of waste dumping causing soil and water pollution in permanent manner.

Recent research works on MSW to energy generation systems suggests that, treated MSW can be utilized as fuel in steam turbine or gas turbine based power plants [3, 4]. Gasification (plasma and fixed bed) of MSW and the usage of product gas in gas engines is also suggested by many researchers [5, 6]. However, gasification of different particle sized waste with blended composition is still being an opportunity to the techno-crafts for the continuous operation and in an efficient way.

External firing of dry and segregated waste in Brayton cycle helps to protect the GT blades from corrosion; particulate deposition related problems as well as recover of energy from waste in an efficient manner [7, 8]. Based on this consideration, present study is aimed at modelling of MSW fired AT cycle based power plant for a typical municipal city; Kandi (23.95°N 88.03°E), located at the Murshidabad district of West Bengal, India [9]. Apart from energy generation, one water heating plant is also integrated with this plant, to meet the hot water requirement. About 20 ton of waste is generated per day basis in this municipal area. After segregation (~50% of raw waste) 10 tons of waste can be fed to the plant for generation electricity. Effect of plant operational parameters on the energy, exergy, environmental and economic performance of this novel configuration is studied and reported during the analysis. In particular, this paper represents the energetic, exergetic and environmental performance of the plant at different considered range of thermodynamic parameters.

2 Plant Configuration

Line diagram of the proposed MSW fired integrated air turbine-municipal hot water generation plant is shown in Fig. 1. Collected solid waste after drying and segregation in the D&S unit enters combustion chamber (CC) the dual combustor-air heating unit (DCAH). Then it gets burnt in the presence of air turbine (AT) exhaust air and as a result of flue gas is generated. Flue gas then enters the shell side of the air heater (AH) of the DCAH unit. Compressed air after flowing through compressor (C) enters the tube bundles of the AH. After then the compressed air stream becomes heated during the heat transfer process at the AH which then enters the AT.

Hot and compressed air after being expanded in the AT, recirculated to the CC to accomplish the combustion of MSW in the DCAH unit. Flue gas, after passing through the water heating unit (WHU), released to atmosphere through the stack. An induced draught compressor (IDC) is placed in among the WHU and stack to compensate the pressure drop in the gas discharge line. A hot water generation plant is also integrated with this MSW fired externally fired plant so that, a unique model

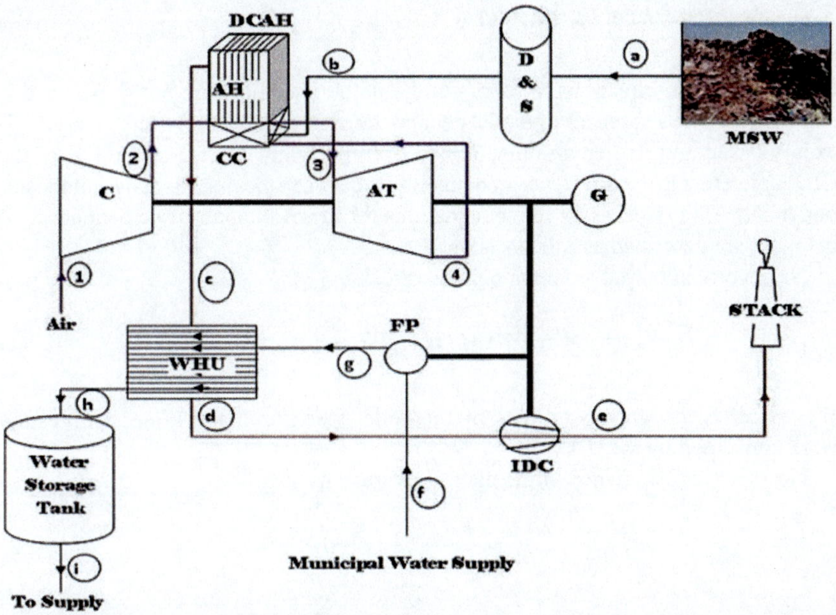

Fig. 1 Schematic presentation of the proposed plant

of integrated MSW to energy plant as well as municipal hot water generation plant can be developed and analyzed.

3 Model Equations

Composition of the combustibles (feed to the gasifier) is C-40%; H-5.5%; O-30%; N-0.6%; M-12.9%; Ash-11%. Ultimate analysis of samples is planned and carried out in *Duttcon Industrial Research and Development Lab, Kolkata* by closely following the International Standards (IS) methods. Calculated LHV of the fuel is 12,000.08 kJ/kg. Assumptions made to accomplish the energetic, exergetic and environmental analyses are as follows:

1. Ash content of the waste is constituted of Al_2O_3, SiO_2 and Fe_2O_3 [10].
2. No heat loss occurs across any component and in the piping. Pressure loss is considered as same as stated in the literature [11].
3. Isentropic efficiencies of air compressor and GT are 87% and 86%, respectively.
4. Operating pressure of the FP is 2 bar and the hot water outlet temperature from the WHU is saturation temperature at 2 bar pressure. Outlet pressure of IDC is 1.1 bar.

3.1 Performance Parameters

Necessary model equations to carry out the numerical study are not shown in this sub-section. The sequences of model equations are arranged as; (i) energy-based, (ii) exergy-based and (iii) environmental-based, respectively. HETD of the HX of the CHR unit refers to the difference of temperature between the flue gas at the combustor outlet and ATIT. Necessary model equations for the evaluation of performances of the plant are described as follows (Eqs. 1–8):

Net electrical output from the plant is calculated as:

$$W_{\mathrm{net}} = (W_{\mathrm{AT}} - W_{\mathrm{C}}) \cdot \eta_{\mathrm{g}} - W_{\mathrm{FP}} \cdot \eta_{\mathrm{m}} - W_{\mathrm{IDC}} \cdot \eta_{\mathrm{m}} \tag{1}$$

W_{FP} signifies the work consumed by the boiler feed pump and W_{IDC} signifies the work consumed by the IDC.

Net electrical efficiency of the plant is expressed as:

$$\eta_{\mathrm{e}} = {}^{W_{\mathrm{net}}}\!/_{m_{\mathrm{w}} \cdot \mathrm{LHV_w}} \tag{2}$$

m_{w} stands for the mass flow of segregated waste (kg/s). Hence, specific electrical waste consumption is presented as:

$$\mathrm{ESWC} = {}^{3600 \cdot m_{\mathrm{w}}}\!/_{W_{\mathrm{net}}} \tag{3}$$

Performance of a combined power and hot water generation plant is evaluated via fuel energy savings ratio (FESR) here. FESR is determined as:

$$\mathrm{FESR} = \frac{\Delta F}{\{(W_{\mathrm{net}}/\eta_{\mathrm{ref,net}}) + (Q_{\mathrm{H}}/\eta_{\mathrm{ref,H}})\}} \tag{4}$$

where W_{net} is net electrical power output of the plant, Q_{H} is the heat output (in terms of hot water generation) of the plant; $\eta_{\mathrm{ref,net}}$ and $\eta_{\mathrm{ref,H}}$ are taken as the reference efficiencies of the separate power and heat plants respectively. ΔF represents the fuel savings of two individual power and process heating plants and calculated as:

$$\Delta F = \left(\frac{W_{\mathrm{net}}}{\eta_{\mathrm{ref,net}}} + \frac{Q_{\mathrm{H}}}{\eta_{\mathrm{ref,H}}}\right) - m_{\mathrm{w}} \cdot \mathrm{LHV_w} \tag{5}$$

Specific thermo-mechanical exergy at any state of a thermodynamic cycle is determined using a generalized equation as follows [12]:

$$\mathrm{ex} = (h_{\mathrm{i}} - h_{\mathrm{o}}) - T_{\mathrm{o}}(s_{\mathrm{i}} - s_{\mathrm{o}}), \tag{6}$$

Exergetic efficiency of the plant as well for the plant components is calculated as:

$$\eta_{ex} = \frac{Ex_{product}}{Ex_{fuel}} \quad (7)$$

$Ex_{product}$ and Ex_{fuel} define product and fuel exergy of the plant as its components.

SI is also a measurement index of thermodynamic and environmental performance of the plant. SI of the plant is calculated as follows:

$$SI = \frac{1}{(1 - \eta_{ex,plant})} \quad (8)$$

It is an important environmental performance index of any plant which helps to identify how much the plant is sustainable in thermodynamic perspective [13].

4 Results and Discussions

Base case performance of the modelled plant is shown in Table 1. It is investigated from Table 1 that, the designed plant can perform well in terms of both power production and hot water generation (2 bar saturated) for typical municipal cities.

In continuation, this plant also environmentally performs well. Such value of thermodynamic state points (as shown in append table) at base case is considered as modern GT plants operates almost with similar conditions.

Exergetic performance of the plant at base case is shown in Fig. 2. Figure 2 reveals that, maximum exergy destruction occurs at the CC and AH of the DCAH unit, together accounting of about 45% of the input exergy. Measureable amount of exergy is destroyed at the WHU. Exergy losses at the other components are found to be traceable, compared to other components.

Furthermore, Fig. 2 is indicative about the fact that, considered range of plant's parameters can influence different plant output (3-E). Therefore, the variations of operational parameters on the different performance parameters are shown and discussed in the succeeding sections.

Variation of net plant's electrical efficiency and FESR with variation is r_p for considered ATIT is shown in Fig. 3. It is noticed that FESR and efficiency attains a

Table 1 Base case performance of the plant (r_p = 4, ATIT = 1100 °C and HETD = 100 °C)

Parameter	Unit	Value
Net electricity delivered	kW_e	1107.09
Plants' electrical efficiency	%	30.75
Steam generation rate	Lt./h	2455.2
FESR	%	63.07
Exergy efficiency of the plant	%	36.11
Sustainability index	–	1.56

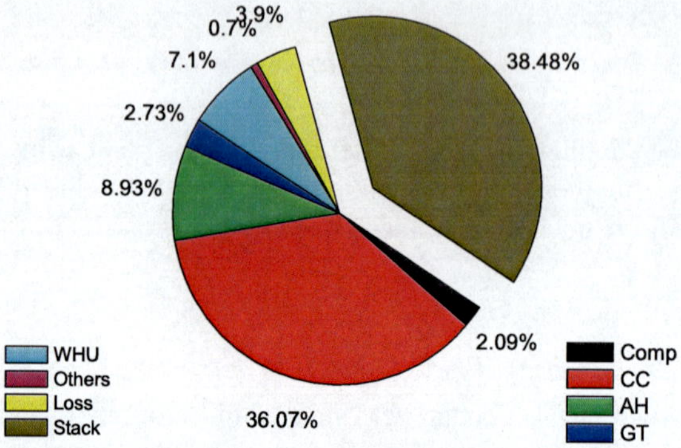

Fig. 2 Exergy balance of the plant at base case

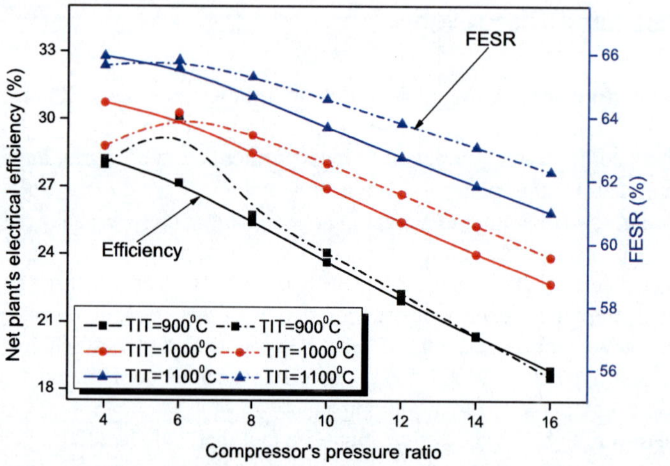

Fig. 3 Electrical efficiency and FESR versus r_p and TIT (fixed HETD = 100 °C)

maxima at $r_p = 6$, for all TITs. Higher TIT offers better performance as observed form the graph. Figure 4 shows the variations in net electrical output from the plant and hot water generation from the plant with variations in r_p and TIT. Except, TIT = 900 °C, electrical output from the plant continuously decreases with increase in r_p. Consequently, hot water generation monotonously increases with increase in r_p. It is also noticed from the graph that higher TIT leads to lower water generation from the plant.

For the graphs, HETD is kept constant at 100 °C. It is also noticed that exergy efficiency of the plant gets maximized at $r_p = 6$ and TIT = 1100 °C. Keeping r_p and TIT fixed at the pre-said thermodynamic co-ordinates, variations of the performance parameters of the plant with variations in HETD of the AH are shown in Figs. 5 and

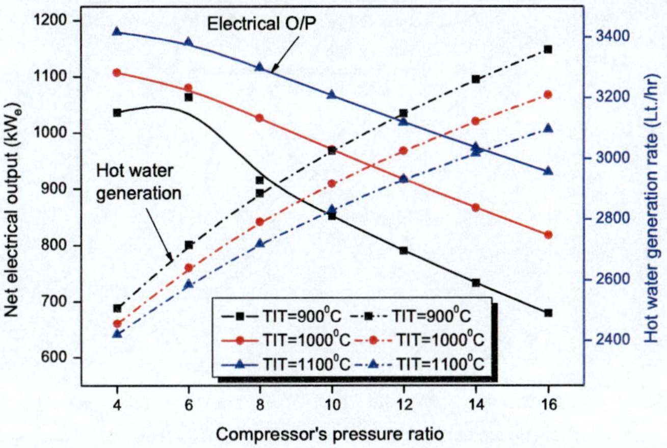

Fig. 4 Electrical output and hot water generation versus r_p and TIT (fixed HETD = 100 °C)

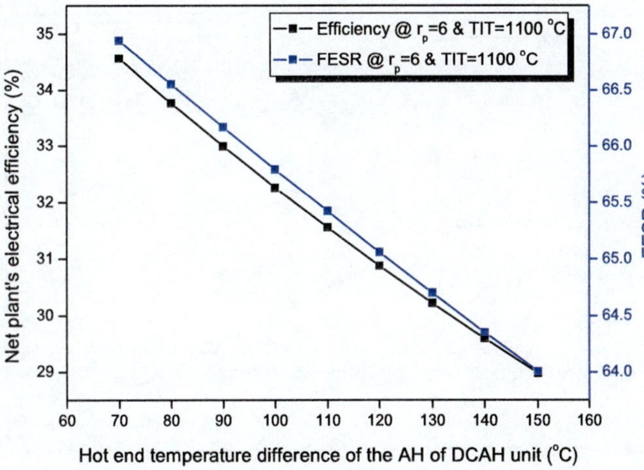

Fig. 5 Electrical efficiency and FESR versus HETD

6, respectively. It is noticed from these two plots that, plant's electrical efficiency, FESR and net power output decrease with increase in HETD. However, hot water generation rate becomes more at higher HETD.

Environmental performance of the plant is measured via evaluating the sustainability index of the plant. Variation in SI of the plant with variation in r_p and TIT of the plant is shown in Fig. 7. It is experienced from Fig. 7 that, SI attains a yield value at $r_p = 6$, for all TITs. Higher SI value is observed at higher TITs.

Figure 8 shows the variation in SI with variation in HETD of the AH of the plant is shown in Fig. 8. The plot reveals that SI of the plant goes on decreasing with

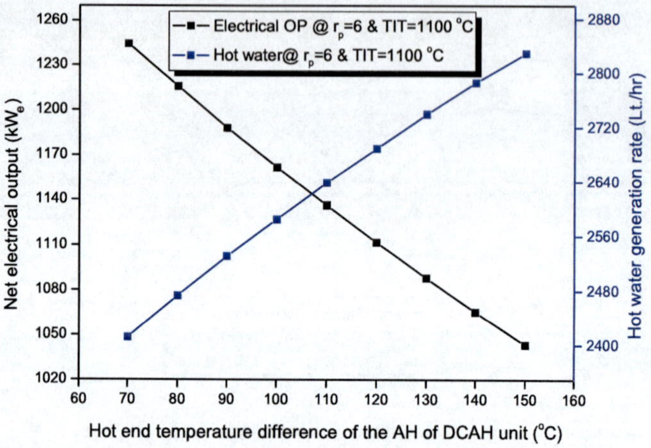

Fig. 6 Electrical output and hot water generation versus HETD

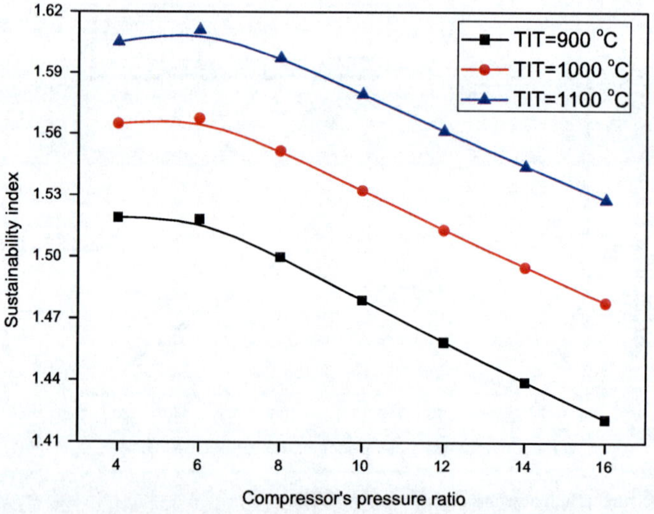

Fig. 7 SI versus r_p and TIT (fixed HETD = 100 °C)

increase in HETD. From the above discussions, it is indicative that, this plant can be very beneficial for simultaneous of power and hot water generation using MSW of Indian cities. Generated power can be used to crater the utility electrical needs (viz. park, street light etc.) and the hot water from the plant can be utilized to meet the hot water demand of any city. This will also solve the ever-emerging problem of MSW management.

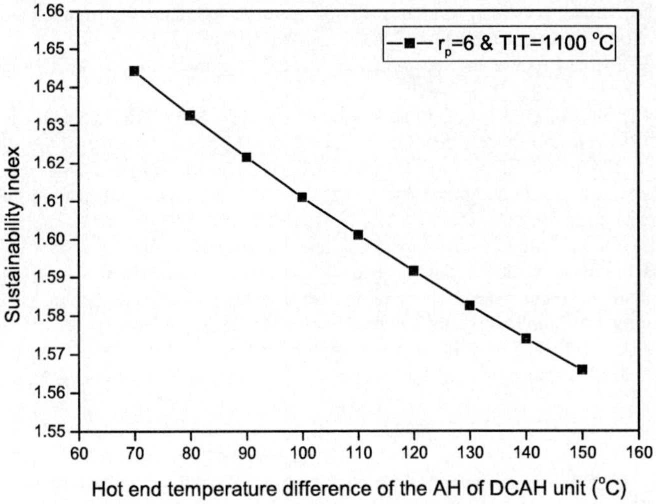

Fig. 8 SI versus HETD

5 Conclusions

The study concludes that the plant needs to be operated at $r_p = 6$; TIT = 1100 °C and HETD = 70 °C to get maximum power output from the plant in efficient and environment-friendly way. However, to get maximum hot water from the plant, it is to be operated at $r_p = 4$; TIT = 900 °C and HETD = 150 °C, slightly compromising to the environment. Economic analysis of the plant also to be carried out further to cross-check the system's economic viability. Government/Administrative support is also needed for such kind of projects to make it more attentive to the environmental Scientists.

References

1. Singh RP, Tyagi VV, Allen TM, Ibrahim H, Kothari R (2011) An overview for exploring the possibilities of energy generation from municipal solid waste (MSW) in Indian scenario. Renew Sustain Energy Rev 15:4797–4808
2. Das S, Bhattacharyya BK (2015) Performance evaluation of the proposed and existing waste management system: economic analysis. In: Proceedings of the 2015 international conference on operations excellence and service engineering, Orlando, pp 267–277
3. Baggio P, Baratieri M, Gasparella A, Longo GA (2008) Energy and environmental analysis of an innovative system based on municipal solid waste (MSW) pyrolysis and combined cycle. Appl Therm Eng 28:136–144
4. Rokni M (2015) Thermodynamic analyses of municipal solid waste gasification plant integrated with solid oxide fuel cell and Stirling hybrid system. Int J Hydrogen Energy 40:7855–7869

5. Arafa HA, Jijakli K (2013) Modeling and comparative assessment of municipal solid waste gasification for energy production. Waste Manage 33:1704–1713
6. Arena U (2012) Process and technological aspects of municipal solid waste gasification. A review. Waste Manage 32:625–639
7. Mondal P, Ghosh S (2017) Exergo-economic analysis of a 1 MW BIGCC plant with indirectly heated GT cycle and supercritical ORC. Clean Technol Environ Policy 19:1475–1486
8. Soltani S, Mahmoudi SMS, Yari M, Rosen MA (2013) Thermodynamic analyses of an externally fired gas turbine combined cycle integrated with a biomass gasification plant. Energy Convers Manage 70:107–115
9. Website of Kandi Municipality. https://kandimunicipality1869.com/
10. Mondal P, Ghosh S (2015) Thermodynamic performance assessment of a bio-gasification based small-scale combined cogeneration plant employing indirectly heated gas turbine for community scale application. Int J Renew Energy Res 5(2):354–366
11. Datta A, Ganguli R, Sarkar L (2010) Energy and exergy analyses of an externally fired air turbine (EGFT) cycle integrated with biomass gasifier for distributed power generation. Energy 35:341–350
12. Ghosh S, De S (2004) First and second law performance variations of coal gasification fuel-cell based combined cogeneration plant with varying load. J Power Energy 218:477–485
13. Jana K, De S (2015) Sustainable polygeneration design and assessment through combined thermodynamic, economic and environmental analysis. Energy 91:540–555

Thermal Performance Evaluation of an Improved Biomass Cookstove for Domestic Applications

Raktimjyoti Barpatragohain, Niyarjyoti Bharali, and Partha Pratim Dutta

Abstract Biomass fuels are normally burnt inefficiently using the traditional cookstoves in rural areas. The uncontrolled harmful emissions released due to the combustion of biomass cause various respiratory problems to the users. Thus in order to overcome such problems related to the traditional cookstove, an improved biomass cookstove is proposed in the present study. An effort is made to design and fabricate an improved biomass cookstove to study various heat losses from it and performance parameters. Water boiling test (WBT) was conducted for understanding the overall performance of the improved stove. Results of WBT were compared with a commercially available biomass cookstove. The average thermal efficiency while boiling 7.5 L water was ~33%, whereas the average thermal efficiency in case of the commercial stove was ~23%. While boiling 7.5 L water in this improved stove, the performance was better than the commercial stove with respect to specific fuel consumption. The maximum level of CO was detected around 20 ppm during the start and it reduced to 5 ppm as combustion progressed. Both the values for CO_2/CO and indoor air quality were within acceptable limit.

Keywords Improved cookstove · Biomass combustion · Thermal performance · Water boiling test · Emission

Nomenclature

w_v	Mass of water vaporized (g)
$P1_i$	Mass of pot of water before test (g)
$P1_f$	Mass of pot of water after test (g)
Δt_c	Time to boil (min)
t_i	Time at start of test (min)

R. Barpatragohain · N. Bharali · P. P. Dutta (✉)
Department of Mechanical Engineering, Tezpur University, Napaam, Sonitpur, Assam 784028, India
e-mail: ppdutta06@gmail.com

S. Revankar et al. (eds.), *Proceedings of International Conference on Thermofluids*, Lecture Notes in Mechanical Engineering,
https://doi.org/10.1007/978-981-15-7831-1_54

t_f	Time at end of test (min)
Δt_c^T	Temperature-corrected time to boil (min)
$T1_i$	Water temperature at start of test (°C)
$T1_f$	Water temperature at end of test (°C)
TE	Thermal efficiency (%)
w_{cv}	Water vaporized (g)
r_b	Burning rate (g/min)
f_c	Mass of biomass combusted (g)
f_{cd}	Energy released (kJ)
h	Heat transfer coefficient (W/m^2 K)
SC_c	Specific fuel consumption (kJ/g)
Re	Reynold number
Pr	Prandtl number

1 Introduction

Biomass is a suitable renewable energy source that would help to overcome fossil fuel dependency partially. Combustion of biomass releases carbon dioxide in the atmosphere but biomass is considered as a carbon-neutral source because it raptures the same amount of carbon dioxide while growing [1]. Biomass energy sources are various domestic wastes, wood, rice husk, etc. These types of various biomass fuels are widely used in industrial furnaces as well as in domestic cooking stoves to produce heat [2]. Bryden and Maccarty [3] observed that biomass fuels deviated in numerous ways from the conventional fossil fuels used in combustion processes, such as coal. They usually have high moisture contents, lower heating values, and a range of trace constituents, such as chlorine, sulfur, phosphorus, nitrogen, and a variety of ash-forming metals. These special properties of biomass fuels pose some challenges; but in many cases it provides advantages to their use in combustion methods. Design of the combustion devices and choice of their operating constraints are very much dependent on the thermophysical properties of the biomass fuels. Fichet et al. [4] reported that biomass had established itself as an energy carrier accomplished with the growing demand of clean and sustainable energy.

2 Literature Review

CFD analysis is an effective approach for evaluating the performance of a biomass furnace complementing conventional combustion analysis in a laboratory. It enables the researchers to study the different combustion reactions involved inside the furnace. By optimizing combustion air (fluid) flow, CFD analysis shows a path to improve the overall efficiency [5–7].

Similarly, water boiling test (WBT) is a very effective methodology for evaluating the biomass stove performance. WBT gives a fair assessment based on the effectiveness and efficiency of different designs. With the help of WBT, the rate of emissions can also be estimated [8–10]. Water boiling test has been applied comprehensively in studying and exploring the combustion characteristics of firewood and recent fuels in traditional and improved biomass cookstoves. Thermogravimetric analysis is used to study the thermal behavior of solid fuels vis-a-vis the ignition behavior of coal, biomass, and coal-biomass blends. Using TGA, one can distinguish the impact of particle size and shape on the thermal behavior of biomass [11–14].

Bhuiyan et al. [1] had executed a 3D numerical modeling of biomass pallet combustion under diverse combustion environments. The effects of overall flame temperatures, species concentration, solid-phase variables, and ignition front spread velocity were reported. Gogoi and Baruah [2] developed a steady-state heat transfer model to forecast the performance of a biomass stove with varying operating and design environments like composition, particle shape and size, shape and material of the combustion chamber. Bryden and Maccarty [3] reviewed the cookstove modeling process for one to three burners, natural draft, wood-fired cookstoves fueled with untreated biomass of different sizes. Fichet et al. [4] had developed a CFD-based model for wood combustion in the domestic stoves with a special emphasis on CO and CO_2 emissions. The CFD-based model was constructed in the framework of RANS approach. They suggested measures for the reduction of emissions and enhancing in overall efficiency. Still et al. [5] did some laboratory experiments regarding the filtration and retained heat for reduction of particulate matter (PM) emissions from biomass cooking. They used WBT test in a closed enclosure. They found that the rates of PM emission were significantly reduced from 7.5 mg/min of PM to 1.5 mg/min of PM.

Quist et al. [7] studied the performance for a basic brick channel cookstove with uncertainty analysis. Grimsby et al. [8] used different biomass fuels like maize stalks, stalks from sunflower in the improved cookstoves and assessed them with (WBT). Djurovic et al. [9] showed that the efficiency of the biomass furnace could be increased and the amount of CO emission could be reduced by proper furnace dimensioning. It also helped in reducing the cost of the furnace. Chen et al. [10] studied the performance of pellet-gasifier stoves, efficiency, and pollutant emissions with the help of WBT. They compared with traditional stoves and different biomass fuels along with the quantity of particulate matter (PM) emitted.

Performance of different types of improved solid biomass cookstoves and the thermal efficiencies were reported in the literature [12]. Shiehnejadhesar et al. [13] had studied about the virtual biomass grate furnace that was comprised of a comprehensive CFD model of all relevant processes for simulation. Lung and Espira [14] presented wood consumption data using kitchen performance tests over an area 3000 km^2 for 7 years on a modified traditional cookstove named *Upesi* ceramic stove. This improved cookstove was meant for communities around a threatened rainforest in western Kenya. They found using the *Upesi* improved cookstove significantly reduced daily wood consumption by 3.87 kg per household, a mean saving of 37.7%. Lombardi et al. [15] performed an investigation on techno-economics prospective of a fully renewable solar micro-grid, ensuring an integrated access to electricity and clean cooking, based on two representative case studies in Tanzania, namely (i) a residential case study and (ii) a community service case study. An inexpensive potassium-based catalyst was incorporated in a chimneyless biomass cookstove reducing harmful emissions through catalytic oxidation [16]. Fine particulate matter and black carbon concentration levels were monitored uninterruptedly inside two types of kitchens that were separated from and attached to the main house under actual cooking practices [17]. Dutta et al. [18–20] studied biomass gasifier improved burners for balck tea drying. Sharma and Jain [21] investigated the impact of increased levels of indoor air pollution caused due to biomass combustion in the rural household of northern India. Gohain and Dutta [22] studied on thermal modeling of an improved biomass cookstove for the rural area. Based on the above literature, the objective of the present work is design and thermal performance testing of an improved biomass cookstove for rural applications.

3 Material and Methodology

3.1 Mathematical Modeling

The important mathematical relationships those are used during the performance testing with the help of water boiling test are as follows [23, 24]:

1. Mass of water vaporized is calculated as:

$$w_v = P1_i - P1_f \tag{1}$$

2. Time to boil the pot:

$$\Delta t_c = t_f - t_i \tag{2}$$

3. The temperature-corrected time to boil is calculated by the following equation:

$$\Delta t_c^T = \Delta t_c \frac{75}{T1_f - T1_i} \tag{3}$$

4. The thermal efficiency is an estimate of the total energy produced by the fire that is used to heat the water in the pot. It can be calculated by the following mathematical relationship:

$$\mathrm{TE} = \frac{\Delta E_{H_2O} + \Delta E_{H_2O,evap}}{E_{released}} \tag{4}$$

where

$$\Delta E_{H_2O} = m_{H_2O}\left(C_p(T_b - T_{fuel}) + \Delta h_{H_2O}\right) \tag{5}$$

$$\Delta h_{H_2O} \approx 2257\ (\text{kJ/kg}) \tag{6}$$

$$T_{fuel} = T_a \tag{7}$$

$$E_{released} = f_c \cdot \text{LHV};\ \Delta E_{H_2O,evap} = w_{cv} \times 2260\ (\text{kJ/kg})$$

5. The burning rate can be calculated by the following expression:

$$r_b = \frac{f_c}{\Delta t_c} \tag{8}$$

6. The specific fuel consumption is calculated by the following expression:

$$\text{SC}_c = \frac{f_{cd}}{f_c} \tag{9}$$

Equations for calculating air velocity and heat loss:

The theoretical velocity of the gas inside the combustion chamber was estimated by the following mathematical relationships [25, 26]:

$$v_{th} = \sqrt{2 \times g \times H \times \left(\frac{T_g}{T_a} - 1\right)} \tag{10}$$

From the continuity equation, the theoretical mass flow rate of the flue gases can be calculated [25, 26]:

$$\dot{m}_{th} = \frac{353}{T_g} \times A \times \sqrt{2 \times g \times h \times \left(\frac{T_g}{T_a} - 1\right)} \tag{11}$$

Heat loss to the surrounding can be calculated by the following expression:

$$Q_{heatloss} = \frac{(T_g - T_a)}{\frac{1}{(h_{ci}+h_{rfl}+h_{rch})} + \frac{\ln\left(\frac{D_o}{D_i}\right)}{2\times\pi\times H\times k_i} + \frac{1}{(h_{co}+h_{ro})A_o}} \times \frac{1}{1000 \times m_f} \tag{12}$$

where, for the fully developed turbulent flow, the following expression can be used [25, 26]

$$h_{ci} = \frac{k}{D} \times 0.023 \times \mathrm{Re}_D^{0.8} \times \mathrm{Pr}^{0.3} \tag{13}$$

From average Nusselt number expression for natural convection, we get the following expression [25, 26]

$$h_{co} = \frac{k}{H} \times 0.59 \times \left(\mathrm{Ra}_H^{0.25}\right) \tag{14}$$

$$h_{rfl} = \frac{\sigma \times \left(\varepsilon_g T_g^4 - \alpha_g T_{wi}^4\right)}{\left(T_g - T_{wi}\right)} \tag{15}$$

$$h_{ro} = \sigma \times \varepsilon_o \times \left(T_{wo}^2 + T_a^2\right) \times \left(T_{wo} + T_a\right) \tag{16}$$

$$h_{rch} = \frac{\sigma \times \left(T_{char}^2 + T_{wi}^2\right) \times \left(T_{char} + T_{wi}\right)}{\frac{(1-\varepsilon_c)}{\varepsilon_c} + \frac{F_{char-wall} + F_{char-pot}}{F_{char-wall} \times \left(F_{char-wall} + 2 \times F_{char-pot}\right)}} \tag{17}$$

4 Experimental Procedure

Two different pots with carrying capacity 5 and 7.5 L were used during the tests. These are the two standard weights taken during the WBT test protocol. The test started with initially taking the weight of water needed for boiling, i.e., whether 5 or 7.5 L. The initial weight of the biomass used for combustion was noted. A calibrated thermocouple was placed in the pot containing water for determining the temperature rise at a regular time interval. A stopwatch was kept ready as soon as the test started for recording the time required for boiling water in every phase.

WBT consists of three phases, i.e., cold start, hot start, and simmering. With the ignition of the biomass as the combustion process starts, the first phase of the test, i.e., cold start, begins. This phase remains till the first water bubble touches the water surface. Another thing noted throughout the test was that, even though water boils at 99/100 °C, this result holds true if the water is boiled at the sea level. However, during the test it was observed that water started boiling at different temperatures varying from 96 to 99 °C. Some of the reasons of this variation may be due to the pressure difference and the impurities present in the water causing the difference of the boiling point. As soon as the water reached the boiling point temperature, the process was stopped. After that, the final weight of the pot with water and the fuel with the container was measured. The amount of time consumed during the first phase was recorded, and the amount of charcoal produced during the first phase was measured. After collecting all the essential data, immediately the pot was emptied and freshly filled it with the same amount of water initially taken during the cold phase. The only difference in this phase is that water is being collected and weighted

Table 1 Technical specifications of instrument used for the water boiling test

S. No.	Name of the instrument	Make/model	Variable ranges	Accuracy/resolution
1	K type thermocouple and digital thermometer	MLABS/HT-9815	−200 to 1372 °C	±1 °C
2	Anemometer	AMV06	0–20 m/s	±1%/0.1 m/s
3	Infrared thermometer	HTC/HTC/MT-4	−50 to 550 °C	±2/0.1 °C
4	Weighing machine	Bijnath	0–10 kg	0.1 g
5	Moisture meter	MD 3G	5–40%	±1%
6	Bomb calorimeter	Saiyam/MD4G	Changsha	
7	CO_2/CO analyzer	5E-1AC/ML TESTO-435	0–10,000/0–500 ppm	±5 ppm/±1 ppm

when the pot is still hot. Hence, the second is called as the hot start. The process is repeated similar to the cold start, and all the necessary data were collected. The third phase or final phase is called simmering. In this phase as soon as the hot start is completed, the pot is kept under observation for 45 min. During this period, the tester has to maintain the temperature between 96 and 99 °C. This step simulates the long cooking of legumes or pulses common throughout the world [8].

5 Experimental Setup

The following instruments were used for performance evaluation of improved biomass stove for domestic application (Table 1).

6 Results and Discussion

The experimental setup of an improved biomass stove which is used for water boiling test has been presented in the Fig. 1a, b. A double-walled cookstove was fabricated, where the inner combustion chamber was made up of stainless steel while the outer cylinder was made up of mild steel. The results of water boiling test of the designed biomass stove are compared with a commercial greenwood cookstove, and they are presented in Table 2.

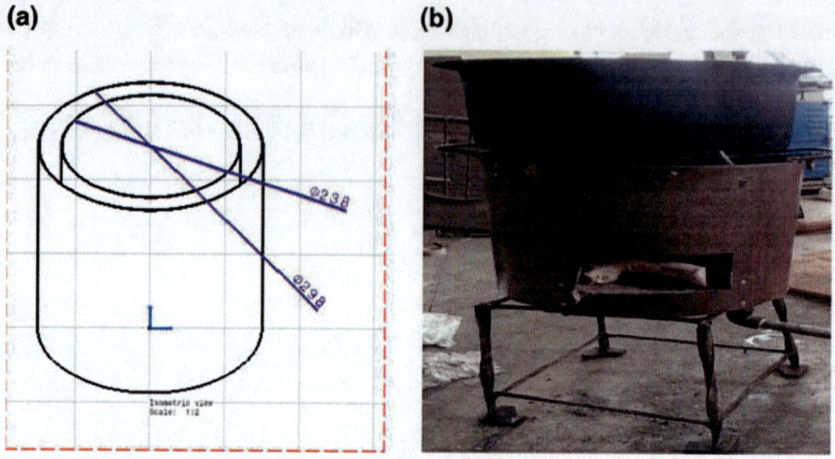

Fig. 1 **a** Double-walled biomass furnace. **b** Biomass furnace water boiling test

Table 2 Performance comparison of designed cookstove with greenwood commercial cookstove

Greenwood commercial stove		Improved biomass cookstove	
Performance parameters		5 L test	7.5 L test
Net change in char (g)	79	56	70
Time to boil (min)	26	16	18
Temperature-corrected time to boil (min)	24	15	16
Burning rate (g/min)	15.53	16	18
Specific fuel consumption (g/L boiled)	86	115	80
Overall thermal efficiency (%)	22	26	31

Table 2 shows comparative results of the important performance parameters of both the commercially available stove and the improved biomass cookstove. Tests were conducted on both the stoves under similar experimental conditions. It is evident from Table 2 that the results in case of the designed and fabricated model were better than the commercial stove. The average time for boiling in case of present model was better than the commercial stove. Other parameters like the specific fuel consumption, net amount of char produced due to the burning of biomass were less in comparison to the commercial stove. As a result, it enhanced the overall thermal efficiency of the improved stove. Thus, the overall efficiency of present model tested with two different weights was found better than the greenwood commercial stove.

Figure 2a represents a fair comparison of thermal efficiencies with water boiling point temperatures. Water boiling test (WBT) was conducted on the designed and fabricated cookstove with two different water weights, and the results were

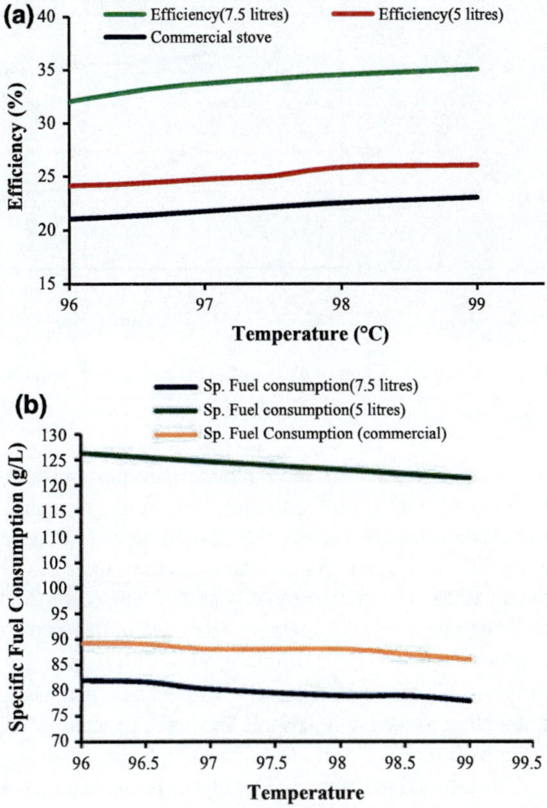

Fig. 2 **a** Variation thermal efficiency with boiling temperature of water. **b** Variation of specific fuel consumption with boiling temperature of water

compared with a commercial cookstove under similar experimental conditions. It can be seen from the plot that the thermal efficiencies of the fabricated cookstove under both the cases, i.e., 5 and 7.5 L, were better than the commercial greenwood cookstove. The average thermal efficiency in case of 7.5 L was around 33%, and in case of 5 L test, it was around 25%, whereas the average thermal efficiency in case of the commercial stove was 23% respectively.

Figure 2b represents the variation of specific fuel consumptions with water boiling point temperatures. WBTs were conducted on both the fabricated model and the commercial cookstove. The lower the specific fuel consumption, the better is the performance of the cookstove. Results showed that in case of 7.5 L when tested in the present designed model, specific fuel consumption (average 80 g/L) was better than the commercial stove, whereas in case of 5 L test, the specific fuel consumption was more comparing to the commercial stove. The reason for such result is because of the spread of flame from the sides of the pot and the full capacity of the stove was not utilized. This resulted in the loss of heat that led for an increment of the fuel consumption in order to provide sufficient heat to the pot's bottom.

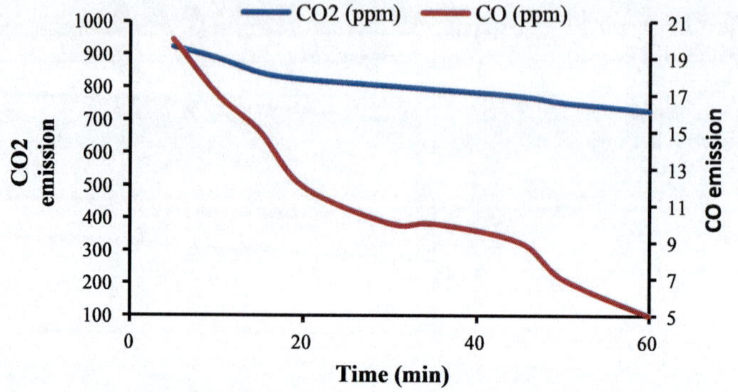

Fig. 3 In door variation of carbon dioxide and carbon monooxide with time during water boiling test

Figure 3 represents the characteristic plots of CO_2 and CO emission (near the cookstove) with time for the improved designed cookstove. CO_2 and CO are very important emission factors that should be kept under consideration while designing a biomass cookstove. As per standard, the values of CO_2 and CO should not increase above 1000 ppm in case of CO_2 and 25 ppm in case of CO when used for domestic cooking purpose. Gas analyzer was used to detect the emission levels while the test was performed. The maximum level of CO_2 detected was 920 ppm which was during the start of combustion. As the combustion process progressed, the CO_2 emission gradually reduced to 820 ppm. So was in the case of CO, which was detected around 20 ppm during the start and reduces to 5 ppm as the combustion progressed. .

The thermal performance results of present improved biomass cookstove are validated with Bhuiyan et al. [1], the normal combustion rate of biomass in an improved cookstove. Gogoi and Baruah [2] observed thermal efficiency 23.63% and time of boiling 16.94 min. The present results with improved biomass cookstove boiling 5 L water 16 min with an average 24.50% thermal efficiency. Therefore, this cookstove may be recommended for domestic applications in rural areas.

7 Conclusions

Thermal performance evaluation of an improved biomass cookstove is demonstrated with water boiling test performed in the laboratory condition. Its performance was compared with a commercially available greenwood cookstove. The average thermal efficiency for boiling of 7.5 L water was around 33% and in case of 5 L water boiling, it was found around 25%. The average thermal efficiency in case of the commercial stove was found 23%. The maximum level of CO_2 detected was 920 ppm which was during the start of combustion. As the combustion process progressed, the CO_2 emission gradually stabilized at 820 ppm. CO was detected around 20 ppm during the start and stablized at 5 ppm as combustion progressed This is well within the tolerable limit of healthy human being.

Acknowledgements AICTE funded RPS project "Design Development and Performance Testing of Solar Dryer for Drying *Garcinia pedunculata* with Thermal energy Storage" 2019–2022.

References

1. Bhuiyan AA, Karim MR, Naser J (2018) Effect of recycled flue gas ratios for pellet type biomass combustion in a packed bed furnace. Int J Heat Mass Transf 120:1031–1043
2. Gogoi B, Baruah DC (2016) Steady state heat transfer modeling of solid fuel biomass stove. Energy 97:283–295
3. Bryden KM, Maccarty NA (2015) Modelling of household biomass cook stoves: a review. Energy Sustain Dev 26:1–13
4. Fichet V, Plion P, Tabet F (2016) A comprehensive CFD based model for domestic biomass heating systems. J Energy Inst 89:199–214
5. Still DK, Bentson S, Murrey N, Andres J, Yue Z, Macarty NA (2018) Laboratory experiments regarding the use of filtration and retained heat to reduce particulate matter emissions from biomass cooking. Energy Sustain Dev 42:129–135
6. Ibarra L, Aragón G, Gomez I, Múgica I, González J, Maroto JR, Rojas E, Sanz D, Ramos R, Escalada R, Borjabad E, Gutierrez-Canas C (2018) Blend design tools for medium combustion plants (MCP) firing biomass wastes. Waste Manage 71:200–214
7. Quist CM, Jones RB, Jones MR, Lewis RS (2016) Uncertainty analysis and design guidelines of biomass cookstove thermal efficiency studies. Energy Sustain Dev 34:54–61
8. Grimsby LK, Rajabu HM, Treiber MU (2016) Multiple biomass fuels and improved cook stoves from Tanzania assessed with the water boiling test. Sustain Energy Technol Assess 14:63–73
9. Djurovic D, Nemoda S, Repic B, Dakic D, Adzic M (2015) Influence of biomass furnace volume change on flue gases burn out process. Renew Energy 76:1–6
10. Chen Y, Shen G, Su S, Du W, Huangfu Y, Liu G, Wang X, Xing B, Smith KR, Tao S (2016) Efficiencies and pollutant emissions from forced-draft biomass-pellet semi-gasifier stoves: comparison of international and Chinese water boiling test protocols. Energy Sustain Dev 31:22–30
11. Kwofie EM, Ngadi M, Sotocinal S (2017) Energy efficiency and emission assessment of a continuous rice husk stove for rice parboiling. Energy 122:340–349
12. Datta A, Malik JK, Pal RC, Singh VK, Suresh R (2016) Evaluation of the performance of improved biomass cooking stoves with different solid biomass fuel types. Biomass Bioenergy 95:27–34
13. Shiehnejadhesar A, Mehrabian R, Hochenauer C, Scharler R (2017) The virtual biomass grate furnace—an overall CFD model for biomass combustion plant. In: INFUB—11th European conference on industrial furnaces and boilers, Austria, vol 120, pp 516–523
14. Lung MA, Espira A (2019) A large-scale, village-level test of wood consumption patterns in a modified traditional cook stove in Kenya. Energy Sustain Dev 49:11–20
15. Lombardi F, Riva F, Sacchi M, Colombo E (2019) Enabling combined access to electricity and clean cooking with PV micro grids: new evidences from a high-resolution model of cooking loads. Energy Sustain Dev 49:78–88
16. Paulsen AD, Kunsa TA, Carpenter AI, Amundsen TJ, Schwartz NR, Harrington J, Reed J, Alcorn B, Gattoni J, Yelvington PE (2019) Gaseous and particulate emissions from a chimneyless biomass cook stove equipped with a potassium catalyst. Appl Energy 235:369–378
17. Rupakheti D, Kimoanh NT, Rupakheti, Sharma RK, Panday AK, Puppala SP, Lawrence MG (2019) Indoor levels of black carbon and particulate matters in relation to cooking activities using different cook stove-fuels in rural Nepal. Energy Sustain Dev 48:25–33

18. Dutta PP, Baruah DC (2014) Possibility of biomass gasification in tea manufacturing industries. Int. J. Renew Energy Technol 5(4):310–322
19. Dutta PP, Das A, Pandey V (2014) Fuel characteristics of some locally available biomass as a potential gasification feed stock for thermal application. Ind Eng Chem Res J 53(51):19806–19813 (American Chemical Society).
20. Dutta PP (2014) Prospect of renewable thermal energy in black tea processing in assam an investigation for energy resources and technology. PhD dissertation Tezpur University, Assam, India
21. Sharma D, Jain S (2019) Impact of intervention of biomass cook stove technologies and kitchen characteristics on indoor air quality and human exposure in rural settings of India. Environ Int 123:240–255
22. Gohain RJB, Dutta PP (2019) Modeling thermal performance of an improved biomass cook stove for rural area. In: 2nd international conference on new frontiers in engineering science and technology (NFEST2019), NIT Kurukshetra
23. Li W, Cao J, Lue L (2017) Study on the ignition behaviour and kinetics of combustion of biomass. In: 9th international conference on applied energy (ICAE2017), Cardiff, vol 142, pp 136–141. Elsevier, Netherlands
24. Cengal Y, Boles AM (2006) Thermodynamics: an engineering approach, 6th edn. McGraw-Hill, Boston
25. Holman JP (2010) Heat transfer, 10th edn. The McGraw-Hill Companies Inc., New York
26. Kshirsagar MP, Kamalkar V (2015) A mathematical tool for predicting thermal performance of natural draft biomass cook stoves and identification of a new operational parameter. Energy 93:188–201

Experimental Investigation of Paddy Drying in Rotating Fluidized Bed in Static Geometry

Rishiraj Purkayastha, Pavitra Singh, Abinash Mahapatro, Alok Kumar, and Pinakeswar Mahanta

Abstract The concept of rotating fluidized bed with static geometry (RFB-SG) is highly beneficial in drying operations. The present work is focused on the drying of granular particles (paddy) in the developed RFB-SG dryer. In this RFB-SG, the tangential injection of fluidizing gas via several inlets induces rotary motion to the solid particles and forms a solid-bed. In the entire process of fluidization, the solid particles experience a radially outward centrifugal force, whereas the gas is forced to move inward towards the chimney. The RFB-SG is expected to operate more efficiently at higher gas flow rates as compared to the conventional fluidized bed reactor. Experiments were carried to dry paddy of initial moisture ranging 23–37% (wb) to 14% (wb) with drying air inlet temperature of 55 and 60 °C and mass flow 250 m^3/h. A better air utilization decreased the drying time by 20% on increasing the air inlet temperature by 9%. Air inlet temperature of 60°C is better for drying paddy as compared to 55°C.

Keywords Fluidized bed drying · Rotating fluidized bed in static geometry · Food drying

1 Introduction

Rice is the primary source of staple food for many Asian countries which can be obtained in the post-harvesting process of paddy drying. Paddy drying is an energy-intensive process. To reduce the energy requirement of drying process farmers, use

R. Purkayastha · P. Singh · P. Mahanta
Centre for Energy, Indian Institute of Technology Guwahati, Guwahati 781039, India

A. Mahapatro (✉) · P. Mahanta
Department of Mechanical Engineering, Indian Institute of Technology Guwahati, Guwahati 781039, India
e-mail: abinashmahapatro@gmail.com

A. Kumar · P. Mahanta
Department of Mechanical Engineering, National Institute of Technology Arunachal Pradesh, Yupia 791112, India

S. Revankar et al. (eds.), *Proceedings of International Conference on Thermofluids*, Lecture Notes in Mechanical Engineering,
https://doi.org/10.1007/978-981-15-7831-1_55

solar, biomass energy or both [1]. In this process, the moisture content of paddy grain is reduced to 14% (wb, approx.) for preventing attack by fungus, pests, and germs. By reduction of moisture content of paddy, microorganism growth, biological activities, i.e. enzymatic reactions are also reduced, and the paddy is less prone to attack by fungus, pests, and germs. Reduction in moisture content prolongs the shelf life of paddy, facilitates storage and transport [2]. Moisture content between 13 and 14% (wb) is required for the safe storage of paddy [3, 4].

Conventionally dryers are categorized as direct and indirect. In direct-type dryers, the source of heat is in direct contact with paddy. In indirect dryers, the heat exchanging medium exchanges the heat and dries the paddy. Forced dryers are often hybridized with solar energy to make it more energy-efficient [5]. In continuous type dryers, paddy is moved slowly and hot air is passed through a thick paddy layer of 5–20 cm thickness. For avoiding reduction in nutritional value and change in colour or texture of paddy the drying air temperature is maintained at 50–60 °C [6]. The residence time for drying typically ranges to several hours and commercial capacities typically ranges from 4 to 12 tons per day. Paddy drying is also done in fluidized bed risers [7]. Air is the most commonly used drying medium. Superheated steam can also be used as a heat supply medium [2]. Due to the relatively high cost and complexity of technology, the commercial application of steam is limited [8, 9].

The drying process can be divided into two distinct regimes namely (a) constant rate period where the gradient of moisture removal rate ($\mathrm{d}M/\mathrm{d}\tau$) remains constant and (b) falling rate period where the drying rate strongly depends on the interfacial heat and mass transfer rates. The second regime starts when the surface vapour pressure of the particle is less than the saturated vapour pressure. In this period the drying rate may be determined by the intra-particle diffusion limitation. The transition between the regimes depends on the structure of drying materials. However, the properties of the product do not affect the constant rate period [10].

The conventional fluidized bed is known for its outstanding heat and mass transfer characteristics. The fluidizing gas is used as a medium for heating and cooling the particle bed. The heat transfer coefficient of gas–solid is restricted by the operational gas–solid slip velocity of conventional fluidized beds. Particle entrainment is observed at higher gas–solid slip velocity. The dense operation at higher gas flowrates is limited by gas–solid slip velocity in conventional fluidized beds. The ratio of breadth and height of the particle bed decides the fluidization gas flow rate per mass which is directly linked to the earth's gravity [11]. The low particle density and non-uniform bed formation are disadvantageous for air-particle contact. Rotating fluidized bed reduces the size and improves the gas handling. Light cohesive particles such as Geldart-C type can also be fluidized in rotating fluidized bed [12, 13].

Fluidized bed performance is restricted by the bed density, non-uniform particle bed, and gas–solid slip velocity but they provide more effective heat and mass transfer [14]. By operating at high-G, in rotating fluidized beds the bed rotates in a cylindrical chamber and can overcome most of the limitations. The gases flow inward and impart drag force whereas particles experience counteracting radially outward centrifugal force [15]. In the case of RFB-SG a uniform and dense particle bed can be observed leading to better interstitial heat, mass, momentum transfer and gas–solid slip velocities [11]. RFBs are considered as promising technologies for developing compact dryers for granular materials in particular biomass.

2 Application of RFB-SG in Drying

Rotating fluidized bed are achieved either through rotation of the chamber or by the use of a static chamber in a fixed geometry. The former faces the problem of sealing, vibration and feeding of particulate material to be dried or cooled [16]. RFB-SGs can be created by the tangential injection of fluidization gas through multiple gas inlet slots. The solid particles experience a radially outward force due to the tangential gas–solid drag force resulting in the rotational motion of the bed. A dense particle bed is formed towards the cylindrical wall. The gas entering the chamber moves radially inward and leaves the chamber through the chimney (diffuser), which is centrally placed [17]. The particles are normally fed and removed through one end of the walls. The particle bed keeps on rotating, whereas the gas exits through the centrally placed chimney. Similarity exists between the RFB-SG, cyclone separator and vibrational fluidized beds [16]. Guidelines for design of the chamber can be found from Kochetov et al. and Dutta et al. [18, 19]. A review on RFB-SG was recently published by de Wilde [20]. Heat and mass transfer intensification have been demonstrated by CFD simulations [11]. The use of RFB-SG for drying of biomass was studied by Eliaers and de Wilde. A comparison of paddy drying in the vortex chamber and bubbling fluidized bed was done by Pati et al. and reported that drying was faster in RFB-SG [16]. The potential for better heat and mass transfer was reported by de Wilde and de Broqueville [11]. The paddy grain used as an inventory for the present experiments is shown in Fig. 1.

Process intensification was much higher in drying with RFB-SG. Limited information was available for scaling up of rotational the beds. Kochetov et al. [18] experimentally found out the ranges of different dimensional ratios that need to be maintained while scaling up of a vortex chamber and are as follows: (a) the ratio of the chimney diameter and chamber diameter should be between 0.3 and 0.5 m, (b) ratios of chamber length and diameter must be maintained between 0.2 and 0.5, (c) the ratio between the inlet slots width and the chamber diameter needs to be maintained between 0.05 and 0.2. A more detailed investigation is needed for the drying of granular materials. In the present work, the concept of rotating fluidized beds in static geometry was experimentally investigated for drying of shell covered paddy. Experiments were performed with hot air having an inlet air temperature of 55 and 60 °C through four inlet ports as shown in Fig. 2.

Fig. 1 The rice containing shell, paddy

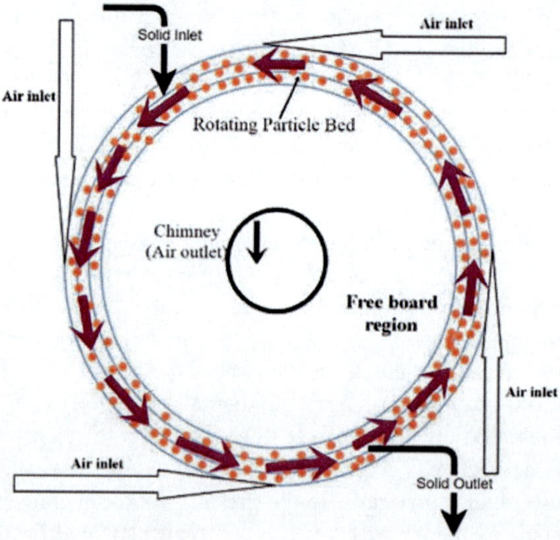

Fig. 2 Schematic of a rotating fluidized bed

3 Experimental Setup

Batch-type paddy experiments were carried out in the vortex chamber. A schematic of the complete experimental setup used for drying of paddy is shown in Fig. 3. It consists of a (a) Blower, (b) Bypass valve, (c) Orifice plate, (d) Manometers, (e)

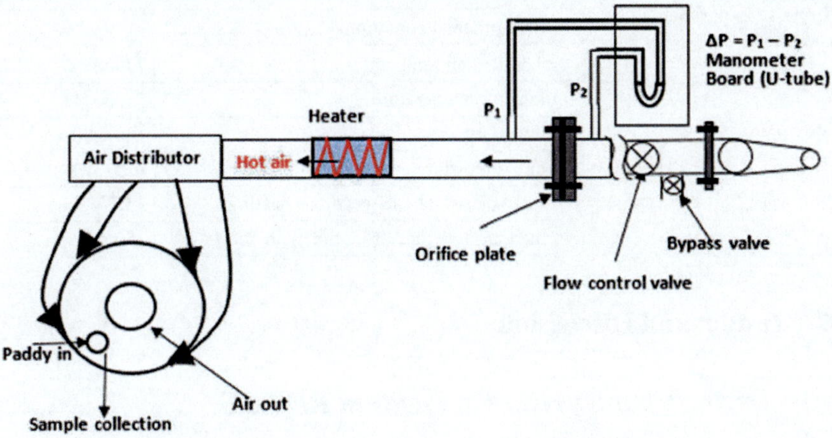

Fig. 3 Schematic of experimental setup

Heater, (f) Air distributor, (g) Vortex chamber/reactor, (h) Feed port/sample collection port, (i) Chimney.

The core of the experimental setup, made of mild steel, consists of a vortex chamber (also called a reactor) with 0.26 m diameter. Air is distributed to the vortex chamber through four holes of an inner diameter of 20 mm. A pictorial view of the air distribution system to the vortex chamber is shown in Fig. 4 and various input parameters for the vortex chamber is shown in Table 1.

Fig. 4 Experimental setup of RFB-SG

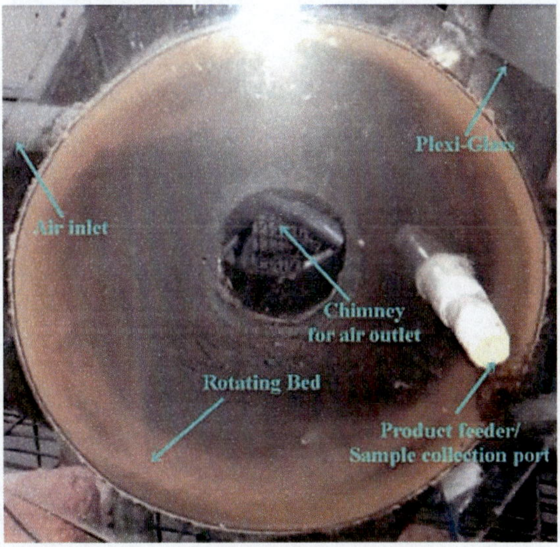

Table 1 Experimental conditions

Experimental conditions	
Ambient air temperature	27–30 °C
Inlet air temperature	55, 60 °C
Inventory	100–200 g
Initial moisture content (wb)	~23–37%
Input electrical energy for air heating	3 kW_e
Inlet air flow rate (from the main duct)	250 m^3/h

4 Results and Discussion

4.1 Drying Characteristics of Paddy in RFB-SG

Experiments were performed for analysis of the drying characteristics of paddy grains in RFB-SG dryer. Air inlet took place through four inlets. Drying characteristics of paddy in RFB-SG are discussed in the following sections. The moisture ratio has been plotted against the time required for drying. The moisture ratio can be calculated from the following formula.

$$MR = \frac{MC - MC_e}{MC_o - MC_e} \tag{1}$$

where MC = Moisture content of the grain at any level, MC_e = Equilibrium moisture content, MC_o = Initial moisture content of wet grain.

4.2 Effect of Air Inlet Temperature

Figures 5, 6 and 7 represents the variation of moisture ratio (MR) present during paddy drying. Experiments were conducted for the air inlet temperature of 55 and 60 °C.

It is observed from Fig. 5 that with the increase in inlet air temperature from 55 to 60 °C the drying time required to reach safe moisture decreases by 50%. With the reduction of drying time from 30 to 15 min for 100 g of paddy. A similar trend is observed in Figs. 6 and 7.

It is observed from Fig. 6 that with 0.9% increase in inlet air temperature, the drying time required to reach safe moisture decreases by 20%. With the reduction of drying time from 25 to 20 min for 150 g of paddy. This can be attributed to the formation of dense bed as compared to lower inventory.

From Fig. 7, it can be observed that with an increase in inlet air temperature by 9%, the drying time required to reach safe moisture decreases by 25% with a reduction

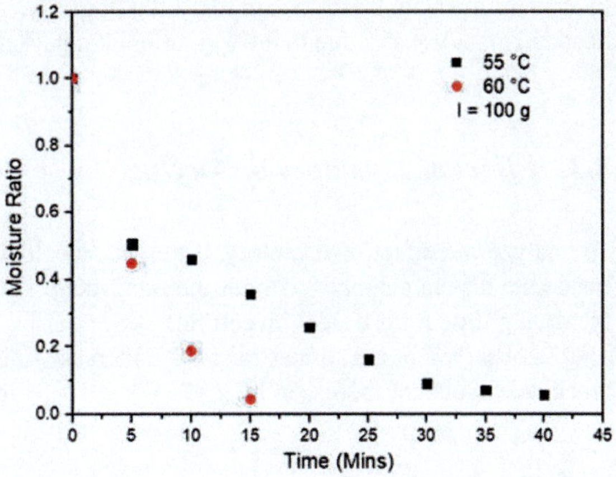

Fig. 5 Variation of MR with time at different air inlet temperature ($I = 100$ g)

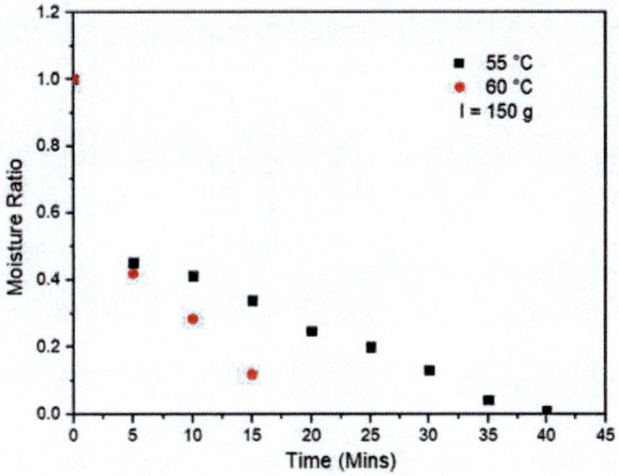

Fig. 6 Variation of MR with time for different air inlet temperature ($I = 150$ g)

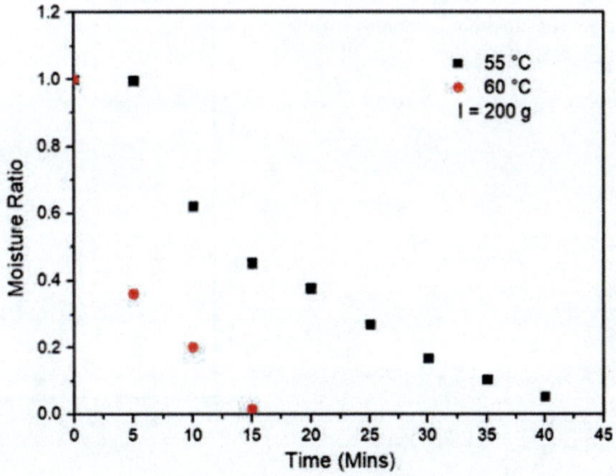

Fig. 7 Variation of MR with time at different air inlet temperature ($I = 200$ g)

in drying time from 20 to 15 min for 200 g of paddy. This can be attributed to better heat and mass transfer due to the formation of a packed bed with a higher inventory.

4.3 *Effect of Inventory on Drying*

To analyze the effect of inventory, different inventory of drying materials was fed inside the drying chamber without the introduction of slits. Figures 8 and 9 shows the drying time for different inventories.

It is observed in Fig. 8 that for a 50% increase in inventory, the time is taken to reach safe moisture increased by 25%. For a 100% increase in inventory, the time

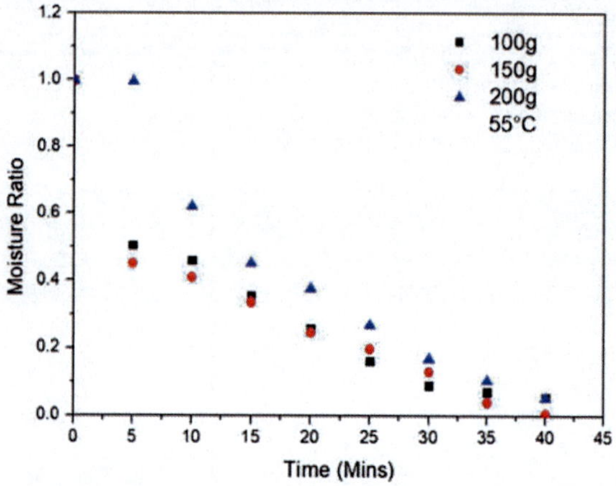

Fig. 8 Variation of MR with time for different inventory (Ta = 55 °C)

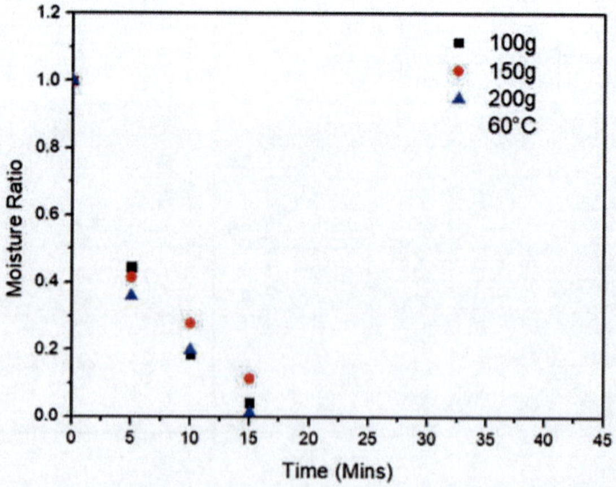

Fig. 9 Variation of MR with time for different inventory (Ta = 60 °C)

taken to reach safe moisture was increased by 25%. This can be attributed to a better density of bed with an inventory of 200 g and better utilization of air.

It is observed in Fig. 9 that for a 50% increase in inventory, the time required to reach safe moisture was the same. This can be attributed to better utilization of air with an increase in inventory. For a 100% increase in inventory, the time required to reach safe moisture was the same. This can be attributed to a better density of bed with an inventory of 200 g and better utilization of air.

5 Conclusion

The present work is carried out to understand the drying characteristics of paddy in rotating fluidized bed in static geometry. The important factors that influence the characteristics of paddy drying are the quantity of inventory, inlet air temperature and the same has been investigated for their influence on drying time. Following was the observation:

- Better utilization of air is observed with an increase in inventory.
- The time required to reach up to safe moisture remains the same with an increase in inventory from 100 to 200 g.
- With an increase in inlet temperature from 55 to 60°C the drying time decreased by 20%.

References

1. Mohapatra SS, Mahanta P (2012) Thermodynamic evaluation of natural convection paddy dryer. In: Proceedings of the IEEE 2nd international conference on the developments in renewable energy technology (ICDRET). IEEE, pp 1–9
2. Mujumdar AS (2007) Book review: handbook of industrial drying. CRC Press, Boca Raton, pp 1133–1134
3. International Rice Research Institute (IRRI) (2013) Determining the physical properties of paddy module 3. Rice Knowledge Bank. https://www.knowledgebank.irri.org/images/docs/training-manual-paddy-drying.pdf. Accessed 6 Sept 2017
4. Wimberly JE (1983) Technical handbook for the paddy rice postharvest industry in developing countries. International Rice Research Institute
5. Sreekumar A, Manikantan PE, Vijayakumar KP (2008) Performance of indirect solar cabinet dryer. Energy Convers Manage 49(6):1388–1395
6. Zhang Q et al (2003) Tracing fissure information by scanning electron microscopy characterization of naturally fissured surfaces of rice kernels. Trans ASAE 46(6):1583
7. Syahrul S, Hamdullahpur F, Dincer I (2002) Energy analysis in fluidized-bed drying of large wet particles. Int J Energy Res 26(6):507–525
8. Smith FJ (1970) Microwave fluidized bed dryer. U.S. patent no. 3,528,179, 15 Sept 1970
9. Kahyaoglu LN, Sahin S, Sumnu G (2012) Spouted bed and microwave-assisted spouted bed drying of parboiled wheat. Food Bioprod Process 90(2):301–308
10. Basunia MA, Abe T (2001) Thin-layer solar drying characteristics of rough rice under natural convection. J Food Eng 47(4):295–301

11. de Broqueville A, de Wilde J (2009) Numerical investigation of gas-solid heat transfer in rotating fluidized beds in a static geometry. Chem Eng Sci 64(6):1232–1248
12. Watano S et al (2004) Fine particle coating by a novel rotating fluidized bed coater. Powder Technol 141(3):172–176
13. Eliaers P et al (2014) High-G, low-temperature coating of cohesive particles in a vortex chamber. Powder Technol 258:242–251
14. de Wilde J, Froment GF (2012) Computational fluid dynamics in chemical reactor analysis and design: application to the ZoneFlow™ reactor for methane steam reforming. Fuel 100:48–56
15. Chen S-S (1987) Flow-induced vibration of circular cylindrical structures
16. Pati JR et al (2016) Experimental study of paddy drying in a vortex chamber. Dry Technol 34(9):1073–1084
17. de Wilde J, de Broqueville A (2007) Rotating fluidized beds in a static geometry: experimental proof of concept. AIChE J 53(4):793–810
18. Kochetov LM, Sazhin BS, Karlik EA (1969) Experimental determination of the optimal ratios of structural dimensions in the whirl chamber for drying granular materials. Chem Petrol Eng 5(2):106–108
19. Dutta A et al (2010) Rotating fluidized bed with a static geometry: guidelines for design and operating conditions. Chem Eng Sci 65(5):1678–1693
20. de Wilde J (2014) Gas–solid fluidized beds in vortex chambers. Chem Eng Process 85:256–290

Conjugate Heat Transfer Analysis of Solar Cooker Cavity Using CFD Approach

Sudhansu Sekhar Sahoo, Prasanta Kumar Satapathy, Pravat Kumar Parhi, Auroshis Rout, and Sanju Thomas

Abstract Conjugate mode of heat transfer experienced in solar cooker cavity is presented in this paper using computational fluid dynamics (CFD) approach. Trapezoidal-shaped box-type cooker has been used to investigate the heat loss from cooker cavity. Steady laminar natural convection and radiation inside the trapezoidal cavity having hot bottom, cold top isothermal wall and adiabatic sidewalls are considered. Forced convection due to wind flow on the top of the glass top cover and radiation is considered as well. The losses due to convection and radiation are considered from the top glass cover as well. Parameters like bottom plate temperature, emissivity of the plate and wind flow over the top cover are found to be more influencing on the heat losses from the cavity. Based on the results using CFD approach, the suitable parameters may be chosen for design of solar cooker cavity better and optimised performance.

Keywords Solar thermal · Solar cooker · Trapezoidal cavity · Heat losses · CFD

1 Introduction

Solar cookers are devices that convert the intensity of solar radiation into heat in order to cook or pasteurize food and beverages. As solar energy is renewable source of energy and it is available in unlimited amount, the operation cost of solar cooker is very low. Heat transfer analysis of a cavity or enclosure has been the topic of

S. S. Sahoo (✉) · P. K. Satapathy
Department of Mechanical Engineering, CET Bhubaneswar, Bhubaneswar, India
e-mail: sahoo.sudhansu@gmail.com

P. K. Parhi
Department of Civil Engineering, CET Bhubaneswar, Bhubaneswar, India

A. Rout
Department of Energy Science and Engineering, IIT Bombay, Mumbai, India

S. Thomas
Department of Mechanical Engineering, COE, CUSAT, Ernakulam, India

S. Revankar et al. (eds.), *Proceedings of International Conference on Thermofluids*, Lecture Notes in Mechanical Engineering,
https://doi.org/10.1007/978-981-15-7831-1_56

research since last two decades because of its various engineering applications. Especially, natural convection has been studied in such cavities quite comprehensively. Combined natural convection and radiation heat transfer analysis in cavity are found in the literature as well. This problem, i.e. heat transfer analysis in cavity, also finds its application in solar field like solar cooker and linear Fresnel reactor system. The following paragraph highlights the brief work carried out in these solar cooker applications.

Different types of solar cookers like box-type cookers, concentrated-type cookers, solar cooker with storage method are available in market [1, 2]. Various methodologies are employed by researchers and manufacturers to analyse the performance of solar cooker having basic principles of solar energy, heat transfer, fluid mechanics and materials. Solar cookers are exposed to sun directly and bottom absorber plate is coated for maximum solar heat gain coming via top glass cover. As the bottom plates get heated up, the internal surface area usually attains uniform temperature at the point of stagnation condition after certain time period. Simultaneously, heat loss also occurs from bottom and top cover through all modes of heat transfer. To reduce conduction heat loss from bottom plate, insulating materials made of glass wool or other kinds are used below the bottom plate. Top loss includes heat loss to atmosphere via the glass cover through convection and radiation modes of heat transfer. Performance analysis and design of solar cookers have widely been done by researchers in the past along through experimentation and analytical approaches [3–7].

From the literature review, it was found that heat loss from the absorber plate is due to many factors which may be classified into three parts, geometrical, operating and materials property. Many researchers focussed on experimentation on heat loss from the cooker cavity. However, due to limitations in experimentation such as cost of experimentation and time required, uncertainty, data obtained are very loss. Hence, this paper is mainly focussed on computational approach to find out the heat losses from the trapezoidal-shaped solar cooker cavity. CFD is less time-consuming, cost-effective approach. Trapezoidal-shaped cooker cavity has been considered in present analysis. Effect of emissivity, temperature ratio, external convective heat loss due to wind flow on the heat losses is studied. Non-dimensional approach has been used for more generic use.

2 Setup Description

Trapezoidal cavity used in solar cooker is considered in the present work (Fig. 1). Bottom plate is hot plate made of black-coated steel sheet. Sidewalls are insulated, and upper glass cover allows sunrays to pass through it. The cavity is filled with air. The fluid zone is taken as stagnant initially. For case 1, the top wall is considered as cold wall having constant temperature. Convective and radiative heat losses on top cover are considered for case 2.

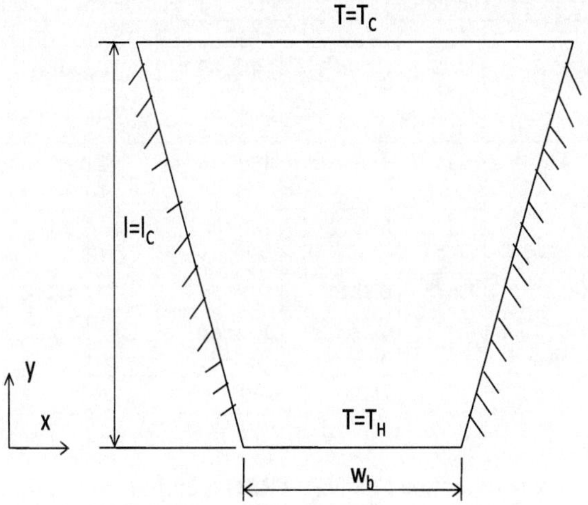

Fig. 1 Solar cooker cavity under consideration

3 Mathematical Modelling and Simulation Approach

The governing equations used for the present study are as follows based on assumptions like two-dimensional, steady, laminar flow, Boussinesq approximation in the cavity with non-participating air in it.

Continuity equation

$$\frac{\partial u}{\partial x} + \frac{\partial v}{\partial y} = 0 \tag{1}$$

X momentum equation

$$u\frac{\partial u}{\partial x} + v\frac{\partial u}{\partial y} = -\frac{1}{\rho}\frac{\partial p}{\partial x} + \nu\left(\frac{\partial^2 u}{\partial x^2} + \frac{\partial^2 u}{\partial u^2}\right) \tag{2}$$

Y momentum equation

$$u\frac{\partial v}{\partial x} + v\frac{\partial v}{\partial y} = -\frac{1}{\rho}\frac{\partial p}{\partial x} + \nu\left(\frac{\partial^2 v}{\partial x^2} + \frac{\partial^2 v}{\partial y^2}\right) + g\beta(T - T_\infty) \tag{3}$$

Energy equation

$$u\frac{\partial T}{\partial x} + v\frac{\partial T}{\partial y} = \alpha\left(\frac{\partial^2 v}{\partial x^2} + \frac{\partial^2 v}{\partial y^2}\right) \tag{4}$$

u and *v* are velocities in *x*- and *y*-direction. Pressure is denoted as *p*. ρ, β and α are density and thermal coefficient and thermal diffusivity, respectively. Discrete

Table 1 Thermo-physical properties of air at different T^* values and that for steel and glass

	T^*	Density (kg/m^3)	Specific heat (J/kg K)	Viscosity (kg/m s)	Thermal conductivity (W/m K)	Thermal expansion coefficient (1/K)
Air	1.33	1.01	1009	2.084×10^{-5}	0.0297	0.002857
	1.5	0.9412	1009.4	2.3277×10^{-5}	0.03154	0.0026667
	1.67	0.8826	1013	2.6042×10^{-5}	0.033325	0.0025
Steel		8030	502.48	Nil	16.27	Nil
Glass		2600	840	Nil	1.05	Nil

transfer radiation model (DTRM) is employed and used to simulate thermal radiation for the present problem [8]. The details are omitted for the sake of brevity.

The bottom wall is taken as hot wall having constant temperature T_H. Sidewalls are taken as adiabatic. The internal emissivity of both the sidewalls is taken as 0.1, while internal emissivity of the top wall is taken as 0.9. Bottom plate emissivity is varied from 0.1 to 0.9. Top plate, which is glass, is considered as constant temperature (T_c) for case 1. The top wall is treated as convective heat loss due to external wind flow (h_e) along with radiation for case 2. The steel, glass and air properties are considered as mentioned in Table 1.

CFD approach has been employed to solve governing equations [8]. Pressure correction method and SIMPLEC algorithm is used as solution methodology. The second-order upwind differencing scheme has been adopted for momentum and energy equations. Convergence criteria of residual considered are 10^{-5} for momentum and mass and 10^{-6} for energy equations.

4 Results and Discussion

The problem under consideration involves five parameters, namely Rayleigh number (Ra), aspect ratio (AR), emissivity (ϵ), ratio of absolute temperature of bottom wall and top wall or atmosphere depending on the case (T^*) and external Nusselt number (Nu_e). Aspect ratio has been fixed as 1 for the current problem. Average Nusselt number (Nu_t) at the top cold wall is obtained with respect to variation of other four parameters. Grid independence study and verification study were done before presenting results. Rayleigh number (Ra) is calculated based on height of the cavity (l), and the required Rayleigh number was obtained by adjusting various parameters. T^* is defined as the ratio of hot wall temperature (T_H) and cold wall temperature (T_C). For cases 2, T^* is defined as the ratio of (T_H) and free stream temperature (T_∞). For case 2, non-dimensional parameter and convection Nusselt Number (Nu_e)

can be defined as

$$\mathrm{Nu_e} = \frac{h_e w_{tc}}{k} \tag{5}$$

where h_{tc} is convective heat transfer coefficient on top cover due to wind, k is the thermal conductivity of air and w_{tc} is width of top cover.

4.1 Temperature Contours and Velocity Vectors

Figure 2 shows the isotherms for $Ra = 10^4$ and $Ra = 10^5$ with $\epsilon_b = 0.1$ and 0.5. In case of the trapezoidal cavity, bicellular flow occurs for all aspect ratio values and emissivity values as shown in Fig. 3. For $\mathrm{Ra} = 10^4$, isotherms are uniformely spaced. In contrast, for higher Rayleigh number (Ra) values, isotherms are more crowded

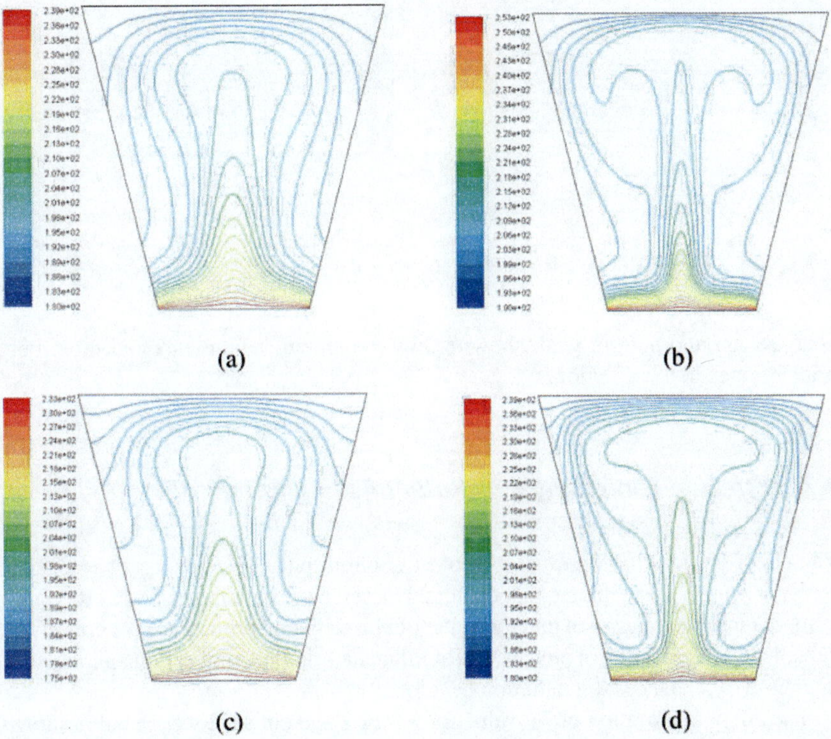

Fig. 2 Isotherms for the trapezoidal cooker cavity (case 1) **a** $\mathrm{Ra} = 10^4$, $\epsilon_b = 0.1$, **b** $\mathrm{Ra} = 10^5$, $\epsilon_b = 0.1$, **c** $\mathrm{Ra} = 10^4$, $\epsilon_b = 0.5$, **d** $\mathrm{Ra} = 10^5$, $\epsilon_b = 0.5$

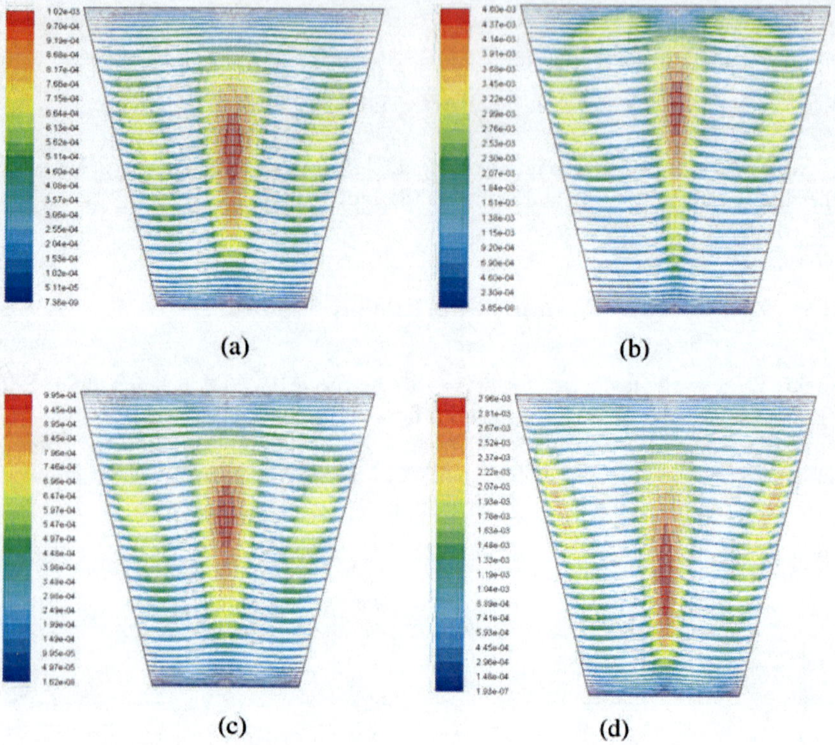

Fig. 3 Velocity vectors for the trapezoidal cooker cavity (case1) **a** Ra $= 10^4$, $\epsilon_b = 0.1$, **b** Ra $= 10^5$, $\epsilon_b = 0.1$, **c** Ra $= 10^4$, $\epsilon_b = 0.5$, **d** Ra $= 10^5$, $\epsilon_b = 0.5$

near the top and bottom wall, showing high temperature gradients and more heat transfer rates in these regions.

4.2 *Effect of Emissivity and Ratio of Temperature (T^*)*

The effect of variation of emissivity of hot bottom plate (ϵ_b) having $T^* = 1.33$ is shown in Fig. 4. It is depicted that the total Nusselt number is increasing as Rayleigh number increases. It is seen that the value of Nusselt number increases as emissivity value increases. It follows physics as Ra influences natural convection and increase in emissivity enhances the radiation heat loss.

The effect of variation of T^* value for AR $= 1$ and emissivity $\epsilon_b = 0.3$ is shown in Fig. 5. As the ratio temperature T^* increases, total Nusselt number increases. As radiation loss is dependent on absolute temperature, the loss would be more in case of increasing the temperature value.

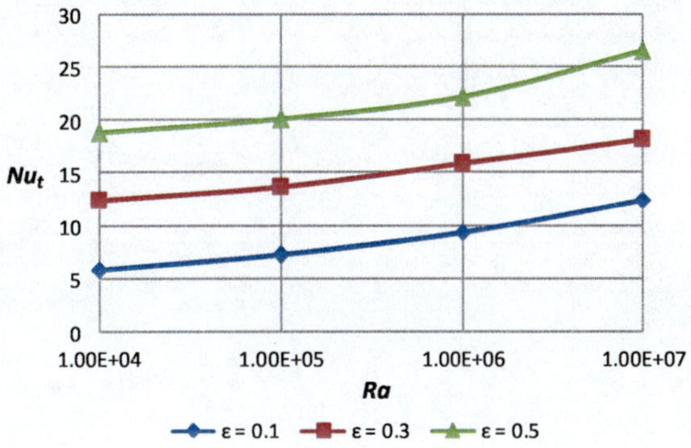

Fig. 4 Variation of $\log(Nu_t)$ with respect to Ra at different emissivities for case 1 with $T^* = 1.33$

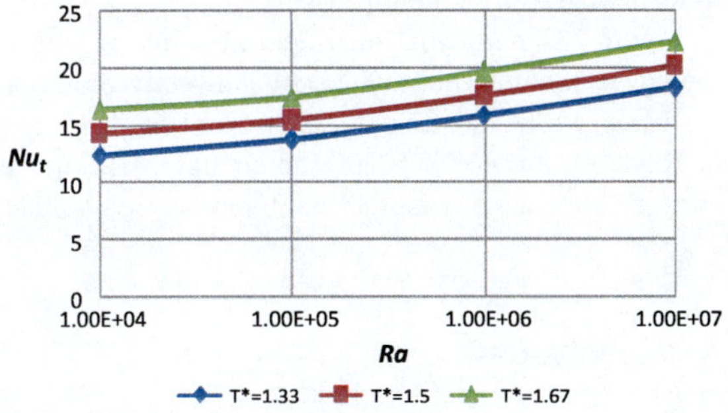

Fig. 5 Variation of $\log(Nu_t)$ with respect to Ra at different T^* for case 1 with $\epsilon_b = 0.3$

4.3 *Effect of External Nusselt Number (Nu_e)*

Due to wind flow over the glass top cover, heat transfer coefficient influences on forced convection. To make it generalised, non-dimensional heat transfer coefficient named as external Nusselt number has been used. Figure 6 depicts variation of total Nusselt number with respect to Rayleigh number at different external Nusselt number values. Nusselt number (Nu_t) seems to be increasing with increase in external Nusselt

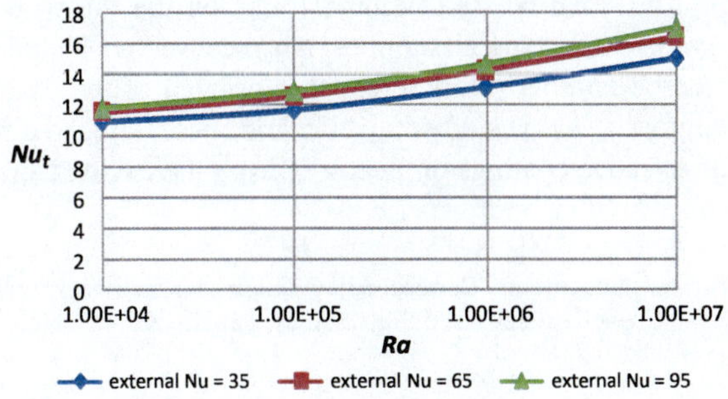

Fig. 6 Variation of $\log(Nu_t)$ with respect to Ra for case 2 at different Nu_e, having $T^* = 1.33$, $\epsilon_b = 0.3$

Fig. 7 Variation of log(Nu) with respect to Ra for case 1 having $T^* = 1.33$, $\epsilon_b = 0.1$

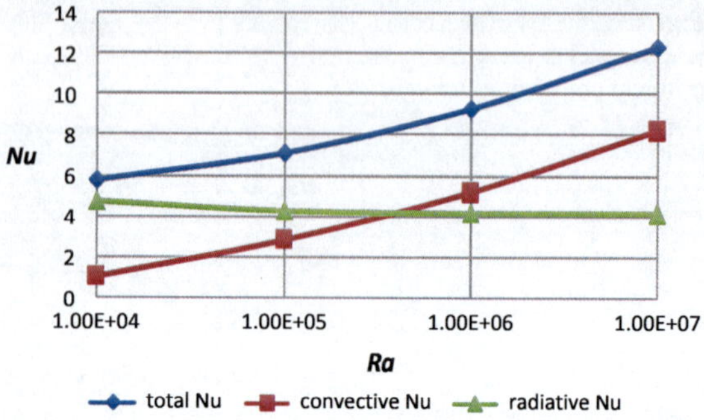

number. This may be due to the more forced convection heat transfer which causes more heat loss from the top cover.

Figure 7 denotes the individual heat losses (convective and radiative) from the top cover. It can be seen that only convective component of Nusselt number (Nu_c) increases while radiative Nusselt number (Nu_r) remains constant with increase in Rayleigh number. The presented figure is depicted for case 1; i.e., top cover temperature is having isothermal temperature boundary condition.

5 Conclusions

CFD analysis to investigate the heat loss in trapezoidal-shaped cooker was carried out in the present work. Conjugate heat transfer, i.e. convection and radiation heat transfer, modes are studied and analysed in solar cooker cavity. Bottom wall was considered hot plate, while top glass wall was cold wall. Sidewalls were considered as adiabatic. Air was considered as cavity medium which is assumed to be non-participating. Steady two-dimensional natural convection along with radiation was carried out in the cooker cavity. Forced convection due to wind flow on the top portion of glass cover and radiation was considered for heat transfer analysis. Trapezoidal cavity having same geometry has been used for all simulations by varying parameters such as absolute temperature ratio (T^*), emissivity of bottom surface (ϵ_b), external Nusselt number (Nu_e) on the top wall and Rayleigh number (Ra). Average total Nusselt number (Nu_t), convective Nusselt number (Nu_c) and radiative Nusselt number (Nu_r) across the top wall of the cavity were obtained using CFD approach. Future study may be carried out related to different aspect ratios and effect of selective coatings on heat loss from the cavity. Correlation development may be carried out also for generic use.

Acknowledgements Corresponding author thankfully acknowledges the financial support from State Council on Science & Technology, Science & Technology Department, and Government of

Odisha under research grant ST-SCST-MISC-0061/2018 2682/dated 24-06-2019 to carry out this project.

References

1. Kumar S (2004) Natural convective heat transfer in trapezoidal enclosure of box-type solar cooker. Renew Energy, 211–222
2. Chen CR, Sharma A, Tyagi SK, Buddhi D (2008) Numerical heat transfer studies of PCMs used in a box-type solar cooker. Renew Energy 33:1121–1129
3. Mirdha US, Dhariwal SR (2008) Design optimization of solar cooker, 33:530–544
4. Balakrishnan M, Claude A, Arun Kumar DR (2012) Engineering design and fabrication of a solar cooker with parabolic conentrator for heating, drying and cooking purposes. Arch Appl Sci Res 4(4):1636–1649
5. Sharma A, Tyagi VV, Chen CR, Buddhi D (2009) Review on thermal energy storage with phase change materials and applications. Renew Sustain Energy Rev 13:318–345
6. Alam TH, Chawdhury SA (2012) Solar cooking and its prospects in Bangladesh. In: IEEE conference, 5–7 Jan 2012
7. Yang DJ, Yin HM (2011) Energy conversion efficiency of a novel hybrid solar system for photovoltaic, thermoelectric, and heat utilization. IEEE Trans Energy Convers 26:662–670
8. ANSYS 16 Manual (2016)

Thermofluids in Materials and Manufacturing

Experimentation with Thermally Stable Mesoporous Silica Powder and Its CO_2 Adsorption

Sanjib Barma and Bishnupada Mandal

Abstract The CO_2 coming out of the power plants at higher temperature. Adsorption is considered as a promising technology for the removal of CO_2 from flue gas. So, the adsorbent material should have the properties which are applicable in high temperature condition. Silica-based materials are available, and they withstand high temperature. This paper describes the experimentation with the ordered mesoporous silica materials obtained by the acidic route as well as the basic route and compares the thermal stability and the CO_2 adsorption capacity. The porosity of OMS materials is characterized by using N_2 adsorption–desorption. The CO_2 isotherms at 30 °C are measured up to 8 bars for OMS samples to find out CO_2 adsorption capacity. CO_2 adsorption studies show promising results for OMS materials obtained by the acidic route. TGA analysis confirms the stability of material upto >200 °C.

Keywords OMS material · CO_2 adsorption · Acidic route · Basic route · Characterization

1 Introduction

CO_2 separation from flue gas or other gaseous mixtures which are generated during combustion of fossil fuels creates interest among the researchers across the globe [1–15]. Adsorption-based separation and purification technology is popular due to low regeneration cost, no corrosion and simplicity of this process [1–15]. In the case of solid state CO_2 storage, the gas is reversibly embedded in a solid matrix. The main challenges for the improvement of this type material are to facilitate controlling the overall treatment apart from the desired selectivity and sorption capacity toward the target gases. A wide range of uniform pore size and the high availability of silanol

S. Barma (✉) · B. Mandal
Department of Chemical Engineering, Indian Institute of Technology Guwahati, Guwahati, Assam 781039, India
e-mail: sanjib.burma@gmail.com

S. Barma
Department of Chemical Technology, University of Calcutta, Kolkata 700009, India

S. Revankar et al. (eds.), *Proceedings of International Conference on Thermofluids*, Lecture Notes in Mechanical Engineering,
https://doi.org/10.1007/978-981-15-7831-1_57

groups in mesoporous silica materials make them attractive for the inclusion of the desired functional groups with high coverage. Ordered mesoporous silica materials (OMS) can fulfill the requirements. The OMS materials include MCM-41, MCM-48, SBA-1, SBA-2, SBA-3, SBA-7 [2–15], etc. Mesoporous silica materials are formed generally by micelle-templating process. There are several literatures which described the synthesis procedure of OMS materials via different pH conditions (pH < 2 and pH > 2). In the acidic medium (pH < 2), the cubic cage type mesoporous silica materials such as SBA-16 [12, 13] or FDU-1, 12 [14, 15] could be synthesized with non-ionic EO_n–PO_m–EO_n triblock copolymer as the template, and with TEOS as the silica source. Huo et al. [16] proposed the synthesis of OMS by different routes involved S^+I^- and $S^+X^-I^+$ pathway where the 'S' stands for surfactant and 'I' stands for the inorganic silicate species. The 'X' refers to anionic species of Cl^-, Br^-, etc. OMS membranes fabricated on different inorganic supports had attracted great attention due to its robust nature in the extreme environment conditions [17–19]. CO_2 coming out of the power plants or vehicles after burning of fossil fuels is in high temperature. Due to the continuous and energy efficient operation, membranes of these materials can be a promising alternative for CO_2 capture and storages. There are huge scopes of research to know the characteristics of the OMS materials synthesized by different routes for the application of CO_2 separation at high thermal condition. An effort has been made in the present work to develop a comparative study for the OMS materials, obtained by the acidic route and the basic route for CO_2 adsorption at extreme conditions.

2 Experimental Section

2.1 Materials

Reagent grade ethanol (EMSURE R ACS, PhEur), sodium hydroxide (NaOH) and HCl (35%) were purchased from MERCK, Germany. Tetraethyl ortho-silicate (TEOS), *N*-hexadecyltrimethyl ammonium bromide (CTAB) were supplied by Sigma Aldrich, India.

2.2 Synthesis of OMS

A requisite mixture of deionized water, NaOH and cetyl-tri-methyl-ammonium-bromide (CTAB) was taken in a beaker and stirred at 50 °C until the CTAB was completely dissolved [20]. Then, the required amount of tetra-ethoxy-silane (TEOS) was added to the cooled solution (at ambient temperature) followed by stirring for 30 min to make the silica sol. The silica sol was then shifted to a teflon container and kept for 3 days at a temperature of 90 °C in an oven. Samples were carefully washed with distilled water in a filter paper and then dried in an oven at 60 °C. Finally, OMS powder was obtained by calcination in the muffle furnace at 550 °C with very slow heating and cooling rate.

Another requisite mixture of TEOS, ethanol and diluted HCl was poured in a glass beaker. The whole mixture was under stirring for one hour at ambient temperature. CTAB of required amount was added to the solution at room temperature with continuous stirring for complete dissolving. The obtained solution was aged for 24 h with continuous stirring [21]. The silica solution was then poured into a flat and open petri dish. The sol was kept for overnight at room temperature in a Fume hood (Dlab Tech, Clean Bench) for drying. Then, the dried silica was removed by a spatula in powder form. Finally, the powder was calcined in the muffle furnace at 550 °C for 6 h amid controlled slow heating and cooling rate.

2.3 CO_2 Adsorption Measurements

CO_2 isotherms were measured following Sieverts method at 30 °C. The Quantachrome-iSorb HP1 unit was used to perform all the measurements. Adsorbent samples were purified by keeping them under the vacuum at 120 °C. The dead volume of sample tube was obtained from helium measurement using non-adsorbing helium assumption.

2.4 Thermal Stability Measurement

Thermo-gravimetric analysis (TGA) was executed using a thermo-gravimetric analyzer (Model 851, Mettler Toledo). The analysis was performed using a nitrogen gas supply. The heating rate was maintained of 10 °C/min. Beckman Coulter surface area analyzer (Coulter SA model 3100) was used for nitrogen physisorption measurements at 77 K. The samples' pores were opened by degassing at 110 °C for 4 h before the N_2 adsorption–desorption measurements. Brunauer–Emmett–Teller (BET) model was used for calculating the specific surface area, and the Barrett–Joyner–Halenda (BJH) approach was used for determining the pore size distribution.

3 Results and Discussions

It has been described [22] before that the silica precursor becomes cationic or anionic species during hydrolysis, depending on the pH of the medium (Fig. 1). So, in the acidic and the basic medium, the silica-surfactant interactions follow the different pathway (Fig. 2) for the synthesis of OMS materials.

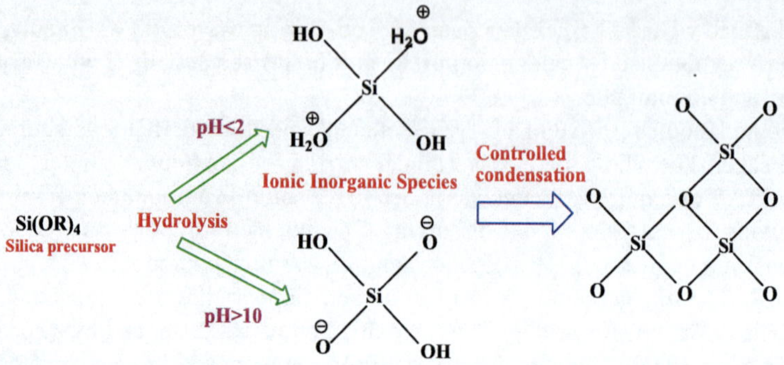

Fig. 1 Silica precursor at acidic and basic condition [22]

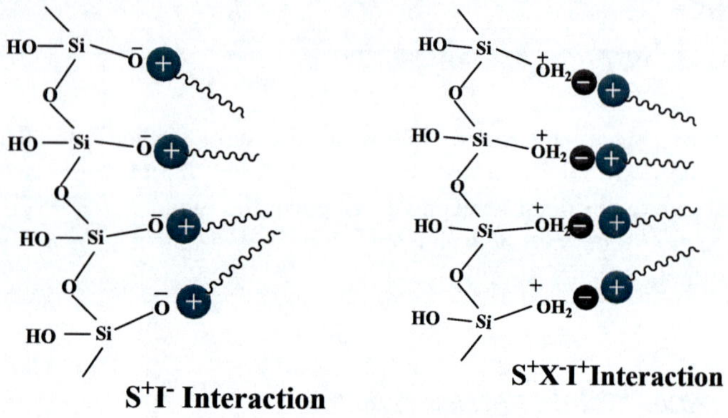

Fig. 2 Different types of silica-surfactant interactions [22]

3.1 Surface Area of OMS

The N_2 adsorption/desorption isotherms of the prepared OMS materials obtained by two different routes are shown in Figs. 3. Both the isotherms observed for the OMS materials show a typical type IV adsorption–desorption isotherm which confirms that the materials are ordered mesoporous in nature. N_2 adsorption–desorption isotherm pattern, surface area and pore volume of the OMS materials are highly comparable with the reported literature [2, 10]. The BET surface area and total pore volume of the OMS materials are determined using the nitrogen adsorption isotherm [23]. Relative pressure varied between 0.01 and 1. All the vales obtained are given in Table 1.

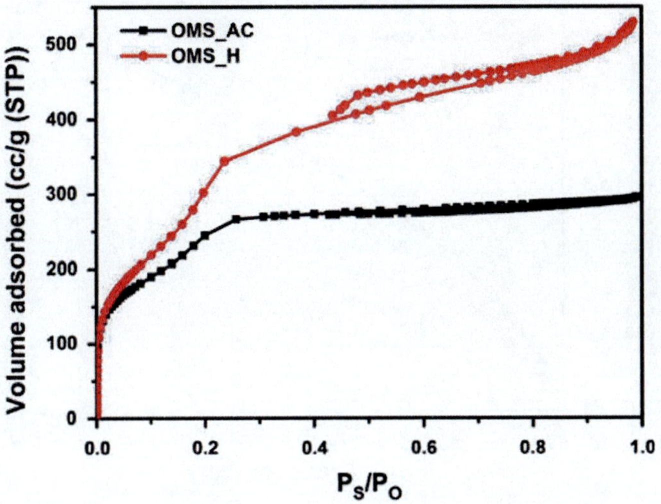

Fig. 3 Adsorption–desorption isotherm of N_2 on different OMS samples at 77 K

Table 1 BET surface area and total pore volume of selected OMS samples

Sample name	BET surface area (m^2/g)	Pore volume (cc/g)
OMS_H	1133.35	0.8114
OMS_AC	890.46	0.4541

3.2 CO_2 Uptake Analysis

The CO_2 uptake capacity measurements on to samples OMS_AC & OMS_H are obtained at room temperature at the pressure range of ~(0–9) bar is shown in Fig. 4. According to the IUPAC classification for adsorption isotherms, type I isotherms were obtained. The adsorption capacities of samples OMS_AC ranging from low pressure to 8 bar give slightly higher value compared to OMS_H materials. At high pressure after about 8 bar, it seems higher CO_2 adsorption capacity for the sample OMS_H compared to OMS_AC.

The high surface area and the high pore volume may be possible reason for higher CO_2 adsorption capacity for OMS_H samples at higher pressures, which is consistent with the literature [24]. At lower pressure ranges, the higher CO_2 adsorption capacity for OMS_AC samples may be due to the higher average pore diameter and uniform pore arrangement or pore structures. The pore network pattern may play a significant role for the higher CO_2 adsorption capacity of the OMS_AC sample [20]. It is seen from Fig. 4 that at about 1 bar pressure, the CO_2 adsorption capacity for OMS_AC and OMS_H samples is observed to be ~0.63 mmol/g and ~0.48 mmol/g, respectively. It is reported [18] that membrane can be synthesized on ceramic support of these materials for the application of CO_2 separation.

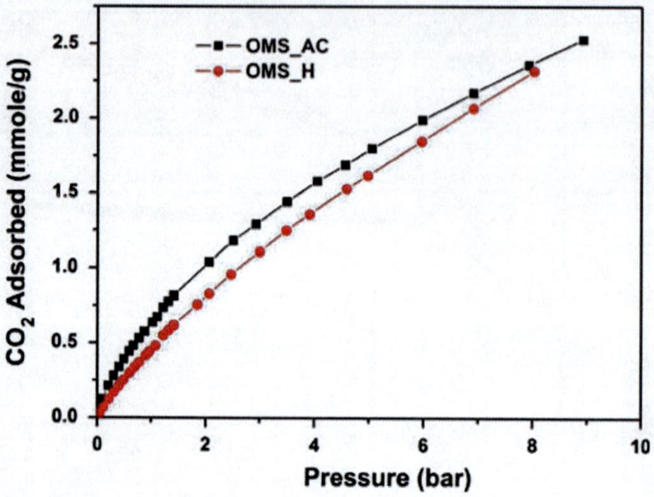

Fig. 4 CO_2 adsorption isotherm of samples OMS_AC & OMS_H at 30 °C

3.3 Thermal Stability of OMS

Thermo-gravimetric (TGA) analysis of the OMS materials is shown in Fig. 5. The OMS sample obtained by acidic route and basic route is denoted as OMS_AC & OMS_B, respectively. The figure exhibits an initial weight loss in the temperature

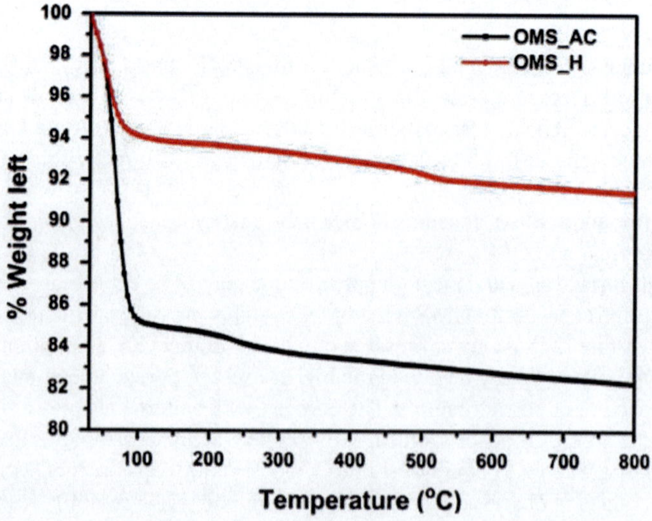

Fig. 5 Thermo-gravimetric analysis for the OMS samples synthesized by various routes

range 30–100 °C for both the samples due to the loss of moisture and hydrated water. The TGA curve shows additional weight loss for pure MCM-48 sample, associated with condensation of surface silanol groups. After the region 200–600 °C, there is a marginal/negligible weight loss. Thereafter, the sample shows constant weight. Thermal stability of samples also seems to be satisfactory in the context of flue gas application (as flue gas temperature typically ranges between 70–120 °C). This also indicates that the surfactant is completely removed from both the OMS materials after calcination at 550 °C.

4 Conclusions

The acidic route is found to be an energy intensive process for synthesizing ordered mesoporous materials and shows a promising alternative for CO_2 capture technologies due to cost effectiveness. Thermal stability of the both samples also seems to be satisfactory in the context of flue gas application (as flue gas temperature typically ranges between 70–120 °C). For membrane applications, these materials can play very promising role due to the continuous process as well as higher thermal and pressure stability.

Acknowledgements The financial support by the CSIR (Grant No. 22(0630)/13/EMR-II), New Delhi, Government of India, is gratefully acknowledged.

References

1. Joss L, Gazzani M, Mazzotti M (2017) Rational design of temperature swing adsorption cycles for post-combustion CO_2 capture. Chem Eng Sci 158:381–394
2. Loganathan S, Tikmani M, Ghoshal AK (2013) Novel pore-expanded MCM-41 for CO_2 capture: synthesis and characterization. Langmuir 29:3491–3499
3. Mortera R, Fiorilli S, Garrone E, Verné E, Onida B (2010) Pores occlusion in MCM- 41 spheres immersed in SBF and the effect on ibuprofen delivery kinetics: a quantitative model. Chem Eng J 156:184–192
4. Pang X, Gao J, Tang F (2005) Controlled preparation of rod- and top-like MCM- 41 mesoporous silica through one-step route. J Non-Cryst Solids 351:1705–1709
5. Frasch J, Lebeau B, Soulard M, Patarin J (2000) In-situ investigations on cetyltrimethylammonium surfactant/silicate systems, precursors of organized mesoporous MCM-41-type siliceous materials. Langmuir 16:9049–9057
6. Angloher S, Kecht J, Bein T (2007) Optimization of reaction conditions for the metal organic modification of MCM-41. Chem Mater 19:3568–3574
7. Chandrasekar G, Son WJ, Ahn WS (2009) Synthesis of mesoporous materials SBA-15 and CMK-3 from fly ash and their application for CO_2 adsorption. J Por Mater 16:545–551
8. Hiyoshi N, Yogo K, Yashima T (2005) Adsorption characteristics of carbon dioxide on organically functionalized SBA-15. Micropor Mesopor Mater 84:357–365
9. Jang HT, Park Y, Ko YS, Lee JY, Margandan B (2009) Highly siliceous MCM-48 from rice husk ash for CO_2 adsorption. Int J Greenh Gas Con 3:545–549

10. Huang HY, Yang RT, Chinn D, Munson CL (2003) Amine-grafted MCM-48 and silica xerogel as superior sorbents for acidic gas removal from natural gas. Ind Eng Chem Res 42:2427–2433
11. Kim S, Ida J, Guliants VV, Lin JYS (2005) Tailoring pore properties of MCM-48 silica for selective adsorption of CO_2. J Phys Chem B 109:6287–6293
12. Yu C, Tian B, Fan J, Stucky GD, Zhao D (2002) Nonionic block copolymer synthesis of large-pore cubic mesoporous single crystals by use of inorganic salts. J Am Chem Soc 124:4556–4557
13. Kleitz F, Kim TW, Ryoo R (2006) Phase domain of the cubic Im 3 m mesoporous silica in the EO106 PO70 EO106-Butanol-H_2O System. Langmuir 22:440–445
14. Fan J, Yu C, Gao F, Lei J, Tian B, Wang L, Luo Q, Tu B, Zhou W, Zhao D (2003) Cubic mesoporous silica with large controllable entrance sizes and advanced adsorption properties. Angew Chem Int Edit 42:3146–3150
15. Matos JR, Mercuri LP, Kruk M, Jaroniec M (2002) Synthesis of large-pore silica with cage-like structure using sodium silicate and triblock copolymer template. Langmuir 18:884–890
16. Huo Q, Margolese DI, Ciesla U, Demuth DG, Feng P, Gier TE, Chmelka BF, Schuth F, Stucky GD (1994) Organization of organic molecules with inorganic molecular species into nano composite biphase arrays. Chem Mater 6:1176–1191
17. Sakamoto Y, Nagata K, Yogo K, Yamada K (2007) Preparation and CO_2 separation properties of amine-modified mesoporous silica membranes. Micropor Mesopor Mater 101:303–311
18. Kumar P, Ida J, Kim S, Guliants VV, Lin JYS (2006) Ordered mesoporous membranes: effects of support and surfactant removal conditions on membrane quality. J Membr Sci 279:539–547
19. Barma S, Mandal B (2015) Synthesis and characterization of ordered mesoporous silica membrane: role of porous support and gas permeation study. Micropor Mesopor Mater 210:10–19
20. Brigante M, Schulz PC (2012) Synthesis of mesoporous silicas in alkaline and acidic media using the systems cetyltrimethylammonium tosylate (CTAT)–Pluronic F127 triblock copolymer and CTAT–Pluronic F68 triblock copolymer as templates. Colloid Interface Sci 369:71–81
21. Besson S, Ricolleau C, Gacoin T, Jacquoid C, Boilot JP (2000) A new 3D organization of mesopores in oriented CTAB silica films. J Phys Chem B 104:12095–12097
22. Pal N, Bhaumik A (2013) Soft templating strategies for the synthesis of mesoporous materials: inorganic, organic–inorganic hybrid and purely organic solids. Adv Colloid Interface Sci 189–190:21–41
23. Sahu D, Mishra P, Edubilli S, Verma A, Gumma S (2013) Hydrogen adsorption on Zn-BDC, Cr-BDC, Ni-DABCO and Mg-DOBDC metal organic frameworks. J Chem Eng Data 58(11):3096–3101
24. Belmabkhout Y, Serna-Guerrero R, Sayari A (2009) Adsorption of CO_2 from dry gases on MCM-41 silica at ambient temperature and high pressure. 1: pure CO_2 adsorption. Chem Eng Sci 64:3721–3728

Experimentation with Thermo-mechanically Stable Epoxy Composite Reinforced with Palm Fiber

Jnanaranjan Kar, Arun Kumar Rout, Priyadarshi Tapas Ranjan Swain, and Alekha Kumar Sutar

Abstract The aim of this current investigation is to evaluate the influence of bio-filler on developing thermo-mechanically stable composites. A new class of low-cost palm fiber-reinforced epoxy composite was fabricated using 0, 5, 10, and 15 wt% rice husk (RH) powder as bio-filler. The influence of filler addition on thermal, dynamic mechanical is investigated. By using thermogravimetric analysis (TGA) and dynamic mechanical analysis (DMA) technique for analysing the thermal stability of the untreated and chemically treated composite materials and find out more stable material. The thermal degradation of palm fiber occurred in the following temperature ranges: hemicelluloses (150–350 °C), cellulose (275–500 °C), and lignin (250–500 °C). Moreover, no significant difference is observed in the thermal degradation of RH filled and unfilled samples beyond 550 °C. However, the storage modulus and loss modulus of the composite are increased with filler addition. A low tan δ value is observed for the unfilled samples. The corrosion property of the composite increased with filler addition.

Keywords Palm fiber · Thermo-mechanically stable composites · Thermogravimetric analysis · Dynamic mechanical analysis

---

J. Kar · A. K. Sutar
Catalysis and Research Lab, Ravenshaw University, Cuttack 753001, India

A. K. Rout (✉)
Department of Production Engineering, Veer Surendra Sai University of Technology, Burla, Sambalpur 768018, India
e-mail: arun.rout.6314@gmail.com

P. T. R. Swain
Department of Mechanical Engineering, Veer Surendra Sai University of Technology, Burla, Sambalpur 768018, India

S. Revankar et al. (eds.), *Proceedings of International Conference on Thermofluids*, Lecture Notes in Mechanical Engineering,
https://doi.org/10.1007/978-981-15-7831-1_58

1 Introduction

Natural fibers are being increasingly considered as an environmentally friendly material in polymer matrices due to their renewable nature. Unlike synthetic fiber, they offer problem-free disposal, and they are non-toxic and non-hazardous to the environment. Moreover, natural fiber has low cost, low density, good specific modulus, easy processing, and high aspect ratio for efficient stress transfer over synthetic fiber [1, 2]. They are widely used in the textile, automobile, construction, and composite manufacturing industries. However, the application of natural fiber has been restricted due to its hydrophilic nature and lower thermal resistance properties. The above limitations can be resolved by employing several chemical treatments of the fiber before any specific use. Chemically treated plant fiber-reinforced composites show improved mechanical and thermal resistance properties compared with the raw fiber [3, 4]. Alkali, silane and different coupling reagents are implemented as a pretreatment in many plant-based fibers.

Better fiber-matrix bond strength along with improved mechanical and thermal properties of the jute/poly (lactic acid) was observed after silane and alkali treatment of jute fiber [5]. Similarly, improved mechanical, thermal, and water absorption properties of coconut and mallow–eucalyptus fiber composites were found after the chemical treatment of fiber surface [6]. Hybridization of natural fiber composites with synthetic fiber or filler yields a positive response towards improvement in the mechanical and thermal resistance of the developed composites [7]. Sisal/glass [8] and bamboo/glass [9] hybrid composites show enhanced thermal and mechanical properties than their individual use as reinforcement.

The dynamic mechanical analysis (DMA) is an essential and effective tool used to optimize the viscoelastic behavior of polymer composites for various engineering applications. In a study of dynamic mechanical and thermal properties of jute/vinyl ester composites, the composite thermal resistance and storage modulus was improved after alkali treatment of the fiber [10]. The flexural, impact, storage modulus, and loss modulus of the treated jute/epoxy are increased compared with the raw fiber [11]. It can be improved by incorporating various synthetic/natural fibers or fillers (micro/nano) to the polymer matrix. Glass fiber addition to the sugar palm fiber and bamboo fiber-reinforced composites improves the DMA [12]. The storage modulus and thermal resistance of oil palm-epoxy composites were improved with the addition of jute fiber into it [13]. The viscoelastic properties of short glass fiber-reinforced polypropylene composites are enhanced when calcite is added [14]. Similarly, various fillers like granite powder [15], alumina [16], fly ash cenosphere [17], nano-ZnO and TiO_2 [18], and cork [19] show a positive response on the thermal behavior of composites.

In the literature, there are reports on the dynamic mechanical analysis and thermal behavior of various plant-based fiber-reinforced composites, but there is no such report available on DMA and thermal analysis of palm-fiber-reinforced epoxy composites, though it is abundantly available. Thus, the objective of this study is to fabricate palm-epoxy composites filled with rice husk (RH) filler in various percentages and to characterize their thermo-mechanical properties for various engineering applications.

2 Experimental Details

2.1 *Materials and Methods*

Epoxy resin LY556 (diglycidyl ether of bisphenol A) and hardener HY-951 were used as the matrix components, and they were purchased from Mohapatra Scientific Supplier, Bhubaneswar, India. The raw leaf was used as the fiber and was collected locally from Baramba village near Bhubaneswar, India. The palm leaf stem stalk fillers were collected locally. Fiber treated with 5% alkali for 6 h shows better mechanical property and has higher cellulose content [20]. Therefore, the fibers treated with 5% NaOH for 6 h were selected as a reinforcement material for composites. The treated fibers were cut into suitable dimensions (200 × 5 × 0.4 mm) and made into a fiber mat.

2.2 *Composite Fabrication*

Palm-epoxy hybrid composites were fabricated by incorporating 0–15 wt% of RH filler. The hand-lay-up technique was implemented for the fabrication of the composite. The detailed designation and composition of the composite are shown in Table 1. Initially, RH and epoxy resin were stirred well for 10 min for homogeneity. The mixture was blended with a hardener in a ratio of (10:1) by weight. The casting of each hybrid composite was cured under a load of 30 kgf for 1 day at room temperature.

Table 1 Designation of composite samples

Designation	Composition		
	Epoxy (wt%)	Palm fiber (wt%)	Filler content (wt%)
EPP-0	70	30	0
EPP-5	65	30	5
EPP-10	60	30	10
EPP-15	55	30	15

2.3 Dynamic Mechanical Analysis (DMA)

The DMA of palm-epoxy composite with and without filler content was performed in a three-point bending mode on Tritec 2000 DMA device (Triton Technology Ltd., Nottinghamshire, UK) per the ASTM D256-08 (2008) standard. The test was carried out at the atmospheric condition in a temperature range of 25–150 °C with a heating rate of 10 °C per minute and a fixed frequency of 1 Hz. The storage modulus (*E*/), loss modulus (*E*//), and tan δ of the specimens were observed during the run.

2.4 Thermogravimetric Analysis (TGA)

The thermal behavior of palm stalk fiber and the RH reinforced palm-epoxy composites were studied by TGA following ASTM E1131-08 (2008) standard. The test was performed using STA7200 supplied by High-tech-Science Corporation (Hitachi, High-Tech Science Corp., Japan). TGA was conducted in a nitrogen atmosphere in the temperature range of 30–550 °C and a heating rate of 10 °C per minute.

2.5 Corrosion Testing

Composite specimens were prepared according to ASTM B117-03 (2003) with dimensions of 12 × 10 × 3 mm. Initially, the edges of the composite specimens were sealed with epoxy matrix, and their initial weight was measured in dry conditions. The test is then conducted by putting the samples in 10 wt% NaCl solution for 24, 48, and 72 h at a temperature of 45 °C. Each specimen was rinsed with distilled water, dried in an oven for 15 h, and final weights were taken.

3 Results and Discussion

Natural fibers are nowadays considered as an environmentally friendly material in developing thermo-mechanically stable materials due to their renewable nature. They are non-toxic and non-hazardous to the environment. Moreover, natural fiber has low cost, low density, good specific modulus, easy processing, and high aspect ratio for efficient stress transfer over synthetic fiber. However, the application of natural fiber has been restricted due to its hydrophilic nature and lower thermal resistance properties. These limitations can be resolved by employing several chemical treatments of the fiber. Chemically treated plant fiber-reinforced composites show improved mechanical and thermal resistance properties.

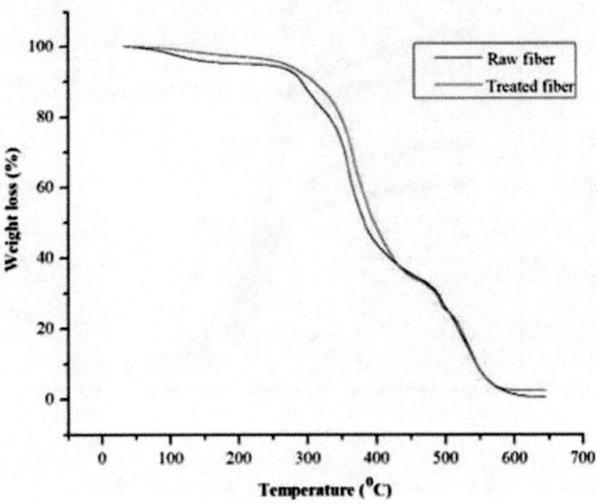

Fig. 1 TGA curve of treated and raw palm stalk fiber

3.1 Thermal Analysis

3.1.1 Thermogravimetric Analysis (TGA) of Fiber

The thermal resistance of raw and 5% NaOH palm fiber treated for 6 h is shown in Fig. 1. Thermal decomposition occurs in three steps. The initial weight loss was observed between 100 and 150 °C indicating the presence of moisture content. A thermal peak at 250–300 °C indicates the degradation of hemicelluloses, and at 340–360 °C, the thermal degradation of cellulose occurs. The thermal degradation occurred slightly earlier in raw fiber than the treated one, which confirms that the thermal resistance of treated fiber improved with alkali treatment. This behavior of the fiber is attributed to the presence of higher cellulose content in treated fiber [21]. Similar results were found for agave and borassus fruit fiber [21, 22].

3.1.2 Dynamic Mechanical Analysis (DMA)

The DMA is generally used to measure the damping capacity of composite material. When a dynamic load is applied to a composite specimen, it absorbs and releases some energy. The energy-absorbing capacity refers to the stiffness and energy dissipation capability refers to the damping capacity of the composite material. DMA measures the stiffness and damping factor by storage modulus (*E*/), loss modulus (*E*//), and tan δ.

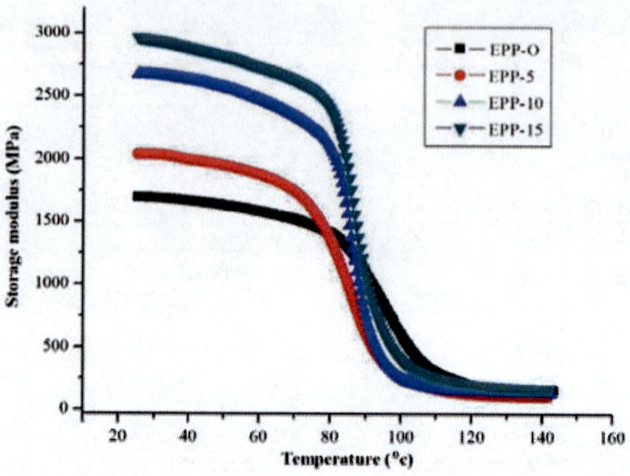

Fig. 2 Storage modulus of raw and RH filled palm-epoxy composite

3.1.3 Storage Modulus (E/)

The variation in storage modulus of palm-epoxy composite with filler addition is shown in Fig. 2. The storage modulus value gradually increased with filler addition. The unfilled palm-epoxy had a storage modulus of 1694 MPa in the glassy region, and this value increased to 2045 MPa, 2690 MPa, and 2962 MPa, respectively, for 5, 10, and 15 wt%. This result indicated that the stiffness of the composite gradually increased to a maximum for the 15 wt% filled composites. Filler addition reduces the molecular mobility and deformation of the matrix resin. Similar results were reported by Patnaik and Nayak [16] for alumina filled jute-epoxy composite and Adak et al. [23] for grapheme oxide filled carbon-epoxy composites. The storage modulus value gradually decreased in the glassy transition region (60–90 °C) with an increase in temperature. This result indicated that the molecular mobility of composite specimens increased. The molecular mobility of the RH filled composites was less than the un-filled one. In the rubbery region (100–140 °C), the value of E/was the maximum for 15 wt% filled composites followed by other composites that have a very close value of *E/*. The improved storage modulus confirmed the superior thermo-mechanical stability of the developed composites. Similar results are reported for kenaf-epoxy composites filled with magnesium hydroxide [24].

3.1.4 Loss Modulus (E//)

Loss modulus represents the viscous response of the material. The variation of the loss modulus of the palm-epoxy composite with and without filler content as a function of temperature is shown in Fig. 3. It is the measure of released heat energy per cycle in a cyclic loading condition for a viscoelastic material. The value of *E//* initially increased, attained a maximum value, and then decreased in the glass-transition

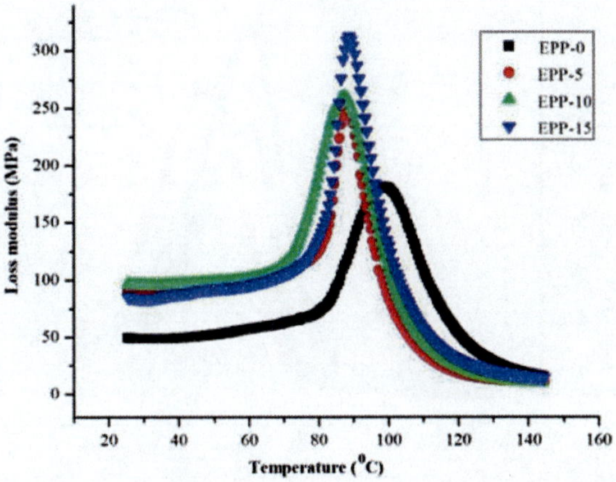

Fig. 3 Loss modulus of palm-epoxy composites with different RH powder

temperature range for all samples. At higher temperature, overlapping of curves indicates that inclusion of RH filler had no significant effect on the viscose dissipation. The inclusion of RH improved the loss modulus of EPP-0 composite from 49.56 MPa to 85.61 MPa, 90.7 MPa, and 96.26 MPa for 5 wt%, 10 wt%, and 15 wt%, respectively. This behavior was attributed to the restriction of polymer chain mobility due to the presence of RH filler. Moreover, RH addition increased the viscose dissipation of palm-epoxy composites. The present results are consistent with previous research on alumina filled jute-epoxy composite [16] and graphene oxide filled carbon-epoxy composite [23].

3.1.5 Tan Delta

Tan δ is expressed as a dimensionless number and represents the mechanical damping factor. It depends on fiber-matrix adhesion. Hence, a weak fiber-matrix adhesion will have higher values of tan δ [25]. Low tan δ also indicates that the composite has good load-bearing capacity [13]. The tan δ values of EPP-0, EPP-5, EPP-10, and EPP-15 composites are shown in Fig. 4. The tan δ was least in the unfilled palm-epoxy composite, and maximum for the 15 wt% RH filler reinforced composite. The peak value of tan δ occurred in the glass transition region. As fiber-matrix interface bonding increases, the mobility of molecular chains decreases, and hence, there is reduction in damping factor. Similar results were reported for jute-epoxy composite filled with granite stone powder [15].

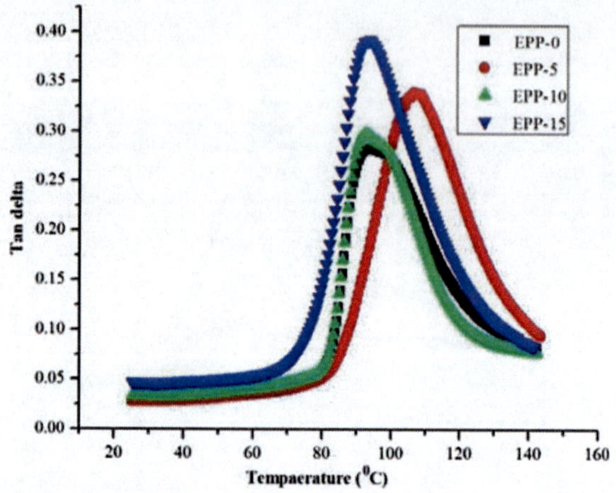

Fig. 4 Tan δ value of palm fiber-reinforced epoxy composites filled with RH filler

3.1.6 Cole–Cole Plot

It is an important tool to observe the relationship between stored and loss modulus for a viscoelastic material [21]. Further, it is generally used to study the change in composite structure at the molecular level after adding micro/nano-sized reinforcements. Cole–Cole Plot is drawn with loss modulus versus storage modulus as shown in Fig. 5. It shows the influence of incorporating RH filler on the Cole–Cole Plot of palm-epoxy composites. The homogeneous or heterogeneous nature of the composite sample could be identified from the shape of the plot [26]. A perfect semicircular arc

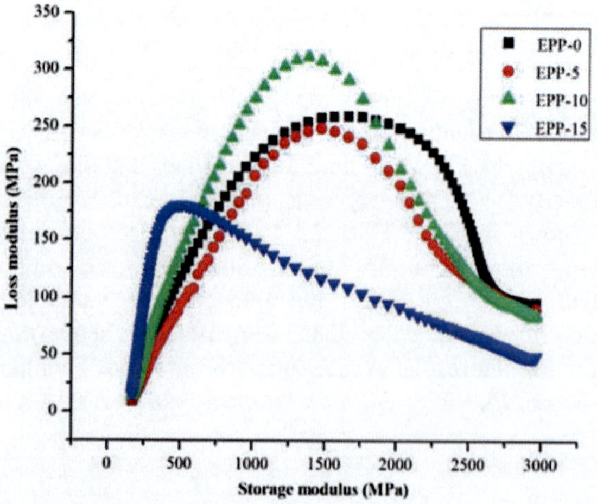

Fig. 5 Cole–Cole Plot for RH filled and unfilled palm-epoxy composites

in the plot shows the homogeneous nature of the system, but an irregular/imperfect semicircular curve represents the heterogeneity of the system. Figure 5 shows that inclusion of RH filler has increased the irregularity of the composite. The unfilled sample (EPP-0) has a less irregular shape compared with other samples as expected. This behavior is attributed to the strong adhesion between the fiber and matrix. A similar trend was found for granite filled jute-epoxy composite [15].

3.1.7 Thermogravimetric Analysis of RH Filled Composites

Thermogravimetric analysis was conducted for palm-epoxy composites filled with RH filler. Figure 6 depicts the effect of RH addition on the thermal stability of palm-epoxy composites. The initial degradation of all the samples occurs at 100–150 °C due to the vaporization of moisture content through dehydration of secondary alcoholic groups. This dehydration forms unsaturated structure that leads to weak C-O and C–N bonds [26]. The second stage of degradation was observed above 300 °C due to the decomposition of epoxy and amine hardener. The thermal degradation of palm fiber occurred in the following temperature ranges: hemicelluloses (150–350 °C), cellulose (275–500 °C), and lignin (250–500 °C). Moreover, no significant difference was observed in the thermal degradation of RH filled and unfilled samples beyond 550 °C. The weight loss of filled samples occurred slightly earlier than unfilled ones in the temperature range of 300–400 °C. This behavior was attributed to the breaking of C–O and C–N bonds, which has accelerated the volatilization. Similar observations have been made for jute and kenaf-epoxy composites filled with granite and magnesium powder, respectively [15, 24].

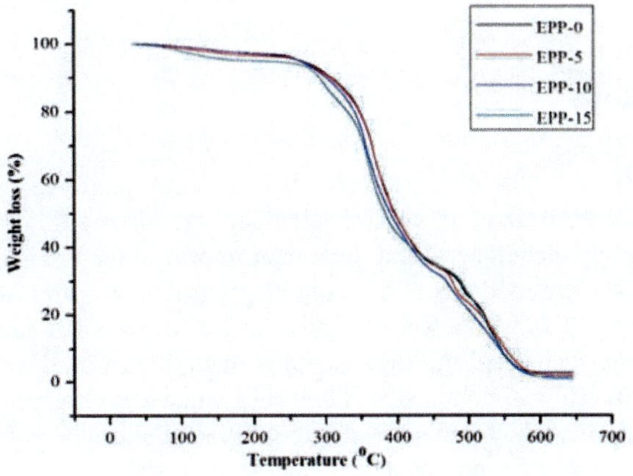

Fig. 6 Thermogravimetric analysis of RH filled and unfilled palm-epoxy composites

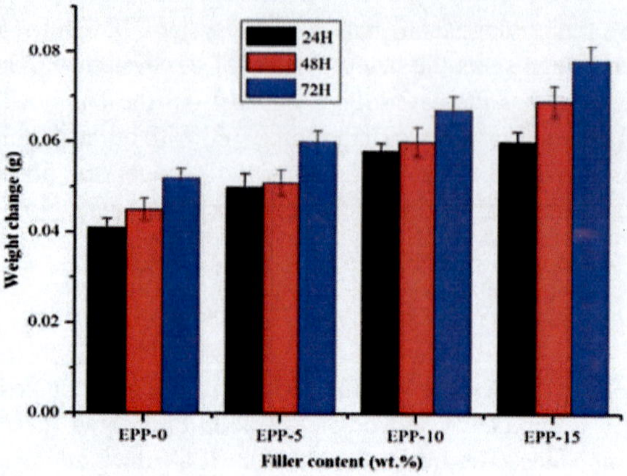

Fig. 7 Corrosive properties of unfilled and RH filled composites

3.1.8 Corrosive Properties

The corrosive tendencies of RH filled palm-epoxy composites are shown in Fig. 7. The weight gain of samples was small even after long hours of sodium chloride (NaCl) treatment. Furthermore, weight gain increased with increasing filler content in the composite samples. Strong interfacial adhesion between the fiber and matrix in the unfilled samples reduced the penetration of NaCl. However, filler reinforcement in the composite increased the void content, which led to the diffusion of NaCl into the matrix. Hence, the weight gain was observed for filled samples. The present result is consistent with previous reports for jute-epoxy hybrid composites [27].

4 Conclusions

A new class of low-cost thermally sable epoxy composites reinforced with palm fiber and filled with 0, 5, 10, and 15 wt% of RH filler was successfully fabricated by a simple hand-lay-up technique. Alkali treatment improved the thermal resistance of raw fiber. The storage modulus of the composite increased with filler addition. A maximum value of 2962 MPa was observed for the 15 wt% RH filled composite. RH filler addition enhanced the loss modulus of unfilled samples from 49.56 to 96.26 MPa for the EPP-15 composite. A low tan δ value was observed for the EPP-0 (unfilled) composite, which indicates strong interfacial adhesion between the fiber and matrix. Filler addition increased the heterogeneity of the composite as observed from the Cole–Cole Plot. The thermogravimetric study of the composite indicated no significant difference in the thermal degradation of RH filled and unfilled samples beyond 550 °C.

References

1. Rout J, Mishra M, Tripathy SS, Nayak SK, Mohanty AK (2001) The influence of fibre treatment on the performance of coir-polyester composites. Compos Sci Technol 61(9):1303–1310
2. Sanjay MR, Yogesha B (2017) Studies on natural/glass fiber reinforced polymer hybrid composites: an evolution. Mater Today Proc 4:2739–2747
3. Asumani OML, Reid RG, Paskaramoorthy R (2012) The effects of alkali-silane treatment on the tensile and flexural properties of short fiber non-woven kenaf reinforced polypropylene composites. Compos Appl Sci Manuf 43:1431–1440
4. Fiore V, Bella GD, Valenza A (2015) The effect of alkaline treatment on mechanical properties of kenaf fibers and their epoxy composites. Compos B Eng 68:14–21
5. Zafar MT, Maiti SN, Ghosh AK (2016) Effects of surface treatments of jute fibers on the microstructural and mechanical responses of poly(lactic acid)/jute fiber bio-composites. RSC Adv 6:73373–73382
6. Araujo RS, Marques MPV, Oliveria PF, Rezende CC (2018) Bio-composites based on poly(lactic acid) containing mallow and eucalyptus surface modified natural fibers. J Polym Environ 26(9):3785–3791
7. Md Shah AU, Sultan MTH, Cardona F, Jawaid M, Talib ARA, Yidris N (2017) Thermal analysis of bamboo fiber and its composites. BioResources 12(2):2394–2406
8. Nayak SK, Mohanty S (2010) Sisal glass fiber reinforced PP hybrid composites: effect of MAPP on the dynamic mechanical and thermal properties. J Reinf Plast Compos 29:1551–1568
9. Samal SK, Mohanty S, Nayak SK (2009) Polypropylene-bamboo/glass fiber hybrid composites: fabrication and analysis of mechanical, morphological, thermal, and dynamic mechanical behavior. J Reinf Plast Compos 28:2729–2747
10. Ray D, Sarkar BK, Das S, Rana AK (2002) Dynamic mechanical and thermal analysis of vinyl ester-resin-matrix composites reinforced with untreated and alkali-treated jute fibers. Compos Sci Technol 62:911–917
11. Jabbar A, Militky J, Wiener J, Karahan M (2015) Static and dynamic mechanical properties of novel treated jute/green epoxy composites. Text Res J 86(9):960–974
12. Atiqah A, Jawaid M, Sapuan SM, Ishak MR, Alothman OY (2018) Thermal properties of sugar palm/glass fiber reinforced thermoplastic polyurethane hybrid composites. Compos Struct 202:954–958
13. Jawaid M, Khalil HA, Hassan A, Dungani R, Hadiyane A (2013) Effect of jute fibre loading on tensile and dynamic mechanical properties of oil palm epoxy composites. Compos B Eng 45(1):619–624
14. Senturk O, Senturk AE, Palabiyik M (2018) Evaluation of hybrid effect on the thermomechanical and mechanical properties of calcite/SGF/PP hybrid composites. Compos B Eng 140:68–77
15. Pawar MJ, Patnaik A, Nagar R (2015) Investigation on mechanical and thermo-mechanical properties of granite powder reinforced epoxy composites. Polym Compos 38(4):736–748
16. Patnaik TP, Nayak SS (2016) Investigation on physico-mechanical and thermo-mechanical analysis of alumina filled needle-punch nonwoven jute epoxy composites. Polym Compos 39:1553–1561
17. Satapathy S, Kothapalli RVS (2017) Mechanical dynamic mechanical and thermal properties of banana fiber/recycled high-density polyethylene biocomposite filled with flyashcenospheres. J Polym Environ 26(1):200–213
18. Prasob A, Sasikumar M (2018) Static and dynamic behavior of jute/epoxy composites with ZnO and TiO_2 fillers at different temperature conditions. Polym Testing 69:52–62
19. Karaduman Y (2018) Experimental investigation on flexural and dynamic mechanical properties of jute fiber/cork-reinforced polyester sandwich composites. Bio Resourc 13(2):3958–3970
20. Rout AK, Kar J, Jesthi DK (2016) Effect of surface treatment on the physical, chemical, and mechanical properties of palm tree leaf stalk fibers. BioResources 11(2):4432–4445
21. Reddy KO, Maheswari CU, Shukla M, Song JI, Rajulu AV (2013) Tensile and structural characterization of alkali treated Borassus fruit fine fibers. Compos B Eng 44(1):433–438

22. Thamae T, Baillie C (2007) Influence of fibre extraction method, alkali and silane treatment on the interface of Agave americana waste HDPE composites as possible roof ceilings in Lesotho. Compos Interfaces 14(7–9):821–836
23. Adak NC, Chhetri S, Kim NH, Murmu NC, Samanta P, Kuila T (2018) Static and dynamic mechanical properties of graphene oxide-incorporated woven carbon fiber/epoxy composite. J Mater Eng Perform 27(3):1138–1147
24. Saba N, Alothman OY, Almutairi Z, Jawaid M (2019) Magnesium hydroxide reinforced kenaf fibers/epoxy hybrid composites: mechanical and thermomechanical properties. Constr Build Mater 201:138–148
25. Martínez-Hernández AL, Velasco-Santos C, De-Icaza M, Castano VM (2007) Dynamical–mechanical and thermal analysis of polymeric composites reinforced with keratin biofibers from chicken feathers. Compos B Eng 38(3):405–410
26. Jesuarockiam N, Jawaid M, Zainudin ES, Thariq Hameed Sultan M, Yahaya R (2019) Enhanced thermal and dynamic mechanical properties of synthetic/natural hybrid composites with graphene nanoplateletes. Polymers 11(7):1085
27. Khalil A, Jawaid HPSA, Abu Bakar M (2011) Woven hybrid composites: water absorption and thickness swelling behaviors. BioResources 6(2):1043–1052

A Review on Materials Used for Combustion in Porous Radiant Burners

Gyan Sagar Sinha, Lav K. Kaushik, and P. Muthukumar

Abstract This paper reviews different materials that are used in porous media combustion. Porous medium combustion is a novel concept which not only increases the thermal efficiency while it also reduces the pollutant emissions. Different thermal and physical properties of porous materials have been discussed in this paper. Different types of ceramic that dominate porous media combustion have been highlighted, which can guide researchers during material selection.

Keywords Thermal efficiency · Thermal and physical properties · NO_x and CO · Porous media combustion

1 Introduction

Combustion has an important role in our daily life that includes transportations and cooking. Earlier cooking was done through biomass (wood, cow dung, etc.). The thermal efficiency of the biomass stove is very low while toxic emissions are very high as compared to kerosene and LPG stoves [1]. The reason behind this is incomplete combustion. LPG combustion is cleaner but due to its low distribution and high cost, it is not affordable for rural peoples as well as low-income homes in urban areas [2]. So people have a choice to use kerosene in which combustion is cleaner than biomass combustion and government of India also giving subsidy to kerosene. So it is mandatory to develop and design a stove related to kerosene as well as LPG to enhance the performance of the same. Porous media combustion is a promising idea that not only increases thermal efficiency while it decreases the pollutant emissions (NO_x and CO). Porous media combustion is a combustion that takes place in a highly emissive and conductive solid porous matrix. Combustion in a solid porous matrix

G. S. Sinha
Kalinga Institute of Industrial Technology, Bhubaneswar, Odisha 751024, India

L. K. Kaushik (✉) · P. Muthukumar
Indian Institute of Technology Guwahati, Guwahati, Assam, India
e-mail: love.kaushik90@gmail.com

S. Revankar et al. (eds.), *Proceedings of International Conference on Thermofluids*, Lecture Notes in Mechanical Engineering,
https://doi.org/10.1007/978-981-15-7831-1_59

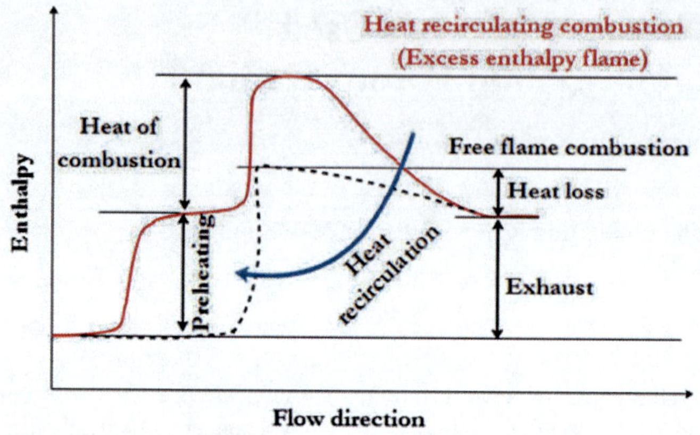

Fig. 1 Concept of heat recirculation [4]

which not only dominants the conduction and radiation heat transfer while it also increases the convection heat transfer due to the increased surface area [3]. Convolute path of porous material helps in mixing the fuel and air properly which leads to a homogeneous mixture of air and fuel. A homogeneous mixture of air-fuel leads to efficient combustion which ultimately results in low CO emission [4]. Temperature distribution in porous burner approximately remains uniform that leads to low NO_x emission [4]. The concept of recirculating combustion heat or excess enthalpy flame prevails in PMC [4], which means that a part of enthalpy of the combustion gases is transferred back to the upstream, into the fresh incoming mixture, via radiation and conduction heat transfer (Fig. 1). Whereas, in the traditional burners, combustion happens in an open environment (gaseous environment). The temperature gradient is more in traditional combustion, which leads to more thermal NO_x formation [4].

Different types of porous materials (or Porous Burner) are available, but their use is based on different operating conditions and applications. The motivation behind this is to use porous material in different types of combustion to enhance combustion stability. In PMC, stabilization of flame at the desired location is of prime concern. Un-stable flame may lead to the adverse conditions of lift-off or flashback, which ultimately extinguish the combustion process and shut down the burner. "Flashback" prevails when the incoming mixture velocity is not sufficient to resist the burning velocity and allows the flame to propagate upstream into the incoming reactants. Similarly, when the mixture velocity is higher than the burning velocity, the flame travels downstream (out of the burner surface), and this phenomenon is termed "Blow-off." Summarizing the development of porous media combustion, it can be said that a variety of materials ranging from metal to ceramic are uncovered. Various burners have been developed for several applications with different materials. The performance of the porous burner is highly dependent on the selection of materials. Therefore, this paper has done a thorough review of different types of porous material and their uses.

2 Porous Burner

2.1 Single and Double Layer

A Porous Burner (PB) can be made of a single material having high porosity (void fraction), or it can be of two different materials having different porosities and thermo-physical properties. The former is known as single-layered and the latter as a double-layered burner. A double-layered burner consists of two distinct zones: preheating zone and combustion zone, and the interface of the two zones serves as a flame holder [6–8]. Figure 2a and b shows the schematic of a single and a double-layered burner. In recent times, the use of a double-layered burner has gained more popularity due to its characteristics of better flame stability compared to single-layered burner attributing to different material properties preventing flash-back [8–10]. Malico and Pereira [10] through their numerical studies, showed the importance of thermal radiation and radiative properties such as extinction coefficient, scattering albedo, optical thickness, etc. on the flame profile.

Research has revealed that the preheating zone of the burner should have shorter length, lower porosity, lower thermal conductivity, lower volumetric heat transfer coefficient, larger albedo, shorter optical path length, higher optical thickness, and higher radiative extinction coefficient to prevent possibility of ignition and flame propagation thereby occurrence of flashback [7, 11, 12]. While combustion zone supporting flame propagation and heat recirculation to the preheating zone should have high conductivity, high volumetric heat transfer coefficient, intermediate extinction coefficient and non-scattering. Porosity decides the optical and thermal properties such as optical thickness, extinction coefficient, heat transfer coefficient, thermal conductivity, etc. The porous bed can be of the same material with different pore sizes [12] or can be of two different materials [3, 14, 15].

Babkin et al. [17] gave the stabilization criteria through the Peclet number. They showed that the stability of the hydrocarbon flame in a double-layered burner as a function of Peclet number (Pe: the ratio of heat flow by convection transport to heat flow by conduction), based on the mean pore diameter. They proposed the following

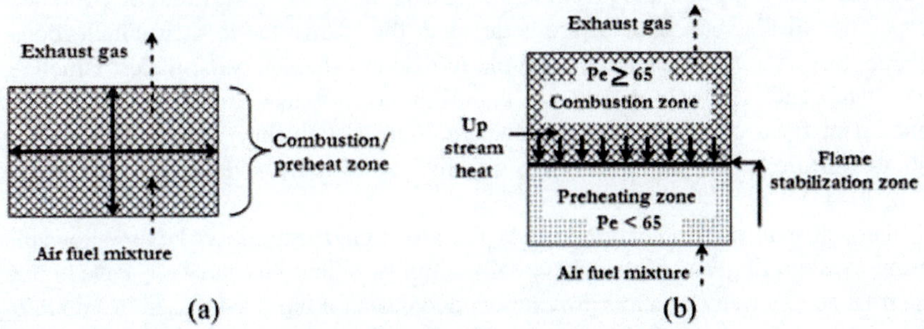

Fig. 2 Schematic of **a** single-layered, **b** double-layered PB [16]

limiting condition for the flame to propagate.

$$\mathrm{Pe} \geq 65$$

Any burner having Pe greater than 65 or equal would allow the flame to propagate and thus, the combustion zone of the burner should be constructed in such a way that the Pe in this region is greater than 65 and the preheating zone is below 65 [16]. For Hydrogen flame (H_2); however, the limit is found to be 37 by Dillon [18]. A careful selection of the pore diameter and thermal conductivities of the two zones allows the flame to stabilize at the interface of the media, and upstream zone (preheat zone) to receive the heat. The size of the pores in this region should be small enough to quench the flame and eliminate flashback while the pores in the downstream section should be large enough to allow significant thermal feedback to the reactants.

3 Materials and Properties of Porous Media

Porous media with certain specific characteristics such as high thermal and chemical stress resistance, large inner surface area, good heat transport properties, high heat capacity, etc. are found suitable for combustion applications, and a range of ceramics such as Aluminum oxide (Al_2O_3), Silicon carbide (SiC), Zirconium dioxide/Zirconia (ZrO_2), Mullite, Cordierite, metal, etc. [19–23] meets the requirement. Metallic materials such as stainless steel, Fe–Cr alloy (nickel–iron chromium), etc. are also found usage [24–28]. However, due to their inadequate thermal stability at high temperatures, and high thermal inertia, ceramics are largely preferred [28]. The notable benefits of ceramics are high melting temperature, high oxidation resistance, high permeability and superior creep resistance [30]. Table 1 enlists a few important thermal and radiative properties of the commonly used ceramics.

Ceramics are available in different forms: reticulated foams, honeycomb, lamellae, pebbles, balls, etc. and due to certain advantages, such as better gas mixing, higher porosity, lesser pressure drop, etc., foams are found to be largely employed [30]. However, they are prone to have larger damage to exposure to high-temperature environments due to their rigid structures. On the other hand, the durability and heat capacities of discrete particles are better than the foams due to their small, robust shape [30]. Al_2O_3 is generally available in discrete shapes; balls, flakes, lamellae, etc., while SiC, ZrO_2 Cordierite, etc. comes in the form of foam. Honeycomb slabs and foam form of Al_2O_3 are not very common though they were used in a few studies, while balls (diameters ranging from 3 to 20 mm) have found wide usages [28, 30–34].

The most attractive features of Al_2O_3 are higher durability, larger allowable temperature rating, and higher chemically stability. While SiC and ZrO_2 have higher thermal conductivity, and this provides a good basis for them to be used in a double-layered burner with Al_2O_3 as one layer. Recently several successful demonstrations are reported where Al_2O_3 is used in upstream and SiC in the downstream to have better

Table 1 Important material properties

Material		A	B	C	D	E	F	G	H	I
Maximum allowable temperature in air (°C)		1600	1900	1800	1400–1750	1380	1700	–	–	–
Thermal conductivity (W/m K)		20–50	5–6	2–4	40–120	110–160	80–145	2.6	0.37	–
		at (1000 °C)			at (20–100 °C)					
Total emissivity at 2000 K		0.9	0.28	0.31	–	–	–	–	–	–
Thermal expansion coefficient from 20 to 1000 °C (10^{-6} K^{-1})		4–5	8	10–13	–	–	–	–	–	–
Heat capacity (25–800 °C) (kJ/kg K)		–	1.05	0.59	0.84	0.84	0.84	0.80	–	–
Optical thickness	10 ppi 65 ppi			28.2 300						
Scattering albedo	10 ppi	0.55–0.88		0.87–0.99		0.7		0.76	0.84	
Extinction coefficient (m^{-1})	10 ppi 20 ppi 25 ppi 30 ppi	234.1 468.1 585.2 702.2	210 300	234.1	100	178 244		270	202	34.1

A SiC: Silicon carbide
B Al_2O_3: Aluminum oxide
C ZrO_2: Zirconium dioxide/Zirconia
D SSiC: sintered silicon carbide
E SiSiC: silicon infiltrated silicon carbide
F HPSiC: hot pressed infiltrated silicon carbide
G Cordierite
H Mullite
I FeCrAl: Fe Cr alloy (nickel–iron chromium)

flame stability due to the differences in conductivity values [15, 35, 36]. However, porosity has a larger impact on the effective thermal conductivity (accounting the combined effect of conduction and radiation) [37, 38] and heat transfer coefficient [38, 39]. At a temperature of about 1000 °C, Hsu et al. [37] reported the thermal conductivity of a 65 ppi PSZ to be 1.6 W/m K, against 0.5 W/m K for a 10 ppi PSZ (partially stabilized zirconia). A few researchers have predicted the effective thermal conductivities of other materials like Al_2O_3 and reticulated vitreous carbon etc. by developing conduction models, however, the findings are still scarce considering different materials having a wide range of porosities [40, 41]. Further, forms of material also affect the conductivity and thereby affecting heat transfer and stability.

SiC has been found to have wide usage in the double-layered burner [42–45] followed by ZrO_2 either in pure or stabilized by other materials such as magnesia or yttria [24, 28, 46–49]. Avdic et al. [50] used ZrO_2 foam in the preheating zone and Al_2O_3 lamellae in the combustion zone while in some studies, ZrO_2 or SiC are used alone [51]. The use of Mullite and cordierite are very few [3, 27, 52–54]. Cordierite exhibits a higher extinction coefficient (Table 1) than SiC or Al_2O_3 having the same pore density (expressed in terms of ppi: pores per inch). Hendricks and Howell [55] measured the transmittance and reflectance of PSZ and SiC having pore sizes of 10, 20, 66 ppi and found that SiC retained its properties at all wavelengths; while PSZ showed a significant change, in the wavelength range 2500–3000 nm.

It can be stated that Al_2O_3, SiC and ZrO_2 are largely used due to their easy availability and low cost. Further, the combinations of Al_2O_3–SiC or Al_2O_3–ZrO_2 are more suitable for constructing a double layer PRB as they offer better flame stability due to the differences in thermal conductivities. Reticulated foams, honeycomb and discrete shapes (balls) are the most common forms of the ceramic matrix used so far, and it has been reported that the heat transfer mechanism varies with the form of the matrix. However, except a few earlier reports [38, 39, 41], in recent times, no study has been reported on the variations of thermal conductivity or heat transfer coefficient as a function of porosity. Prediction of accurate thermal behavior of the burner, however, demands a good knowledge of the thermo-physical properties. Not enough experimental studies are available on the influence of these thermo-physical properties on the burner behavior.

4 Conclusions

Summarizing the development of materials for PMC (porous media combustion), it can be said that a variety of materials ranging from metal to ceramics are under consideration. Metal has lower temperature resistance while ceramics have received more attention due to its higher resistance to thermal and chemical shock. The SiC and Al_2O_3 are found to be largely used material due to their better thermal and radiative properties and easy availability. Double-layered burners are found to be constructed with Al_2O_3 and SiC as a combination. Al_2O_3 is reported to be used mostly in the form of balls while other ceramics such as SiC, ZrO_2 as a reticulated structure. Discrete particles (like balls, wires) are found to provide better stability at high temperatures while reticulated ceramics provide good mixing of fuel and air. Long term testing is needed to establish the structural stability of the porous material operated at high temperature and also to know their abilities to sustain heat cycle (cooling and heating) before disintegration. For economic consideration, these findings would be important.

References

1. Dendukuri G, Mittal JP (1993) Some field experiences with improved Chulhas in introduced in rural households of Andhra Pradesh. India. Energy Convers Manag 34:457–464
2. Dixit CSB, Paul PJ, Mukunda HS (2006) Experimental studies on a pulverized fuel stove. Biomass Bioenergy 30:673–683
3. Mital R, Gore JP, Viskanta R (1997) A study of the structure of submerged reaction zone in porous ceramic radiant burners. Combustion Flame 111:175–184
4. Weinberg FJ (1971) Combustion temperatures: the future. Nature 233:239–241
5. Howell J, Hall M, Ellzey J (1996) Combustion of hydrocarbon fuels within porous inert media. Prog Energy Combust Sci 22:121–145
6. Viskanta R (1995) Interaction of combustion and heat transfer in porous inert media. In: 8th international symposium on transport phenomena in combustion, San Francisco
7. Barra AJ, Ellzey JL (2004) Heat recirculation and heat transfer in porous burners. Combust Flame 137:230–241
8. Trimis D, Durst F (1996) Combustion in a porous medium-advances and applications. Combust Sci Technol 121:153–168
9. Hsu PF, Evan WD, Howell JR (1993) Experimental and numerical study of premixed combustion within nonhomogeneous porous ceramics. Combust Sci Technol 90:149–172
10. Malico I, Pereira JCF (2001) Numerical study on the influence of radiative properties in porous media combustion. J Heat Transf 123:951–957
11. Kulkarni M, Peck R (1996) Analysis of a bilayered porous radiant burner. Int J Comput Methodol 30:219–232
12. Barra AJ, Diepvens G, Ellzey JL, Henneke MR (2003) Numerical study of the effects of material properties on flame stabilization in a porous burner. Combust Flame 134:369–379
13. Bubnovich V, Zhdanok S, Dobrego K (2006) Analytical study of the combustion waves propagation under filtration of methane-air mixture in a packed bed. Int J Heat Mass Transf 49:2578–2586
14. Pantangi V, Kumar AK, Mishra SC, Sahoo N (2007) Performance analysis of domestic LPG cooking stoves with porous media. Int Energy J 8:139–144
15. Mishra NK, Mishra SC, Muthukumar P (2015) Performance characterization of a medium-scale liquefied petroleum gas cooking stove with a two-layer porous radiant burner. Appl Therm Eng 89:44–50
16. Sinha GS (2018) Development and performance analysis of self-aspirated porous radiant burners for kerosene pressure stove. Ph.D. thesis, IIT Guwahati, Guwahati, India
17. Babkin V, Korzhavin AA, Bunev VA (1991) Propagation of premixed gaseous explosion flames in porous media. Combust Flame 87:182–190
18. Dillon J (1999) Combustion in porous media. Los Alamos National Lab, Final report, California Institute of Technology
19. Weclas M (2005) Porous media in internal combustion engines, cellular ceramics: structure, manufacturing. Propert Appl, 580–595
20. Mößbauer S, Pickenäcker O, Pickenäcker K, Trimis D (2002) Application of the porous burner technology in energy and heat-engineering. Clean Air 3:185–198

21. Pickenäcker O, Pickenäcker K, Wawrzinek K, Trimis D, Pritzkow WEC, Müller C, Goedtke P, Papenburg U, Adler J, Standke G, Heymer H, Tauscher W, Jansen F (1999) Innovative ceramic materials for porous-medium Burners II. Interceram 48:424–433
22. Cookson EJ, Floyd I, Donald E (2007) Application of reticulated metal foam to gas fired infrared burners. High Temp Mater Processes (Lond) 26:269–274
23. Tong TW, Li W (1995) Enhancement of thermal emission from porous radiant burners. J Quant Spectrosc Radiat Transf 53:235–248
24. Dhamrat RS, Ellzey JL (2006) Numerical and experimental study of the conversion of methane to hydrogen in a porous media reactor. Combust Flame 144:698–709
25. Bakry A (2008) Stabilized premixed combustion within atmospheric gas porous inert medium (PIM) burner. Proc Inst Mech Eng Part J Power Energy 222:781–789
26. Krittacom B, Kamiuto K (2009) Radiation emission characteristics of an open-cellular porous burner. J Therm Sci Technol 4:13–24
27. Yu B, Kum SM, Lee CE, Lee S (2013) Combustion characteristics and thermal efficiency for premixed porous-media types of burners. Energy 53:343–350
28. Gao H, Qu Z, Feng X, Tao W (2014) Methane/air premixed combustion in a two-layer porous burner with different foam materials. Fuel 115:154–161
29. Tiwari A, Alenezi MR, Jun SC (2016) Advanced composite materials. Wiley
30. Gao HB, Qu ZG, He Y, Tao WQ (2012) Experimental study of combustion in a double-layer burner packed with alumina pellets of different diameters. Appl Energy 100:295–302
31. Gao H, Qu Z, Tao W, He Y, Zhou J (2011) Experimental study of biogas combustion in a two-layer packed bed burner. Energy Fuels 25:2887–2895
32. Gao HB, Qu ZG, Tao WQ, He YL (2013) Experimental investigation of methane/(Ar, N_2, CO_2)–air mixture combustion in a two-layer packed bed burner. Exp Thermal Fluid Sci 44:599–606
33. Yoksenakul W, Jugjai S (2011) Design and development of a SPMB (self-aspirating, porous medium burner) with a submerged flame. Energy 36:3092–3100
34. Arrieta CE, Amell AA (2014) Combustion analysis of an equimolar mixture of methane and syngas in a surface-stabilized combustion burner for household appliances. Fuel 137:11–20
35. Mujeebu MA, Abdullah MZ, Bakar MZA, Mohamad AA (2011) A mesoscale premixed LPG burner with surface combustion in porous ceramic foam. Energy Sourc Part Recov Utilization Environ Effects 34:9–18
36. Wongwatcharaphon K, Tongtem P, Jugjai S (2013) Numerical and experimental study of a nonsprayed porous burner for liquid kerosene. J Porous Media 16:227–239
37. Hsu P, Howell JR (1992) Measurements of thermal conductivity and optical properties of porous partially stabilized zirconia. Exp Heat Transf Int J 5:293–313
38. Fu X, Viskanta R, Gore JP (1998) Prediction of effective thermal conductivity of cellular ceramics. Int Commun Heat Mass Transfer 25:151–160
39. Younis L, Viskanta R (1993) Experimental determination of the volumetric heat transfer coefficient between stream of air and ceramic foam. Int J Heat Mass Transf 36:1425–1434
40. Zumbrunnen DA, Viskanta R, Incropera FP (1986) Heat transfer through porous solids with complex internal geometries. Int J Heat Mass Transf 29:275–284
41. Singh R, Kasana HS (2004) Computational aspects of effective thermal conductivity of highly porous metal foams. Appl Therm Eng 24:1841–1849
42. Pantangi V, Mishra SC, Muthukumar P, Reddy R (2011) Studies on porous radiant burners for LPG (liquefied petroleum gas) cooking applications. Energy 36:6074–6080
43. Muthukumar P, Anand P, Sachdeva P (2011) Performance analysis of porous radiant burners used in LPG cooking stove. Int J Energy Environ 2:367–374
44. Keramiotis C, Stelzner B, Trimis D, Founti M (2012) Porous burners for low emission combustion: an experimental investigation. Energy 45:213–219
45. Herrera B, Cacua K, Villalba LO (2015) Combustion stability and thermal efficiency in a porous media burner for LPG cooking in the food industry using Al_2O_3 particles coming from grinding wastes. Appl Therm Eng 91:1127–1133
46. Mathis WM, Ellzey JL (2003) Flame stabilization, operating range, and emissions for a methane/air porous burner. Combust Sci Technol 175:825–839

47. Djordjevic N, Habisreuther P, Zarzalis N (2012) Experimental study on the basic phenomena of flame stabilization mechanism in a porous burner for premixed combustion application. Energy Fuels 26:6705–6719
48. Smucker MT, Ellzey JL (2004) Computational and experimental study of a two-section porous burner. Combust Sci Technol 176:1171–1189
49. Ellzey JL, Goel R (1995) Emissions of CO and NO from a two stage porous media burner. Combust Sci Technol 107:81–91
50. Avdic F (2004) Application of the porous medium gas combustion technique to household heating systems with additional energy sources. Ph.D. Thesis, University of Erlangen-Nuremberg
51. Kaplan M, Hall MJ (1995) The combustion of liquid fuels within a porous media radiant burner. Exp Thermal Fluid Sci 11:13–20
52. Takami H, Suzuki T, Itaya Y, Hasatani M (1998) Performance of flammability of kerosene and NOx emission in the porous burner. Fuel 77:165–171
53. Mital R, Gore JP, Viskanta R (1998) A radiation efficiency measurement procedure for gas-fired radiant burners. Exp Heat Transf Int J 11:3–21
54. Abdelaal M, El-Riedy M, El-Nahas A (2013) Effect of oxygen enriched air on porous radiant burner performance and NO emissions. Exp Thermal Fluid Sci 45:163–168
55. Hendricks T, Howell J (1996) Absorption/scattering coefficients and scattering phase functions in reticulated porous ceramics. J Heat Transfer 118:79–87
56. Hale MJ, Bohn MS (1993) Measurement of the radiative transport properties of reticulated alumina foams. In: ASME/ASES joint solar energy conference, New York, 1–9
57. Orenstein RM, Green DJ (1992) Thermal shock behavior of open-cell ceramic Foams. J Am Ceram Soc 75:1899–1905
58. Delalic N, Mulahasanovic D, Ganic EN (2004) Porous media compact heat exchanger unit-experimental analysis. Exp Thermal Fluid Sci 28:185–192

Thermal Analysis of Al 7075-T651 During High-Speed Machining

P. Chandrasekhar, Sudipta Chand, Radha Kanta Sarangi, Satya Praksah Kar, and Abhilas Swain

Abstract In the present study, finite element analysis during machining of Al7075-T651 was conducted and chip–tool interface temperature has been evaluated and validated with the available literature. The design of experiment model based on Taguchi's orthogonal array (L_{16}) was used to simulate the cutting experiments considering three factors: cutting speed, feed, and depth of cut with 4 levels. %. For optimizing the process and controlling the chip–tool interface temperature, cutting speed and feed need to be optimized appropriately. ANOVA test identified cutting speed and feed were the significant process parameters for the chip–tool interface temperature for which cutting speed contributes 19.44% and feed contributes 65%.

Keywords Chip–tool temperature · Finite element analysis · ABAQUS · High-speed machining

1 Introduction

Machining is a common material removal process where excess material is removed from workpiece in the form of chips using various cutting tools. Machining is essential for applications where parts have complex geometry or when dimensional accuracy is an important performance characteristic of the manufactured part. Even with the use of near-net-shape manufacturing processes, machining is preferred to achieve the final desired dimension. High-speed machining yields better part quality along with increased productivity. Major advantages of high-speed machining compared to conventional machining are reduced power consumption due to reduced cutting force, high material removable rate, and reduced machining time. Al7075 is replacing Al6061 in high-performance applications due to higher hardness values, tensile strength, high fatigue strength, and high stress-corrosion cracking resistance. Al7075-T651 is equally popular in adventure sports equipment, lightweight ammunition, and

P. Chandrasekhar (✉) · S. Chand · R. K. Sarangi · S. P. Kar · A. Swain
School of Mechanical Engineering, Kalinga Institute of Industrial Technology Deemed To Be University, Bhubaneswar, Odisha 751024, India
e-mail: csekharfme@kiit.ac.in

S. Revankar et al. (eds.), *Proceedings of International Conference on Thermofluids*, Lecture Notes in Mechanical Engineering,
https://doi.org/10.1007/978-981-15-7831-1_60

many more. In conventional machining process, cutting forces involved and temperature at the chip–tool interface are critical responses for high-speed machining. It is always desirable to have minimal cutting force and minimal chip–tool interface temperature.

An increase in chip–tool interface temperature leads to higher residual stress in the workpiece, higher wear, and tear and sticking of the workpiece to the chip. So, it is important to reduce this temperature. Chip–tool interface temperature depends on various parameters like cutting force, stress distribution, and plastic strain during machining. Finite element analysis can produce multiple outputs, such as cutting force, temperature distribution, stress distribution, plastic strain from a single model with reasonable accuracy which gives finite element analysis a competitive edge over experimental work. Daoud et al. [1] applied the inverse approach method based on response surface methodology for determining J-C constitutive parameters using orthogonal cutting experiments. Shet and Deng [2] studied the effect of rake angle and friction condition defined in finite element model on plastic strain distribution in orthogonal machining with fully coupled thermo-mechanical model developed in general-purpose finite element code, ABAQUS. Mohammadpour et al. [3] developed a 2D finite element model of high-speed machining of AISI1045 steel for studying the effect of machining parameters; cutting speed, feed and depth of cut, on the residual stress induced in machined parts due to machining. Tamizharasan and Senthil [4] modeled oblique turning of AISI1045 with uncoated carbide tool in Deform-3D software. During the literature review, most machinability studies of aluminum alloys used experimental analysis and focused on conventional machining. Most common commercial codes used for machining are Abaqus; [5], Deform; [6], AvantEdge [7], Forge; [8], MSC. Marc; [9]. Abaqus was selected for this study, as it is a general-purpose finite element code and provides greater flexibility and detailing while modeling the machining process. ABAQUS/Explicit 6.14 was used for 2D modeling of orthogonal machining of Al7075-T651 with uncoated carbide tool insert.

2 Model Geometry and Material Modeling

2.1 Model Geometry

A small section of workpiece with length 5 mm and height 0.5 mm, were 0.35 mm was base of workpiece and 0.15 mm was the uncut chip thickness; 45° inclination was provided to uncut chip thickness for reducing excessive mesh distortion. Section of uncoated carbide tool with height 2 and 1 mm width with −5° rake angle and 5° relief angle was considered for this model. A two-dimensional assembly of the model which includes workpiece and tool geometry is shown in Fig. 1.

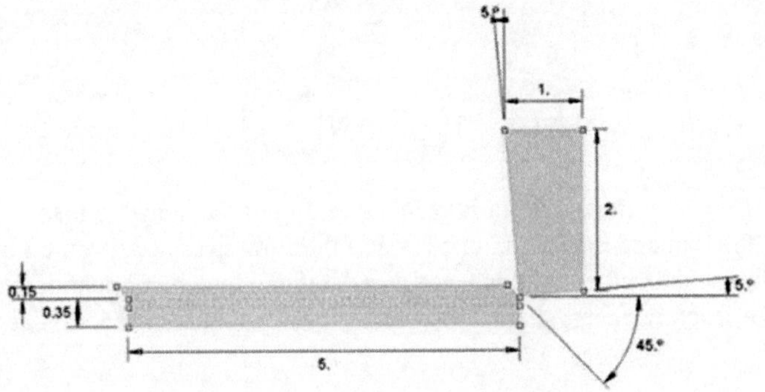

Fig. 1 Workpiece and tool geometry

2.2 *Material Modeling*

Machining involves large plastic deformations with high temperature along primary and secondary shear zones and plastic strain rates often in the order of 10^4/s, hence the effect of strain rate and temperature essential to material model. Johnson–Cook constitutive model defines material as a thermo-elastic visco-plastic material as shown in Eq. 1, where T = temperature, A = yield strength coefficient, B = hardening modulus, n = strain hardening coefficient, C = strain rate sensitivity coefficient, m = thermal softening coefficient, (ε°) = reference strain. Johnson–Cook model constants for Al7075-T651 [10] are being presented in Table 1.

$$\sigma = \left(A + B\varepsilon^{n}\right)\left(1 + C\ln\left(\frac{\varepsilon}{\varepsilon_0}\right)\right)\left(1 - \left(\frac{T - T_0}{T_{\mathrm{m}} - T_0}\right)^{m}\right) \tag{1}$$

The Johnson–Cook failure criterion in ABAQUS/Explicit is implemented by cumulative damage law shown in Eq. 2, where D = damage parameter, $\Delta\varepsilon$ = increment of the equivalent plastic strain, which is updated at every step and ε_{f} = equivalent plastic strain at failure calculated by Eq. 3. In Eq. 3, P denotes pressure stress, σ is misses stress, and d_1, d_2, d_3, d_4, d_5 are known as Johnson–Cook damage constants. Table 2 shows Johnson–Cook damage constants for Al7075-T651.

Table 1 Johnson–Cook material constants for Al7075-T651

A	B	n	C	m	ε
527	575	0.72	0.017	1.61	1

Table 2 Johnson–Cook damage parameters for Al7075-T651 [11]

d_1	d_2	d_3	d_4	d_5	ε
0.11	0.572	−3.446	0.016	1.099	1

$$D = \sum \frac{\Delta\varepsilon}{\varepsilon_f} \tag{2}$$

$$\varepsilon_f = \left(d_1 + d_2 \exp\left(d_3 \frac{P}{\sigma}\right)\right)\left(1 + d_4 \ln\left(\frac{\varepsilon}{\varepsilon_0}\right)\right)\left(1 + d_5\left(\frac{T - T_0}{T_m - T_0}\right)\right) \tag{3}$$

When $D = 1$, failure criteria is reached and material starts to lose its stiffness based on the damage evaluation criteria and the flow stress is governed by Eq. (4) where d is a scalar value between 0 and 1 and defined by either linear or exponential function of fracture energy, here d was defined by exponential function as shown in Eq. (5).

$$\sigma_d = (1 - d)\sigma \tag{4}$$

$$d = 1 - \exp\left(\int_0^{u_p} \sigma / G_f^{II} du_p\right) \tag{5}$$

In Eq. (5), u_p is equivalent to plastic displacement after damage initiation and G_f^{II} is fracture energy for Mode II fracture in-plane strain condition. G_f^{II} is given by Eq. (6). For Al7075-T651, $k_{IIC} = 40\ \text{MPa}\sqrt{m}$, $G_f^{II} = 19.88$ MN/m physical properties of workpiece and tool [11] are listed in Table 3.

$$G_f^{II} = \left(\frac{1 - \upsilon^2}{E}\right) k_{IIC}^2 \tag{6}$$

Table 3 Physical properties of workpiece and tool

Properties	Al 7075-T651	WC tool
Density, ρ (kg/m^3)	2810	11,900
Elastic modulus, E (GPa)	71.7	612
Poisson's ratio, v	0.33	0.22
Specific heat, C_p (J/kgK)	$C_p = 0.786\ T + 839.2$	$C_p = 0.12\ T + 839.01$
Thermal conductivity, K (W/mK)	$K = 0.0002T^2 + 0.1366\ T + 128.63$	$K = 0.042\ T + 35.95$
Thermal expansion, α (µm/mK)	25.2	4.9
Melting temperature, (°K)	908	
Room temperature (°K)	298	

3 Friction Modeling and Heat Generation

Interaction between rake face and internal chip surface is critical to the thermal analysis model considered two regions, sliding and sticking region, existing simultaneously along the chip–tool interface. Model can be represented by simple Columbus law as shown in Eqs. 7 and 8 where τ_f, τ_{max}, μ are frictional stress, equivalent shear stress, friction coefficient, respectively. Surface to surface contact type with penalty formulation available in ABAQUS (6.14) was used to implement the friction model.

$$\tau_f = \mu P, \mu P < \tau_{max} \tag{7}$$

$$\tau_f = \tau_{max}, \mu P \geq \tau_{max} \tag{8}$$

In thermo-coupled problems, the amount of heat generated during inelastic deformation has to be balanced. Plastic deformation during the machining process generates heat, resulting in a rise in temperature; heat flux generated is treated as the fraction of plastic work, usually 0.9 for metals. The same value was considered for this study. Another major source of heat in machining is the friction between tool and workpiece, fraction of friction work converted into heat is assumed to be 1 and 0.63 is the fraction of this heat transferred to chip and rest in tool. The same assumptions were made by Jomaa et al. [11].

4 Boundary Condition and Mesh Generation

Boundary conditions in machining problems are of two types: mechanical and thermal. Figure 2 and Table 4 show boundary conditions applied on tool and workpiece. For simplification, a tool was treated as a mechanically rigid body and fixed along both directions. The bottom and right edges of the workpiece were constrained to zero displacement in the vertical direction, and cutting velocity was provided in

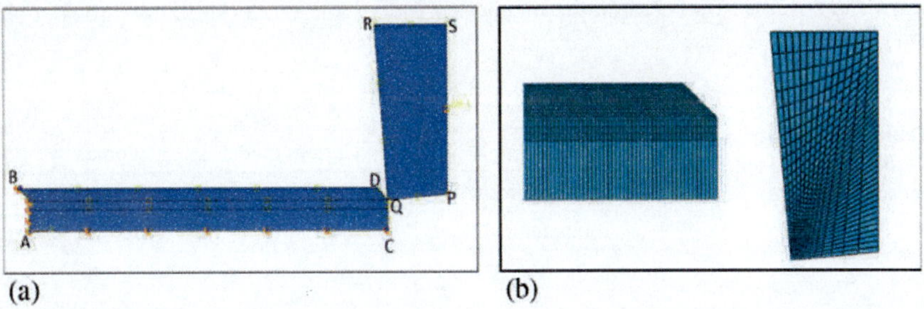

Fig. 2 **a** Boundary conditions on tool and workpiece; **b** mesh in workpiece and tool

Table 4 Boundary conditions on tool and workpiece

$u_x, u_y = 0$	On tool PQRS
$u_y = 0$	On the workpiece edges, AB and BC
$u_x = 1144$ m/min	On the edge of the workpiece, AB and BC
$T = 25$ °C	On the tool PQRS and workpiece ABCD
$Q = k_{t-c}(T_A - T_B)$	On tool–chip interface

horizontal direction. Room temperature was considered as 25 °C, and the initial temperature of both tool and workpiece was 25 °C. High heat transfer coefficient of 100 kW/m^2 K was provided between tool and chip contact to shorten time to reach a steady-state [11].

Both workpiece and tool were modeled with a 4-node plane strain quadrilateral elements, CPE4RT, with bilinear displacement and temperature. CPE4RT uses reduced integration; hence, hourglass control becomes necessary to prevent excessive distortion and disintegration of the model due to the hourglassing effect. Uncut chip thickness meshed with element height 10 μm inclined at 45° and up to 100 μm depth below uncut chip thickness elements of average height 10 μm. The rest of the workpiece was meshed with single element thickness, similarly. The tool consisted of 570 elements with high element concentration at the tip of the tool.

5 Model Validation

The output of the developed model was compared with the results of the model [11] at cutting speed 1144 m/min, feed rate 0.15 mm/rev, and depth of cut 4.3 mm. It took 82 min to complete the simulation on a computer with 2.2 GHz quad-core processor and a 16 Gb RAM. Deviation of predicted cutting force and maximum temperature in temperature profile was 8.14% and 0.58%, respectively. Since the model was able to predict the cutting force and temperature distribution with a reasonable accuracy, the model was used for simulation of machining experiments. Table 5 and Fig. 3 show the comparison between the developed model and model used by Jomaa et al. [11].

Table 5 Comparison of cutting force and temperature between the developed model and reported values of Jomaa et al. [11]

	Current model	Jomaa
Cutting force (*N*)	638	590
Temperature (°C)	493	496

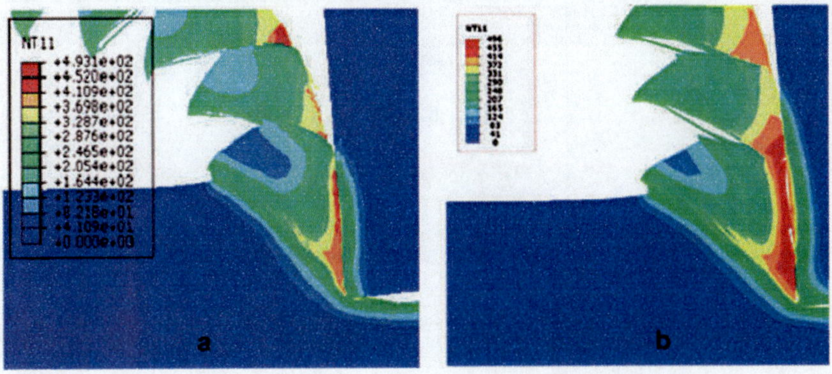

Fig. 3 Temperature profile in **a** current model, **b** Jomma et al. at 0.056 μs [11]

6 Simulation Runs

The design of the experiment model based on Taguchi's orthogonal array (L_{16}) was used to simulate the cutting experiments considering 3 control factors: cutting speed, feed, and depth of cut with 4 levels. The average cutting force and chip–tool interface temperature were taken as output. Table 6 shows the control factors and their levels used for experiments and Taguchi's L_{16} orthogonal array was generated using Minitab software. The machining experiments were simulated using the previously validated model up to 4.6 mm in length.

Stress profiles for different runs are shown in Fig. 4a–c, and the temperature profiles for different runs are shown in Fig. 4d–f. Cutting force response was measured as the average of the forces generated after the stabilization of machining process, output graphs of cutting forces generated in different runs are presented in Fig. 5a, b. In order to determine the chip–tool interface, temperature node on the chip–tool interface with the highest temperature along the interface was identified and the average of the nodal temperature when it reaches thermally stable condition was calculated. Output graphs of chip-interface temperature generated in different runs are presented in Fig. 5c–f. The results of simulation runs are presented in Table 7.

Table 6 Control factors and their levels

Sl. No	Control factors	Units	Level 1	Level 2	Level 3	Level 4
1	Cutting speed	m/min	600	800	1000	1200
2	Feed	mm/rev	0.05	0.1	0.15	0.2
3	Depth of cut	mm	1.5	2	2.5	3

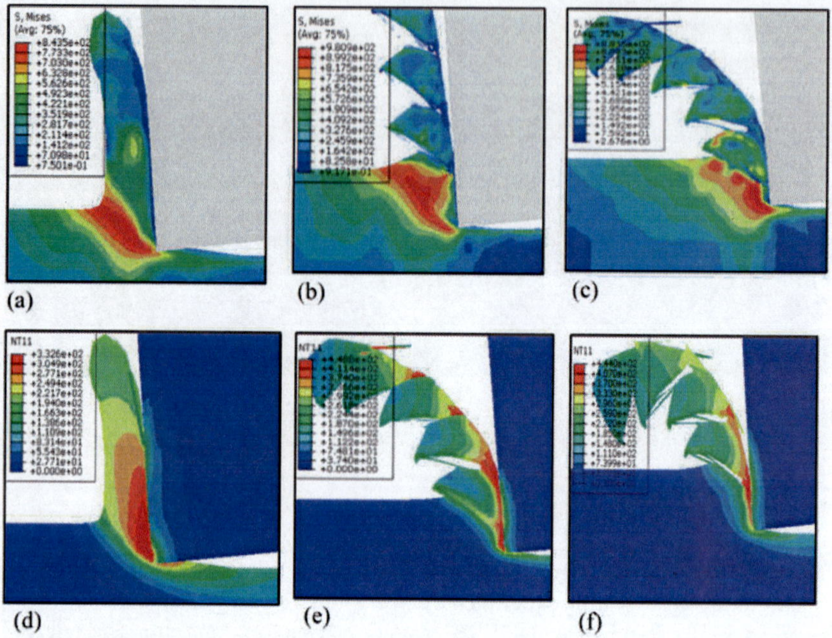

Fig. 4 **a–c** Stress profile; **d–f** temperature profile

7 Analysis of Variance

Tables 8 and 9 show the ANOVA results for cutting force and chip–tool interface temperature. The *p*-value for all the control variables is less than 0.05. The contribution of feed-in cutting force was 65% while cutting with 10.473% was the least contributing factor for cutting force. *P*-value for depth of cut is greater than 0.05 and no significance for chip–tool interface temperature, whereas *p*-value for feed and cutting speed is well below 0.05 and contribution of cutting speed and feed was 19.4% and 65%, respectively.

8 Conclusion

This work presents the multi-objective optimization of high-speed machining Al7075-T651 using the finite element modeling for predicting the chip–tool interface temperature. A fully thermo-mechanical coupled finite element model for orthogonal machining of Al7075-T651 was prepared. Simulation runs were conducted based on Taguchi's orthogonal array (L_{16}) with 3 control factors: speed feed and depth of cut and 4 levels. ANOVA test identified cutting speed and feed were the significant process parameters for the chip–tool interface temperature for which cutting speed

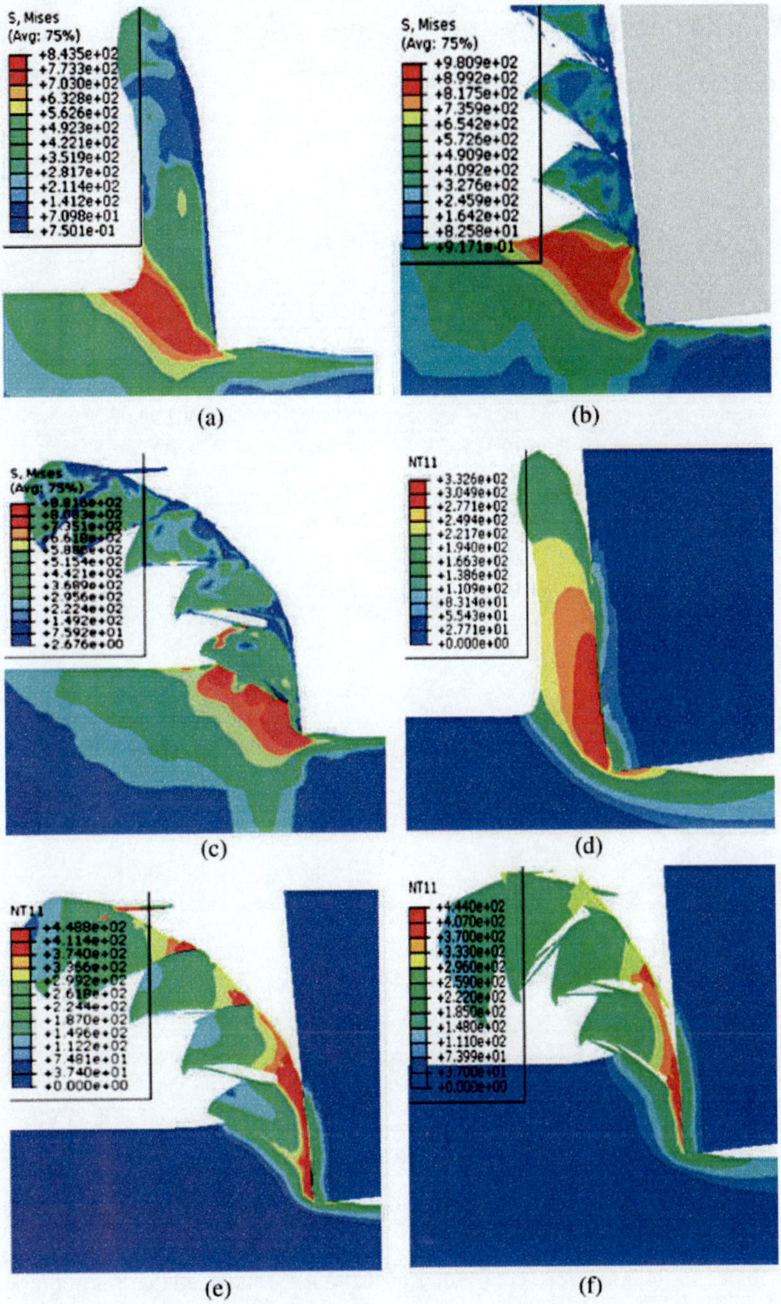

Fig. 5 **a**, **b** Output graphs of cutting forces generated in different runs; **c–f** output graphs of chip-interface temperature generated in different runs

Table 7 Simulation results

Run No	Force (N)	Temperature (°C)
1	124.537	246.89
2	288.341	321.88
3	475.503	381.17
4	656.851	316.61
5	144.800	246.46
6	188.660	364.64
7	395.158	401.24
8	485.592	391.38
9	182.196	248.07
10	377.095	385.23
11	252.725	407.90
12	407.862	416.73
13	197.760	253.47
14	280.098	409.22
15	264.020	426.58
16	276.073	432.48

Table 8 ANOVA for F ($R^2 = 98.08\%$)

Source	DF	SS	MS	F	P	%C
Speed	3	35,799	11,932.9	11.94	0.006	11.44
Feed	3	181,338	60,445.9	60.50	0.000	57.93
Depth	3	89,894	29,964.6	29.99	0.001	28.72
Error	6	5995	999.1			1.92
Total	15	313,025				

Table 9 ANOVA for temp $R^2 = 95.90\%$, $R^2{}_{\text{adj}} = 89.76\%$

Source	DF	SS	MS	F	P	%C
Speed	3	8845	2948.3	5.87	0.032	19.40
Feed	3	60,451	2015.4	40.12	0.000	73.96
Depth	3	1246	415.4	0.83	0.525	1.74
Error	6	3014	502.3			4.90
Total	15	73,556				

contributes 19.44% and feed contributes 65%. So it is concluded that for optimizing the process and controlling the chip–tool interface temperature, cutting speed and feed need to be optimized appropriately.

References

1. Daoud M, Jomaa W, Chatelain JF, Bouzid A (2015) A machining-based methodology to identify material constitutive law for finite element simulation. Int J Adv Manuf Technol 77(9–12):2019–2033
2. Shet C&Deng Xn (2000) Finite element analysis of the orthogonal metal cutting process. J Mater Process Technol 105:95–109
3. Mohammadpour M, Razfar MR, Saffar RJ (2010) Numerical investigating the effect of machining parameters on residual stresses in orthogonal cutting. Simul Model Pract Theory 18:378–389
4. Tamizharasan T, Senthil NK (2012) Optimization of cutting insert geometry using DEFORM-3D: numerical simulation and experimental validation. Int J Simul Model 11(2):65–76
5. Mabrouki T, Rigal JF (2006) A contribution to a qualitative understanding of thermo-mechanical effects during chip formation in hard turning. J Mater Process Technol 176:214–221
6. Ozel T (2006) The influence of friction models on finite element simulations of machining. Int J Mach Tools Manuf 46(5):518–530
7. Grzesik W (2006) Determination of temperature distribution in the cutting zone using hybrid analytical-FEM technique. Int J Mach Tools Manuf 46:651–658
8. Ng EG, Aspinwall DK, Brazil D, Monaghan J (1999) Modeling of temperature and forces when orthogonally machining hardened steel. Int J Mach Tools Manuf 36(6):885–903
9. Mamalis AG, Horvath M, Branis AS, Manolakos DE (2001) Finite element simulation of chip formation in orthogonal metal cutting. J Mater Process Technol 110(1):19–27
10. Barzani MM, Zalnezhad E, Sarhan AAD, Farahany S, Singh R (2015) Fuzzy logic based model for predicting surface roughness of machined Al–Si–Cu–Fe die casting alloy using different additives-turning. Measurement 61:150–161
11. Jomaa W, Mechri O, Levesque J, Songmene V, Bocher P (2017) Gakwaya: A Finite element simulation and analysis of serrated chip formation during high–speed machining of AA7075–T651. J Manuf Process 26:446–458

Analysis of Heat Transfer Coefficient in Turning Process

Ramanuj Kumar, Amlana Panda, Ashok Kumar Sahoo, and Deepak Singhal

Abstract Heat development during the metal machining process is a major challenging issue for the machinist. In the turning process, heat is distributed among chip, tool, workpiece, and cutting surrounding. This heat dissemination significantly influenced the complete cutting action. Further, heat transfer directly influenced the cutting attributes like wear growth rate, cutting life of the tool, surface finish, and workpiece dimensional precision achieved. In the current work, the stepwise procedure is deliberated to estimate the convection heat transfer coefficient for the workpiece into the surrounding and tool–workpiece interface into the surrounding. Further, the influence of input cutting terms (axial feed and cutting speed) onto the convection heat transfer coefficient is studied with the help of the main effect plot and surface plot. Highest heat transfer coefficient ($h_c = 32.2$ W/m^2 K) value (from workpiece into surrounding) is obtained at highest cutting speed (75 m/min) with the lowest feed (0.25 mm/rev) cutting conditions, while greatest heat transfer coefficient ($h_{in} = 3.1$ W/m^2 K) value (from the interface of tool–workpiece into surrounding) is noticed at a moderate speed (60 m/min) with axial feed (0.30 mm/rev) cutting conditions. Heat transfer coefficient (h_c) for workpiece into surrounding is improving with accelerating cutting speed, while it reduces sharply till 0.30 mm/rev of axial feed but further slowly increases. Heat transfer coefficient (h_{in}) for the interface of tool–workpiece into surrounding is retarding with axial feed, whereas it is sharply increasing up to 60 m/min of machining speed beyond this it is almost constant.

Keywords Convection heat transfer coefficient · Cutting speed · Axial feed · Turning

---

R. Kumar (✉) · A. Panda · A. K. Sahoo · D. Singhal
School of Mechanical Engineering, Kalinga Institute of Industrial Technology, Deemed To Be University, Bhubaneswar, India
e-mail: ramanujkumar22@gmail.com

S. Revankar et al. (eds.), *Proceedings of International Conference on Thermofluids*, Lecture Notes in Mechanical Engineering,
https://doi.org/10.1007/978-981-15-7831-1_61

1 Introduction

Metal machining is a thermo-mechanical practice due to the appearance of plastic deformation which leads to the generation of heat [1]. Basically, there are three zones where heat is generated in the turning process as shown in Fig. 1 [2, 3]. According to literature, a higher amount of heat is occurred in the primary shear zone (due to plastic deformation of material) and secondary shear zone, i.e., interface of tool-chip zone (due to rubbing action between rake of tool and chip) [4]. During machining, the maximum amount of heat is carried by chip (80%) [5]. The remaining amount of heat is sufficient to arise many thermo-mechanical issues in the processing of the metallic pieces. Many works reported the undesirable influence of high heat generation on the characteristics of the work–tool materials, dimensional precision, quality of surface produced, and life of tool [6, 7]. Also, heat intensity is proportional to input cutting terms (speed, feed, and depth of cut). Larger speed supports the formation of higher heat which leads to failure or premature failure of tools. Further, the generated heat is transferred into surrounding, metal-piece, tool, and chip through either by convection, conduction, or radiation. By conduction mode, heat is transmitted into a tool, workpiece, and chip. Convection and radiation modes of heat transformation are responsible to dissipate the generated cutting heat into the surrounding. The radiation mode of heat transformation is very less, but convection mode of heat transformation is dominant. However, in the current work, the role of various input parameters on the convection heat transformation coefficient is studied with the help of previously published experimental data of Kumar et al. [8]. Here, convection heat transformation from workpiece and tool–work interface into surrounding are studied.

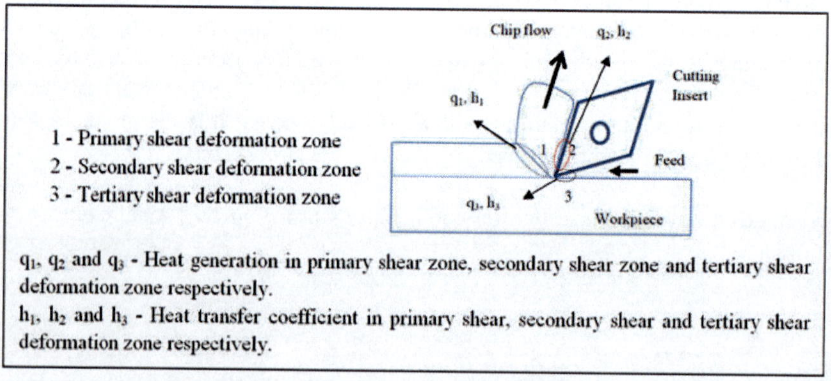

Fig. 1 Heat generation zones in turning process [2, 3]

2 Experimental Methodology

Ti-6Al-4V titanium grade 5 alloys offer a combination of low weight, high strength, and elevated corrosion resistance, which made it attractive for aerospace and medical applications. Machining of this alloy comes under difficult to cut due to its high chemical reactivity, low thermal conductivity, and elevated hardness. However, current work focused on convection heat transfer estimation from machined surface based on cutting temperature generated during cutting. Also, interface heat transfer coefficient is estimated considering heat generated at tool–work interface. The cutting temperature (T) and cutting force (F_c) data were taken from the work carried by Kumar et al. [7]. *L*9 experiments were conducted using different combination of feed values (0.25, 0.30, and 0.35 mm/rev) and cutting speed values (45, 60 and 75 m/min) parameters with constant depth of cut (0.5 mm).

2.1 *Convection Heat Transfer Coefficient*

In the process of turning of Ti-6Al-4V alloy, the temperature of work surface increases due to heat generation during cutting, and thus, heat from the workpiece is transmitted into the surrounding in the mode of convection and radiation. As the heat directly affects the properties of the workpiece, so it must be removed rapidly. To identify the heat removal rate, the heat transfer coefficient is an important term to quantify the transfer of heat between the work surface and the environment. Therefore, calculating the convection coefficient between the work surface and environment correctly is necessary for understanding the betterment of cutting process. During estimating the convection coefficient, some assumptions have been made as follows: The workpiece is homogeneous, the constant heat flux throughout the workpiece surface, neglecting radiation, i.e., heat loss from the work surface to the environment by convection only, consider the combined free and force convection, consider the thermal physical property of the air at mean temperature. The following equations (Eqs. 1–7) have been utilized to estimate the convection coefficient [9, 10]:

Step 1: Select air properties at mean temperature.

$$T_m = \frac{(T_w + T_f)}{2} \tag{1}$$

where T_m denotes average temperature (°C), T_w is the surface temperature of the workpiece (°C), T_f is ambient or fluid (air) temperature (°C).

Step 2: Reynolds number

$$R_e = \frac{\omega d^2}{\upsilon} \tag{2}$$

where R_e symbolized as a Reynolds number, ω represents the angular speed of test specimen (rad/s), d is diameter of test specimen, υ shows kinematic viscosity (m^2/s).

If $R_e < 5 \times 10^4$, then it is the condition where heat transfer happened by combined free and forced convection (mixed convection), otherwise check, if $R_e < (G_r/P_r)^{0.5}$, then it is free convection. Where, G_r and P_r are Grashof number and Prandtl number correspondingly.

Step 3: Coefficient of volumetric coefficient (β) where, T_w is in Kelvin.

$$\beta = \frac{1}{T_w} \tag{3}$$

Step 4: Grashof number (G_r):

$$G_r = \frac{\beta g d^3 (T_w - T_f)}{\upsilon^2} \tag{4}$$

Step 5: Nusselt number (N_u):

Case 1: If heat transfer by combined free and forced convection, then

$$N_u = 0.18[(0.5R_e^2 + G_r)P_r]^{0.315} \tag{5}$$

Case 2: If heat transfer by free convection, then

$$N_u = 0.456(G_r P_r)^{0.25} \tag{6}$$

Step 6: Convection coefficient (h_c)

$$h_c = \frac{N_u K}{L_c} \tag{7}$$

where h_c is convection coefficient (W/m^2 K), L_c is characteristic length (L_c = diameter of test specimen d (mm), K is thermal conductivity (W/m K).

2.2 Estimation of Tool–Work Interface Heat Transfer Coefficient

In turning, friction between tool and workpiece attributes the higher heat generation which directly affects the quality of work surface and tooling cost. This heat is dissipated in the surrounding by convection and radiation. However, it is very important to quantify the transfer of heat at the tool–workpiece interface and to accurately predict the temperature distribution within the cutting tool. In the current work, only the convection mode of heat transfer is considered. Various assumptions have been made to establish the model to calculate the convection heat transfer coefficient.

The following assumptions are the workpiece is homogeneous, all the mechanical works done in the machining process are converted into heat. The heat is calculated on the primary deformation zone, neglecting the effect of axial and radial forces for calculating the heat, neglecting radiation, i.e., heat loss from the tool–workpiece interface to surrounding by convection only, uniform temperature distributions in job, i.e., consider lumped body condition, consider unit area at tool–workpiece interface, and assume perfect contact between tool and workpiece. The following steps are mentioned in Eqs. (8–9):

Step 1—Heat flow (average) across the interface of tool–workpiece [11]:

$$q = F_c \cdot V_c \tag{8}$$

where q denotes the heat flow (average) across the interface of tool–work (W/m^2), F_c represents the cutting force (N), V_c denotes cutting speed in m/s.

Step 2—Heat transfer coefficient [12]:

$$h_{in} = \frac{q}{\Delta T} \tag{9}$$

where, h_{in} is the tool–work interface heat transfer coefficient (W/m^2 K), and ΔT is the amount of net temperature fall (°C).

3 Results Analysis and Discussion

From the result (Table 1), convection heat transfer coefficient (h_c) varies in between 14.1 and 32.2 W/m^2 K. In machining of high carbon steel, heat transfer coefficient was found in range of 10–30 W/m^2 K [13]. Highest h_c value is obtained at the highest cutting speed (75 m/min) with the lowest feed (0.25 mm/rev) cutting conditions. Due to this combination of speed–feed, the least magnitude of cutting force is noticed during experimentation as shown in Table. The lowest speed with the highest axial feed attributed the lowest h_c, and as a result, highest cutting force was reported in Table 1. Convection heat transfer coefficient (h_c) is leading with accelerating cutting speed, while it reduces sharply till 0.30 mm/rev of axial feed but further slowly increases as displayed in the main effects graph (Fig. 2a). Kopas et al. [14] also agreed with the results obtained here. It was stated that the workpiece cooled slower at lower cutting speed, hence the heat transfer rate was lower, i.e., heat transfer rate was increasing with cutting speed. From surface plot (Fig. 2b–c), h_c is almost constant up to 60 m/min of speed with a larger feed rate (above 0.3 mm/rev), while it is sharply lifted when speed is beyond 60 m/min with lower ranges of axial feed (up to 0.3 mm/rev). Thus, favorable h_c can be achieved when machining can be done at beyond 60 m/min with lower feed values (up to 0.3 mm/rev).

Similarly, from result data (Table 1), interface heat transfer coefficient (h_{in}) ranges from 1.92 to 3.10 W/m^2 K. Highest interface heat transfer coefficient (h_{in}) 3.10 W/m^2

Table 1 Estimated convection heat transfer coefficient

Sequence of test	Axial feed (mm/rev) [8]	Cutting speed, (m/min) [8]	Max temp., °C [8]	Cutting force (F_c) N [8]	Convective heat transfer coefficient of workpiece, h_c (W/m^2 K)	Convective heat transfer coefficient at tool–work interface, h_{in} (W/m^2 K)
1	0.25	45	188	445	14.3	2.1
2	0.30	45	197	453	14.2	1.98
3	0.35	45	209	479	14.1	1.92
4	0.25	60	192	426	16.6	3.0
5	0.30	60	204	442	17.6	3.10
6	0.35	60	216	463	17.2	2.37
7	0.25	75	198	395	32.2	3.06
8	0.30	75	210	414	18	2.70
9	0.35	75	225	425	19.2	2.71

K was achieved at moderate speed (60 m/min)-feed (0.30 mm/rev) cutting conditions, while lowest one is obtained at the lowest speed (45 m/min) with highest feed (0.35 mm/rev) cutting scenario [14]. According to Fig. 3a, interface heat transfer coefficient (h_{in}) is in decrease mode when feed value is increasing, while it is sharply increasing up to 60 m/min of machining speed, and beyond this, h_{in} is almost constant. From the surface plot (Fig. 3b–c), the slope of the surface is leading to 60 m/min of speed but with growing feed slope is in decreasing in nature. However, up to 60 m/min of speed with lower ranges of feed (up to 0.3 mm/rev) is recommended to get a better interface heat transfer coefficient.

Overall, it is recommended that up to 60 m/min of speed with lower ranges of feed values (up to 0.3 mm/rev) is best combinations of input terms to get a better work surface convection heat transfer coefficient and work–tool interface heat transfer coefficient during machining of titanium alloys.

4 Conclusion

Current work emphasized on the estimation of the convection heat transfer coefficient in the turning process. A stepwise procedure is formulated to estimate the convection heat transfer coefficient (h_c) between the workpiece and surrounding. Also, the convection heat transfer coefficient (h_{in}) between the interface of tool–workpiece and surrounding is estimated. The following conclusions have been made

- Heat transfer coefficient between workpiece and surrounding is varied from 14.1 to 32.2 W/m^2 K. It is increasing with increasing cutting speed, while it is retarding sharply till 0.3 mm/rev of axial feed but further slowly increases.

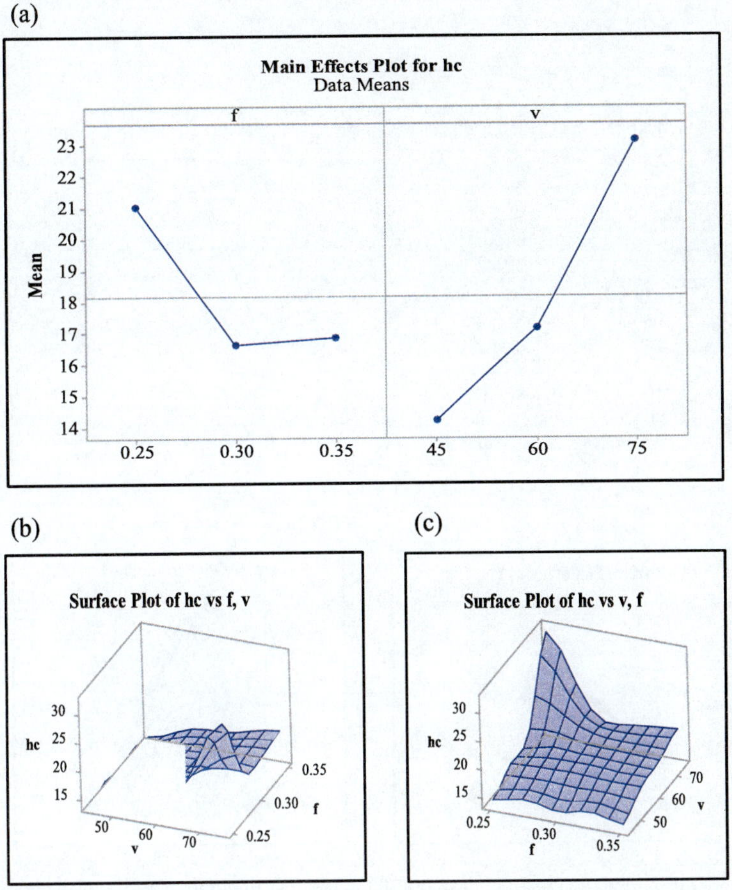

Fig. 2 Graphical plot of convection heat transfer coefficient from workpiece to surrounding **a** main effect plot, **b–c** surface plot

- Most favorable heat transfer coefficient between workpiece and surrounding can be achieved with greater speed of 60 m/min with lower feed values (up to 0.3 mm/rev).
- Heat transfer coefficient between the interface of tool–workpiece and surrounding is varied from 1.92 to 3.10 W/m^2 K. It is decreasing with rising axial feed, while it is sharply increasing up to 60 m/min of machining speed, but beyond this, it is almost constant.
- Up to 60 m/min of speed with lower ranges of feed (up to 0.3 mm/rev) are recommended to get better heat transfer coefficient between the interface of tool–workpiece and surrounding.

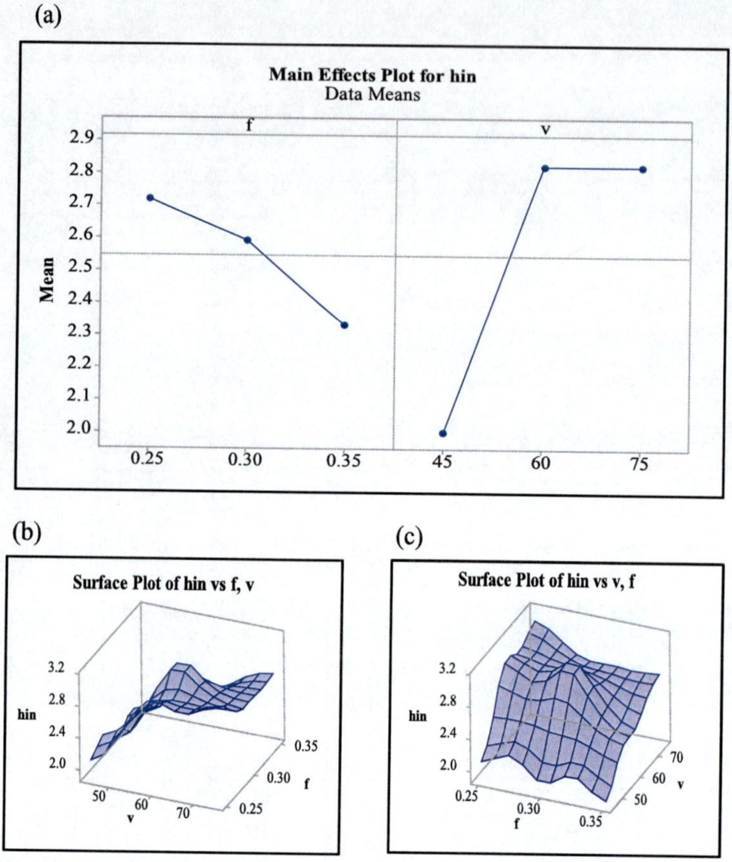

Fig. 3 Graphical plot of convection heat transfer coefficient from interface of tool–workpiece to surrounding **a** main effect plot, **b–c** surface plot

Optimizing the cutting parameters is needed to select the best combination of input terms to get the optimum convection heat transfer coefficient for both cases. Also, effects of heat transfer coefficient on tool wear, tool life, and surface finish can be studied.

References

1. Danish M, Ginta TL, Habib K, Rani AMA, Saha BB (2019) Effect of cryogenic cooling on the heat transfer during turning of AZ31C magnesium alloy. Heat Transfer Eng 40(12):1023–1032
2. Abukhshim NA, Mativenga PT, Sheikh MA (2005) Investigation of heat partition in high speed turning of high strength alloy steel. Int J Mach Tools Manuf 45(15):1687–1695

3. Haddag B, Atlati S, Nouari M, Zenasni M (2015) Analysis of the heat transfer at the tool–workpiece interface in machining: determination of heat generation and heat transfer coefficients. Heat Mass Transf 51:1355–1370
4. Pervaiz S, Deiab I, Wahba E, Rashid A, Nicolescu M (2018) A numerical and experimental study to investigate convective heat transfer and associated cutting temperature distribution in single point turning. Int J Adv Manuf Technol 94:897–910
5. Kus A, Isik Y, Cakir MC, Coskun S, Özdemir K (2014) Thermocouple and infrared sensor-based measurement of temperature distribution in metal cutting. Sensors 15(1):1274–1291
6. Brito RF, Carvalho SR, Lima E, Silva SMM (2015) Experimental investigation of thermal aspects in a cutting tool using comsol and inverse problem. Appl Therm Eng 86:60–68
7. Ng EG, Aspinwall DK, Brazil D, Monaghan J (1999) Modelling of temperature and forces when orthogonally machining hardened steel. Int J Mach Tools Manuf 39(6):885–903
8. Kumar R, Sahoo AK, Satyanarayana K, Rao GV (2013) Some studies on cutting force and temperature in machining Ti-6Al-4V alloy using regression analysis and ANOVA. Int J Ind Eng Comput 4:427–436
9. Domkundwar, A. V., Domkundwar, V. M.: Heat and mass transfer data book, 3rd edn. Dhanpar Rai & Co. (P) Limited, Delhi (2004)
10. Kothandaraman CP, Subramanyan S (2007) Heat and mass transfer data book, 6th edn. New Delhi, New Age International (P) Limited
11. Muller, B., Renz, U.: Thermal analysis of high-speed metal cutting process. In: 6th ASME-JSME Thermal Engineering Joint Conference, Hawaii Island (2003)
12. Sofyani SA, Marinescu ID (2017) Analytical modeling of the thermal aspects of metalworking fluids in the milling process. Int J Adv Manuf Technol 92:3953–3966
13. Santos MR, Lima e Silva SMM, Machado ÁR, Silva MB, Guimarães G, Carvalho SR (2014) Analyses of effects of cutting parameters on cutting edge temperature using inverse heat conduction technique. Math Probl Eng 1–11
14. Kops L, Arenson M (1999) Convective heat transfer coefficients in turning. In: Proceedings of the 15th Brazilian Congress of Mechanical Engineering, Sao Paulo, Brazil, (1999)

Embedded Heat Pipe-Assisted Cooling in Machining Process: A Comprehensive Review

Shailesh Kumar Sharma, Amlana Panda, Ramanuj Kumar, Ashok Kumar Sahoo, and Bharat Chandra Routara

Abstract During the machining process, the utilization of minimal cooling application generates difficulties with the cutting fluid. Very minimum quantity of cutting fluid is applied to the machining zone during the process. Considering the correlation amid tool-chip contact and the sources of heat from tool taken into account, the influences of the embedded heat pipe are of prime limelight in the technologically advanced machining sector. Since the quantity of fluid is considered to be low, therefore, some extra supporting internal cooling systems in terms of the heat pipe is the topic of research interest. Notably, this is innovative solution against dry and cooling machining to improve heat dissipation, and thus improving the cutting performance in advanced machining technology. Keeping this in view this innovative procedure with the backdrop of heat generation sources has been focused for the betterment of heat pipe embedded technology. This review study articulates various types of cutting tool and heat pipe materials and explores the relevance of the cutting performance. Considering the future outlook, this review will converge internal cooling strategies particularly embedded heat pipe using industry implemented factors for machining sector to bring the technological changes.

Keywords Embedded heat pipe · Dry turning · Sources of heat generation

1 Introduction

More meaningful attention has been paid toward the internal cooling strategy of cutting tool to overcome the predicament occurred during the machining lubrication approach. Moreover, during the dry machining, the coolant absence decreases the tooling life and supportive surface trait. Heat pipe-assisted lubrication scheme has been initiated as a substitution for direct coolant application in the machining region for investigation [1]. The dry machining needs abundantly rigid cutting condition that is difficult to be accomplished in contemporary machining shop floors. The

S. K. Sharma · A. Panda (✉) · R. Kumar · A. K. Sahoo · B. C. Routara
Kalinga Institute of Industrial Technology (KIIT) Deemed to be University, Bhubaneswar, India
e-mail: amlanpanda@gmail.com

S. Revankar et al. (eds.), *Proceedings of International Conference on Thermofluids*, Lecture Notes in Mechanical Engineering,
https://doi.org/10.1007/978-981-15-7831-1_62

application of cutting fluids has become a widespread choice for enhancing the machining characteristics of metals. Moreover, the role of cutting fluids is recognized for their pivotal contribution in extracting heat from the tip, cooling the tool-chip interface and conveying the metal chips carried away from the machining section. Though, traditional cutting fluids can cause serious problems related to the ecology and operator's health. The difficulties with traditional cutting fluids and dry cutting that tend to scientific study on innovative approaches for lubrication in cutting. The problems related to machining with cutting fluids and dry cutting tend to research for cooling in cutting operations. In a dry environment, the friction amid tool and chip tends to be serious that causes higher temperature, wearing rates and accordingly shorter longevity of the insert, especially in the cutting of hard-to-cut materials [2–4].

With advance technological machining system, the effectiveness of the cutting tool and machined material plays a significant pillar. Heat production in the tool-workpiece interface is a function of flank face tool wear [5].

A scientific study has been exposed that approximately an input energy of 99% is switched into the heat. The radical rapport of thermodynamics heat transfer performance in metal machining is taken into consideration for newly developed heat pipe to carry away the produced heat at the cutting tool. By and large, the pollution of cutting fluid and the health-related issue can be significantly reduced [6]. Tool wear is one of the machinability indicators for assessing tool life. Wearing of the tool signifies some degradation of mechanical characteristics. The pulsating-assisted heat pipe is another approach for indirect cooling at dry machining [7].

Due to the more work, material hardness the higher cutting temperature is generated during hard turning [8]. Cutting temperature has more negative consequences on the cutting tool as it monitors pointedly the tool lifespan with fast wear [9]. Because of the thermal softening repercussion persuaded by comparatively higher tool temperature tools are more inclined toward worn-out. Moreover, the tool declines the precision accuracy of the machined component. Interaction between tool and workpiece finds out the heat generation during the machining process. Because for the sharp edge of the tool, it is regarded as that the majority of the heat that performs to the tool is extracted from the tool-formed chip friction [10, 11]. Basically, a heat pipe is heat transfer equipment that is competent in conveying huge amounts of energy by evaporation–condensation cyclic processes. The interior surface of the heat pipe container is wrapped with a wick [12]. The topic of temperature and heat distribution in metal machining has gained significant attention among the researchers. Various types of tool-workpiece combination have been used by various researchers. Thermal problem in metal machining with tribological phenomenon is of great concern. More cutting temperature has a strong effect on tooling wear, tool span life and chip formation mechanism. Thermal deformation of the cutting tool is considered as main heat source error. In this review study, a general overview of the cutting performance enhancement using heat pipe embedded in the cutting tool during machining has been presented. The main contributions of this study are: (i) it presents a general review on heat pipe and its working tenet, (ii) application of embedded heat pipe in various machining sector, (iii) issues and (iv) it offers research outlook to the machinist.

2 Review of Selected Literature

More research focus is shifted toward the internal cooling to achieve environmental conscious manufacturing in the region of cooling technology. Moreover, the application of heat pipe becomes a new research thrust due to its better thermal conductivity and draws significant attention in dry cutting [13]. Indirect lubrication of tools needs the elevated heat produced on the rake tool face to be carried by a particular end that is eradicated from the machining region. The position and intensity of the cooling resource with regard to machining region plays a great importance in indirect cooling [14]. More attention has been paid to internal cooling to overcome the difficulties of cutting fluids [15]. The prominent types of internal cooling strategies are presented in Fig. 1.

At the tool-chip contacts, the contact pressure is a multitude that the supplied cutting fluid does not perform well for minimizing the crater wear. Internal cooling suggests for internally cooled tool that is very effective for precision [16]. The newly innovated censored cutting tool with changeable inside cooling for dry machining near cutting tip has been focused. In practical real-time cutting, this approach minimize the cutting temperature, thus, escalate the tooling life [17].

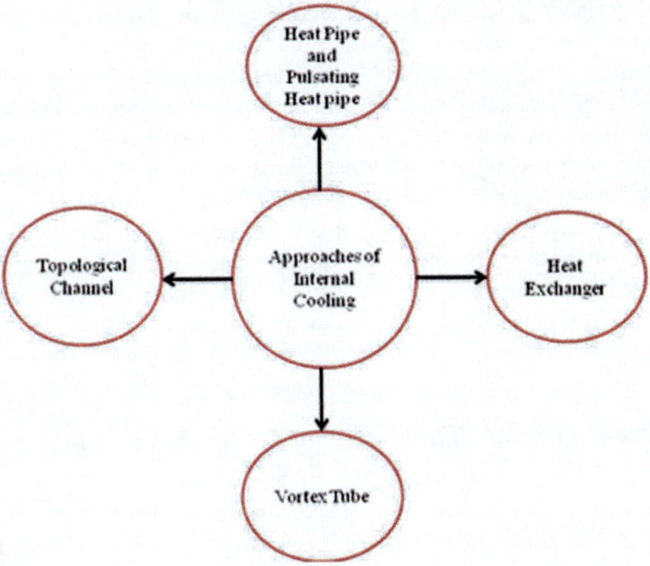

Fig. 1 Classification of internal cooling [13, 14]

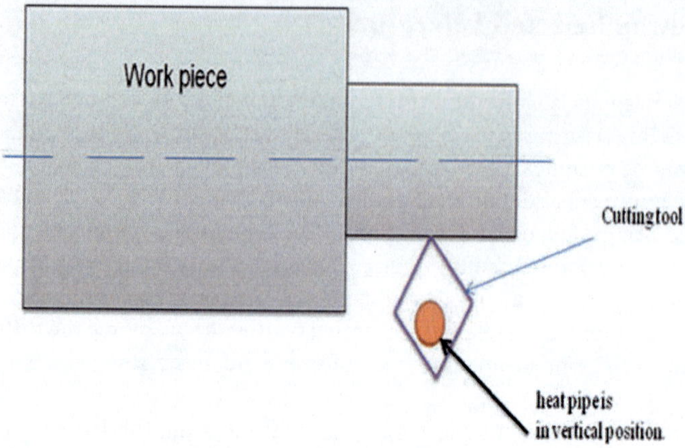

Fig. 2 Scheme of experimental set up with heat pipe [19]

2.1 Framework on Heat Pipe-Assisted Cooling

In general, the use of heat pipe in metal cutting procedure as a substitute approach for cooling is a research thrust area. Some scientific researchers have confirmed the efficiency of the heat pipe in enhancing the machining characteristics of metals in various cutting practices. The heat pipe is the equipment that moves energy with comparatively lower temperature dissimilarity with no requirement for an outer power. The heat pipe is rigidly fixed in a vertical place in connection with the tool for extorting heat. During the machining process, the application of embedded heat pipe in tool is very significant lubrication approaches that can fastly transfer part of produced in machining section. Generally, the heat pipe is cheap, compact and environmentally friendly and conforms to the needs of green manufacturing without consuming the energy [18]. The investigational set up for machining with heat pipe is presented in Fig. 2.

2.2 Regions of Heat Pipe

Heat pipe is the equipment used in transferring the heat from one region to another. Various types of the heat pipe, broadly used in micro-grooved, sintered wick and oscillating, etc. [20]. Heat pipe used to convey epic amounts of heat over comparatively more distances, without moving components. It maintains consistent temperature with both regions (heated and cooled). The working fluid survives both in liquid and vapor phases. The appropriate choice of the fluid decides the thermal behavior of the pipe [21]. There are three divisions namely evaporator, adiabatic and condenser. The outer heat on the evaporation region bases on the working fluid to vaporize. The vapor pressure markedly compels the vapor during the adiabatic zone to the condenser region. Further, the condensation of vapor takes place in the condensing section. The fluid at condensed condition is further back by capillary pressure produced in the wick. The returns to the evaporator zone due to gravity [19]. The various zone of the heat pipe is presented in Fig. 3 and principle in Fig. 4, respectively.

In recent years, pulsating heat pipe is an effective and new heat transfer equipment-assisted cutting tools in minimizing the cutting temperature and increasing the tool wear resistance was markedly proved in dry machining of titanium alloys [2]. The heat is extracted at the tool edge contact by a pipe embedded in the tool having more thermal conductance [25]. Authors have used tool with 0° inclination and 0° lead angle with cutting depth large adequately than the tool feed rate and tool nose radius. The investigational outcome demonstrates that there is a drop in temperature of 25 °C when heat pipe is used [1]. Use of heat pipe in metal cutting as substitute to traditional approach of extracting heat from the cutting region is a promising characteristic of attention amid researchers. Cutting tool cooling prunes cutting temperature and enhances the insert lifetime by minimizing tool wear [26]. Introduction of heat pipe is signified as passive method that combines evaporation and condensation of fluid [27].

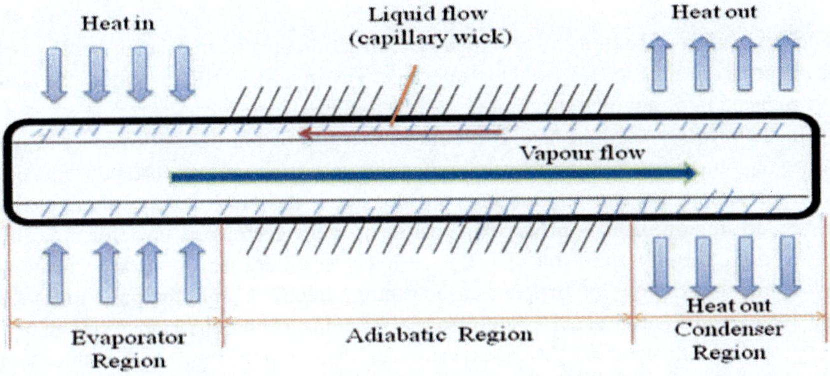

Fig. 3 Regions of heat pipe [22–24]

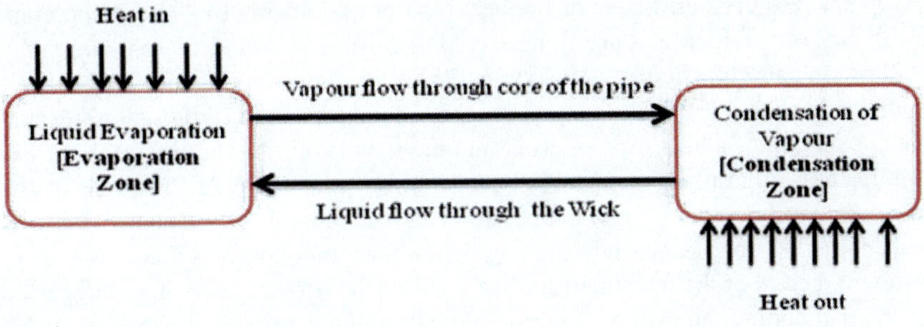

Fig. 4 Working principle of heat pipe [24]

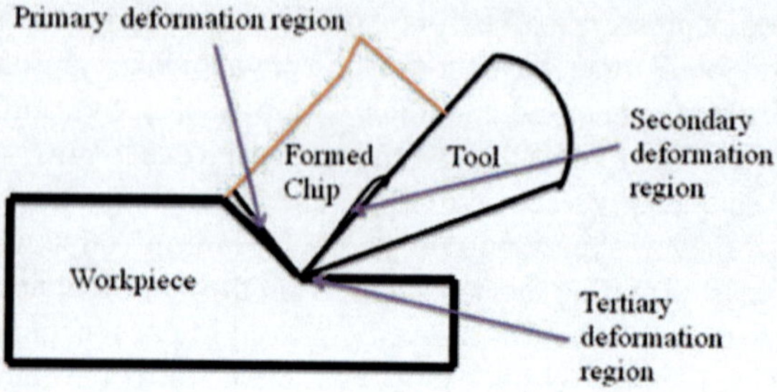

Fig. 5 Heat generation sources in the orthogonal cutting [5]

Circumferentially, a wick is positioned about the inside face of the pipe that extracts the condensed liquid return back to the evaporator. The vapor and the liquid pass throughout an adiabatic zone [24]. In heat pipe, the thermal energy is significantly changed to kinetic energy is owing to latent heat of evaporation and vaporization.

2.3 *Heat Generation Sources*

The generation of heat at the secondary deformation area is significant because of the friction among tool-formed chip interface. The important section where the heat is produced in orthogonal machining is shown in Fig. 5.

In the primary section, heat is produced because of plastic deformation in the shear region. In this zone, due to local heating very high temperature is produced, therefore, softening the material and further allow the more deformation. In the secondary region, heat is produced because of work performed in deforming the chip. In the tertiary zone (tool-workpiece interface), because of work perform to overcome the friction that happens at the rubbing interface amid the flank tool face and freshly work surface [15]. Computational domain and boundary must be specified for proper tool distribution [24]. Heat can be dissipated to the tool, work surface and chip by convection process through the fluid application. The multitude of heat flows through the formed chip, the other one is directed to work material [28]. Authors have experimented with the efficiency of pipe in end milling from dynamic and thermal viewpoint [29]. The attributes of the heat pipe in minimizing the tool temperature were verified by critical examinations of cutting tool materials and microphotography of chips. In particular, it is used to extract the heat from the cutting tool-chip contact in the carbide tool [30]. The sundry quantity of experimental studies signifies that the application of a heat pipe inserted into the tool of milling, turning tools assists significantly minimizing temperature and tooling wear [26]. Performance of the tool influenced notably by tool temperatures in machining [10]. Advanced coolants and cooling approaches are far-reaching for the machining process that influence of different factors of the machined surface quality [31]. Investigators have suggested an internal cooling approach adopting internal fluid channels inside the holder for putting on heat dissipation through dry machining practices [32]. In dry cutting

Table 1 Cutting tool and tool holder details

Author [Ref.]	Tool characteristics	Heat pipe parameters	Machining type
Zhu et al. [23]	Three-flute tool Dia head: 20 mm Dia end: 18 mm Long: 125 mm	Diameter: 5.5 mm Length: 110 mm	End milling
Gnanadurai and Varadarajan [19]	Tungsten carbide inserts	Air cooled heat pipe wall thickness: 1 mm Working fluid: water Outside radius: 6.3 mm Material: electrolytic copper	Hard turning
Chiou et al. [6]	Carbide inserts	Flat copper heat pipe, steel heat pipe Length: 10 mm Width: 5 mm Thickness: 1 mm	Dry machining
Quan and Mai [18]	Ordinary cylindrical-cutter (no heat pipe) Cylindrical-cutter (embedded heat pipe)	Diameter: 6 mm Thickness: 0.4 mm Length: 150 mm Working medium used: water Material: copper	Turning
Noorul Haq and Tamizharasan [1]	CBN cutting tool	Length = 20, 30, 40 mm Diameter = 5, 7, 10 mm Material: brass, steel copper	Hard turning
Liang and Quan [26]	WC–15%TiC–6%Co P10 carbide insert	Round copper–water diameter = 6 mm Length = 150 mm Thickness = 0.65 mm	Dry turning
Jen et al. [22]	Overall length of the drill tool: 216 mm	Water-based heat pipe diameter = 6.35 mm	Drilling

procedures, the adhesion and friction amid formed chip and tool tend to be more than grounds more wear rates, temperatures, and as a result, less tooling lives. The detail characteristics of the tool and heat pipe are presented in Table 1.

3 Issues Related to Embedded Heat Pipe

The multifarious researchers have worked upon many facets of heat pipe embedded cutting tool technology and came up with their own suggestions about this innovative process. Various structural parameters, like length of the evaporator, adiabatic section, condenser and outer radius of the heat pipe, play an important role in the cutting of 4340-grade steel. Metal cutting process is mostly a deformation process where deformation is extremely concentrated in a very small region.

From the cutting area, the extraction of huge amount of heat by the occurrence of heat pipes tends to the minimization of cutting temperature, thermal degradation of cutting fluids [19]. Intercooling method is a significant substitute to minimize cutting temperature and augment tooling life. However, the heat pipe may likely enhance heat portioning to the cutting tool at the tool [33]. Heat pipe-based lubrication is an environment-friendly cleaner technology for the required check of temperature. The flow of heat to the tool holder may lead to the irregular expansion. The orientation of the heat pipe and tool holder center line should be ideal arrangements for carrying away the heat transferred to the tool [24].

Topological channel has been used by authors [34] to enhance the mechanical characteristics, the heat transfer and wearability of the tool. More temperature significantly run down the tool span life. The hefty amount of heat accumulated in the workpiece and tool can significantly contribute to the poor dimensional of the work surface [22]. The direction of coolant application through rake, flank and clearance face may be single or multi is of prime significance to responses. Also, the influences of cooling approaches on various types of tools (multi-coated, ceramic, diamond) play a more expansive pillar in machining [35]. Authors [36] demonstrated that the heat pipe lubrication in drilling performs efficiently thermal performance offering prolong of tool service life. In this backdrop, the research repercussions have been depicted in Table 2.

The efficiency of the internal lubrication mechanism to curtail the temperature and tooling wear in turning of high chrome white cast iron [35]. In addition, in composite machining, a heat dissipation heat pipe has been introduced. The heat pipe embedded in a cutting tool is able to reduce the tooling of wear [33]. The impact of heat pipe on the thermal characteristic of the tool through dry cutting procedure has been numerically examined [16]. The outcomes from the experiment demonstrated that using a heat pipe can perform thermal behavior compared to flood cooling used widely in the machining industry improving sizeable tool lifetime [38]. Generated heat at tool-chip interface at the rake face corner of tool holder and insert of some segment of it flow into tool [6].

4 Research Summary

The heat pipe cooling can improve the cutting temperature at the cutting tool edge and particularly the dissimilarities of temperature amid the tool tip and the size of the tool tip by improving heat dissipation [23]. The traditional heat pipe is a single one. However, the imperative characteristic of the pulsating heat pipe is that wick is not needed to come back the condensate to the evaporation region [39]. Moreover, the research needs to be executed for better analysis of machining situations under which the effectiveness of the heat pipe embedded tool can be developed in both rough and

Table 2 Research outcomes by various researchers using heat pipe

Investigators and reference	Research outcomes
Gnanadurai and Varadarajan [19]	The proper utilization of heat pipe in minimum coolant application reduced tooling wear and cutting temperature to maximum 22% and 15%, respectively
Wu et al. [2]	More cutting temperatures of the pulsating heat assisted a tool holder could be minimized by 10% as so contrasted to that of the traditional one
Chiou et al. [37]	The newly embedded heat pipe efficiently carry the generated heat at tool-chip contact in machining of hardened AISI 1040 steel to minimize the tool wear
Quan and Mai [18]	The cooling consequence of the heat pipe efficiently embedded into the tool was more than the ordinary one
Haq and Tamizharasan [1]	The tool flank face wear and crater wear of the tools are considerably minimized by concerning 6 and 9%, respectively, by minimizing the temperature of the machining region well under the softening temperature of tools using installed heat pipe
Jen et al. [34]	The temperature has been reduced significantly in the order of 41% in vertical place and 31% in the horizontal place for comparison of the solid drill with heat pipe drill
Liu and Chou [33]	The modification of heat participation by the heat pipe-assisted cooling in cutting so as to evaluate and paradigm the influences on the minimization of tool temperature

finish machining. To control the process temperatures in metal cutting, the application of heat pipe is of prime concern related to manufacturing. Heat pipe-assisted lubrication system can minimize the application of cutting fluids. Therein, various analyses revealed that heat pipe-based cooling partly obviates heat so the effectiveness and efficiency of this innovative technology need more interest. Subsequently, for real practical application, the machining scheme of embedded heat pipe should be more experimented. Even so, parametric and numerical cumulative analysis of heat pipe embedded tool design should be taken into consideration for the betterment of cooing thermal performance. To ameliorate the machining characteristics, newer tool holder designs consisting of cooling approaches have paid significant interest to the machining sphere.

5 Conclusion

This review paper collectively provides the various response variables obtained by the investigators. However, the particular focus in on cutting temperature. The cutting temperature is generally influenced on the machined surface, cutting tool service life. In recent times, pulsating assisted heat pipe is used in metal cutting. The major focus of this review was to study the heat pipe, its principle and its application in the metal cutting. Moreover, improved cutting tool design leads to better promising cutting performance in turning of different metals. Mechanism of heat transfer needs more investigation in heat pipe embedded tool during cutting. Cost-effective cooling technology attracted many researchers. In addition, the use of embedded heat pipes with its tribological behavior will provide a better solution in the field of machining.

References

1. Haq AN, Tamizharasan T (2006) Investigation of the effects of cooling in hard turning operations. Int J Adv Manuf Technol 30(9–10):808–816
2. Wu Z, Yang Y, Luo C (2016) Design, fabrication and dry cutting performance of pulsating heat pipe self-cooling tools. J Clean Prod 124:276–282
3. Das RK, Sahoo AK, Mishra PC, Kumar R, Panda A (2018) Comparative machinability performance of heat treated 4340 Steel under dry and minimum quantity lubrication surroundings. Procedia Manuf 20:377–385
4. Tiwari H, Kumar R, Panda A, Sahoo AK, Roy S (2019) A brief review on different lubricants used in MQL process during hard turning. In: Advances in industrial and production engineering. Springer, Singapore, pp 853–860
5. Abukhshim NA, Mativenga PT, Sheikh MA (2005) Investigation of heat partition in high speed turning of high strength alloy steel. Int J Mach Tools Manuf 45(15):1687–1695
6. Chiou RY, Chen JS, Lu L, Cole I (2002) Prediction of heat transfer behavior of carbide inserts with embedded heat pipes for dry machining. In: ASME 2002 international mechanical engineering congress and exposition. American society of mechanical engineers digital collection, pp 83–90
7. Gupta S, Venkatesan K, Devendiran S, Mathew AT (2019) Experimental investigation of IN725 under different cooling environments using new tool holder. Mater Manuf Process 34(6):637–647
8. Panda A, Sahoo AK, Kumar R, Das RK (2019) A review on machinability aspects for AISI 52100 bearing steel. Mater Today Proc. https://doi.org/10.1016/j.matpr.2019.05.422
9. Bartarya G, Choudhury SK (2012) State of the art in hard turning. Int J Mach Tools Manuf 53(1):1–14
10. Cui X, Guo J (2019) Effects of bioinspired microscopic geometry of ceramic tool surface on tool temperatures in continuous and interrupted turning. Int J Mech Sci 157:715–725
11. Sun M, Appel E, Kovalev A, Gorb E, Liang A, Gorb SN (2017) The influence of the topography and physico-chemical properties of the cuticle surface on the wettability and adhesive properties of the elytra of the dung beetle *Geotrupes stercorarius* (Coleoptera, Scarabaeidae). Bioinspir Biomim 13(1):016008
12. Mahdavi M, Tiari S, De Schampheleire S, Qiu S (2018) Experimental study of the thermal characteristics of a heat pipe. Exp Therm Fluid Sci 93:292–304
13. Wu Z, Yang Y, Su C, Cai X, Luo C (2017) Development and prospect of cooling technology for dry cutting tools. Int J Adv Manuf Technol 88(5–8):1567–1577
14. Minton T, Ghani S, Sammler F, Bateman R, Fürstmann P, Roeder M (2013) Temperature of internally-cooled diamond-coated tools for dry-cutting titanium. Int J Mach Tools Manuf 75:27–35
15. Zhao H, Barber GC, Zou Q (2002) A study of flank wear in orthogonal cutting with internal cooling. Wear 253(9–10):957–962
16. Al-Odat MQ (2010) Numerical analysis of cutting tool temperature in dry machining processes with embedded heat pipe. Eng Comput

17. Sun X, Bateman R, Cheng K, Ghani SC (2012) Design and analysis of an internally cooled smart cutting tool for dry cutting. Proc Inst Mech Eng Part B J Eng Manuf 226(4):585–591
18. Quan Y, Mai Q (2015) Investigation of the cooling effect of heat pipe-embedded cutter in dry machining with different thermal conductivities of cutter/workpiece materials and different cutting parameters. Int J Adv Manuf Technol 79(5–8):1161–1169
19. Gnanadurai RR, Varadarajan AS (2016) Investigation on the effect of cooling of the tool using heat pipe during hard turning with minimal fluid application. Eng Sci Technol Int J 19(3):1190–1198
20. Gupta NK, Tiwari AK, Ghosh SK (2018) Heat transfer mechanisms in heat pipes using nanofluids—a review. Exp Therm Fluid Sci 90:84–100
21. Jouhara H, Chauhan A, Nannou T, Almahmoud S, Delpech B, Wrobel LC (2017) Heat pipe based systems—advances and applications. Energy 128:729–754
22. Jen TC, Gutierrez G, Eapen S, Barber G, Zhao H, Szuba PS, Labataille J, Manjunathaiah J (2002) Investigation of heat pipe cooling in drilling applications: part I: preliminary numerical analysis and verification. Int J Mach Tools Manuf 42(5):643–652
23. Zhu L, Jen TC, Yin CL, Yen YH, Zhu M, Zhang J (2009) Investigation of heat pipe cooling in drilling applications: part 2—thermal, structural static, and dynamic analyses. In: ASME 2009 international mechanical engineering
24. Judd RL, MacKenzie HS, Elbestawi MA (1995) An investigation of a heat pipe cooling system for use in turning on a lathe. Int J Adv Manuf Technol 10(6):357–366
25. Chiou RY, Chen JS, Lu L, North MT (2003) The effect of an embedded heat pipe in a cutting tool on temperature and wear. In: ASME 2003 international mechanical engineering congress and exposition. American society of mechanical engineers digital collection, pp 369–376
26. Liang L, Quan Y (2013) Investigation of heat partition in dry turning assisted by heat pipe cooling. Int J Adv Manuf Technol 66(9–12):1931–1941
27. Pour MG, Ghaedi E (2015) Experimental investigation of performance of heat pipe for drilling purposes. Int J Mater Chem Phys 1:343–346
28. Sharif MN, Pervaiz S, Deiab I (2017) Potential of alternative lubrication strategies for metal cutting processes: a review. Int J Adv Manuf Technol 89(5–8):2447–2479
29. Wu Z, Deng J, Su C, Luo C, Xia D (2014) Performance of the micro-texture self-lubricating and pulsating heat pipe self-cooling tools in dry cutting process. Int J Refract Metal Hard Mater 45:238–248
30. Chiou RY, Lu L, Chen JS, North MT (2007) Investigation of dry machining with embedded heat pipe cooling by finite element analysis and experiments. Int J Adv Manuf Technol 31(9–10):905–914
31. Anton S, Andreas S, Friedrich B (2015) Heat dissipation in turning operations by means of internal cooling. Procedia Eng 100:1116–1123
32. Yao B, Sun W, Chen B, Yu X, He Y, Feng W, Wang S (2017) An independent internal cooling system for promoting heat dissipation during dry cutting with numerical and experimental verification. Appl Sci 7(4):332
33. Liu J, Chou YK (2007) Cutting tool temperature analysis in heat-pipe assisted composite machining. J Manuf Sci Eng 129(5):902–910
34. Jen TC, Chen YM, Gutierrez G (2003) Thermal performance of heat pipe drill: experimental study. In: ASME 2003 heat transfer summer conference. American society of mechanical engineers digital collection, pp 87–93
35. Ravi AM, Murigendrappa SM (2018) Experimental study on internal cooling system in hard turning of HCWCI using CBN tools. AIP Conf Proc 1943(1):020053
36. Zhu L, Jen TC, Liu YB, Zhao JW, Liu WL, Yen YH (2015) Cutting tool life analysis in heat-pipe assisted drilling operations. J Manuf Sci Eng 137(1):011008
37. Chiou RY, Lu L, Chen JS, North MT (2004) Dry machining characterization simulations and experiments for new cutting tool design with embedded heat pipe. Intell Syst Des Manuf Int Soc Opt Photonics 5605:154–165
38. Zhu L, Jen TC, Yin CL, Kong XL, Yen YH (2012) Experimental analyses to investigate the feasibility and effectiveness in using heat pipe-embedded drills. Int J Adv Manuf Technol 58(9 12):861–868

39. Wu Z, Xing Y, Liu L, Huang P, Zhao G (2020) Design, fabrication and performance evaluation of pulsating heat pipe assisted tool holder. J Manuf Process 50:224–233